T0189552

Lecture Notes in Computer Science 1294

Edited by G. Goos, J. Hartmanis and J. van Leeuwen

Advisory Board: W. Brauer D. Gries J. Stoer

Springer
Berlin
Heidelberg
New York
Barcelona
Budapest
Hong Kong
London
Milan
Paris
Santa Clara
Singapore
Tokyo

Burton S. Kaliski Jr. (Ed.)

Advances in Cryptology – CRYPTO '97

17th Annual International
Cryptology Conference
Santa Barbara, California, USA
August 17-21, 1997
Proceedings

 Springer

Series Editors

Gerhard Goos, Karlsruhe University, Germany

Juris Hartmanis, Cornell University, NY, USA

Jan van Leeuwen, Utrecht University, The Netherlands

Volume Editor

Burton S. Kaliski Jr.
RSA Laboratories
20 Crosby Drive, Bedford, MA 01730-1402, USA
E-mail: burt@rsa.com

Cataloging-in-Publication data applied for

Die Deutsche Bibliothek - CIP-Einheitsaufnahme

Advances in cryptology : proceedings / CRYPTO '97, 17th Annual
International Cryptology Conference, Santa Barbara, California, USA,
August 17 - 21, 1997. Burton S. Kaliski (ed.). [IACR]. - Berlin ;
Heidelberg ; New York ; Barcelona ; Budapest ; Hong Kong ;
London ; Milan ; Paris ; Santa Clara ; Singapore ; Tokyo : Springer,
1997
  (Lecture notes in computer science ; Vol. 1294)
  ISBN 3-540-63384-7

CR Subject Classification (1991): E.3, G.2.1, D.4.6, K.6.5,F.2.1-2, C.2, J.1,
E.4

ISSN 0302-9743
ISBN 3-540-63384-7 Springer-Verlag Berlin Heidelberg New York

Typesetting: Camera-ready by author
SPIN 10546375   06/3142 – 5 4 3 2 1 0   Printed on acid-free paper

Preface

Crypto '97, the Seventeenth Annual Crypto conference organized by the International Association for Cryptologic Research (IACR) in cooperation with the IEEE Computer Society Technical Committee on Security and Privacy and the Computer Science Department of the University of California, Santa Barbara, represents another step forward in the steady progression of the science of cryptology. There is both a tremendous need for and a great amount of work on securing information with cryptologic technology. As one of the two annual meetings held by the IACR, the Crypto conference provides a focal point for presentation and discussion of research on all aspects of this science.

It is thus a privilege to coordinate the efforts of this community in focusing on its steps forward. Crypto '97 is a conference for its community, and to the researchers who have contributed to it — those whose papers appear in the proceedings, those whose submissions were not accepted, and those who have laid the foundation for the work — the community owes a debt of gratitude.

The process of developing a conference program is a challenging one, and this year's committee made the process both enjoyable and effective. My thanks go to Antoon Bosselaers, Gilles Brassard, Johannes Buchmann, Ivan Damgård, Donald Davies, Alfredo de Santis, Susan Langford, James L. Massey, Moni Naor, David Naccache, Tatsuaki Okamoto, Douglas Stinson, Michael J. Wiener, Rebecca Wright, and Yuliang Zheng for many hours of reviewing submissions and presenting their comments to the committee.

My thanks also to the committee's two advisory members, Neal Koblitz and Hugo Krawcyzk, the program chairs of Crypto '96 and '98. Neal's experience from a year ago and Hugo's perspective on the year ahead have helped to make this year's conference what it is, and should provide continuity to the next one.

Continuing a recent tradition, the review process for Crypto '97 was conducted entirely by e-mail and fax, without a program committee meeting. Each submission was assigned anonymously to three committee members (though many submissions were reviewed by more than three people), and decisions were made through several rounds of e-mail discussions. Of the 160 submissions received, the committee accepted 36, of which 35 appear in final form in these proceedings. Except for the papers themselves, nearly all correspondence with authors was also conducted by e-mail.

Gilles Brassard and Oded Goldreich complete this year's program with their invited lectures on quantum information processing and the theoretical foundations of cryptology. My appreciation to both of them, as well as to Stuart Haber, who chairs the conference's informal rump session (whose papers, due to logistics, cannot be included in these proceedings).

The program committee benefited from the expertise of many colleagues: Carlisle Adams, Carlo Blundo, Dan Boneh, Jørgen Brandt, Ran Canetti, Don Coppersmith, Erik De Win, Giovanni Di Crescenzo, Matthew Franklin, Atsushi Fujioka, Eiichiro Fujisaki, Rosario Gennaro, Helena Handschuh, Michael Jacobson Jr., Markus Jakobsson, Joe Kilian, Lars Knudsen, Tetsutaro Kobayashi, Françoise Levy-dit-Vehel, Keith Martin, Markus Maurer, Andreas Meyer, David M'raïhi, Volker Mueller, Stefan Neis, Kobbi Nissim, Kazuo Ohta, Pascal Paillier, Sachar Paulus, Giuseppe Persiano, Erez Petrank, Benny Pinkas, Bart Preneel, Tal Rabin, Omer Reingold, Mike Reiter, Pankaj Rohatgi, Taiichi Saitoh, Berry Schoenmakers, Martin Strauss, Edlyn Teske, Shigenori Uchiyama, Paul Van Oorschot, Susanne Wetzel, and Hugh Williams. My thanks to each one, as well as to any others whom I have inadvertently omitted.

The successful organization of this year's conference is due to its general chair, Bruce Schneier. The functions of general chair and program chair are for the most part independent, but at those times where collaboration was required, Bruce was very helpful, and I appreciate the opportunity to have worked with him. On behalf of Bruce, I would also like to extend my thanks to Raphael Carter and Karen Cooper for their assistance in the organization of Crypto '97.

My work was also not without assistance, and I would like to thank Ari Juels and Gerri Sireen for their participation in administrative aspects of the program.

In the Proverbs, it is written, "It is the glory of God to conceal a thing; but the honour of kings is to search out a matter." The search for knowledge about cryptology — itself the science of secrets — is an essential part of protecting information in today's increasingly open world. Another step in this search is expressed in these proceedings. May the search of such matters, and the search for knowledge about cryptology, continue for many years to come.

Burt Kaliski   June 16, 1997
                                                                                                    Bedford, Massachusetts

CRYPTO '97

August 17–21, 1997, Santa Barbara, California, USA

Sponsored by the

International Association for Cryptologic Research (IACR)

in cooperation with

IEEE Computer Society Technical Committee on Security and Privacy
Computer Science Department, University of California, Santa Barbara

General Chair

Bruce Schneier, Counterpane Systems, USA

Program Chair

Burt Kaliski, RSA Laboratories, USA

Program Committee

Antoon Bosselaers ....Katholieke Universiteit Leuven, Belgium
Gilles Brassard...Université de Montréal, Canada
Johannes Buchmann....................Techniche Hochschule Darmstadt, Germany
Ivan Damgård...Aarhus University, Denmark
Donald Davies...................Royal Holloway College London, United Kingdom
Alfredo de Santis.. Università di Salerno, Italy
Susan Langford ....Atalla Corporation, USA
James L. Massey ....Swiss Federal Institute of Technology, Switzerland
Moni Naor.. Weizmann Institute, Israel
David Naccache .... Gemplus, France
Tatsuaki Okamoto.. NTT Laboratories, Japan
Douglas Stinson .... University of Nebraska, USA
Michael J. Wiener ....Entrust Technologies, Canada
Rebecca Wright... AT&T Labs, USA
Yuliang Zheng.. Monash University, Australia

Advisory Members

Neal Koblitz *(Crypto '96 program chair)* .... University of Washington, USA
Hugo Krawczyk *(Crypto '98 program chair)* IBM T.J. Watson Research Center, USA
....and Technion, Israel

Contents

Contents

Elliptic Curve Implementation

Number-Theoretic Systems

Distributed Cryptography

Hash Functions

Cryptanalysis of Secret-Key Cryptosystems

The Complexity of Computing Hard Core Predicates

Mikael Goldmann and Mats Näslund

Royal Institute of Technology,
Dept. of Numerical Analysis and Computing Science,
S–100 44 Stockholm, Sweden
e-mail: {migo,matsn}@nada.kth.se

Abstract. We prove that a general family of hard core predicates requires circuits of depth $(1-o(1))\frac{\log n}{\log\log n}$ or super-polynomial size to be realized. This lower bound is essentially tight. For constant depth circuits, an exponential lower bound on the size is obtained. Assuming the existence of one-way functions, we explicitly construct a one-way function $f(x)$ such that for any circuit c from a family of circuits as above, $c(x)$ is almost always predictable from $f(x)$.

Keywords: pseudo-randomness, small-depth circuit, one-way function

1 Introduction

One of the most useful cryptographic primitives is the *pseudo-random generator*. This is a function that deterministically expands a short random seed to a longer, random "looking" string. Blum and Micali [3], and Yao [13], showed a simple way to construct such a generator, using the following method. Assume that f is a permutation and that we have a set of boolean functions C. Choose c at random in C, a random x, and output $f(x), c(x)$, and (the description of) c. Clearly it suffices that $c(x)$ is unpredictable given $f(x)$ and c. Since computing $c(x)$ is no harder than inverting $f(x)$, this in turn implies that $f(x)$ must be a so called *one-way function*. If we indeed have this situation, C is said to be a *family of hard core predicates for f*.

For efficiency reasons, we would like the functions in C to be very simple to compute, so that an essential question is: *How simple can they be?* It seems natural to consider functions such as $c(x) = $ "some bit of x". Although there are examples of certain conjectured one-way functions f for which such simple c are hard core predicates, e.g. see [1,3,7], these constructions are too simple to work in a general setting without any assumptions on f. The reason for this is that a one-way function may depend on a relatively small number of its input bits and output the rest of them unchanged. The knowledge of these latter bits may be enough to deduce the value of such a "hard core". Since the case when no assumptions on f (except that it's one-way) are needed clearly is the most attractive, this is the case that we study in this paper.

Yao observed in [13] that although a one-way function may reveal many bits of the input, it *must* hide at least *some* bits. We may in general not know which these bits are, so a good candidate for a hard core should depend on all (or almost all) of its input bits. The first construction of a hard core predicate for any one-way function is due to Goldreich and Levin, [5], and uses the inner product modulo 2 of x and a random binary string r. Two more constructions, affine functions in GF[2^n] and $\mathbb{Z}_p$, are due to Näslund, [11,12]. These are functions depending on all bits in x. Though simple, it is not obvious that we cannot use even simpler functions as long as they depend on all bits.

In this paper we shall prove that the existing constructions are basically the simplest possible. To measure the simplicity/complexity we will use the computational model of circuits, i.e. how large/deep a circuit of boolean AND/OR/NOT-gates that is needed to compute the hard core predicate. All the three general constructions mentioned above can be computed by circuits of logarithmic depth, polynomial size, and constant fan-in, that is, NC^1-circuits. So, the next natural step-down in complexity would be to consider AC^0-circuits; circuits of constant depth, polynomial size, and unbounded fan-in. There are numerous results indicating that this class of circuits is not very powerful. For instance, it is known from [10] by Mansour, Nisan, and Tiwari that universal hash functions (in general good candidates for hard core predicates) can not be computed by such simple circuits. A similar negative result on the existence of so called *pseudo-random functions* was given by Linial, Mansour, and Nisan in [9].

The widely used technique for showing computational limitations of small-depth circuits is the application of the Håstad switching lemma, see [6]. This proves to be useful here too since the lemma basically says that knowing some of the inputs to a small-depth circuit is very likely to be enough to deduce the output value of the circuit. This method is probabilistic and will give non-uniform results. However, we show that it is possible to obtain uniform results as well.

The paper is organized as follows. First we give some basic definitions and a proof outline in Section 2. Section 3 describes some tools from the theory of circuit complexity. Although perhaps known as a "folklore theorem", we prove in Section 4 that no family of constant depth, constant fan-in circuits can be a family of hard core predicates. We choose to do this since it illustrates the basic techniques. In Section 5 we then prove that not even polynomial size, constant depth, and unbounded fan-in circuits can be hard core predicates.

2 Preliminaries

If x is a binary string, $|x|$ is the length of x (if S is a set $|S|$ is the cardinality). By $y \in_{\mathcal{D}} S$, we mean a y chosen from S according to the distribution $\mathcal{D}$. Here, $\mathcal{U}$ will denote the uniform distribution on S. For two binary strings x, y, $x \circ y$ denotes the concatenation of the strings. If $x = x_1 x_2 \cdots x_n \in \{0,1\}^n$ and $I \subseteq \{1, 2, \ldots, n\}$ let $x_I = x_{i_1} x_{i_2} \cdots x_{i_{|I|}}$, $i_j \in I$, $i_1 < i_2 < \cdots < i_{|I|}$. $x_{\bar{I}}$ is defined analogously by taking the complement of I.

Let $\mathcal{B} = \{b : \{0,1\}^* \mapsto \{0,1\}\}$, $\mathcal{B}_n = \{b : \{0,1\}^n \mapsto \{0,1\}\}$. A *circuit* is a directed acyclic graph having *gates* as vertices. A gate can be of type OR, AND, or NOT and computes the corresponding boolean function of its incoming edges, the incoming edges being outputs of other gates or one of the inputs, x_i, $i = 1, 2, \ldots, n$ or the negation of an input $\bar{x_i}$. The *fan-in* of a gate is the number of incoming edges. There is a unique gate the output of which is the output of the whole circuit. The *size* of the circuit is the number of gates. By modifying the circuit (and making it slightly bigger), we can assume that NOT-gates only appear at the inputs and that the circuit is leveled with the gates at level i taking their inputs from gates at level $i - 1$ and that all gates at a given level are of the same type (AND/OR), types alternating from level to level. Hence all inputs x_i are at level 0. The *depth* of the circuit is the number of levels.

A circuit c computing $b \in \mathcal{B}_n$ is said to *depend* on m bits if there is a fixed $I \subseteq \{1, 2, \ldots, n\}$, $|I| = m$, so that for all x, $|x| = n$, $c(x)$ is uniquely determined by x_I. Notice that a circuit c can be evaluated on input x by an algorithm whose running time is polynomial in the size of c by simply traversing c's gates.

By NC^0 we mean the set of $b \in \mathcal{B}$ so that for some $c, d, k \in O(1)$, for all n and $x \in \{0, 1\}^n$, $b(x)$ is computable by a circuit with size, depth, and fan-in bounded by n^c, d, and k respectively. AC^0 is defined similarly but without the restriction on the fan-in.

An *ensemble of circuits* is a sequence, $\mathfrak{C} = \{\mathfrak{C}_n\}_{n \geq 1}$, where each $\mathfrak{C}_n$ is a probability distribution on circuits computing functions in $\mathcal{B}_n$. If there is a probabilistic polynomial time Turing machine (pptm) that on input 1^n outputs a c according to $\mathfrak{C}_n$, we shall say that we have a *polynomial ensemble of circuits*. An ensemble of *functions*, $\mathfrak{F} = \{\mathfrak{F}_n\}_{n \geq 1}$, is defined analogously, but with each $\mathfrak{F}_n$ being a distribution on functions mapping $\{0, 1\}^n \mapsto \{0, 1\}^*$. An ensemble of circuits, $\{\mathfrak{C}_n\}_{n \geq 1}$, is said to be $(s(n), d(n), k(n))$-*bounded* if for all n, $\mathfrak{C}_n$ has support only on circuits c with $\text{size}(c) \leq s(n)$, $\text{depth}(c) \leq d(n)$, and fan-in bounded by $k(n)$. If one of the three parameters, e.g. the fan-in, is unbounded we shall omit it and write $(s(n), d(n), \cdot)$-bounded etc.

A function $\xi(n)$ is *negligible* if for every constant $a > 0$ and for every sufficiently large n, $\xi(n) < n^{-a}$. A *one-way function* is a deterministic poly-time computable function f such that for every pptm, M, the probability that $M(f(x)) \in f^{-1}(x)$ is negligible. The probability is taken over $x \in_{\mathcal{U}} \{0, 1\}^n$ and M's random choices. Referring to a simple padding argument, we shall assume that all one-way functions are *length-preserving*, $|f(x)| = |x|$.

Let $\mathfrak{C} = \{\mathfrak{C}_n\}_{n \geq 1}$ be a polynomial ensemble of circuits and let f be a one-way function. An $\varepsilon(n)$-*adversary* for $\mathfrak{C}$ is a pptm A such that $\Pr[A(f(x), c) = c(x)] \geq 1/2 + \varepsilon(n)$, the probability taken over $x \in_{\mathcal{U}} \{0, 1\}^n$, c chosen according to $\mathfrak{C}_n$, and A's random choices. We call $\mathfrak{C}$ a *hard core predicate for* f if no $\varepsilon(n)$-adversary exists for non-negligible $\varepsilon(n)$. Normally, $\mathfrak{C}_n$ is the uniform distribution on some set of circuits, but we shall here allow other distributions. If $\mathfrak{C}$ is a hard core predicate for *any* one-way function, we simply call $\mathfrak{C}$ a (general) hard core predicate.

2.1 General Proof Outline

Assume that we have a one-way function[1] f of the form $f(x) = g(x_I) \circ x_{\bar{I}}$ where $I \subset \{1, 2, \ldots, n\}$ and where g is another one-way function. In other words, f is defined by applying g to a part of x and output the rest of x unchanged. (It may be the case that g itself outputs some bits unchanged, but we shall see that this can only help us.)

Suppose now that we are the adversary A. Given $f(x)$ and a circuit c, we want to compute $c(x)$. How would we go about this? Since we know the bits in $x_{\bar{I}}$, a natural approach would be to try to make a partial evaluation of c using only these bits. If, for instance, we know that one of the inputs to an AND-gate in c is a zero, we can simplify the circuit by deleting this gate and replacing it by the constant zero and so on. If we are able to make enough simplifications from the information in $x_{\bar{I}}$, the circuit

---

[1] Without this assumption, the notion of hard core predicate is, of course, meaningless.

4

will be a constant, determined by $x_{\bar{I}}$, and independent of x_I. It is not clear how to do this simplification/evaluation in polynomial time, nor is it clear how to tell *if* the circuit indeed is independent of x_I. However, if it *almost always* is the case that c "collapses" in this way, we can always act as if the value x_I *is* unimportant, substitute an arbitrary value z for x_I, and then evaluate the circuit using $z, x_{\bar{I}}$. In the case where c doesn't depend on x_I, this strategy will give a correct value for $c(x)$. Also, this is only a simple evaluation, and can be done in time polynomial in $\text{size}(c)$. (This is polynomial in A's input, $(f(x), c)$, but it is not polynomial in $n = |x|$ unless $\text{size}(c)$ is. We point out this difference since we shall include larger circuits later in our study.)

The hardness of inverting f is now reduced to the hardness of inverting g and the length of the argument of g (g's security parameter) is decreased. Hence, we lower the security of f correspondingly. As long as this length reduction is within a polynomial factor though, this is, at least from a theoretical standpoint, of no importance.

If we for the moment accept this idea, there remains one big concern. How should we choose the set I that g is applied to? Surely, we cannot hope that a fixed I will work as it seems likely that we could find a circuit that only uses the bits in x_I that are hidden to us and thus the circuit output would be unpredictable. We should therefore use a random I each time. This randomness must be taken somewhere and there are two ways of doing this; either we "hardwire" the randomness into f and we have a non-uniform construction or, to get uniformity, we "borrow" randomness from x since x is assumed to be random. This second approach can be realized as follows. Writing x as $x = x' \circ x''$ (we shall determine the lengths of x', x'' later), we now interpret (in some way) x' as an encoding of a subset I of the bits in x''. We then compute f as $f(x) = f(x' \circ x'') = g(x''_I) \circ x''_{\bar{I}} \circ x'$. Since all information on I is available in x' which is supplied to the circuit c, we must choose this encoding carefully to avoid c "figuring out" which bits it should use, namely those in I, hidden by g.

These are the main ingredients and the bulk of the paper basically concerns three things. 1. Find an encoding that circumvents the problem just mentioned. 2. Quantify how many bits in x we (the adversary) will need to know (the size of I). 3. Analyze how likely it is that the circuit indeed "collapses" given the bits in I.

3 Random Restrictions

The notion of "knowing" bits in x is formalized by *random restrictions*, introduced in [4].

Definition 1. A *restriction* is a partial assignment to the inputs of a circuit c, assigned inputs are given values in $\{0,1\}$ and the rest are assigned the symbol $*$ to denote that they remain variables. By $R^{(n,p)}$ we mean the set of restrictions assigning $*$ to some pn-subset of $x_1, \ldots, x_n$ and values in $\{0,1\}$ to the other $n(1-p)$ x_is. A *random restriction* in $R^{(n,p)}$ then, assigns $*$ to a random pn-subset of the inputs and values in $\{0,1\}$ independently, and with equal probability, to the other $n(1-p)$ inputs.

For a circuit c and a restriction ρ, $c \restriction_\rho$ denotes the circuit computing the restricted function of the remaining $*$ after ρ is applied. For $\rho \in R^{(n,p)}$, let $*(\rho)$ be the np-subset (of indices) that is assigned $*$ by ρ. Since $*(\rho)$ and the assignment to the other bits in x uniquely determines ρ, we can by setting $I = *(\rho)$ specify ρ by the notation $[I; x_{\bar{I}}]$.

3.1 Encoding Restrictions as Integers

Consider restrictions in $R^{(n,p)}$. We will encode these as integers. For a $\rho = [*(\rho); z]$, it is trivial to encode z as a binary string, so the only possible problem is how to encode the set $*(\rho)$. We show how to do this. Note that there are $\binom{n}{np}$ possibilities for $*(\rho)$.

Lemma 2. Let $u(n) = \binom{n}{np}$. For every $v \geq 0$ there is a polynomial time computable surjective function

$$Q_v : \{0,1\}^{\lceil \log u(n) \rceil + v} \mapsto J = \{I \mid I \subset \{1, 2, \ldots, n\}, |I| = np\}$$

such that for $h \in_u \{0,1\}^{\lceil \log u(n) \rceil + v}$, for every $I \in J$:

$$\left(1 - \frac{1}{2^v}\right) \frac{1}{u(n)} \leq \Pr_h[Q_v(h) = I] \leq \left(1 + \frac{1}{2^v}\right) \frac{1}{u(n)}.$$

The proof is straightforward and therefore omitted. The main idea is to interpret the integer $q = h \mod u(n)$ as "the lexicographically qth np-subset".

As noted, there could still be a problem with how we perform the encoding, since the circuits could gain information on the restriction. We will take care of this when computing the value h that Q_v is applied to and we return to this later. In the remainder of this paper we abuse notation slightly and refer to the value h as coding a restriction rather than $Q_v(h)$. As long as h is uniformly distributed in $\{0,1\}^{\lceil \log u(n) \rceil + v}$, this is by the Lemma above basically the same thing.

4 There are no Hard Core Predicates in NC^0

The situation for NC^0 circuits is quite simple. Since such circuits have fan-in bounded by $k \in O(1)$, they can depend on only $k^{\mathrm{depth}(c)} \in O(1)$ of the inputs x_i. The proofs in this case are simple combinatorial arguments.

Proposition 3. Let c be a circuit of depth d, fan-in k, $d, k \in O(1)$, computing some function $b \in \mathcal{B}_n$ and let $\rho \in_u R^{(n,p)}$, $p \leq k^{-d}$. Then

$$\Pr_{*(\rho)}[c \upharpoonright_\rho \text{ is a constant function}] \geq 1 - k^d p.$$

Furthermore, if this is the case then given ρ we can deterministically in time polynomial in $\mathrm{size}(c)$ decide what this constant is.

Proof. Since c can depend on at most k^d of its inputs, the probability that c depends on an input x_i such that $i \in *(\rho)$ is at most $k^d \frac{np}{n}$.

If c in this way collapses under ρ, we can simply evaluate the circuit by assigning an arbitrary (even fixed) value to the x_is for $i \in *(\rho)$ since c does not depend on these. $\square$

4.1 Non-uniform Case

Theorem 4. *For any* $\delta \in (0,1)$ *there is a polynomial ensemble of one-way functions* $\mathfrak{F}(\delta) = \{\mathfrak{F}_n\}_{n \geq 1}$ *and a deterministic polynomial time algorithm A such that for all constants* d,k, *for any* $(\cdot,d,k)$*-bounded ensemble of circuits* $\{\mathfrak{C}_n\}_{n \geq 1}$, *for every c supported by* $\mathfrak{C}_n$, *and for all* $x \in \{0,1\}^n$,

$$\Pr_{f \in \mathfrak{F}_n}[A(f(x),c) = c(x)] \geq 1 - O\left(n^{-(1-\delta)}\right).$$

Proof. Let g be a one-way function and let $\mathfrak{F}_n$ be the uniform distribution on the following set of one-way functions:

$$\{f_I(x) = g(x_I) \circ x_{\bar{I}} \circ I \mid I \subseteq \{1,2,\ldots,n\}, |I| = n^\delta\}.$$

For any c chosen according to $\mathfrak{C}_n$ and any x, having a value of the form $f_I(x)$ for random I, corresponds in a natural way to having a restriction $\rho = [I; x_{\bar{I}}]$ in $R^{(n,n^{-(1-\delta)})}$ on the input of c. The result now follows directly from Proposition 3. $\qquad \square$

By standard probabilistic arguments we get the Corollary below.

Corollary 5. *For all constants* d,k,δ, $\delta \in (0,1)$, *for any* $(\cdot,d,k)$*-bounded ensemble of circuits* $\{\mathfrak{C}_n\}_{n \geq 1}$, *there is a non-uniform one-way function f and a deterministic polynomial time algorithm A so that for all* $x \in \{0,1\}^n$,

$$\Pr_{c \in \mathfrak{C}_n}[A(f(x),c) = c(x)] \geq 1 - O\left(n^{-(1-\delta)}\right).$$

4.2 Uniform Case

In the construction of the one-way functions $\{f_I\}$ in the previous subsection we used extra randomness when selecting I. To get a uniform result we must somehow eliminate this. As mentioned in the outline, we cannot use a fixed subset I.

The idea is that since x, the argument of $f(x)$, is supposed to be a random string, we will "borrow" a few random bits from x itself to "point out" which subset I to use when computing $g(x_I)$ (and thus also which subset to output unaffected). We will therefore need a mapping from, say the first $l(n)$ bits of x to the set of all n^δ-subsets of $\{l(n)+1, l(n)+2, \ldots, n\}$. If we split x as $x = x' \circ x''$, we would like to interpret x' as an encoding of a random subset of the bits in x''. But we know how to do this from Lemma 2. We need roughly $l(n) = \lceil \log \binom{n}{n^\delta} \rceil \leq n^\delta \log n$ bits to encode all n^δ-subsets. To be more precise we should have $l(n) = |x'| = \lceil \log \binom{|x''|}{n^\delta} \rceil$, and since $|x''| = |x| - |x'| = n - l(n)$, $l(n)$ should in fact satisfy the equation $l(n) = \lceil \log \binom{n-l(n)}{n^\delta} \rceil$. Instead of solving this equation we can cheat slightly and simply choose $l(n)$ "large enough". This also means that we will use a v-value greater than zero when referring to Lemma 2 and this will give a more uniform distribution on the restrictions.

We must also be slightly careful, since the subset we are to compute g on is now supplied to the circuit c via x', and c might use that information to correlate itself to that subset and maybe even become dependent on some of the bits hidden by g. To avoid this we will use a slightly more elaborate encoding of the n^δ-subsets as described in the proof below.

Theorem 6. *For any* $\delta \in (0,1)$ *there is a one-way function f and a deterministic poly-nomial time algorithm A such that for all constants d,k, for any $(\cdot,d,k)$-bounded ensemble of circuits $\{\mathfrak{C}_n\}_{n\geq 1}$, for all sufficiently large n, for every c supported by $\mathfrak{C}_n$,*

$$\Pr_{x \in_\mathcal{U} \{0,1\}^n}[A(f(x),c) = c(x)] \geq 1 - O\left(n^{-(1-\delta)}\right).$$

Consequently, there are no hard core predicates in NC^0.

Proof: Let $0 < \tau < 1 - \delta$, let g be a one-way function and write $x = x' \circ x''$ where x' is the first $l(n) = n^\tau n^\delta \log n$ bits of x and x'' the last $n - l(n)$ bits. Define

$$h^{(i)}(x') = \sum_{j=(i-1)n^\tau+1}^{in^\tau} x'_j \bmod 2$$

(the exclusive-or over the ith n^τ-bit segment of x') and set

$$h(x') = h^{(1)}(x') \circ h^{(2)}(x') \circ \cdots \circ h^{(n^\delta \log n)}(x').$$

Observe that $h(x')$ is an $n^\delta \log n$-bit string where, for sufficiently large n, each individual bit is totally random to the circuits we are considering. This holds since each bit is an exclusive-or of n^τ bits, and the circuits we study can depend on no more than $k^d \in O(1)$ bits. Hence, if x and therefore x' is random, any such circuit is completely uncorrelated with $h(x')$. Now use $h = h(x')$ as described in Lemma 2 to encode an n^δ-subset of x''. (To simplify notation, we abuse it slightly by writing $h(x')$ rather than $Q_v(h(x'))$.) The encoding can be viewed as in the figure above. Now define the one-way function

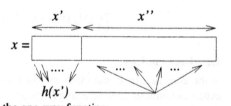

$$f(x) = f(x' \circ x'') = g(x''_{h(x')}) \circ x''_{\overline{h(x')}} \circ x'.$$

For any c from $\mathfrak{C}_n$, given $f(x)$, $c(x)$ is now equivalent to a circuit $c'(x'') = c(x' \circ x'')$ upon which we have an (almost) random restriction $\rho = [h(x');x''_{\overline{h(x')}}]$ in $R^{(n-l(n),p)}$, $p = n^\delta/(n-l(n)) = (n^{1-\delta} - n^\tau \log n)^{-1}$. The distribution on the restrictions is not *exactly* the uniform distribution, but by Lemma 2, no "bad" restriction (one that does not force c' to a constant) is chosen by more than twice the probability it is chosen by the uniform distribution. (And this can also be made arbitrarily close to uniform distribution by Lemma 2.) Hence, by Proposition 3 the output of this circuit is completely determined by ρ with probability at least $1 - 2k^d(n^{1-\delta} - n^\tau \log n)^{-1}$, and if so, we can determine the output in polynomial time. $\qquad\Box$

5   There are no Hard Core Predicates in AC^0

For AC^0 we cannot use the same simple counting arguments since these circuits may very well depend on all n bits in x. We therefore need some more powerful tools from the theory of circuit complexity.

5.1 The Switching Lemma

The Håstad switching lemma, see [6], quantifies how much and how likely it is that a circuit is simplified under a restriction. Similar results are known from [4] and [14]. We have the following version of the switching lemma, derived from [2].

Lemma 7 (The Switching Lemma). *Let G be an AND-gate whose inputs are OR-gates all of fan-in at most r and let $\rho \in_{\mathcal{U}} R^{(n,p)}$, where $p \leq 1/7$. Then the probability that $G \upharpoonright_{\rho}$ can be written as an OR of ANDs, each AND having fan-in strictly less than t is at least $1 - (7pr)^t$.*

A dual lemma holds by replacing AND by OR and vice versa. We can prove the following powerful result.

Lemma 8. *For any $\delta \in (0,1)$, for all sufficiently large n the following holds. If c is a circuit of depth $d(n) \leq \frac{\log n}{\log \log n}$ and size $s(n) \leq n^{-(1-\delta)/d(n)} 2^{\frac{1}{14} n^{(1-\delta)/d(n)}}$, computing some function in $\mathcal{B}_n$, then with $p = n^{-(1-\delta)}$,*

$$\Pr_{\rho \in_{\mathcal{U}} R^{(n,p)}} [c \upharpoonright_{\rho} \text{ is a constant function}] \geq 1 - 8n^{-(1-\delta)/d(n)}.$$

If this is the case, we can given ρ find this constant in time polynomial in $\text{size}(c)$.

Proof. Let c be a circuit as mentioned in the lemma. We will choose $\rho \in R^{(n,p)}$ in $d(n)$ steps, each step consisting of picking a random restriction ρ_i on the remaining unset inputs. Our ρ will be the composition of all these restrictions. Let $t = \frac{1}{14} n^{(1-\delta)/d(n)}$.

The first step is needed to get fan-in at most t at level 1. Assume it consists of AND-gates. For the purposes of the switching lemma, we view these gates as ANDs of ORs where each OR has fan-in 1 (a variable x_i or its negation). We pick a random restriction from $R^{(n,p_0)}$, where $p_0 = 1/14$. By the Switching Lemma we know that each AND of fan-in-1 ORs can be replaced by an OR of ANDs of fan-in t with probability at least $1 - (7p_0)^t = 1 - 2^{-t}$. The OR can now be "collapsed" into the level above, as that level contains OR-gates.

In steps 2 through $d(n) - 1$ we reduce the circuit by applying the switching lemma to the bottom two levels of the circuit, switch ANDs of ORs to ORs of ANDs (or vice versa) and collapsing adjacent levels of OR-gates (or AND-gates) maintaining the bound t on the bottom fan-in. This is done as follows.

Let $p_1 = (14t)^{-1}$, and let $n_i = p_0(p_1)^{i-1} n$. At step i we pick a random restriction $\rho_i \in_{\mathcal{U}} R^{(n_{i-1}, p_1)}$, where the domain of ρ_i is the input variables that have not been set by ρ_1 through ρ_{i-1}. Notice that after step i there are n_i variables that remain unset.

For every AND of ORs (or OR of ANDs) that we consider, the probability that the restriction doesn't allow us to switch is at most $(7p_1t)^t = 2^{-t}$. Over steps 1 through $d(n) - 1$ we invoke the Switching Lemma once for each gate in the circuit (except the top gate), and each time the probability of failure is at most 2^{-t}. So with probability at least $1 - s(n)2^{-t}$ the entire circuit has been collapsed to a single AND of ORs of fan-in $\leq t$ (or to an OR of ANDs of fan-in $\leq t$), and there are still $n_{d(n)-1} = p_0(p_1)^{d(n)-2} n$ variables unset.

Finally, in step $d(n)$ we do as follows. Assume we have been successful in steps 1 through $d(n) - 1$, and that we are left with an AND of ORs, where each OR has fan-in at most t. Let $p_2 = (14t^2)^{-1}$ and pick $\rho_d \in_\mathcal{U} R^{(n_{d(n)-1}, p_2)}$. By the Switching Lemma, the probability that the AND of ORs can be written as an OR of ANDs, each AND of fan-in strictly less than 1 (and must thus be a constant) is at least $1 - (7p_2t) = 1 - (2t)^{-1}$.

The probability that *all* the ρ_i are successful is at least $1 - s(n)2^{-t} - (2t)^{-1} \geq 1 - 8n^{-(1-\delta)/d(n)}$. Notice also that $p = p_{d(n)} = p_0(p_1)^{d(n)-2}p_2 = n^{-(1-\delta)}$.

Finally, to find the constant we substitute arbitrary values for x_i, $i \in *(\rho)$, and evaluate the circuit like before. □

Notice that for the circuit depths covered by the Lemma, the failure probability, $8n^{-(1-\delta)/d(n)} \in o(1)$. Hence, almost surely, for such circuits, $c \restriction_\rho$ *will be a constant.*

5.2 Non-uniform Case

Theorem 9. *For any* $\delta \in (0,1)$ *there is a polynomial ensemble of one-way functions* $\{\mathfrak{F}_n\}_{n\geq 1}$, *and a deterministic polynomial time algorithm A for which the following hold. For any* $(s(n), d(n), \cdot)$-*bounded ensemble of circuits* $\{\mathfrak{C}_n\}_{n\geq 1}$, *where* $d(n) \leq \frac{\log n}{\log\log n}$ *and* $s(n) \leq n^{-(1-\delta)/d(n)}2^{\frac{1}{14}n^{(1-\delta)/d(n)}}$, *for all sufficiently large n, for every c supported by* $\mathfrak{C}_n$,

$$\Pr[A(f(x), c) = c(x)] = 1 - O\left(n^{-(1-\delta)/d(n)}\right),$$

the probability taken over $x \in_\mathcal{U} \{0,1\}^n$, *and f chosen according to* $\mathfrak{F}_n$.

Proof. Let $p = n^{-(1-\delta)}$, assume that g is a one-way function and let $\mathfrak{F}_n$ be the uniform distribution on the one-way functions

$$\{f_I(x) = g(x_I) \circ x_{\bar{I}} \circ I \mid I \subset \{1, 2, \ldots, n\}, |I| = np\}.$$

Note that random for random x and I, $f_I(x)$ corresponds to the random restriction $\rho = [I; x_{\bar{I}}] \in R^{(n,p)}$ on the input of c.

The result now follows, since for any c chosen from $\mathfrak{C}_n$, the probability that $c \restriction_\rho$ is a constant is by Lemma 8 at least $1 - 8n^{-(1-\delta)/d(n)}$ and this constant can be found in polynomial time using $f_I(x)$. □

Again, by standard "averaging" arguments we have as an immediate Corollary:

Corollary 10. *Let* $\delta \in (0,1)$ *and let* $\{\mathfrak{C}_n\}_{n\geq 1}$ *be an* $(s(n), d(n), \cdot)$-*bounded ensemble of circuits where* $d(n) \leq \frac{\log n}{\log\log n}$ *and* $s(n) \leq n^{-(1-\delta)/d(n)}2^{\frac{1}{14}n^{(1-\delta)/d(n)}}$. *Then, there is a non-uniform one-way function f and deterministic polynomial time algorithm A such that for all sufficiently large n,*

$$\Pr[A(f(x), c) = c(x)] = 1 - O\left(n^{-(1-\delta)/d(n)}\right),$$

the probability taken over $x \in_\mathcal{U} \{0,1\}^n$, *and c chosen according to* $\mathfrak{C}_n$.

5.3 Uniform Case

In the bounded fan-in case we could derandomize our proofs by encoding the restriction as a part of x, the argument to the one-way function. We had to choose this encoding so that the circuit was completely uncorrelated with the restriction and this could be done by observing that bounded fan-in circuits cannot "see" all bits in x. For unbounded fan-in circuits however, the situation is more difficult, since theoretically at least, the circuit can have full information on the restriction. We will still use the same principal encoding of the restrictions, but we have to be more careful in the analysis.

We will now consider restrictions in $R^{(n,n^{-(1-\varepsilon)})}$, i.e. leaving $n^\varepsilon *$ for some $\varepsilon > 0$. We will encode them lexicographically like before. Let $x = x' \circ x''$ where x' is the first $L(n) = n^\alpha n^\varepsilon \log n$ bits in x and x'' is the $n - L(n)$ last bits. The constants ε and α will be determined later. Now let

$$H_\alpha^{(i)}(x') = \sum_{j=(i-1)n^\alpha+1}^{in^\alpha} x'_j \mod 2,$$

i.e. the XOR over the ith n^α-bit segment of x', and let

$$H_{\alpha,\varepsilon}(x') = H_\alpha^{(1)}(x') \circ H_\alpha^{(2)}(x') \circ \cdots \circ H_\alpha^{(n^\varepsilon \log n)}(x')$$

which we by setting $h = H_{\alpha,\varepsilon}(x')$ like in Lemma 2 interpret as an encoding of a restriction on the bits in x''. (We simplify, writing $H_{\alpha,\varepsilon}(x')$ instead of $Q_v(H_{\alpha,\varepsilon}(x'))$.)

We now get what in a natural way corresponds to restrictions on $x = x' \circ x''$ of the form $\rho = [H_{\alpha,\varepsilon}(x'); x' \circ x''_{\overline{H_{\alpha,\varepsilon}(x')}}]$. What we would like to do is to analyze the probability that a circuit collapses when subjected to such a restriction. However, we now clearly do not have the uniform distribution on restrictions, in particular we have all $*$ concentrated to the x''-part of x. The simple combinatorial arguments applicable to NC^0-circuits cannot be applied here. We must make a closer analysis of the induced distribution on the restrictions to be able to apply the switching lemma.

Below we describe three distributions $\mathcal{D}_1(\alpha)$, $\mathcal{D}_2(\alpha)$, and $\mathcal{D}_3(p,\alpha)$ on $R^{(n,n^{-(1-\varepsilon)})}$. The plan is to show that for suitable choices of p, α, ε,

(A) A random restriction from $\mathcal{D}_3(p,\alpha)$ will collapse our circuits with probability close to 1. (This will be Lemma 13.)

(B) $\mathcal{D}_3(p,\alpha)$ is equal to $\mathcal{D}_2(\alpha)$. (See Lemma 14.)

(C) $\mathcal{D}_2(\alpha)$ is very close to $\mathcal{D}_1(\alpha)$. (Lemma 15.)

The distribution $\mathcal{D}_1(\alpha)$ will be the one we actually have, and $\mathcal{D}_3(p,\alpha)$ is the one we will analyze. Where will this lead? Like before, we shall construct a one way function,

$$f(x) = f(x' \circ x'') = g(x''_{H_{\alpha,\varepsilon}(x')}) \circ x''_{\overline{H_{\alpha,\varepsilon}(x')}} \circ x'$$

(where again, g is another one-way function). For $x \in_{\mathcal{U}} \{0,1\}^n$, $f(x)$ will correspond in a natural way to a random restriction $\rho \in_{\mathcal{D}_1(\alpha)} R^{(n,n^{-(1-\varepsilon)})}$. Hence, using our previous strategy for evaluating circuits under restrictions (substituting arbitrary values for unknown inputs), we have an algorithm A such that by (A), (B), and (C) above,

$$\Pr_{x}[A(f(x),c) = c(x)] \geq \Pr_{\mathcal{D}_1(\alpha)}[c \lceil_\rho \text{ is a constant}]$$

$$\approx \Pr_{\mathcal{D}_3(p.\alpha)}[c \lceil_\rho \text{ is a constant}] \quad \text{(by (B),(C))}$$

$$= 1 - o(1) \quad \text{(by (A))}.$$

With this program in mind, we now define the distributions.

DISTRIBUTION $\mathcal{D}_1(\alpha)$
1. Choose x' uniformly at random in $\{0,1\}^{L(n)}$, and set $I = H_{\alpha,\varepsilon}(x')$.
2. Assign 0/1 with equal probability to the bits in $x''_{\bar{I}}$.
3. Assign $*$ to all of x''_I.

Let $\rho = [I; x' \circ x''_{\bar{I}}] \in R^{(n, n^{-(1-\varepsilon)})}$ be the induced restriction on x.

DISTRIBUTION $\mathcal{D}_2(\alpha)$
1. Choose I, a random n^ε-subset of the bits in x'' and assign $*$ to the bits in x''_I.
2. Assign 0/1 with equal probability to the bits in $x''_{\bar{I}}$
3. Choose x' uniformly at random in $H_{\alpha,\varepsilon}^{-1}(I)$.

Let $\rho = [I; x' \circ x''_{\bar{I}}] \in R^{(n, n^{-(1-\varepsilon)})}$ be the induced restriction on x.

The difference between $\mathcal{D}_1(\alpha)$ and $\mathcal{D}_2(\alpha)$ is that in $\mathcal{D}_1(\alpha)$ we choose the argument of $H_{\alpha,\varepsilon}$ and compute I from this, in $\mathcal{D}_2(\alpha)$ we reverse the procedure. Intuitively, since the "hash-function" $H_{\alpha,\varepsilon}$ is well behaved, it should not matter too much in which order we do these operations.

Finally we define a last distribution, the one that we will analyze. This last distribution is constructed in four steps where we first apply a random restriction from $R^{(n,p)}$ to all of x, i.e. *both* to x' and x''. For $\rho \in R^{(n,p)}$, write $\rho = \rho' \circ \rho''$ where ρ' and ρ'' are the parts of ρ assigning values to x' and x'' respectively.

DISTRIBUTION $\mathcal{D}_3(p,\alpha)$
1. Fix some bits in x', x'' by choosing a random $\rho = \rho' \circ \rho'' \in_{\mathcal{U}} R^{(n,p)}$, such that
   (a) $|*(\rho'')| \geq n^\varepsilon$ and such that
   (b) for $i = 1, \ldots, n^\varepsilon \log n$, $H_\alpha^{(i)} \lceil_\rho = *$. (I.e. each $H_\alpha^{(i)}(x')$ is undetermined by the restriction ρ' on x'.)
2. Choose I, a random n^ε-subset of $*(\rho'')$ and let these remain as $*$ in x''.
3. Assign 0/1 with equal probability to the remaining $*$ in $x''_{\bar{I}}$.
4. Assign 0/1 with equal probability to the remaining $*$ in x', but assert that $x' \in H_{\alpha,\varepsilon}^{-1}(I)$.

Let $\rho = [I; x' \circ x''_{\bar{I}}] \in R^{(n, n^{-(1-\varepsilon)})}$ be the induced restriction on x.

Let us give some motivation for studying this distribution. In step 1, we will fix some of the bits in x', but by condition 1b, they are not many enough to determine any of the $H_\alpha^{(i)}$ components of $H_{\alpha,\varepsilon}$ at all. To be able to later fix x' so that $H_{\alpha,\varepsilon}(x')$ can take any n^ε-subset of the indices in x'' as a value, we need at least n^ε $*$ in the x''-part and this is asserted by condition 1a. Therefore, after step 1, all possibilities for I (determined in step 2) are still at this stage equally likely. The point is now that *if* our circuit c has collapsed after step 1, it will have done so without having any chance of obtaining information concerning I. Later steps, 2, 3 and 4, will only decrease the number of $*$ and thus c will remain collapsed. There is some hope to apply Lemma 8 to the restriction we have at step 1 *if* we can show that it "almost" random. Finally, steps 2, 3, and 4 will assert that the final restriction is consistent with distribution $\mathcal{D}_1(\alpha)$.

We start by showing that the restriction obtained after step 1 above is "just as good" as a uniformly distributed restriction. We first need two preparatory propositions.

Proposition 11. *Let* $p = n^{-(1-\delta)}$, $\rho \in R^{(n,p)}$. *For any* $J \subset \{1,2,\dots,n\}$, $|J| = n^{\alpha}$ *with* $1 - \delta < \alpha < 1$, *then*

$$\Pr_{\rho \in \mathcal{U} R^{(n,p)}} [|*(\rho) \cap J| = 0] \leq e^{-pn^{\alpha}}$$

for all sufficiently large n.

We omit the elementary proof.

Proposition 12. *If* $I \subset \{1,2,\dots,n\}$, $|I| = n - n^{\alpha+\varepsilon}\log n$, $0 < \varepsilon < \delta < 1$, $\alpha + \varepsilon < 1$, *and* $p = n^{-(1-\delta)}$, *then for all sufficiently large n,*

$$\Pr_{\rho \in \mathcal{U} R^{(n,p)}} [|*(\rho) \cap I| < n^{\varepsilon}] \leq e^{-\frac{1}{2}(n^{\delta-\varepsilon}-1)}$$

Proof. The proposition follows from simple combinatorial arguments, approximating the binomial coefficients involved by Stirling's formula. $\square$

Lemma 13. *Assume* $0 < \varepsilon < \delta < 1$, $\alpha \in (1-\delta, 1-\varepsilon)$, *and let c be a circuit of depth* $d(n) \leq \frac{\log n}{\log\log n}$ *and size* $s(n) \leq n^{-(1-\delta)/d(n)}2^{\frac{1}{14}n^{(1-\delta)/d(n)}}$. *Then with* $p = n^{-(1-\delta)}$, *for all sufficiently large n,*

$$\Pr_{\rho \in \mathcal{D}_3(p,\alpha) R^{(n,n^{-(1-\varepsilon)})}} [c\restriction_{\rho} \text{ is a constant function}] \geq 1 - O\left(n^{-(1-\delta)/d(n)}\right),$$

and if so, given ρ, *this constant can be found in time polynomial in* $s(n)$.

Proof. By Lemma 8, if we for the moment consider all of $R^{(n,p)}$ (i.e. regardless of whether or not it passes the constraints in step 1 in the definition of $\mathcal{D}_3(p,\alpha)$), we have

$$\Pr_{\rho \in \mathcal{U} R^{(n,p)}} [c\restriction_{\rho} \text{ is a constant function}] \geq 1 - 8n^{-(1-\delta)/d(n)}.$$

Let us call a restriction in $R^{(n,p)}$ *bad* if it does not satisfy the additional constraints in step 1. Write $\rho = \rho' \circ \rho''$ as defined above. First we see that the probability that at least one of the $H_{\alpha}^{(i)}$s (there are $n^{\varepsilon}\log n$ of them) becomes determined by ρ' (a violation of constraint 1b) is for large n by Proposition 11 bounded by

$$(n^{\varepsilon}\log n)e^{-pn^{\alpha}} = (n^{\varepsilon}\log n)e^{-n^{\alpha-(1-\delta)}}$$

which is negligible since $\alpha > 1 - \delta$.

Furthermore, the probability that we get fewer than n^ε $*$ in ρ'' (violation of constraint 1a) is by Proposition 12 bounded by $e^{-\frac{1}{2}(n^{\delta-\varepsilon}-1)}$ since $\alpha + \varepsilon < 1$. Thus, for large n

$$\Pr_{\rho \in _{\mathcal{U}} R^{(n,p)}} [\rho \text{ is bad}] \leq (n^\varepsilon \log n) e^{-n^{\alpha-(1-\delta)}} + e^{-\frac{1}{2}(n^{\delta-\varepsilon}-1)}.$$

The lemma now follows since

$$\Pr_{\rho \in \mathcal{D}_3(p,\alpha) R^{(n,n^{-(1-\varepsilon)})}} [c\lceil_\rho \text{ is a constant}] \geq 1 - 8n^{-(1-\delta)/d(n)} - \Pr_{\rho \in _{\mathcal{U}} R^{(n,p)}} [\rho \text{ is bad}].$$

This is actually the probability the circuit has collapsed already after step 1 in forming $\mathcal{D}_3(p,\alpha)$, but as noted, the circuit will then surely remain collapsed. $\qquad\square$

Next, notice that the distributions $\mathcal{D}_3(p,\alpha), \mathcal{D}_2(\alpha)$ are the same and furthermore, that $\mathcal{D}_1(\alpha), \mathcal{D}_2(\alpha)$ are close.

Lemma 14. *With $p = n^{-(1-\delta)}$, $0 < \varepsilon < \delta < 1$, $\alpha \in (1-\delta, 1-\varepsilon)$, the two distributions $\mathcal{D}_3(p,\alpha), \mathcal{D}_2(\alpha)$ on $R^{(n,n^{-(1-\varepsilon)})}$ are equal.*

Proof. We prove that for both distributions: *(i)* The location of $*$ in ρ'', i.e. the set I, have the same distribution. *(ii)* Non-$*$ in ρ'' are uniformly distributed in $\{0,1\}$. *(iii)* The ρ'-part is a uniformly distributed value consistent with $H_{\alpha,\varepsilon}(x') = I$. This will establish the claim.

(i) In $\mathcal{D}_2(\alpha)$ we first choose a random n^ε-subset as $*(\rho'')$ so each such is chosen with probability $\binom{|x''|}{n^\varepsilon}^{-1}$. In $\mathcal{D}_3(p,\alpha)$ after step 1 in forming the distribution, let R_1 be the random variable corresponding to $*(\rho'')$ at this stage and let R_2 similarly be the value of $*(\rho'')$ after step 2. Since the final $*(\rho'')$ is obtained by first choosing R_1 at random and then R_2 as a subset of R_1, we have by symmetry that for any n^ε-subset I:

$$\Pr_{\rho \in \mathcal{D}_3(p,\alpha) R^{(n,n^{-(1-\varepsilon)})}} [*(\rho'') = I] = \binom{|x''|}{n^\varepsilon}^{-1} \Pr[|R_1| \geq n^\varepsilon].$$

But in $\mathcal{D}_3(p,\alpha)$ we discard precisely those ρ for which $|R_1| < n^\varepsilon$, and thus each n^ε-set is chosen with the same probability as by $\mathcal{D}_2(\alpha)$.

(ii) Note that there is no difference in the distribution on non-$*$ in ρ'' in the two distributions since they are in both cases assigned 0/1 with equal probability.

(iii) Lastly, when assigning 0/1 to ρ' in $\mathcal{D}_2(\alpha)$, we choose at random a $x' \in H_{\alpha,\varepsilon}^{-1}(I)$. By the XOR-construction of each $H_\alpha^{(i)}$, we can in each n^α-bit segment of x' choose any set of $n^\alpha - 1$ indices uniformly at random in $\{0,1\}$ and the last bit in each block will have to be assigned a unique value determined by the condition $H_{\alpha,\varepsilon}(x') = I$. In $\mathcal{D}_3(p,\alpha)$ we also choose I before choosing x'. Since we in $\mathcal{D}_3(p,\alpha)$ by constraint 1b have at least one $*$ in each n^α-bit segment of ρ' and all $*$ were selected uniformly at random, we can there also fix all but one of the remaining $*$ as 0/1 at random and the last bit is determined uniquely by $H_{\alpha,\varepsilon}(x') = I$. $\qquad\square$

Lemma 15. *For any fixed $\rho_0 \in R^{(n,n^{-(1-\varepsilon)})}$,*

$$2^{-1} \Pr_{\mathcal{D}_2(\alpha)} [\rho = \rho_0] \leq \Pr_{\mathcal{D}_1(\alpha)} [\rho = \rho_0] \leq 2 \Pr_{\mathcal{D}_2(\alpha)} [\rho = \rho_0].$$

Proof. Follows immediately from Lemma 2. □

Corollary 16. *Let $0 < \varepsilon < \delta < 1$, $\alpha \in (1 - \delta, 1 - \varepsilon)$, and let c be a circuit of depth $d(n) \leq \frac{\log n}{\log\log n}$ and size $s(n) \leq n^{-(1-\delta)/d(n)} 2^{\frac{1}{14} n^{(1-\delta)/d(n)}}$. Then, for sufficiently large n,*

$$\Pr_{\rho \in \mathcal{D}_1(\alpha) R^{(n,n^{-(1-\varepsilon)})}} [c \restriction_\rho \text{ is a constant function}] \geq 1 - O\left(n^{-(1-\delta)/d(n)}\right),$$

and if so, this constant can from ρ be computed in time polynomial in $s(n)$.

Proof. With $p = n^{-(1-\delta)}$, the result follows immediately from lemmas 13, 14, and 15. □

Theorem 17. *For any $\delta \in (0,1)$ there is a one-way function f and a deterministic polynomial time algorithm A for which the following hold. For all sufficiently large n, for any $(s(n), d(n), \cdot)$-bounded ensemble of circuits $\{\mathcal{C}_n\}_{n \geq 1}$, where $d(n) \leq \frac{\log n}{\log\log n}$, $s(n) \leq n^{-(1-\delta)/d(n)} 2^{\frac{1}{14} n^{(1-\delta)/d(n)}}$, for every c supported by $\mathcal{C}_n$,*

$$\Pr_{x \in \mathcal{U}\{0,1\}^n} [A(f(x), c) = c(x)] = 1 - O\left(n^{-(1-\delta)/d(n)}\right).$$

Proof. Choose $\varepsilon \in (0, \delta)$, $\alpha \in (1 - \delta, 1 - \varepsilon)$, let g be a one-way function and define the one-way function

$$f(x) = f(x' \circ x'') = g(x''_{H_{\alpha,\varepsilon}(x')}) \circ x''_{\overline{H_{\alpha,\varepsilon}(x')}} \circ x'.$$

For any $c \in \mathcal{C}_n$, for random x, $f(x)$ gives a restriction $\rho = [H_{\alpha,\varepsilon}(x'); x' \circ x''_{\overline{H_{\alpha,\varepsilon}(x')}}] \in R^{(n,n^{-(1-\varepsilon)})}$ on c, and this ρ is chosen according to the distribution $\mathcal{D}_1(\alpha)$. It follows from Corollary 16 that $c \restriction_\rho$ will be a constant with probability $1 - O(n^{-(1-\delta)/d(n)})$, and if this is the case, we can use ρ (i.e. $f(x)$) to determine what this constant is in time polynomial in size(c). □

We immediately get the following corollary.

Corollary 18. *An $(s(n), d(n), \cdot)$-bounded ensemble of circuits, computing a (general) hard core predicate, requires depth $d(n) \geq \frac{\log n}{(1+\varepsilon)\log\log n}$ for every $\varepsilon > 0$, or otherwise, requires size $s(n) \geq 2^{\omega(\log n)}$. If $d(n)$ is a constant, size $s(n) \geq 2^{n^{\Omega(1)}}$ is required. In particular, there are no hard core predicates computable in AC^0.*

The existing constructions of hard core predicates such as [5] can be computed by polynomial size circuits of depth $\frac{\log n}{\log\log n}$ and hence, the lower bound in the Corollary is essentially tight.

6 Summary and Open Problems

This paper does not rule out the possibility of generating pseudo-random sequences in AC^0, but it does tell us that "generic" constructions based on an arbitrary one-way function and a hard core predicate does not work. For instance, the construction in [8] *could* still be a pseudo-random generator since it is based on a *particular* conjectured one-way function.

We have found an essentially tight lower bound on the complexity of computing (general) hard core predicates. Note that with respect to AC^0 circuits, the results obtained are uniform in a very strong sense: We have a fixed one-way function that has no hard core predicates computable by any circuit family of constant depth d and size n^r regardless of d, r and the distribution on the circuits.

Together with existing constructions of hard core predicates, we now have a good characterization of them with respect to computational complexity. The next step would therefore be to give, if possible, a more functional characterization of them.

Acknowledgment. We would like to thank Johan Håstad for fruitful discussions and suggestions. We also thank Alex Russell and Jean-Pierre Seifert.

References

1. W. Alexi, B. Chor, O. Goldreich, and C. P. Schnorr: *RSA and Rabin Functions: Certain Parts Are as Hard as the Whole.* SIAM J. on Computing **17** (1988), no 2, pp. 194–209.
2. P. Beame: *A Switching Lemma Primer.* Manuscript, 1994.
3. M. Blum and S. Micali: *How to Generate Cryptographically Strong Sequences of Pseudo-random Bits.* SIAM J. on Computing **13** (1986), no 4, pp. 850–864.
4. M. Furst, J. Saxe, and M. Sipser: *Parity, Circuits, and the Polynomial Time Hierarchy.* Proc. 22nd Symposium on Foundations of Computer Science, IEEE, 1981, pp. 260–270.
5. O. Goldreich and L. A. Levin: *A Hard Core Predicate for all One Way Functions.* Proc. 21st Symposium on Theory of Computing, ACM, 1989, pp. 25–32.
6. J. Håstad: *Computational Limitations of Small-Depth Circuits.* ACM doctoral dissertation award, 1986. MIT Press 1987.
7. J. Håstad, A. W. Schrift, and A. Shamir: *The Discrete Logarithm Modulo a Composite Hides $O(n)$ Bits.* J. of Computer and System Sciences **47** (1993), pp. 376–403.
8. R. Impagliazzo and M. Naor: *Efficient Cryptographic Schemes Provably as Secure as Subset Sum.* J. of Cryptology **9** (1996), no 4, pp. 199–216.
9. N. Linial, Y. Mansour, and N. Nisan: *Constant Depth Circuits, Fourier Transform, and Learnability.* J. of the ACM **40** (1993), no 3, pp. 607–620.
10. Y. Mansour, N. Nisan, and P. Tiwari: *The Computational Complexity of Universal Hashing.* Theoretical Computer Science **107** (1993), pp. 121–133.
11. M. Näslund: *Universal Hash Functions & Hard Core Bits.* Proc. Eurocrypt 1995, LNCS 921, Springer Verlag, pp. 356–366.
12. M. Näslund: *All Bits in $ax + b$ mod p are Hard.* Proc. Crypto 1996, LNCS 1109, Springer Verlag, pp. 114–128.
13. A. C. Yao: *Theory and Applications of Trapdoor Functions.* Proc. 23rd Symposium on Foundations of Computer Science, IEEE, 1982, pp. 80–91.
14. A. C. Yao: *Separating the Polynomial-Time Hierarchy by Oracles.* Proc. 26th Symposium on Foundations of Computer Science, IEEE, 1985, pp. 1–10.

Statistical Zero Knowledge Protocols to Prove Modular Polynomial Relations

Eiichiro FUJISAKI and Tatsuaki OKAMOTO

NTT Laboratories,
1-1 Hikarinooka, Yokosuka-shi, 239 Japan
Email: {fujisaki, okamoto}@sucaba.isl.ntt.co.jp

Abstract. This paper proposes a bit commitment scheme, $BC(\cdot)$, and *efficient* statistical zero knowledge (in short, SZK) protocols in which, for any given multi-variable polynomial $f(X_1, .., X_t)$ and any given modulus n, prover $\mathcal{P}$ gives $(I_1, .., I_t)$ to verifier $\mathcal{V}$ and can convince $\mathcal{V}$ that $\mathcal{P}$ knows $(x_1, .., x_t)$ satisfying $f(x_1, .., x_t) \equiv 0 \pmod{n}$ and $I_i = BC(x_i)$, $(i = 1, .., t)$. The proposed protocols are $O(|n|)$ times more efficient than the corresponding previous ones [Dam93, Dam95, Oka95]. The (knowledge) soundness of our protocols holds under a computational assumption, the intractability of a modified RSA problem (see Def.3), while the (statistical) zero-knowledgeness of the protocols needs no computational assumption. The protocols can be employed to construct various practical cryptographic protocols, such as fair exchange, untraceable electronic cash and verifiable secret sharing protocols.

1 Introduction

1.1 Problem

In many cryptographic protocols, a party often wants to prove something related to his secret while concealing his secret from the others. Such relations are often specified by modular polynomials and bit commitments are very useful in such protocols. This paper focuses on the following problem: for given multi-variable polynomial $f(X_1, .., X_t)$ and modulus n, a party (prover) $\mathcal{P}$ gives $(I_1, .., I_t)$ to another party (verifier) $\mathcal{V}$ and convinces $\mathcal{V}$ that $\mathcal{P}$ knows $(x_1, .., x_t)$ satisfying $f(x_1, .., x_t) \equiv 0 \pmod{n}$ and $I_i = BC(x_i)$, $(i = 1, .., t)$, without revealing the values, $x_1, .., x_t$.

This problem is indeed raised on many cryptographic protocols. In fair exchange and contract signing protocols based on RSA signatures [Dam93, Dam95], (n, e) is the public-key of the RSA scheme, $f(x) = x^e - m$ and $I = BC(x)$. After proving that $\mathcal{P}$ knows x satisfying the relations, $\mathcal{P}$ releases x bit by bit using I. In untraceable off-line electronic cash protocols, restricted blind signatures [Bra95, Oka95] play an important role, where, for instance, $f(x_1, x_2, x_3) = (x_1 x_2)x_3^e$ and $I_1 = BC(x_1)$ and $I_2 = BC(x_2)$. After proving that $\mathcal{P}$ knows (x_1, x_2, x_3) satisfying the relations, $\mathcal{V}$ stores I_1 and I_2. If $\mathcal{P}$ double-spends a coin, $\mathcal{V}$ can get (x_1, x_2) from $x_1 x_2$ as evidence of double-spending (See [Oka95] for more details). If f is a polynomial and n is a prime for Shamir's secret sharing scheme, some protocols related to secret sharing such as (publicly) verifiable secret sharing [CGMA85, Ped91, Sta96] can be interpreted as this type of problem.

1.2 Previous Works

Theoretically, the general construction of protocols to solve the above-mentioned problem has been already given assuming that a secure bit commitment scheme exists. This is derived from the results of zero knowledge proof for NP-language [GMRa89, GMW86, BCC86] and converting them to proof of knowledge [FFS88, TW87, BG92]. Depending on the types of the underlying bit commitment schemes, there exist two different results: namely, computational ZK (CZK) for interactive proof (IP) and perfect ZK (PZK) for argument (computationally-sound proof). However, those protocols are very inefficient in general.

In 1993, Damgård proposed the first efficient protocol to solve the problem with a specific form for constructing a *fair exchange and contract signing* protocol [Dam93, Dam95]. He proposed the protocols in which prover $\mathcal{P}$ can convince verifier $\mathcal{V}$ that he knows s of bit commitment $BC(s)$ and that it is a Rabin signature ($s = m^{1/2} \bmod n$) or a RSA signature ($s = m^{1/e} \bmod n$), for a message m. The protocols are PZK computationally-sound proof of knowledge systems (PZK arguments of *knowledge*). Those protocols essentially consist of some primitives: a bit commitment scheme and three protocols, which correspond to the *basic, comparing,* and *mod-multi* protocols in this paper. His *basic* protocol is the protocol in which $\mathcal{P}$ proves to $\mathcal{V}$ that secret s is in a given range $[a, b]$ and the *comparing* and *mod-multi* protocols are compositions of the *basic* protocol. It is easy to construct a PZK argument of knowledge for any multi-variable modular polynomial (f, n) based on these primitives.

In 1995, Okamoto showed another application of the problem above. He constructed an RSA-type *restricted blind signature* for his *untraceable off-line electronic cash* [Oka95] by using similar primitives: a bit commitment scheme and three protocols, which are essentially equivalent to those of Damgård's except for the bit commitment scheme.

Unfortunately, both of their protocols are not so efficient, because $\mathcal{V}$, in their *basic* protocols, needs to request $\mathcal{P}$ to open one of the commitments, BC(t) or BC(t+s), *many times* (the so called *cut-and-choose* method).

1.3 Results

This paper gives a more efficient solution to the problem above than previous ones. We first propose primitives, a bit commitment scheme and four (statistical) *witness indistinguishable* (WI) protocols (See [FS90] for WI). Then we construct, by using these primitives, statistical zero-knowledge protocols (SZK argument of knowledge) in which, for any given multi-variable polynomial $f(X_1, .., X_t)$ and any given modulus n, prover $\mathcal{P}$ gives $(I_1, .., I_t)$ to verifier $\mathcal{V}$ and can convince $\mathcal{V}$ that $\mathcal{P}$ knows $(x_1, .., x_t)$ satisfying $f(x_1, .., x_t) \equiv 0 \pmod{n}$ and $I_i = BC(x_i)$, $(i = 1, .., t)$ without revealing any additional information. The proposed protocols are $O(|n|)$ times more efficient than the corresponding protocols in [Dam93, Dam95, Oka95], because our protocols do not need to confirm that a secret is in any range nor to execute any (single-bit based) *cut-and-choose* method. At the same time, the communication complexity of our protocols is $O(|n|)$ times less than those of [Dam93, Dam95] and [Oka95]. Although a set-up procedure for the parameter of the underlying bit-commitment is necessary and plays an essential role to satisfy the zero-knowledgeness of our protocols, the procedure can be done separately before the main parts of the protocols in pre-processing and can be shared by repeated execution of the main parts. (Similarly, a set-up procedure is also necessary in [Dam93, Dam95].)

A computational assumption, the intractability of a modified RSA problem (defined in Def.3), is necessary to prove the (knowledge) soundness regarding $(x_1, .., x_t)$ in our protocols, while no computational assumption is required for (statistical) zero-knowledgeness. In addition, any poly-time bounded prover $\mathcal{P}^*$ can open the bit commitment in any different ways with negligible probability under the factoring assumption.

These protocols can be employed to construct various practical cryptographic protocols such as fair exchange, untraceable electronic cash and some protocols regarding secret sharing. In Section 5, we demonstrate how to employ the proposed protocols to construct the *fair exchange and contract signing* protocol.

2 Definitions and Assumptions

This section mainly defines the factoring assumption and the modified RSA problem and its assumption; the modified RSA problem is a little different from the well-known RSA problem at the point that a cracking algorithm, A, can on input (N, Y) choose a convenient exponent, e (≥ 2), to output (X, e) such that $X \equiv \sqrt[e]{Y} \pmod{N}$ (Of course, it is less intractable than the *factoring* problem since a cracking algorithm, A, which can factor N, can solve the modified RSA problem of N). The validity (soundness) of the *whole* protocols against $\mathcal{P}$ can be guaranteed under Assumption 4 while the validity of the commitment against $\mathcal{P}$ can be guaranteed under Assumption 2.

Definition 1. $f(n)$ is **negligible** in n if, for any constant c, there exists a constant, N, such that $f(n) < (1/n)^c$ for any $n > N$. $f(n)$ is **non-negligible** in n if, there exits constants c and N such that $f(n) > (1/n)^c$ for any $n > N$. f(n) is **overwhelming** in n if, for any constant c, there exists a constant, N, such that $f(n) > 1 - (1/n)^c$ for any $n > N$.

Assumption 2. (Factoring Assumption) A probabilistic polynomial-time generator Δ_1 exists which on input $1^{|N|}$ outputs composite N where N is a composite of two prime numbers, P and Q, such that for any probabilistic polynomial-size circuit family, A, the probability that A can factor N is negligible. The probability is taken over the random choices of Δ_1 and A.

Definition 3. Modified RSA problem is, for given (N, Y), finding X and e ($e \geq 2$), such that $Y \equiv X^e \pmod{N}$, where N is the composite of two prime numbers, P and Q.

Assumption 4. (Modified RSA Assumption) A probabilistic polynomial-time generator Δ_2 exists which on input $1^{|N|}$ outputs (N, Y) such that for any probabilistic polynomial-size circuit family A, the probability that A can solve the modified RSA problem is negligible. The probability is taken over the random choices of Δ_2 and A.

In this paper, we use the following symbols. "$\alpha \in_R S$" means uniformly choosing a random element, α, from a set, S. Let Z_N be a residue class ring modulo N, and Z_N^* the reduced residue class group. Other symbols and definitions will be set as needed.

3 Bit Commitment and WI protocols

In this section, we propose a bit commitment scheme, and four WI protocols. The suitable parallel executions of those protocols, for any multi-variable polynomial $f(X_1, .., X_t)$ and any modulus n, can construct an WI protocol over the relation ($(I_1, .., I_t, I_{t+1})$, $(x_1, r_1, .., x_t, r_t, y, r_{t+1})$) such that $y \equiv f(x_1, .., x_t)$ (mod n) (For simplicity, we often call the protocol WI protocol to confirm $y \equiv f(x_1, .., x_t)$ (mod n)). We show later, as an example, a WI protocol to confirm $y \equiv ax^5 + b \bmod n$.

3.1 Bit Commitment Scheme

Our proposed commitment statistically reveals to the verifier no information of secret s in $BC(s)$ and holds computational validity against the prover. The validity of the commitment is guaranteed if the factoring assumption (Assumption 2) holds true. The commitments are given by

$$BC_{b_0}(s, r) = b_0{}^s b_1{}^r \bmod N \text{ or } BC_{b_0}(s, r_1, r_2) = b_0{}^s b_1{}^{r_1} b_2{}^{r_2} \bmod N.$$

Here, (N, b_0, b_1, b_2) is a set of system parameters given by verifier $\mathcal{V}$ or authority (i.e. trusted third party).

To set the system parameters, verifier $\mathcal{V}$ (or authority) executes the following procedure:

[Set-up procedure]

1. $\mathcal{V}$ generates large primes, P and Q, including odd prime divisors, p and q, such that $p = (P-1)/2$, $q = (Q-1)/2$, and $p \neq q$).
2. $\mathcal{V}$ finds at random $g_p \in G_p \backslash \{1\}$, and $g_q \in G_q \backslash \{1\}$, where G_p, G_q are subgroups of the order p, q in Z_P^*, Z_Q^* respectively (The complexity of finding g_p and g_q is comparable to that of finding generator elements of Z_P^* and Z_Q^*).
3. $\mathcal{V}$ computes, $b_0 \in Z_N^*$, by using the Chinese Remainder Theorem, such that $b_0 = g_p \bmod P$ and $b_0 = g_q \bmod Q$ (b_0 is a generator element of G_{pq}).
4. $\mathcal{V}$ finds at random $\alpha, \beta \in Z_{pq}^*$ and sets $b_1 = b_0{}^\alpha \bmod N$ and $b_2 = b_0{}^\beta \bmod N$.
5. $\mathcal{V}$ sends (N, b_0, b_1, b_2) to prover $\mathcal{P}$. Then $\mathcal{V}$ proves that he knows α, α^{-1}, β, and β^{-1} such that $b_1 = b_0{}^\alpha \bmod N$, and $b_2 = b_0{}^\beta \bmod N$ in the zero knowledge manner (that is, the orders of b_0, b_1, and b_2 are equivalent).

In the bit-commitment phase, $\mathcal{P}$ sends to $\mathcal{V}$, $BC_{b_0}(x, r) = b_0{}^x b_1{}^r \bmod N$ or $BC_{b_0}(x, r_1, r_2) = b_0{}^x b_1{}^{r_1} b_2{}^{r_2} \bmod N$ where $x \in [0, N)$ is a secret and $r, r_1, r_2 \in [0, 2^m N)$ are auxiliary random numbers.

Lemma 5. (Indistinguishability) *If $m = O(|N|)$, $BC_{b_0}(x, r)$ and $BC_{b_0}(x, r_1, r_2)$ statistically reveal no information of x to $\mathcal{V}$.*

The following results show that the validity (security) of these commitments are guaranteed if the factoring assumption (Assumption 2) holds true.

Lemma 6. (Miller) *Let $N = p_1^{v_1} \cdots p_m^{v_m}$ be the prime factorization of the odd integer N. Let $\lambda(N) = \mathrm{lcm}\{p_1^{v_1-1}(p_1 - 1), .., p_m^{v_m-1}(p_m - 1)\}$ (the Carmichael λ-function) and L be a multiple of $\lambda(N)$ (i.e., $\lambda(N)|L$). There exists a probabilistic polynomial-time algorithm M which, on input (N, L), can output the factorization of N with non-negligible probability in $|N|$. (Note: N is given by Δ_1 and the probability is taken over the coin tosses of Δ_1 and M.)*

The proof of Lemma 6 is implied by Theorem 2 in [Mil76].

Definition 7. (Generator Δ_{BC}) Let Δ_{BC} be a probabilistic polynomial-time algorithm which on input $1^{|N|}$ outputs (N, b_0, b_1, b_2) where the distribution of N is equal to that of Δ_2 and (b_0, b_1, b_2) is generated by the **Set-up procedure** of the bit commitment scheme.

Theorem 8. (Validity against $\mathcal{P}^*$) *If Assumption 2 holds true, there exists no probabilistic polynomial-time algorithm $\mathcal{P}^*$ which, on input (N, b_0, b_1), given by Δ_{BC}, can output (s_1, r_1) and (s_2, r_2), with non-negligible probability in $|N|$, where $(s_1, r_1) \neq (s_2, r_2)$ and $b_0^{s_1} b_1^{r_1} \equiv b_0^{s_2} b_1^{r_2} \pmod{N}$. (Note: the probability is taken over the coin tosses of Δ_{BC} and $\mathcal{P}^*$.)*

Sketch of Proof:

The proof is by contradiction. Assuming that a probabilistic polynomial-time algorithm $\mathcal{P}^*$ can output (s_1, r_1) and (s_2, r_2), with non-negligible probability, then we can construct a probabilistic poly-time algorithm M which can factor N with non-negligible probability. Let $s = s_1 - s_2$, and $r = r_2 - r_1$. The algorithm $\mathcal{P}^*$ above can be replaced by the algorithm which, on input (N, b_0, b_1), can output

$$b_0^{s} b_1^{r} \equiv 1 \pmod{N}, \tag{1}$$

where $(s, r) \neq (0, 0)$. In addition, by Lemma 6, the algorithm M can be replaced by the algorithm which on input N outputs L' such that $\lambda(N)|L'$.

The strategy of M is the following:

Algorithm M

1. Input N generated by Δ_{BC} to M.
2. M picks $b_0 \in_R Z_N$ and $\alpha \in_R (0, 2^k N)$ ($k = O(|N|)$), then computes $b_1 = b_0^{\alpha} \bmod N$.
3. M inputs (N, b_0, b_1) to $\mathcal{P}^*$.
4. If $\mathcal{P}^*$ returns (s, r), go the next step, otherwise M halts.
5. M outputs $L = 2(r - \alpha s)$ if $L \neq 0$, otherwise halts.

The algorithm M can output L with non-negligible probability.

When M picks b_0 uniformly in Z_N in Step 2, the probability that the order of b_0 is pq is non-negligible because $\frac{\varphi(pq)}{\#Z_N} = \frac{(p-1)(q-1)}{(2p+1)(2q+1)} \approx \frac{1}{4}$, where $\varphi(\cdot)$ is the Eulerian function and $\varphi(pq)$ is the number of generators of G_{pq}. This means that the distribution of a non-negligible fraction (about $1/4$) of (N, b, b_1)'s picked by M is indistinguishable from those generated by Δ_{BC}. $\mathcal{P}^*$ therefore outputs (s, r) with non-negligible probability in Step 4. In Step 5, the probability of $L \neq 0$ is non-negligible. This is because even infinite power $\mathcal{P}^*$ can only know $\alpha_0 = \alpha \bmod pq$. Therefore, if α is uniformly picked in $[0, 2^k N)$, the probability of $L \neq 0$ is non-negligible. From equation (1), $L \equiv 0 \pmod{pq}$. This means that

$$L = 2kpq = k\lambda(N), \tag{2}$$

where $k \neq 0$, $\lambda(N) = \text{lcm}(P - 1, Q - 1)$. $\qquad \square$

21

Corollary 9. (Validity against $\mathcal{P}^*$) *If Assumption 2 holds true, there exists no probabilistic polynomial-time algorithm $\mathcal{P}^*$ which, on input (N, b_0, b_1, b_2), given by Δ_{BC}, can output $(t, u, v) \neq (0, 0, 0)$ such that $b_0^t b_1^u b_2^v \equiv 1 \pmod{N}$, with non-negligible probability.*

If base b_0 is clear, we use the expressions $BC(s, r)$ and $BC(s, r_1, r_2)$. If auxiliary parameters are not important, we use just $BC(s)$.

3.2 Basic Protocol

Let $R^{(1)}_{(N, b_0, b_1)} := \{(I, (x, r)) | I = BC_{(N, b_0, b_1)}(x, r)\}$. The basic protocol is (statistical) *witness indistinguishable* (WI) over the relation $R^{(1)}_{(N, b_0, b_1)}$ and convinces $\mathcal{V}$ that $\mathcal{P}$ knows (x, r) such that $I = BC_{(N, b_0, b_1)}(x, r)$.

[Basic Protocol]

1. $\mathcal{V}$ executes with $\mathcal{P}$ the *set-up procedure* for parameter (N, b_0, b_1, b_2).
2. $\mathcal{P}$ sets $I = BC_{(N, b_0, b_1)}(x, r)$ and sends it to $\mathcal{V}$.
3. $\mathcal{P}$ chooses $w_1^0, w_1^1 \in_R [0, 2^{2m}N)$ and sets w_2^0, w_2^1 by $w_2^0 = w_1^0 - 2^{2m}N$ and $w_2^1 = w_1^1 - 2^{2m}N$. $\mathcal{P}$ picks four elements, $w_{i,j}^2$'s $\in_R [0, 2^m N)$, then computes $t_{i,j} = BC(w_i^0, w_j^1, w_{ij}^2)$, where $1 \leq i, j \leq 2$.
4. $\mathcal{P}$ sends to $\mathcal{V}$, four unordered commitments, $t_{i,j}$'s.
5. $\mathcal{V}$ picks a challenge $c \in_R [0, 2^m)$ and sends it to $\mathcal{P}$.
6. $\mathcal{P}$ sets $X = cx + w_i^0$ and $R = cr + w_j^1$ such that $X, R \in [0, 2^{2m}N)$, and sends to $\mathcal{V}$, the pair, $(X, R, w_{i,j}^2)$.
7. $\mathcal{V}$ checks there exists a t_{ij} such that $BC(X, R, w_{ij}^2) \equiv t_{i,j} I^c \pmod{N}$.

The completeness is obvious, since when $X = cx + w_i^0$ and $R = cr + w_j^1$ (if $\mathcal{P}$ is honest, there exists $X, R \in [0, 2^{2m}N)$), the left-hand side in the verification equation is equal to the right-hand, because

$$b_0^X b_1^R b_2^{w_{i,j}^2} \equiv b_0^{cs + w_i^0} b_1^{cr + w_j^1} b_2^{w_{i,j}^2} \equiv b_0^{w_i^0} b_1^{w_j^1} b_2^{w_{i,j}^2} I^c \pmod{N}.$$

Lemma 10. (Soundness) *Under Assumption 4, there exists a probabilistic poly-time algorithm M such that, for any probabilistic poly-time algorithm $\mathcal{P}^*$, if probabilistic interactive algorithm $(\mathcal{P}^*, \mathcal{V})$ accepts with non-negligible probability in $|N|$, then M with Δ_{BC} and $\mathcal{P}^*$ as oracles can extract (x, r) satisfying $I = BC_{(N, b_0, b_1)}(x, r)$ with overwhelming probability in $|N|$ where I is given by $\mathcal{P}^*$ as output. The success probability of $(\mathcal{P}^*, \mathcal{V})$ is taken over the coin tosses of $\mathcal{P}^*$ and $\mathcal{V}$ (including Δ_{BC}), while the success probability of M over those of Δ_{BC}, $\mathcal{P}^*$ and M.*

The proof of the soundness is given in Appendix A.

Lemma 11. (Witness Indistinguishable) *If $m = O(|N|)$ and $x, r \in [0, 2^m N)$, the basic protocol is statistically witness indistinguishable over $R^{(1)}_{(N, b_0, b_1)}$.*

3.3 Checking Protocol

The following protocol is considered as a kind of the basic protocol. However, since it is also utilized in the mod-multi protocol and in Subsection 3.6, we state it as a different one. Let $R^{(2)}_{(N,b_0)} := \{(I, \gamma)| I = b_0^\gamma \bmod N\}$. The checking protocol is WI over the relation $R^{(2)}_{(N,b_0)}$ and convinces $\mathcal{V}$ that $\mathcal{P}$ can know γ such that $I = b_0{}^\gamma \bmod N$.

[Checking Protocol]

1. $\mathcal{V}$ executes with $\mathcal{P}$ a *set-up procedure* for (N, b_0, b_1).
2. $\mathcal{P}$ sets $I = b_0{}^\gamma \bmod N$ and sends it to $\mathcal{V}$.
3. $\mathcal{P}$ chooses $w_1^0 \in_R [\,0, 2^{2m}l)$ and sets w_2^0 by $w_2^0 = w_1^0 - 2^{2m}l$. $\mathcal{P}$ picks two elements, w_i^1's $\in_R [\,0, 2^m N)$, then computes $t_i = BC_{b_0}(w_i^0, w_i^1)$, where $1 \leq i \leq 2$ and $l := \max[b - a, N]$.
4. $\mathcal{P}$ sends to $\mathcal{V}$, two unordered commitments, t_i's.
5. $\mathcal{V}$ picks a challenge $c \in_R [\,0, 2^m)$ and sends it to $\mathcal{P}$.
6. $\mathcal{P}$ sets $X := c(\gamma - a) + w_i^0 \in [\,0, 2^{2m}l)$, and sends to $\mathcal{V}$, the pair, (X, w_i^1).
7. $\mathcal{V}$ checks there exists a t_{ij} such that $BC(X, w_i^1) \equiv t_i(Ib_0{}^{-a})^c \pmod{N}$.

The following results are easily obtained by the properties of the basic protocol.

Lemma 12. (Soundness) *Under Assumption 4, there exists a probabilistic algorithm M such that, for any probabilistic poly-time algorithm $\mathcal{P}^*$, if probabilistic interactive algorithm $(\mathcal{P}^*, \mathcal{V})$ accepts with non-negligible probability in $|N|$, then M with Δ_{BC} and $\mathcal{P}^*$ as oracles can extract γ satisfying $I = b_0{}^\gamma \bmod N$ with overwhelming probability in $|N|$ where I is given by $\mathcal{P}^*$ as output. The success probability of $(\mathcal{P}^*, \mathcal{V})$ is taken over the coin tosses of $\mathcal{P}^*$ and $\mathcal{V}$ (including Δ_{BC}), while the success probability of M over those of Δ_{BC}, $\mathcal{P}^*$ and M.*

Lemma 13. (Witness Indistinguishable) *If $m = O(|N|)$ and $\gamma \in [\,a, b\,)$, the checking protocol is statistically witness indistinguishable over $R^{(2)}_{(N,b_0)}$.*

3.4 Comparing Protocol

Let $R^{(3)}_{(N,b_0,b_1,a)} := \{((I_1, I_2), (x, r_1, r_2))| I_1 = BC_{b_0}(x, r_1), I_2 = BC_a(x, r_2)\}$. The comparing protocol is WI over the relation $R^{(3)}_{(N,b_0,b_1,a)}$, in which $\mathcal{P}$ can convince $\mathcal{V}$ that he knows (x, r_1, r_2) such that $I_1 = BC_{b_0}(x, r_1)$ and $I_2 = BC_a(x, r_2)$.

[Comparing Protocol]

1. $\mathcal{V}$ executes with $\mathcal{P}$ the *set-up procedure* for parameters (N, b_0, b_1, b_2).
2. $\mathcal{P}$ sets $I_1 = BC_{b_0}(x, r_1)$ and $I_2 = BC_a(x, r_2)$, and sends them to $\mathcal{V}$.
3. $\mathcal{P}$ computes, for $1 \leq i, j \leq 2, t_{ij} = BC_{b_0}(w_i^0, w_j^1, w_{ij}^2)$ and $u_{ij} = BC_a(w_i^0, \eta_j^1, \eta_{ij}^2)$.
4. $\mathcal{P}$ sends to $\mathcal{V}$, four unordered pairs, $(t_{i,j}, u_{i,j})$'s.
5. $\mathcal{V}$ picks a $c \in_R [\,0, 2^m)$ and sends it to $\mathcal{P}$.
6. $\mathcal{P}$ sets $X := cx + w_i^0$, $R_1 := cr_1 + w_j^1$, and $R_2 := cr_2 + w_k^2$ such that $X, R_1, R_2 \in [\,0, 2^{2m}N)$. $\mathcal{P}$ then sends to $\mathcal{V}$, the pair, $(X, R_1, R_2, w_{i,j}^2, \eta_{i,k}^2)$.

7. $\mathcal{V}$ checks that there exists a pair ($t_{i,j}$, $u_{i,k}$) such that

$$BC_{b_0}(X, R_1, w_{ij}^2) \equiv t_{i,j} I_1^c \pmod{N} \text{ and } BC_a(X, R_2, \eta_{ij}^2) \equiv u_{i,k} I_2^c \pmod{N}.$$

If $\mathcal{V}$ sets a new base a, he has to convince $\mathcal{P}$ that there exists an α such that $a = b_0{}^\alpha \bmod N$ before executing this protocol, but in many cases, a is set by $\mathcal{P}$ as $a := I_1$. Note that in the case of $a := I_1$, $\mathcal{P}$ can show $\mathcal{V}$ $I_2 = BC_{b_0}(x^2, r_1 x_1 + r_2)$. This means $\mathcal{P}$ can convince $\mathcal{V}$ that commitments, $BC_{b_0}(x, r_1)$ and $BC_{b_0}(y, r_2)$, satisfy $y = x^2$.

The following results are easily obtained by the properties of the basic protocol.

Lemma 14. (Soundness) *Under Assumption 4, there exists a probabilistic algorithm M such that, for any probabilistic poly-time algorithm $\mathcal{P}^*$, if probabilistic interactive algorithm $(\mathcal{P}^*, \mathcal{V})$ accepts with non-negligible probability in $|N|$, then M, with Δ_{BC} and $\mathcal{P}^*$ as oracles, can extract (x, r_1, r_2) with overwhelming probability in $|N|$, where (I_1, I_2) is given by $\mathcal{P}^*$ as output, and $I_1 = b_0^x b_1^{r_1} \bmod N$, $I_2 = a^x b_1^{r_2} \bmod N$. The success probability of $(\mathcal{P}^*, \mathcal{V})$ is taken over the coin tosses of $\mathcal{P}^*$ and $\mathcal{V}$ (including Δ_{BC}), while the success probability of M over those of Δ_{BC}, $\mathcal{P}^*$ and M.*

Lemma 15. (Witness Indistinguishable) *If $m = O(|N|)$ and $x_1, r_1, x_2, r_2 \in [\, 0, \, 2^m N)$, the comparing protocol is statistically witness indistinguishable over $R_{(N, b_0, b_1)}^{(3)}$.*

3.5 Mod-Multi Protocol

Let $R_{(N, b_0, b_1)}^{(4)} := \{\ ((I_1, I_2, I_3), (x_1, r_1, .., x_3, r_3)) \mid I_i = BC(x_i, r_i),\ x_3 \equiv x_1 x_2$ (mod n) }. The mod-multi protocol is WI over the relation $R_{(N, b_0, b_1)}^{(4)}$ (We call it a mod-multi protocol to confirm $x_3 \equiv x_1 x_2 \pmod{n}$). In the mod-multi protocol, $\mathcal{P}$ can convince $\mathcal{V}$ that he knows $(x_1, x_2, x_3, r_1, r_2, r_3)$ such that $x_3 \equiv x_1 x_2$ (mod n), where $I_1 = BC_{b_0}(x_1, r_1)$, $I_2 = BC_{b_0}(x_2, r_2)$ and $I_3 = BC_{b_0}(x_3, r_3)$.

[Mod-Multi Protocol]

1. $\mathcal{V}$ executes with $\mathcal{P}$ the *set-up procedure* and sends to $\mathcal{P}$, parameters (N, b_0, b_1, b_2).
2. $\mathcal{P}$ sets $I_1 = BC_{b_0}(x_1, r_1)$, $I_2 = BC_{b_0}(x_2, r_2)$, and $I_3 = BC_{b_0}(x_3, r_3)$, and sends them to $\mathcal{V}$.
3. $\mathcal{P}$ sets $I_2^1 = BC_{I_2}(x_1, r_4) = BC_{b_0}(x_1 x_2, r_2 x_1 + r_4)$, and $I_d = BC_{b_0}(d, r_d)$ where $d = (x_3 - x_1 x_2)/n$.
4. $\mathcal{P}$ executes in parallel with $\mathcal{V}$ the *comparing* protocol for (I_1, I_2^1) and the three *basic* protocols for I_2, I_3, and I_d.
5. $\mathcal{P}$ computes $\gamma = (r_2 x_1 + r_4 + r_d n) - r_3$, and executes with $\mathcal{V}$ the *checking* protocol for $b_1^\gamma = I_3 (I_2^1 I_d^n)^{-1} \bmod N$ and range $[\, -2^m N, \, 2^{2m+1} N)$ over the relation $R_{(N, b_1)}^{(2)}$.

In this protocol, $\mathcal{P}$ executes one comparing protocol, three basic protocols, and one checking protocol for b_1^γ in parallel. (in the case of $x_1 = x_2$ the number of the basic protocols is reduced to two). This protocol is also WI.

Lemma 16. (Soundness) *Under Assumption 4, there exists a probabilistic algorithm M such that, for any probabilistic poly-time algorithm $\mathcal{P}^*$, if probabilistic interactive algorithm $(\mathcal{P}^*,\mathcal{V})$ accepts with non-negligible probability in $|N|$, then M, with Δ_{BC} and $\mathcal{P}^*$ as oracles, can extract $(x_1, r_1, .., x_3, r_3)$ with overwhelming probability in $|N|$, where (I_1, I_2, I_3) is given by $\mathcal{P}^*$ as output, $I_i = BC(x_i, r_i)(i = 1, \ldots, 3)$ and $x_3 \equiv x_1 x_2 \pmod{n}$. The success probability of $(\mathcal{P}^*, \mathcal{V})$ is taken over the coin tosses of $\mathcal{P}^*$ and $\mathcal{V}$ (including Δ_{BC}), while the success probability of M over those of Δ_{BC}, $\mathcal{P}^*$ and M.*

Sketch of Proof:

From lemma 10, if $(\mathcal{P}^*, \mathcal{V})$ has non-negligible success probability, we can construct a probabilistic poly-time knowledge extractor M, which extracts $(x_1, r_1, .., x_3, r_3)$ and d such that $I_i = BC(x_i, r_i)$ and $x_3 \equiv x_1 x_2 + dn \pmod{pq}$. Then the probability of $x_3 \neq x_1 x_2 + dn$ is negligible. If it is non-negligible, we can construct an algorithm M', with poly-time bounded $\mathcal{P}^*$ as an oracle, which can factor N given by Δ_{BC} with non-negligible probability in $|N|$. This is a contradiction. M' indeed extract L such that $L = 2(x_3 - x_1 x_2 - dn)$ ($= 2kpq = k\lambda(N)$) where $\lambda(N) = \mathrm{lcm}(P - 1, Q - 1)$. By Lemma 6, this contradicts Assumption 2 and thereby contradicts Assumption 4. Consequently, M extracts (x_1, x_2, x_3, d) such that $x_3 \equiv x_1 x_2 \pmod{n}$ with overwhelming probability in $|N|$. $\quad\square$

Lemma 17. (Witness Indistinguishable) *If $m = O(|N|)$ and $x_1, r_1, .., x_3, r_3 \in [\, 0, \, 2^m N)$, the mod-multi protocol is statistically witness indistinguishable over $R^{(4)}_{(N, b_0, b_1)}$.*

3.6 WI protocol to Confirm $y \equiv ax^5 + b \pmod{n}$

We show, as an example, a WI protocol to confirm $y \equiv ax^5 + b \bmod n$.

Let $[x_1, x_2; x_3]$ be the mod-multi protocol to confirm $x_3 \equiv x_1 x_2 \pmod{n}$ and let $[x_1; x_2]$ be the mod-multi protocol to confirm $x_2 \equiv x_1^2 \pmod{n}$.

Prover $\mathcal{P}$ sets $(I_y, I_x, I_d, I_1, I_2, I_3)$ as $(BC_{b_0}(y, r_y), BC_{b_0}(x, r), BC_{b_0}(d, r_d), BC_{b_0}(t_1), BC_{b_0}(t_2), BC_{b_0}(t_3, r_3))$ where $d = \frac{y - (a t_3 + b)}{n}$, $t_1 = x^2 \bmod n$, $t_2 = x^4 \bmod n$ and $t_3 = x^5 \bmod n$. $\mathcal{P}$ executes with $\mathcal{V}$ the two basic protocols for I_y and I_d and the three mod-multi protocols, $[x; t_1]$, $[t_1; t_2]$, and $[x, t_2; t_3]$, in parallel. $\mathcal{P}$ then executes with $\mathcal{V}$ a checking protocol for b_1^γ and range $[\, -2^m N, \, 2^{2m+1} N)$, where $\gamma = a r_3 + r_d n - r_y$ and $b_1^\gamma \equiv I_3^a b_0^{\,b} (I_y I_d)^{-1} \pmod{N}$.

4 Statistical Zero Knowledge Protocol

In this section, we state the main results of this paper. As mentioned above in Section 3, for any multi-variable polynomial $f(X_1, .., X_t)$ and any modulus n, we can construct a statistical WI protocol to prove that $\mathcal{P}$ knows $(x_1, r_1, .., x_t, r_t, y, r_{t+1})$ such that $I_i = BC(x_i, r_i)$ $(i = 1, .., t)$, $I_{t+1} = BC(y, r_{t+1})$, and $y \equiv f(x_1, .., x_t) \pmod{n}$. This WI protocol can be transformed to the following statistical zero knowledge (SZK) protocol.

Here, we define some terminology. Let f be a multi-variable polynomial. Let $\mathcal{S} := \{(f, n) \mid \exists (x_1, .., x_t) \in Z_n^t \text{ s.t. } f(x_1, .., x_t) \equiv 0 \bmod n\}$. We can assume, without loss of generality, coefficients of f, number of variables in f, i.e. t, and parameters (N, b_0, b_1, b_2) are related to modulus n regarding their size (that is, the size of them is $O(|n|)$).

The SZK protocol is constructed as follows:

[SZK Protocol]
common input: (f, n).
output: $I_1, .., I_t, N, b_0, b_1$, and the remaining conversation of $(\mathcal{P}, \mathcal{V})$.
knowledge of $\mathcal{P}$: $(x_1, r_1, .., x_t, r_t)$ such that $I_i = BC_{(N, b_0, b_1)}(x_i, r_i)$ $(i = 1, .., t)$
  and $f(x_1, .., x_t) \equiv 0 \pmod n$.

1. $\mathcal{V}$ executes with $\mathcal{P}$ a *set-up procedure* for (N, b_0, b_1, b_2).
2. $\mathcal{P}$ sets $I_i := BC_{b_0}(x_i, r_i)$ $(i = 1, .., t)$, $I_{t+1} := BC_{b_0}(0, r_{t+1})$, and sends them to $\mathcal{V}$.
3. $\mathcal{P}$ executes, with $\mathcal{V}$, the WI protocol mentioned above to prove that $\mathcal{P}$ knows $x_1, r_1, .., x_t, r_t$, and y, r_{t+1} such that $I_i = BC(x_i, r_i)$ $(i = 1, .., t)$, $I_{t+1} = BC(y, r_{t+1})$, and $y \equiv f(x_1, .., x_t) \pmod n$ where $y = 0$.
4. $\mathcal{P}$ sends r_{t+1} to $\mathcal{V}$.
5. $\mathcal{V}$ checks that $I_{t+1} \equiv BC(0, r_{t+1}) \pmod N$.

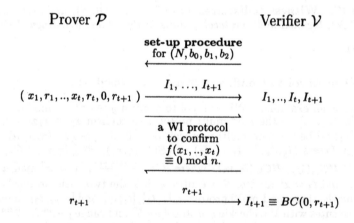

Prover $\mathcal{P}$ Verifier $\mathcal{V}$

set-up procedure
for (N, b_0, b_1, b_2)
$\longleftarrow$

$I_1, \ldots, I_{t+1}$
$(x_1, r_1, .., x_t, r_t, 0, r_{t+1})$ $\xrightarrow{\hspace{2cm}}$ $I_1, .., I_t, I_{t+1}$

$\xrightarrow{\hspace{2cm}}$
a WI protocol
to confirm
$f(x_1, .., x_t)$
$\equiv 0 \bmod n$.
$\longleftarrow$

r_{t+1}
r_{t+1} $\xrightarrow{\hspace{2cm}}$ $I_{t+1} \equiv BC(0, r_{t+1})$

Fig. 1. The SZK protocol that convinces $\mathcal{V}$ that $\mathcal{P}$ knows $(x_1, .., x_t)$ satisfying $f(x_1, .., x_t) \equiv 0 \pmod n$ and $I_i = BC(x_i)$ $(i = 1, .., t)$.

Theorem 18. (Soundness) *Under Assumption 4, there exists a probabilistic poly-time algorithm M such that, for any probabilistic poly-time algorithm $\mathcal{P}^*$ and for any input $(f, n) \in S$, if probabilistic interactive algorithm $(\mathcal{P}^*, \mathcal{V})$ accepts on input (f, n) with non-negligible probability in $|n|$, then M, with Δ_{BC} and $\mathcal{P}^*$ as oracles, can extract $(x_1, r_1, \ldots, x_t, r_t)$ with overwhelming probability in $|n|$, where $I_i = BC(x_i, r_i)$ and $f_t(x_1, .., x_t) \equiv 0 \bmod n$. The success probability of $(\mathcal{P}^*, \mathcal{V})$ is taken over the coin tosses of $\mathcal{P}^*$ and $\mathcal{V}$ (including Δ_{BC}), and the success probability of M over those of Δ_{BC}, $\mathcal{P}^*$ and M.*

Sketch of Proof:

Assume that $(\mathcal{P}^*, \mathcal{V})$ has non-negligible success probability. The sketch of the proof is as follows: M executes, with $\mathcal{P}^*$, the set-up procedure for parameters (N, b_0, b_1, b_2) given by Δ_{BC}, in which M should convince $\mathcal{P}^*$ that he knows α, α^{-1}, β, and β^{-1} such that $b_1 = b_0{}^{\alpha} \bmod N$, and $b_2 = b_0{}^{\beta} \bmod N$. Instead of using the values, α, α^{-1}, β, and β^{-1}, M can execute the set-up procedure using the resetable simulation technique for $\mathcal{P}^*$ because the set-up procedure is a zero-knowledge system of $(M, \mathcal{P}^*)$. After M completes the set-up procedure, $\mathcal{P}^*$ sends to M, $I_1, .., I_t$, and I_{t+1} to start a WI protocol. Note that this protocol has (knowledge) soundness over the relation $((I_1, .., I_t, I_{t+1}), (x_1, r_1, .., x_t, r_t, 0, r_{t+1}))$ such that $I_i = BC(x_i, r_i)$ $(i = 1, .., t)$, $I_{t+1} = BC(0, r_{t+1})$, and $f(x_1, .., x_t) \equiv 0 \bmod n$. Therfore M can extract from $\mathcal{P}^*$ desirable witnesses, $(x_1, r_1, ..., x_t, r_t, 0, r_{t+1})$. $\qquad\square$

Theorem 19. (Zero Knowledge) *Let $m = O(|n|)$. There exists a probabilistic algorithm M which runs in expected polynomial time such that, for any $\mathcal{V}^*$, and for any common input $(f, n) \in S$, the view of $\mathcal{V}^*$ is statistically indistinguishable from the output of M with $\mathcal{V}^*$ as an oracle.*

Sketch of Proof:

Let M be an expected poly-time algorithm allowed to use $\mathcal{V}^*$ as an oracle. M can extract α and α^{-1} from $\mathcal{V}^*$ in the set-up procedure. Let $L := \alpha\alpha^{-1} - 1$. Note that the order of b_0, b_1, and b_2 divides L. Next, M chooses $x_1', r_1', .., x_t', r_t', r_{t+1} \in_R [\, 0, 2^m N)$ and sets $I_1', ..., I_t'$ and $I_{t+1} := BC(0, r_{t+1})$. M computes $y := f(x_1', .., x_t') \bmod n$ and $r_{t+1}' := r_{t+1} - \alpha^{-1}y \bmod L$. Note that $I_{t+1} = BC(0, r_{t+1}) = BC(y, r_{t+1}')$. M executes with $\mathcal{V}^*$ a (statistical) WI protocol over the relation $((I_1', .., I_t', I_{t+1}), (x_1', r_1', .., x_t', r_t', y, r_{t+1}'))$ such that $I_i' = BC(x_i', r_i')$ $(i = 1, .., t)$, $I_{t+1} = BC(y, r_{t+1}')$, and $y \equiv f(x_1, .., x_t) \pmod{n}$. Finally, M sends r_{t+1} to $\mathcal{V}^*$.

Here the distribution of $(I_1, ..., I_t, I_{t+1})$ such that $I_i = BC(x_i, r_i)$ $(i = 1, .., t)$, $I_{t+1} = BC(0, r_{t+1})$ and $f(x_1, .., x_t) \equiv 0 \pmod{n}$ and that of $(I_1', ..., I_t', I_{t+1})$ such that $I_i = BC(x_i', r_i')$ $(i = 1, .., t)$, $I_{t+1} = BC(y, r_{t+1}')$ and $y \equiv f(x_1, .., x_t) \pmod{n}$ are statistically indistinguishable. In addition, for common input $(I_1', ..., I_t', I_{t+1})$ the protocols with witness $(x_1, r_1, ..., x_t, r_t, 0, r_{t+1})$ and with witness $(x_1', r_1', ..., x_t', r_t', y, r_{t+1}')$ are statistically indistinguishable. Therefore the view of $\mathcal{V}^*$ is also statistically indistinguishable from the output of $M^{\mathcal{V}^*}$. $\qquad\square$

Example 1. Suppose that $f(X) \equiv X^e - m \pmod{n}$. $\mathcal{P}$ can prove, in the statistical zero knowledge manner, that he knows s such that $f(s) \equiv 0 \pmod{n}$ and $BC(s)$.

Remark. Although the set-up procedure is described in the first step of the proposed SZK protocol, the procedure can be executed in an off-line manner before the remaining protocol begins. In addition, the set-up procedure can be shared by repeated execution of the main protocol. The zero-knowledgeness is still guranteed even if the set-up procedure is shared by repeated execution of the main protocol between $\mathcal{P}$ and $\mathcal{V}$.

5 Application to *Fair Exchange and Contract Signing*

We propose a gradual release protocol to realize *fair exchange and contract signing*. We modify our commitments into *bit releaseable* commitments like those of [Dam93, Dam95] for our gradual release protocol. The protocol is as follows:

V executes with P a *set-up procedure* and they hold parameters (N, b_0, b_1, b_2) in common. P and V set l such that $|s| < l$ and compute $b_1' := b_1^{2^l} \bmod N$ (V should prove that he knows $(2^l)^{-1} \bmod pq$ in the ZK manner to show that b_1 and b_1' have the same order) (**set-up phase**). P sends to V, $(m, BC_{b_0,b_1'}(s, r_1), BC_{b_0,b_1'}(0, r_2))$. For parameters (N, b_0, b_1', b_2), P executes, with V, the protocol to confirm that $BC_{b,b_1'}(s, r_1)$ and $BC_{b,b_1'}(0, r_2)$ satisfy the relation $s^e - m \equiv 0 \pmod{n}$, where (e, n) are RSA (or Rabin) public-key. P then open the commitment $BC_{b,b_1'}(0, r)$ (**confirming phase**). P releases the secret s bit by bit from the least-significant bit (LSB). Let s_k be the remaining secret of s after k bit release. P opens the LSB of s_k by revealing $X_{k+1} = b_0^{\frac{s_k - i}{2}} b_1'^{2^{l-k-1}} \bmod N$. V can know the LSB of s_k by checking $X_k \equiv X_{k+1}^2 b_0^i \pmod{N}$ (**bit by bit release phase**).

6 Efficiency

We compare our protocols with those in [Dam95] from the view points of computational and communication complexity. In [Dam95], the commitment is defined by the form $BC(s, r) = g^s r^{2^l} \bmod N$. As our *comparing* and *mod-multi* protocols are constructed in a similar manner to those in [Dam95], it is enough to compare our *basic* protocol with that in [Dam95]. Our *comparing* protocol is composed of at most two *basic* protocols and our *mod-multi* protocol consists of three *basic*, a *comparing*, and a *checking* protocols. Therefore those in [Dam95] have nearly the same construction. We assume below that $m = |N| = |n| = |c|$.

We estimate the computational complexity of the both *basic* protocols from the number of modular multiplications. In our *basic* protocol, P needs to compute four auxiliary parameters, t_{ij}'s ($t_{ij} = b_0^{w_i^1} b_1^{w_j^1} b_2^{w_{ij}^2}$), and V needs to check the verification $t_{ij} I^c = b_0^X b_1^R b_2^{w_{ij}^2}$, where $|w_i^0| = |w_j^1| = |X| = |R| = |2^{2m} N| = 3m$ and $|w_{ij}^2| = |2^m N| = 2m$. P and V both need $O(m)$ modular multiplications of N (about $32m$, $9m$ respectively). In Damgård's *basic* protocol, P needs to compute $2m$ auxiliary parameters, t_i's ($t_i = g^{w_i^0} w_i^{1^{2^l}}$), and V needs to check m verifications, $t_i I = g^X R^{2^l}$, where $|w_i^0| = |w_i^1| = |X| = 3m$ and $l = 2m$. P and V both need $O(m^2)$ modular multiplications of N (about $6m^2$, $3m^2$ respectively). Accordingly, our protocol is about $O(m)$ times more efficient than Damgård's.

The amount of communication in our *basic* protocol is $O(m)$ bits since $4|t_{ij}| + |c| + |X| + |R| + |w_{ij}^2| = 8m$ while that of [Dam95] is $O(m^2)$ bits since $m \cdot (2|t_i| + |X| + |R|) = 4m^2$. Hence, the communication complexity of ours is also $O(m)$ times less than that of Damgård's.

Comparing our protocols with those in [Oka95], the modulus size of Okamoto's bit commitment, $BC(s, r) = g^s G^r \bmod p$, should be at least twice ours. Hence, our protocol is about $O(m^3)$ times more efficient than [Oka95]. The communication complexity of ours is also $O(m)$ times less than that of [Oka95].

7 Conclusions

We have proposed a bit commitment scheme, $BC(\cdot)$, and related statistical zero knowledge (SZK) protocols in which, for any given multi-variable polynomial $f(X_1, .., X_t)$ and any given modulus n, prover $\mathcal{P}$ gives $(I_1, .., I_t)$ to verifier $\mathcal{V}$ and can convince $\mathcal{V}$ that $\mathcal{P}$ knows $(x_1, .., x_t)$ satisfying $f(x_1, .., x_t) \equiv 0$ (mod n) and $I_i = BC(x_i)$, $(i = 1, .., t)$. The proposed protocols are $O(|n|)$ times more efficient than the corresponding previous ones [Dam93, Dam95, Oka95]. The (knowledge) soundness of our protocols holds under a computational assumption, the intractability of the modified RSA problem, while the (statistical) zero-knowledgeness of the protocols needs no computational assumption. We have also shown the applications of fair exchange and contract signing by using the proposed protocol.

References

[BCC86]   G.Brassard, D.Chaum, and C.Crépeau, "Minimum Disclosure Proofs of Knowledge," Journal of Computer and System Sciences, Vol.37, pp.156-189 (1988)

[BG92]   Bellare, M. and Goldreich, O., "On Defining Proofs of Knowledge", Proceedings of Crypto 92, pp.390–420 (1992).

[Bra95]   Brands, S., "Restrictive Blinding of Secret-Key Certificates", Proceedings of Eurocrypt 95, pp.231–247 (1995).

[CDS94]   Cramer, R., Damgård, I. and Schoenmakers, B., "Proofs of Partial Knowledge and Simplified Design of Witness Hiding Protocols", Proc. of Crypto'94, LNCS, Springer, pp.174-187 (1994)

[CGMA85]   Chor, B., Goldwasser, S., Micali, S. and Awerbuch, B., "Verifiable Secret Sharing and Achieving Simultaneity in the Presence of Faults", Proc. of FOCS, pp.383-395 (1985).

[Dam93]   Damgård, I., "Practical and Provably Secure Release of a Secret and Exchange of Signatures," Proceedings of Eurocrypt 93 (1993).

[Dam95]   Damgård, I., "Practical and Provably Secure Release of a Secret and Exchange of Signatures," vol. 8 pp.201–222, Journal of CRYPTOLOGY(1995).

[FFS88]   U.Feige, A.Fiat and A.Shamir, "Zero Knowledge Proofs of Identity," Journal of Cryptology, Vol. 1, pp.77-94 (1988).

[FS90]   U.Feige, and A.Shamir, "Witness Indistinguishable and Witness Hiding Protocols," Proc. of STOC90.

[GMRa89]   Goldwasser, S., Micali, S., and Rackoff, C., "The knowledge complexity of interactive proof systems", SIAM J. Comput., vol.18, pp.186-208 (1989).

[GMW86]   O.Goldreich, S.Micali, and A.Wigderson, "Proofs that Yield Nothing But their Validity and a Methodology of Cryptographic Protocol Design," Proc. FOCS, pp.174-187 (1986)

[Mil76]   Miller, G.L., "Riemann's Hypothesis and Tests for Primality", Journal of Computer and System Sciences 13, 300-317 (1976).

[Oka95]   Okamoto, T., "An Efficient Divisible Electronic Cash Scheme", Proceedings of Crypto 95, pp.438-451 (1995).

[Ped91]   Pedersen, T. P., "Non-Interactive and Information-Theoretic Secure Verifiable Secret Sharing", Proceedings of Crypto 91, pp. 129–140 (1992).

[Sta96]   Stadler, M., "Publicly Verifiable Secret Sharing", Proc. of Eurocrypt'96, LNCS 1070, Springer, pp.190-199 (1996)

[TW87]   Tompa, M., and Woll, H., "Random Self-Reducibility and Zero-Knowledge Interactive-Proofs of Possession of Information", Proc. FOCS, pp 472–482 (1987).

A Proof of Lemma 10

Sketch of Proof:

The top level strategy of knowledge extractor M is as follows:

Protocol:

Step 1 M inputs $1^{|N|}$ to Δ_{BC} and gets parameter (N, b_0, b_1, b_2).

Step 2 M executes, with $\mathcal{P}^*$, the set-up procedure for parameters (N, b_0, b_1, b_2), in which M should convince $\mathcal{P}^*$ that he knows α, α^{-1}, β, and β^{-1} such that $b_1 = b_0{}^\alpha \bmod N$, and $b_2 = b_0{}^\beta \bmod N$. Instead of using the values, α, α^{-1}, β, and β^{-1}, M can execute the set-up procedure using the resetable simulation technique for $\mathcal{P}^*$ because the set-up procedure is a zero-knowledge system of $(M, \mathcal{P}^*)$.

Step 3 M can extract $(t_{i,j}, c, X, R, w_{i,j}^2)$ and $(t_{i,j}, c', X', R', w_{i,j}^2)$ for the same $t_{i,j}$, by using $\mathcal{P}^*$ as an oracle.

Step 4 M outputs $(\frac{\Delta X}{\Delta c}, \frac{\Delta X}{\Delta c})$ as a witness of I, where $\Delta c := c - c'$, $\Delta X := X - X'$ and $\Delta R := R - R'$.

We explain Step 3 and Step 4.

Consider Step 3. Let $\epsilon_{i,j}$ be the success probability of $(\mathcal{P}^*, \mathcal{V})$ with the conversation, $(t_{i,j}, c, X, R, w_{i,j}^2)$. Note that at least one of $\epsilon_{i,j}$'s is non-negligible if $(\mathcal{P}^*, \mathcal{V})$ accepts with non-negligible probability. Then M can find two different pairs for a $t_{i,j}$ in expected polynomial time in $|N|$. Indeed, the following strategy succeeds with overwhelming probability (See also [FFS88]):

1. For any (i, j), do the following steps.
2. Probe $O(1/\epsilon)$ random entries in $H_{i,j}$ (Here $H_{i,j}$'s are boolean matrices and each $H_{i,j}$'s rows corresponds to all possible states α of RP and its columns correspond to all possible choices c of RV, where the RP is $\mathcal{P}^*$'s random tape, and the RV is $\mathcal{V}$'s random tape.).
3. If find the first $(t_{i,j}, c, X, R, w_{i,j}^2)$ which $(\mathcal{P}^*, \mathcal{V})$ accepts, then probe $O(1/\epsilon)$ random entries along the same row in order to find $(t_{i,j}, c', X', R', w_{i,j}^2)$ which $(\mathcal{P}^*, \mathcal{V})$ accepts.

In Step 4, $(t_{i,j}, c, X, R, w_{i,j}^2)$ and $(t_{i,j}, c', X', R', w_{i,j}^2)$ satisfy that $X \equiv cx + w_i^0 \bmod pq$, $X' \equiv c'x + w_i^0 \bmod pq$, $R \equiv cr + w_j^1 \bmod pq$, and $R' \equiv e'R + w_j^1 \bmod pq$. Therefore,

$$\Delta X \equiv \Delta c \cdot x \pmod{pq} \quad \text{and} \quad \Delta R \equiv \Delta c \cdot r \pmod{pq}. \tag{3}$$

M can obtain x and r only ΔX and ΔR dividing by Δc respectively, with overwhelming probability in $|N|$ under Assumption 4.

Let $\alpha_0 \in Z_{pq}$ such that $b_1 = b_0^{\alpha_0} \bmod N$. Let $d := \gcd(\Delta c, \Delta X + \Delta R\alpha_0)$. From (3), the following relation holds

$$\Delta X + \Delta R\alpha_0 \equiv \Delta c(x + r\alpha_0) \pmod{pq}. \tag{4}$$

Here we replace, without loss of generality, $\mathcal{P}^*$ with the poly-time bounded machine which, on input (N, b_0, b_1, b_2) given by Δ_{BC}, outputs $(I, \Delta c, \Delta X, \Delta R)$ with overwhelming probability in $|N|$. We consider a poly-time bounded algorithm M' using $\mathcal{P}^*$ as an oracle in the following:

Algorithm M'

1. inputs (N, C) generated by Δ_2 to M'.
2. M' picks $b_2 \in_R Z_N$, $\alpha \in_R (0, 2^k N)$ (k is a constant.), then computes $b_1 = C^\alpha \bmod N$.
3. M' inputs (N, C, b_1, b_2) to $\mathcal{P}^*$.
4. If $\mathcal{P}^*$ returns $(I, \Delta c, \Delta X, \Delta R)$, go to the next step, otherwise M halts.
5. M' outputs $(I^Y C^Z \bmod N, \frac{\Delta c}{d})$ and halts, where Y and Z are integers such that
$$\frac{\Delta X + \Delta R\alpha}{d}Y + \frac{\Delta c}{d}Z = 1.$$

Note that $C \equiv C^{\frac{\Delta X + \Delta R\alpha}{d}Y + \frac{\Delta c}{d}Z} \equiv I^{\frac{\Delta c}{d}Y} C^{\frac{\Delta c}{d}Z} \equiv (I^Y C^Z)^{\frac{\Delta c}{d}} \pmod{N}$.

If $d \neq \Delta c$, M' is a machine, with $\mathcal{P}^*$ as an oracle, which can solve the modified RSA problem with non-negligible probability. It contradicts Assumption 4. Therefore $d = \Delta c$, namely $\Delta c | (\Delta X + \Delta R\alpha)$. Moreover, $(\Delta c, \Delta X, \Delta R)$ must satisfy that $\Delta c | \Delta X$ and $\Delta c | \Delta R$ to hold $d = \Delta c$. Let $\alpha = \alpha_0 + \xi pq$. From $d = \Delta c$,
$$\frac{\Delta X + \Delta R\alpha}{\Delta c} = \frac{\Delta X + \Delta R\alpha_0 + \Delta R\xi}{\Delta c}.$$

As even an infinite power $\mathcal{P}^*$ can never know ξ, The condition of $\Delta c | \Delta X$ and $\Delta c | \Delta R$ has to be held to satisfy that of $d = \Delta c$.

Thus, M can extract (x, r) with overwhelming probability in $|N|$. $\qquad\square$

Keeping the SZK-Verifier Honest Unconditionally

Giovanni Di Crescenzo*   Tatsuaki Okamoto**   Moti Yung***

Abstract. This paper shows that using direct properties of a zero-knowledge protocol itself, one may impose a honest behavior on the verifier (without additional cryptographic tools). The main technical contribution is showing that if a language L has an Arthur-Merlin (i.e. public coins) honest-verifier statistical SZK proof system then L has an (any-verifier) SZK proof system when we use a non-uniform simulation model of SZK (where the simulation view and protocol view can be made statistically closer than any given polynomial given as a parameter). Three basic questions regarding statistical zero-knowledge (SZK) are solved in this model:

- If L has a honest-verifier SZK proof then L has an any-verifier non-uniform simulation SZK proof.
- If L has an SZK proof then $\overline{L}$ has an non-uniform simulation SZK proof.
- If L has a private-coin SZK proof then L has a public-coin non-uniform simulation SZK proof.

1 Introduction

Statistical zero-knowledge proofs (SZK), introduced by Goldwasser, Micali and Rackoff [13], are an important notion with practical as well as theoretical relevance. In practice, SZK proofs have proved very useful in the design of cryptographic protocols, such as identification schemes [8]. From a theoretical point of view, SZK proofs seem to capture the intrinsic properties of the zero-knowledge concept, since they do not need further cryptographic assumptions, as it is the case for computational zero-knowledge (CZK) proofs. For CZK, all languages in NP [11] and in IP (=PSPACE) [17] (also [4]) are known to have a CZK proof system, while a precise characterization for the languages having SZK proof systems is not known. It is known that the class SZK is in AM ∩ co-AM [9, 1], and that NP-complete languages do not have such proofs unless the polynomial hierarchy collapses. Nevertheless, very few properties of SZK have been proved and for many years the problem of establishing unconditional relations among, and properties of SZK proofs, has attracted much attention (see, e.g., [2, 3, 7, 20] and the results below).

* Computer Science and Engineering Dep., University of California San Diego, La Jolla, CA, 92093-0114. E-mail: giovanni@cs.ucsd.edu

** NTT Laboratories, Nippon Telegraph and Telephone Corporation, 1-1 Hikarinooka, Yokosuka-shi, Kanagawa-ken, 239 Japan. E-mail: okamoto@sucaba.isl.ntt.co.jp

*** CertCo, New York NY, E-mail: moti@certco.com, moti@cs.columbia.edu

The notion of zero-knowledge achieved: non-uniform simulation. A few notions of zero-knowledge have been defined in the literature (see [13, 12, 10]). One notion, called *auxiliary-input* zero-knowledge, requires security with respect to any polynomial-time adversary, having an additional auxiliary-input, which models information obtained by the verifier in his past history (which was not required originally). Most protocols in the literature achieve a stronger notion, called *black-box simulation* zero-knowledge, where a simulator treats the verifier as a black box. The simulator may be characterized by an additional parameter, e.g. a polynomial (or a constant). In [10] a black-box simulation was considered where the additional parameter quantifies the random bits used by the simulator. We employ a simulator which uses a sampling technique to assess the bias of the verifier, thus the simulator gets a parameter indicating the sampling bias (which can be made non uniformly smaller than any given polynomial and relaxes the simulation notion which makes the simulator's view smaller than all inverse polynomials). The technique builds a simulation based on simulation of an underlying honest-verifier protocol and preserves the "black-box" property. We call this model, used throughout, *non-uniform (black-box) simulation* statistical zero-knowledge.

Public-coin honest-verifier vs. any-verifier. We show that for any honest-verifier public-coin SZK proof system for a language L there exists an "any-verifier" non-uniform simulation SZK proof system for L. The first unconditional construction for this result was given in [2] and worked for random self-reducible languages. Another such result good only for constant-round proof systems was given in [5] (based on techniques in [19, 22]). Later, two transformations were shown in [6]: one unconditional, for constant-round proof systems, which improved the round-complexity of the transformation in [5], the other for unbounded-round proof systems, assuming one-way functions.

Honest-verifier vs. any-verifier statistical zero-knowledge. Combining our theorem with results in [20], we show a transformation between a proof system which is non-uniform simulation SZK wrt the honest-verifier into one which is SZK wrt any verifier for the same language. This problem was first posed by [2] who solved it under the intractability of discrete log. Later, this problem has been solved in [22], assuming one-way permutations, and, recently, in [20], assuming one-way functions.

SZK for the complemented language. Combining our theorem with results in [20], we show a transformation between a proof system which is SZK for L into one which is non-uniform simulation SZK for the complemented language $\overline{L}$. The first result along these lines was the one in [9], who constructed a (non zero-knowledge) proof system for $\overline{L}$, assuming an SZK proof system for L (a full proof of this was given in [1]). The result then followed from [17, 4] for the case of CZK, assuming one-way functions. Later, this problem was solved by [2] in the case of SZK, assuming the intractability of discrete log. Recently, in [20], an SZK proof system for $\overline{L}$ was given both assuming one-way functions and in the honest-verifier case.

Private-coin vs. public-coin statistical zero-knowledge. Combining our theorem with results in [20], we show a transformation between a proof system which is private-coin SZK into one which is public-coin non-uniform simulation SZK for the same language. The first result along these lines is due to [14], who proved this transformation between interactive proof systems (i.e., not zero-knowledge). The result then followed from [17, 4] for the case of CZK, assuming one-way functions. Recently, in [20], the result is shown both assuming one-way function, and in the honest-verifier case, for SZK.

2 Non-uniform simulation SZK proof systems

Interactive protocols. Let a pair (A,B) denote an interactive protocol between two interacting probabilistic machines A and B [13]. We denote by x an input common to A and B, by R the content of B's random tape and by y B's auxiliary-input (if any). The *transcript* of an execution of protocol (A,B) on input x, denoted by $tr_{(A,B)}(x)$, is the messages written by A and B during such execution. By $Out_B(tr_{(A,B)}(x)) \in \{accept, reject\}$ we denote B's *output* at the end of the execution of protocol (A,B) on common input x. We define $View_{B(y)}(x)$, B's view of the interaction with A on input x, as the probability space that assigns to pairs $(R; tr_{(A,B(y;R))}(x))$ the probability that R is the content of B's random tape and that $tr_{(A,B(y;R))}(x)$ is the transcript of an execution of protocol (A,B) on common input x given that R is B's random tape and y is B's auxiliary input.

Zero-knowledge proof systems. A zero-knowledge proof system of membership in L is an interactive protocol where the prover convinces a poly-bounded verifier that $x \in L$, without giving additional computational advantage; formally:

Definition 1. *Let P be a probabilistic interactive Turing machine and V a probabilistic poly-time interactive Turing machine sharing input x. Let L be a language. (P,V) is a* SZK PROOF SYSTEM *for L if*

1. *(Completeness) For all $x \in L$, $|x| = n$, for all sufficiently large n, and all constants c, $\text{Prob}(Out_V(tr_{(P,V)}(x)) = accept) = 1 - |x|^c$.*

2. *(Soundness) For any machine P', for all $x \notin L$, $|x| = n$, for all sufficiently large n, and all constants c, $\text{Prob}(Out_V(tr_{(P,V)}(x)) = accept) \leq |x|^{-c}$.*

3. *(Non-uniform Simulation Statistical Zero-Knowledge)*
For any probabilistic polynomial-time Turing machine V' There exists a probabilistic Turing machine $S_{V'}$ such that for all $x \in L$, any auxiliary-input y (of size polynomial in $|x|$), and any constant c, it holds that

- *$S_{V'}$ runs in expected polynomial time (which may depend on c);*
- *$\sum_\alpha |\text{Pr}(View_{V'(y)}(x) = \alpha) - \text{Pr}(S_{V'(y)}(x) = \alpha)| \leq |x|^{-c}$.*

We notice that in our definition, the running time of the simulator is expected polynomial time, and the statistical difference between the two spaces is smaller than the inverse of any polynomial, as in the usual definition of black-box simulation SZK. However, the running time of the simulator and the statistical difference between the two spaces depends also on the constant c (it is true for any constant so it can be made arbitrarily hard to distinguish differences). We will call this relaxed notion, achieving a non-uniform statistical bias based on an input, *non-uniform simulation SZK.*

3 Auxiliary-input cryptographic primitives

In this section we present definitions for auxiliary-input cryptographic primitives with point-wise security arguments (unlike the usual eventually secure notion). They will include distributionally one-way functions, one-way functions, pseudo-random generators, and bit-commitment schemes.

Auxiliary-input one-way functions. Let n be an integer, $aux \in \{0,1\}^n$ be a string, and $f_{aux} : \{0,1\}^n \to \{0,1\}^n$ be an auxiliary-input function. Also, let D_n be a distribution over $\{0,1\}^n$, and U_n be the uniform distribution over $\{0,1\}^n$. We formalize the notion of (locally) breaking and distributionally-breaking an auxiliary-input function.

Definition 2. *We say that an algorithm A_{aux} $(t(n)\epsilon(n))$-breaks the auxiliary-input function f_{aux} on point x if*

$$\mathrm{prob}(y = f_{aux}(x);\ x' \leftarrow A_{aux}(y) \wedge f_{aux}(x') = y) = \epsilon(n),$$

and A_{aux} runs in time $t(n)$ on input 1^n, where the probability is taken over the random coins used by A_{aux}.

Definition 3. *We say that an algorithm A_{aux} $(t(n), \epsilon(n))$-distributionally-breaks the auxiliary-input function f_{aux} on x if*

$$\sum_{\alpha} |\, prob(\, A_{aux}(f_{aux}(x)) \circ f_{aux}(x) = \alpha\,) - prob(\, x \circ f_{aux}(x) = \alpha\,)\,| = \epsilon(n),$$

where by $a \circ b$ we denote concatenation of strings a, b, A_{aux} runs in time $t(n)$ on input 1^n, and the probability is taken over any random coins used by A_{aux}.

Auxiliary-input pseudo-random generators. Let n be an integer, $aux \in \{0,1\}^n$ be a string, and $g_{aux} : \{0,1\}^n \to \{0,1\}^m$, for $m > n$, be an auxiliary-input generator. We formalize the notion of breaking an auxiliary-input generator at a point x.

Definition 4. *We say that algorithm A_{aux} $(t(n), \epsilon(n))$-breaks generator g_{aux} if*

$$\mathrm{prob}(y_1 = g_{aux}(x); y_2 \leftarrow U_{l(n)} :\ A_{aux}(1^n, y_1) \oplus A_{aux}(1^n, y_2) = 1) = \epsilon(n),$$

and A_{aux} runs in time $t(n)$ on input 1^n, where the probability is taken over any random coins used by A_{aux}.

Auxiliary-input bit-commitment schemes.

Definition 5. *Let a pair of interacting Turing machines be Alice which can be an infinitely powerful machine and Bob which is poly-time bounded. An auxiliary-input bit-commitment scheme* $\{(Alice_{aux}, Bob_{aux}(t(n)\epsilon(n)))\}$ *for Alice's input* x *of size* n, *is a two-phase interactive protocols (commit and reveal):*

- *After the commit phase, any probabilistic polynomial time* $Bob'_{aux}(t(n), \epsilon(n))$ *which works* $t(n)$ *time can compute the bit committed by* $Alice_{aux}$ *only with probability* $\leq 1/2 + \epsilon(n)$.
- *In the reveal phase,* $Alice_{aux}$ *can reveal one value (by releasing* x). *For any* $Alice'_{aux}$, *if it tries to reveal a bit different from the one committed in the first phase, then* Bob_{aux} *rejects with overwhelming probability.*

3.1 Extensions to the auxiliary-input case

The following lemmas are simple adaptations of [21, 23, 16, 15, 18] to the auxiliary-input case. They employ reductions between the primitives that relates breaking an input in a given time with a certain probability one primitive to breaking another primitive on a related input with polynomially related time and probability. Lemma 6, due to [21], relates SZK proofs to distributionally one-way functions. It shows that for an input which does not give a distributionally one-way function, the prover's function can be performed efficiently.

Lemma 6. *[21] Let* L *be a language,* (A,B) *a (honest-verifier) SZK proof system for* L *where* M *is its simulator. Let* $x \in \{0,1\}^n$; *for all sufficiently large* n, *there exists an auxiliary-input distributionally one-way function* f_{aux} *such that: If there exists an algorithm* A *which, on input* $x, pref_i$, $(t(n), \epsilon(n))$-*distributionally-breaks* f_{aux}, *where* $aux = x$, *then there exists an algorithm* A' *which, on input* x *of size* n, *runs in time* $t(n) \cdot poly(n)$ *and with overwhelming probability computes* r *such that* $M(r, x) = (pref \circ suff)$, *and* $B(x, pref \circ suff) = 1$, *where* pref *and* suff *are prefix and suffix of* M's *output, respectively.*

Lemma 7 gives an auxiliary-input one-way function evaluated at a point assuming the existence of an auxiliary-input distributionally one-way function. It follows from a result by Impagliazzo and Luby.

Lemma 7. *[16] Let* $aux \in \{0,1\}^n$ *and let* $\{DF_{aux}\}$ *be an auxiliary-input distributionally one-way functions. Then there exists an auxiliary-input one-way functions* $\{F_{aux}\}$ *such that the following holds. For all sufficiently large* n, *if there exists an algorithm* A_{aux} *which* $(t(n), \epsilon(n))$-*breaks* F_{aux} *on input* x *of size* n, *then there exists an algorithm* B_{aux} *which* $(t'(n), \epsilon'(n))$-*distributionally-breaks* DF_{aux} *on* x, *where* $t'(n) = poly(t(n))$ *and* $\epsilon'(n) = poly(\epsilon(n))$.

Lemma 8 gives an auxiliary-input pseudo-random generators assuming the existence of an auxiliary-input one-way function with pointwise translation of hardness. It follows from a result by Håstad, Impagliazzo, Levin and Luby.

Lemma 8. [15] *Let* $aux \in \{0,1\}^n$ *and let* $\{F_{aux}\}$ *be an auxiliary-input one-way functions. Then there exists an auxiliary-input pseudo-random generators* $\{g_{aux}\}$ *such that the following holds. For all sufficiently large n, if there exists an algorithm A_{aux} which $(t(n), \epsilon(n))$-breaks g_{aux} on input x of size n, then there exists an algorithm B_{aux} which $(t'(n), \epsilon'(n))$-breaks F_{aux} on x, where $t'(n) = poly(t(poly(n)))$ and $\epsilon'(n) = poly(\epsilon(poly(n)))$.*

Lemma 9 gives an auxiliary-input bit-commitment schemes assuming the existence of an auxiliary-input pseudo-random generators, which maintains pointwise hardness up to a given polynomial. It follows from a result by Naor.

Lemma 9. [18] *Let* $aux \in \{0,1\}^n$ *and let* $\{g_{aux}\}$ *be an auxiliary-input pseudorandom generators. Then there exists an auxiliary-input bit commitment scheme $BC_{aux} = (Alice_{aux}, Bob_{aux})$ such that for all sufficiently large n for Alice's input x of size n, the following two conditions hold:*

1. *If there exists an algorithm Bob'_{aux} which guesses in time $t(n)$ and probability $\epsilon(n)$ the bit committed by $Alice_{aux}$ in the first phase of $(Alice_{aux}, Bob_{aux})$ then there exists an algorithm D_{aux} which $(t'(n), \epsilon'(n))$-breaks g_{aux} on input x, where $t'(n) = poly(t(n))$ and $\epsilon'(n) = poly(\epsilon(n))$.*
2. *The probability that there exists an algorithm $Alice'_{aux}$ which in the second phase reveals a bit different from the one committed in the first phase is negligible.*

Let us try to summarize the above lemmas. Assuming that L has a honest-verifier SZK proofs, Lemma 6 constructs a collection of auxiliary-input distributionally one-way functions. The sequence of Lemmas 7, 8, 9 transforms any auxiliary-input distributionally one-way function into an auxiliary-input one-way functions, which is in turn transformed into an auxiliary-input pseudo-random generators, which in turn transformed into an auxiliary-input strong-to-weak bit-commitment schemes. The hardness of the various primitive is polynomially related locally (from input to one to the same input of the other). Moreover, combining Lemma 6 and Lemma 7, we have that an algorithm which pointwise guesses (given some resources) a bit committed by BC_x can be used to compute an almost uniformly distributed preimage for the prefix of the output of simulator M for (A,B) (for related polynomial resources). We obtain:

Theorem 10. [21, 16, 15, 18] *Let L be a language, let (A,B) be a (honest-verifier) SZK proof system for L. For all sufficiently large n, let $aux \in \{0,1\}^n$. there exists an auxiliary-input commitment schemes $BC_{aux} = (Alice_{aux}, Bob_{aux})$ such that the following two conditions hold:*

1. *If there exists an algorithm Bob'_{aux} which guesses in time $t(n)$ and with probability $\epsilon(n)$ the bit committed by $Alice_{aux}$ in the commit phase of $(Alice_{aux}, Bob_{aux})$ where x of size n is Alice's input, then there exists an algorithm D_{aux} which, on input a prefix $pref_i$ of the output of simulator M ($pref_i$ is a related function of the transcript of the first phase of the*

commitment scheme), runs in time $(poly(t(poly(n))))$ and with probability $poly(\epsilon(poly(n)))$ computes r such that $M(r,x) = (pref_i \circ suff_i)$, and $B(x, pref_i \circ suff_i) = 1$. Moreover, the distribution of string r is statistically close to the uniform distribution over all strings r such that $M(r,x) = (pref_i \circ suff_i)$ and $B(x, pref_i \circ suff_i) = 1$

2. *The probability that there exists an algorithm $Alice'_{aux}$ which in the second phase reveals a bit different from the one committed in the commit phase is negligible.*

4 The transformation

In this section we will prove the following result.

Theorem 11. *Let L be a language, let (A,B) be a public-coin honest-verifier SZK proof system for L, and let (P,V) be the protocol constructed in Section 4.1. Then (P,V) is an any-verifier non-uniform simulation SZK proof system for L.*

4.1 The protocol (P,V)

Let L be a language and let (A,B) be a public-coin honest-verifier SZK proof system for L. A first step of our transformation is to construct a protocol (C,D) as the parallel execution of n copies of (A,B). Clearly, (C,D) is public-coin and honest-verifier SZK, and we call M the associated simulator (which is the parallel execution of the simulator for (A,B)). We denote by $c(n)$ be the maximum (polynomial) length of a conversation of (C,D), and by $s(n)$ the length of the random string used by M on inputs of size n. In order to construct protocol (P,V), we will construct two functions DF_x, F_x, a generator G_x and a bit-commitment scheme BC_x, where string x is the common input to protocol (P,V). At a very high level, protocol (P,V) can be considered as a way of compiling protocol (C,D) using the bit commitment scheme BC_x where x is the common input. In fact, at any time, protocol (P,V) will use a (single) accepting conversation of protocol (C,D); which we will call the *inner conversation* for protocol (P,V). At the end of the protocol (P,V), V will verify that the inner conversation sent by P is accepting for D. The definition of scheme BC_x, and, more precisely, the definitions of functions F_x and DF_x, will depend on the prefix so far obtained of the inner conversation.

The function F_x. We will define a function F_x in two steps. Informally, in the first step, we would like to construct a function DF_x which, given as input a $s(n)$-bit string R and an index $i \in \{0,1\}^{c(n)}$, returns an i-bit long prefix of the output of the simulator M on input (R,x) concatenated to some padding string $ps = 1 \circ 0^{c(n)-i-1}$. Formally, $DF_x : \{0,1\}^{s(n)} \times \{0,1\}^{\lceil \log c(n) \rceil} \rightarrow \{0,1\}^{c(n)}$ is defined as $DF_x(r,i) = (M(r,x)_{1,...,i}) \circ ps$, where $a_{1,...,i}$ denotes the first i bits of string a. By now, we may think of DF_x as a distributionally one-way function.

Now, we would like to transform function DF_x into a one-way function using the transformation from distributionally one-way functions to one-way functions given in [16]. Precisely, we obtain function F_x by applying a slightly modified version of this transformation, as follows. Let us consider the mentioned transformation from [16]: it takes a function f as input and returns another function g; now, observe that function g requires multiple independent applications of f. Formally, we define function F_x as the function obtained by applying this transformation to function DF_x, with the exception that one of the applications of DF_x (precisely, a uniformly chosen one) is replaced with the input-output pair $((r, i); (pref \circ ps))$, where $pref$ is the prefix so far in the inner conversation for protocol (P,V), $i = |pref|$, ps is some padding string, and r is uniformly distributed among the strings r such that $M(r, x) = (x, pref \circ suff)$ and $D(x, pref \circ suff) = 1$. Since the inner conversation is defined for protocol (C,D), a parallel repetition of (A,B), we will still be able to use the result in [16] for F_x.

The bit commitment scheme BC_x. In order to construct a generator G_x and a bit commitment scheme BC_x, we adapt to the auxiliary-input case results from [15], [18]. Namely, let G_x be the generator that can be constructed starting by the given values of function F_x using the transformation from a one-way function to a pseudo-random generator given in [15]. Finally, let BC_x be the generator that can be constructed starting by generator G_x using the transformation from a pseudo-random generator to a bit-commitment scheme given in [18].

Constructing protocol (P,V). Starting from the assumed public-coin protocol (A,B) for language L, and using scheme BC_x, we construct a protocol (P,V) and show that it is any-verifier SZK.

Previous approaches. The main difficulty with proving that (A,B) is any-verifier SZK is that a dishonest B′ might send messages having a distribution quite different from the uniform one. Given a bit commitment scheme, this problem can be overcome using an idea of [2]. That is, by compiling each step in (A,B) in which B uniformly chooses a bit b and sends it to A, with the following flipping coin protocol. First P commits to a random bit a; then V replies with some possibly biased bit c, and finally the resulting bit b is set equal to $a \oplus c$. Clearly, no matter how V behaves, if P behaves honestly the distribution of bit $a \oplus c$ will be uniform. Now, an idea in [6] is to implement the bit commitment as follows: first, assume that the language L is not in AVBPP (see [23] for definitions), then use the assumption that L has a honest-verifier SZK proof, and sequentially apply known results in the literature to obtain a bit-commitment scheme. Specifically, the results in [21, 23, 15, 18] give a bit commitment scheme.

Our approach. In our protocol we do not make any assumption on the language L. Instead, we adapt results in [21, 24, 16, 15, 18] to the auxiliary-input setting and construct the following auxiliary-input primitives: distributionally one-way function, one-way function, pseudo-random generator and a bit commitment scheme, respectively. One implication of using auxiliary-input primitives in the context of zero-knowledge proofs is that there may exist some x's for which the constructed primitives are not secure. This case considerably complicates the proof of the zero-knowledge property of our protocol. Roughly speaking, we can think of some 'hard' x's for which all primitives are secure, and some 'easy' x's for which they are not, for various level of 'easiness'. In particular, the commitment scheme performed by P seems meaningless in the case of the 'easy'

x's, since it becomes easy for V' to compute the committed bit. We will overcome this problem with the following strategy where the simulator assesses whether the verifier is cheating or not! More precisely, it will make a close estimate of the probability that the verifier V' sends a certain random bit, given that he has received a commitment to a certain bit. By looking at this probability, the simulator will be able to compute if the dishonest V', influences the outcome of the flipping-coin protocol depending on the committed bit (for a given level of influence, where non-uniformity of bias determined by the simulator's input is used for the sampling procedure to work). If the bias of V' is greater than the given polynomial in n, then the simulator will use V' as a black box to break the commitment scheme, invert the "one-way" function, and, finally, run the program of the prover. Here, the fact that the construction of the function uses a prefix of the inner conversation so far will help in keeping the following invariant: the simulator always knows a random string that generates the current inner conversation, no matter what is the cheating behavior of the verifier. On the other hand, if the dishonest V' does not influence the outcome of the flipping-coin protocol (i.e., its level of influence is unnoticeable), then the output bit will be "close enough" to the uniform distribution (it will non-uniformly be smaller than any given polynomial), and therefore the simulator can simulate an execution of the flipping-coin protocol using the usual rewinding technique. Note that the simulation strategy employs the verifier as a black box and thus maintains "black-box simulation". Now, we give more details.

A more formal description of (P,V). Let x, $|x| = n$, be the common input to (P,V). Recall that (A,B) is a public-coin honest-verifier SZK for L, and (C,D) is the parallel execution of n copies of (A,B), and M is the simulator associated with (C,D). Moreover, we assume wlog that the first message in (C,D) is sent by D, and that (C,D) has $r(n)$ rounds; also, each message sent by D has length $k(n)$, and the simulator M uses a random string of length $s(n)$.

The Protocol (P,V)

1. For $i = 1, \ldots, r(n)$,
   P and V set $pref_{i-1} = (a_1, d_1, \ldots, a_{i-1}, d_{i-1})$;
   P computes a string $r_i \in \{0,1\}^{s(n)}$ such that $M(r_i, x) = (pref_{i-1}, a_i, suff)$,
   for some a_i and $B(x, pref_{i-1}, a_i, suff) = 1$;
   P sends a_i to V;
   for $j = 1, \ldots, k(n)$,
      P uniformly chooses $a_{i,j} \in \{0,1\}$ and a seed $s_{i,j} \in \{0,1\}^{g(n)}$;
      P and V run the commit phase of BC_x,
         where P uses bit $a_{i,j}$ and seed $s_{i,j}$ as his private input;
      V uniformly chooses $c_{i,j} \in \{0,1\}$ and sends it to P;
      P and V run the reveal phase of BC_x and set $d_{i,j} = a_{i,j} \oplus c_{i,j}$;
   P and V set $d_i = d_{i,1} \circ \cdots \circ d_{i,k(n)}$;
2. V accepts if all verifications are satisfied and
   if D accepts on input transcript $(x, pref_{r(n)-1}, a_{r(n)})$.

4.2 Proof of correctness for (P,V)

This subsection is devoted to show the following

Lemma 12. *If (A,B) is a public-coin honest-verifier SZK for L, then (P,V) is any-verifier non-uniform simulation SZK for L.*

Completeness. Clearly V runs in polynomial time since so do D and the receiver of scheme BC_x. Now, assume $x \in L$, and P,V are honest. In order to show that P can perform his program with high probability it will be enough to show that he can find a string $r_i \in \{0,1\}^{r(n)}$ satisfying $M(r_i, x) = (x, pref_{i-1}, a_i, suff)$, for some a_i such that transcript $(x, pref_{i-1}, a_i, suff)$ is accepting for (C,D). We observe that since (A,B) is honest-verifier SZK, so is (C,D), and by definition of simulator M, P can always find such a string r_i. Finally, we observe that given that P can successfully perform his program, then the acceptance probability of (P,V) is at least as in (A,B), which is overwhelming.

Soundness. Assume that $x \notin L$. Then notice that the bit $d_{i,j}$ resulting from the output of the j-th execution of the flipping coin protocol compiling round i of (C,D) plays the same role of bit $b_{i,j}$ in the i-th message from D in protocol (C,D). Now, we show that for any P^*, bit $d_{i,j}$ is almost uniformly distributed. To see this, notice that bit $d_{i,j}$ is computed as $c_{i,j} \oplus a_{i,j}$, where bit $c_{i,j}$ is uniformly chosen by V and bit $a_{i,j}$ is the bit decommitted by P', using the scheme in [18]. Now, from property 2 of Lemma 9, we obtain that for any x, P' can decommit two possible values for bit $a_{i,j}$ only with negligible probability, for any x (both in L or not, when not in L the commitment may be non-concealing– but we do not care). Following the analysis done in Claim 3.1 of [18], one can show that the probability that there exist two seeds $s_{i,j}, s'_{i,j}$ that can be used as decommitments of two distinct bits is at most $2^{-g(n)}$, which is negligible. This implies that the distribution of bit $a_{i,j}$ is almost uniform; namely, the probability that $a_{i,j} = b$ is different from $1/2$ only by at most a negligible factor, for $b = 0,1$, and the same holds for bit $d_{i,j}$. Now, since there are at most a polynomial number of bits $d_{i,j}$, the probability that all bits $d_{i,j}$ are not independently and uniformly distributed is negligible. Thus, with probability $1-$ a negligible factor, bits $d_{i,j}$ are distributed exactly as bits $b_{i,j}$ in (C,D). This implies that, if $x \notin L$, and for any P', the probability that V accepts is equal to a negligible factor plus the error probability in the soundness of (C,D), which is negligible.

Any-verifier non-uniform simulation statistical zero-knowledge. We show that for (P,V) for any probabilistic polynomial-time verifier V', there exists a simulator $S_{V'}$, such that, for any $x \in L$, and for any constant c, the statistical distance between $S_{V'}(x)$ and $View_{V'}(x)$ is at most $|x|^{-c}$. The simulation is black-box whenever the one of (A,B) is. Informally, our simulator S will try to simulate an accepting conversation of (P,V), as follows. First, S will generate a conversation of (C,D), and then will try to force the outcomes of the flipping-coin subprotocols executed by (P,V) consistently with the messages of the verifier D, in the inner conversation. In this process, the simulator S will use the rewinding simulation technique to obtain the desired outcome of any flipping-coin protocol. Contrarily to what usually happens, this strategy may not be successful in this case, especially if the cheating verifier V' somehow is able to guess the bit committed by P. We will show that such a cheating verifier can be used as a black box to break the commitment scheme BC_x. Now, notice that by Lemma 10 an algorithm breaking scheme BC_x can be transformed into an algorithm that inverts a prefix of the output of the simulator M for (C,D). The construction of our proof system (P,V) will guarantee that the preimage thus obtained has distance smaller than any given (input) polynomial from the uniform distribution among those strings generating the given prefix of the inner conversation. This will allow us to generate a random string for the simulator which generates the inner conversation obtained so far. Thus, for any cheating verifier V', the simulation will always make some progress, either thanks to the rewinding technique, or because the cheating behavior of V' allows us to compute a random string which generates the inner conversation so far. In this way, $r(n)k(n)$ phases are sufficient to guarantee that the simulator outputs an accepting conversation of protocol (P,V). Now we proceed more formally.

The procedures Estimate, Guess and Invert. The algorithm S will use three procedures. The first procedure, called Estimate will be run by S to evaluate the cheating behavior of V' in each execution of a flipping-coin protocol. Specifically, notice that for some ('easy' or 'hard') x's, V' might be able to correctly guess the committed bit $a_{i,j}$ with some probability bounded away from $1/2$, and then influence the distribution of the bit $d_{i,j}$ output of the flipping-coin protocol. Using procedure Estimate for each execution of the flipping-coin protocol, the simulator will estimate the bias $\delta_{i,j}$ caused by V' on bit $d_{i,j}$. The estimate will be done as follows: on input a constant c let $r(n) = n^{r'}, k(n) = n^{k'}$, and let $q = \max\{100, 2(c+r'k')+1\}$, S will try to estimate $d_{i,j}$ with a number $e_{i,j}$ such that $e_{i,j}$ is near $1/2$: $e_{i,j} \in [1/2 - n^{-q}, 1/2 + n^{-q}]$ (i.e., be unnoticeable) with exponentially small probability of errors, or not (i.e., $e_{i,j}$ is bounded away from $1/2$, e.g. $e_{i,j} \in [0, 1/2 - 3n^{-q}] \cup [1/2 + 3n^{-q}, 1]$) Using a large enough polynomial-size sample n^s, extreme assumption on the bias probability, and Chernoff bounds, an exponentially small error in the estimate is possible (using an appropriate ϵ in the estimation test below). Thus the accuracy error of this estimate will be less than n^{-q}, which is enough for us since we would like the distance between the output of the simulator and the output of the protocol to be smaller than n^{-c}.

Now, on input the common input to the proof system x, the constant q, the number i of round and the number j of bit in the i-th round of (C,D), procedure Estimate runs the following steps:

Procedure Estimate:

1. set $count_{0,0} = count_{0,1} = count_{1,0} = count_{1,1} = 0$;
2. for $l = 1, \ldots, n^s$,
   uniformly choose bit a_l and commit to a_l using scheme BC_x;
   get bit c_l from V';
   set $count_{a_l,c_l} = count_{a_l,c_l} + 1$;
   reveal bit a_l to V' using scheme BC_x;
   rewind V' to the state just before running the j-th execution of protocol BC_x
      in the i-th round;
3. set $p_{h,k} = count_{h,k}/n^{2q+1}$, for $h, k = 0, 1$, and $\delta_{i,j} = |p_{0,1} + p_{1,0} - p_{1,1} - p_{0,0}|$;

4. if $\delta_{i,j} > \epsilon$ then set $bias = yes$ else set $bias = no$;
5. output $(bias, p_{0,0}, p_{1,0}, p_{0,1}, p_{1,1})$ and return.

Procedure **Estimate** will be used by the simulator $S_{V'}$ to distinguish whether the bias added by the verifier V' to the distribution of the output of a flipping coin protocol is at least n^{-q} or not. In fact, the procedure satisfies the following.

Fact 13. Let $bias$ be output by procedure **Estimate** on input x, i, j. The bias $\delta_{i,j}$ of bit $b_{i,j}$ in an execution of (P,V') is at least n^{-q} if $bias = yes$ or smaller than n^{-q} if $bias = no$.

In the case $bias = no$, Fact 13 guarantees that the bias $\delta_{i,j}$ is smaller than n^{-q}. Then the simulator $S_{V'}$ is able to successfully simulate the j-th execution in the i-th round of the bit commitment scheme BC_x, by using the rewinding technique until it holds that $d_{i,j} = b_{i,j}$. This happens in an expected number of steps that is at most $2n^q$. In the second case, namely, when $bias = yes$, the simulator $S_{V'}$ is *not* able to successfully simulate the j-th execution in the i-th round of the bit commitment scheme BC_x. However, since the distribution of bit $b_{i,j}$ is far from uniform, with sufficiently high probability the value of $c_{i,j}$ is chosen by V' depending on that of $a_{i,j}$, and thus V' can break the commitment scheme with high probability. Formally, on input x, and given the transcript $tr_{i,j}$ of the commit phase of an execution of scheme BC_x, procedure **Guess** does the following:

Procedure Guess:

1. Let $tr_{i,j}$ be the transcript of the commit phase of an execution of BC_x with V';
2. get bit $c_{i,j}$ from V';
3. if $p_{1,1} + p_{0,0} > p_{0,1} + p_{1,0}$ then set $a_{i,j} = c_{i,j}$ else set $a_{i,j} = 1 - c_{i,j}$;
4. output $a_{i,j}$ and return.

Procedure **Guess** assumed (for exposition) uniform behavior on cheating on commitment of 1 and commitment of 0. A refinement leading to a guess in other behaviors is possible. The procedure satisfies the following fact

Fact 14. The bit $a_{i,j}$ output by procedure **Guess** on input $x, i, j, tr_{i,j}$ is equal to the bit committed by P in transcript $tr_{i,j}$ with probability at least $\delta_{i,j}$.

From Fact 14 it follows that there exists an algorithm which breaks the commitment scheme BC_x. Then, using the fact that protocol (C,D) is the parallel execution of protocol (A,B), together with the reduction in Theorem 10, we obtain that there exists an algorithm **Invert** which (using appropriate polynomial resources and success probability), given input $x, pref$, returns a randomly chosen string which allows the simulator M to generate the current prefix of the inner conversation. Formally, we obtain the following

Fact 15. Let r be the output of algorithm **Invert** on input $x, pref_i$. Then the distribution of r is statistically close to the uniform distribution on the strings r such that $M(r, x) = (x, pref_i \circ suff_i)$, and $D(x, pref_i \circ suff_i) = 1$.

The simulator $S_{V'}$. On input x, and constant c, algorithm $S_{V'}$ uses procedures Estimate and Invert, as follows.

1. Uniformly choose an $s(n)$-bit string R.
2. Run M on input R, x thus obtaining conversation
   $$conv = ((b_{1,1}, \ldots, b_{1,tk}), a_1, \ldots, a_{r-1}, (b_{r,1}, \ldots, b_{r,tk}), a_r) \text{ as output.}$$
3. For $i = 1, \cdots, r$,
       for $j = 1, \ldots, tk$,
           run procedure Estimate on input x, i, j, c, and let $bias$ be its output;
           if $bias = no$ then
               repeat
                   rewind V' until after message a_{i-1} was sent to him;
                   run a flipping coin protocol interacting with V'
                     and let $d_{i,j}$ be the resulting output;
               until $d_{i,j} = b_{i,j}$;
           if $bias = yes$ then
               run a flipping coin protocol interacting with V'
                   and let $b_{i,j}$ be the resulting output;
               rewind V' until after message a_{i-1} was sent to him;
               set $pref_i = (\mathbf{b}_1, a_1, \ldots, \mathbf{b}_{i-1}, a_{i-1}, (b_{i,1}, \ldots, b_{i,j}))$;
               run algorithm Invert on input $x, pref_i$, and let R_i be its output;
               set $conv = (\mathbf{b}_1, a_1, \ldots, a_{r-1}, \mathbf{b}_r, a_r) = M(R_i, x)$;
           send message a_i to V'.
4. Output: $conv$ and halt.

We observe that the expected running time of simulator S is $poly(n) \cdot n^{O(q)}$, and thus it is expected polynomial time, for any given q derived from any given constant c. Now we need to show that the output of the simulator is statistically close to the view of the verifier.

In the next lemma we prove that the output of the simulator is statistically close to the view of the verifier.

Lemma 16. *For any common input x and auxiliary input y, for any constant c, the statistical distance between the output of $S_{V'}$ on input x, c and the view of V' in protocol (P, V') on input x is at most $|x|^{-c}$.*

Proof. All messages from V' are equally distributed in both spaces since they are computed in the same way. Now, we consider the messages from the prover in both spaces. The messages sent by the prover in the commitment phase of BC_x are computed almost in the same way in both spaces; here, the only difference is that the output of each flipping coin protocol is uniformly distributed in the simulation (since it is equal to the random bit output by M) while it has a bias smaller than n^{-q} in the view of V'. However, this contributes a factor smaller than n^{-c} to the statistical difference between the two spaces. Now, we consider the messages a_i from the prover, for $i = 1, \ldots, r$. We observe that in the output of S they are all computed in two ways: either they are taken from the output of M on input the uniformly distributed string R and x, or they are computed by first using procedure Invert to compute a random string R_i for M and then they are taken from the output of M on input R_i and x. In the first case it holds that in both spaces the message a_i is distributed as a message output from M on a random string. In the second case, the distribution of this message in the output of the protocol is as in the first case. In the output of the simulator S, instead, the procedure Invert is executed. Now, notice that, given prefix $pref_i$ of a conversation, procedure Invert computes a string R_i which, by Fact 15, is chosen uniformly enough among those satisfying $M(R_i, x) = (pref_i, suff_i)$, and $D(x, pref_i, suff_i) = 1$. In other words, message a_{i+1} is computed with a distribution statistically close to that of the simulator $S_{V'}$, conditioned by the conversation so far, which is statistically close to the prover's distribution. Therefore, the overall statistical distance between the two spaces is less than n^{-c}. $\square$

5 Implications on non-uniform simulation SZK proofs

We combine the result in Theorem 11 with some results in the literature. This will show results regarding non-uniform simulation SZK.

Theorem 17. *Let L be a language. If L has a honest-verifier SZK proof then L has a non-uniform sim. SZK proof.*

Proof. Assume L has a honest-verifier SZK proof system (A,B). If (A,B) is public-coin, applying Theorem 11 will prove the result. If (A,B) is private-coin, then a result in [20] allows to obtain a public-coin honest-verifier SZK proof system, and, then, applying again Theorem 11 will prove the result. $\square$

Theorem 18. *Let L be a language. If L has a private-coin SZK proof then L has a non-uniform sim. public-coin SZK proof.*

Proof. Assume L has a private-coin SZK proof system (A,B). Using a result in [20], (A,B) can be transformed into a honest-verifier public-coin SZK proof system (C,D) for L. Then, using Theorem 11, (C,D) can be transformed into an SZK proof system for L. $\square$

Theorem 19. *Let L be a language. If L has a SZK proof then $\overline{L}$ has a non-uniform sim. SZK proof.*

Proof. Assume L has a SZK proof system (A,B). Using a result in [20], (A,B) can be transformed into a honest-verifier SZK proof system (C,) for $\overline{L}$. Then, using Theorem 17, (C,D) can be transformed into a SZK proof system for $\overline{L}$. $\square$

Acknowledgements. Many thanks go to Alfredo De Santis, Oded Goldreich and Russell Impagliazzo for valuable discussions and remarks.

References

1. W. Aiello and J. Håstad. *Statistical Zero Knowledge Can Be Recognized in Two Rounds*, Journal of Computer and System Sciences, vol. 42, 1991, pp. 327–345.
2. M. Bellare, S. Micali, and R. Ostrovsky, *The (True) Complexity of Statistical Zero-Knowledge Proofs*, in STOC 90.
3. M. Bellare, and E. Petrank, *Making Zero-Knowledge Provers Efficient*, STOC 92.
4. M. Ben-Or, O. Goldreich, S. Goldwasser, J. Håstad, J. Kilian, S. Micali, and P. Rogaway, *Everything Provable is Provable in Zero Knowledge*, in CRYPTO 88.
5. I. Damgård, *Interactive Hashing can Simplify Zero-Knowledge Design without Complexity assumptions*, in CRYPTO 92.
6. I. Damgård, O. Goldreich, T. Okamoto, and A. Wigderson, *Honest-Verifier vs. Dishonest-Verifier in Public-Coin Zero-Knowledge Proofs*, in CRYPTO 95.
7. A. De Santis, G. Di Crescenzo, P. Persiano, and M. Yung, *On Monotone Formula Closure of SZK*, in FOCS 94.
8. U. Feige, A. Fiat, and A. Shamir, *Zero-Knowledge Proofs of Identity*, Journal of Cryptology, vol. 1, 1988, pp. 77–94.
9. L. Fortnow, *The Complexity of Perfect Zero Knowledge*, in STOC 87.
10. O. Goldreich and H. Krawczyk, *On the Composition of Zero-Knowledge Proof Systems*, SIAM Journal on Computing, 1996.
11. O. Goldreich, S. Micali, and A. Wigderson, *Proofs that Yield Nothing but their Validity or All Languages in NP Have Zero-Knowledge Proof Systems*, Journal of the ACM, vol. 38, n. 1, 1991, pp. 691–729.
12. O. Goldreich and Y. Oren, *Definitions and Properties of Zero-Knowledge Proof Systems*, Journal of Cryptology, v. 7, n. 1, 1994.
13. S. Goldwasser, S. Micali, and C. Rackoff, *The Knowledge Complexity of Interactive Proof-Systems*, SIAM Journal on Computing, vol. 18, n. 1, February 1989.
14. S. Goldwasser and M. Sipser, *Private Coins versus Public Coins in Interactive Proof-Systems*, in STOC 1986.
15. J. Håstad, R. Impagliazzo, L. Levin, and M. Luby, *Construction of a Pseudo-Random Generator from One-Way Function*, to appear in SIAM Journal on Computing, previous versions: FOCS 89 and STOC 90.
16. R. Impagliazzo and M. Luby, *One-Way Functions are Necessary for Complexity-Based Cryptography*, in FOCS 90.
17. R. Impagliazzo and M. Yung, *Direct Minimum Knowledge Computations*, in CRYPTO 87.
18. M. Naor, *Bit-Commitment Using Pseudo-Randomness*, in CRYPTO 89.
19. M. Naor, R. Ostrovsky, R. Venkatesan, and M. Yung, *Perfectly-Secure Zero-Knowledge Arguments Can be Based on General Complexity Assumptions*, in CRYPTO 92.
20. T. Okamoto, *On Relations Between Statistical Zero-Knowledge Proofs*, STOC 96.
21. R. Ostrovsky, *One-Way Functions, Hard on Average Problems and Statistical Zero-Knowledge Proofs*, in Structures 91.
22. R. Ostrovsky, R. Venkatesan, and M. Yung, *Interactive Hashing Simplifies Zero-Knowledge Protocol Design*, in EUROCRYPT '93.
23. R. Ostrovsky, and A. Wigderson, *One-way Functions are Essential for Non-Trivial Zero-Knowledge*, in ISTCS 93.
24. A. Yao, *Theory and Applications of Trapdoor Functions*, in FOCS 81.

On the Foundations of Modern Cryptography

Oded Goldreich

Department of Computer Science and Applied Mathematics, Weizmann Institute of Science,
Rehovot, ISRAEL. Email: oded@wisdom.weizmann.ac.il.

Abstract. In our opinion, the Foundations of Cryptography are the paradigms, approaches and techniques used to conceptualize, define and provide solutions to natural cryptographic problems. In this essay, we survey some of these paradigms, approaches and techniques as well as some of the fundamental results obtained using them. Special effort is made in attempt to dissolve common misconceptions regarding these paradigms and results.

> *It is possible to build a cabin with no foundations,*
> *but not a lasting building.*
>
> Eng. Isidor Goldreich (1906–1995)

1 Introduction

Cryptography is concerned with the construction of schemes which are robust against malicious attempts to make these schemes deviate from their prescribed functionality. Given a desired functionality, a cryptographer should design a scheme which not only satisfies the desired functionality under "normal operation", but also maintains this functionality in face of adversarial attempts which are devised after the cryptographer has completed his/her work. The fact that an adversary will devise its attack after the scheme has been specified makes the design of such schemes very hard. In particular, the adversary will try to take actions other than the ones the designer had envisioned. Thus, our approach is that it makes little sense to make assumptions regarding the specific *strategy* that the adversary may use. The only assumptions which can be justified refer to the computational *abilities* of the adversary. Furthermore, it is our opinion that the design of cryptographic systems has to be based on *firm foundations*; whereas ad-hoc approaches and heuristics are a very dangerous way to go. A heuristic may make sense when the designer has a very good idea about the environment in which a scheme is to operate, yet a cryptographic scheme has to operate in a maliciously selected environment which typically transcends the designer's view.

Providing firm foundations to Cryptography has been a major research direction in the last two decades. Indeed, the pioneering paper of Diffie and Hellman [46] should be considered the initiator of this direction. Two major (interleaved) activities have been:

1. Definitional Activity: The identification, conceptualization and rigorous definition of cryptographic tasks which capture natural security concerns; and
2. Constructive Activity: The study and design of cryptographic schemes satisfying definitions as in (1).

The definitional activity provided a definition of secure encryption [73]. The reader may be surprised: *what is there to define* (beyond the basic setting formulated in [46])? Let us answer with a question (posed by [73]): *should an encryption scheme which leaks the first bit of the plaintext be considered secure?* Clearly, the answer is negative and so some naive conceptions regarding secure encryption (e.g., "a scheme is secure if it is infeasible to obtain the plaintext from the ciphertext when not given the decryption key") turn out to be unsatisfactory. The lesson is that even when a natural concern (e.g., "secure communication over insecure channels") has been identified, work still needs to be done towards a satisfactory (rigorous) definition of the underlying concept. The definitional activity also undertook the treatment of unforgeable signature schemes [75]: One result of the treatment was the refutation of a "folklore theorem" (attributed to Ron Rivest) by which "a signature scheme that is robust against chosen message attack cannot have a proof of security". The lesson here is that unclear/unsound formulations (i.e., those underlying the above folklore paradox) lead to false conclusions.

Another existing concept which was re-examined is the then-fuzzy notion of a "pseudo-random generator". Although ad-hoc "pseudorandom generators" which pass some ad-hoc statistical tests may be adequate for some statistical samplings, they are certainly inadequate for use in Cryptography: For example, sequences generated by linear congruential generators are easy to predict [24, 58] and endanger cryptographic applications even when not given in the clear [8]. The alternative suggested in [22, 73, 119] is a robust notion of pseudorandom generators – such a generator produces sequences which are *computationally indistinguishable* from truly random sequences, and thus, can replace truly random sequences in any practical application. We mention that the notion of computational indistinguishability has played a central role in the formulation of other cryptographic concepts (such as secure encryption and zero-knowledge).

The definitional activity has identified concepts which were not known before. One well-known example is the introduction of zero-knowledge proofs [74]. A key paradigm crystallized in making the latter definition is the *simulation paradigm*: A party is said to have gained nothing from some extra information given to it if it can generate (i.e., simulate the receipt of) essentially the same information by itself (i.e., without being given this information). The simulation paradigm plays a central role in the related definitions of secure multi-party computations (with respect to varying settings such as in [96, 2, 72, 27]). However, it has been employed also in different settings such as in [13, 14, 31].

The definitional activity is an on-going process. Its more recent targets have included mobile adversaries (aka "proactive security") [107, 32, 80], Electronic Cash [36], Coercibility [30, 29], Threshold Cryptography [45], and more.

The constructive activity. As new definitions of cryptographic tasks emerged, the first challenge was to demonstrate that they can be achieved. Thus, the first goal of the constructive activity is to *demonstrate the plausibility* of obtaining certain goals. Thus, standard assumptions such as that the RSA is hard to invert were used to construct secure public-key encryption schemes [73, 119] and unforgeable digital schemes [75]. We stress that assuming that RSA is hard to invert is different from assuming that RSA is a secure encryption scheme. Furthermore, plain RSA (alike any deterministic public-key encryption scheme) is not secure (as one can easily distinguish the encryption of one *predetermined* message from the encryption of another). Yet, RSA can be easily transformed into a secure public-key encryption scheme by using a construction which is reminiscent of a common practice (of padding the message with random noise). We stress that the resulting scheme is not merely believed to be secure but rather its security is linked to a much simpler assumption (i.e., the assumption that RSA is hard to invert). Likewise, although plain RSA signing is vulnerable to "existential forgery" (and other attacks), RSA can be transformed into a signature scheme which is unforgeable (provided RSA is hard to invert). Using the assumption that RSA is hard to invert, one can construct pseudorandom generators [22, 119], zero-knowledge proofs for any NP-statement [68], and multi-party protocols for securely computing any multi-variant function [120, 69].

A major misconception regarding theoretical work in Cryptography stems from not distinguishing work aimed at demonstrating the plausibility of obtaining certain goals from work aimed at suggesting paradigms and/or constructions which can be used in practice. For example, the general results concerning zero-knowledge proofs and multi-party protocols [68, 120, 69] mentioned above are merely *claims of plausibility*: What they say is that any problem of the above type (i.e., any protocol problem as discussed in Section 7) can be solved in principle. This is a very valuable piece of information. Thus, if you have a specific problem which falls into the above category then you should know that the problem is solvable in principle. However, if you need to construct a real system then you should probably construct a solution from scratch (rather than employing the above general results). Typically, *some* tools developed towards solving the general problem may be useful in solving the specific problem. Thus, we distinguish three types of results:

1. *Plausibility results:* Here we refer to mere statements of the type "any NP-language has a zero-knowledge proof system" (cf., [68]).
2. *Introduction of paradigms and techniques which may be applicable in practice:* Typical examples include construction paradigms as the "choose n out of $2n$ technique" of [109], the "authentication tree" of [92, 94], the "randomized encryption" paradigm of [73], proof techniques as the "hybrid argument" of [73] (cf., [62, Sec. 3.2.3]), and many others.
3. *Presentation of schemes which are suitable for practical applications:* Typical examples include the public-key encryption schemes of [21], the digital signature schemes of [50, 49], the session-key protocols of [13, 14], and many others.

Typically, it is quite easy to determine to which of the above categories a specific technical contribution belongs. Unfortunately, the classification is not always stated in the paper; however, it is typically evident from the construction. We stress that all results we are aware of (and in particular all results cited in this essay), come with an explicit construction. Furthermore, the security of the resulting construction is explicitly related to the complexity of certain intractable tasks. In contrast to some uninformed beliefs, for each of these results there is an explicit translation of concrete intractability assumptions (on which the scheme is based) into lower bounds on the amount of work required to violate the security of the resulting scheme.[1] We stress that this translation can be invoked for any value of the security parameter. Doing so determines whether a specific construction is adequate for a specific application under specific reasonable intractability assumptions. In many cases the answer

---

[1] The only exception to the latter statement is Levin's observation regarding the existence of a *universal one-way function* (cf., [89] and [62, Sec. 2.4.1]).

is in the affirmative, but in general this does depend on the specific construction as well as on the specific value of the security parameter and on what is reasonable to assume for this value. When we say that a result is suitable for practical applications (i.e., belongs to Type 3 above), we mean that it offers reasonable security for reasonable implementation values of the security parameter and reasonable assumptions.

Other activities. This essay is focused on the definitional and constructive activities mentioned above. Other activities in the foundations of cryptography include the exploration of new directions and the marking of limitations. For example, we mention novel modes of operation such as split-entities [16, 45, 95], batching operations [55], off-line/on-line signing [50] and Incremental Cryptography [6, 7]. On the limitation side, we mention [83, 66]. In particular, [83] indicates that certain tasks (e.g., secret key exchange) are unlikely to be achieved by using a one-way function in a "black-box manner".

Organization: Although encryption, signatures and secure protocols are the primary tasks of Cryptography, we start our presentation with basic paradigms and tools such as computational difficulty (Section 2), pseudorandomness (Section 3) and zero-knowledge (Section 4). Once these are presented, we turn to encryption (Section 5), signatures (Section 6) and secure protocols (Section 7). We conclude with some notes (Section 8), two suggestions for future research (Section 9) and some suggestions for further reading (Section 10).

This essay has been written under close to impossible time constraints[2] and the result is likely to have faults of various types. I intend to revise the essay in the future and make the revision available from http://theory.lcs.mit.edu/~oded/tfoc.html.

2 Central Paradigms

Modern Cryptography, as surveyed here, is concerned with the construction of *efficient* schemes for which it is *infeasible* to violate the security feature. Thus, we need a notion of efficient computations as well as a notion of infeasible ones. The computations of the legitimate users of the scheme ought be efficient; whereas violating the security features (via an adversary) ought to be infeasible. Our notions of efficient and infeasible computations are "asymptotic": They refer to the running time as a function of the security parameter. This is done in order to avoid cumbersome formulations which refer to the actual running-time on a specific model for specific values of the security parameter. As discussed above one can easily derive such specific statements from the asymptotic treatment. Actually, the term "asymptotic" is misleading since, from the functional treatment of the running-time (as a function of the security parameter), one can derive statements for ANY value of the security parameter.

Efficient computations are commonly modeled by computations which are polynomial-time in the security parameter. The polynomial bounding the running-time of the legitimate user's strategy is fixed and typically explicit and small (still in some cases it is indeed a valuable goal to make it even smaller). Here (i.e., when referring to the complexity of the legitimate user) we are in the same situation as in any algorithmic research. Things are different when referring to our assumptions regarding the computational resources of

---

[2] Specifically, I was invited to write this essay less than a month before the deadline.

the adversary. A common approach is to postulate that the latter are polynomial-time too, where the polynomial is NOT a-priori specified. In other words, the adversary is restricted to the class of efficient computations and anything beyond this is considered to be infeasible. Although many definitions explicitly refer to this convention, this convention is INESSENTIAL to any of the results known in the area. In all cases, a more general (and yet more cumbersome) statement can be made by referring to adversaries of running-time bounded by any function (or class of functions). For example, for any function $T : \mathbb{N} \mapsto \mathbb{N}$ (e.g., $T(n) = 2^{\sqrt[3]{n}}$), we may consider adversaries which on security parameter n run for at most $T(n)$ steps. Doing so we (implicitly) define as infeasible any computation which (on security parameter n) requires more than $T(n)$ steps. A typical result has the form[3]

> If RSA with n-bit moduli cannot be inverted in time $T(n)$ then the following construction (using security parameter n) is secure against adversaries operating in time $T'(n) = T(g(n))/f(n)$, where f and g^{-1} are explicitly given polynomials.

However, most papers prefer to present a simplified statement of the form "if RSA cannot be inverted in polynomial-time then the following construction is secure against polynomial-time adversaries". This is unfortunate since it is the specific functions f and g, which are (sometimes explicit and) always implicit in the proof, that determine the practicality of the construction.[4] The smaller f and g^{-1}, the better. Our rule of thumb is that results with $g^{-1}(n) = O(n)$ (e.g., $g(n) = n/2$) are practical, whereas results with, say, $g^{-1}(n) = n^4$ (i.e., $g(n) = \sqrt[4]{n}$) are to be considered merely plausibility results.

Lastly we consider the notion of a negligible probability. The idea behind this notion is to have a robust notion of rareness: A rare event should occur rarely even if we repeat the experiment for a feasible number of times. That is, if we consider any polynomial-time computation to be feasible then any function $f : \mathbb{N} \mapsto \mathbb{N}$ so that $(1 - f(n))^{p(n)} < 0.01$, for any polynomial p, is considered negligible (i.e., f is negligible if for any polynomial p the function $f(\cdot)$ is bounded above by $1/p(\cdot)$). However, if we consider the function $T(n)$ to provide our notion of infeasible computation then functions bounded above by $1/T(n)$ are considered negligible (in n).

In the rest of this essay we adopt the simpler convention of defining infeasible computations as ones which cannot be conducted in polynomial-time. (However, we explicitly state the level of practicality of each of the results presented.) The interested reader is referred to [90] for a more general treatment.

2.1 Computational Difficulty

Modern Cryptography is concerned with the construction of schemes which are easy to operate (properly) but hard to foil. Thus, a complexity gap (i.e., between the complexity of proper usage and the complexity of defeating the prescribed functionality) lies in the heart of Modern Cryptography. However, gaps as required for Modern Cryptography are

---

[3] Actually, the form below is over-simplified. The actual statement refers also to the success probabilities of both attacks, and relates the two (time,probability)-pairs of measures.

[4] The importance of *explicitly* relating the security of the resulting scheme to the quantified intractability assumption has been advocated (and practiced) in a sequence of recent works by Bellare and Rogaway (cf., [10, p. 343]).

not known to exist – they are only widely believed to exist. Indeed, almost all of Modern Cryptography raises or falls with the question of whether one-way functions exist (e.g., see [78, 63, 114, 98, 68] for positive results and [89, 114, 106] for negative ones). One-way functions are functions which are easy to evaluate but hard (on the average) to invert.

Definition 1 (one-way functions [46]): *A function* $f : \{0, 1\}^* \mapsto \{0, 1\}^*$ *is called one-way if*

- easy direction: *there is an efficient algorithm which on input x outputs $f(x)$.*
- hard direction: *given $f(x)$, where x is uniformly selected, it is infeasible to find, with non-negligible probability, a preimage of $f(x)$. That is, any feasible algorithm which tries to do invert f may succeed only with negligible probability, where the probability is taken over the choices of x and the algorithm's coin tosses.*

Warning: the above definition, as well as all other definitions in this essay, avoids some technicalities and so is imprecise. The interested reader is referred to other texts (see Section 10).

2.2 Computational Indistinguishability

A central notion in Modern Cryptography is that of "effective similarity". The underlying idea is that we do not care if objects are equal or not – all we care is whether a difference between the objects can be observed by a feasible computation. In case the answer is negative, we may say that the two objects are equivalent as far as any practical application is concerned. Indeed, it will be our common practice to interchange such (computationally indistinguishable) objects.

Definition 2 (computational indistinguishability [73, 119]): *Let* $X = \{X_n\}_{n \in \mathbb{N}}$ *and* $Y = \{Y_n\}_{n \in \mathbb{N}}$ *be probability ensembles such that each X_n and Y_n ranges over strings of length n. We say that X and Y are* computationally indistinguishable *if for every feasible algorithm A the difference*

$$d_A(n) \stackrel{\text{def}}{=} |\Pr(A(X_n)=1) - \Pr(A(Y_n)=1)|$$

is a negligible function in n.

2.3 The Simulation Paradigm

A key question regarding the modeling of security concerns is how to express the intuitive requirement that an adversary "gains nothing substantial" by deviating from the prescribed behavior of an honest user. The approach initiated in [73, 74] is that the adversary *gains nothing* if whatever it can obtain by deviating from the prescribed honest behavior can also be obtained in an appropriately defined "ideal model". The definition of the "ideal model" captures what we want to achieve in terms of security, and so is specific to the security concern to be addressed. For example, an encryption scheme is considered secure (against eavesdropping) if an adversary which eavesdrops on a channel on which messages are sent, using this encryption scheme, gains nothing over a user which does not tap this channel. Thus, the encryption scheme "simulates" an ideal private channel between parties.

A notable property of the above simulation paradigm, as well as of the entire approach surveyed here, is that this approach is very liberal with respect to its view of the abilities of the adversary as well as to what might constitute a gain for the adversary. For example, we consider an encryption scheme to be secure only if it can simulate a private channel. Indeed, failure to provide such a simulation does NOT necessarily mean that the encryption scheme can be "broken" in some intuitively harmful sense. Thus, it seems that our approach to defining security is overly cautious. However, it seems impossible to come up with definitions of security which distinguish "breaking the scheme in a harmful sense" from "breaking it in a non-harmful sense": What is harmful is application-dependent, whereas a good definition of security ought to be application independent (as otherwise using the scheme in any new application will require a full re-evaluation of its security). Furthermore, since we are interested in secure schemes, there is no harm in employing overly cautious definitions, provided that this does not prevent us (or even disturb us) from constructing "good" schemes. We claim that this has been the case in the past. In most cases it has been possible to construct schemes which meet the overly cautious definitions (of security), and in other cases the difficulty to construct such schemes has demonstrated an inherent problem (e.g., [83, 66]).

3 Pseudorandomness

In practice "pseudorandom" sequences are used instead of truly random sequences in many applications. The underlying belief is that if an (efficient) application performs well when using a truly random sequence then it will perform essentially as well when using a "pseudorandom" sequence. However, this belief is not supported by previous characterizations of "pseudorandomness" (e.g., such as passing the statistical tests in Knuth's book or having large linear-complexity). In contrast, the above belief is an easy corollary of defining pseudorandom distributions as ones which are computationally indistinguishable from uniform distributions. We are interested in pseudorandom sequences which can be generated and determined by short random seeds. That is,

Definition 3 (pseudorandom generator [22, 119]): *Let $\ell: \mathbb{N} \mapsto \mathbb{N}$ be so that $\ell(n) > n$, $\forall n$. A* pseudorandom generator, *with* stretch function ℓ, *is an* efficient *(deterministic)* algorithm which on input a random n-bit *seed* outputs a $\ell(n)$-bit sequence which is computationally indistinguishable from a uniformly chosen $\ell(n)$-bit sequence.

We stress that pseudorandom sequences can replace truly random sequences not only in "ordinary" computations but also in cryptographic ones. That is, ANY cryptographic application which is secure when the legitimate parties use truly random sequences, is also secure when the legitimate parties use pseudorandom sequences. Various cryptographic applications of pseudorandom generators will be presented in the sequel, but first let us consider the construction of pseudorandom generators. A key paradigm is presented next. It uses the notion of a *hard-core* predicate [22] of a (one-way) function: The predicate b is a hard-core of the function f if b is easy to evaluate but $b(x)$ is hard to predict from $f(x)$. That is, it is infeasible, given $f(x)$ when x is uniformly chosen, to predict $b(x)$ substantially better than with probability $1/2$. Intuitively, b "inherits *in a concentrated sense*" the difficulty of inverting f. (Note that if b is a hard-core of an efficiently computable 1-1 function f then f must be one-way.)

The iteration paradigm [22]: Let f be a 1-1 function which is length-preserving and efficiently computable, and b be a hard-core predicate of f. Then

$$G(s) = b(s) \cdot b(f(s)) \cdots b(f^{\ell(|s|)-1}(s))$$

is a pseudorandom generator (with stretch function ℓ), where $f^{i+1}(x) \stackrel{\text{def}}{=} f(f^i(x))$ and $f^0(x) \stackrel{\text{def}}{=} x$. As a concrete example, consider the permutation $x \mapsto x^2 \bmod N$, where N is the product of two primes each congruent to 3 (mod 4). We have $G_N(s) = \text{lsb}(s) \cdot \text{lsb}(s^2 \bmod N) \cdots \text{lsb}(s^{2^{\ell(|s|)-1}} \bmod N)$, where $\text{lsb}(x)$ is the least significant bit of x (which by [1, 117] is a hard-core of the modular squaring function). We note that for any one-way permutation f', the inner-product mod 2 of x and r is a hard-core of $f(x, r) = (f'(x), r)$ [67]. Thus, using any one-way permutation, we can easily construct pseudorandom generators.

The iteration paradigm is even more beneficial when one has a hard-core function rather than a hard-core predicate: h is called a *hard-core function* of f if h is easy to evaluate but, for a random $x \in \{0, 1\}^*$, the distribution $f(x) \cdot h(x)$ is pseudorandom. (Note that a hard-core predicate is a special case.) Using a hard-core function h for f, we obtain the pseudorandom generator $G'(s) = h(s) \cdot h(f(s)) \cdot h(f^2(s)) \cdots$. In particular, assuming the intractability of the subset sum problem (for suitable densities) this paradigm was used in [82] to construct very efficient pseudorandom generators. Alternatively, encouraged by the results in [1, 79], we conjecture that the first $n/2$ least significant bits of the argument constitute a hard-core function of the modular squaring function for n-bit long moduli. This conjecture yields an efficient pseudorandom generator: $G'_N(s) = \text{LSB}_N(s) \cdot \text{LSB}_N(s^2 \bmod N) \cdot \text{LSB}_N(s^4 \bmod N) \cdots$, where $\text{LSB}_N(x)$ denotes the $0.5 \log_2 N$ least significant bits of x.

A plausibility result: By [78], pseudorandom generators exist if one-way functions exist. Unlike the construction of pseudorandom generators from one-way permutations, the known construction of pseudorandom generators from *arbitrary* one-way functions has no practical significance. It is indeed an important open problem to provide an alternative construction which may be practical and still utilize an *arbitrary* one-way function.

Pseudorandom Functions

Pseudorandom generators allow to efficiently generate long pseudorandom sequences from short random seeds. Pseudorandom functions (defined below) are even more powerful: They allow efficient direct access to a huge pseudorandom sequence (which is not feasible to scan bit-by-bit). Put in other words, pseudorandom functions can replace truly random functions in any application where the function is used in a black-box fashion (i.e., the adversary may obtain the value of the function at arguments of its choice but is not able to evaluate the function by itself).[5]

Definition 4 (pseudorandom functions [63]): *A pseudorandom function is an efficient* (deterministic) *algorithm which given an n-bit seed, s, and an n-bit argument, x, returns an n-bit string, denoted $f_s(x)$, so that it is infeasible to distinguish the responses of f_s, for a uniformly chosen s, from the responses of a truly random function.*

---

[5] This is different from the *Random Oracle Model* of [12].

That is, the distinguisher is given access to a function and is required to distinguish a random function $f : \{0,1\}^n \mapsto \{0,1\}^n$ from a function chosen uniformly in $\{f_s : s \in \{0,1\}^n\}$. We stress that in the latter case the distinguisher is NOT given the description of the function f_s (i.e., the seed s), but rather may obtain the value of f_s on any n-bit string of its choice.[6]

Pseudorandom functions are a very useful cryptographic tool (cf., [64, 60] and Section 5): One may first design a cryptographic scheme assuming that the legitimate users have black-box access to a random function, and next implement the random function using a pseudorandom function.

From pseudorandom generators to pseudorandom functions [63]: Let G be a pseudorandom generator with stretching function $\ell(n) = 2n$, and let $G_0(s)$ (resp., $G_1(s)$) denote the first (resp., last) n bits in $G(s)$ where $s \in \{0,1\}^n$. We define the function ensemble $\{f_s : \{0,1\}^{|s|} \mapsto \{0,1\}^{|s|}\}$, where $f_s(\sigma_{|s|} \cdots \sigma_2 \sigma_1) = G_{\sigma_{|s|}}(\cdots G_{\sigma_2}(G_{\sigma_1}(s)) \cdots)$. This ensemble is pseudorandom.

An alternative construction of pseudorandom functions has been suggested in [101].

4 Zero-Knowledge

Loosely speaking, zero-knowledge proofs are proofs which yield nothing beyond the validity of the assertion. That is, a verifier obtaining such a proof only gains conviction in the validity of the assertion. Using the simulation paradigm this requirement is stated by postulating that anything that is feasibly computable from a zero-knowledge proof is also feasibly computable from the valid assertion alone. All the above refers to proofs as to interactive and randomized processes. That is, here a proof is a (multi-round) protocol for two parties, called verifier and prover, in which the prover wishes to convince the verifier of the validity of a given assertion. Such an *interactive proof* should allow the prover to convince the verifier of the validity of any true assertion, whereas NO prover strategy may fool the verifier to accept false assertions. Both the above *completeness* and *soundness* conditions should hold with high probability (i.e., a negligible error probability is allowed). The prescribed verifier strategy is required to be efficient. No such requirement is made with respect to the prover strategy; yet we will be interested in "relatively efficient" prover strategies (see below). Zero-knowledge is a property of some prover-strategies. More generally, we consider interactive machines which yield no knowledge while interacting with an arbitrary feasible adversary on a common input taken from a predetermined set (in our case the set of valid assertions).

Definition 5 (zero-knowledge [74]): *A strategy A is* zero-knowledge *on inputs from S if, for every feasible strategy B^*, there exists a feasible computation C^* so that the following two probability ensembles are computationally indistinguishable:*

---

[6] Typically, the distinguisher stands for an adversary that attacks a system which uses a pseudorandom function. The values of the function on arguments of the adversary's choice are obtained from the legitimate users of the system who, *unlike the adversary*, know the seed s. The definition implies that the adversary will not be more successful in its attack than it could have been if the system was to use a truly random function. Needless to say that the latter system is merely a *Gedanken Experiment* (it cannot be implemented since it is infeasible to even store a truly random function).

1. $\{(A, B^*)(x)\}_{x \in S} \stackrel{\text{def}}{=}$ *the output of B^* when interacting with A on common input $x \in S$; and*

2. $\{C^*(x)\}_{x \in S} \stackrel{\text{def}}{=}$ *the output of C^* on input $x \in S$.*

Note that whereas A and B^* above are interactive strategies, C^* is a non-interactive computation. The above definition does NOT account for auxiliary information which an adversary may have prior to entering the interaction. Accounting for such auxiliary information is essential for using zero-knowledge proofs as subprotocols inside larger protocols (see [66, 70]). Another concern is that we prefer that the complexity of C^* be bounded as a function of the complexity of B^*. Both concerns are taken care of by a more strict notion of zero-knowledge presented next.

Definition 6 (zero-knowledge, revisited [70]): *A strategy A is* black-box zero-knowledge *on inputs from S if there exists an efficient* (universal) *subroutine-calling algorithm U so that for every feasible strategy B^*, the probability ensembles $\{(A, B^*)(x)\}_{x \in S}$ and $\{U^{B^*}(x)\}_{x \in S}$ are computationally indistinguishable, where U^{B^*} is algorithm U using strategy B^* as a subroutine.*

Note that the running time of U^{B^*} is at most the running-time of U times the running-time of B^*. Actually, the first term may be replaced by the number of times U invokes the subroutine. All known zero-knowledge proofs are in fact black-box zero-knowledge.

A general plausibility result: By [68], assuming the existence of commitment schemes, there exist (black-box) zero-knowledge proofs for membership in any NP-language.[7] Furthermore, the prescribed prover strategy is efficient provided it is given an NP-witness to the assertion to be proven. This makes zero-knowledge a very powerful tool in the design of cryptographic schemes and protocols.

Zero-knowledge as a tool: In a typical cryptographic setting, a user, referred to as A, has a secret and is supposed to take some steps depending on its secret. The question is how can other users verify that A indeed took the correct steps (as determined by A's secret and the publicly known information). Indeed, if A discloses its secret then anybody can verify that it took the correct steps. However, A does not want to reveal its secret. Using zero-knowledge proofs we can satisfy both conflicting requirements. That is, A can prove in zero-knowledge that it took the correct steps. Note that A's claim to having taken the correct steps is an NP-assertion and that A has an NP-witness to its validity (i.e., its secret!). Thus, by the above result, it is possible for A to efficiently prove the correctness of its actions without yielding anything about its secret. (However, in practice one may want to design a specific zero-knowledge proof, tailored to the specific application and so being more efficient, rather than invoking the general result above. Thus, the development of techniques for the construction of such proofs is still of interest – see, for example, [43].)

---

[7] NP is the class of languages having efficiently verifiable (and short) proofs of membership. That is, L is in NP if there exists a polynomial-time recognizable binary relation R_L and a polynomial ℓ so that $x \in L$ if and only if there exists y so that $|y| \leq \ell(|x|)$ and $(x, y) \in R_L$.

Some Variants

Perfect zero-knowledge arguments: This term captures two deviations from the above definition; the first being a strengthening and the second being a weakening. Perfect zero-knowledge strategies are such for which the ensembles in Definition 5 are identically distributed (rather than computationally indistinguishable). This means that the zero-knowledge clause holds regardless of the computational abilities of the adversary. However, *arguments* (aka *computationally sound proofs*) differ from interactive proofs in having a weaker soundness clause: it is infeasible (rather than impossible) to fool the verifier to accept false assertion (except with negligible probability) [25]. Perfect zero-knowledge arguments for NP were constructed using any one-way permutation [99].

Non-Interactive zero-knowledge proofs [20, 52]: Here the interaction between the prover and the verifier consists of the prover sending a single message to the verifier (as in "classical proofs"). In addition, both players have access to a "random reference string" which is postulated to be uniformly selected. Non-interactive zero-knowledge proofs are useful in applications where one of the parties may be trusted to select the abovementioned reference string (e.g., see Section 5.3). Non-interactive zero-knowledge arguments for NP were constructed using any trapdoor permutation [52, 87].

Zero-knowledge proofs of knowledge [74, 56, 5]: Loosely speaking, a system for proofs of knowledge guarantees that whenever the verifier is convinced that the prover knows X, this X can be efficiently extracted from the prover's strategy. One natural application of (zero-knowledge) proofs of knowledge is for *identification* [56, 51].

Relaxations of Zero-knowledge: Important relaxations of zero-knowledge were presented in [53]. Specifically, in *witness indistinguishable* proofs it is infeasible to tell which NP-witness to the assertion the prover is using. Unlike zero-knowledge proofs, this notion is closed under parallel composition. Furthermore, this relaxation suffices for some applications in which one may originally think of using zero-knowledge proofs.

5 Encryption

Both Private-Key and Public-Key encryption schemes consists of three efficient algorithms: *key generation, encryption* and *decryption*. The difference between the two types is reflected in the definition of security – the security of a public-key encryption scheme should hold also when the adversary is given the encryption key, whereas this is not required for private-key encryption scheme. Thus, public-key encryption schemes allow each user to broadcast its encryption key so that any user may send it encrypted messages (without needing to first agree on a private encryption-key with the receiver). Below we present definitions of security for private-key encryption schemes. The public-key analogies can be easily derived by considering adversaries which get the encryption key as additional input. (For private-key encryption schemes we may assume, without loss of generality, that the encryption key is identical to the decryption key.)

5.1 Definitions

For simplicity we consider only the encryption of a single message; however this message may be longer than the key (which rules out information-theoretic secrecy [115]). We present two equivalent definitions of security. The first, called *semantic security*, is a computational analogue of Shannon's definition of *perfect secrecy* [115]. The second definition views secure encryption schemes as ones for which it is infeasible to distinguish encryptions of any (known) pair of messages (e.g., the all-zeros message and the all-ones message). The latter definition is technical in nature and is referred to as *indistinguishability of encryptions*.

We stress that the definitions presented below go way beyond saying that it is infeasible to recover the plaintext from the ciphertext. The latter statement is indeed a minimal requirement from a secure encryption scheme, but we claim that it is way too weak a requirement: An encryption scheme is typically used in applications where obtaining specific partial information on the plaintext endangers the security of the application. When designing an application-independent encryption scheme, we do not know which partial information endangers the application and which does not. Furthermore, even if one wants to design an encryption scheme tailored to one's own specific applications, it is rare (to say the least) that one has a precise characterization of all possible partial information which endanger these applications. Thus, we require that it is infeasible to obtain any information about the plaintext from the ciphertext. Furthermore, in most applications the plaintext may not be uniformly distributed and some a-priori information regarding it is available to the adversary. We require that the secrecy of all partial information is preserved also in such a case. That is, even in presence of a-priori information on the plaintext, it is infeasible to obtain any (new) information about the plaintext from the ciphertext (beyond what is feasible to obtain from the a-priori information on the plaintext). The definition of semantic security postulates all of this. The equivalent definition of indistinguishability of encryptions is useful in demonstrating the security of candidate constructions as well as for arguing about their usage as part of larger protocols.

The actual definitions: In both definitions we consider (feasible) adversaries which obtain, in addition to the ciphertext, also auxiliary information which may depend on the potential plaintexts (but not on the key). By $E(x)$ we denote the distribution of encryptions of x, when the key is selected at random. To simplify the exposition, let us assume that on security parameter n the key generation produces a key of length n, whereas the scheme is used to encrypt messages of length n^2.

Definition 7 (semantic security (following [73])): *An encryption scheme is* semantically secure *if for every feasible algorithm, A, there exists a feasible algorithm B so that for every two functions* $f, h : \{0,1\}^* \mapsto \{0,1\}^*$ *and all sequences of pairs,* $(X_n, z_n)_{n \in \mathbb{N}}$, *where* X_n *is a random variable ranging over* $\{0,1\}^{n^2}$ *and* $|z_n|$ *is of feasible (in n) length,*

$$\Pr(A(E(X_n), h(X_n), z_n) = f(X_n)) < \Pr(B(h(X_n), z_n) = f(X_n)) + \mu(n)$$

where μ *is a negligible function. Furthermore, the complexity of B should be related to that of A.*

What this definition says is that a feasible adversary does not gain anything by looking at the ciphertext. That is, whatever information (captured by the function f) it tries to compute from the ciphertext, can be essentially computed as efficiently from the available a-priori information (captured by the function h). In particular, the ciphertext does not help in (feasibly) computing the least significant bit of the plaintext or any other information regarding the plaintext. This holds for any distribution of plaintexts (captured by the random variable X_n).

Definition 8 (indistinguishability of encryptions (following [73])): *An encryption scheme has indistinguishable encryptions if for every feasible algorithm, A, and all sequences of triples, $(x_n, y_n, z_n)_{n \in \mathbb{N}}$, where $|x_n| = |y_n| = n^2$ and $|z_n|$ is of feasible (in n) length, the difference*

$$d_A(n) \stackrel{\text{def}}{=} |\Pr(A(E(x_n), z_n) = 1) - \Pr(A(E(y_n), z_n) = 1)|$$

is a negligible function in n.

In particular, z_n may equal (x_n, y_n). Thus, it is infeasible to distinguish the encryptions of any two fix messages such as the all-zero message and the all-ones message.

Probabilistic Encryption: It is easy to see that a secure *public-key* encryption scheme must employ a probabilistic (i.e., randomized) encryption algorithm. Otherwise, given the encryption key as (additional) input, it is easy to distinguish the encryption of the all-zero message from the encryption of the all-ones message. The same holds for *private-key* encryption schemes when considering the security of encrypting several messages (rather than a single message as done above).[8] This explains the linkage between the above robust security definitions and the *randomization paradigm* (discussed below).

5.2 Constructions

It is common practice to use "pseudorandom generators" as a basis for private-key stream ciphers. We stress that this is a very dangerous practice when the "pseudorandom generator" is easy to predict (such as the linear congruential generator or some modifications of it which output a constant fraction of the bits of each resulting number – see [24, 58]). However, this common practice becomes sound provided one uses pseudorandom generators (as defined in Section 3).

Private-Key Encryption Scheme based on Pseudorandom Functions: The key generation algorithm consists of selecting a seed, denoted s, for such a function, denoted f_s. To encrypt a message $x \in \{0, 1\}^n$ (using key s), the encryption algorithm uniformly selects a string $r \in \{0, 1\}^n$ and produces the ciphertext $(r, x \oplus f_s(r))$. To decrypt the ciphertext (r, y) (using key s), the decryption algorithm just computes $y \oplus f_s(r)$. The proof of security of this encryption scheme consists of two steps (suggested as a general methodology in Section 3):

---

[8] Here, for example, using a deterministic encryption algorithm allows the adversary to distinguish two encryptions of the same message from the encryptions of a pair of different messages.

1. Prove that an idealized version of the scheme, in which one uses a uniformly selected function $f: \{0,1\}^n \mapsto \{0,1\}^n$, rather than the pseudorandom function f_s, is secure.
2. Conclude that the real scheme (as presented above) is secure (since otherwise one could distinguish a pseudorandom function from a truly random one).

Note that we could have gotten rid of the randomization if we had allowed the encryption algorithm to be history dependent (e.g., use a counter in the role of r). Furthermore, if the encryption scheme is used for FIFO communication between the parties and both can maintain the counter value then there is no need for the sender to send the counter value.

The randomization paradigm [73]: To demonstrate this paradigm suppose we have a trapdoor one-way permutation, $\{p_\alpha\}_\alpha$, and a hard-core predicate, b, for it.[9] The key generation algorithm consists of selecting at random a permutation p_α together with a trapdoor for it: The permutation (or rather its description) serves as the public-key, whereas the trapdoor serves as the private-key. To encrypt a single bit σ (using public key p_α), the encryption algorithm uniformly selects an element, r, in the domain of p_α and produces the ciphertext $(p_\alpha(r), \sigma \oplus b(r))$. To decrypt the ciphertext (y, τ) (using the private key), the decryption algorithm just computes $\tau \oplus b(p_\alpha^{-1}(y))$ (where the inverse is computed using the trapdoor (i.e., private-key)). The above scheme is quite wasteful in bandwidth; however, the paradigm underlying its construction is valuable in practice. For example, it is certainly better to randomly pad messages (say using padding equal in length to the message) before encrypting them using RSA than to employ RSA on the plain message. Such a heuristic could be placed on firm grounds if a conjecture analogous to the one mentioned in Section 3 is supported. That is, assume that the first $n/2$ least significant bits of the argument constitute a hard-core function of RSA with n-bit long moduli. Then, encrypting $n/2$-bit messages by padding the message with $n/2$ random bits and applying RSA (with an n-bit moduli) on the result constitutes a secure public-key encryption system, hereafter referred to as Randomized RSA.

An alternative public-key encryption scheme is presented in [21]. The encryption scheme augments the construction of a pseudorandom generator, given in Section 3, as follows. The key-generation algorithm consists of selecting at random a permutation p_α together with a trapdoor. To encrypt the n-bit string x (using public key p_α), the encryption algorithm uniformly selects an element, s, in the domain of p_α and produces the ciphertext $(p_\alpha^n(s), x \oplus G_\alpha(s))$, where $G_\alpha(s) = b(s) \cdot b(p_\alpha(s)) \cdots b(p_\alpha^{n-1}(s))$. (We use the notation $p_\alpha^{i+1}(x) = p_\alpha(p_\alpha^i(x))$ and $p_\alpha^{-(i+1)}(x) = p_\alpha^{-1}(p_\alpha^{-i}(x))$.) To decrypt the ciphertext (y, z) (using the private key), the decryption algorithm first recovers $s = p_\alpha^{-n}(y)$ and then outputs $z \oplus G_\alpha(s)$.

Assuming that factoring Blum Integers (i.e., products of two primes each congruent to 3 (mod 4)) is hard, one may use the modular squaring function in role of the trapdoor permutation above (see [21, 1, 117, 57]). This yields a secure public-key encryption scheme with efficiency comparable to that of RSA. Recall that RSA itself is not secure (as it employs a deterministic encryption algorithm), whereas Randomized RSA (defined above) is not known to be secure under standard assumption such as intractability of factoring (or of inverting the RSA function).[10]

---

[9] Hard-core predicates are defined in Section 3. Recall that by [67], every trapdoor permutation can be modified into one having a hard-core predicate.

[10] Recall that Randomized RSA is secure assuming that the $n/2$ least significant bits constitute a hard-core function for n-bit RSA moduli. We only know that the $O(\log n)$ least significant bits constitute a hard-core function for n-bit moduli [1].

5.3 Beyond eavesdropping security

The above definitions refer only to a "passive" attack in which the adversary merely eavesdrops on the line over which ciphertexts are being sent. Stronger types of attacks, culminating in the so-called Chosen Ciphertext Attack, may be possible in various applications. In such an attack, the adversary may obtain the plaintexts of ciphertexts of its choice (as well as ciphertexts of plaintexts of its choice) and its task is to obtain information about the plaintext of a different ciphertext. Clearly, the private-key encryption scheme based on pseudorandom functions (described above) is secure also against such attacks. Public-key encryption schemes secure against Chosen Ciphertext Attacks can be constructed, assuming the existence of trapdoor permutations and utilizing non-interactive zero-knowledge proofs [105] (which can be constructed under this assumption [52]).

Another issue is the *non-malleability* of the encryption scheme, considered in [47]. Here one requires that it should be infeasible for an adversary, given a ciphertext, to produce a valid ciphertext for a related plaintext. For example, given a ciphertext of a plaintext of the form $1x$, it should be infeasible to produce a ciphertext to the plaintext $0x$. It is easy to turn a private-key encryption scheme into a non-malleable one, by using a message authentication scheme on top. Non-malleable public-key encryption schemes are known to exist assuming the existence of trapdoor permutation [47].

6 Signatures

Again, there are private-key and public-key versions both consisting of three efficient algorithms: *key generation*, *signing* and *verification*. (Private-key signature schemes are commonly referred to as *message authentication schemes* or *codes* (MAC).) The difference between the two types is again reflected in the definition of security. This difference yields different functionality (even more than in the case of encryption): Public-key signature schemes (hereafter referred to as signature schemes) may be used to produce signatures which are *universally verifiable* (given access to the public-key of the signer). Private-key signature schemes (hereafter referred to as message authentication schemes) are only used to authenticate messages sent among a small set of *mutually trusting* parties (since ability to verify signatures is linked to the ability to produce them). Put in other words, message authentication schemes are used to authenticate information sent between (typically two) parties, and the purpose is to convince *the receiver* that the information was indeed sent by the legitimate sender. In particular, message authentication schemes cannot convince *a third party* that the sender has indeed sent the information (rather than the receiver having generated it by itself). In contrast, public-key signatures can be used to convince third parties: A signature to a document is typically sent to a second party so that in the future this party may (by merely presenting the signed document) convince third parties that the document was indeed generated/sent/approved by the signer.

6.1 Definitions

We consider very powerful attacks on the signature scheme as well as a very liberal notion of breaking it. Specifically, the attacker is allowed to obtain signatures to any message of its choice. One may argue that in many applications such a general attack is not possible (as messages to be signed must have a specific format). Yet, our view is that it is impossible to define a general (i.e., application-independent) notion of admissible messages, and thus a general/robust definition of an attack seems to have to be formulated as suggested here. (Note that at worst, our approach is overly cautious.) Likewise, the adversary is said to be successful if it can produce a valid signature to ANY message for which it has not asked for a signature during its attack. Again, this defines the ability to form signatures to possibly "nonsensical" messages as a breaking of the scheme. Yet, again, we see no way to have a general (i.e., application-independent) notion of "meaningful" messages (so that only forging signatures to them will be consider a breaking of the scheme).

Definition 9 (unforgeable signatures [75]):

- *A* chosen message attack *is a process which on input a verification-key can obtain signatures* (relative to the corresponding signing-key) *to messages of its choice.*
- *Such an attack is said to* succeeds (in existential forgery) *if it outputs a valid signature to a message for which it has* NOT *requested a signature during the attack.*
- *A signature scheme is* secure (or unforgeable) *if every* feasible *chosen message attack succeeds with at most negligible probability.*

We stress that *plain* RSA (alike plain versions of Rabin's scheme [110] and DSS [97]) is not secure under the above definition. However, it may be secure if the message is "randomized" before RSA (or the other schemes) is applied (cf., [15]). Thus, the randomization paradigm (see Section 5) seems pivotal here too.

6.2 Constructions

Message authentication schemes can be constructed using pseudorandom functions (see [64] or the better constructions in [10, 9, 3]). However, as noted in [4], an *extensive* usage of pseudorandom functions seem an overkill for achieving message authentication, and more efficient schemes may be obtained based on other cryptographic primitives. We mention two approaches:

1. *Fingerprinting* the message using a scheme which is *secure against forgery provided that the adversary does not have access to the scheme's outcome* (e.g., using Universal Hashing [33]), and *"hiding"* the result using a *non-malleable* scheme (e.g., a private-key encryption or a pseudorandom function). (Non-malleability is not required in certain cases; see [118].)
2. *Hashing* the message *using a collision-free scheme* (cf., [41, 42]), and *authenticating* the result using a MAC which operates on (short) fixed-length strings [4].

Three central paradigms in the construction of *signature schemes* are the "refreshing" of the "effective" signing-key, the usage of an "authentication tree" and the "hashing paradigm".

The refreshing paradigm [75]: To demonstrate this paradigm, suppose we have a signature scheme which is robust against a "random message attack" (i.e., an attack in which the adversary only obtains signatures to randomly chosen messages). Further suppose that we have a *one-time* signature scheme (i.e., a signature scheme which is secure against an attack in which the adversary obtains a signature to a single message of its choice). Then, we can obtain a secure signature scheme as follows: When a new message is to be signed, we generate a new random signing-key for the one-time signature scheme, use it to sign the message, and sign the corresponding (one-time) verification-key using the fixed signing-key of the main signature scheme[11] (which is robust against a "random message attack") [50]. We note that one-time signature schemes (as utilized here) are easy to construct (see, for example [93]).

The tree paradigm [92, 75]: To demonstrate this paradigm, we show how to construct a general signature scheme using only a one-time signature scheme (alas one where an $2n$-bit string can be signed w.r.t an n-bit long verification-key). The idea is to use the initial singing-key (i.e., the one corresponding to the public verification-key) in order to sign/authenticate two new/random verification keys. The corresponding signing keys are used to sign/authenticate four new/random verification keys (two per a signing key), and so on. Stopping after d such steps, this process forms a binary tree with 2^d leaves where each leaf corresponds to an instance of the one-time signature scheme. The signing-keys at the leaves can be used to sign the actual messages, and the corresponding verification-keys may be authenticated using the path from the root. Pseudorandom functions may be used to eliminate the need to store the values of intermediate vertices used in previous signatures [60]. Employing this paradigm and assuming that the RSA function is infeasible to invert, one obtains a secure signature scheme [75, 60] in which the ith message can be signed/verified in time $2 \log_2 i$ slower than plain RSA. Using a tree of large fan-in and assuming that RSA is infeasible to invert, one may obtain a secure signature scheme [49] which for reasonable parameters is only 5 times slower than plain RSA (alas uses a much bigger key).[12] We stress that plain RSA is not a secure signature scheme, whereas the security of its randomized version (mentioned above) is not known to be reducible to the assumption that RSA is hard to invert.

The hashing paradigm: A common practice is to sign real documents via a two stage process: First the document is hashed into a (relatively) short bit string, and next the basic signature scheme is applied to the resulting string. We note that this heuristic becomes sound provided the hashing function is *collision-free* (as defined in [41]). Collision-free functions can be constructed assuming the intractability of factoring [41]. One may indeed postulate that certain off-the-shelve products (as MD5 or SHA) are collision-free, but such assumptions need to be tested (and indeed may turn out false). We stress that using a hashing scheme in the above two-stage process without evaluating whether it is collision-free is a very dangerous practice.

---

[11] Alternatively, one may generate the one-time key-pair and the signature to its verification-key ahead of time, leading to an "off-line/on-line" signature scheme [50].

[12] This figure refers to signing up-to 1,000,000,000 messages. The scheme requires a universal set of system parameters consisting of 1000–2000 integers of the size of the moduli. We believe that in *some* applications the storage/time trade-off provided by [49] may be preferred over [75, 60].

A useful variant on the above paradigm is the use of *Universal One-Way Hash Functions* (as defined in [104]), rather than the collision-free hashing used above. In such a case a new hash function is selected per each application of the scheme, and the basic signature scheme is applied to both the (succinct) description of the hash function and to the resulting (hashed) string. (In contrast, when using a collision-free hashing function, the same function – the description of which is part of the signer's public-key – is used in all applications.) The advantage of using Universal One-Way Hash Functions is that their security requirement seems weaker than the collision-free condition.

A plausibility result: By [104, 114] signature schemes exist if and only if one-way functions exist. Unlike the constructions of signature schemes described above, the known construction of signature schemes from *arbitrary* one-way functions has no practical significance [114]. It is indeed an important open problem to provide an alternative construction which may be practical and still utilize an *arbitrary* one-way function.

7 Cryptographic Protocols

A general framework for casting cryptographic (protocol) problems consists of specifying a random process which maps n inputs to n outputs. The inputs to the process are to be thought of as local inputs of n parties, and the n outputs are their corresponding local outputs. The random process describes the desired functionality. That is, if the n parties were to trust each other (or trust some outside party), then they could each send their local input to the trusted party, who would compute the outcome of the process and send each party the corresponding output. The question addressed in this section is to what extent can this trusted party be "simulated" by the mutually distrustful parties themselves.

7.1 Definitions

For simplicity we consider the special case where the specified process is deterministic and the n outputs are identical. That is, we consider an arbitrary n-ary function and n parties which wish to obtain the value of the function on their n corresponding inputs. Each party wishes to obtain the correct value of the function and prevent any other party from gaining anything else (i.e., anything beyond the value of the function and what is implied by it).

We first observe that (one thing which is unavoidable is that) each party may change its local input before entering the protocol. However, this is unavoidable also when the parties utilize a trusted party. In general, the basic paradigm underlying the definitions of *secure multi-party computations* amounts to saying that situations which may occur in the real protocol, can be simulated in the ideal model (where the parties may employ a trusted party). Thus, the "effective malfunctioning" of parties in secure protocols is restricted to what is postulated in the corresponding ideal model. The specific definitions differ in the specific restrictions and/or requirements placed on the parties in the real computation. This is typically reflected in the definition of the corresponding ideal model – see examples below.

An example – computations with honest majority: Here we consider an ideal model in which any minority group (of the parties) may collude as follows. Firstly this minority shares its original inputs and decided together on replaced inputs[13] to be sent to the trusted party. (The other parties send their respective original inputs to the trusted party.) When the trusted party returns the output, each majority player outputs it locally, whereas the colluding minority may compute outputs based on all they know (i.e., the output and all the local inputs of these parties). A *secure multi-party computation with honest majority* is required to simulate this ideal model. That is, the effect of any feasible adversary which controls a minority of the players in the actual protocol, can be essentially simulated by a (different) feasible adversary which controls the corresponding players in the ideal model. This means that in a secure protocol the effect of each minority group is "essentially restricted" to replacing its own local inputs (independently of the local inputs of the majority players) before the protocol starts, and replacing its own local outputs (depending only on its local inputs and outputs) after the protocol terminates. (We stress that in the real execution the minority players do obtain additional pieces of information; yet in a secure protocol they gain nothing from these additional pieces of information.)

Secure protocols according to the above definition may even tolerate a situation where a minority of the parties aborts the execution. An aborted party (in the real protocol) is simulated by a party (in the ideal model) which aborts the execution either before supplying its input to the trusted party (in which case a default input is used) or after supplying its input. In either case, the majority players (in the real protocol) are able to compute the output although a minority aborted the execution. This cannot be expected to happen when there is no honest majority (e.g., in a two-party computation) [40].

Another example – two-party computations: In light of the above, we consider an ideal model where each of the two parties may "shut-down" the trusted (third) party at any point in time. In particular, this may happen after the trusted party has supplied the outcome of the computation to one party but before it has supplied it to the second. A *secure multi-party computation allowing abort* is required to simulate this ideal model. That is, each party's "effective malfunctioning" in a secure protocol is restricted to supplying an initial input of its choice and aborting the computation at any point in time. We stress that, as above, the choice of the initial input of each party may NOT depend on the input of the other party.

7.2 Constructions

General plausibility results: Assuming the existence of trapdoor permutations, one may provide secure protocols for any two-party computation (allowing abort) [120] as well as for any multi-party computations with honest majority [69]. Thus, a host of cryptographic problems are solvable assuming the existence of trapdoor permutations. As stressed in the case of zero-knowledge proofs, we view these results as asserting that very wide classes of problems are solvable in principle. However, we do not recommend using the solutions

---

[13] Such replacement may be avoided if the local inputs of parties are verifiable by the other parties. In such a case, a party (in the ideal model) has the choice of either joining the execution of the protocol with its correct local input or not join the execution at all (but it cannot join with a replaced local input). Secure protocols simulating this ideal model can be constructed as well.

derived by these general results in practice. For example, although Threshold Cryptography (cf., [45, 59]) is merely a special case of multi-party computation, it is indeed beneficial to focus on its specifics.

Analogous plausibility results were obtained in a variety of models. In particular, we mention secure computations in the private channels model [17, 35] and in the presence of mobile adversaries [107].

8 Some Notes

On information theoretic secrecy: Most of Modern Cryptography aims at achieving *computational* security; that is, making it infeasible (rather than impossible) for an adversary to break the system. The departure from *information theoretic* secrecy was suggested by Shannon in the very paper which introduced the notion [115]: In an information theoretic secure encryption scheme the private-key must be longer than the total entropy of the plaintexts to be sent using this key. This drastically restricts the applicability of (information-theoretic secure) private-key encryption schemes. Furthermore, notions such as public-key cryptography, pseudorandom generators, and most known cryptographic protocols cannot exist in an information theoretic sense.

On the need for and choice of assumptions: As stated in Section 2, most of Modern Cryptography is based on computational difficulty. Intuitively, this is an immediate consequence of the fact that Modern Cryptography wish to capitalize on the difference between feasible attacks and possible-but-infeasible attacks. Formally, the existence of one-way functions has been shown to be a necessary condition for the existence of secure private-key encryption [81], pseudorandom generators [89], digital signatures [114], "non-trivial" zero-knowledge proofs [106], and various basic protocols [81].

As we need assumptions anyhow, why not assume what we want? Well, first we need to know what we want. This calls for a clear definition of complex security concerns – an non-trivial issue which is discussed at length in previous sections. However, once a definition is derived how can we know that it can at all be met? The way to demonstrate that a definition is viable (and so the intuitive security concern can be satisfied at all) is to construct a solution based on a *better understood* assumption. For example, looking at the definition of zero-knowledge proofs [74], it is not a-priori clear that such proofs exists in a non-trivial sense. The non-triviality of the notion was demonstrated in [74] by presenting a zero-knowledge proof system for statements, regarding Quadratic Residuosity, which are believed to be hard to verify (without extra information). Furthermore, in contrary to prior beliefs, it was shown in [68] that the existence of commitment schemes[14] implies that any NP-statement can be proven in zero-knowledge. Thus, statements, which were not known at all to hold (and even believed to be false), where shown to hold by reduction to widely believed assumptions (without which most of Modern Cryptography collapses anyhow). Furthermore, reducing the solution of a new task to the assumed security of a well-known primitive typically means providing a construction which using the known primitive solves

---

[14] Consequently, it was shown how to construct commitment schemes based on any pseudorandom generator [98], and that the latter exists if one-way functions exist [78].

the new task. This means that we do not only know (or assume) that the new task is solvable but rather have a solution based on a primitive which, being well-known, typically has several candidate implementations. More on this subject below.

On the meaning of asymptotic results: Asymptotic analysis is a major simplifying convention. It allows to disregard specifics like the model of computation and to focus on the essentials of the problem at hand. Further simplification is achieved by identifying efficient computations with polynomial-time computations, and more importantly by identifying infeasible computations with ones which are not implementable in polynomial-time. However, none of these conventions is really essential for the theory discussed in this essay.[15]

As stated in Section 2, all know results (referring to computational complexity) consists of an explicit construction in which a complex primitive is implemented based on a simpler one. The claim of security in many papers merely states that if the resulting (complex) primitive can be broken in polynomial-time then so can the original (simpler) primitive. However, all papers provide an explicit construction showing how to use any breaking algorithm for the resulting primitive in order to obtain a breaking algorithm for the original primitive. This transformation does not depend on the running-time of the first algorithm; it typically uses the first algorithm as a black-box. Thus, the running-time of the resulting breaking algorithm (for the simpler primitive) is explicitly bounded in terms of the running-time of the given breaking algorithm (for the complex primitive). This means that for each of these results, one can instantiate the resulting (complex) scheme for any desired value of the security parameter, make a concrete assumption regarding the security of the underlying (simpler) primitive, and derive a concrete estimate of the security of the proposed implementation of the complex primitive.

The applicability of a specific theoretical result depends on the complexity of the construction and the relation between the security of the resulting scheme and the quantified intractability assumption. Some of these results seem applicable in practice, some only offer useful paradigm/techniques, and other only state the plausibility of certain results. In the latter cases it is indeed the task of the theory community to work towards the improvement of these results. In fact, many improvements of this type have been achieved in the past (and we hope to see more in the future). Following are some examples:

- A plausibility result of Yao (commonly attributed to [119]) on the existence of hard-core predicates, assuming the existence of one-way permutations, was replaced by a practical construction of hard-core predicates for any one-way functions [67].
- A plausibility result of Yao (commonly attributed to [119]) by which any weak one-way permutation can be transformed into an ordinary one-way permutation was replaced by an efficient transformation of weak one-way permutation into ordinary one-way permutation [65].
- A plausibility result of [68] by which one may construct Verifiable Secret Sharing schemes (cf., [39]), using any one-way function, was replaced by an efficient construction the security of which is based on DLP [54]. In general, many concrete problems

---

[15] As long as the notions of efficient and feasible computation are sufficiently robust and rich. For example, they should be closed under various functional compositions and should allow computations such as RSA.

which are solvable in principle (by the plausibility results of [68, 120, 69]) were given efficient solutions.

Forget the result, use its ideas: As stated above, some theoretical results are not directly applicable in practice. Still, in many cases these results utilize ideas which may be of value in practice. Thus, if you know (by a theoretical result) that a problem is solvable in principle, but the known construction is not applicable for your purposes, you may try to utilize some of its underlying ideas when trying to come-up with an alternative solution tailored for your own purposes. We note that the underling ideas are at least as likely to appear in the proof of security as in the construction itself.

The choice of assumptions, revisited. When constructing a solution to a cryptographic problem one may have a choice of which building blocks to use (e.g., one-way functions or pseudorandom functions). In a coarse sense these tools may look equivalent (e.g., one exists if and only if the other exists), but when deciding which to use in practice one should consider the actual level of security attributed to each of them and the "cost" of using each of them as a building block in a particular construction. In the latter term ("cost") we mean the relationship of the security of the building block to the security of the resulting solution. For further discussion the reader is referred to [3, Sec. 1.5]. *Turning the table around*, if we note that a specific primitive provides good security, when used as a building block in many constructions, then this may serve as incentive to focus attention on the implementation of this primitive. The last statement should be understood both as referring to the theory and practice of cryptography. For example, it is our opinion that the industry should focus on constructing fixed-length-key pseudorandom functions rather than on constructing fixed-length-key pseudorandom permutations (or, equivalently, private-key block ciphers).[16]

Security as a quantity rather than a quality: From the above it should be clear that our notions of security are quantitative in nature. They refers to the minimal amount of work required to break the system (as a function of the security parameter). Thus alternative constructions for the same task may (and need to) be compared based on the security they provide. This can be done whenever the underlying assumption are compareable.

"Too cautious" definitions: As stated in Sections 5 and 6, our definitions seem "too cautious" in the sense that they also imply things which may not matter in practice. This is an artifact of our approach to security which requires that the adversary gains *nothing* (rather than "gains nothing we care about") by its malicious actions. We stress two advantages of our approach. First it yields application-independent notions of security (since the notion of a "gain we care about" is application-dependent). Secondly, even when having a specific application in mind, it is close to impossible to come-up with a precise characterization of the set of "gains we care about". Thus, even in the latter case, our approach of depriving the adversary from *any* gain seems to be the best way to go.

"Provable Security": Some of the papers discussed in this essay use the term "provable security". The term is supposed to reflect the fact that these papers only make well-defined technical claims and that proofs of these claims are given or known to the authors.

---

[16] Not to mention that the latter can be efficiently constructed from the former [91, 102].

Specifically, whenever a term such as "security" is used, the paper offers or refers to a rigorous definition of the term (and the authors wish to stress this fact in contrast to prior papers where the term was used as an undefined intuitive phrase). We personally object to this terminology since it suggests the possibility that there can be technical claims[17] which are well-defined and others which are not, and among the former some can be stated even when no proof is known. This view is wrong: A technical claim must always be well-defined, and it must always have a proof (otherwise it is a conjecture – not a claim). There is room for non-technical claims, but these claims should be stated as opinions and such. In particular, a technical claim referring to security must always refer to a rigorous definition of security and the person making this claim must always know a proof (or state the claim as a conjecture).

Still, do consider specific attack (but as a last resort). We do realize that sometimes one is faced with a situation where all the paradigms described above offer no help. A typical example occurs when designing an "atomic" cryptographic primitive (e.g., a one-way function). The first thing we suggest in such a case is to formulate precise specifications/assumptions regarding the security of this primitive. Once this is done, one may need to turn to ad-hoc methods for trying to test these assumptions (i.e., if the known attack schemes fail then one gains some confidence in the validity of the assumptions). For example, if we were to invent RSA today then we would have postulated that it is a trapdoor permutation. To evaluate the validity of our conjecture, we would have noted (as Rivest, Shamir and Adleman did in [113, Sec. IX]) that known algorithms for factoring are infeasible for reasonable values of the security parameter, and that there seems to be no other way to invert the function.

9 Two Suggestions for Future Research

A very important direction for future research consists of trying to "upgrade" the utility of some of the constructions mentioned above. In particular, we have mentioned four plausibility results: two referring to the construction of pseudorandom generators and signature schemes and two referring to the construction of zero-knowledge proofs and multi-party protocols. For the former two results, we see no fundamental reason why the corresponding constructions can not be replaced by reasonable ones (i.e., providing very efficient constructions of pseudorandom generators and signature schemes based on *arbitrary* one-way functions). Furthermore, we believe that working towards this goal may yield new and useful paradigms (which may be applicable in practice regardless of these results). As for the latter general plausibility results (i.e., the construction of zero-knowledge proofs and multi-party protocols), here there seem to be little hope for a result which may both maintain the generality of the results in [68, 120, 69] and yield practical solutions for each specific task. However, we believe that there is work to be done towards the development of additional paradigms and techniques which may be useful in the construction of schemes for specific tasks.

Another important direction is to provide results and/or develop techniques for guaranteeing that individually-secure protocols remain secure when many copies of them are run

---

[17] We refer to theorems, lemmas, propositions and such.

in parallel and, furthermore, obliviously of one another. Although some negative results are known [66], they only rule out specific approaches (such as the naive false conjecture that ANY zero-knowledge proof maintains its security when executed twice in parallel).

10 Some Suggestions for Further Reading

The intention of these suggestions is NOT to provide a scholarly account of the due credits but rather to provide sources for further reading. Thus, our main criteria is the readability of the text (not its novelty). The recommendations are arranged by subjects.

Computational Hardness, One-Way Functions, Pseudorandom Generators and Zero-Knowledge: For these, our favorite source is our own text [62].

Encryption Schemes: A good motivating discussion appears in [73]. The definitional treatment in [61] is the one we prefer, although it can be substantially simplified if one adopts non-uniform complexity measures (as done above). Further details on the constructions of public-key encryption schemes (sketched above) can be found in [73, 61] and [21, 1], respectively. For discussion of Non-Malleable Cryptography, which actually transcends the domain of encryption, see [47].

Signature Schemes: For a definitional treatment of *signature schemes* the reader is referred to [75] and [108]. Easy to understand constructions appear in [11, 50, 49]. Variants on the basic model are discussed in [108] and in [34, 85]. For discussion of *message authentication schemes* (MACs) the reader in referred to [4].

General Cryptographic Protocols: This area is both most complex and most lacking good expositions. Our own preference is to refer to [27] for the definitions and to [61] for the constructions.

New Directions: Incremental Cryptography [6, 7], Realizing the Random Oracle Model [28], Coercibility [30, 29], sharing of cryptographic objects [45, 44, 59], Private Information Retrieval [38, 37, 88], Cryptanalysis by induced faults [23], Visual Cryptography [103, 100] and many others.

Acknowledgments

I wish to thank Ran Canetti, Shafi Goldwasser and Hugo Krawczyk for helpful discussions. Special thanks to Hugo for carefully reading and commenting on an early draft.

Bibliographic Abbreviations

- STOC is *ACM Symposium on the Theory of Computing.*
- FOCS is *IEEE Symposium on Foundations of Computer Science.*

References

1. W. Alexi, B. Chor, O. Goldreich and C.P. Schnorr. RSA/Rabin Functions: Certain Parts are As Hard As the Whole. *SIAM J. on Comput.*, Vol. 17, April 1988, pages 194–209.
2. D. Beaver. Foundations of Secure Interactive Computing. In *Crypto91*, Springer-Verlag LNCS (Vol. 576), pages 377–391.
3. M. Bellare, R. Canetti and H. Krawczyk. Pseudorandom functions Revisited: The Cascade Construction and its Concrete Security. In *37th FOCS*, pages 514–523, 1996.
4. M. Bellare, R. Canetti and H. Krawczyk. Keying Hash Functions for Message Authentication. In Crypto96, Springer LNCS (Vol. 1109), pages 1–15.
5. M. Bellare and O. Goldreich. On Defining Proofs of Knowledge. In *Crypto92*, Springer-Verlag LNCS (Vol. 740), pages 390–420.
6. M. Bellare, O. Goldreich and S. Goldwasser. Incremental Cryptography: the Case of Hashing and Signing. In *Crypto94*, Springer-Verlag LNCS (Vol. 839), pages 216–233, 1994.
7. M. Bellare, O. Goldreich and S. Goldwasser. Incremental Cryptography and Application to Virus Protection. In *27th STOC*, pages 45–56, 1995.
8. M. Bellare, S. Goldwasser and D. Micciancio. "Pseudo-random" Number Generation within Cryptographic Algorithms: the DSS Case. These proceedings.
9. M. Bellare, R. Guerin and P. Rogaway. XOR MACs: New Methods for Message Authentication using Finite Pseudorandom Functions. In *Crypto95*, Springer-Verlag LNCS (Vol. 963), pages 15–28.
10. M. Bellare, J. Kilian and P. Rogaway. The Security of Cipher Block Chaining. In *Crypto94*, Springer-Verlag LNCS (Vol. 839), pages 341–358.
11. M. Bellare and S. Micali. How to Sign Given Any Trapdoor Function. *J. of the ACM*, Vol. 39, pages 214–233, 1992.
12. M. Bellare and P. Rogaway. Random Oracles are Practical: a Paradigm for Designing Efficient Protocols. In *1st Conf. on Computer and Communications Security*, ACM, pages 62–73, 1993.
13. M. Bellare and P. Rogaway. Entity Authentication and Key Distribution. In *Crypto93*, Springer-Verlag LNCS (Vol. 773), pages 232–249, 1994.
14. M. Bellare and P. Rogaway. Provably Secure Session Key Distribution: The Three Party Case. In *27th STOC*, pages 57–66, 1995.
15. M. Bellare and P. Rogaway. The Exact Security of Digital Signatures: How to Sign with RSA and Rabin. In *EuroCrypt96*, Springer LNCS (Vol. 1070).
16. M. Ben-Or, S. Goldwasser, J. Kilian and A. Wigderson. Multi-Prover Interactive Proofs: How to Remove Intractability. In *20th STOC*, pages 113–131, 1988.
17. M. Ben-Or, S. Goldwasser and A. Wigderson. Completeness Theorems for Non-Cryptographic Fault-Tolerant Distributed Computation. In *20th STOC*, pages 1–10, 1988.
18. L. Blum, M. Blum and M. Shub. A Simple Secure Unpredictable Pseudo-Random Number Generator. *SIAM J. on Comput.*, Vol. 15, 1986, pages 364–383.
19. M. Blum, A. De Santis, S. Micali, and G. Persiano. Non-Interactive Zero-Knowledge Proof Systems. *SIAM J. on Comput.*, Vol. 20, No. 6, pages 1084–1118, 1991. (Considered the journal version of [20].)
20. M. Blum, P. Feldman and S. Micali. Non-Interactive Zero-Knowledge and its Applications. In *20th STOC*, pages 103–112, 1988. See [19].
21. M. Blum and S. Goldwasser. An Efficient Probabilistic Public-Key Encryption Scheme which hides all partial information. In *Crypto84*, LNCS (Vol. 196) Springer-Verlag, pages 289–302.
22. M. Blum and S. Micali. How to Generate Cryptographically Strong Sequences of Pseudo-Random Bits. *SIAM J. on Comput.*, Vol. 13, pages 850–864, 1984.
23. D. Boneh, R. DeMillo and R. Lipton. On the Importance of Checking Cryptographic Protocols for Faults. In *EuroCrypt97*, Springer LNCS (Vol. 1233), pages 37–51, 1997.

24. J.B. Boyar. Inferring Sequences Produced by Pseudo-Random Number Generators. *J. of the ACM*, Vol. 36, pages 129–141, 1989.

25. G. Brassard, D. Chaum and C. Crépeau. Minimum Disclosure Proofs of Knowledge. *J. of Comp. and Sys. Sci.*, Vol. 37, No. 2, pages 156–189, 1988.

26. G. Brassard and C. Crépeau. Zero-Knowledge Simulation of Boolean Circuits. In *Crypto86*, Springer-Verlag LNCS (Vol. 263), pages 223–233, 1987.

27. R. Canetti. *Studies in Secure Multi-Party Computation and Applications*. Ph.D. Thesis, Department of Computer Science and Applied Mathematics, Weizmann Institute of Science, Rehovot, Israel, June 1995.
   Available from from http://theory.lcs.mit.edu/~tcryptol/BOOKS/ran-phd.html.

28. R. Canetti. Towards Realizing Random Oracles: Hash Functions that Hide All Partial Information. These proceedings.

29. R. Canetti, C. Dwork, M. Naor and R. Ostrovsky. Deniable Encryption. These proceedings.

30. R. Canetti and R. Gennaro. Incoercible Multiparty Computation. In *37th FOCS*, pages 504–513, 1996.

31. R. Canetti, S. Halevi and A. Herzberg. How to Maintain Authenticated Communication in the Presence of Break-Ins. In *16th Symp. on Principles of Distributed Computing*, 1997.

32. R. Canetti and A. Herzberg. Maintaining Security in the Presence of Transient Faults. In *Crypto94*, Springer-Verlag LNCS (Vol. 839), pages 425–439.

33. L. Carter and M. Wegman. Universal Hash Functions. *J. of Comp. and Sys. Sci.*, Vol. 18, 1979, pages 143–154.

34. D. Chaum. Blind Signatures for Untraceable Payments. In *Crypto82*, Plenum Press, pages 199–203, 1983.

35. D. Chaum, C. Crépeau and I. Damgård. Multi-party unconditionally Secure Protocols. In *20th STOC*, pages 11–19, 1988.

36. D. Chaum, A. Fiat and M. Naor. Untraceable Electronic Cash. In *Crypto88*, Springer-Verlag LNCS (Vol. 403), pages 319–327.

37. B. Chor and N. Gilboa. Computationally Private Information Retrieval. In *29th STOC*, pages 304–313, 1997.

38. B. Chor, O. Goldreich, E. Kushilevitz and M. Sudan, Private Information Retrieval. In *36th FOCS*, pages 41–50, 1995.

39. B. Chor, S. Goldwasser, S. Micali and B. Awerbuch. Verifiable Secret Sharing and Achieving Simultaneity in the Presence of Faults. In *26th FOCS*, pages 383–395, 1985.

40. R. Cleve. Limits on the Security of Coin Flips when Half the Processors are Faulty. In *18th STOC*, pages 364–369, 1986.

41. I. Damgård. Collision Free Hash Functions and Public Key Signature Schemes. In *Euro-Crypt87*, Springer-Verlag, LNCS (Vol. 304), pages 203–216.

42. I. Damgård. A Design Principle for Hash Functions. In *Crypto89*, Springer-Verlag LNCS (Vol. 435), pages 416–427.

43. I. Damgård, O. Goldreich, T. Okamoto and A. Wigderson. Honest Verifier vs Dishonest Verifier in Public Coin Zero-Knowledge Proofs. In *Crypto95*, Springer-Verlag LNCS (Vol. 963), pages 325–338, 1995.

44. A. De-Santis, Y. Desmedt, Y. Frankel and M. Yung. How to Share a Function Securely. In *26th STOC*, pages 522–533, 1994.

45. Y. Desmedt and Y. Frankel. Threshold Cryptosystems. In *Crypto89*, Springer-Verlag LNCS (Vol. 435), pages 307–315.

46. W. Diffie, and M.E. Hellman. New Directions in Cryptography. *IEEE Trans. on Info. Theory*, IT-22 (Nov. 1976), pages 644–654.

47. D. Dolev, C. Dwork, and M. Naor. Non-Malleable Cryptography. In *23rd STOC*, pages 542–552, 1991.

48. C. Dwork, and M. Naor. Pricing via Processing or Combatting Junk Mail. In *Crypto92*, Springer-Verlag LNCS (Vol. 740), pages 139–147.
49. C. Dwork, and M. Naor. An Efficient Existentially Unforgeable Signature Scheme and its Application. To appear in *J. of Crypto.*. Preliminary version in *Crypto94*.
50. S. Even, O. Goldreich and S. Micali. On-line/Off-line Digital signatures. *J. of Crypto.*, Vol. 9, 1996, pages 35–67.
51. U. Feige, A. Fiat and A. Shamir. Zero-Knowledge Proofs of Identity. *J. of Crypto.*, Vol. 1, 1988, pages 77–94.
52. U. Feige, D. Lapidot, and A. Shamir. Multiple Non-Interactive Zero-Knowledge Proofs Based on a Single Random String. In *31th FOCS*, pages 308–317, 1990. To appear in *SIAM J. on Comput.*.
53. U. Feige and A. Shamir. Witness Indistinguishability and Witness Hiding Protocols. In *22nd STOC*, pages 416–426, 1990.
54. P. Feldman. A Practical Scheme for Non-interactive Verifiable Secret Sharing. In *28th FOCS*, pages 427–437, 1987.
55. A. Fiat. Batch RSA. *J. of Crypto.*, Vol. 10, 1997, pages 75–88.
56. A. Fiat and A. Shamir. How to Prove Yourself: Practical Solution to Identification and Signature Problems. In *Crypto86*, Springer-Verlag LNCS (Vol. 263), pages 186–189, 1987.
57. R. Fischlin and C.P. Schnorr. Stronger Security Proofs for RSA and Rabin Bits. In *Euro-Crypt97*, Springer LNCS (Vol. 1233), pages 267–279, 1997.
58. A.M. Frieze, J. Håstad, R. Kannan, J.C. Lagarias, and A. Shamir. Reconstructing Truncated Integer Variables Satisfying Linear Congruences. *SIAM J. on Comput.*, Vol. 17, pages 262–280, 1988.
59. P.S. Gemmell. An Introduction to Threshold Cryptography. In *CryptoBytes*, RSA Lab., Vol. 2, No. 3, 1997.
60. O. Goldreich. Two Remarks Concerning the GMR Signature Scheme. In *Crypto86*, Springer-Verlag LNCS (Vol. 263), pages 104–110, 1987.
61. O. Goldreich. *Lecture Notes on Encryption, Signatures and Cryptographic Protocol.* Spring 1989. Available from http://theory.lcs.mit.edu/~oded/ln89.html
62. O. Goldreich. *Foundation of Cryptography – Fragments of a Book.* February 1995. Available from http://theory.lcs.mit.edu/~oded/frag.html
63. O. Goldreich, S. Goldwasser, and S. Micali. How to Construct Random Functions. *J. of the ACM*, Vol. 33, No. 4, pages 792–807, 1986.
64. O. Goldreich, S. Goldwasser, and S. Micali. On the Cryptographic Applications of Random Functions. In *Crypto84*, Springer-Verlag LNCS (Vol. 263), pages 276–288, 1985.
65. O. Goldreich, R. Impagliazzo, L.A. Levin, R. Venkatesan, and D. Zuckerman. Security Preserving Amplification of Hardness. In *31st FOCS*, pages 318–326, 1990.
66. O. Goldreich and H. Krawczyk. On the Composition of Zero-Knowledge Proof Systems. *SIAM J. on Comput.*, Vol. 25, No. 1, February 1996, pages 169–192.
67. O. Goldreich and L.A. Levin. Hard-core Predicates for any One-Way Function. In *21st STOC*, pages 25–32, 1989.
68. O. Goldreich, S. Micali and A. Wigderson. Proofs that Yield Nothing but their Validity or All Languages in NP Have Zero-Knowledge Proof Systems. *J. of the ACM*, Vol. 38, No. 1, pages 691–729, 1991. See also preliminary version in *27th FOCS*, 1986.
69. O. Goldreich, S. Micali and A. Wigderson. How to Play any Mental Game – A Completeness Theorem for Protocols with Honest Majority. In *19th STOC*, pages 218–229, 1987.
70. O. Goldreich and Y. Oren. Definitions and Properties of Zero-Knowledge Proof Systems. *J. of Crypto.*, Vol. 7, No. 1, pages 1–32, 1994.
71. O. Goldreich and R. Ostrovsky. Software Protection and Simulation on Oblivious RAMs. *J. of the ACM*, Vol. 43, 1996, pages 431–473.

72. S. Goldwasser and L.A. Levin. Fair Computation of General Functions in Presence of Immoral Majority. In *Crypto90*, Springer-Verlag LNCS (Vol. 537), pages 77-93.

73. S. Goldwasser and S. Micali. Probabilistic Encryption. *J. of Comp. and Sys. Sci.*, Vol. 28, No. 2, pages 270–299, 1984. See also preliminary version in *14th STOC*, 1982.

74. S. Goldwasser, S. Micali and C. Rackoff. The Knowledge Complexity of Interactive Proof Systems. *SIAM J. on Comput.*, Vol. 18, pages 186–208, 1989.

75. S. Goldwasser, S. Micali, and R.L. Rivest. A Digital Signature Scheme Secure Against Adaptive Chosen-Message Attacks. *SIAM J. on Comput.*, April 1988, pages 281–308.

76. S. Goldwasser, S. Micali and P. Tong. Why and How to Establish a Private Code in a Public Network. In *23rd FOCS*, 1982, pages 134–144.

77. S. Goldwasser, S. Micali and A.C. Yao. Strong Signature Schemes. In *15th STOC*, pages 431–439, 1983.

78. J. Håstad, R. Impagliazzo, L.A. Levin and M. Luby. Construction of Pseudorandom Generator from any One-Way Function. To appear in *SIAM J. on Comput.*. Preliminary versions by Impagliazzo et. al. in *21st STOC* (1989) and Håstad in *22nd STOC* (1990).

79. J. Håstad, A. Schrift and A. Shamir. The Discrete Logarithm Modulo a Composite Hides $O(n)$ Bits. *J. of Comp. and Sys. Sci.*, Vol. 47, pages 376–404, 1993.

80. A. Herzberg, S. Jarecki, H. Krawczyk and M. Yu. Proactive Secret Sharing, or How to Cope with Perpetual Leakage. In *Crypto95*, Springer-Verlag LNCS (Vol. 963), pages 339–352.

81. R. Impagliazzo and M. Luby. One-Way Functions are Essential for Complexity Based Cryptography. In *30th FOCS*, pages 230-235, 1989.

82. R. Impagliazzo and M. Naor. Efficient Cryptographic Schemes Provable as Secure as Subset Sum. *J. of Crypto.*, Vol. 9, 1996, pages 199–216.

83. R. Impagliazzo and S. Rudich. Limits on the Provable Consequences of One-Way Permutations. In *21st STOC*, pages 44–61, 1989.

84. R. Impagliazzo and M. Yung. Direct Zero-Knowledge Computations. In *Crypto87*, Springer-Verlag LNCS (Vol. 293), pages 40–51, 1987.

85. A. Juels, M. Luby and R. Ostrovsky. Security of Blind Digital Signatures. These proceedings.

86. J. Kilian. A Note on Efficient Zero-Knowledge Proofs and Arguments. In *24th STOC*, pages 723–732, 1992.

87. J. Kilian and E. Petrank. An Efficient Non-Interactive Zero-Knowledge Proof System for NP with General Assumptions. To appear in *J. of Crypto.*.

88. E. Kushilevitz and R. Ostrovsky. Replication Is NOT Needed: A SINGLE Database, Computational PIR. TR CS0906, Department of Computer Science, Technion, May 1997.

89. L.A. Levin. One-Way Function and Pseudorandom Generators. *Combinatorica*, Vol. 7, pages 357–363, 1987.

90. M. Luby. *Pseudorandomness and Cryptographic Applications*. Princeton University Press, 1996.

91. M. Luby and C. Rackoff. How to Construct Pseudorandom Permutations from Pseudorandom Functions. *SIAM J. on Comput.*, Vol. 17, 1988, pages 373–386.

92. R.C. Merkle. Protocols for public key cryptosystems. In *Proc. of the 1980 Symposium on Security and Privacy*.

93. R.C. Merkle. A Digital Signature Based on a Conventional Encryption Function. In *Crypto87*, Springer-Verlag LNCS (Vol. 293), 1987, pages 369-378.

94. R.C. Merkle. A Certified Digital Signature Scheme. In *Crypto89*, Springer-Verlag LNCS (Vol. 435), pages 218–238.

95. S. Micali. Fair Public-Key Cryptosystems. In *Crypto92*, Springer-Verlag LNCS (Vol. 740), pages 113–138.

96. S. Micali and P. Rogaway. Secure Computation. In *Crypto91*, Springer-Verlag LNCS (Vol. 576), pages 392–404.

97. National Institute for Standards and Technology. Digital Signature Standard (DSS), *Federal Register*, Vol. 56, No. 169, August 1991.

98. M. Naor. Bit Commitment using Pseudorandom Generators. *J. of Crypto.*, Vol. 4, pages 151–158, 1991.

99. M. Naor, R. Ostrovsky, R. Venkatesan and M. Yung. Zero-Knowledge Arguments for NP can be Based on General Assumptions. In *Crypto92*, Springer-Verlag LNCS (Vol. 740), pages 196–214.

100. M. Naor and B. Pinkas. Visual Authentication and Identification. These proceedings.

101. M. Naor and O. Reingold. Synthesizers and their Application to the Parallel Construction of Pseudo-Random Functions. In *36th FOCS*, pages 170–181, 1995.

102. M. Naor and O. Reingold. On the Construction of Pseudo-Random Permutations: Luby-Rackoff Revisited. In *29th STOC*, pages 189–199, 1997.

103. M. Naor and A. Shamir. Visual Cryptography. In *EuroCrypt94*, Springer-Verlag LNCS (Vol. 950), 1995, pages 1–12.

104. M. Naor and M. Yung. Universal One-Way Hash Functions and their Cryptographic Application. *21st STOC*, 1989, pp. 33-43.

105. M. Naor and M. Yung. Public-Key Cryptosystems Provably Secure Against Chosen Ciphertext Attacks. In *22nd STOC*, pages 427-437, 1990.

106. R. Ostrovsky and A. Wigderson. One-Way Functions are essential for Non-Trivial Zero-Knowledge. In *2nd Israel Symp. on Theory of Computing and Systems*, IEEE Comp. Soc. Press, pages 3–17, 1993.

107. R. Ostrovsky and M. Yung. How to Withstand Mobile Virus Attacks. In *10th Symp. on Principles of Distributed Computing*, pages 51–59, 1991.

108. B. Pfitzmann. *Digital Signature Schemes (General Framework and Fail-Stop Signatures)*. Springer LNCS (Vol. 1100), 1996.

109. M.O. Rabin. Digitalized Signatures. In *Foundations of Secure Computation* (R.A. DeMillo et. al. eds.), Academic Press, 1977.

110. M.O. Rabin. Digitalized Signatures and Public Key Functions as Intractable as Factoring. MIT/LCS/TR-212, 1979.

111. M.O. Rabin. How to Exchange Secrets by Oblivious Transfer. Tech. Memo TR-81, Aiken Computation Laboratory, Harvard U., 1981.

112. T. Rabin and M. Ben-Or. Verifiable Secret Sharing and Multi-party Protocols with Honest Majority. In *21st STOC*, pages 73–85, 1989.

113. R. Rivest, A. Shamir and L. Adleman. A Method for Obtaining Digital Signatures and Public Key Cryptosystems. *CACM*, Vol. 21, Feb. 1978, pages 120–126.

114. J. Rompel. One-way Functions are Necessary and Sufficient for Secure Signatures. In *22nd STOC*, 1990, pages 387–394.

115. C.E. Shannon. Communication Theory of Secrecy Systems. *Bell Sys. Tech. J.*, Vol. 28, pages 656–715, 1949.

116. A. Shamir. How to Share a Secret. *CACM*, Vol. 22, Nov. 1979, pages 612–613.

117. U.V. Vazirani and V.V. Vazirani. Efficient and Secure Pseudo-Random Number Generation. *25th FOCS*, pages 458–463, 1984.

118. M. Wegman and L. Carter. New Hash Functions and their Use in Authentication and Set Equality. *J. of Comp. and Sys. Sci.*, Vol. 22, 1981, pages 265–279.

119. A.C. Yao. Theory and Application of Trapdoor Functions. In *23rd FOCS*, pages 80–91, 1982.

120. A.C. Yao. How to Generate and Exchange Secrets. In *27th FOCS*, pages 162–167, 1986.

Plug and Play Encryption

Donald Beaver *

IBM/Transarc

Abstract. We present a novel protocol for secret key exchange that is provably secure against attacks by an adversary that is free to attack zero, one, or both parties in an adaptive fashion, at any time. This high degree of robustness enables larger, multiparty interactions (including multiparty secure computations) to substitute our protocol for secure private channels in a simple, plug-and-play fashion, without simultaneously limiting security analysis to attacks by static adversaries, *i.e.* adversaries whose corruption choices are fixed in advance.

No reliance on the assistance of third parties or on erasing partial computations is required. In addition to providing order-of-magnitude speedups over alternative approaches, the simplicity of our protocols lends itself to simple demonstrations of security. We present constructions that are based on a novel and counterintuitive use of the Diffie-Hellman key exchange protocol; our methods extend to other standard cryptographic assumptions as well.

1 Introduction

Historically, the pressing theme of cryptology has been to convey a private message securely, as though an absolutely secure private channel were available. Encryption schemes of ever increasing robustness have been proposed, from private-key methods that assume an initially-secure exchange, to public-key schemes that enable key exchange over public lines without prior communication [DH76]. Although not always explicit, one goal is common to all such efforts: to be able to plug in a replacement for an absolutely secure channel without compromising security, *i.e.* while maintaining the level of security of the original (even if imaginary) channel.

The construction of large systems is guided by several natural motivations for pursuing a component-based approach. First, the design and construction of large systems is simplified, as is the analysis of their properties. Second, the ability to replace costly or idealized components reduces the overall costs of implementing a system. Third, the system flexibly accommodates advances in quality and technology – as well as newly uncovered disadvantages or flaws. The goal of component-based design is thus not merely aesthetic: functionality, utility, affordability, and flexibility tend to decrease as complexity increases.

---

* Transarc Corp., Pittsburgh, PA 15219; 412-338-4365; beaver@transarc.com, http://www.transarc.com/~beaver.

Security is no exception. Yet security analyses often make implicit assumptions about the context in which a component is used, at the risk of compromising overall security.

In particular, when a cryptosystem is analyzed in isolation, *e.g.* as a system involving just two parties and an eavesdropper, many constraints that apply to its use in a larger system are easily overlooked, or misleadingly treated as moot. Even if the result is not catastrophic failure, there is a price: the deceptively simple verification of security in isolation provides no formal guarantee about security after installation.

The goal of this work is to provide *plug-and-play encryption systems* that are robust, sufficiently efficient and (most importantly) sufficiently analyzed to merit simple installation in large-scale network interactions, whether tightly coupled (as in multiparty computations) or loosely (as on internets). In either case, we demand and achieve privacy in the face of adaptive attacks.

Adaptive Attacks. One subtle sticking point that has received relatively little attention is the distinction between static and adaptive attacks. An adversary mounting a static attack is required to choose which players it wants to corrupt before the protocol begins (although its later behavior, *viz* substitution of messages, can be freely adapted). Adaptive attacks permit the adversary to choose whom to corrupt (typically, up to some limit) at any time.

Clearly, security against adaptive attacks is the stronger, more realistic, and more desirable achievement. But analysis of static attacks is simpler, and far more common. There are indeed apparently good reasons to discard the more complex (hence more risky) analysis needed to assure adaptive security. For example, if a sender or receiver is attacked, then any messages are compromised *a fortiori*, so further analysis seems unnecessary. Worse, if a sender is attacked, then what happens when the receiver is attacked later should be moot, since any messages are likewise already compromised.

Unfortunately, this sort of reasoning takes place in a cryptographic vacuum: there exists nothing apart from sender, receiver, and eavesdropper. It does not extend to situations in which there are multiple parties employing multiple cryptosystems along with diverse other protocols and interactions. In such a setting, the behavior of the network is not made moot by the failure of two parties, hence an overall analysis *does* depend on the behavior of a component protocol when one or both of its participants are overrun.

Thus – subtly – the security of a system that has been verified only against static attacks may remain in question if it is plugged into a large-scale interaction in which attackers pick and choose victims at will.

While obvious when stated thus, this simple observation is easily overlooked when the simpler component (cryptosystem, zero-knowledge proof, *etc.*) is analyzed outside the context of the larger system.

The natural questions, then, are whether the static analysis is sufficient anyway (it is not), whether existing cryptosystems can hope to enjoy a demonstration of security in dynamic, adversarial environments (most cannot), and whether there are alternatives that can indeed be shown robust (there are).

Obstacles to Adaptive Security. The primary technical difficulty in developing cryptosystems that can be proven secure against adaptive attacks is the fact that most cryptosystems bind the sender and receiver to the cleartext message, even though the cleartext itself may be hard to calculate from the ciphertext alone.

An adversary (and anyone else) can simply observe $E(m, r)$ over a public line (where m is the cleartext and r represents random bits). Not only are the sender and receiver unable to later pretend that another message m' was sent, but the adversary can expect to discover a message m and list of random bits r that is consistent with the string $E(m, r)$, if it decides to corrupt the sender or receiver later on.

This concern arises even in the simple application of key-exchange protocols. Consider Diffie-Hellman key exchange (DH), in which Alice sends some $g^a \bmod p$, Bob responds with some $g^b \bmod p$, and the two use $K = g^{ab} \equiv (g^a)^b \equiv (g^b)^a \bmod p$ as their secret key [DH76].[2]

Unless computing discrete logarithms is feasible, an eavesdropper will likely find it hard to compute g^{ab}. But it is impossible to supply a different a', b' and $g^{a'b'}$ that are consistent with the public values g^a and g^b. Thus, it is impossible to pretend later that a value other than g^{ab} was used as the key, if the value of a or b is obtained (e.g. via corruption of one of the parties). Messages encrypted with g^{ab} (or some derivative) are private but immutable.

Simulation. This property, although apparently just technical, is critical in any approach that measures knowledge through Turing tests, i.e. through simulations [GMR89]. In the domain of computational security, zero-knowledge approaches are a common standard for demonstrating interactive security ([GMR89, GMW86, B91, MR91, B95]). Typically, one must find a simulator that presents a convincing but faked conversation without having access to the private information that normally may play a role in generating the actual conversation. If the fake conversation is indistinguishable from a real one, then we may infer that the real one leaks no "knowledge" about the sensitive private information.

In the case of encryption, this reduces to being able to simulate a ciphertext (or a key-exchange conversation) without having access to the messages and secret bits held by the sender and receiver. The static case is simple to analyze: if either party is corrupt, then the simulator is entitled to the secret message m, and can easily form a valid encryption $E(m, r)$ as necessary; if neither is corrupt, then the simulator can typically offer up $E(0, r)$ as the fake ciphertext. Because the key and random bits will never be obtained by the static adversary, distinguishing between $E(0, r)$ and $E(1, r')$ (for example) is tantamount to breaking a cryptographic assumption [GM84].

It might be argued that this simulation-based approach is unnecessarily burdensome for cryptosystems. Yet it is often the case that a larger protocol employs other convenient modules, such as zero-knowledge proofs or oblivious transfers

---

[2] Here, p is a large prime, g a generator of $\mathbf{Z}_p^*$, and a and b are chosen at random $\bmod (p - 1)$ by the respective parties.

or bit committals, and as a result demands an overall simulator-based approach. To facilitate plug-and-play usage of cryptosystems, then, a simulator for the component cryptosystem is a minimum requirement.

Equivocation. For encryption systems, the central technical problem lies in message equivocation. Although a ciphertext may be infeasible to break, it is also likely to be unequivocal: that is, it may be impossible to make it appear as the encryption of two different messages. In some sense, this is intuitively unavoidable, since a receiver must be able to decide on a cleartext interpretation. But therein lies the central problem:

> **Equivocation Paradox.** If a ciphertext can be decrypted to more than one cleartext, then a receiver cannot be sure what message it received. If a ciphertext cannot be decrypted to more than one cleartext, then a simulator cannot demonstrate that it is secure.

Advances: Past and Present. Beaver and Haber presented a direct and efficient cryptosystem that requires each party to erase certain internal records [BH92]. Without access to these records, a later attacker will not find sufficient information to determine that a simulated ciphertext is different from an actual ciphertext. A similar construction has been reported to have been developed by Feldman (see [CFGN96]).

Although it is good security practice to erase keys as soon as possible, it is certainly preferable to avoid basing security on such demands. Canetti, Feige, Goldreich, and Naor broke through this barrier with an ingenious method that requires no erasing [CFGN96].

Neither of these methods is fully satisfactory: one requires erasing and careful attention regarding automatic backups, while the other is complex and expensive in both its design and its verification.

This paper capitalizes on important concepts from each and extends them to broader and simpler techniques, achieving quasi-practical (as opposed to carefully-managed or merely theoretic) performance.

Using an unusual twist on Diffie-Hellman key exchange, we present a novel method for adaptively-secure key exchange:

Theorem 1. *There exists a non-erasing implementation of secret key exchange, using expected $O(1)$ invocations of Diffie-Hellman key exchange per bit, that is secure against adaptive 2-adversaries, if the Diffie-Hellman Assumption holds.*

In comparison, the protocol of [BH92] is more efficient but requires erasing, while the protocol of [CFGN96] achieved the same goal already but with $\Omega(k)$ invocations per key bit of a similar underlying primitive, where k is a security parameter.

We employ DH-based solutions for the purpose of exposition. These results generalize naturally to factoring, discrete logarithm, and, if the cost of involving network computations is permissible, to any one-way trapdoor permutation (as in [CFGN96]).

These results also provide the most efficient open-channel replacements available for the private channels used in the information-theoretically secure multiparty computation protocols of Ben-Or, Goldwasser and Wigderson [BGW88] and Chaum, Crépeau, and Damgard [CCD88].

Notably, the proof of security against adaptive attacks is radically simplified, as well. This is particularly important in light of the apparent increase in complexity (and hence risk) of extending formal analysis from static to adaptive scenarios. That is, even though one might expect to encounter more complicated analyses and therefore enjoy less confidence in the results, we demonstrate that this is unnecessary.

Further Remarks: Commitment and Deniability. We have avoided the term "non-committing" [CFGN96] because of certain ambiguities. In particular, all of the above results are "committing" in the sense that an honest sender cannot later pretend that an alternate message was sent. That is, these cryptosystems are non-committing for the *simulator*, and they are non-committing in that they do not immediately serve as a bit commitment scheme as a typical cryptosystem does. These caveats said, our cryptosystem shares the same such properties as [CFGN96].

The protocol presented here is committing yet "equivocable": real ciphertexts are unequivocal and bind an honest sender to the cleartext, but they can be made to appear equivocal using specially-crafted facsimiles.

In separate work, the methodology presented here has been extended to be *non-committal* in a different sense, *viz* an honest sender and receiver can convince outside inspectors that an arbitrary alternative cleartext was sent [B96], enabling them thus to deny that a particular cleartext was sent. The solution presented in [B96] uses a more complicated and less efficient mechanism, and it obscures the primary solution presented here. Moreover, this paper presents a full proof of security.

2 Formalities

Attacks: Static or Adaptive. An adversary is a probabilistic poly-time TM (PPTM) that issues two sorts of messages: "*corrupt i*," "*send m* from i to j." It receives two sorts of responses: "*view of i*," "*receive m* from j to i."

A **static t-adversary** is an adversary who issues up to t *corrupt* requests before the protocol starts. An **adaptive t-adversary** may issue up to t such requests at any time.

Encryption. The **specification protocol for secure channels** is a two-party protocol consisting of $(\hat{A}, \hat{B})$, in which $\hat{A}$ inputs a bit m which is transferred securely to $\hat{B}$. An eavesdropper knows only that a bit was sent, or that one or the other party decided to abort.

To simplify analysis, we consider an implementation network that provides authenticated, service-undeniable, point-to-point connections. Our protocols can otherwise be extended, though such strengthenings are uninstructive, here. The traffic over the lines is public, of course.

Simulation-based security. In the adaptive case, there is a single interface/simulator, $\mathcal{I}$, who receives requests from and delivers responses to the attacker, $\mathcal{A}$, creating an environment for $\mathcal{A}$ as though $\mathcal{A}$ were attacking a given implementation. $\mathcal{I}$ is itself an attacker acting within the specification protocol, which is run with players $\hat{\imath}$ following the specification's programs on inputs x_i. When $\mathcal{A}$ corrupts player i, $\mathcal{I}$ issues a corruption request and is given $\hat{\imath}$'s information.[3] $\mathcal{I}$ responds to $\mathcal{A}$ with a facsimile of the *"view of i"* response that $\mathcal{A}$ expects. $\mathcal{I}$ receives all of $\mathcal{A}$'s *"send m"* requests and provides $\mathcal{A}$ with facsimiles of *"receive m"* responses. Finally, $\mathcal{A}$ (or $\mathcal{I}$ on $\mathcal{A}$'s behalf) writes its output, $y_{\mathcal{A}}$.

In the case of secure channels, let $\mathcal{A}$, with auxiliary input $x_{\mathcal{A}}$, attack a given implementation in which Alice holds input m. The execution induces a distribution $(A(m), B, \mathcal{A}(x_{\mathcal{A}}))$ on output triples, $(y_A, y_B, y_{\mathcal{A}})$.

Let $\mathcal{I}(\mathcal{A}(x_{\mathcal{A}}))$ attack the specification (described above). The execution induces a distribution $(\hat{A}(b), \hat{B}, \mathcal{I}(\mathcal{A}(x_{\mathcal{A}})))$ on output triples, $(y_{\hat{A}}, y_{\hat{B}}, y_{\mathcal{I}})$.

An extra, "security" parameter k may be considered. This provides a sequence of distributions on output triples in each scenario. Let $\approx$ denote *computational indistinguishability*, a notion whose formal definition is omitted for reasons of space (*cf.* [GMR89]).

An **encryption scheme secure against adaptive t-adversaries** is a (two-party) protocol such that, for any adaptive t-adversary $\mathcal{A}$, there is a PPTM simulator $\mathcal{I}$ such that for any m, $(A(m), B, \mathcal{A}(x_{\mathcal{A}})) \approx (\hat{A}(m), \hat{B}, \mathcal{I}(\mathcal{A}(x_{\mathcal{A}})))$.[4]

Notation. Let $\$(S)$ denote the uniformly random distribution over finite set S. For a prime p, let $\mathbf{Z}_p^* = \{1, 2, \ldots, p-1\}$ and $\mathbf{Z}_{p-1} = \{0, 1, 2, \ldots, p-2\}$.

Assumptions. Let $p - 1 = 2p'$, where p and p' are prime. Let $\hat{g}$ be a generator of $\mathbf{Z}_p^*$, and define $g = \hat{g}^2 \bmod p$. Then g generates a subgroup $\langle g \rangle$ (of quadratic residues mod p). Define the **Diffie-Hellman distribution** D_p as the triple of random variables (A, B, C) obtained through

$$a \leftarrow \$(\mathbf{Z}_{p-1}), b \leftarrow \$(\mathbf{Z}_{p-1}), A \leftarrow g^a \bmod p, B \leftarrow g^b \bmod p, C \leftarrow g^{ab} \bmod p.$$

The **Decision Diffie-Hellman Assumption** (DDHA) can be described as follows:

(DDHA) Let p be a prime and g a subgroup generator selected as described above. Then D_p is computationally indistinguishable from $(\$(\langle g \rangle), \$(\langle g \rangle), \$(\langle g \rangle))$.

Note that without the precaution of moving to a subgroup, typical Diffie-Hellman triples can be distinguished from three random elements. The quadratic residuosity of g^{ab} can be deduced from that of g^a and g^b, hence a random element would be distinguishable from g^{ab}.

---

[3] $\hat{\imath}$ is a player in the specification protocol and is unaware of messages being passed in a given implementation. In particular, $\hat{\imath}$ knows only its input x_i and the messages it sends and receives over channels supported in the specification.

[4] A multiparty implementation is also possible; the formalities are similar.

3 Adaptively Secure Key Exchange

The novel idea (and surprising twist) behind our protocols is two-fold: although Alice and Bob engage in a classical, statically-secure key exchange protocol, they (1) sometimes garble their strings and (2) always reveal the secret key!

More specifically, Alice and Bob perform two parallel DH exchanges (indexed 0 and 1, say). They independently choose to garble computations and strings on precisely one index. Bob responds not only with (possibly garbled) g^b but with the (possibly garbled) key in each exchange.

If they chose identical indices to garble, then the key will be recognizably correct in the ungarbled instance. This provides them with a common secret bit: the index of the ungarbled instance. Of course, if they chose opposite indices, then they detect garbage on both instances and try again (sequentially or in parallel).

An eavesdropper, Eve, learns only whether Alice and Bob made the same guess in a given attempt, but not what that guess was. It remains to determine whether the details ensure three constraints: (1) Eve cannot tell the difference between garbage and a valid exchange; (2) Alice and Bob can agree on which index is garbled; and critically, (3) a simulator can construct conversations that can later be made consistent with (valid,garble) or (garble,valid) as needed.

3.1 Creating Garbage

The "garbling" of the Diffie-Hellman protocol occurs in one of two ways. Instead of choosing an exponent e and computing $r = g^e$, a player can choose $r \in \langle g \rangle$ directly without knowing its discrete logarithm. Naturally, that player will be unable to calculate or verify the final DH key, g^{ab}, but this is unimportant to Alice and Bob. (It is *extremely* important to the proof of security, however!) Second, a player can garble g^{ab} by likewise choosing a uniformly random residue whose discrete logarithm is unknown.

Fig. 1 describes the full protocol for a single attempt to transmit a one-bit message, m. Alice and Bob each conclude either fail or succeed:m. Clearly, they can trivially establish a one-time pad bit instead by using $m = 0$.

Note that we have no need of implicit zero-knowledge proofs of behavior or knowledge, or even simple verification beyond parsability: a corrupt Alice or Bob is fully permitted to learn all available logarithms and to force the agreed-upon bit to be anything they like.

If Alice's choice, c, matches Bob's choice, d, then the relevant variables describe a normal execution of Diffie-Hellman key exchange. In particular, the Diffie-Hellman key would be $x_c^b = y_c^a = z_c$. Thus, Alice can simply check whether $y_c^a = z_c$.[5] If they are equal, Alice knows $c = d$ and uses c to mask m.

---

[5] We are concerned only with sharing a single, equivocal random bit, thus revealing the value z_d is not an issue. Alternatively, the key can be hashed or used to encrypt a known or redundant message in order to detect whether the chosen index "makes sense" or whether it is garbled.

Send-Bit-Attempt(m)

0.	Public:	prime p, subgroup generator g	3.1. A:
1.1.	A:	$c \leftarrow \$(\{0,1\})$, $a \leftarrow \$(Z_{p-1})$,	
		$x_c \leftarrow g^a \bmod p$, $x_{1-c} \leftarrow \$(\langle g \rangle)$	
1.2.	A$\rightarrow$B:	x_0, x_1	
2.1.	B:	$d \leftarrow \$(\{0,1\})$, $b \leftarrow \$(Z_{p-1})$,	3.2. A$\rightarrow$B:
		$y_d \leftarrow g^b \bmod p$, $y_{1-d} \leftarrow \$(\langle g \rangle)$,	4.1. B:
		$z_d \leftarrow x_d^b \bmod p$, $z_{1-d} \leftarrow \$(\langle g \rangle)$	
2.2.	B$\rightarrow$A:	y_0, y_1, z_0, z_1	

3.1. A: if $y_c^a = z_d$, then $s \leftarrow 0$
 else $s \leftarrow 1$
 if $s = 0$, then $f \leftarrow m \oplus c$
 else $f \leftarrow 0$
3.2. A$\rightarrow$B: (s, f)
4.1. B: if $s = 0$, then
 $m \leftarrow f \oplus d$
 conclude succeed:m
 else conclude fail

Fig. 1. Three-pass attempt to transmit one bit, m.

If, however, their choices differ, then Bob has chosen both y_c and z_c as random residues, thus with high probability, $y_c^b \neq z_c$. Alice therefore informs Bob of failure.

It may seem possible to utilize even the failed attempts, since Alice could nevertheless calculate $d = 1 - c$ and use d as a one-time pad bit. This unfortunately disables equivocation by the simulator.

3.2 Three Passes or Four?

To establish a 1-bit shared secret key (or exchange a 1-bit message) with high probability, it suffices to use k parallel attempts. To establish a k-bit shared secret key, $3k$ parallel attempts clearly suffice.

(Note that, unlike the case of parallel zero-knowledge, we do not face the issue of mutually antagonistic parties attempting to withold information from one another. Even if one of the parties is malicious, there are no challenges which a simulator needs to overcome, despite lacking a secret proof (or message): here, the simulator would have *all* knowledge, including any desired m, and it trivially simulates an honest party.)

Sending a k-bit message requires a touch more thought, however. Clearly, a fourth pass will suffice. The fourth pass could be avoided by using k invocations of 1-message-bit exchange, resulting in $O(k^2)$ attempts. Instead, applying the linear codes of Sipser and Spielman [SiSp94], a k-bit message can be transmitted in 3 passes using only $O(k)$ attempts.

4 Proof of Security

The motivation for our counterintuitive disposal of DH is to enable a simulator to produce a fake conversation that can be explained as representing either a 0 or a 1, without yet knowing which explanation will be required. Normally, this is impossible, since a standard key exchange will uniquely define a secret. (Indeed, even this protocol commits an honest Alice and Bob to the secret bit.)

Interface-No-Corruption

0. $\mathcal{I} \to \mathcal{A}$: prime p, subgroup generator g (2.1 cont.)   else

1.1. Internal: $a_0 \leftarrow \$(\mathbb{Z}_{p-1})$, $a_1 \leftarrow \$(\mathbb{Z}_{p-1})$
         $x_0 \leftarrow g^{a_0} \bmod p$, $x_1 \leftarrow g^{a_1} \bmod p$

1.2. $\mathcal{I} \to \mathcal{A}$: "A$\to$B: x_0, x_1"

2.1. Internal: $s \leftarrow \$(\{0,1\})$
         $b_0 \leftarrow \$(\{0,1\})$, $b_1 \leftarrow \$(\{0,1\})$
         $y_0 \leftarrow g^{b_0} \bmod p$, $y_1 \leftarrow g^{b_1} \bmod p$
         if $s = 0$ then

                 $z_0 \leftarrow g^{a_0 b_0} \bmod p$,
                 $z_1 \leftarrow g^{a_1 b_1} \bmod p$

         (cont.)

else
         $c \leftarrow \$(\{0,1\})$,
         $a \leftarrow a_c$,
         $d \leftarrow 1 - c$,
         $b \leftarrow b_d$,
         $z_d \leftarrow y_d^{b_d} \bmod p$,
         $z_{1-d} \leftarrow \$(\langle g \rangle)$

2.2. $\mathcal{I} \to \mathcal{A}$: "B$\to$A: y_0, y_1, z_0, z_1"

3.1. Internal: if $s = 0$ then $f \leftarrow \$(\{0,1\})$
         else $f \leftarrow 0$

3.2. $\mathcal{I} \to \mathcal{A}$: "A$\to$B: (s, f)"

Fig. 2. Single-attempt interface/simulator, without corruption.

The simulator uses clean garbage, however. It constructs a conversation containing two *valid* DH exchanges, for which it knows *all* discrete logarithms. It can later pretend that both parties chose index c by witholding the discrete logs for exchange $1 - c$. Even though the resulting conversation is information-theoretically distinct from a real conversation distribution, a poly-bounded judge cannot detect the difference without breaking DDHA.

Our solution expands the intuition implicit in [CFGN96], in which the receiver is helped to avoid learning full information. Here, we arrange for *both* sender and receiver to avoid learning full information.

To prove security against adaptive adversaries, we present an interface that, when attacking an interaction between Alice and Bob over an absolutely secure channel, provides an adversary $\mathcal{A}$ with a fake view that is computationally indistinguishable from a real one. In the past, the great difficulty lay in patching a partially-committed view to accommodate $m_i = 0$ or $m_i = 1$ flexibly, particularly when one or both parties may be corrupted much later on.

For the moment, let us focus on a single execution of Send-Bit-Attempt. We first describe the action of the interface $\mathcal{I}$ when $\mathcal{A}$ makes no corruption requests. Unlike Alice and Bob, $\mathcal{I}$ "cheats" by discovering *both* keys. Instead of setting x_{1-c} and y_{1-d} to random values with unknown discrete logs, as an honest Alice or Bob would do, $\mathcal{I}$ knowingly selects their logarithms at random. In doing so, $\mathcal{I}$ retains the ability to "open" x_0 or x_1 consistently with $c = 0$ or $c = 1$. Moreover, $\mathcal{I}$ does not garble z_{1-d}, but improperly uses $x_{1-d}^{\log_g y_{1-d}}$. See Fig. 2 for details.

The key is that *the logarithms of the critical values are never explicitly represented anywhere in the network* – not even as a shared secret. Thus even an adversary who gains *both* Alice's and Bob's *complete, unerased* internal histories cannot calculate the discrete logs or even distinguish the critical residues from random values. (Naturally, a computationally-*unbounded* adversary will be able to distinguish $\mathcal{I}$'s fake view from a real-life view.)

In sum, $\mathcal{I}$ generates three fake messages:

$$(x_0, x_1), (y_0, y_1, z_0, z_1), (s, f)$$

where, if $s = 0$, then $z_i = y_i^{\log_g x_i}$.

Let us call x_{1-d}, y_{1-d}, and z_{1-d} "critical variables." In an execution of the Send-Message protocol in which Alice's message is generated honestly, the critical variables are uniformly random and independent of all other variables. The significant and sole difference between an actual transcript and $\mathcal{I}$'s generated version is that in $\mathcal{I}$'s version, when $s = 1$, the critical variables are *not independent* – although they are individually uniformly random. Specifically, $z_{1-d} = x_{1-d}^{\log_g y_{1-d}}$ – which is a relationship that does not hold when honest Alice generates the three values. This relationship is, however, infeasible to detect.

4.1 Corruption Requests

We now turn to how the interface handles corruption requests. There are four cases, depending on when $\mathcal{A}$ makes its first corruption request.

Case 0: $\mathcal{A}$ makes its first corruption request before any messages are sent.
Case 1: $\mathcal{A}$ makes its first corruption request after Alice sends her message, but before Bob sends his message.
Case 2: $\mathcal{A}$ makes its first corruption request after Alice and Bob have each sent one message.
Case 3: $\mathcal{A}$ makes its first corruption request after Alice and Bob have sent all their messages.

When the first corruption request is made, $\mathcal{I}$ will "patch" the views of *both* players, hand over the view of the corrupted player to $\mathcal{A}$, and then assume the role of the uncorrupted player based on the patched view. Note that the message bit m plays no role until Alice's second message to Bob.

Patching fake views. We turn first to how $\mathcal{I}$ patches the fake views. For clarity, the handling of auxiliary inputs (such as histories from previous protocols) is left implicit.

We further subdivide the cases into 0A, 0B, 1A, 1B, 2A, 2B, 3A and 3B, according to whether $\mathcal{A}$ selects Alice or Bob to corrupt first.
Case 0.A: $\mathcal{I}$ obtains $\hat{A}$'s input m from corrupting $\hat{A}$ in the specification protocol and reports it to $\mathcal{A}$.
Case 0.B: Nothing to patch.

Case 1.A: $\mathcal{I}$ obtains $\hat{A}$'s input m from the specification protocol. $\mathcal{I}$ performs $c \leftarrow \$(\{0,1\})$; $a \leftarrow a_d$. $\mathcal{I}$ patches A's view with m, c, a, (x_0, x_1) and reports it to $\mathcal{A}$. $\mathcal{I}$ patches B's view with (x_0, x_1).
Case 1.B: $\mathcal{I}$ patches B's view with (x_0, x_1).

Case 2.A: $\mathcal{I}$ obtains $\hat{A}$'s input m from the specification protocol. $\mathcal{I}$ performs $c \leftarrow \$(\{0,1\})$; $a \leftarrow a_d$; $d \leftarrow \$(\{0,1\})$; $b \leftarrow b_d$. $\mathcal{I}$ patches A's view with

m, c, a, (x_0, x_1), (y_0, y_1, z_0, z_1), and reports it to $\mathcal{A}$. $\mathcal{I}$ patches B's view with (x_0, x_1), d, b_c, (y_0, y_1, z_0, z_1).

Case 2.B: $\mathcal{I}$ performs $c \leftarrow \$(\{0, 1\})$; $a \leftarrow a_d$; $d \leftarrow \$(\{0, 1\})$; $b \leftarrow b_d$. $\mathcal{I}$ patches A's view with c, a, (x_0, x_1), (y_0, y_1, z_0, z_1). $\mathcal{I}$ patches B's view with (x_0, x_1), d, b_c, (y_0, y_1, z_0, z_1) and reports it to $\mathcal{A}$.

Case 3.A: $\mathcal{I}$ obtains $\hat{A}$'s input m from the specification protocol. At this point, $\mathcal{I}$ has irrevocably decided whether this attempt will succeed ($s = 0$) or fail ($s = 1$).

If $s = 0$, perform the following: $c \leftarrow m \oplus f$; $a \leftarrow a_c$; $d \leftarrow c$; $b \leftarrow b_d$.

If $s = 1$, $\mathcal{I}$ has already chosen values for c, a_c, d, and b_d.

$\mathcal{I}$ patches A's view with c, a, (x_0, x_1), (y_0, y_1, z_0, z_1), (s, f), and reports it to $\mathcal{A}$. $\mathcal{I}$ patches B's view with (x_0, x_1), d, b_c, (y_0, y_1, z_0, z_1), (s, f).

Case 3.B: $\mathcal{I}$ obtains message bit m from the secure channel (after honest $\hat{A}$ sends it to now-corrupt $\hat{B}$). As in case 3.A., $\mathcal{I}$ has irrevocably decided whether this attempt will succeed ($s = 0$) or fail ($s = 1$).

If $s = 0$, perform the following: $c \leftarrow m \oplus f$; $a \leftarrow a_c$; $d \leftarrow c$; $b \leftarrow b_d$.

If $s = 1$, $\mathcal{I}$ has already chosen values for c, a_c, d, and b_d.

$\mathcal{I}$ patches A's view with c, a, (x_0, x_1), (y_0, y_1, z_0, z_1), (s, f). $\mathcal{I}$ patches B's view with (x_0, x_1), d, b_c, (y_0, y_1, z_0, z_1), (s, f), and reports it to $\mathcal{A}$.

Assuming the role of the remaining, honest player. Generally speaking, $\mathcal{I}$ merely runs an internal copy of Alice or Bob, having set Alice's or Bob's state according to the given view. Should $\mathcal{A}$ then corrupt the remaining honest player, $\mathcal{I}$ simply hands over the current view. If the remaining honest player is Bob, this procedure is straightforward, since Bob has no special input.

If the honest player is Alice, however, $\mathcal{I}$ plays Alice's role without knowing m, at least until step 3 of Send-Bit-Attempt. Since Alice's computations do not depend on m until then, this presents no problem. Once $\mathcal{I}$ has received $\mathcal{A}$'s message to honest Alice on behalf of corrupt Bob, $\mathcal{I}$ waits for corrupt $\hat{B}$ to receive m along the secure channel. It then resumes its internal simulation of honest Alice.

Deciding what to send. If Alice is corrupted before she sends her second message, or equivalently, before $\hat{A}$ has sent the message m, then $\mathcal{I}$ must decide what to send on the secure channel on behalf of the now-corrupt $\hat{A}$. In this case, $\mathcal{I}$ did not yet commit a fake second message to $\mathcal{A}$; that is, $\mathcal{I}$ obtains Alice's message (s, f) from $\mathcal{A}$. $\mathcal{I}$ then continues to run the honest, internal copy of Bob, deriving Bob's effective result, either **fail** or **succeed**:m. In the former case, $\mathcal{I}$ does not send a bit (or append a bit to a longer secure message, in the context of a k-bit protocol). In the latter case, $\mathcal{I}$ sends the bit (resp., appends the bit m to the secure message from corrupt $\hat{A}$).

Syntactic errors. The response of an honest player to a syntactic error (messages that cannot be parsed, *etc.*) is to abort the protocol (resp., place the ideal secure channel in a publicly aborted state). The necessary refinements to the preceding discussion are obvious, tedious, and omitted.

4.2 Reduction to DDHA

By inspection, the distribution that $\mathcal{I}$ hands to $\mathcal{A}$ is identical to that obtained in an actual execution, *except for* the lack of independence among the critical variables, when transmission is successful. It remains to show that this lack of independence is unnoticeable to $\mathcal{A}$. In particular, if it were detectable (to polynomial-bounded observers), then DDHA would fail.

Note that when Alice or Bob is corrupted before Alice sends her second message, the critical variables raise no concerns: they are either properly independent (because they follow honest Alice's program) or generated by an $\mathcal{A}$-controlled Alice. In particular, the results obtained when Alice or Bob is corrupt by the time Alice sends her second message are identical whether obtained through the protocol or through $\mathcal{I}$. The interesting case occurs when neither Alice nor Bob is corrupt at the time Alice sends her message – that is to say, at the time $\mathcal{I}$ supplies Alice's second message to $\mathcal{A}$.

Suppose that the implementation were insecure, namely that there were a poly-time machine D that distinguishes $(A, B, \mathcal{A})$ from $(\hat{A}, \hat{B}, \mathcal{I}(\mathcal{A}))$ with success probability $1/2 + k^{-c}$, for some fixed $m \in \{0,1\}$, $c > 0$ and infinitely many k. Call any such k "vulnerable."

We describe an algorithm, **Break**, that violates the DDHA. The input to **Break** consists of three values (in addition to g and p):

$$g^\alpha, g^\beta, \gamma.$$

Let distribution δ_0 generate these as follows: $\alpha \leftarrow \$(\mathbf{Z}_{p-1})$; $\beta \leftarrow \$(\mathbf{Z}_{p-1})$; $\gamma \leftarrow g^{\alpha\beta}$. In contrast, distribution δ_1 applies the following: $\alpha \leftarrow \$(\mathbf{Z}_{p-1})$; $\beta \leftarrow \$(\mathbf{Z}_{p-1})$; $\gamma \leftarrow \$(\langle g \rangle)$. Our goal is to distinguish δ_0 from δ_1, thereby violating the DDHA.

Observe particularly that in our earlier construction, $\mathcal{I}$ needs to know all four values a_0, a_1, b_0, and b_1, because it does not know what m will be, but it may have to equivocate later on. The converse is also true: *knowing the result* **fail** *or* **succeed**$:m$, *one does not need to know all four logarithms in order to duplicate the behavior of* $\mathcal{I}$. The **Break** routine takes advantage of this fact.

Intuitively, $\mathcal{I}$ must stick with the s and f it selected without knowing m, and (when $s = 0$) it later adapts d to its discovery of m, so that $d = m \oplus f$. On the other hand, **Break** can select d and s, and when $s = 0$, knowing m already, it *calculates* $f = m \oplus d$ instead of choosing f independently at random.

In more detail, **Break** runs an internal execution of the specification protocol, permitting a built-in interface $\mathcal{I}_{brk}$ to interact with $\mathcal{A}$. **Break** supplies its input (sampled from δ_0 or δ_1) to $\mathcal{I}_{brk}$. **Break** knows m and can therefore operate internal copies of $\hat{A}$ and $\hat{B}$ as well. Depending on the distribution given to **Break**, the final distribution is identical to either (0) an attack by $\mathcal{A}$ assisted by $\mathcal{I}$ against the specification protocol, or (1) an attack by $\mathcal{A}$ on the **Send-Message** protocol. Once **Break** has obtained the final results, it passes them to an internal copy of distinguisher D and simply reports whatever D reports.

The built-in interface $\mathcal{I}_{brk}$ follows the general outline of $\mathcal{I}$'s program, except that $\mathcal{I}_{brk}$ commits to d and s *without* knowing the discrete logarithms of all the

$\mathcal{I}_{brk}(g^\alpha, g^\beta, \gamma)$

0. Given: prime p, subgroup generator g
1.1. Internal: $c \leftarrow \$(\{0,1\})$,
  $a_c \leftarrow \$(\mathbb{Z}_{p-1})$,
  $x_c \leftarrow g^{a_c} \bmod p$,
  $x_{1-c} \leftarrow g^\alpha \bmod p$
1.2. $\mathcal{I} \to \mathcal{A}$: "A→B: x_0, x_1"
2.1. Internal: $s \leftarrow \$(\{0,1\})$
  if $s = 0$ then
    $d \leftarrow c$,
    $b_d \leftarrow \$(\mathbb{Z}_{p-1})$,
    $y_d \leftarrow g^{b_d} \bmod p$,
    $z_d \leftarrow g^{a_d b_d} \bmod p$,
    $y_{1-d} \leftarrow g^\beta \bmod p$,
    $z_{1-d} \leftarrow \gamma \bmod p$

(2.1 cont.) else
    $d \leftarrow 1 - c$,
    $b_d \leftarrow \$(\mathbb{Z}_{p-1})$,
    $y_d \leftarrow g^{b_d} \bmod p$,
    $z_d \leftarrow g^{a b_d} \bmod p$,
    $y_{1-d} \leftarrow \$(\langle g \rangle)$,
    $z_{1-d} \leftarrow \$(\langle g \rangle)$
2.2. $\mathcal{I} \to \mathcal{A}$: "B→A: y_0, y_1, z_0, z_1"
3.1. Internal: if $s = 0$ then
    $f \leftarrow c \oplus m$,
  else
    $f \leftarrow 0$
3.2. $\mathcal{I} \to \mathcal{A}$: "A→B: (s, f)"

Fig. 3. Behavior of $\mathcal{I}_{brk}$, used as a subroutine to violate DDHA.

values. It does not get caught trying to equivocate, because it already knows m. Fig. 3 describes the details.

Fig. 4 illustrates the results induced by $\mathcal{I}_{brk}$. If $\gamma = g^{\alpha\beta}$, these tables correspond to the results of $\mathcal{I}$. If γ is chosen independently at random, these tables correspond to the results of executing Send-Bit-Attempt.

The following two observations are straightforward. If $(g^\alpha, g^\beta, \gamma) \leftarrow \delta_0$, then

$$(\hat{A}(m,k), \hat{B}(k), \mathcal{I}_{brk}(g^\alpha, g^\beta, \gamma; \mathcal{A}(k), k, m)) = (\hat{A}(m,k), \hat{B}(k), \mathcal{I}(\mathcal{A}(k), k)).$$

If $(g^\alpha, g^\beta, \gamma) \leftarrow \delta_1$, then

$$(\hat{A}(m,k), \hat{B}(k), \mathcal{I}_{brk}(g^\alpha, g^\beta, \gamma; \mathcal{A}(k), k, m)) = (A(m,k), B(k), \mathcal{A}(k)).$$

By the construction of Break,

$$\text{Break}(\delta_0) = D((\hat{A}(m,k), \hat{B}(k), \mathcal{I}(\mathcal{A}(k), k)))$$
$$\text{Break}(\delta_1) = D((A(m,k), B(k), \mathcal{A}(k))).$$

Thus, $|\text{Break}(\delta_0) - \text{Break}(\delta_1)| \geq 1/2 + k^{-c}$ at all vulnerable k, contradicting DDHA as desired.

4.3 Full Key Exchange

When establishing a k-bit key K using $3k$ parallel attempts, we randomly select $i \leftarrow \$(\{1, \ldots, 3k\})$ and use the above Break routine in the ith parallel iteration. We generate the portions of Alice's second message in the locations $1..(i-1)$ using $\mathcal{I}$. We generate the portions of Alice's second message in the locations $(i+1)..3k$ according to Send-Bit-Attempt.

Through standard arguments, we conclude that our algorithm distinguishes δ_0 from δ_1 with advantage $1/2 + (k^{-c}/3k)$ at all vulnerable k. This suffices to contradict the DDHA, extending the proof to k-bit exchanges.

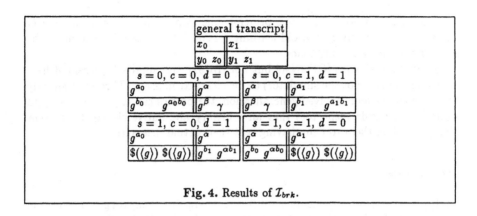

Fig. 4. Results of $\mathcal{I}_{brk}$.

5 Concluding Remarks

Although it appears that Alice could calculate and use d in all cases (a mismatch informs her that $d = 1 - c$), this simple optimization fails for technical reasons. It is not hard to show that $\mathcal{I}$'s fake encryptions must contain two DH triples as before, but in this case they are easily unveiled by detecting whether $z_{1-d} = y_{1-d}^{a_c}$.

Generalizations. To use other intractability assumptions, such as RSA or factoring, a suitable key-exchange construction suffices. In particular, the dense secure public-key cryptosystems of DeSantis and Persiano are appropriate [DP92].

When third parties are available, one-way trapdoor permutations suffice. Unlike [CFGN96], *both* Alice and Bob obtain one of two composed trapdoors; instead of applying large-scale permutations of multiple encryptions, they then detect a matched choice in the simple manner suggested here.

Committal and Equivocability. We identify two key properties useful for analyzing and designing encryption systems robust enough for plug-and-play usage:

1. *Equivocability:* the ability to make convincing fake ciphertexts that are equivocal, even if the real ciphertexts are not.
2. *Selective Ignorance:* arranging for a party to avoid learning full information.

The results of [BH92] apply the first approach, destroying sufficient information to allow ciphertexts to be made consistent with different cleartexts. Note that the *real* ciphertexts are ultimately equivocal; the cryptosystem cannot be used indirectly as a committal.

The methods in [CFGN96] identified and applied the second approach as well. In their protocol, a clever if costly interaction enables the sender to obtain partial information from the receiver and/or third parties. The lack of full information permits a simulator to create equivocal fake ciphertexts, even though real ciphertexts are unequivocal. Certain roots can be seen in DeSantis' and Persiano's work on non-interactive zero-knowledge proofs [DP92].

The current work applies this principle one step further, arranging for *both* the sender and receiver to avoid learning certain information. The result is far greater simplicity and efficiency.

As is the case with [CFGN96] but not [BH92], the approach presented here binds the parties (not the simulator) to the cleartext message. Through a slightly more expensive and involved generalization, the parties can enjoy the ability to pretend to outside inspectors that an arbitrary alternative cleartext was sent [B96], enabling them to deny having sent their actual messages.

References

[B91]   D. Beaver. "Foundations of Secure Interactive Computing." *Advances in Cryptology – Crypto '91 Proceedings,* Springer–Verlag LNCS 576, 1992, 377–391.

[B95]   D. Beaver. "Adaptive Zero Knowledge and Computational Equivocation." *Proceedings of the 28th STOC,* ACM, 1996, 629–638.

[B96]   D. Beaver. "Plausible Deniability." *Advances in Cryptology – Pragocrypt '96 Proceedings,* CTU Publishing House, 1996, 272–288.

[BH92]   D. Beaver, S. Haber. "Cryptographic Protocols Provably Secure Against Dynamic Adversaries." *Advances in Cryptology – Eurocrypt '92 Proceedings,* Springer–Verlag LNCS 658, 1993, 307–323.

[BGW88]   M. Ben-Or, S. Goldwasser, A. Wigderson. "Completeness Theorems for Non-Cryptographic Fault-Tolerant Distributed Computation." *Proceedings of the 20th STOC,* ACM, 1988, 1–10.

[CFGN96]   R. Canetti, U. Feige, O. Goldreich, M. Naor. "Adaptively Secure Multiparty Computation." *Proceedings of the 28th STOC,* ACM, 1996, 639–648.

[CCD88]   D. Chaum, C. Crépeau, I. Damgård. "Multiparty Unconditionally Secure Protocols." *Proceedings of the 20th STOC,* ACM, 1988, 11–19.

[DP92]   A. DeSantis, G. Persiano. "Zero-Knowledge Proofs of Knowledge Without Interaction." *Proceedings of the 33rd FOCS,* IEEE, 1992, 427–436.

[DH76]   W. Diffie, M. Hellman. "New Directions in Cryptography." *IEEE Transactions on Information Theory* **IT-22**, November 1976, 644–654.

[GM84]   S. Goldwasser, S. Micali. "Probabilistic Encryption." *J. Comput. Systems Sci.* **28**, 1984, 270–299.

[GMR89]   S. Goldwasser, S. Micali, C. Rackoff. "The Knowledge Complexity of Interactive Proof Systems." *SIAM J. on Computing* **18**:1, 1989, 186–208.

[GMW86]   O. Goldreich, S. Micali, A. Wigderson. "Proofs that Yield Nothing but Their Validity and a Methodology of Cryptographic Protocol Design." *Proceedings of the 27th FOCS,* IEEE, 1986, 174–187.

[MR91]   S. Micali, P. Rogaway. "Secure Computation." *Advances in Cryptology – Crypto '91 Proceedings,* Springer–Verlag LNCS 576, 1992, 392–404.

[RSA78]   R. Rivest, A. Shamir, L. Adleman. "A Method for Obtaining Digital Signatures and Public Key Cryptosystems." *Communications of the ACM* **21**:2, 1978, 120–126.

[SiSp94]   M. Sipser, D. Spielman. "Expander Codes." *Proceedings of the 35th FOCS,* IEEE, 1994, 566–576.

Deniable Encryption[*]

Ran Canetti[1]   Cynthia Dwork[2]   Moni Naor[3]   Rafail Ostrovsky[4]

[1] IBM T.J. Watson Research Center. *email:* canetti@watson.ibm.com
[2] IBM Almaden Research Center. *email:* dwork@almaden.ibm.com
[3] Dept. of Computer Science, the Weizmann Institute. *email:*
naor@wisdom.weizmann.ac.il
[4] Bell Communications Research, MCC 1C-365B, Morristown, N.J. *email:*
rafail@bellcore.com

Abstract. Consider a situation in which the transmission of encrypted messages is intercepted by an adversary who can later ask the sender to reveal the random choices (and also the secret key, if one exists) used in generating the ciphertext, thereby exposing the cleartext. An encryption scheme is deniable if the sender can generate 'fake random choices' that will make the ciphertext 'look like' an encryption of a different cleartext, thus keeping the real cleartext private. Analogous requirements can be formulated with respect to attacking the receiver and with respect to attacking both parties.

In this paper we introduce deniable encryption and propose constructions of schemes with polynomial deniability. In addition to being interesting by itself, and having several applications, deniable encryption provides a simplified and elegant construction of *adaptively secure* multiparty computation.

1 Introduction

The traditional goal of encryption is to maintain the privacy of communicated data against passive eavesdroppers. That is, assume that Alice wants to communicate private information to Bob over a channel where Eve can eavesdrop. Alice obtains Bob's (public) encryption key of an asymmetric encryption scheme and uses it, together with local randomness, to encrypt her messages. Now only Bob, who possesses the decryption key, should be able to decrypt. Semantic security [15] captures the security requirements that this setting imposes on the encryption function. Basically, semantic security means that Eve learns nothing from the ciphertexts she hears: whatever she can compute having heard the ciphertexts she can also compute from scratch. It follows that Alice must use local randomness in order to achieve semantic security.

While (passive) semantic security appropriately captures the security needed against passive eavesdroppers, there are settings in which it falls short of providing the desired degree of protection. Such settings include protection against

---

[*] Research on this paper was supported by BSF Grant 32-00032.

chosen ciphertext attacks (e.g., [17, 18]), non-malleable encryption [8], and protection against adaptive adversaries [7].

We investigate the additional properties required to protect the privacy of transmitted data in yet another hostile setting. Assume that the adversary Eve now has the power to approach Alice (or Bob, or both) *after* the ciphertext was transmitted, and demand to see all the private information: the cleartext, the random bits used for encryption and any private keys Alice (or Bob) have. Once Alice hands over this information, Eve can verify that the cleartext and randomness provided by Alice indeed match the transmitted ciphertext. Can the privacy of the communicated data be still somehow maintained, in face of such an attack?

We first concentrate on the case where Eve attacks only Alice in the above way. Certainly, if Alice must hand Eve the *real* cleartext and random bits then no protection is possible. Also if Eve approaches Alice *before* the transmission and requires Alice to send specific messages there is no way to hide information. However, in case Eve has no direct physical access to Alice's memory, and Alice is allowed to hand Eve *fake* cleartext and random bits, is it possible for Alice to maintain the privacy of the transmitted data? That is, we ask the following question. Assume Alice sent a ciphertext $c = E(m_1, r)$, where m_1 is some message, E is the public encryption algorithm and r is Alice's local random input. Can Alice now come up with a fake random input r' that will make c 'look like' an encryption of a different message m_2? We call encryption schemes that have this property deniable.

The following valid question may arise at this point: if Eve has no physical access to Alice's memory, then why should Alice present Eve with any data at all? That is, why not have Alice tell Eve: 'Sorry, I erased the cleartext and the random bits used'. Indeed, if Eve will be willing to accept such an answer, then deniable encryption is not needed. But there may well exist cases where being able to provide Eve with convincing fake randomness will be valuable to Alice. (Presenting convincing data is almost always more credible than saying 'I erased', or 'I forgot'.) In fact, there may be cases where Alice is required to record all her history including the randomness used, and can be punished/prosecuted if she claims to have destroyed the "evidence", *i.e.* any part of her history. Furthermore, the mere fact that Alice is able to 'open' any ciphertext in many ways makes it impossible for Alice to convince Eve in the authenticity of *any* opening. This holds even if Alice *wishes* to present Eve with the real data. In this sense, the privacy of Alice's data is protected even from the *future behavior of Alice herself*.

Standard encryption schemes do not guarantee deniability. Indeed, typically there do not *exist* two different messages that may result in the same ciphertext (with *any* random input). In fact, encryption is often conceived of as a *committing* process, in the sense that the ciphertext may serve as a commitment to the cleartext. (This is a common use for encryption schemes, e.g. in [13, 14].) Deniable encryption radically diverges from this concept.

Deniable encryption may seem impossible at first glance: consider a ciphertext c sent from Alice to Bob. If, using two different random choices, Alice could have generated c both as an encryption of a message m_1 and as an encryption of a different message, m_2, then how can Bob correctly decide, from c alone, whether Alice meant to send m_1 or m_2? A more careful inspection shows that such schemes can indeed be constructed, based on trapdoor information unavailable to Eve.

Deniable encryption has applications to the prevention of vote-buying in electronic voting schemes [4, 10, 11, 19], storing encrypted data in a deniable way, and uncoercible multiparty computation [5]; it also yields an alternative solution to the adaptive security problem [7]. We elaborate on these applications in the sequel.

We classify deniable encryption schemes according to which parties may be coerced: a sender-deniable scheme is resilient against coercing (i.e., demanding to see the secret data) of the sender of the ciphertext; receiver-deniable and sender-and-receiver-deniable schemes are defined analogously. We also distinguish between shared-key schemes, in which the sender and receiver initially share some information, and public-key deniable encryption schemes, in which no prior communication is assumed. Another issue is the time at which the coerced party must decide on the fake message: at time of attack (preferable) or at time of encryption.

Let us informally sketch the requirements for a one-round public-key, sender-deniable, bit-by-bit encryption scheme (Section 2 contains a more general definition). Let E_k be the sender's encryption algorithm with public key k. First, a deniable encryption scheme should be semantically secure in the sense of [15]. In addition we require that the sender have a (publicly known) faking algorithm. Given a bit b, a random input r, and the resulting ciphertext $c = E_k(b, r)$, the faking algorithm generates a fake random input $\rho = \phi(b, r, c)$ that 'makes c look like an encryption of $\bar{b}$'. That is, given b, ρ, c, the adversary should be unable to distinguish between the following cases:
(a) ρ is uniformly chosen and $c = E_k(b, \rho)$
(b) c was generated as $c = E_k(\bar{b}, r)$ where r is independently and uniformly chosen, and $\rho = \phi(\bar{b}, r, c)$.
We say that a scheme is δ-deniable if the adversary can distinguish between cases (a) and (b) with probability at most δ.

We construct a sender-deniable public-key encryption scheme based on any trapdoor permutation (Section 3). However, our scheme falls short of achieving the desired level of deniability. That is, while we can construct a δ-deniable scheme for arbitrarily small δ, the length of the ciphertext is *linear* in $1/\delta$. Consequently, if we want δ to be negligible, we end up with ciphertexts of super-polynomial length. (The semantic security of our scheme against passive eavesdroppers holds in the usual sense.) We present evidence that constructing substantially better one-round schemes requires a different approach (Section 4).

We also consider a more flexible notion of deniability than the one sketched above. An encryption scheme for encrypting a single bit can be generally viewed as defining two distributions on ciphertexts: a distribution T_0 of encryptions of

0, and a distribution T_1 of encryptions of 1. Here, in contrast, the sender chooses the ciphertext according to one of *four* distributions, T_0, T_1, C_0, C_1. Distribution T_b is used by a sender who wishes to send the binary value b and does not wish to have the ability to open dishonestly when attacked. Distribution C_b is also used to send the bit value b, but by a sender who wishes to preserve both the ability to open "honestly" and the ability to open dishonestly when attacked. (This choice can be made at time of attack.) In particular, if the sender encrypts according to distribution C_b then, when attacked, the sender can appear to have chosen either from T_0 or T_1. This alternative notion allows us to construct efficient deniable schemes with negligible δ.

Section 6 shows, via simple constructions, how to transform any sender-deniable encryption scheme into a receiver-deniable scheme, and vice-versa. We also show how a scheme resilient against corrupting both the sender and the receiver can be constructed based on a scheme resilient against corrupting the sender. This last construction requires the help of other parties in a network, and works as long as at least one other party remains unattacked. In Section 5 we review some shared-key deniable schemes.

APPLICATIONS AND RELATED WORK   A natural application of deniable encryption is to prevent coercion in electronic secret voting schemes [10]: a coercer may offer bribe in exchange for proof of a person's vote, after hearing the corresponding ciphertext. The coercion problem in the context of voting has been studied in the past [4, 19, 11]. However, these previous works assume that, for a crucial part of the conversation, the communicating parties share a physically secure channel; thus, the coercer hears no ciphertext and the 'deniability problem' disappears.[5] Deniable encryptions may be incorporated in these works to replace these physical security assumptions. (One still has to make sure, as before, that the voters are not coerced *prior* to the elections.)

Based on the public-key, sender-deniable construction presented here, [5] describe a general multiparty protocol permitting a set of parties to compute a common function of their inputs while keeping their internal data private even in the presence of a coercer.

Finally, our work on deniable encryption provides a conceptually simple and elegant alternative solution to the problem of general secure multiparty computation in the presence of an *adaptive* adversary – one that chooses whom to corrupt *during* the course of the computation, based on the information seen as the execution unfolds. Protocols for securely computing any function in a multiparty scenario in the presence of a non-adaptive adversary were shown in [14]. Almost a decade passed before the restriction to non-adaptive adversaries was lifted [7].[6] These protocols are based on another type of encryption protocol,

---

[5] In [11] a slightly different physical security assumption is made, namely that the random choices used for encryption are physically unavailable. The result is the same: the 'deniability problem' disappears.

[6] [9, 3] obtain solutions for this problem under the assumption that the parties are trusted to keep *erasing* past information. Such solutions are unsatisfactory in a setting where parties aren't trusted since erasing cannot be externally verified. Furthermore, the physical design of computer systems makes erasing information difficult and unreliable [16].

called non-committing encryption. Non-committing encryptions have the same
flavor as deniable encryptions, in that there exist ciphertexts that can be opened
as encryptions of, say, both '1' and '0'. However, non-committing encryptions are
strictly weaker than deniable ones. For example, in non-committing encryptions
the parties *using* the scheme are, in general, *not* able to generate ciphertexts
that can be opened both ways; such ciphertexts can only be generated by a *simulator* (which is an artifact of the [7] model). In contrast, in deniable encryption
each ciphertext generated by parties using the scheme has unique decryption,
and at the same time can be opened in several ways for an adversary (thus,
the non-committing encryption scheme in [7] is not deniable). The key insight is
that any deniable encryption scheme resilient against attacking both the sender
and the receiver is non-committing. Indeed, applying the transformation of Section 6 to the basic scheme described in Section 3 yields a complete solution to
the adaptive security problem. See [6] for more details.

2 Definitions

Let us first recall the definition of computational distance of distributions. Here
and in the sequel a function $\delta : \mathrm{N} \to [0, 1]$ is negligible if it approaches zero faster
than any polynomial (when its argument approaches infinity).

Definition 1. Let $\mathcal{A} = \{A_n\}_{n\in\mathrm{N}}$ and $\mathcal{B} = \{B_n\}_{n\in\mathrm{N}}$ be two ensembles of probability distributions, and let $\delta : \mathrm{N} \to [0, 1]$. We say that $\mathcal{A}$ and $\mathcal{B}$ are $\delta(n)$-
close if for every polynomial time distinguisher D and for all large enough n,
$|\mathrm{Prob}(D(A_n) = 1) - \mathrm{Prob}(D(B_n) = 1)| < \delta(n)$.

If $\delta(n)$ is negligible then we say that $\mathcal{A}$ and $\mathcal{B}$ are computationally indistinguishable and write $\mathcal{A} \stackrel{c}{\approx} \mathcal{B}$.

2.1 Public-key encryption

Consider a sender S and a receiver R that, *a priori*, have no shared secret information. They engage in some protocol in order to transmit a message from
S to R. (If a standard public key encryption scheme is used then this protocol may consist of the receiver sending his public encryption key to the sender,
who responds with the encrypted message.) Intuitively, we desire: (1) the receiver should be able to decrypt the correct value (except, perhaps, with negligible probability of error); (2) the protocol should be semantically secure against
eavesdroppers; and (3) the sender should have a faking algorithm ϕ such that,
given (m_1, r_S, c, m_2) (where m_1 is the transmitted message, r_S is the sender's
random input, c is a transcript of the conversation between S and R for transmitting m_1, and m_2 is the required fake message), ϕ generates a fake random
input for the sender, that makes c look like a conversation for transmitting m_2.

More precisely, let M be the set of all possible messages to be sent from S to R (M can be $\{0,1\}^s$ for some s). Let π be a protocol for transmitting a message $m \in M$ from S to R. Let $\text{COM}_\pi(m, r_S, r_R)$ denote the communication between S and R for transmitting m, when S has random input r_S and R has random input r_R. Let $\text{COM}_\pi(m)$ denote the random variable describing $\text{COM}_\pi(m, r_S, r_R)$ when r_S and r_R are uniformly and independently chosen.

Definition 2. A protocol π with sender S and receiver R, and with security parameter n, is a $\delta(n)$-sender-deniable encryption protocol if:

Correctness: The probability that R's output is different than S's input is negligible (as a function of n).

Security: For any $m_1, m_2 \in M$ we have $\text{COM}_\pi(m_1) \stackrel{c}{\approx} \text{COM}_\pi(m_2)$.

Deniability: There exists an efficient faking algorithm ϕ having the following property with respect to any $m_1, m_2 \in M$. Let r_S, r_R be uniformly and independently chosen random inputs of S and R, respectively, let $c = \text{COM}_\pi(m_1, r_S, r_R)$, and let $\tilde{r}_S = \phi(m_1, r_S, c, m_2)$. Then, the random variables

$$(m_2, \tilde{r}_S, \text{COM}_\pi(m_1, r_S, r_R)) \text{ and } (m_2, r_S, \text{COM}_\pi(m_2, r_S, r_R)) \qquad (1)$$

are $\delta(n)$-close.

The right hand side of (1) describes the adversary's view of an honest encryption of m_2 according to protocol π. The left hand side of (1) describes the adversary's view when c was generated while transmitting m_1, and the sender falsely claims that c is an encryption of m_2. The definition requires that the adversary cannot distinguish between the two cases with probability more than $\delta(n)$.

REMARKS: 1. When the domain of messages is $M = \{0,1\}$ the definition may be simplified. In the sequel we concentrate on such schemes, encrypting one bit at a time.

2. Definition 2 requires the parties to choose new public keys for each message transmitted. The definition can be modified in a natural way to capture schemes where a 'long-lived' public key is used to encrypt several messages, requiring the sender to be able to 'fake' each message independently of the other messages encrypted with the same public key. The scheme described in the sequel indeed enjoys this additional property.

3. Schemes in which the coerced party chooses the fake message m_2 at time of encryption are called *plan-ahead* deniable encryption schemes. Some modifications of the constructions described below yield plan-ahead deniable encryption schemes with negligible $\delta(n)$.

Next we define a somewhat weaker notion of deniability, called flexible deniability.

Definition 3. A protocol π with sender S and receiver R, binary Preserve parameter P, and security parameter n, is a $\delta(n)$-flexible-sender-deniable encryption protocol if:

Correctness: The probability that R's output is different than S's input is negligible (as a function of n).

Security: For any $m_1, m_2 \in M$ and for any $P \in \{T, C\}$ we have $\text{COM}_\pi(P, m_1) \stackrel{c}{\approx} \text{COM}_\pi(P, m_2)$

(encryptions of m_1 and m_2 are indistinguishable independent of P).

Weak Deniability: There exists an efficient 'faking' algorithm ϕ having the following property with respect to any $m_1, m_2 \in M$. Let r_S, r_R be uniformly chosen random inputs of S and R, respectively, let $c = \text{COM}_\pi(C, m_1, r_S, r_R)$, and let $\tilde{r}_S = \phi(m_1, r_S, c, m_2)$. Then, the random variables

$$(m_2, \tilde{r}_S, c) \text{ and } (m_2, r'_S, \text{COM}_\pi(T, m_2, r'_S, r'_R)) \tag{2}$$

are $\delta(n)$-close, where r'_S, r'_R are independent, uniformly chosen random inputs of S and R, respectively.

The left-hand side of Equation 2 describes the view of the adversary when the sender, having preserved the ability to open dishonestly when sending m_1, opens with value m_2 (which might or might not equal m_1). The right-hand side of Equation 2 describes the adversary's view when the sender, not having preserved the ability to open dishonestly, opens an encryption of m_2.

Schemes resilient against attacking the receiver, or simultaneous attack of both the sender and the receiver, are defined analogously. They appear in [6].

2.2 Shared-key encryption

In a shared-key scenario, the sender and receiver share a random, secret key about which the adversary is assumed to have no *a priori* information. Consequently, here the parties can also present the adversary with a fake shared key, on top of presenting fake random inputs. This is captured as follows. The communication between the parties now depends also on a shared key k, and is denoted $\text{COM}_\pi(m, k, r_S, r_R)$ (where m, r_S, r_R are the same as before). Below we define sender-deniability.

Definition 4. A protocol π with sender S and receiver R, and with security parameter n, is a shared-key $\delta(n)$-sender-deniable encryption protocol if:

Correctness: The probability that R's output is different than S's input is negligible (as a function of n).

Security: For any $m_1, m_2 \in M$ and for a shared-key k chosen at random, we have $\text{COM}_\pi(m_1, k) \stackrel{c}{\approx} \text{COM}_\pi(m_2, k)$.

Deniability: There exists an efficient 'faking' algorithm ϕ having the following property with respect to any $m_1, m_2 \in M$. Let k, r_S, r_R be uniformly chosen shared-key and random inputs of S and R, respectively, let $c = \text{COM}_\pi(m_1, k, r_S, r_R)$, and let $(\tilde{k}, \tilde{r}_S) = \phi(m_1, k, r_S, c, m_2)$. Then, the random variables

$$(m_2, \tilde{k}, \tilde{r}_S, c) \text{ and } (m_2, k, r_S, \text{COM}_\pi(m_2, k, r_S, r_R))$$

are $\delta(n)$-close.

Note that Definition 4 also covers the case where the same key is used to encrypt several messages: let m_1 (resp., m_2) in the definition denote the concatenation of all real (resp., fake) messages. In Section 5 we mention some shared-key schemes.

3 Public-key Deniable Encryption

OVERVIEW. We describe two public-key deniable encryption schemes. The first, called the basic scheme, is only a partial solution to the problem. We use it as a building-block to construct our main scheme. (It can also be used to construct a non-committing encryption scheme, as described in [6].) Our main scheme, called the Parity Scheme, is $\frac{4}{n}$-sender-deniable according to Definition 2. Roughly speaking, this means that the probability of successful attack vanishes *linearly* in the security parameter. (By a simple renaming of parameters this scheme can be regarded as $\frac{1}{n^c}$-sender-deniable for any $c > 0$. Yet, the probability of successful attack vanishes only linearly in the amount of work invested in encryption and decryption.)

The schemes are sender-deniable. Receiver-deniable and Sender-and-receiver-deniable schemes can be constructed from these using the techniques of Section 6. Our schemes encrypt one bit at a time. Here they are described in the standard terms of encryption and decryption algorithms. In terms of Definition 2, the interaction consists of the receiver sending the public encryption key to the sender, who responds with the encrypted message.

THE BASIC APPROACH. Our schemes are based on the following simple idea. Assume that the sender can pick an element in some domain either *randomly,* or according to some *pseudorandom* distribution. Assume further that the receiver, having some secret information, can tell whether the element was chosen randomly or pseudorandomly; other parties cannot tell the difference. Then, the sender can proceed as follows: to encrypt a 1 (resp., 0) send a pseudorandom (resp., random) element. The receiver will be able to decrypt correctly; but if a *pseudorandom* element e was transmitted, then when attacked the sender can claim that e was *randomly* chosen — and the adversary will not be able to tell the difference.

Here the sender could fake its message only in one direction (from 1 to 0). Using simple tricks one can come up with schemes that allow faking in both directions. We now describe the schemes in detail.

TRANSLUCENT SETS. Our schemes are based on a construct that can be informally described as follows. (Formal definitions can be extracted from this description.) We assume that there exists a family $\{S_t\}_{t \in \mathbb{N}}$ of sets, where $S_t \subset \{0,1\}^t$, together with secret 'trapdoor information' d_t, such that:

1. S_t is small: $|S_t| \leq 2^{t-k}$ for some sufficiently large $k(t)$.
2. It is easy to generate random elements $x \in S_t$, even without the secret d_t.
3. Given $x \in \{0,1\}^t$ and d_t it is easy to decide whether $x \in S_t$.
4. Without d_t, values chosen uniformly from S_t are indistinguishable from values chosen uniformly from $\{0,1\}^t$.

We first present two simple constructions of translucent sets. Both use a trapdoor permutation $f : \{0,1\}^s \rightarrow \{0,1\}^s$, and its hard-core predicate $B : \{0,1\}^s \rightarrow \{0,1\}$ (say, use the Goldreich-Levin predicate [12]).

Construction I: Let $t = sk$. Represent each $x \in \{0,1\}^t$ as a vector $x = x_1...x_k$ where each $x_i \in \{0,1\}^s$. Then let $S_t = \{x_1...x_k \in \{0,1\}^{sk} \mid \forall i = 1..k, B(f^{-1}(x_i)) = 0\}$. Here $|S_t| \approx 2^{(s-1)k} = 2^{t-k}$.

Construction II: Let $t = s + k$. Represent each $x \in \{0,1\}^t$ as $x = x_0, b_1...b_k$ where $x_0 \in \{0,1\}^s$ and for $i \geq 1$ each $b_i \in \{0,1\}$. Then let $S_t = \{x_0, b_1...b_k \in \{0,1\}^{s+k} \mid \forall i = 1..k, B(f^{-i}(x_0)) = b_i\}$. Here $|S_t| = 2^s = 2^{t-k}$.

It is easy to verify that both constructions satisfy requirements 1-4. Construction II is more efficient in that, given a trapdoor permutation on $\{0,1\}^s$, the length of x is only $t = s + k$ instead of $t = sk$.

A third construction relies on the latticed-based public-key cryptosystem described in [2]. Roughly speaking, the secret information is an n-dimensional vector u of length at most 1. Let $\mathcal{K}$ denote the cube $2^{n \log n} U^{(n)}$, where $U^{(n)}$ is the n-dimensional unit cube. The vector u induces a collection of $(n-1)$-dimensional hyperplanes as follows: for integer i the ith hyperplane is the set of all vectors v whose inner product with u is equal to i. Let X be the intersection of the hyperplanes with $\mathcal{K}$. The public key consists of a collection of $m = n^c$ points $v_1, \ldots, v_m$, each of which is a small perturbation of a randomly chosen point in X. The encryption procedure makes use of a certain parallelepiped $\mathcal{P}$, computable from the public key. An encryption of zero is a point chosen uniformly at random from $\mathcal{K} \cap 2^{-n} \mathbf{Z}^n$. An encryption of one is $\sum_{i=1}^{m} \delta_i v_i \mod \mathcal{P}$, where each $\delta_i \in_R \{0,1\}$. Thus, encryptions of one are close to hyperplanes in X, while encryptions of zero, typically, are not. Decryption of the ciphertext is performed by computing the distance of the ciphertext from the nearest hyperplane in X: if the distance is sufficiently small the ciphertext is decrypted as one (there is a polynomial probability of error). This construction yields a translucent set in which t is the length of a ciphertext ($t \approx n^2$), and, once the public key has been chosen, S_t is the set of encryptions of one, and $\mathcal{R}_t$ is the set of encryptions of zero.

THE BASIC SCHEME. The public encryption key is a method for generating uniformly at random a member of a translucent set $S_t \subset \{0,1\}^t$. The private decryption key is the corresponding secret d.

Encryption: *To encrypt 1, send a random element of S_t. To encrypt 0, send a random element in $\{0,1\}^t$.*

Decryption: *If the ciphertext x is in S_t then output 1. Else output 0.*

Opening an encryption honestly: *reveal the true random choices used.*

Opening an encryption dishonestly: *If the encrypted bit is 1, i.e., the ciphertext x is a random element in S_t, then claim that x was chosen at random from $\{0,1\}^t$ and thus x is an encryption of 0. If the encrypted bit is 0 then lying will be infeasible since the ciphertext x is in S_t only with negligible probability 2^{-k}.* Analysis: Correctness: An encryption of 1 is always decrypted correctly. An encryption of 0 may be decrypted as 1 with probability 2^{-k}. Standard security against eavesdroppers is straightforward. Deniability: the faking algorithm ϕ and its validity are described above. Since lying is possible only in one direction, this is only a partial solution to the problem. Next we describe a scheme where lying is possible in both directions.

THE PARITY SCHEME. Let $S_t \subset \{0,1\}^t$ be a translucent set. We call elements drawn uniformly from S (resp., from $\{0,1\}^t$) S-elements (resp., $\mathcal{R}$-elements).

Encryption: *To encrypt 0 (resp., 1), choose a random even (resp., odd) number $i \in 0, \ldots, n$. Construct a ciphertext consisting of i S-elements followed by $n - i$ $\mathcal{R}$-elements.*

Decryption: *Output the parity of the number of elements in the received ciphertext that belong to S.*

Opening an encryption honestly: *Reveal the real random choices used in generating the ciphertext.*

Opening an encryption dishonestly: *Let i be the number chosen by the sender. The sender claims that she has chosen $i - 1$ rather than i. (Consequently, the parity of i flips.) For this, she claims that the ith element in the ciphertext is an $\mathcal{R}$-element (whereas it was chosen as an S-element). If there are no S-elements (i.e., $i = 0$) then cheating fails.*

Theorem 5. *Assume trapdoor permutations exist. Then the Parity Scheme is a $4/n$-sender-deniable encryption scheme.*

Proof (Sketch): The probability of erroneous decryption is at most $n2^{-k}$. Security of the Parity Scheme against eavesdroppers that see only the ciphertext is straightforward. We show deniability. Assume that n is odd, and let c be an encryption of 1. Let i be the number chosen for generating c. Then, i was chosen at random from $1, 3, \ldots n$. Consequently, the value $i - 1$ is uniformly distributed over $0, 2, \ldots, n - 1$. Thus, when the sender claims that she has chosen $i - 1$, she demonstrates the correct distribution of i for encrypting 0. Thus, cheating in this direction is undetectable (as long as S-elements cannot be distinguished from $\mathcal{R}$-elements). Assume now that c is an encryption of 0. Thus i is chosen uniformly from $0, 2, \ldots, n - 1$. Now, $i - 1$ is distributed uniformly in $-1, 1, 3, \ldots, n - 2$ (where -1 is interpreted as "cheating impossible"). It is easy to verify that the statistical distance between the distribution of i in the case of an honest opening (i.e., uniform on $1, 3, \ldots, n$) and the distribution of i in the case of fake opening (i.e., uniform on $-1, 1, 3, \ldots, n - 2$) is $4/n$. It follows that, as long as S-elements cannot be distinguished from $\mathcal{R}$-elements, cheating is detectable with probability at most $4/n$. □

The Parity Scheme can be modified to let the sender first choose a vector v uniformly out of all vectors in $\{0,1\}^n$ with the parity of the bit to be encrypted. Next the ciphertext is constructed by replacing each 1 entry in v with an S-element, and replacing each 0 with an $\mathcal{R}$-element. Here the probability of $i = 0$ (i.e., the probability of the case where cheating is impossible) is negligible. Now, however, the statistical distance between i's distribution in honest and fake openings grows to $\Omega(\sqrt{\frac{1}{n}})$. A 'hybrid' scheme, omitted from this abstract, achieves both negligible probability of impossible cheating and probability $O(1/n)$ of detection.

The *unique shortest vector* problem for lattices is: "Find the shortest nonzero vector in an n dimensional lattice L where the shortest vector v is unique in the sense that any other vector whose length is at most $n^c \|v\|$ is parallel to v." The unique shortest vector problem is one of the three famous problems listed in [1]. There, a random method is given to generate hard instances of a particular lattice problem so that if it has a polynomial time solution then all of the three worst-case problems (including the unique-shortest vector problem) has a solution. The cryptosystem in [2] outlined above is secure provided the unique shortest vector problem is hard in the worst case. From this and the proof of Theorem 5 we have:

Theorem 6. *Assume that the unique shortest vector problem is hard in the worst case. Then the Parity Scheme is a $4/n$-sender-deniable encryption scheme.*

A FLEXIBLY DENIABLE SCHEME.   Let $T_0 = \{\mathcal{R}, \mathcal{R}\}$, $T_1 = C_1 = \{\mathcal{S}, \mathcal{R}\}$, and $C_0 = \{\mathcal{S}, \mathcal{S}\}$.
Encryption: *To encrypt b without preserving the ability to open dishonestly (that is, if Preserve $= 0$), send $V \in_R T_b$. To encrypt b preserving the ability to open dishonestly (Preserve $= 1$), send $V \in_R C_b$.*
Decryption: *Output the parity of the number of elements in V that belong to $\mathcal{S}$.*
Opening an encryption drawn from T_b: *Reveal the true random choices used.*
Opening an encryption drawn from C_0 as value v: *Let b be the number of elements in the ciphertext drawn from $\mathcal{S}$. If $v = 0$ then claim that V was chosen as $\{\mathcal{R}, \mathcal{R}\}$. If $v = 1$ then claim that V was chosen as $\{\mathcal{S}, \mathcal{R}\}$.*
Opening an encryption drawn from C_1 as value v: *If $v = 0$ then claim that V was chosen as $\{\mathcal{R}, \mathcal{R}\}$. If $v = 1$ then reveal the real random choices used in generating V.*
Analysis: Security against eavesdroppers seeing only the ciphertext is straightforward. The probability of erroneous decryption is at most 2^{-k}. Weak deniability with negligible $\delta(n)$ is immediate by inspection, assuming polynomial time indistinguishability of $\mathcal{S}$ and $\mathcal{R}$.

4   Efficiency Vs. Deniability

In this section we describe an attack suggesting that no one-round scheme of the type presented above can enjoy negligible $\delta(n)$. The attack works against all schemes that we describe as separable (the reason for the name will become clear shortly). Roughly, in a separable scheme the decryption key is the trapdoor of some translucent set $\mathcal{S} \subset \{0,1\}^t$; a ciphertext consists of a sequence of elements $y_1....y_m$ in $\{0,1\}^t$. The sender chooses some of the y_i's at random, and the rest at random from S. The encrypted bit is encoded in the number and placement of the y_i's that are in the translucent set S. To fake the value of the cleartext the sender claims that one (or more) of the y_i's was randomly chosen, whereas this y_i was chosen from S.

For any separable scheme, and for each value $b \in \{0, 1\}$, one can compute the expected number of y_i's in S in an encryption of b. Denote this number by E_b. Now, since the faking algorithm always *decreases* the number of y_i's for which the sender claims to know the preimage, the adversary decides that the sender is lying if the sender claims to have sent b but the number of y_i's which the sender claims to have chosen from S is less than E_b. It is shown below that this strategy succeeds with probability at least $\Omega(\frac{1}{m})$.

A more precise (and somewhat more general) description follows.

Definition 7. A $\frac{1}{k}$-sender-deniable public key encryption scheme π is m-separable if there exists an efficient, deterministic classification algorithm C that, on any input ρ (interpreted as a claimed random input of the sender), outputs a number $C(\rho) \in 1, \ldots, m$. Furthermore:

1. For a value ρ (interpreted as a random input for the sender), let $\rho^{(b)}$ be the random variable describing $\phi(b, \rho, c)$, where ϕ is the sender's faking algorithm, $b \in \{0, 1\}$, r_R is the receiver's random input, and and $c = \text{COM}_\pi(b, \rho, r_R)$ is the resulting communication. Let $EC^{(b)}(\rho)$ denote the expected value (over the choices of r_R) of $C(\rho^{(b)})$.
   Then for any value ρ such that $C(\rho) > 1$, either $EC^{(0)}(\rho) \le C(\rho) - 1$ or $EC^{(1)}(\rho) \le C(\rho) - 1$.
2. If the sender's random input ρ satisfies $C(\rho) = 1$ then the faking algorithm fails, i.e. it outputs a special symbol denoting that no suitable fake random input was found.

Claim 8 *For any m-separable, $\frac{1}{k}$-sender-deniable public key encryption scheme we have $2m \ge k$.*

REMARKS:

- Using the terminology of the above informal description of separable schemes, the coercer will use the classification algorithm that outputs the number of y_i's which the sender claims to have chosen as S-elements. It follows that any such scheme with only m y_i's is m-separable.
- In all the m-separable schemes that we know of, the length of the ciphertext grows linearly with m. This seems to be inherent in our approach for constructing deniable schemes.

Proof. Consider an m-separable deniable scheme π with faking algorithm ϕ. We show an algorithm A that for some $b \in \{0, 1\}$ distinguishes between

$$(\bar{b}, r_S^{(b)}, \text{COM}_\pi(b, r_S, r_R)) \text{ and } (\bar{b}, r_S, \text{COM}_\pi(\bar{b}, r_S, r_R)) \qquad (3)$$

with probability $\frac{1}{2m}$, where r_S, r_R are random inputs for the sender and the receiver respectively, and $r_S^{(b)} = \phi(b, r_S, \mathrm{COM}_\pi(b, r_S, r_R))$.

Let C be the classification algorithm. For $b \in \{0, 1\}$, let DC denote the distribution of $C(r_S)$ where r_S is chosen at random from the domain of random inputs of the sender, and let $DC^{(b)}$ denote the distribution of $C(r_S^{(b)})$ when r_R is chosen at random. Let $EC, EC^{(b)}$ denote the expected values of $DC, DC^{(b)}$, respectively. It follows from Definition 7 that either $EC - EC^{(0)} \geq \frac{1}{2}$ or $EC - EC^{(1)} \geq \frac{1}{2}$.

Let $\mathrm{SD}(D_1, D_2)$ denote the statistical distance between two distributions D_1, D_2 over $1, \ldots, m$,[7] and let E_1, E_2 denote the corresponding expected values. It can be verified that $|E_1 - E_2| \leq m \cdot \mathrm{SD}(D_1, D_2)$. In our case this implies that either $\mathrm{SD}(DC, DC^{(0)}) > \frac{1}{2m}$ or $\mathrm{SD}(DC, DC^{(1)}) > \frac{1}{2m}$.

The distinguisher A is now straightforward. Assume that $\mathrm{SD}(DC, DC^{(0)}) > \frac{1}{2m}$. Then A distinguishes between $(0, r_S^{(1)}, \mathrm{COM}_\pi(1, r_S, r_R))$ and $(0, r_S, \mathrm{COM}_\pi(0, r_S, r_R))$ as follows. Let $Z \subset 1...m$ be the set of numbers that have higher probability under $DC^{(0)}$ than under DC. Then, given a triplet $(0, \rho, c)$, first check that the ciphertext c is consistent with 0 and ρ. Next, if $C(\rho) = 1$ then by Definition 7 above A can distinguish between the two distributions of (3). Otherwise, say that the triplet describes an honest encryption of 0 iff $C(\rho) \in Z$. By definition of statistical distance, A distinguishes correctly with probability at least $\frac{1}{2m}$. (Since Z is a subset of $1...m$, it can be found by sampling.)

5 Shared-key deniable encryption

In this section we briefly remark on some shared-key deniable schemes. Clearly, a public-key deniable scheme is also deniable in the shared-key setting. Thus the public key constructions described in previous sections apply here as well. Yet better shared-key deniable schemes may be easier to find than public-key ones.

A one-time-pad is a shared-key deniable encryption scheme: Assume that the sender and the receiver share a sufficiently long random string, and each message m is encrypted by bitwise xoring it with the next unused $|m|$ bits of the key. Let k denote the part of the random key used to encrypt m, and let $c = m \oplus k$ denote the corresponding ciphertext. Then, in order to claim that c is an encryption of a message $m' \neq m$, the parties claim that the shared key is $k' = c \oplus m'$. It is easy to verify that this trivial scheme satisfies Definition 4. Here the message m' can be chosen as late as *at time of attack*. However, using a one-time pad is generally impractical, since the key has to be as long as all the communication between the parties.

Recall that in plan-ahead sender-deniability the sender chooses the fake message(s) *at time of encryption*. Although restrictive, this notion can be useful, *e.g.* for maintaining 'deniable records' of data, such as a private diary, that may be publicly accessible but is kept private using a deniable encryption scheme (alternative examples include a psychiatrist's or lawyer's notes.) The records are

---

[7] That is, $\mathrm{SD}(D_1, D_2) = \sum_{i \in 1, \ldots, m} |\mathrm{Prob}_{D_1}(i) = \mathrm{Prob}_{D_2}(i)|$.

deniable if, when coerced to reveal the cleartext and the secret key used for encryption and decryption, the owner of the record can instead "reveal" a variety of fake cleartexts of her choice.

Plan-ahead shared-key deniability is trivially solved: given l alternative messages to encrypt, use l different keys, and construct the ciphertext as the concatenation of the encryptions of all messages, where the ith message is encrypted using the ith key. When coerced, the party simply claims that the key he used is the one that corresponds to the message he wishes to open.

One problem with this simple scheme is that the size of the ciphertext grows linearly in the number of different messages to be encrypted. It is possible (details omitted for lack of space) to transform any given shared key encryption to a deniable one, without any increase in the message length, and with a key of length $1 - \frac{1}{l}$ times the length of the message. The shared-key deniable schemes can also be used to make public-key deniable schemes more efficient by way of first sending (using public-key deniable scheme) a deniable shared key and then switching to a private-key (deniable) scheme. We omit the details from this abstract.

6 Coercing the Sender vs. Coercing the Receiver

We describe simple constructions that transform sender-deniable schemes into receiver-deniable schemes and vice-versa. If there are other parties that can help in transmitting the data, we also construct a sender-and-receiver-deniable scheme from any sender-deniable scheme. We describe the constructions with respect to schemes that encrypt only one bit at a time. Generalizing these constructions to schemes that encrypt arbitrarily long messages is straightforward. These constructions apply to both shared-key and public-key settings.

RECEIVER-DENIABILITY FROM SENDER-DENIABILITY.   Assume a sender-deniable encryption scheme $\mathcal{A}$, and construct the following scheme $\mathcal{B}$. Let b denote the bit to be transmitted from S to R. First R chooses a random bit r, and invokes the scheme $\mathcal{A}$ to send r to S. (That is, with respect to scheme $\mathcal{A}$, R is the sender and S is the receiver.) Next, S sends $b \oplus r$ to R, in the clear.

If scheme $\mathcal{A}$ is sender-deniable then, when attacked, R can convincingly claim that the value of r was either 0 or 1, as desired. Consequently R can claim that the bit b was either 0 or 1, at wish, and scheme $\mathcal{B}$ is receiver-deniable.

SENDER-DENIABILITY FROM RECEIVER-DENIABILITY.   We use the exact same construction. It is easy to verify that if $\mathcal{A}$ is receiver-deniable then $\mathcal{B}$ is sender-deniable.

SENDER-AND-RECEIVER-DENIABILITY.   Assume that S and R can use other parties $I_1, ..., I_n$ as intermediaries in their communication. The following scheme is resilient against attacking the sender, the receiver and some intermediaries, as long as at least one intermediary remains unattacked.

In order to transmit a bit b to R, S first chooses n bits $b_1...b_n$ such that $\oplus_i b_i = b$. Next, S transmits b_i to each intermediary I_i, using a sender-deniable scheme. Next, each I_i transmits b_i to R using a receiver-deniable scheme. Finally R computes $\oplus_i b_i = b$.

When an intermediary I_i is attacked, it reveals the true value of b_i. However, as long as one intermediary I_j remains unattacked, both S and R can convincingly claim, when attacked, that the value of b_j (and consequently the value of b) is either 0 or 1.

Note that this scheme works only if the parties can 'coordinate their stories'. (This is further treated in [5].)

References

1. M. Ajtai, Generating Hard Instances of Lattice Problems, STOC'96
2. M. Ajtai, C. Dwork, A Public-Key Cryptosystem with Average-Case/Worst-Case Equivalence, *STOC'97*; see also *Electronic Colloquium on Computational Complexity TR96-065*, http://www.eccc.uni-trier.de/eccc-local/Lists/TR-1996.html
3. D. Beaver and S. Haber, Cryptographic Protocols Provably Secure Against Dynamic Adversaries, *Eurocrypt*, 1992.
4. J. Benaloh and D. Tunistra, Receipt-Free Secret-Ballot Elections, *26th STOC*, 1994, pp. 544-552.
5. R. Canetti and R. Gennaro, Incoercible multiparty computation, *FOCS'96*
6. R. Canetti, C. Dwork, M. Naor and R. Ostrovsky, Deniable Encryption, *Theory of Cryptology Library*, http://theory.lcs.mit.edu/ tcryptol, 1996.
7. R. Canetti, U. Feige, O. Goldreich and M. Naor, Adaptively secure computation, *28th STOC*, 1996.
8. D. Dolev, C. Dwork and M. Naor, Non-malleable cryptography, *STOC'91*
9. P. Feldman, *Private Communication*, 1986.
10. A. Herzberg, Rump-Session presentation at CRYPTO 1991.
11. R. Gennaro, unpublished manuscript.
12. O. Goldreich and L. Levin, A Hard-Core Predicate to any One-Way Function, *21st STOC*, 1989, pp. 25-32.
13. O. Goldreich, S. Micali and A. Wigderson, Proofs that Yield Nothing but the Validity of the Assertion, and a Methodology of Cryptographic Protocol Design, *27th FOCS*, 174-187, 1986.
14. O. Goldreich, S. Micali and A. Wigderson, How to Play any Mental Game, *19th STOC*, pp. 218-229, 1987.
15. S. Goldwasser and S. Micali, Probabilistic encryption, *JCSS*, Vol. 28, No 2, April 1984, pp. 270-299.
16. P. Gutman, Secure Deletion of Data from Magnetic and Solid-State Memory, Sixth USENIX Security Symposium Proceedings, San Jose, California, July 22-25, 1996, pp. 77-89.
17. M. Naor and M. Yung " Public key cryptosystems provably secure against chosen ciphertext attacks", Proc. 22nd ACM Annual Symposium on the Theory of Computing, 1990, pp. 427-437.
18. C. Rackoff and D. Simon, Non-interactive zero-knowledge proof of knowledge and chosen ciphertext attack, *CRYPTO'91, (LNCS 576)*, 1991.
19. K. Sako and J. Kilian, Receipt-Free Mix-Type Voting Scheme, *Eurocrypt* 1995, pp. 393-403.

Eliminating Decryption Errors in the Ajtai-Dwork Cryptosystem

Oded Goldreich[1]   Shafi Goldwasser[1,2]   Shai Halevi[2]

[1] Department of Computer Science and Applied Mathematics,
Weizmann Institute of Science, Rehovot, ISRAEL.
[2] Laboratory for Computer Science, MIT, Cambridge, MA 02139.
E-mail: {oded, shafi, shaih}@theory.lcs.mit.edu

Abstract. Following Ajtai's lead, Ajtai and Dwork have recently introduced a public-key encryption scheme which is secure under the assumption that a certain computational problem on lattices is hard on the worst-case. Their encryption method may cause decryption errors, though with small probability (i.e., inversely proportional to the security parameter). In this paper we modify the encryption method of Ajtai and Dwork so that the legitimate receiver always recovers the message sent. That is, we make the Ajtai-Dwork Cryptosystem error-free.

Keywords: Public-key Encryption Schemes, Computational Problems in Lattices.

1 Introduction

A major project of our field is to find concrete hard problems which can be used for "doing Cryptography" (e.g., constructing encryption schemes, message-authentication codes and digital signatures). As current state of the art in Complexity Theory does not allow to prove that such (cryptographically-useful) problems are hard, one has to rely on unproven and yet plausible assumptions. It is thus important to have as many alternative/unrelated assumption as possible, so that Cryptography can be based on any one of them. So far there are very few alternatives; and so Ajtai's work [1], which suggests a new domain out of which adequately-hard problems can be found, marks an important day for Cryptography.

In particular, Ajtai constructed a one-way function based on the assumption that Lattice Reduction is hard in the worst-case. Following his lead, Ajtai and Dwork have recently introduced a public-key encryption scheme which is secure, provided that the following (worst-case complexity) assumption holds [2]:

Assumption ISVP (Infeasibility of Shortest Vector Problem): There exists no polynomial-time algorithm, which given an arbitrary basis for an n-dimensional lattice, having a "unique poly(n)-shortest vector", finds the shortest (non-zero) vector in the lattice. By having a *unique poly(n)-shortest vector* we mean that any vector of length at most poly(n) times bigger than the shortest vector is an integer multiple of the shortest vector.

The encryption method of Ajtai and Dwork [2], has a non-zero decryption-error probability. Specifically, when working with security parameter n, the ciphertext of the message bit '1' is decrypted to be a '0' with probability $\frac{1}{n}$. (The ciphertext corresponding to the message bit '0' is always decrypted as '0'.)

In this paper we modify the encryption method of Ajtai and Dwork so that every message is always decrypted correctly. Thus, we obtain a error-free encryption scheme which is secure under the same assumption used by Ajtai and Dwork.

2 The Encryption Scheme

In this section we recall the construction of Ajtai and Dwork [2] and describe our modification of it. We start by introducing a few notations which are used throughout the paper.

2.1 Notations

We denote the set of integers by $\mathcal{Z}$, and the set of real numbers by $\mathcal{R}$. For any number ϵ between 0 and $\frac{1}{2}$, we denote by $\mathcal{Z} \pm \epsilon$ the set of real numbers for which the distance to the nearest integer is at most ϵ.

The n-dimensional Euclidean space is denoted by $\mathcal{R}^n$. For two vectors $x, y \in \mathcal{R}^n$, we denote the inner-product of x and y by $\langle x, y \rangle$. Given a set of n linearly independent vectors $w_1, \ldots, w_n \in \mathcal{R}^n$, the *parallelepiped which is spanned by the* w_i's is the set

$$P(w_1, \ldots, w_n) \stackrel{\text{def}}{=} \left\{ \sum_i \alpha_i w_i \ : \ \alpha_i \in [0,1), \ i = 1, \ldots, n \right\}$$

The *width* of $P(w_1, \ldots, w_n)$ is the minimum over i of the Euclidean distance between w_i and the subspace spanned by the other w_j's.

Given a parallelepiped $P = P(w_1, \ldots, w_n)$ and a vector v, we *reduce v modulo P* by obtaining a vector $v' \in P$ so that $v' = v + \sum_i c_i w_i$, where the c_i are all integers. We denote this process by $v' = v \mod P$.

2.2 The Ajtai-Dwork Construction

Let us recall the Ajtai-Dwork construction.[3] To simplify the exposition we present the scheme in terms of real numbers, but we always mean numbers with some fixed finite precision. (Following [2], one should use n-bit binary expansion of real numbers when working with security parameter n).

---

[3] The scheme which we describe below is slightly different than the original scheme in [2]. The difference between these schemes is insignificant, however (this is mostly a matter of presentation style).

Common Parameters. Given security parameter n, we let $m \overset{\text{def}}{=} n^3$, and $\rho_n \overset{\text{def}}{=}$ $2^{n \log n}$. We denote by B_n (for Big or cuBe) the n-dimensional cube of side-length ρ_n. Also, we denote by S_n (for Small or Sphere) the n-dimensional sphere of radius n^{-8}. Namely, we have

$$B_n \overset{\text{def}}{=} \{x \in \mathcal{R}^n : 0 \le x_i < \rho_n, \ i = 1, \ldots, n\} \quad \text{and} \quad S_n \overset{\text{def}}{=} \{x \in \mathcal{R}^n : \|x\| \le n^{-8}\}$$

Private-key. Given security parameter n, the private-key is a uniformly chosen vector in the n-dimensional unit sphere. We denote this vector by u.

Public-key. For a private key u, denote by $\mathcal{H}_u$ the distribution on points in B_n which is induced by the following process.

1. Pick a point a uniformly at random from the set $\{x \in B_n : \langle x, u \rangle \in \mathcal{Z}\}$.
2. For $i = 1, \ldots, n$, select $\delta_1, \ldots, \delta_n$ uniformly at random from S_n.
3. Output the point $v = a + \sum_i \delta_i$.

Using this notation, the public key which correspond to the private key u is obtained by picking the points $w_1, \ldots, w_n, v_1, \ldots, v_m$ independently at random from the distribution $\mathcal{H}_u$, subject to the constraint that the width of the parallelepiped $P(w_1, \ldots, w_n)$ is at least $n^{-2}\rho_n$. In the sequel, we often use the notations $\mathbf{w} \overset{\text{def}}{=} (w_1, \ldots, w_n)$, $\mathbf{v} \overset{\text{def}}{=} (v_1, \ldots, v_m)$, and $\mathbf{e} \overset{\text{def}}{=} (\mathbf{w}, \mathbf{v})$.
(Remark: It is shown in [2] that of we pick $w_1, \ldots, w_n$ uniformly in $\mathcal{H}_u$, then the width of $P(w_1, \ldots, w_n)$ will be large enough, with probability at least $1 - n^{-1/2}$.)

Encryption. The encryption works in a bit-by-bit fashion. Namely, to encrypt a string $s = \sigma_1 \sigma_2 \ldots \sigma_\ell$, each bit σ_i is encrypted separately.

To encrypt a '0', we uniformly select $b_1, \ldots, b_m$ in $\{0, 1\}$, and reduce the vector $\sum_{i=1}^{m} b_i \cdot v_i$ modulo the parallelepiped $P(\mathbf{w})$. The vector $x = (\sum_{i=1}^{m} b_i \cdot v_i) \bmod P(\mathbf{w})$ is the ciphertext which correspond to the bit '0'.

To encrypt a '1' we uniformly select a vector x in the parallelepiped $P(\mathbf{w})$. This vector is the ciphertext which correspond to the bit '1'.

Decryption. Given a ciphertext, x, and the private-key u, we compute $\tau = \langle x, u \rangle$. We decrypt the ciphertext as a '0' if τ is within $1/n$ of some integer and decrypt it as a '1' otherwise.

Decryption errors. It is easy to see that if x is an encryption of '1', then the fractional part of $\langle x, u \rangle$ is distributed almost uniformly in $[0, 1)$. On the other hand, a simple argument show that if x is an encryption of '0' then the fractional part of $\langle x, u \rangle$ is always less than $1/n$ in absolute value. Thus, an encryption of '0' will always be decrypted as '0', and an encryption of '1' has a probability of $2/n$ to be decrypted as '0'.

2.3 An Error-free Construction

We proceed now to describe our modification which eliminates the decryption errors from the construction above. In this modified scheme, just like in the original Ajtai-Dwork scheme, encrypting a '0' results in a ciphertext x such that $\langle x, u \rangle$ is close to an integer. However, in our scheme we also make sure that encrypting a '1' results in a ciphertext x such that $\langle x, u \rangle$ is far from any integer. The modified scheme is as follows:

Common Parameters and private-key. The common parameters n, m, ρ_n, B_n and S_n, and the private key u, are set in exactly the same manner as in the original scheme.

Public-key (modified). The vectors $w_1, \ldots, w_n, v_1, \ldots, v_m$ are chosen in exactly the same manner as in the original scheme.

In addition, we pick i_1 uniformly at random from all the indices i for which $\langle a_i, u \rangle \in 2\mathcal{Z} + 1$, where a_i is the large vector used to generate v_i (i.e., $v_i = a_i + \sum_j \delta_j$). That is, i_1 is selected so that $\langle a_{i_1}, u \rangle$ is an odd integer. We note that with probability $1 - 2^{-\Omega(m)}$ such an index exists.[4] The public-key consists of the sequence of points $(w_1, \ldots, w_n, v_1, \ldots, v_m)$ and the integer i_1.

Encryption (modified). We encrypt a '0' just like in the original scheme, by uniformly selecting $b_1, \ldots, b_m \in \{0, 1\}$, and reducing the vector $\sum_{i=1}^{m} b_i \cdot v_i$ modulo the parallelepiped $P(\mathbf{w})$. The vector $x = \left(\sum_{i=1}^{m} b_i \cdot v_i \right) \bmod P(\mathbf{w})$ is the ciphertext which correspond to the bit '0'.

The difference is in the encryption of a '1'. We do that by uniformly selecting $b_1, \ldots, b_m \in \{0, 1\}$, and reducing the vector $\frac{1}{2} v_{i_1} + \sum_{i=1}^{m} b_i \cdot v_i$ modulo the parallelepiped $P(\mathbf{w})$. The vector $x = \left(\frac{1}{2} v_{i_1} + \sum_{i=1}^{m} b_i \cdot v_i \right) \bmod P(\mathbf{w})$ is the ciphertext which correspond to the bit '1'.

Decryption (modified)*:* Given a ciphertext, x, and the private-key u, we compute $\tau = \langle v, u \rangle$. We decrypt the ciphertext as a '0' if τ is within $1/4$ of some integer and decrypt it as a '1' otherwise.

In contrast to the encryption scheme in [2], we can show that in our scheme there is no decryption error. Specifically, we have:

Proposition 1 (error-free decryption): *For every* $\sigma \in \{0, 1\}$, *every choice of the private and public keys, and every choice of b_i's by the encryption algorithm, the ciphertext, x, satisfies* $\langle x, u \rangle \in \mathcal{Z} + \frac{\sigma}{2} \pm \frac{1}{n}$.

Proof (sketch): The case of $\sigma = 0$ is the same as for the original Ajtai-Dwork scheme. The case of $\sigma = 1$ follows from the same arguments, using the fact that $\langle \frac{1}{2} v_{i_1}, u \rangle \in \mathcal{Z} + \frac{1}{2} \pm n^{-7}$. $\qquad\square$

---

[4] Otherwise, we may simply use the identity function for encryption/decryption.

3   Security of the Modified Scheme

To prove the security of the modified scheme, we start by invoking the main result of Ajtai and Dwork [2]:

Theorem 2 [2, Thm 7.1]: *Under Assumption ISVP, it is infeasible to distinguish the encryption of $\sigma = 0$ from a uniformly distributed point in $P(\mathbf{w})$, when given $\mathbf{w}, \mathbf{v}$. (We stress that $\mathbf{w}, \mathbf{v}$ and the encryption of '0' are distributed as described above.)*

Note that this theorem establishes the security (as defined in [3]) of the encryption scheme of Ajtai and Dwork [2], since in that scheme $\sigma = 1$ is encrypted as a uniformly chosen point in $P(\mathbf{w})$. To establish the security of our (modified) encryption scheme (under the same assumption), we need to prove

Theorem 3 (security): *Under Assumption ISVP, it is infeasible to distinguish the encryption of $\sigma = 0$ from the encryption of $\sigma = 1$, when given $\mathbf{w}, \mathbf{v}$ and i_1. (We stress that $\mathbf{w}, \mathbf{v}, i_1$ and the encryptions are distributed as described in the modified scheme.)*

Proof: Recall our notations $\mathbf{w} \stackrel{\text{def}}{=} (w_1, \ldots, w_n)$, $\mathbf{v} \stackrel{\text{def}}{=} (v_1, \ldots, v_m)$ and $\mathbf{e} \stackrel{\text{def}}{=} (\mathbf{w}, \mathbf{v})$. For a bit $\sigma \in \{0, 1\}$, and an encryption key $(\mathbf{e}, i)$, let us denote by $E_{\mathbf{e},i}(\sigma)$ the probabilistic encryption of σ using $(\mathbf{e}, i)$. Also, let us denote by $\Pi_{\mathbf{w}}$ the uniform distribution over $P(\mathbf{w})$. Assuming ISVP, we will show that for both $\sigma = 0$ and $\sigma = 1$, it is infeasible to distinguish $(\mathbf{e}, i, E_{\mathbf{e},i}(\sigma))$ from $(\mathbf{e}, i, \Pi_{\mathbf{w}})$.

First we show that this holds for $\sigma = 0$. Note that this claim is not identical to Theorem 2, as here the distinguisher is given i (for which $\langle v_i, u \rangle \in 2\mathcal{Z} + 1 \pm n^{-7}$ holds) as extra information. Still, Theorem 2 does imply the following

Lemma 4 *Under Assumption ISVP, it is infeasible to distinguish $(\mathbf{e}, i, E_{\mathbf{e},i}(0))$ from $(\mathbf{e}, i, \Pi_{\mathbf{w}})$, where $(\mathbf{e}, i)$ are selected as above and $\Pi_{\mathbf{w}}$ is uniformly distributed in $P(\mathbf{w})$.*

Proof. Suppose towards the contradiction that there exists a distinguisher, D, of running-time $t(n)$ and distinguishing gap $\epsilon(n)$ (between $(\mathbf{e}, i, E_{\mathbf{e},i}(0))$ and $(\mathbf{e}, i, \Pi_{\mathbf{w}})$ as in the claim). We construct a new distinguisher, D', which violates Theorem 2. D' works as follows:

input: $\mathbf{e} = (w_1, \ldots, w_n, v_1, \ldots, v_m)$ and x.

preprocessing: Using D, we find an index j which approximately maximizes the distinguishing gap of D on inputs of the form $(\mathbf{e}, j, \cdot)$. This is done by estimating, for every $j = 1, \ldots, m$, the value

$$\text{Prob}[D(\mathbf{e}, j, E_{\mathbf{e},j}(0)) = 1] - \text{Prob}[D(\mathbf{e}, j, \Pi_{\mathbf{w}}) = 1]$$

where the probability is taken over the internal coin tosses of both the encryption algorithm (i.e., choice of b_i's) and D. Invoking D for $\text{poly}(n)/\epsilon(n)^2$ times we may obtain, with overwhelmingly high probability, an approximation of the above upto $\epsilon(n)/4$. Let $\tau \in \{\pm 1\}$ denote the sign of the approximated difference for the best j.

decision: Using j and τ, found in the preprocessing, we invoke D on input $(\mathbf{e}, j, x)$. Let $\sigma \in \{\pm 1\}$ denote the output of D. Then D' outputs $\tau \cdot \sigma$.

Clearly, D' has running time $\mathrm{poly}(n, t(n), \epsilon(n)^{-1})$, which is polynomial in n as long as $t(n)/\epsilon(n)$ is polynomial in n. It is easy to see that

$$|\mathrm{Prob}[D'(\mathbf{e}, E_{\mathbf{e}}(0)) = 1] - \mathrm{Prob}[D'(\mathbf{e}, \Pi_{\mathbf{w}}) = 1]| > \frac{\epsilon(n)}{2} - 2^{-n}$$

(The second term is due to the case where we made some wrong approximation in the preprocessing stage.) Thus, we have a distinguisher violating the conclusion of Theorem 2, and so contradiction follows. □

Using Lemma 4, we easily derive

Lemma 5 *Under Assumption ISVP, it is infeasible to distinguish* $(\mathbf{e}, i, E_{\mathbf{e},i}(1))$ *from* $(\mathbf{e}, i, \Pi_{\mathbf{w}})$, *where* $(\mathbf{e}, i)$ *and* $\Pi_{\mathbf{w}}$ *are as in Lemma 4.*

Proof. Suppose towards the contradiction that there exists a distinguisher, D, of running-time $t(n)$ and distinguishing gap $\epsilon(n)$ (between $(\mathbf{e}, i, E_{\mathbf{e},i}(1))$ and $(\mathbf{e}, i, \Pi_{\mathbf{w}})$ as in the claim). We construct a new distinguisher, D', as follows

input: $\mathbf{e} = (w_1, \ldots, w_n, v_1, \ldots, v_m)$, i and x.
decision: Algorithm D' computes $x' = (x - \frac{1}{2}v_i) \bmod P(\mathbf{w})$, and outputs $D(x')$.

Observe that $E_{\mathbf{e},i}(0)$ and $E_{\mathbf{e},i}(1) - \frac{1}{2}v_i$ (reduced mod $P(\mathbf{w})$) are identically distributed. Similarly, $\Pi_{\mathbf{w}}$ and $\Pi_{\mathbf{w}} - \frac{1}{2}v_i$ (reduced mod $P(\mathbf{w})$) are identically distributed. Thus, D' distinguishes $(\mathbf{e}, i, E_{\mathbf{e},i}(0))$ from $(\mathbf{e}, i, \Pi_{\mathbf{w}})$, in contradiction to the claim of Lemma 4. The current lemma follows. □

Combining Lemmas 4 and 5, we have established Theorem 3. ■

Comment – An alternative proof of Theorem 3. The security of the encryption scheme in [2] is established via a sequence of reductions. The first reduction assumes an algorithm D which distinguishes between encryptions of 0's and 1's. It then constructs another algorithm D' which distinguishes between sequences of vectors $(\mathbf{w}, \mathbf{v})$ which constitute a public-key, and sequences uniformly distributed points in the big cube B_n (See [2, Lemma 8.1]). On a high level, this is done as follows: Algorithm D' uses the input vectors, $(\mathbf{w}, \mathbf{v})$, to encrypt 0's and 1's as if they constitute a public-key. If D is able to distinguish between encryptions of 0's and 1's, then D' concludes that these vectors indeed constitute a public-key. Otherwise, D concludes that they are just uniformly distributed points.

One can easily verify the argument in [2] holds also for distinguishers of encryptions under our modified scheme. Specifically, one needs to verify that when applying our encryption scheme using m uniformly distributed vectors, the result is distributed almost uniformly in the parallelepiped $P(\mathbf{w})$, regardless of whether a '0' or a '1' was encrypted. □

Acknowledgments

This research was supported by DARPA grant DABT63-96-C-0018.

References

1. Miklos Ajtai. Generating Hard Instances of Lattice Problems. In *28th ACM Symposium on Theory of Computing*, pages 99–108, Philadelphia, 1996.
2. Miklos Ajtai and Cynthia Dwork. A Public-Key Cryptosystem with Worst-Case/Average-Case Equivalence, In *29th ACM Symposium on Theory of Computing*, pages 284-293, 1997.
3. Shafi Goldwasser and Silvio Micali. Probabilistic Encryption, *JCSS*, Vol. 28, No. 2, pages 270–299, 1984.

Public-Key Cryptosystems
from Lattice Reduction Problems

Oded Goldreich*   Shafi Goldwasser**   Shai Halevi
Weizmann Institute of Science, ISRAEL   MIT, Laboratory for Computer Science

{oded, shafi, shaih}@theory.lcs.mit.edu

Abstract. We present a new proposal for a trapdoor one-way function, from which we derive public-key encryption and digital signatures. The security of the new construction is based on the conjectured computational difficulty of lattice-reduction problems, providing a possible alternative to existing public-key encryption algorithms and digital signatures such as RSA and DSS.

Keywords: Public-Key Cryptosystems, Lattice Reduction Problems

1 Introduction

The need for public-key encryption and digital signatures is spreading rapidly today as more people use computer networks to exchange confidential documents, buy products and access sensitive data. In fact, several of these tasks are impossible to achieve without the availability of secure and efficient public-key cryptography.

In light of the importance of public key cryptography, it is surprising that there are relatively few proposals of public key cryptosystems which have received any attention. Moreover, the source of security of these proposals almost always relies on the (apparent) computational intractability of problems in finite integer rings, specifically integer factorization (e.g., [20, 19, etc.]) and discrete logarithm computations (e.g.,[8, 9, 7, etc]).

In this paper we propose a new trapdoor one-way function relying on the computational difficulty of lattice reduction problems, in particular the problem of finding closest vectors in a lattice to a given point (CVP). From this trapdoor function, we then derive a public-key encryption and digital signature methods.

These methods are asymptotically more efficient than the RSA and ElGamal encryption schemes, in that the computation time for encryption, decryption, signing, and verifying are all quadratic in the natural security parameter. The size of the public key, however, is longer than for these systems. Specifically, for security parameter k, the new system has public key of size $O(k^2)$ and computation time of $O(k^2)$, compared to public key of size $O(k)$ and computation time of $O(k^3)$ for the RSA and ElGamal systems. We believe that, given today's technologies, the increase in size of the keys is more than compensated by the decrease in computation time.

---

* This research was done while visiting in the Laboratory for Computer Science, MIT.
** This research was supported by DARPA grant DABT63-96-C-0018.

113

Our trapdoor function. The idea underling our construction is that, given *any* basis for a lattice, it is easy to generate a vector which is close to a lattice point (i.e., by taking a lattice point and adding a small error vector to it). However it seems hard to return from this "close-to-lattice" vector to the original lattice point (given an arbitrary lattice basis). Thus, the operation of adding a small error vector to a lattice point can be thought of as a one-way computation.

To introduce a trapdoor mechanism into this one-way computation, we use the fact that different bases of the same lattice seems to yield a difference in the ability to find close lattice points to arbitrary vectors in $\mathcal{R}^n$. Therefore the trapdoor information may be a basis of a lattice which allows very good approximation of the closest lattice point problem. Thus, we use two different bases of the same lattice. One basis is chosen to allows computing the function but not inverting it, while the other basis is chosen to allow computing the inverse function by permitting good approximation to the closet lattice vector problem (CVP). For the sake of the introduction, we simply call such a basis a *reduced basis*. Below we give an informal description of our trapdoor one-way function which uses the above ideas.

The parameters of the system includes the security parameter n (which is the dimension of the lattices that we work with) and a "threshold" parameter σ which determines the size of the error-vectors which we add to the lattice points.

A particular function and its trapdoor information are specified by a pair of bases of the same (full rank) lattice in $\mathcal{R}^n$: A "non-reduced" basis B which is used to compute the function and a reduced basis R which serves as the trapdoor information and is used for inversion. The "reduced" basis is selected "uniformly" and the "non-reduced" basis is derived from it using a randomized unimodular transformation.

The input to the function is a lattice point (which is specified by an integral linear combination of the columns of B) and an error vector whose size is bounded by σ. The value of the function on this input is just the vector sum of the two points. To invert the function, we use a reduced basis R in one of Babai's nearest-vector approximation algorithms [4] to find a lattice point which is at most σ away from the given vector.

The cryptanalytic problem underlying our scheme is to approximate the closest vector problem (CVP) in a lattice, given a "non-reduced" basis for that lattice. A related problem is the problem of reducing the given public basis (since one obvious attack is to reduce the given basis and then use the result for inverting the function). See Section 2.1 for a description of these computational problems in lattices.

From trapdoor function to encryption scheme. In order to use the above trapdoor function for public-key encryption, we need a way to embed the message in the arguments to this function, in such a way that no "partial information" about the message is leaked by the ciphertext (cf., [13]).

There are several ways to do that, and we discuss some of them in Section 4. One generic way is to use hard core bits of the trapdoor function to embed the bits of the message (e.g., [12]). This approach has the advantage of ensuring that the encryption scheme is as secure as the underlying trapdoor function, but it is inefficient in terms of message expansion.

Another plausible way, which may be more efficient, is to map the message to a lattice point by taking the integer combinations of the public basis vectors which is "specified" by the message bits, and then add to the lattice point a "small error vector" chosen at random. To decrypt, we look for a lattice point which is close to the ciphertext. By using the private basis, which is a reduced basis, the correct decryption is obtained with high probability. We remark that our encryption algorithm is similar in its algorithmic nature to a scheme based on algebraic coding that was suggested by McEliece's in [18].

A signature scheme. It is also possible to construct a signature scheme along similar lines: Regard the message as a n-dimensional vector over the reals. Then, a signature of such vector, is a lattice point which is "close" to it (where closeness is defined by a published threshold). The private basis is reduced so that finding "close" points is possible. Verifying correctness amounts to checking that a signature is indeed a lattice point and that the message is close to the signature.

It is important to remark at the outset, that messages which are close to each other will have the same signature. When applying the method in a setting where this property is desirable (e.g., signing analog signals which may change a little in time), this feature may be of great benefit. However, to get secure signatures in the sense of [14], this property pause a significant problem. When applying the method to a message space where such property is undesirable, we propose to first hash the message and only then sign it. This is good practice also in case that the scheme is subject to a chosen message attack, as otherwise being able to obtain different signatures of two messages which are close to each other when viewed as points in $\mathcal{R}^n$ will imply the ability to compute a small basis for the lattice which in turn will enable the attacker to find close vectors in a lattice and break the scheme. (Interestingly, a family of collision-free hash functions can be constructed assuming that Lattice-Reduction is hard on the worst-case, see [10]). Due to lack of space, we do not discuss that construction in this extended abstract.

1.1 Discussion

Our work was inspired by a remarkable result of Ajtai [1] who introduced a function which is provably a one-way function if approximating the shortest non-zero vector (SVP) in a lattice is hard *on the worst case*. Ajtai's work may be viewed as exhibiting a samplable distribution on lattices and proving that approximating the shortest non-zero vector in lattices chosen according to this distribution is as hard as the worst case instance of approximating the shortest non-zero vector in a lattice. Ajtai's construction, however, does not provide a trapdoor function and thus does not provide a way of doing public-key encryption based on lattice problems. Constructing such a trapdoor function is the novelty and focus of our work.

Independently of our work, Ajtai and Dwork [2] suggested a public-key encryption scheme whose security is reducible to a variant of SVP. Although exhibiting a trapdoor Boolean predicate (which is sufficient for public-key encryption – see [13]), the Ajtai-Dwork construction does not provide a trapdoor function. That is, given the trapdoor information it is possible to decide whether the predicate evaluates to 0 or 1 but not known how to find an inverse. Also, the variant of SVP used in the security proof of [2], called the "poly(n)-unique shortest vector problem" seems to be considerably easier than the general SVP. Finally, we note that the Ajtai-Dwork construction is less efficient than ours, both in terms of the key-size and in terms of encryption time ($O(n^4)$ vs. $O(n^2)$ for both measures). Thus, it seems that their current construction is not really practical.

In retrospect, our encryption scheme bears much similarity to McEliece's scheme [18]. His scheme utilizes a pair of matrices over GF(2), which corresponds to two representations of the same linear code. The encryption method is probabilistic: one multiplies the public matrix by the message vector and adds a random noise vector to the resulting codeword. Thus in both McEliece and our encryption scheme, encryption amounts to a matrix-by-vector multiplication and the addition of a suitable random vector to the result. However, the domains in which these operations take place are vastly different and so is the algebra. Another difference is in the way the private-key is generated. In McEliece's scheme the private-key is a random Goppa code and has structure essential for legitimate decoding. In our scheme the private-key can be chosen uniformly and thus is "structure-less" – legitimate decoding merely depends on a property of such random choices. In both schemes the public-key is obtained by a suitable random linear transformation of the private-key; however, in our scheme the choice of this transformation seems richer. In general, we believe that McEliece's suggestion as well as ours deserve further investigation, especially due to the difference in computational complexity required from the legal sender and receiver in these schemes as compared with the factoring/DLP based schemes.

1.2 Evaluation of Security

To provide some feeling for the security of our construction, we analyzed a few plausible attacks against it and evaluated their effectiveness. Our analysis, combined with extensive testing, indicate that the work-load of these attacks grows exponentially with the dimension of the lattice. In particular, according to our estimates these attacks should be intractable in practice for dimension 300 or so.

1.3 Organization

In Section 2 we review necessary material about lattices and lattice problems. In Section 3 we describe our construction of a trapdoor function and discuss various parameters and attacks, and in Section 4 we describes our encryption scheme. In Section 5 we describe our experimental results.

2 Lattices and Lattice Reduction Problems

In the sequel we use the following conventions: We denote the set of real numbers by $\mathcal{R}$ and the set of integers by $\mathcal{Z}$. We denote real numbers by small Greek letters (e.g., β, ρ, τ etc.) and integers by one of the letters i, j, k, l, m, n. We denote vectors by bold-face lowercase letters (e.g., $\mathbf{b}, \mathbf{c}, \mathbf{r}$ etc.). We use capital letters (e.g., B, C, R, etc.) to denote matrices or sets of vectors. If β is a real number, we denote the integer closest to β by $\lfloor \beta \rceil$ and the smallest integer which is $\geq \beta$ by $\lceil \beta \rceil$. If $\mathbf{b}$ is a vector in $\mathcal{R}^n$, then $\lfloor \mathbf{b} \rceil$ denotes the vector in $\mathcal{Z}^n$ which is obtained by rounding each entry in $\mathbf{b}$ to the nearest integer. In this paper we only care about lattices of full rank, so the definitions below only deal with those.

Definition 1. Given a set of n linearly independent vectors in $\mathcal{R}^n$, $B = \{\mathbf{b}_1, \cdots, \mathbf{b}_n\}$, we define the lattice spanned by B as the set of all possible linear combinations of the $\mathbf{b}_i$'s with integral coefficients, namely $L(B) \stackrel{\text{def}}{=} \{\sum_i k_i \mathbf{b}_i : k_i \in \mathcal{Z} \text{ for all } i\}$

We call B a *basis* of the lattice $L(B)$. If the vector $\mathbf{v}$ belongs to the lattice L, then we say that $\mathbf{v}$ is a lattice-vector (or a lattice point). In the sequel we view a basis for a lattice in $\mathcal{R}^n$ as an $n \times n$ non-singular matrix B whose columns are the basis vectors. Viewed this way, the lattice spanned by B is the set $L(B) = \{B\mathbf{v} : \mathbf{v}$ is an integral vector$\}$. Below we briefly mention a few well-known facts about lattices. We note that there are many different bases for any lattice L. In fact, if the set $B = \{\mathbf{b}_1, \cdots, \mathbf{b}_n\}$ spans some lattice then by taking any vector $\mathbf{b}_i \in B$ and adding to it any integral linear combination of the other vectors we obtain a different basis for the same lattice. An important fact about lattices is that all the bases of a given lattice have the same determinant (up to the sign). This fact follows since there is an integer matrix T such that $BT = C$ and another integer matrix T^{-1} such that $CT^{-1} = B$. The notion of of the *orthogonality defect* of a basis, which was introduced by Schnorr in [21], plays a crucial role in the security of our schemes.

Definition 2. Let B be a real non-singular $n \times n$ matrix. The orthogonality defect of B is defined as orth-defect$(B) \stackrel{\text{def}}{=} \frac{\prod_i \|\mathbf{b}_i\|}{|det(B)|}$, where $\|\mathbf{b}_i\|$ is the Euclidean norm of the i'th column in B.

Clearly, orth-defect$(B) = 1$ if and only if the columns of B are orthogonal to one another, and orth-defect$(B) > 1$ otherwise. When comparing different bases of the same lattice in $\mathcal{R}^n$, we really only care about the product of the $\|\mathbf{b}_i\|$'s, since $det(B)$ is the same for all of them (and serves just as a normalization factor).

   Another important notion in our scheme is the dual lattice. If $B = \mathbf{b}_1, \cdots, \mathbf{b}_n$ is a basis for some lattice in $\mathcal{R}^n$ (where we think of B as an $n \times n$ matrix whose columns are the $\mathbf{b}_i$'s) then the dual lattice of $L(B)$ is the lattice which is spanned by the rows of the matrix B^{-1}. In Section 3.4 we show that when we use a basis B for a lattice $L = L(B)$ for our trapdoor function, the work-load which is associated with some natural attacks on the scheme is proportional to the orthogonality defect *of the corresponding basis for the dual lattice*. It would therefore be convenient for us to define the *dual orthogonality defect* for a matrix.

Definition 3. Let B be a real non-singular $n \times n$ matrix. The dual orthogonality defect of B is defined as orth-defect$^*(B) \stackrel{\text{def}}{=} \prod_i \|\hat{\mathbf{b}}_i\|/|det(B^{-1})| = |det(B)| \cdot \prod_i \|\hat{\mathbf{b}}_i\|$, where $\hat{\mathbf{b}}_i$ is the i'th row in B^{-1}.

2.1 Hard problems in lattices

The security of our constructions is related to the (conjectured) intractability of a few computational problems in lattices.

The Closest Vector Problem (CVP). In this problem we are given a basis B for a lattice in $\mathcal{R}^n$ and another vector $\mathbf{v} \in \mathcal{R}^n$, and our task is to find the vector in $L(B)$ which is closest to $\mathbf{v}$ (in some norm). The CVP was shown by van Emde Boas [6] to be $\mathcal{NP}$-hard for any l_p norm. Also, Arora et al. [3] proved that approximating the CVP to within any constant factor is also NP-hard.

   No polynomial-time algorithm is known for approximating the CVP in $\mathcal{R}^n$ to within a polynomial factor in n. The best polynomial time algorithms for approximating CVP are based on the LLL algorithm [17] and its variants. Babai [4] proved that the CVP in $\mathcal{R}^n$ can be approximated in polynomial time to within a factor of $2^{n/2}$. This was later improved by Schnorr [21] to a factor of $(1 + \varepsilon)^n$ for any $\varepsilon > 0$. We note, however, that these bounds refer to worst-case instances, and these algorithms "typically" perform much better than the above upper-bounds.

As we explain in Section 3, an attack against our trapdoor function amounts to finding an exact solution for some random instance of CVP.

The Smallest Basis Problem (SBP). In this problem, we are given a basis B for a lattice in $\mathcal{R}^n$ and our goal is to find the "smallest" basis B' for the same lattice. There are many variants of this problem, depending on the exact meaning of "smallest". In the context of this paper, we care about bases with small orthogonality defect. Thus, we consider the version in which we look for the basis B' of $L(B)$ which has smallest orthogonality defect. For this problem too there are no known polynomial-time algorithms, and the best polynomial-time approximation algorithms for it are variants of the LLL algorithms, which achieve an approximation ratio of $2^{O(n^2)}$ in the worst case for SBP instances in $\mathcal{R}^n$.

In our public key constructions, finding the private-key from the public-key requires solving some random SBP instances.

3   A Candidate Trapdoor Function

In this section we define our candidate trapdoor function and analyze a few possible attacks against it. Informally, a collection of trapdoor functions consists of four algorithms, GENERATE, SAMPLE, EVALUATE and INVERT, where GENERATE outputs a description of a function and the associate trapdoor information, SAMPLE picks an element in the domain of the function, EVALUATE evaluates the function on that element and INVERT uses the trapdoor information to inverts the function. Below we describe our construction.

GENERATE.   On input 1^n, we generate two bases B and R of the same full-rank lattice in $\mathcal{Z}^n$ and a positive real number σ. We generate these bases so that R has a low dual-orthogonality-defect and B has a high dual-orthogonality-defect. We describe the generation process in details in Section 3.2. The bases B, R are represented by $n \times n$ matrices where the basis-vectors are the columns of these matrices. In the sequel we call B the "public basis" and R the "private basis". We view (B, σ) as the description of a function $f_{B,\sigma}$ and R as the trapdoor information. The domain of $f_{B,\sigma}$ consists of some pairs of vectors $\mathbf{v}, \mathbf{e} \in \mathcal{R}^n$ (see below).

SAMPLE.   Given (B, σ), we output vectors $\mathbf{v}, \mathbf{e} \in \mathcal{R}^n$ as follows: The vector $\mathbf{v}$ is chosen at random from a "large enough" cube in $\mathcal{Z}^n$. For example, we can pick each entry in $\mathbf{v}$ uniformly at random[3] from the range $\{-n, \ldots, +n\}$. The vector $\mathbf{e}$ is chosen by setting each entry in it to either $+\sigma$ or $-\sigma$, each with probability $\frac{1}{2}$. (Alternatively, if we want $\mathbf{e}$ to have integral entries we can pick each entry as equal to $\pm \lceil \sigma \rceil$ each with probability $p_\sigma = \frac{\sigma^2}{2\lceil \sigma \rceil^2}$, and 0 with probability $1 - 2p_\sigma$.)

EVALUATE.   Given $B, \sigma, \mathbf{v}, \mathbf{e}$, we compute $\mathbf{c} = f_{B,\sigma}(\mathbf{v}, \mathbf{e}) = B\mathbf{v} + \mathbf{e}$.

---

[3] We do not know if the size of this range has any influence on the security of the construction. The value n is rather arbitrary, and was only chosen to get integers of about 8 bits for the parameters which we work with.

INVERT. Given R and c, we use Babai's Round-off algorithm [4] to invert the function. Namely, we represent c as a linear combination on the columns of R and then round the coefficients in this linear combination to the nearest integers to get a lattice point. The representation of this lattice point as a linear combination on the columns of B is the vector v. Once we have v we can compute e. More precisely, denote $T \overset{\text{def}}{=} B^{-1}R$, so we compute $v \leftarrow T \lceil R^{-1}c \rfloor$ and $e \leftarrow c - Bv$.

3.1 The Inversion Algorithm

In this section we show how σ can be chosen so that the inversion algorithm is successful with high probability. Recall that the inversion algorithm succeeds in inverting the function on c if using the private basis R in Babai's Round-off algorithm results in finding the closest lattice-point to c. Below we suggest two different ways to bound the value of σ, based on the L_1 norm and L_∞ norm of rows in R^{-1}. Both bounds uses the following lemma.

Lemma 4. *Let R be the private basis used in the inversion of $f_{B,\sigma}(v, e)$. Then an inversion error occurs if and only if $\lceil R^{-1}e \rfloor \neq \bar{0}$.*

Proof. Let T be the unimodular transformation matrix $T = B^{-1}R$. Then the inversion algorithm is $v = T \lceil R^{-1}c \rfloor$ and $e = c - Bv$. Obviously, if v is computed correctly then so is e. Thus, let us examine the conditions under which this algorithm finds the correct vector v. Recall that c was computed as $c = Bv + e$, so

$$T \lceil R^{-1}c \rfloor = T \lceil R^{-1}(Bv + e) \rfloor$$
$$= T \lceil R^{-1}Bv + R^{-1}e \rfloor = T \lceil (BT)^{-1}Bv + R^{-1}e \rfloor = T \lceil T^{-1}v + R^{-1}e \rfloor$$

But since T is a unimodular matrix (and therefore, so it T^{-1}) and since v is an integral vector, then $T^{-1}v$ is also an integral vector. Hence we have $\lceil T^{-1}v + R^{-1}e \rfloor = T^{-1}v + \lceil R^{-1}e \rfloor$, and therefore

$$T \lceil R^{-1}c \rfloor = T(T^{-1}v + \lceil R^{-1}e \rfloor) = v + T \lceil R^{-1}e \rfloor$$

Thus the inversion algorithm succeeds if and only if $\lceil R^{-1}e \rfloor = \bar{0}$. □

Theorem 5. *Let R be the private basis used in the inversion of $f_{B,\sigma}$, and denote the maximum L_1 norm of the rows in R^{-1} by ρ. Then as long as $\sigma < 1/(2\rho)$, no inversion errors can occur.*

Proof omitted.

Although Theorem 5 gives a sufficient condition to get the error-probability down to 0, we may choose to set a higher value for σ in order to get better security. The next theorem asserts a different bound on σ, which guarantee a low error probability.

Theorem 6. *Let R be the private basis used in the inversion of $f_{B,\sigma}$, and denote the maximum L_∞ norm of the rows in R^{-1} by $\frac{\gamma}{\sqrt{n}}$. Then the probability of inversion errors is bounded by*

$$\Pr[\text{ inversion error using } R] \leq 2n \cdot \exp\left(-\frac{1}{8\sigma^2\gamma^2}\right) \qquad (1)$$

Proof. We first introduce a few notations. We denote $\mathbf{d} \overset{\text{def}}{=} R^{-1}\mathbf{e}$ and denote the i'th entry in $\mathbf{d}$ and $\mathbf{e}$ by δ_i and ϵ_i respectively. Also, we denote the i'th row in R^{-1} by $\hat{\mathbf{r}}_i$ and the i, j'th element in R^{-1} by ρ_{ij}. We fix some i and evaluate $\Pr[|\delta_i| \geq \frac{1}{2}]$. Recall that $\delta_i = \hat{\mathbf{r}}_i \circ \mathbf{e} = \sum_j \rho_{ij}\epsilon_j$. Since for all j, $|\rho_{ij}| \leq \gamma/\sqrt{n}$ and $\epsilon_j = \pm\sigma$, each with probability $\frac{1}{2}$, then all the random variables $\rho_{ij}\epsilon_j$ have zero mean and they are all limited to the interval $[-\frac{\sigma\gamma}{\sqrt{n}}, +\frac{\sigma\gamma}{\sqrt{n}}]$. Therefore we can use Hoeffding bound to conclude that

$$\Pr\left[|\delta_i| > \frac{1}{2}\right] = \Pr\left[|\sum_j \rho_{ij}\epsilon_j| > \frac{1}{2}\right] < 2\exp\left(-\frac{1}{8\sigma^2\gamma^2}\right)$$

Using the union bound to bound the probability that any such i exists completes the proof. □

Remark. The last theorem implies that to get the error probability below ε it is sufficient to choose $\sigma \leq \left(\gamma\sqrt{8\ln(2n/\varepsilon)}\right)^{-1}$. In fact, the above bound is overly pessimistic in that it only looks at the largest entry in R^{-1}. A more refined bound can be obtained by considering the few largest entries in each row separately and applying the above argument to the rest of the entries.

Alternatively, we can get an estimate (rather that a bound) of the error probability by using Equation 1 as if all the entries in each row of R^{-1} have the same absolute value. In this case γ is the maximum Euclidean norm of the rows in R^{-1} so we get an estimate of the error-probability in terms of the Euclidean norm of the rows in R^{-1}. This estimate is about the same as the one which we get by viewing each of the δ_i's as a zero-mean Gaussian random variable with variance $(\sigma\|\hat{\mathbf{r}}_i\|)^2$ (where $\|\hat{\mathbf{r}}_i\|$ is the Euclidean norm of the i'th row in R^{-1}).

To get a feeling for the size of the parameters involved, consider the parameters $n = 120$, $\varepsilon = 10^{-5}$. For a certain setting of the parameters which we tested (in which the entries in R were chosen from the range ± 4), the maximal Euclidean norm of the rows in R^{-1} is about $1/30$. Evaluating the expression above for $\gamma = 1/30$ yields

$$\sigma \leq \left(\frac{1}{30}\sqrt{8\ln\left(\frac{2\cdot120}{10^{-5}}\right)}\right)^{-1} \approx \frac{30}{11.9} \approx 2.5.$$

3.2 The GENERATE Algorithm

In this section we discuss various aspects of the GENERATE algorithm. We described in Section 3.1 how the value of σ can be computed once we have the private basis R. Now we suggest a few ways to pick R and B. Recall that R, B are two bases for some lattice in $\mathcal{Z}^n$, where R has small dual orthogonality defect and B has a large dual orthogonality defect. Our high-level approach for generating the private and public bases is to choose at random n vectors in $\mathcal{Z}^n$ to get the private basis and then to "mix" them so as to get the public one. There are two distributions to consider in this process

- The choice of the private basis R induces a distribution on the lattices in $\mathcal{Z}^n$.
- For any private basis R, the process of "mixing" R to get the public basis B induces some distribution on the bases of $L(R)$.

To guide us through the choices of the various parameters, we relied on experimental results. Below we briefly discuss the various parameters which are involved in this process.

Lattice dimension. The first parameter we need to set is the dimension of the lattice (the value of n). Clearly, the larger n is, we expect that our schemes will be more secure. On the other hand, both the space needed for the key pair and the running-time of function-evaluation and function-inversion grow (at least) as $\Omega(n^2)$.

The lattice-reduction algorithm which we used for our experiments is capable of finding a basis with very small orthogonality defect as long as the lattice dimension is no more than 60-80 (depending on other parameters). Beyond this point, the quality of the bases we get from this lattice reduction algorithm degrades rapidly with the dimensions. In particular, we found that in dimension 100, the bases we obtained had a high dial-orthogonality-defect. At the present time, the best "practical lattice-reduction algorithm" which we are aware of is Schnorr's block-reduction scheme (which was used to attack the Chor-Rivest cryptosystem, see [22]). We speculate that working in dimensions about 250-300 should be good enough with respect to this algorithm.

Distribution of the private bases. We considered two possible distributions for choosing the private basis.

Choosing a "random lattice": We choose a matrix R which is uniformly distributed in $\{-l, \cdots, +l\}^{n \times n}$ for some integer bound l. In our experiments, the value of l had almost no effect on the quality of the bases which we got. Therefore we chose to work with small integers (e.g., between ± 4).

Choosing an "almost rectangular lattice": We start from the box $k \cdot I$ in $\mathcal{R}^n$ (for some number k), and add "noise" to each of the box vectors. Namely, we pick a matrix R' which is uniformly distributed in $\{-l, \cdots, +l\}^{n \times n}$, and then compute $R \leftarrow R' + kI$. The larger the value of k is, this process generates a basis with smaller dual orthogonality factor, so it may be possible to choose a larger value of σ. On the other hand, it may also allow an attacker to obtain a basis with smaller dual orthogonality factor by reducing the public basis. Our experiments show that we get the best parameters when k is about $\sqrt{n} \cdot l$.

Generating the public basis. Once we have the private basis R, we should pick the public basis B according to some distribution on the bases of the lattice $L(R)$. We tested two methods for generating B from R:

In the *first method*, we transform R into B via a sequence of many "mixing steps", in which we take one basis vector and add to it a random integer linear combination of the other vectors.

In our experiments, we went through the basis vectors one at a time, to make sure that we replace them all. The coefficients in the linear combination were chose at random from $\{-1, 0, 1\}$ with a bias towards 0 (specifically, we used $\Pr[1] = \Pr[-1] = 1/7$). This was done so that the size of the numbers in the public basis will not grow too fast. Our experiments indicate that using $2n$ mixing steps was sufficient to prevent LLL from recovering the original basis.

In the *second method* we multiply R by a few "random" unimodular matrices to get B, namely $B = R \cdot T_1 \cdot T_2 \cdots$. Each of these unimodular transformation matrices is chosen as a product of and upper- and lower-triangular matrices, $T_i = L_i U_i$, where the diagonal entries in L_i, U_i are ± 1. In our experiments, we chose the other non-zero entries in L_i, R_i from $\{-1, 0, 1\}$. We found that we need to multiply R by at least four transformation matrices to prevent LLL from recovering the original basis. Also, our experiments show that this process generates public matrices with larger entries than using $2n$ mixing steps according to the previous method. Thus, we chose to use the first method for most of our experiments.

3.3 Bases representation

To make evaluating and inverting the function more efficient, we chose the following representation for the private and public bases. The public bases is represented by the integer matrix B whose columns are the basis-vectors, so that evaluating $f_{B,\sigma}(\mathbf{v}, \mathbf{e}) = B\mathbf{v} + \mathbf{e}$ can be done in quadratic time. To invert $f_{B,\sigma}$ efficiently, however, we do not store the private basis R itself. Instead, we store the matrix R^{-1} and the unimodular matrix $T = B^{-1}R$. Then, to compute $f_{B,\sigma}^{-1}(\mathbf{c})$ we set $\mathbf{v} = T \lceil R^{-1}\mathbf{c} \rfloor$ and $\mathbf{e} = \mathbf{c} - B\mathbf{v}$, both of which can be done in quadratic time.

Representing B, T is easy since they are integral matrices, but R^{-1} is not an integral matrix, so we need to consider how it should be represented. Although it is possible to store the exact values for R^{-1}, the entries in R^{-1} may have hundreds of bits of precision, which makes working with them rather inefficient. A different approach is to only keep a few bits of each entry in R^{-1}. This, of course, may introduces errors. If we only keep ℓ bits per entry then we get an error of at most $2^{-\ell}$ in each entry of R^{-1}.

Clearly, this has no effect on the security of the system (since it only effects the operations done using the private basis), but it may increase the probability of inversion errors. Since we only perform linear operations on R^{-1}, it is rather straightforward to evaluate the effect of adding small errors to its entries. Denote the "error matrix" by $E = (\epsilon_{ij})$. That is, ϵ_{ij} is the difference between the value which is stored for $(R^{-1})_{ij}$ and the real value of that entry. Then we have $|\epsilon_{ij}| < 2^{-\ell}$ for all i, j. When inverting the function, we apply the same procedure as above, but uses the matrix $R' \stackrel{\text{def}}{=} R^{-1} + E$ instead of the matrix R^{-1} itself.

Recall that the value of the function is $\mathbf{c} = B\mathbf{v} + \mathbf{e}$, where $\mathbf{v}$ is an integer vector and $\mathbf{e}$ is the "error vector". Thus the vector $\mathbf{v}'$ computed by the (modified) inversion routine is

$$\mathbf{v}' = T \lfloor R'\mathbf{c} \rceil = T \lceil (R^{-1} + E)(B\mathbf{v} + \mathbf{e}) \rfloor = \mathbf{v} + T \lceil R^{-1}\mathbf{e} + E(B\mathbf{v} + \mathbf{e}) \rfloor$$

where the last equality follows since $R^{-1}B\mathbf{v}$ is an integral vector so we can take it out of the rounding operation and then we have $TR^{-1}B\mathbf{v} = \mathbf{v}$. Therefore, we invert correctly if and only if $\lceil R^{-1}\mathbf{e} + E(B\mathbf{v} + \mathbf{e}) \rfloor = \bar{\mathbf{0}}$, which means that all the entries in $R^{-1}\mathbf{e} + E(B\mathbf{v} + \mathbf{e})$ are less than a $\frac{1}{2}$ in absolute value. The size of the entries in the vector $R^{-1}\mathbf{e}$ is analyzed in Section 3.1, so here we only consider the vector $E(B\mathbf{v} + \mathbf{e})$.

Recall that all the entries in E are less than $2^{-\ell}$ in absolute value, and that the entries of error vector $\mathbf{e}$ are all $\pm\sigma$ (for our choice of parameters, we have $\sigma \approx 3$). Thus the contribution of the vector $E\mathbf{e}$ can be ignored. To evaluate the entries in $EB\mathbf{v}$, assume that we represent each entry in the matrix B using k bits, and each entry in the vector $\mathbf{v}$ using

m bits. Then, each entry in the vector $EB\mathbf{v}$ must be smaller than $n \cdot 2^{k+m-\ell}$ in absolute value.

For example, if we work in dimension 200, use 16 bits for each entry in B and 8 bits for each entry in $\mathbf{v}$, and keep only the 64 most significant bits of each entry in R^{-1} then the entries in $EB\mathbf{v}$ will be bounded by $200 \cdot 2^{16+8-64} \approx 2^{-32}$. Thus, a sufficient condition for correct inversion is that each entry in $R^{-1}\mathbf{e}$ is less than $\frac{1}{2} - 2^{-32}$ in absolute value (as opposed to less than $\frac{1}{2}$ which we get when we store the exact values for R^{-1}). Clearly, this has almost no effect on the probability of inversion errors.

3.4 Security Analysis

In this section we provide some initial analysis for the security of the suggested trapdoor function by considering several possible attacks and trying to analyze their work-load. An obvious pre-processing step in just about every attack on our construction is to reduce the public basis B to get a better basis B' which can then be used for the attack. For the sake of simplicity, we therefore assume that the public basis itself is already reduced via a "good lattice-reduction algorithm".

Our numerical estimates for the work-load of the various attacks are based on experiments reported in Section 5. In these experiments we used the implementation of the LLL lattice-reduction algorithm from the LiDIA project [16]. The bottom line of our experiments is that all the attacks below become infeasible in dimensions above 150. We do not have data about the performance of these attack using better lattice-reduction algorithms (such as the ones described in [22]. We speculate that when using these better algorithms, the attacks will become infeasible in dimensions about 250-300.

The Round-off Attack The most obvious attack on our scheme (other than a brute-force search for the error vector e) is to try and use the public basis B for inverting the function in the same manner as we use the private basis R. Namely, given the output of the function $\mathbf{c} = B\mathbf{v} + \mathbf{e}$, we compute $B^{-1}\mathbf{c} = \mathbf{v} + B^{-1}\mathbf{e}$. Then we can do an exhaustive search for the vector $\mathbf{d} \overset{\text{def}}{=} B^{-1}\mathbf{e}$. Below we give an approximate analysis for the size of the search space that the attacker needs to go through before it finds the correct vector $\mathbf{d}$.

Denote the i'th entry in $\mathbf{d}$ and $\mathbf{e}$ by δ_i and ϵ_i respectively, the i'th row of B^{-1} by $\hat{\mathbf{b}}_i$ and the (i, j)'th element in B^{-1} by β_{ij}. Using these notations we can write $\delta_i = \hat{\mathbf{b}}_i \circ \mathbf{e} = \sum_j \beta_{ij} \epsilon_j$, and therefore $E[\delta_i] = 0$ and $\text{Var}[\delta_i] = \sum_j \beta_{ij}^2 E[\epsilon_j^2] = (\sigma \|\hat{\mathbf{b}}_i\|)^2$, where $\|\hat{\mathbf{b}}_i\|$ is the Euclidean norm of the i'th row of B^{-1}.

To evaluate the size of this search space for $\mathbf{d}$, we make the simplifying assumptions that each entry δ_i in $\mathbf{d}$ is Gaussian, and that the entries are independent. Based on these simplifying assumptions, the size of the effective search space is exponential in the differential entropy of the Gaussian random vector $\mathbf{d}$. Recall that the differential entropy of a Gaussian random variable x with variance σ^2 is $h(x) = \frac{1}{2}\log(\pi e \sigma^2)$. Since we assume that the δ_i's are independent, then the differential entropy of the vector $\mathbf{d}$ equals the sum of the differential entropies of the entries, so we get

$$h(\mathbf{d}) = \frac{1}{2}\sum_i \log(\pi e \sigma^2 \|\hat{\mathbf{b}}_i\|^2) = \frac{n}{2}\log(\pi e \sigma^2) + \sum_i \log\|\hat{\mathbf{b}}_i\|$$

so the size of the search space is

$$2^{h(\mathbf{d})} = (\pi e)^{n/2} \cdot \sigma^n \cdot \prod_i \|\hat{\mathbf{b}}_i\| = (\pi e)^{n/2} \cdot \sigma^n \cdot \frac{\text{orth-defect}^*(B)}{|det(B)|}$$

Note that the term $det(B)$ in the last expression depends only on the lattice and is independent of the actual basis B.

Typical numeric values. In Subsection 5.2 we describe experiments which we performed in dimension 80 through 160. Upto dimension 80, the LLL algorithm is capable of reconstructing a "good" basis, so that the work-load of this attack is essentially 1. In higher dimensions, however, LLL fails to provide a good basis, and consequently the work load of the Round-off attack grows by a factor of about 8000 per dimension. Thus, already in dimension 100 this attack is worse than the trivial brute-force search for the error vector **e**.

The Nearest-plane Attack One rather obvious improvement to the Round-off attack from above is to use a better approximation algorithm for the CVP. In particular, instead of using Babai's Round-off algorithm we can use the Nearest-plane algorithm which was also described in [4]. On a high-level, the difference between the Round-off and the Nearest-plane algorithms is that in the Nearest-plane, the rounding in the different entries are done adaptively (rather that all at once). More precisely, the Nearest-plane algorithm works as follows: It is given a point **c** and an LLL reduced basis $B = \{\mathbf{b}_1, \ldots, \mathbf{b}_n\}$ (in the order induced by the LLL reduction). It then considers all the affine spaces

$$H_k = \left\{ k\mathbf{b}_n + \sum_{i=1}^{n-1} \alpha_i \mathbf{b}_i \ : \alpha_i \in \mathcal{R} \right\}$$

for all $k \in \mathcal{Z}$, finds the hyperplane H_k which is closest to the point **c**, and projects the point $c - k\mathbf{b}_n$ onto the $(n-1)$-dimensional space which is spanned by $\{\mathbf{b}_1, \ldots, \mathbf{b}_{n-1}\}$. This yields a new point c' and a new basis $B' = \{\mathbf{b}_1, \ldots, \mathbf{b}_{n-1}\}$, and the algorithm now proceeds recursively to find a point p' in this $(n-1)$-dimensional lattice which is close to c'. Finally, the algorithm sets $p = p' + k\mathbf{b}_n$.

It was pointed to us by Don Coppersmith that the Nearest-plane attack can be improved in practice in several different ways:

- Instead of picking the vectors by the order which was induced by LLL, we can pick them by the size of the Euclidean norm in the corresponding rows of B^{-1}. An analysis similar to Subsection 3.1 shows that this choice locally maximizes the probability that the hyperplane H_k is the correct one (this analysis is omitted from this extended abstract).
- We can "peel off" more than one vector in each level of the recursion, if there are several vectors for which the corresponding rows of B^{-1} have small norm.
- We can apply a lattice-reduction procedure to the remaining basis vectors in each level of the recursion. This improvement is particularly useful since the performance of the lattice-reduction algorithm improves rapidly as the dimension decreases.
- If all the rows in B^{-1} have a large Euclidean norm, we can apply an exhaustive search to the few entries which has the smallest Euclidean norm. That is, instead of just trying the closest H_k, we can also try the second closest one, etc.

The work-load of the Nearest-plane attack can be analyzed and tested in a similar manner to that of the Round-off attack: We can describe this attack as consisting of an off-line phase, in which we construct from the public basis B another matrix $\hat{B}$, and an on-line phase in which $\hat{B}$ used in a manner similar to the way B^{-1} is used in the Round-off attack. An estimate for the work-load of this attack can be computed from the Euclidean norm of the rows in $\hat{B}$. Due to lack of space, the analysis is omitted from this extended abstract.

Experiments reported in Subsection 5.3 indicate that the Nearest-plane attack has a much lower work-load than the Round-off attack. Nonetheless, its work-load also grows exponentially with the dimension of the lattice. Our experiments show that when using LLL as our lattice-reduction algorithm, some amount of search is needed starting from dimensions 110-120, and the attack becomes infeasible in dimensions 140-150.

The Embedding Attack Finally, another heuristic which is often used to approximate CVP (and which was brought to our attention by Clause Schnorr and Don Coppersmith) is to embed the n basis-vectors and the point c for which we want to find a close lattice point in an $(n + 1)$-dimensional lattice like so

$$B' = \begin{pmatrix} | & | & | & & | \\ c & b_1 & b_2 & \cdots & b_n \\ | & | & | & & | \\ 1 & 0 & 0 & & 0 \end{pmatrix}$$

Then we use a lattice reduction algorithm to search for the shortest non-zero vector in $L(B')$, in the hope that the first n entries in this vector will be the closest point to c. As opposed to the other attacks, we do not know how to use the output of this attack as a starting point for an exhaustive search (in the case where the output is not the "right lattice point"). Thus the only thing that we can measure about this attack is whether it works or not. Some experiments which we made with this attack (using LLL as our tool for finding shortest vectors) indicate that this heuristic works up to dimensions about 110-120. Recall that the Round-off attack becomes worse than the simple exhaustive search already in dimension 100.

4 Encryption Scheme

Our public-key Encryption scheme is based on our candidate one-way trapdoor function in the usual way. That is, to encrypt a message we embed it inside the argument to the function, compute the function and the result is the ciphertext. To decrypt, we use the trapdoor information to invert the function and extract the message from the argument.

Recall from Section 3 that our one-way trapdoor function takes a lattice vector and adds to it a small error vector. In the context of an encryption scheme, we can think of this process as 'encrypting a lattice vector' by adding to it a small error vector, and we can think of the resulting vector in $\mathcal{R}^n$ as the ciphertext. To encrypt arbitrary messages, we must specify an (easily invertible) encoding which maps messages into lattice vectors which are then encrypted as above. Describing such an encoding is the focus of this section.

To obtain a semantically secure encryption scheme [13], we need an encoding scheme such that seeing the ciphertext does not help a polynomial time adversary in getting "any information" about the message. Other parameters which need to be considered (besides security) are the efficiency of encoding and decoding, and the message expansion. Below we describe two possible encoding methods.

4.1 A Generic Encoding

The first method is a generic one. Since we have a candidate for a trapdoor one-way function, we may use hard-core bits of this function as the message bits. In particular, we can use the general construction of Goldreich-Levin, [12]) which shows how and where to hide hard core bits in a pre-image of any one-way function. (This construction enables hiding $\log n$ bits in one function evaluation.)

This approach has the advantage of being able to prove that it is impossible to even distinguish in polynomial time between any two messages, under the assumption that we started with a trapdoor function. The major drawback of this scheme is the message expansion, since we can only send $\log n$ bits at a time for one function evaluation. Moreover, since this approach is generic, it doesn't provide us with any insight which we may exploit to increase the bandwidth.

4.2 Encoding via the low-order bits in v

Another approach is to embed the bits of the message directly in an integer vector $\mathbf{v}$, and then compute the ciphertext as $\mathbf{c} = B\mathbf{v} + \mathbf{e}$, where B is the public basis and $\mathbf{e}$ is an error vector.

The main problem with this approach is that the adversary can in fact use $\mathbf{c}$ to obtain an estimate on each entry in $\mathbf{v}$. To see that, notice that $B^{-1}\mathbf{c} = \mathbf{v} + B^{-1}\mathbf{e}$, and so each entry in $B^{-1}\mathbf{c}$ is equal to the corresponding entry in $\mathbf{v}$ plus some "noise" from $B^{-1}\mathbf{e}$. Below we denote $\mathbf{d} \stackrel{\text{def}}{=} B^{-1}\mathbf{e}$. Also, the i'th entry in $\mathbf{d}$ is denoted by δ_i and the i'th entry of $\mathbf{v}$ is denoted ν_i.

We saw in Section 3.1 that if the Euclidean norm of the row $\hat{\mathbf{b}}_i$ in B^{-1} is small, then the variance of δ_i will also be small (notice that the dual-orthogonality-defect of B may still be large because of other rows in B^{-1} that have much larger Euclidean norm). In particular, if $\sigma \cdot \|\hat{\mathbf{b}}_i\| < 1$ then there is a reasonable probability that $|\delta_i| < 1/2$, in which case ν_i can be obtained simply by rounding the i'th entry in $B^{-1}\mathbf{c}$ to the nearest integer. Thus, an attacker could focus on the rows of B^{-1} which have low Euclidean norm, and compute the corresponding entries in $\mathbf{v}$. More generally, the adversary may view the i'th entry of $B^{-1}\mathbf{c}$ as an estimate for ν_i (which is probably accurate up to $\sigma\|\hat{\mathbf{b}}_i\|$).

Remark. Somewhat surprisingly, for the purpose of this attack - reducing the basis B does not seem to help (of course, as long as the resulting basis is not "reduced enough" to break the underlying trapdoor function). To see why, consider the unimodular transformation T' between the original basis B and the reduced basis B' ($T' = (B')^{-1}B$). Since $\mathbf{c}$ is computed using the original matrix B, then when trying to extract partial information using B' we compute

$$\mathbf{v}' = (B')^{-1}\mathbf{c} = (B')^{-1}(B\mathbf{v} + \mathbf{e}) = (B')^{-1}B\mathbf{v} + (B')^{-1}\mathbf{e} = T'\mathbf{v} + (B')^{-1}\mathbf{e}$$

If $(B')^{-1}$ has rows with small Euclidean norm, then the attacker may be able to learn the corresponding entries in $T'\mathbf{v}$, but this still does not seem to yield an estimate about any entry in $\mathbf{v}$. This suggests that in this encryption scheme, it may be useful to publish public basis which is not LLL reduced.

Embedding the message in the vector $\mathbf{v}$. From the above discussion, it is clear that if we are to embed the message in the vector $\mathbf{v}$ itself, then it should be embedded in the least significant bits of $\mathbf{v}$'s entries. Also, we should not put any bits of the message in entries of $\mathbf{v}$ which correspond to rows with small Euclidean norm in B^{-1}. We start by examining the simple case in which we only use the least-significant-bit of each entry (except for the "weak entries"), and pick all the other bits at random. Then, given an estimate $\tilde{\nu}_i = \nu_i + \delta_i$ for the entry ν_i, the attacker should decide whether the number in that entry was even or odd (that is, whether the message bit is a 0 or 1).

If we assume that each entry in $\tilde{\nu}_i$ can be approximated by a Gaussian random variable with mean ν_i and variance $\sigma^2 \|\hat{\mathbf{b}}_i\|^2$ (which is reasonable since $\tilde{\nu}_i$ is a sum of n independent random variable which are all "more or less the same"), then given the experimental value $\tilde{\nu}_i$, the statistical advantage $|\Pr[\nu_i \text{ is even} \mid \tilde{\nu}_i] - \Pr[\nu_i \text{ is odd} \mid \tilde{\nu}_i]|$ is exponentially small in $\sigma \|\hat{\mathbf{b}}_i\|$. If the Euclidean norm of $\hat{\mathbf{b}}_i$ is large enough, then the attacker, who knows $\tilde{\nu}_i$, gets only a small statistical advantage in guessing the corresponding bit of the message. If we have a row of B^{-1} with very high Euclidean norm, we may be able to use the corresponding entry of $\mathbf{v}$ for ℓ message-bits. It can be shown that the statistical advantage in guessing any of these bits is exponentially small in $\sigma \|\hat{\mathbf{b}}_i\|/2^\ell$. If the Euclidean norm of each individual row in B^{-1} is too small, we can represent each bit of s using several entries by making that bit the XOR of the least significant bit in all those entries. The statistical advantage then is exponentially small in $\sigma \cdot \sum_i \|\hat{\mathbf{b}}_i\|$ (where the sum is taken over the XOR'ed entries).

4.3 Additional Properties

Detecting decryption errors. One property of the above decryption procedure is that although there is a probability of error, it is still possible to verify when the message is decrypted correctly. This enables the legitimate user to identify decryption errors, so that it can take measures to correct them. Recall that we encrypt the lattice point $\mathbf{p}$ by adding to it a small error vector $\mathbf{e}$, thus obtaining the ciphertext $\mathbf{c} = \mathbf{p} + \mathbf{e}$. When we decrypt $\mathbf{c}$ and find a lattice point $\mathbf{p}'$ (which we hope is the same as $\mathbf{p}$), we can verify that this is the right lattice point by checking that all the entries in the error vector $\mathbf{e}' = \mathbf{c} - \mathbf{p}'$ are $\pm\sigma$. Thus as long as the lattice does not contain a point in which all the entries are exactly $\pm 2\sigma$, decryption errors can always be detected.

Plaintext Awareness. It seems that our scheme enjoys some weak notion of "plaintext awareness" in that there is no obvious way to generate from scratch a valid ciphertext (i.e., one which the decryption algorithm can decrypt) without knowing the corresponding lattice point. Still this plaintext awareness is limited, since after seeing one valid ciphertext c, it is possible to generate other valid ciphertexts without knowing the corresponding lattice-points (simply by adding any lattice point to c).

5 Experimental Results

Throughout this work, we used experimental data to guide us through the choices of various parameters in our construction, and to help us evaluate the effectiveness of some of the attacks. In this section we describe our testing methods and sketch a few of the main results. A full report on these tests will be available in the full version of this paper. For these experiments we used the implementation of the LLL lattice-reduction algorithm from the LiDIA project [16].

5.1 Choosing Parameters for the Key-Generation

The tests which we performed to determine the parameters of the key-generation process are omitted from this extended abstract. These tests are described in the TR version of this work [11].

5.2 Evaluation of the Round-Off Attack

We used the analysis from Subsection 3.4, in conjunction with our tests, to evaluate the performance of the Round-off attack in dimensions 80 through 160 (in increments of 10). We performed the following experiments:

1. In each dimension n we generated *five private bases*. Each basis was chosen as $R_i = 4 \lceil \sqrt{n} \rceil \cdot I + \text{rand}(\pm 4)$, where I is the identity matrix and $\text{rand}(\pm 4)$ is a square matrix whose entries are selected uniformly from the range $\{-4, \ldots, +4\}$.

For each basis R_i we computed the value σ_i (which is used for the error vector in our construction) as $\sigma_i = (\gamma_i \sqrt{8 \ln(2n/\varepsilon)})^{-1}$, where γ_i is the Euclidean norm of the largest row in R_i^{-1}, and $\varepsilon = 10^{-5}$. (By the Remark at the end of Subsection 3.1, we estimate that the probability of decryption errors using the private basis R_i and this value of σ_i is about 10^{-5}.)

2. For each private basis R_i we generated *five public bases*. Each public basis B_{ij} was generated by first applying $2n$ "mixing steps" to R_i and then LLL-reducing the resulting basis. As explained in Subsection 3.4, we evaluated the work-load of the Round-off attack using the public basis B_{ij} as

$$\text{work-load}_{ij} = (\pi e)^{n/2} \cdot \sigma_i^n \cdot \frac{\text{orth-defect}^*(B_{ij})}{|det(B_{ij})|}$$

3. We evaluated the work-load of the Round-off attack against the private basis R_i by the minimum of the work-loads for the corresponding private bases B_{i*}. Namely we set $\text{work-load}_i = \min_j \text{work-load}_{ij}$.

4. We evaluated the "typical work-load" in dimension n, by the median of the work-loads for the five private bases in this dimension.

The results of these tests in dimensions 80-160 are summarized in Figure 1. It can be seen in the figure that this attack falls apart once the dimension grows above 90, where the work-load increases by an amazing multiplicative factor of about 8000 per dimension (!!). Clearly, in dimensions 100 and above it is already easier to perform an exhaustive search for the value of the error vector e than to use the Round-off attack.

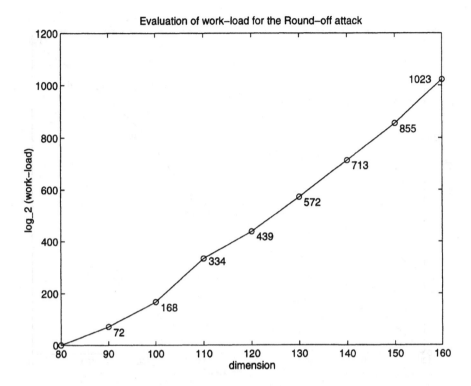

Fig. 1. Evaluation of the work-load of a Round-off attack in dimensions 80-160.

5.3 Evaluation of the Nearest-plane Attack

To evaluate the Nearest-plane attack, we used the same private bases R_i and public bases B_{ij} as for the Round-off attack. For each of the public bases B_{ij}, we carried out the off-line phase in the Nearest-plane attack, thereby generating the transformed matrices $\hat{B}_{ij}$. We then used the $\hat{B}_{ij}$'s to evaluate the work-load of the attack.

As before, the work-load for a private basis R_i is the minimum work-load for all the $\hat{B}_{ij}$'s, and the "typical" work-load for dimension n is the median work-load of the private bases in this dimension.

The results of our tests in dimensions 100-170 are summarized in Figure 2. As the figure clearly demonstrates, this attack is far better than the Round-off attack. Nonetheless, once the dimension grows above 110, the work-load monotonically increases by a multiplicative factor of about 4 per dimension. In dimensions 140-150 this attack is already infeasible. Extrapolating from this line, we estimate that in dimensions higher than 200, it would be easier to perform an exhaustive search for the value of the error vector e than to use the Nearest-plane attack.

129

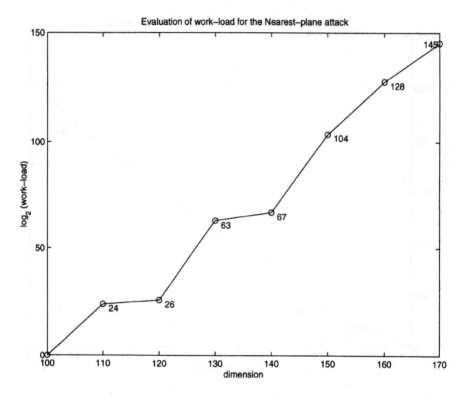

Fig. 2. Evaluation of the work-load of a Nearest-plane attack in dimensions 100-170.

5.4 Evaluation of the Embedding Attack

As we said in Subsection 3.4, we do not know how to turn a failed run of the Embedding attack into a starting point for some exhaustive search, and so we cannot talk about the "work-load" of this attack. Instead, we only measured what is the maximum value of σ (the bound on the error vector) for which this attack works.

For these experiments we used the same private bases R_i and public bases B_{ij} as for the previous two attacks. We then used each public basis to evaluate the function on a few points using a few different values of σ, and tested whether the Embedding attack recovers the encrypted message.

In our experiments we tested several values of σ between 1 and 3. For each setting of σ, we encrypted five messages and declared the attack successful if it recovered at least one of them. For each private basis R_i we computed the highest value of σ for which one of the B_{ij} was successful. For any dimension n we then computed the median among the σ values of the private bases in this dimension.

In Figure 3 we draw these values of σ for dimensions 80-130. These value are compared to the values of σ which we suggest to use in our construction to obtain a probability of 10^{-5} for decryption errors. It can be seen from this figure that for this choice of σ, the Embedding attack stops working around dimensions 110-120.

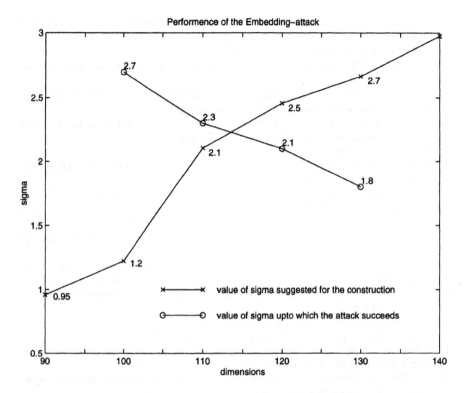

Fig. 3. Performance of the Embedding attack in dimensions 100-130

Acknowledgments.

We thank Svetoslav Tzvetkov for participating in the implementation of the experiments reported in Section 5. We also thank Dan Boneh, Don Coppersmith, Claus Schnorr and Jacques Stren for several very helpful conversations.

References

1. M. Ajtai. Generating hard instances of lattice problems. In *Proceedings of the 28th Annual ACM Symposium on Theory of Computing*, pages 99-108, 1996.
2. M. Ajtai and C. Dwork. A Public-Key Cryptosystem with Worst-Case/Average-Case Equivalence, In *29th ACM Symposium on Theory of Computing*, pages 284-293, 1997.
3. S. Arora, L. Babai, J. Stern, and Z. Sweedyk. The hardness of approximate optimia in lattices, codes, and systems of linear equations. In *Journal of Computer and System Sciences*, 54(2), pages 317-331, 1997.
4. L. Babai, On Lovász lattice reduction and the nearest lattice point problem. in *Combinatorica*, vol. 6, 1986, pp. 1-13.
5. M. Blum and S. Goldwasser. An Efficient Probabilistic Public-Key Encryption Scheme which Hides All Partial Information. in *Proceedings of CRYPTO '84*, Springer-Verlag, 1985, pp. 289-299.

6. P. van Emde Boas, Another $\mathcal{NP}$-complete problem and the complexity of computing short vectors in a lattice. Reprot 81-04, Mathematische Instituut, University of Amsterdam, 1981.

7. Digital Signature Standard (DSS). FIPS PUB 186, 1994.

8. W. Diffie and M.E. Hellman. New Directions In Cryptography. *IEEE Transactions on Information Theory*, Vol IT-22, 1976, pp. 644-654.

9. T. El-Gamal. A Public Key Cryptosystem and a Signature Scheme Based on Discrete Logarithms. *IEEE Trans. Information Theory*, vol. 31, 1985, pp. 469-472

10. O. Goldreich, S. Goldwasser and S. Halevi Collision-Free Hashing from Lattice Problems. Theory of Cryptography Library: Record 96-09. Available from
    http://theory.lcs.mit.edu/~tcryptol/1996/96-09.html

11. O. Goldreich, S. Goldwasser and S. Halevi Public-Key Cryptosystems from Lattice Reductions Problems. ECCC Report TR96-056. Available from
    http://www.eccc.uni-trier.de/eccc-local/Lists/TR-1996.html

12. O. Goldreich and L.A. Levin A Hard-Core Predicate for All One-Way Functions *Proceedings of the 21st ACM Symposium on Theory of Computing*, 1989, pp. 25-32

13. S. Goldwasser and S. Micali, Probabilistic Encryption. *Journal of Computer and System Sciences*, Vol. 28, 1984, pp. 270-299.

14. S. Goldwasser, S. Micali and R.L. Rivest. A Digital Signature Scheme Secure Against Adaptive Chosen Message Attack. *SIAM Journal on Computing*, Vol. 17, no. 2, 1988, pp. 281-308.

15. R. Kannan. Algorithmic Geometry of Numbers. in *Annual Review of Computer Science*, vol. 2, 1987, Annual Reviews Inc.

16. The LiDIA project software-package and user-manual.
    Available from http://www.informatik.th-darmstadt.de/TI/LiDIA/

17. A.K. Lenstra, H.W. Lenstra, L. Lovász. Factoring polynomials with rational coefficients. *Mathematische Annalen* 261, 515-534 (1982).

18. R.J. McEliece, A Public-Key Cryptosystem Based on Algebraic Coding Theory. DSN Progress Report 42-44, Jet Propulsion Laboratory

19. M.O. Rabin, Digital Signatures and Public-Key Functions as Intractable as Factorization. Technical Report MIT/LCS/TR-212, M.I.T., 1978.

20. R.L. Rivest, A. Shamir and L. Adleman. A Method for Obtaining Digital Signatures and Public-Key Cryptosystems. *Communications of the ACM*, Vol. 21, 1978, pp. 120-126.

21. C.P. Schnorr. A hierarchy of polynomial time lattice basis reduction algorithms. in *Theoretical Computer Science*, vol. 53, 1987, pp. 201-224

22. C.P. Schnorr and H.H. Horner, Attacking the Chor-Rivest Cryptosystem by Improved Lattice Reduction. in *Proceedings of EUROCRYPT '95*, Louis C. Guillou and Jean-Jacques Quisquater, editors. Lecture Notes in Computer Science, volume 921, Springer-Verlag, 1995. pp. 1-12

RSA-Based Undeniable Signatures*

Rosario Gennaro**, Hugo Krawczyk*** and Tal Rabin**

Abstract. We present the first undeniable signatures scheme based on RSA. Since their introduction in 1989 a significant amount of work has been devoted to the investigation of undeniable signatures. So far, this work has been based on discrete log systems. In contrast, our scheme uses regular RSA signatures to generate undeniable signatures. In this new setting, both the signature and verification exponents of RSA are kept secret by the signer, while the public key consists of a composite modulus and a sample RSA signature on a single public message.

Our scheme possesses several attractive properties. First of all, provable security, as forging the undeniable signatures is as hard as forging regular RSA signatures. Second, both the confirmation and denial protocols are zero-knowledge. In addition, these protocols are efficient (particularly, the confirmation protocol involves only two rounds of communication and a small number of exponentiations). Furthermore the RSA-based structure of our scheme provides with simple and elegant solutions to add several of the more advanced properties of undeniable signatures found in the literature, including convertibility of the undeniable signatures (into publicly verifiable ones), the possibility to delegate the ability to confirm and deny signatures to a third party without giving up the power to sign, and the existence of distributed (threshold) versions of the signing and confirmation operations.

Due to the above properties and the fact that our undeniable signatures are identical in form to *standard* RSA signatures, the scheme we present becomes a very attractive candidate for practical implementations.

1   Introduction

The central role of digital signatures in the commercial and legal aspects of the evolving electronic commerce world is well recognized. Digital signatures bind signers to the contents of the documents they sign. The ability for any third party to verify the validity of a signature is usually seen as the basis for the "non-repudiation" aspect of digital signatures, and their main source of attractiveness. However, this universal verifiability (or self-authenticating) property of digital signatures is not always a desirable property. Such is the case of a signature binding

---

* Extended Abstract. A complete version of the paper is available from
http://www.research.ibm.com/security/papers1997.html

** IBM T.J. Watson Research Center, PO Box 704, Yorktown Heights, New York 10598, USA
Email: rosario,talr@watson.ibm.com.

*** IBM T.J. Watson Research Center and Department of Electrical Engineering, Technion, Haifa 32000, Israel. Email: hugo@ee.technion.ac.il.

parties to a confidential agreement, or of a signature on documents carrying private or personal information. In these cases limiting the ability of third parties to verify the validity of a signature is an important goal. However, if we limit the verification to such an extent that it cannot be verified by, say, a judge in case of a dispute then the whole value of such signatures is seriously questioned. Thus, the question is how to generate signatures which limit the verification capabilities yet without giving up on the central property of non-repudiation.

An answer to this problem was provided by Chaum and van Antwerpen [CA90] who introduced *undeniable signatures*. Such signatures are characterized by the property that verification can only be achieved by interacting with the legitimate signer (through a *confirmation protocol*). On the other hand, the signer can prove that a forgery is such by engaging in a *denial protocol*. It is required that the following property be satisfied: if on a specific message and signature the confirmation protocol outputs that it is a valid signature then on the same input the denial protocol would not output that it is a forgery. The combination of these two protocols, confirmation and denial, protects both the recipient of the signature and the signer, and preserves the non-repudiation property found in traditional digital signatures. The protection of the recipient is established through the required property. Indeed, the ability of a signer to confirm a signature means that at no later point will the signer be able to deny the signature. For example, in the case of an eventual dispute, the recipient of the signature can resort to a designated authority (e.g., a judge) in order to demonstrate the signature's validity. In this case the signer will be required to confirm or deny the signature. If the signer does not succeed in denying (in particular, if it refuses to cooperate) then the signer remains legally bound to the signature (such will be the case if the alleged signature was a correct one). On the other hand the signer is protected by the fact that his signatures cannot be verified by unauthorized third parties without his own cooperation and the denial protocol protects him from false claims.

The protection of signatures from universal verifiability is not only justified by confidentiality and privacy concerns but it also opens a wide range of applications where verifying a signature is a valuable operation by itself. A typical example presented in the undeniable signatures literature is the case of a software company (or for this matter any other form of electronic publisher) that uses signature confirmation as a means to provides a proof of authenticity of their software to authorized (e.g., paying) customers only. This example illustrates the core observation on which the notion of undeniable signatures stands: verification of signatures, and not only their generation, is a valuable resource to be protected.

1.1 Components and security of undeniable signatures schemes

There are three main components to undeniable signature schemes. The signature generation algorithm (including the details of private and public information), the confirmation protocol, and the denial protocol. Signature generation is much like a regular signature generation, namely, an operation performed by the signer on the message which results in a string that is provided to the requester of the signature. The confirmation protocol is usually modeled after an interactive proof where the signer acts as the prover and the holder of the signature as the verifier. The input to the protocol is a message and its alleged signature (as well as the public key information associated with the signer). In case that the input pair is formed by a message and its legitimate signature then the prover can convince the verifier that this is the case, while if the signature does not correspond to the message then the probability of the prover to convince the verifier is negligible. Similarly, the denial protocol is an interactive proof designed to prove that a given input pair does *not* correspond to a message and its signature. In particular, if the alleged input signature does correspond to the input message then the probability of the prover to convince the verifier of the contrary is negligible. Note, that engaging in the confirmation protocol and having it fail is not an indication that the signature is invalid, this can only be established through the denial protocol. That is the confirmation protocol only establishes validity, and the denial – invalidity.

In addition to the above properties required from the confirmation and denial protocol, there are two basic security requirements on undeniable signatures. The first is unforgeability, namely, without access to the private key of the signer no one should be able to produce legitimate signatures by himself. This is similar to the unforgeability requirement in the case of regular digital signatures, but here the modeling of the attacker is somewhat more complex. In addition to having access to chosen messages signed by the legitimate signer, the attacker may also get to interact with the signer on different instances of the above protocols, possibly on input pairs of his own choice. The second requirement is non-transferability of the signature, namely, no attacker (under the above model) should be able to convince any other party, without the cooperation of the legitimate signer, of the validity or invalidity of a given message and signature. Both of these requirements induce necessary properties on the components of an undeniable signature scheme. In particular, the confirmation and denial protocols should not leak any information that can be used by an attacker to forge or transfer a signature. As a consequence it is desirable that these protocols be zero-knowledge. As for the strings representing signatures, they should provide no information that could help a party to get convinced of the validity (or invalidity) of signatures. Somewhat more formally, it is required that the legitimate signature(s) corresponding to a given message be *simulatable*, namely, they should be indistinguishable from strings that can be efficiently generated without knowledge of the secret signing key.

1.2 Advanced properties of undeniable signatures

Much of the work on undeniable signatures has been motivated by the search for schemes that provide all of the above properties but that, in addition, enjoy some additional attractive properties. These include *convertibility* (the possibility to transform undeniable signatures into regular, i.e. self-authenticating, signatures by just publishing a short piece of information, [BCDP91]), *delegation* (enabling selected third parties to confirm/deny signatures but not to sign), *distribution of power* (threshold version of the signature and confirmation protocols, [Ped91]), *designated confirmer* schemes (in which the recipient of the signature is assured that a specific third party will be able to confirm the signature at a later time, [Cha94]), and *designated verifier* schemes (in which the prover can make sure that only a specified verifier benefits from interacting with the prover on the confirmation of a signature, [JSI96]). More details are provided in Section 5.

1.3 Previous work on undeniable signatures

Since their introduction in 1989, undeniable signatures have received a significant attention in the cryptographic research community [CA90, Cha90, BCDP91, DY91, FOO91, Ped91, CHP92, Cha94, Jak94, Oka94, M96, DP96, JSI96, JY96]. These works have provided a variety of different schemes for undeniable signatures with variable degrees of security, provability, and additional features. Interestingly, all these works are discrete logarithm based. In [BCDP91] the problem of constructing schemes based on different assumptions, in particular RSA, was suggested as a possible research direction.

Most influential are the works of Chaum and van Antwerpen [CA90] and Chaum [Cha90]. The first work introduces the notion of undeniable signatures and provides protocols which are the basis for many of the subsequent works. The second improves significantly on the initial solution by providing zero-knowledge versions of these protocols. The formalization of the basic notions behind undeniable signatures was mainly carried out in the works by Boyar, Chaum, Damgard and Pedersen [BCDP91] and by Damgard and Pedersen [DP96]. In [BCDP91] the notion of *convertible* schemes was introduced. In such schemes the signer can publish a short string that converts the scheme into a regular signature scheme. However the scheme presented in [BCDP91] was recently broken in [M96]. The repaired solution presented therein however does not come with a proof of security. [DP96] present the first convertible schemes with proven security (based on cryptographic assumptions).

1.4 Our contribution

Our work is the first to present undeniable schemes based on RSA[1] Our undeniable signature scheme produces signatures that are *identical in form* to RSA signatures. The essential difference from traditional RSA signatures is that in our case both the signature and verification exponents of RSA are kept secret by the signer, while the public key consists of a composite modulus and a sample RSA signature on a single public message.

---

[1] Chaum in [Cha94] uses RSA signatures *on top* of regular undeniable signatures to provide "designated confirmer signatures"; however the underlying undeniable signatures are still discrete log-based.

Not only does our solution expand the list of available number-theoretic assumptions that suffice to build undeniable signatures, but it achieves and improves, as we show below, in a simple and elegant way several of the desirable properties of undeniable signatures.

Unforgeability: Our construction allows us to prove in a simple way that security of these signatures against forging is equivalent to the unforgeability of RSA signatures[2]. Provable unforgeability of undeniable signatures was presented for the first time in the recent paper by [DP96] where forgery of the proposed scheme is proven equivalent to forgery of the ElGamal scheme.

Simulatability: Non-transferability of an RSA signature is a non-standard requirement in the context of traditional RSA. We prove this property under the assumption that deciding on the equality of discrete logarithms under different bases is intractable. This assumption is required in previous works as well[3] although by itself is not always sufficient to prove simulatability of the undeniable signatures. For example in [DP96] the simulatability property is only conjectured to follow from such assumptions.

Zero-Knowledge: Our confirmation and denial protocols have the interactive proof properties as explained above and are also zero-knowledge. Therefore they do not leak any information that could otherwise be used for forging signatures. The soundness of our protocols (i.e. the guarantee that the prover/signer cannot cheat) relies on the use of composite numbers of a special form (specifically, with "safe prime" factors), which are secure moduli for RSA. A signer who chooses a modulus of a different form may have some way to cheat in our protocols. To force the signer to choose a "proper" modulus we require that he prove the correct choice of primes at the time he registers his public key with a certification authority. A discussion of this issue is presented in Section 4.

Efficiency: Our protocols are efficient (comparable to the most efficient alternatives found in the undeniable signatures literature). The confirmation protocol takes two rounds of communication (which is minimal for zero-knowledge protocols) and involves a small number of exponentiations. The denial protocol is somewhat more expensive as it consists of a basic two-round protocol with small, but not negligible, probability of error (e.g., 1/1000) which needs to be repeated

---

[2] As with regular RSA, the use of a strong one-way hash function is assumed to provide unforgeability against chosen message attacks.

[3] However, in our case the discrete logarithms are computed modulo a composite number while in previous works they are modulo a prime. In both cases, the problem is related to the problem of computing discrete logarithms which is considered to be hard (in the case of a composite modulus that difficulty is implied by the hardness of factoring and also directly by the assumed security of RSA). However, while the feasibility of computing discrete logarithm implies the feasibility of the above decision problem, the reverse direction is not known to hold.

sequentially in order to further reduce the error probability. Its performance is still significantly better (by a factor of 10) than alternative protocols that only achieve probability 1/2 in each execution. We also note that in typical uses of undeniable signature schemes one expects to apply more frequently confirmation than denial. The latter is mainly needed to settle legal disputes.

Advanced Properties: In addition to the above security and efficiency properties, our solution naturally achieves several of the advanced features of undeniable signatures mentioned above. Once again it is the structure of RSA, in particular the presence of a secret verification exponent, that allows to achieve such properties very elegantly. Convertibility is achieved by publishing the verification exponent, thus converting the signatures into regular RSA signatures; delegation is achieved by providing the verification exponent to the delegated party which can then run the confirmation and denial protocols but cannot sign messages or forge signatures; distribution of the signature operation builds on the existing threshold solutions for RSA signatures; distribution of confirmation can be also achieved by an adaptation of the regular threshold RSA solutions. We can also adapt existing techniques for the construction of *designated confirmer* and *designated verifier* undeniable signatures, thus obtaining these variants also for our scheme. More details are provided in Section 5.

Standard RSA compatibility: An important practical advantage of our RSA-based undeniable scheme is that the signatures themselves are identical in form to standard RSA signatures. In particular, this means that they fit directly into existing standardized communication protocols that use (regular) RSA signatures.

Technically, our work builds on previous ideas and protocols which we adapt to the RSA case. These previous solutions are designed to exploit the algebraic properties of cyclic groups like Z_p^* (and its subgroups). This is probably the main reason that subsequent work concentrated on these structures as well. Here we show that many of these ideas can be used in the context of RSA, thus answering in the affirmative a question suggested in [BCDP91]. In doing so we use ideas from the work of Gennaro et al. [GJKR96].

2 Preliminaries

Notation. Throughout the paper we use the following notations:

For a positive integer k we denote $[k] \stackrel{\text{def}}{=} \{1, \cdots, k\}$. Z_n^* denotes the multiplicative group of integers modulo n, and $\phi(n) = (p-1)(q-1)$ the order of this group. For an element $w \in Z_n^*$ we denote by $ord(w)$ the order of w in Z_n^*. The subgroup generated by an element $w \in Z_n^*$ is denoted by $<w>$.

The following lemmas are needed in our proofs in Section 3.

Lemma 1. *Let* $n = pq$, *where* $p < q$, $p = 2p' + 1$, $q = 2q' + 1$, *and* p, q, p', q' *are all prime numbers. The order of elements in* Z_n^* *is one of the set* $\{1, 2, p', q', 2p', 2q', p'q', 2p'q'\}$. *Given an element* $w \in Z_n^* \setminus \{-1, 1\}$, *such that* $ord(w) < p'q'$ *then* $gcd(w - 1, n)$ *is a prime factor of* n.

As a consequence of the above lemma we can assume in our protocols that any value found by a party that does not know the factorization of n must be of order at least $p'q'$ in Z_n^* (except for 1,-1).

Lemma 2. *Let* n *be as in Lemma 1. Given an element* w *such that* $ord(w) \in \{p'q', 2p'q'\}$ *then for every* $m \in Z_n^*$ *it holds that* $m^4 \in <w>$.

3 Our Undeniable Signature Scheme

In this section we give the details of our scheme. We start by defining the following set:

$$\mathcal{N} = \{n \mid n = pq, \ p < q, \ p = 2p' + 1, \ q = 2q' + 1,$$
$$\text{and } p, q, p', q' \text{ are all prime numbers}\}$$

The system is set up by the signer in the following manner: chooses an element $n \in \mathcal{N}$; selects elements $e, d \in \phi(n)$ such that $ed \equiv 1 \mod \phi(n)$; chooses a pair (w, S_w) with $w \in Z_n^*, w \neq 1, S_w = w^d \mod n$; sets the public key parameters to the tuple (n, w, S_w); sets the private key to (e, d).

We shall denote by $\mathcal{PK}$ the set of all tuples (n, w, S_w) generated as above. We refer the reader to Section 4.3 for a discussion on the form of the public key and how to verify its correctness. In particular, we state that the value of w can always be set to a fixed number, e.g. $w = 2$. This simplifies the public key system and adds to the efficiency of computing exponentiations with base w.

3.1 Generating a Signature

To generate a signature on a message m the signer carries out a regular RSA signing operation, i.e. he computes $S_m = m^d \mod n$, outputting the pair (m, S_m). More precisely, the message m is first processed through a suitable encoding (e.g., via one-way hashing) before applying the exponentiation such that the resultant signature scheme can be assumed to be unforgeable even against chosen message attacks (plain RSA does not have this property). Given a message m we will denote by $\bar{m}$ the output of such an encoding of m (we do not specify any encoding in particular)[4]. Thus, the resultant signature of m will be $S_m \stackrel{\text{def}}{=} \bar{m}^d \mod n$. In the case of the pair (w, S_w) we will slightly abuse the notation and write $S_w = w^d \mod n$ (without applying the encoding $\bar{w}$).

---

[4] For simplicity we will assume a deterministic encoding; however randomized encodings, e.g. [BR96], can be used as well but then, in our case, the random bits used for the encoding need to be attached to the signature.

3.2 Confirmation Protocol

In Figure 1 we present a protocol for confirming a signature. It is carried out by two players a prover and a verifier. The public input to the protocol are the public key parameters, namely $(n, w, S_w) \in \mathcal{PK}$, and a pair $(m, \hat{S}_m)$. For the case that $\hat{S}_m$ is a valid signature of m, then P will be able to convince V of this fact, while if the signature is invalid then no prover (even a computationally unbounded one) will be able to convince V to the contrary except for a negligible probability.

This protocol is basically the same as the protocol of Gennaro et al. [GJKR96] (based on [Cha90]) where it is used in a different application, namely, threshold RSA. Our variation on this protocol uses the verification key e rather than the signature key d as originally used in [GJKR96] (in their case, the signer knows only d but not e). Still the basic proof given in that paper applies to our case due to the symmetry that exists between d and e when both exponents are kept secret. This modification allows us to provide solutions where the ability to confirm signatures can be delegated to third parties while keeping the ability to sign new messages only for the original signer (it also allows for a distributed prover solution). See Section 5 for the details.

An interesting aspect of this protocol is that a prover could succeed in convincing the verifier to accept a signature on m even when this signature is not $\bar{m}^d \bmod n$ but $\alpha \bar{m}^d \bmod n$ where α is an element of order 2 (in Z_n^*). [GJKR96] solve this problem through the assumption (valid in their case) that the prover cannot factor n and thus cannot find such an element α. In our case, this assumption does not hold. We deal with this problem by accepting as valid signatures also these multiples of $\bar{m}^d$. On the other hand, when designing the denial protocol we make sure that the signer cannot deny a signature of this extended form. That is, we define the set of valid signatures for a message m as $\mathcal{SIG}(m) \stackrel{\text{def}}{=} \{S_m : S_m = \alpha \bar{m}^d, \ \text{ord}(\alpha) \leq 2\}$.

Signature Confirmation Protocol

Input: Prover: Secret key $(d, e) \in [\phi(n)]^2$
Common: Public key $(n, w, S_w) \in \mathcal{PK}$,
$m \in Z_n^*$ and alleged $\hat{S}_m$

1. V chooses $i, j \in_R [n]$ and computes $Q \stackrel{\text{def}}{=} \hat{S}_m^i S_w^j \bmod n$
   $V \longrightarrow P : Q$
2. P computes $A \stackrel{\text{def}}{=} Q^e \bmod n$
   $P \longrightarrow V : A$
3. V verifies that $A = \bar{m}^i w^j \bmod n$.
   If equality holds then V accepts $\hat{S}_m$ as the signature on m, otherwise "undetermined".

Fig. 1. Proving that $\hat{S}_m \in \mathcal{SIG}(m)$ (ZK steps omitted)

For ease of exposition the protocol in Figure 1 appears in a non zero-knowledge format. However, there are well-known techniques [GMW86, BCC88, Gol95] to add the zero-knowledge property to the above protocol using the notion of a *commitment function*: Instead of P sending A in Step 2, he sends a commitment $commit(A)$, after which V reveals to P the values of i and j. After checking that $Q \stackrel{\text{def}}{=} \hat{S}_m^i S_w^j \bmod n$, P sends A to V. The verifier checks that A corresponds to the value committed by P and then performs the test of Step 3 above.

The zero-knowledge condition is achieved through the properties of the commitment function, namely, (I) $commit(x)$ reveals no information on x, and (II) P cannot find x' such that $commit(x) = commit(x')$. Commitment functions can be implemented in many ways. For example, in the above protocol $commit(A)$ can be implemented as a probabilistic (semantically secure) RSA encryption of A using a public key for which the private key is not known to V (and possibly, not even known to P). To open the commitment, P reveals both A and the string r used for the probabilistic encryption. This implementation of a commitment function is very efficient as it does not involve long exponentiations (and is secure since we assume our adversary, the verifier in this case, is unable to break RSA). A proof of the theorem below can be found in [GJKR96].

Theorem 3. Confirmation Theorem. *Let* $(n, w, S_w) \in \mathcal{PK}$.

Completeness. *If P and V follow the Signature Confirmation protocol then V always accepts.*

Soundness. *A cheating prover P^*, even computationally unbounded, cannot convince V to accept $\hat{S}_m \notin \mathcal{SIG}(m)$ with probability greater than $\frac{O(1)}{n}$.*

Zero-knowledge. *The protocol is zero-knowledge, namely, on input a message and its valid signature, any (possibly cheating) verifier V^* interacting with prover P does not learn any information aside from the validity of the signature.*

3.3 Denial Protocol

Figure 2 exhibits the Denial Protocol. The public input to the protocol are the public key parameters, namely $(n, w, S_w) \in \mathcal{PK}$, and a pair $(m, \hat{S}_m)$. In the case that $\hat{S}_m \notin \mathcal{SIG}(m)$, then P will be able to convince V of this fact, while if $\hat{S}_m \in \mathcal{SIG}(m)$ then no prover (even a computationally unbounded one) will be able to convince V that the signature is invalid except with negligible probability. Our solution is based on a protocol due to Chaum [Cha90], designed to prove in zero-knowledge the inequality of the discrete logarithms of two elements over a prime field Z_p relative to two different bases. The protocol and proof presented in the above paper do not work over Z_n^* for a composite n as required here, in particular, since they strongly rely on the existence of a generator for the multiplicative group Z_p^*. However, a careful adaptation of that protocol and a more involved proof can be shown to solve our problem over Z_n^*.

The protocol has probability of error $\frac{1}{k}$, where $k = O(\log n)$ is a parameter chosen by the system. Due to an elegant observation of Chaum [Cha90] the desired probability of error can be achieved while incurring only a constant number of exponentiations. He notes that while carrying out k consecutive multiplications, which is equivalent in computation to a single exponentiation, we can compute all the powers in a range $[k]$. If we take $k = 1024$ we can repeat the protocol ten times in order to achieve a security of $\frac{1}{2^{100}}$. As stated in the introduction this allows for a ten fold increase in efficency relative to alternative protocols.

The protocol as presented in Figure 2 omits the steps that make it zero-knowledge. This is similar to the case of the confirmation protocol. Yet, in this protocol special care needs to be taken in Step 2. If the (honest) prover does not find a value i that satisfies the equation, which means that V is cheating, P aborts the execution of the protocol. Though aborting the protocol does not reveal much information it does reveal some, and in the zero-knowledge version we do not want even this much information to leak. Thus, P should continue the execution of the protocol by committing to the value 0, in a "dummy commitment" this will conceal the information of whether a value i was found or not. Note that in the case where no i was found, the verifier will be exposed later as a cheater and the commitment of 0 will never be revealed.

Denial Protocol

Input: Prover: Secret key $(d, e) \in [\phi(n)]^2$
       Common: Public key $(n, w, S_w) \in \mathcal{PK}$,
                  $m \in Z_n^*$ and alleged non-signature $\hat{S}_m$

1. V chooses $i = 4b$, $b \in_R [k]$ and $j \in_R [n]$.
   Sets $Q_1 = \bar{m}^i w^j \bmod n$ and $Q_2 = \hat{S}_m^i S_w^{\ j} \bmod n$
   $V \longrightarrow P : (Q_1, Q_2)$
2. P computes $Q_1/Q_2^e = \left(\frac{m}{\hat{S}_m^e}\right)^i$ and computes i by testing all possible values of $i \in [k]$.
   If such a value was found then P sets $A = i$, otherwise abort.
   $P \longrightarrow V : A$
3. V verifies that $A = i$. If equality holds then V rejects $\hat{S}_m$ as a signature of m, otherwise, undetermined.

Fig. 2. Proving that $\hat{S}_m \notin \mathcal{SIG}(m)$ (ZK steps omitted)

Theorem 4. Denial Protocol *Let $(n, w, S_w) \in \mathcal{PK}$.*
Completeness. *Assuming that $\hat{S}_m \notin \mathcal{SIG}(m)$, and if P and V follow the protocol then V always accepts that $\hat{S}_m$ is not a valid signature of m.*
Soundness. *Assuming that $\hat{S}_m \in \mathcal{SIG}(m)$ then a cheating prover P^*, even computationally unbounded, cannot convince V to reject the signature with probability greater than $\frac{1}{k} + \frac{1}{n}$.*

Zero-knowledge. *The protocol is zero-knowledge, namely, on input a message and a non-valid signature, any (possibly cheating) verifier V^* interacting with prover P does not learn any information aside from the fact that $\hat{S}_m$ is in fact not a valid signature for the message m.*

4 Security Analysis

We do not present here a formal treatment of the notion of undeniable signatures and its security requirements. For such a formal and complete treatment we refer the reader to the paper by Damgard and Pedersen [DP96]; an outline of these notions can be found above in our Introduction (in particular, in Section 1.1). Here we argue the security properties of our solution in an informal way based on this outline.

4.1 Unforgeability of signatures

We consider an attacker that cannot forge regular RSA signatures. When attacking our undeniable signatures scheme this attacker may request signatures (and their confirmation) on any messages of its choice. The attacker can also choose pairs of messages and alleged signatures and engage in confirmation or denial protocols with the signer on these inputs (whether it engages in a confirmation or denial protocol depends on the validity or invalidity, respectively, of the input pair). The goal of the attacker is to *forge* a signature, namely, to generate a valid signature on a message not previously signed by the legitimate signer.

We first note that since both confirmation and denial protocols are zero-knowledge then the information provided to the attacker by these protocols is useless for attacking the signatures (in the sense that the same information can be generated by the attacker alone). Therefore, an attacker could essentially try to forge signatures based on the public keys and a (possibly chosen) list of messages and their valid signatures. However, since our signatures are equivalent to regular RSA signatures (except for the fact that the verification exponent is secret which can only make it harder for the attacker) then the ability to forge our undeniable signatures would translate into forging regular RSA signatures which we assume infeasible. (As noted before, RSA is not directly immune against chosen message attacks but we assume this to be countered by additional means, e.g. by the appropriate encoding of the message prior to the exponentiation – see Section 3.1.)

Formalizing the above arguments is quite straightforward and standard. Such a formal proof would show how to transform any given forging attacker against our undeniable signatures into a forging attacker against regular RSA signatures; the transformation would make use of the simulators for our zero-knowledge protocols (both confirmation and denial). We summarize this discussion in the following theorem.

Theorem 5. *Assuming that the underlying RSA signatures are unforgeable (against known and/or chosen message attacks) then our undeniable signatures are unforgeable (against the same attacks).*

4.2 Indistinguishability of signatures

A basic goal of undeniable signatures is that no one should be able to verify the validity (or invalidity) of a message and its (alleged) signature without interacting with the legitimate signer in a confirmation (or denial) protocol. Following [DP96] we need to show that given the public key information and any message m (but not the signature exponent d) one can efficiently generate a *simulated signature* $s(m)$ of m, in the sense that the distribution of simulated signatures cannot be distinguished (efficiently) from the distribution of true signatures on m. We achieve this property in the following way. Given any message m, we apply to it the encoding $\bar{m}$ as determined by the underlying RSA scheme and then raise the result $\bar{m}$ to a random exponent modulo n (i.e., $s(m) = \bar{m}^r \bmod n$, for $r \in_R [n]$). Notice that distinguishing $s(m)$ from the signature $\bar{m}^d \bmod n$ on m is equivalent to deciding whether

$$\log_m(s(m)) \stackrel{?}{=} \log_w(S_w) \qquad (1)$$

where the discrete logarithm operation is taken in Z_n^*. This problem has no known efficient solution, though its equivalence to RSA, factoring, or the discrete logarithm problems has not been established. We thus require the following intractability assumption in order to claim the hardness of distinguishing between valid and simulated signatures.

Assumption EDL: For values $n, w, S_w, \bar{m}$, and $s(w)$ as defined above it is infeasible to decide the validity of equation 1 over Z_n^*.

We stress that the analogous assumption modulo a prime number is necessary for claiming the security of previous undeniable signature schemes as well (see [DP96]). However in the case of [DP96] the EDL assumption is not sufficient to prove simulatability, which in that paper is indeed simply conjectured.

Theorem 6. *Under the above EDL assumption, our signatures are simulatable and hence cannot be verified without the signer's (or its delegated confirmers) cooperation.*

Remark: The above theorem does not concern itself with a general problem of undeniable signatures pointed out first by Desmedt and Yung [DY91]. It is possible that the signer is fooled into proving a signature to several (mutually distrustful) verifiers while he is convinced of proving the signature to only one of them. We will address this problem in Section 5.

4.3 Choosing the signer's keys

In Section 3 we defined what the public and private parameters for the signer should be. Our analysis of the (soundness of the) confirmation and denial protocols depends on these parameters being selected correctly. Typically, the verification of this public key will be done whenever the signer registers it with a trusted party (e.g., a certification authority). Here we outline protocols to check the right composition of the modulus n, the sample element w, and the fact that S_w is chosen as a power of w (the latter serves as the "commitment" of the signer to the signature exponent d). Notice that these protocols are executed only once at registration time and not during the more common signing/verification operations. We denote by V the entity that acts as the verifier of these parameters, and by P the signer that proves its correct choices.

VERIFICATION THAT w IS OF HIGH ORDER. Specifically, we use in our analysis the assumption that w is an element of order at least $p'q'$. By virtue of Lemma 1 all that V needs to verify is that $w \notin \{-1, 1\}$ and that $gcd(w - 1, n)$ is not a factor of n. Actually, the value w can be chosen as a constant, e.g. $w = 2$, for *all* the undeniable signatures public keys. Such a value must always pass the verification (or otherwise factoring is trivial).

VERIFICATION THAT $S_w \in <w>$. The following protocol is essentially the protocol for proving possession of discrete logarithms as presented in [CEG87], once again modified in order to work with composite moduli. The signer P chooses a value $r \in_R [\phi(n)]$ and sends to V the value $w' = w^r$. The verifier V answers with a random bit b. If $b = 0$, P returns the value r, otherwise it returns the value $d + r \mod \phi(n)$. In the first case, V checks whether $w^r = w'$, and in the second, whether $w^{(r+d)} = w'S_w$. If $w \notin <w>$ then the probability that P passes this test is $1/2$. By repeating this procedure k times the probability that the dealer can cheat reduces to 2^{-k}. The protocol is statistical zero-knowledge as the simulator does not know $\phi(n)$, but can use the uniform distribution on $[1..n]$ to statistically approximate the one on $[1..\phi(n)]$. As a practical matter, we observe that this protocol can be performed non-interactively if one assumes the existence of an ideal hash function (a la Fiat-Shamir [FS86]).

VERIFICATION OF THE PRIME FACTORS. We need to check that the signer chooses the modulus n of the right form, i.e. $n = pq$ with $p = 2p' + 1$ and $q = 2q' + 1$ and p, q, p', q' are all prime numbers. We have three alternative solutions for this problem. The first is to use a generic zero-knowledge proof of the above property using the general results of [GMW86]; although the resultant solution would be highly inefficient this task is performed only once at system initialization. A more efficient (but less secure) solution to this problem is to let the signer generate a large set of moduli $n_1, n_2, \cdots, n_k$ from which V chooses a random element, say n_i. Next, P shows the factorization into primes of all the other moduli in the set. If all are of the right form then n_i is chosen as the modulus n, otherwise P

is disqualified. The drawback of this solution is that the probability of cheating, i.e. $1/k$, reduces only linearly with the amount of work in the protocol. Yet once again, the protocol needs to be performed only at initialization of the modulus and thus a relatively large number k of moduli can be produced. (Although this gives only "linear security" we stress that under the appropriate legal circumstances a probability of, say $999/1000$, to be caught cheating can be a significant deterrent for anyone to register an invalid key.)

Finally, there is a solution [Dam] that allows for a trade-off between the error probability at the key registration stage and the performance cost of the undeniable signature scheme. Initially, we let P generate $2k$ moduli. V chooses at random k of them of which the signer must reveal the factorizations. If the factorization of those moduli was of the correct form, we run our basic scheme in parallel for all the remaining k moduli. A confirmation or denial is accepted only if it works for all k moduli. The signer can only cheat if all the opened moduli were good, and all the remaining bad, but for any given set of moduli, this will only happen with exponentially small probability in k. In practice, one can choose parameters more appropriately. For example it doesn't have to be $2k$ and k moduli since with a total of 100 moduli, V choosing 95 of them and keeping only 5 to do the scheme in parallel, the error probability is close to 10^{-9}.

5 Extensions

Our protocols lend themselves to many of the existing extensions in the literature for undeniable signatures.

Convertible Undeniable Signatures. This variation appeared first in [BCDP91], and secure schemes based on ElGamal signatures have been recently presented in [DP96]. Convertible undeniable signatures enable the signer to publish a value which transforms the undeniable signature into a regular (i.e., self-authenticating) digital signature. In our scheme conversion can be easily achieved by simply publishing the value $e = d^{-1} \bmod \phi(n)$. Doing so the signer will transform the undeniable signatures into regular RSA signatures with public key (n, e). Notice that this will automatically imply the security (i.e., unforgeability) of the converted scheme, based on the security of regular RSA signatures.[5]

---

[5] Notice that this holds if the signer issued for the message m its *intended* signature $S_m = \bar{m}^d \bmod n$. If, instead, the signer generated a signature of the form $S_m = \alpha \bar{m}^d$, where α is an element of order 2, then when e is made public it is easy to recover α (and then the factorization of n) from a triple $(m, S_m = \alpha \bar{m}^d, e)$ since e is odd. We stress that although we consider as valid also signatures of that form (see Section 3.2), it is in the interest of the prover not to generate them in that way.

In some applications it may be desirable to convert only a subset of the past signatures (*selective conversion* [BCDP91]). For this scenario we can make use of a non-interactive zero-knowledge confirmation proof for those messages. Such an efficient scheme is described in the final paper.

Delegation. The idea is for the signer to delegate the ability to confirm and deny to a third party without providing that party the capabilities to generate signatures. In the literature this notion is usually treated in the context of convertibility of signatures. However the two notions are conceptually different. Clearly the information used in order to delegate confirmation/denial authority to a third party if made public would basically convert undeniable signatures into universally verifiable ones. However the converse is not necessarily true. It may be that the information used to convert signatures, if given secretly to a third party, would still not allow that party to prove *in a non–transferable way* the validity/invalidity of a signature. In our setting the signer can simply give the third party the key e which is the only needed information in order to carry out successfully the denial and confirmation protocols. Clearly, the recipient of e cannot sign by itself as this is the basic assumption behind regular RSA signatures.

Distributed Provers (and signers). Distributed Provers for undeniable signatures were introduced by Pedersen [Ped91]. With distributed provers the signer can delegate the capability to confirm/deny signatures, without needing to trust a single party. This is obtained by sharing the key, used to verify signatures, using a (verifiable) secret sharing scheme among the provers. This way only if t out of the n provers cooperate it is possible to verify or deny a signature. The existing solutions for threshold RSA signatures [DDFY94, GJKR96] can then be used to obtain an efficient distributed scheme as the only operation needed during confirmation or denial protocols is RSA exponentiations. The fault-tolerance of the protocol in [GJKR96] guarantees the security of the scheme even in the presence of t (out of n) maliciously behaving provers.

As Pedersen pointed out in [Ped91], undeniable signatures with distributed provers present some difficulties. Indeed when the provers are presented with a message and its alleged signature, they have to decide which protocol (either the denial or the confirmation) to use. They can do this by first distributively checking for themselves if the claimed signature is correct or not. But this in turn means that a dishonest prover can use the other provers as an oracle to the verification key at his will. The problem applies to our schemes as well. Several ways of dealing with the problem have been suggested in the literature [Ped91, JY96] some of which easily extend to our scenario.

Also solutions for threshold RSA allow to share the power to sign (in addition to the power to verify/deny signatures) among several servers. Once again in case of possibly maliciously behaving signers a fault-tolerant scheme as [GJKR96] must be used.

Designated Verifier. The following problem of undeniable signatures has been pointed out (see [DY91, Jak94]): in general a mutually suspicious group of verifiers can get simultaneously convinced of the validity of a signature by interacting with the signer in a single execution of the confirmation protocol (in other words, the signer may believe that it is providing the signature confirmation to a single verifier while in actuality several of them are getting convinced at once). This is possible by having the "official" verifier act as the intermediary (or man in the middle) between the prover and the larger set of verifiers. While this is not always a problem, in some cases this may defeat the purpose of undeniable signatures (e.g., if the signer wants to receive payment from each verifier that gets a signature confirmation).

Jakobsson et al. [JSI96] present a solution to this problem through the notion of *designated verifiers proofs* that is readily applicable to our scheme. All that is required is for the verifier to have a public key. Then when the prover commits to his answer during the zero-knowledge steps of our protocols he will use a trapdoor commitment scheme (as in [BCC88]) which the verifier can open in any way. This will prevent the verifier from "transferring" the proof (see [JSI96] for the details).

Designated Confirmer. Designated confirmer undeniable signatures were introduced by Chaum in [Cha94] and further studied by Okamoto in [Oka94]. This variant of undeniable signature is used to provide the recipient of a signature with a guarantee that a specified third party (called a "designated confirmer") will later be able to confirm that signature. Notice the difference between this variant and the delegation property described above. Indeed in the present case the signature is specifically bound at time of generation to a particular confirmer.

The techniques of [Cha94, Oka94] easily extend to our scheme.

Acknowledgments. We would like to thank Ivan Damgård for useful suggestions.

References

[BCC88] G. Brassard, D. Chaum, and C. Crépeau. Minimum disclosure proofs of knowledge. *JCSS*, 37(2):156–189, 1988.

[BCDP91] J. Boyar, D. Chaum, I. Damgård, and T. Pedersen. Convertible undeniable signatures. In A.J. Menezes and S. A. Vanstone, editors, *Proc. CRYPTO 90*, pages 189–205. Springer-Verlag, 1991. Lecture Notes in Computer Science No. 537.

[BR96] M. Bellare and P. Rogaway. The exact security of digital signatures, how to sign with RSA and Rabin. In U. Maurer, editor, *Advances in Cryptology: EUROCRYPT'96*, volume 1070 of *Lecture Notes in Computer Science*, pages 399–416. Springer-Verlag, 1996.

[CA90] David Chaum and Hans Van Antwerpen. Undeniable signatures. In G. Brassard, editor, *Proc. CRYPTO 89*, pages 212–217. Springer-Verlag, 1990. Lecture Notes in Computer Science No. 435.

[CEG87] D. Chaum, J.-H. Evertse, and J. van der Graaf. An improved protocol for demonstrating possession of a discrete logarithm and some generalizations. In *EUROCRYPT'87*, pages 127–141, 1987.

[Cha90] D. Chaum. Zero–knowledge undeniable signatures. In *Proc. EURO-CRYPT 90*, pages 458–464. Springer-Verlag, 1990. Lecture Notes in Computer Science No. 473.

[Cha94] David Chaum. Designated confirmer signatures. In *EUROCRYPT'94*, pages 86–91, 1994.

[CP93] D. Chaum and T. Pedersen. Wallet databases with observers. In *CRYPTO'92*, pages 89–105. Springer-Verlag, 1993. Lecture Notes in Computer Science No. 740.

[CHP92] D. Chaum, E. van Heijst, and B. Pfitzmann. Cryptographically strong undeniable signatures, unconditionally secure for the signer. In J. Feigenbaum, editor, *Proc. CRYPTO 91*, pages 470–484. Springer, 1992. Lecture Notes in Computer Science No. 576.

[Dam] I. Damgård. Personal communication. November, 1996.

[DDFY94] Alfredo De Santis, Yvo Desmedt, Yair Frankel, and Moti Yung. How to share a function securely. In *Proc. 26th ACM Symp. on Theory of Computing*, pages 522–533, Santa Fe, 1994. IEEE.

[DP96] I. Damgard and T. Pedersen. New convertible undeniable signature schemes. In *Eurocrypt'96*, pages 372–386. Springer-Verlag, 1996. Lecture Notes in Computer Science No. 1070.

[DY91] Y Desmedt and M. Yung. Weaknesses of undeniable signature schemes. In *Eurocrypt'91*, pages 205–220, 1991.

[FOO91] A. Fujioka, T. Okamoto, and K. Ohta. Interactive bi-proof systems and undeniable signature schemes. In *Eurocrypt'91*, pages 243–256, 1991.

[FS86] Fiat, A. and Shamir, A. How to Prove Yourself: Practical Solutions to Identification and Signature Problems. In *Crypto'86*, pages 186–194. Springer-Verlag, 1986. Lecture Notes in Computer Science No. 263.

[GJKR96] R. Gennaro, S. Jarecki, H. Krawczyk, and T. Rabin. Robust and efficient sharing of RSA functions. In *Crypto'96*, pages 157–172. Springer-Verlag, 1996. Lecture Notes in Computer Science No. 1109. Complete version available from http://www.research.ibm.com/security/papers1997.html

[GMW86] O. Goldreich, S. Micali, and A. Wigderson. Proofs that Yield Nothing but the Validity of the Assertion, and a Methodology of Cryptographic Protocol Design. In *Proceeding 27th Annual Symposium on the Foundations of Computer Science*, pages 174–187. ACM, 1986.

[Gol95] Oded Goldreich. *Foundation of Cryptography—Fragments of a Book.* Electronic Colloquium on Computational Complexity, February 1995. Available online from *http://www.eccc.uni-trier.de/eccc/.*

[Jak94] M. Jakobsson. Blackmailing using undeniable signatures. In *EURO-CRYPT'94*, pages 425–427, 1994.

[JSI96] M. Jakobsson, K. Sako, and R. Impagliazzo. Designated verifier proofs and their applications. In U. Maurer, editor, *Advances in Cryptology: EUROCRYPT'96*, volume 1070 of *Lecture Notes in Computer Science*, pages 143–154. Springer-Verlag, 1996.

[JY96] M. Jakobsson and M. Yung. Proving without knowing: On oblivious, agnostic and blindfolded provers. In *Crypto'96*, pages 201–215. Springer-Verlag, 1996. Lecture Notes in Computer Science No. 1109.

[M96] M. Michels. Breaking and Repairing a Convertible Undeniable Signature Scheme. In Proceedings of the 1996 ACM Conference on Computer and Communications Security, 1996.

[Oka94] Tatsuaki Okamoto. Designated confirmer signatures and public-key encryption are equivalent. In Yvo G. Desmedt, editor, *Advances in Cryptology: CRYPTO '94*, volume 839 of *Lecture Notes in Computer Science*, pages 61–74. Springer-Verlag, 1994.

[Ped91] T. Pedersen. Distributed provers with applications to undeniable signatures. In *Eurocrypt'91*, pages 221–242, 1991.

Security of Blind Digital Signatures

(EXTENDED ABSTRACT)

Ari Juels[1]*   Michael Luby[2]   Rafail Ostrovsky[3]

[1] RSA Laboratories. Email: ari@rsa.com.
[2] Digital Equipment Corporation, 130 Lytton Avenue, Palo Alto, CA 94301-1044.
Email: luby@pa.dec.com.
[3] Bell Communications Research, 445 South St., MCC 1C-365B, Morristown, NJ
07960-6438, USA. Email: rafail@bellcore.com.

Abstract. Blind digital signatures were introduced by Chaum. In this
paper, we show how *security* and *blindness* properties for blind digital
signatures, can be simultaneously defined and satisfied, assuming an arbitrary one-way trapdoor permutation family. Thus, this paper presents
the first complexity-based proof of security for blind signatures.

1 Introduction

A digital signature scheme allows one to "sign" documents in such a way that
everyone can verify the validity of authentic signatures, but no one can forge
signatures of new documents. The strongest definition of security for a digital
signature scheme was put forth by Goldwasser, Micali, and Rivest [15]. Several
schemes, based on both specific and general complexity assumptions, were subsequently shown to satisfy this strongest definition. A variation on basic digital
signatures, known as *blind digital signatures*, was proposed by Chaum. Blind
digital signature schemes include the additional requirement that a signer can
"sign" a document (which is given to him in some "encrypted" form) without
knowing what the document contains. Blind digital signatures play a central role
in anonymous electronic cash applications. In this paper, we show how security
and blindness properties in digital signatures can be simultaneously defined and
satisfied, assuming an arbitrary one-way trapdoor permutation family.

While our construction achieves the strongest guarantees under general complexity assumptions and runs in polynomial time (in all the parameters), it is
quite complicated and inefficient. The contribution of this paper is therefore
twofold: (1) we show that the notions of blindness and security can be simultaneously formalized and (2) we exhibit a "constructive proof of existence" of

---

* Part of this work was done while this author was at U.C. Berkeley under NSF Grant
CCR-9505448.

a blind digital signature scheme which satisfies these strong requirements. The current paper leaves open the question of an efficient implementation. We stress, though, that it was previously not clear whether the strong security guarantees for blind digital signatures could be satisfied under *any* complexity assumptions.

We preface definitions and our main result with some background.

Digital Signatures: Informally, a signature scheme allows a user with a public key and a corresponding private key to sign a document in such a way that everyone can verify the signature of the document (using her public key) but no one else can forge the signature of another document. Digital signatures were originally defined by Diffie and Hellman [9], and the first implementation was based on the RSA trapdoor function [23]. Goldwasser, Micali, and Rivest [15] defined the strongest known "existential adaptive chosen-message attack" against digital signature schemes. They also demonstrated the first scheme which is secure against such an attack[4] assuming the existence of claw-free permutations, which in turn may be based on the hardness of factoring. Subsequently, signatures secure against existential adaptive chosen-message attacks were shown assuming the existence of trapdoor permutations [2], one-way permutations [19], and general one-way functions [24]. More efficient schemes secure against such an attack were shown in [8], and schemes with additional properties were considered in [11, 3, 16].

Blind Signatures: Chaum [6] proposed the notion of "blind digital signatures" as a key tool for constructing various anonymous electronic cash instruments. These are instruments for which the bank cannot trace where (and hence for what purpose) a user spends her electronic currency. In this paper we do not address the broad issues of electronic commerce, but concentrate our attention solely on blind signatures. Informally, a blind digital signature scheme may be thought of as an abstract game between a "user" and a "bank". The user has a secret document for which she needs to get the signature from the bank. She should be able to obtain this signature without revealing to the bank anything about her document except its length. On the other hand, the security of the signature scheme should guarantee that it is difficult for the user to forge a signature of any additional document, even after getting from the bank a number of blind signatures. Blind/untraceable signatures have attracted considerable attention in the literature (see, for example, [7, 20, 1, 22] and references therein), and are used in several proposed electronic digital cash systems. Researchers use two

---

[4] We remark that [23] is not secure against existential adaptive chosen-message attacks since there are signatures that can be forged under this attack.

different approaches for proving the security of signature schemes: complexity-based proofs of security [9, 15, 2, 19, 24, 3, 16, 8] and random-oracle model proofs of security [10, 4, 21, 22]. Let us elaborate on these two notions of security:

Two Notions of Security for Digital Signatures:

- **Complexity-based proofs:** The complexity-based approach was put forth by Diffie and Hellman [9]. They suggested that the security of a cryptographic primitive could be *reduced* to a hardness assumptions of certain fundamental problems, such as the existence of one-way functions. The approach proved very successful, as a large number of cryptographic primitives, including pseudo-random generators, signatures and secure protocols were shown to exist based on general complexity assumptions.

- **Proofs based on random oracle model:** In the case when complexity-based proofs seem to be difficult to attain, the approach used, for example in [10, 4, 21, 22], is to assume that a cryptographic primitive (such as DES or MD5) behaves like a truly random function. The security of the scheme is then shown under the assumption that the underlying primitive behaves in a near ideal fashion. Such proofs are weaker than complexity-based proofs. (For a related discussion see [5]).

Clearly, the complexity-based proofs of security are preferable to random-oracle model proofs of security. Until now, however, the only proofs of security for blind digital signature schemes have been in the random oracle model. This paper presents the first blind signature scheme with complexity-based proof of security.

Pointcheval and Stern [22] address the security of several blind digital signatures schemes, including blind variants of the Okamoto [20], Schnorr [25], and Guillou-Quisquater [17] signature schemes. In particular, [22] proves the security of Okamoto-Schnorr and Okamoto-Guillou-Quisquater blind signatures in the random oracle model. Thus, Pointcheval and Stern consider blind signatures which rely on number-theoretic assumptions and show proofs of security only in the random-oracle model. In addition, their security proofs, while polynomial in the size of the cryptographic keys, are exponential in the number of blind digital signatures obtained before the break (i.e. if the number of signatures that are required before the break is greater than logarithmic, then the reduction is not polynomial.) The authors pose as an open problem the question of whether one can achieve a scheme where the security of the reduction can be made polynomial both in the number of signatures obtained by the adversary before the break and in the size of the keys.

Our Result: In the next section, we formally define the notion of security of a blind digital signature scheme. Informally, a blind digital signature scheme is secure if it satisfies both a *blindness* and a *non-forgeability* property. The blindness property was formulated in the original paper of Chaum [6], and non-forgeability was considered in the paper of Pointeval and Stern [22] (where it is called called "one more" forgery). Again, informally, (see the next section for formal definitions) blindness means that a signer can not distinguish, except with negligible probability, the order in which she issued signatures, and non-forgeability means that after getting ℓ signatures, it is infeasible for the receiver to compute $\ell + 1$ signatures. We consider a non-forgeability requirement where the forger is allowed to run many parallel protocol executions for many blind signatures, in an arbitrarily interleaved and adaptive fashion, and to abort many such executions in the middle of the protocol, without having to count them as signatures. We call such an attack an *adaptive interleaved chosen-message attack*. We demonstrate a blind digital signature scheme which is secure against this attack, and which can be implemented based on any one-way trapdoor permutation.

MAIN THEOREM: *Assume that one-way trapdoor permutations exist. Then there exists polynomial-time blind digital signature scheme, secure against an adaptive interleaved chosen-message attack.*

Our scheme has both advantages and disadvantages. We list them below.

Advantages:

- We give the first complexity-theoretic proof of security for blind digital signatures; our scheme is shown to be secure against the adaptive interleaved chosen-message attack. (All previous proofs of security for blind digital signatures were in the random-oracle model only and were not fully polynomial.)
- We show how to achieve our protocol based on any one-way trapdoor permutation. (All previous blind digital signatures schemes were based on number-theoretic assumptions only).
- Our scheme and proof of security are fully polynomial in all suitable parameters, including the number of blind signatures requested before the break. (We thus resolve in the affirmative the open question posed by Pointcheval and Stern [22].)

Disadvantages:

- Our scheme, while polynomial in all suitable parameters, is inefficient. Thus, it should be viewed merely as a proof of existence which should pave the way for efficient future implementations.

Organization of the Paper: The remainder of this paper is organized as follows. In section 2, we present the definitions of blindness and security to be used in this paper. We discuss some of the complications and solutions involved in constructing a blind signature scheme in section 3. We present our blind signature scheme in section 4 and sketch a proof of its security in section 5. We conclude in section 6 with a brief discussion of the significance of our result to the area of anonymous electronic cash.

2 Definitions

In the proof and the construction of blind digital signatures, we will use the security of standard digital signatures, as defined by Goldwasser, Micali, and Rivest [15]. Hence, before we give the definition of blind digital signatures, we remind the reader of the standard signature definitions.

Signature schemes: The standard signature scheme is a triple of algorithms, $(Gen, Sign, Verify)$, where $Gen(1^k)$ is a probabilistic polynomial time key-generation algorithm, which takes as an input a security parameter 1^k and outputs a pair (pk, sk) of public and secret keys. The signing algorithm $Sign(pk, sk, m)$ is a probabilistic polynomial time algorithm which takes as an input a public key pk a secret key sk a message m to be signed and outputs a signature of a message $\sigma(m)$ as well as a new (i.e., updated) secret key sk'. (In a *memoryless* signature scheme, the secret key sk stays the same throughout.) A verification algorithm $Verify(pk, m, \sigma(m))$ is a deterministic polynomial time algorithm which takes as an input a public key pk a message m and a purported signature $\sigma(m)$ and outputs *accept/reject*. We require, of course, that for all signatures computed by first executing a key generation algorithm and then signing a sequence of messages according to the above process, the verification algorithm always output *accept*.

As mentioned above, security against the existential adaptive chosen-message attack of Goldwasser, Micali, and Rivest is the strongest known security measure for signatures [15].

Security of Signature Schemes: In this attack, an adversary A, which is a probabilistic polynomial-time machine, is given a public key pk generated by the key-generation algorithm. The adversary A can request in an adaptive fashion a polynomial number of signatures of his choice. A must then produce a valid signature on a document for which he has not yet seen a signature. If he can produce any such document/signature pair which is accepted by the verification algorithm, then the attack is successful. A signature scheme is defined to be *secure* if for all constants c, and for all probabilistic polynomial-time A, there exists a security parameter $k_{c,A}$

such that for all $k > k_{c,A}$ the probability (taken over coin-flips of the adversary) that A is successful is less then $1/k^c$.

We shall use the term *polynomially bounded* in this paper to refer to a quantity which is polynomial in the security parameter. Similarly, we shall denote by $1/poly$ the inverse of a polynomially bounded quantity.

We are now ready to give a formal definition of a blind signature scheme and its security. In the definition below, digital signatures are treated as an interactive protocols between two players: a *Signer* (who "blindly" signs a document m) and the *User* (who obtains the signature of her document m). We rely on the formalism of Interactive Turing machines, defined by Goldwasser, Micali and Rackoff [13]. The security of a blind digital signature consists of two requirements: the **blindness** property and the **non-forgeability** of additional signatures. We say the blind digital signature scheme is *secure* if it satisfies both properties, as defined below. (We remark that our non-forgeability definition follows the definition of "one-more" forgery by Pointcheval and Stern [22])

Blind Digital Signatures: A blind digital signature scheme is a four-tuple, consisting of two Interactive Turing machines *(Signer, User)* and two algorithms *(Gen, Verify)*. $Gen(1^k)$ is a probabilistic polynomial time key-generation algorithm which takes as an input a security parameter 1^k and outputs a pair (pk, sk) of public and secret keys. The *Signer(pk,sk)* and *User(pk,m)* are a pair of polynomially-bounded probabilistic Interactive Turing machines, where both machines have the following (separate) tapes: read-only input tape, write-only output tape, a read/write work tape, a read-only random tape, and two communication tapes, a read-only and a write-only tape. They are both given (on their input tapes) as a common input a pk produced by a key generation algorithm. Additionally, the *Signer* is given on her input tape a corresponding secret key sk and the *User* is given on her input tape a message m, where the length of all inputs must be polynomial in the security parameter 1^k of the key generation algorithm. The User and Signer engage in the interactive protocol of some polynomial (in the security parameter) number of rounds. At the end of this protocol the Signer outputs either *completed* or *not-completed* and the User outputs either *fail* or $\sigma(m)$. The *Verify(pk, m, $\sigma(m)$)* is a deterministic polynomial-time algorithm, which outputs *accept/reject* with the requirement that for any message m, and for all random choices of key generation algorithm, if both Signer and User follow the protocol then the Signer always outputs *completed*, and the output of the user is always *accepted* by the verification algorithm.

We now describe the security of blind signatures.

The Security of Blind Digital Signature: a blind digital signature scheme is *secure* if for all constants c, and for all probabilistic polynomial-time algorithms A, there exists a security parameter $k_{c,A}$ such that for all $k > k_{c,A}$ the following two considerations hold:

- **Blindness property**: Let $b \in_R \{0,1\}$ (i.e. b is a random bit which is kept secret from A). A executes the following experiment (where A controls the "signer", but not the "user", and tries to predict b):
  - **(Step 1)**: $(pk, sk) \leftarrow Gen(1^k)$
  - **(Step 2)**: $\{m_0, m_1\} \leftarrow A(1^k, pk, sk)$ (i.e. A produces two documents, polynomial in 1^k, where $\{m_0, m_1\}$ are by convention lexicographically ordered and may even depend on pk and sk).
  - **(Step 3)**: We denote by $\{m_b, m_{1-b}\}$ the same two documents $\{m_0, m_1\}$, ordered according to the value of bit b, where the value of b is hidden from A. $A(1^k, pk, sk, m_0, m_1)$ engages in two parallel (and arbitrarily interleaved) interactive protocols, the first with $User(pk, m_b)$ and the second with $User(pk, m_{1-b})$.
  - **(Step 4)**: If the first User outputs on her private tape $\sigma(m_b)$ (i.e. does not output *fail*) and the second user outputs on her private tape $\sigma(m_{1-b})$ (i.e., also does not output *fail*) then A is given as an additional input $\{\sigma(m_b), \sigma(m_{1-b})\}$ ordered according to the corresponding (m_0, m_1) order. (We remark that we do not insist that this happens, and either one or both users may output *fail*)
  - **(Step 5)**: A outputs a bit $\tilde{b}$ (given her view of steps 1 through 3, and if conditions are satisfied of step 4 as well).

  Then the probability, taken over the choice of b, over coin-flips of key-generation algorithm, the coin-flips of A, and (private) coin-flips of both users (from step 3), that $\tilde{b} = b$ is at most $\frac{1}{2} + \frac{1}{k^c}$.

- **Non-forgeability property**: A executes the following experiment (where A controls the "user", but not the "signer", and tries to get "one-more" signature):
  - **(Step 1)**: $(pk, sk) \leftarrow Gen(1^k)$
  - **(Step 2)**: $A(pk)$ engages in polynomially many (in k) adaptive, parallel and arbitrarily interleaved interactive protocols with polynomially many copies of $Signer(pk, sk)$, where A decides in an adaptive fashion when to stop. Let ℓ denote the number of executions, where the Signer outputted *completed* in the end of Step 2.
  - **(Step 3)**: A outputs a collection $\{(m_1, \sigma(m_1)), \ldots (m_j, \sigma(m_j))\}$ subject to the constraint the all $(m_i, \sigma(m_i))$ for $1 \leq i \leq j$ are all *accepted* by $Verify(pk, m_i, \sigma(m_i))$.

  Then the probability, taken over coin-flips of key-generation algorithm, the coin-flips of A, and over the (private) coin-flips of the Signer, that $j > \ell$ is at most $\frac{1}{k^c}$.

Remarks on Blindness Property:

- We stress that we do not require the adversary to follow the signing protocol, nor do we require the protocol to terminate with the valid signature. Moreover, we require that the probability bound holds even if the protocol is aborted in the middle of execution.
- By standard hybrid arguments, the above definition is as general as the definition in which polynomially many signatures are obtained and then recalled, leaving A to distinguish between the last two signatures.
- Finally, we note that since the User does not have any special ID or other special identification (or else embeds such information in the message to be signed), we restrict our view to a single user program.

3   Towards Our Scheme

As mentioned in the introduction, our scheme is somewhat complicated. Instead of presenting it immediately, we shall offer a sequence of refinements which in the end yields a correct scheme. Our aim is twofold: (1) to explain why the complications in our the scheme are necessary and (2) to elaborate on the subtleties of the problem, even when using general completeness results.

Basic Ingredients: The two basic ingredients we start with are the secure signature scheme of Naor-Yung [19], and the two-party completeness theorem of Yao and Goldreich, Micali and Widgerson [26, 14]. Let us briefly recall both ingredients.

- The signature scheme of Naor-Yung is secure against existential adaptive chosen-message attack and can be built based on any one-way permutation f [19] (we remark that we do not need the result of [24] which is based on weaker assumptions since other tools in our protocol require one-way permutations anyway.)

- The two-party completeness theorem of Yao and Goldreich, Micali and Wigderson [26, 14] basically says that for any two parties A, and B, where A is given a secret input x and B is given a secret input y, and for any polynomial-time computable function $g(\cdot, \cdot)$ there exists a protocol for computing $g(x, y)$ such that nothing except the output of the function is revealed to the players. Moreover, the schemes could be easily extended to require that only one player learns $g(x, y)$, while for the other player learns nothing (i.e. all interactions are computationally indistinguishable.) Furthermore, the value of $g(x, y)$ can be learned by one of the players only as the last message of the protocol, with the condition that if the protocol is aborted before this last message, then again no information is revealed (i.e. all interactions are computationally indistinguishable.) In fact, we use a stronger definition, used by [26, 14]: that there exists a polynomial-time *simulator* which can simulate the views of the players, even in the case of Byzantine (i.e. malicious) faults. (For details see the above references.) Furthermore, the two-party protocol can be augmented to leave part of the input of one of the players unspecified, and allow this player to set this value in an arbitrary fashion during the actual protocol execution.

A first simple idea would be to use these two general theorems in order to construct blind signatures in the following way: instead of having the User request that the Signer sign the message in the clear, engage in the two-party private protocol, at the end of which the User learns the signature of the document, and the Signer learns nothing. This "solution" suffers from several problems, which we now elaborate upon.

Problem 1: The scheme of Naor-Yung is not "memoryless", and future signatures reveal previous signatures, which violates the blindness property.

Solution to Problem 1: Goldreich [11] showed how to make any signature scheme (including the signature scheme of Naor and Yung) "memoryless" [11], using pseudo-random functions of [12]. In our setting, the key-generation algorithm can add to the secret key a seed s for pseudo-random function and add to a public key a *commitment* [18] of this seed. Then, during secure two-party computation, the Signer must generate all of her random choices (and a random tree of [19, 11]) using an agreed-upon pseudo-random function with the committed seed.

Problem 2: Let us take a closer look at the proof of security of Naor-Yung scheme [19]. Their scheme takes as its basis a tree; messages are inserted in the leaves of this tree, and a signature involves the construction of a path from the root of the tree to the appropriate leaf. Naor and Yung show that if there exists a Forger that can replace the User and forge the signature of a new document, then this Forger can be used as a subroutine to invert a one-way permutation on a random input in this tree. The key idea of their proof is to replace the Signer with an Inverter which is able to set a "trap" in this tree as follows: in order to forge a signature, the Forger must diverge from the path of previous signatures in the tree (see, for example, [19, 11, 8]), and if the Inverter can guess where in the path this divergence takes place (which she can do with $1/poly$ probability) then it can place an output of a one-way permutation at this point and force the forger to invert. The problem is that for this proof to work, the Inverter must know all the previous signatures, in order to know where to set a "trap". But the knowledge of previous signatures on the part of the Signer is exactly what blind signatures are trying to prevent! These would seem to be contradictory requirements.

Solution to Problem 2: Since the Inverter is deployed in a simulation of the signature process, the Inverter is allowed to "reset" the Forger. So how can we assure that the Signer (who can not "reset") does not know which documents she signs while the Inverter (which is allowed to "reset") has full information? The idea is to use a variant of a proof of knowledge procedure. The User first commits to a random string r and to her message exclusive-ored with r. The Signer requests to see the decommitment of either one or the other commitment (but not both). The Inverter will be able to retrieve the message by first requesting to see one commitment, then resetting the state of the Forger, and then requesting to see the other commitment. We call such a commitment an *extractable* commitment.

We should point out that since both commitments (and their decommitments) are now part of the input (public and private) of the secure two-party completeness protocol, they are included in the execution of the two-party completeness protocol and hence force correct behavior of both players (see [26, 14]).

Problem 3: In the scheme of [11] for rendering the signature scheme memoryless, it was not necessary for the Signer to prove that she is only using coin-flips that come from a pseudo-random function. In order to achieve the blindness property, however, we must insist that this is always the case. (This is done through use of the completeness theorem in conjunction with a published commitment of the pseudo-random seed S, as we shall see.) The memoryless property of the signature guarantees that the Signer can not "mark" the signatures in any way, an absolutely necessary property for blind signatures! In the proof of security, though – i.e., when dealing with a forger – the Inverter must be able to replace a pseudo-random string by a "trap". This trap is a completely random input (on which the forger will invert with $1/poly$ probability). Again, these would seem to be contradictory requirements, since if the Signer can insert new random bits into the singing process, then it can "mark" the signature and violate blindness property.

Solution to Problem 3: Again, the ability of the Inverter to "reset" the Forger is vital to the resolution of the above somewhat paradoxical issue. The idea is again to have the Signer commit (in an *extractable* form – see above) to some poly-long string X. The Signer picks a secret input Y of the same length as X; both X and Y are used as private inputs for the secure protocol guaranteed by the two-party completeness theorem. We modify our secure function evaluation protocol to allow the Signer to deviate from the above pseudo-random choices and insert other inputs, but only in case when $X = Y$. If $X \neq Y$ we demand that the Signer follow the protocol as before. The chances that the Signer can correctly "guess" X are negligible, so the signature scheme remains blind with overwhelming probability. On the other hand, the Inverter, by resetting the Forger, can find out what X is, set her guess Y to the same value, and then set a trap.

Problem 4: Since the definition of the non-forgeability property allows the Inverter to run many parallel sessions interleaved in an arbitrary fashion, we must be assured that it can insert a "trap" (on which the Forger will invert during forging of a "one-more signature") in a consistent manner in all the runs. The Inverter must therefore be able to specify a point in the exposed sub-tree of signatures (see [19, 11]) at which to insert her trap. But how can this be consistently specified, not knowing the order or the interleaving nature of the adversary?

Solution to Problem 4: The solution is as follows: if $X = Y$ the Signer/ Inverter can insert arbitrary values at an arbitrary point (i.e. it does not commit where to insert the trap) and thus can consistently do so during parallel interleaving sessions in the same fashion as before, i.e. consistently at some point in an exposed sub-tree of signatures (see [19, 11]) We now give details how this can be done.

Recall that we use a secure computation protocol in such a way that the User/Forger receives no information about the signature prior to the last round from the signer. We refer to this as the *atomic signature* property. Recall that the Forger may request at most a polynomial number of signatures, say $p(k)$, before producing her forgery. The Inverter therefore chooses a number r uniformly at random from $[1, p(k)]$. This specifies the interaction with the Forger in which she will try to plant her trap. The Inverter also chooses a height a of a tree uniformly at random at which to plant her trap. The Inverter specifies in interaction r that trap w will be planted at height a. Once the message m in interaction r has been specified, the Inverter may determine the node v in which she has chosen to plant her trap. With probability $1/poly$, the Inverter will have chosen to plant her trap in such a way that no previously issued signature has yet made use of the node v; thus planting of the trap will not invalidate signatures issued previous to interaction r. We say in this case that the trap choice has been successful: the Inverter plants her trap with impunity. On the other hand, if the Inverter has chosen an address for her trap such that previous signatures would be invalidated, then we say that the trap choice has been unsuccessful.

In this case, the Inverter does not plant the trap in node v. By the atomic signature property, no information about signatures has been divulged to the User/Forger in any other interaction. Therefore, the Inverter may continue to plant her trap in node v in a consistent fashion for all incomplete interactions. Since the simulation is successful with probability $1/poly$, a trap is planted as in Naor and Yung's scheme with probability $1/poly$. It follows that the Inverter causes the Forger to invert w with $1/poly$ probability.

4 The Blind Signature Scheme

We shall now assemble all of the above and describe our blind signature scheme. We shall denote by $c(z)$ the secure commitment of a string z. We shall denote by $c^*(z)$ an extractable commitment of z. (Recall from above that such a commitment reveals nothing about z to the Signer, but enables an Inverter, by rewinding a Forger, to extract z.)

The scheme works as follows. The Signer publishes $c(s)$, that is, a commitment of her secret pseudo-random key s, along with her public key pk, and the one-way permutation f used in the Naor and Yung [19] scheme. (Also made public are the pseudo-random generation function g, as well as a set of public hash functions required by the scheme of Naor and Yung.)

Each time a signature is to be issued, the Signer and User engage in a secure two-party computation. The User provides as input to the computation the message m to be signed, as well as a random string X. In addition, the User provides extractable commitments $c^*(X)$ and $c^*(m)$. Through a variation on the standard secure two-party computation protocol, these two commitments are passed in the clear to the Signer. (Recall that in the Inverter/Forger scenario, these commitments enable the Inverter, by rewinding the Forger, to learn X and m, thereby effectively circumventing the blindness of the scheme.)

The Signer provides to the secure computation (of [26, 14]) her private information as well as information respecting the trap she may wish to plant (when she plays the role of the Inverter). In particular, the Signer provides to the computation her secret signing key sk and her secret pseudorandom seed s. She also provides a string Y constituting her guess of X. Finally, the Signer provides to the computation a specification of the trap she wishes to have inserted. More precisely, the Signer specifies w, the value she wishes to have planted in the signature tree, and either a node v in a tree where she wishes to put w (in case v is already known from other sessions) or a boolean value indicating that in the current signature, on its way to the leaf, at height a in the tree at which trap w should be inserted.

The memoryless property [11] is incorporated into our our scheme as follows. The secure two-party computation protocol produces a choice of leaf in which to insert the message m; this is computed to be the output of the pseudo-random generation function g of [12] with secret seed s and index m (truncated appropriately to yield a uniform selection of leaves). If the Signer's guess Y is successful, i.e., if $Y = X$, then the signer can deviate from $g_s(m)$ path and insert instead w at a node v as specified above. (If the current signature does not use v, no trap is planted and $g_s(m)$ is followed.) On successful completion of the protocol (i.e., if cheating during secure computation was not detected) the decodinbg of signature $\sigma(m)$ (with or without trap) is sent to the User.

5 Security of our scheme

The blindness of the scheme follows from the properties of two-party secure computation of [26, 14]. The security of the computation is violated only when the guess Y of the Signer is correct, and consequently $X = Y$. This happens with negligible probability.

It now remains to be seen that if there exists a successful Forger for this scheme, then this Forger may be used by the Inverter in a polynomial-time algorithm Q capable of inverting the one-way permutation f on an arbitrary value with probability $1/poly$.

Since the Forger makes extractable commitments of X and m, the Forger can be used by the Inverter to rewind the protocol and extract X and m. By setting $X = Y$ (which is indistinguishable for any polynomial-time Forger from the case $X \neq Y$), the Inverter can now plant a trap in a consistent manner.

When signatures are issued sequentially, therefore, by making use of its knowledge of the history of issued signatures, the Inverter may set a "trap" in exactly the way that this was done in a memoryless analog of Naor and Yung's scheme. The ability of algorithm Q to invert f now follows from the security of the memoryless version of Naor and Yung's memoryless analog [11, 19].

When signatures are issued over the course of multiple, interleaved executions of the blind signature protocol, the same "trap" may be planted consistently over many executions using the method described in Section 3 (in response to Problem 4). Thus, the Inverter remains capable of inverting with probability $1/poly$ even over interleaved protocol executions.

6 Conclusion: Anonymous Electronic Cash

As mentioned in the introduction to this paper, blind digital signatures are principally of interest to the cryptographic community for their importance in anonymous electronic cash schemes. In many of these schemes, a coin consists of a pair $(d, \sigma(d))$, where d is selected from a suitable message space, and $\sigma(d)$ represents a blind signature of d or of a digest of d.

An early example of an electronic cash scheme of this sort is a paper by Chaum, Naor, and Fiat [7]. (Their system has in fact been deployed with some additional apparatus in a real-world implementation.) Here a coin assumes the form $(d, f^{1/3}(d))$, where f is a suitable hash function, such as MD5. Computations are performed in $\mathcal{Z}_N$ for some product of primes $N = pq$, where N is published, and p and q are held in secret by the Signer (the bank). A coin is issued as follows. The User generates the value d and a random blindness factor r, and sends the quantity $r^3 f(d)$ to the Signer. The Signer computes $r f^{1/3}(d)$, and sends it to the User. On dividing out r from this last quantity, the User computes $f^{1/3}(d)$, and has therefore obtained a valid coin pair $(d, f^{1/3}(d))$. It is easy to see that the described scheme is blind. (It is, in fact, blind in an information theoretic sense.) The scheme would also appear at first glance to be secure since, given d, it is hard to compute $f^{1/3}(d)$, and vice versa. We wish to point out, however, that this rationale does not give a proof of security, and is in fact deceptive: there might nonetheless be some computationally feasible way of generating the pair of values constituting the coin *simultaneously*.

This and similar weaknesses appear to vex many implementations of anonymous digital cash. Although a proof of security of several blind digital signature schemes based on the random oracle model was given by Pointcheval and Stern [22], the current paper gives the first complexity-based proof for this important primitive. We have therefore shown that secure anonymous digital cash is possible to achieve in a complexity-based sense, i.e. we have shown that it may be as secure as, say, factoring. As mentioned above, however, our protocol is inefficient. Combining the requirements of efficiency and provable security to create a new blind digital signature scheme is an interesting open problem.

References

1. S.A. Brands. Untraceable Off-line Electronic Cash Based on Secret-key Certificates. Latin 95.
2. M. Bellare and S. Micali. "How to Sign Given Any Trapdoor Function". *STOC 88.*
3. M. Bellare and S. Goldwasser. "New Paradigms for Digital Signatures and Message Authentication Based on Non-Interactive Zero Knowledge Proofs". Crypto 89.
4. M. Bellare and P. Rogaway. "The Exact Security of Digital Signatures – How to Sign with RSA and Rabin". Eurocrypt-96.
5. R. Canetti "De-mystifying Random Oracles" CRYPTO-97 (this proceedings).
6. D. Chaum. "Blind Signatures for Untraceable Payments". Crypto-82.
7. D. Chaum, A. Fiat, and M. Naor. "Untraceable Electronic Cash", Crypto-89.
8. C. Dwork and M. Naor. "An Efficient Existentially Unforgeable Signature Scheme and its Applications". Crypto 94.

9. W. Diffie and M. Hellman. "New Directions in Cryptography". *IEEE Trans. on Inf. Theory*, IT-22, pp. 644–654, 1976.

10. A. Fiat and A. Shamir. "How to Prove Yourself: Practical Solutions of Identification and Signature Problems, CRYPTO 86.

11. O. Goldreich. "Two Remarks Concerning the GMR Signature Scheme" MIT Tech. Report 715, 1986. CRYPTO 86.

12. O. Goldreich, S. Goldwasser, and S. Micali. "How to Construct Random Functions". *JASM* V. 33 No 4. (October 1986) pp. 792-807.

13. S. Goldwasser, S. Micali and C. Rackoff, "The Knowledge Complexity of Interactive Proof-Systems". *SIAM J. Comput.* 18 (1989), pp. 186-208; (also in STOC 85, pp. 291-304.)

14. O. Goldreich, S. Micali, and A. Wigderson. "How to Play Any Mental Game". Proc. of 19th STOC, pp. 218-229, 1987.

15. S. Goldwasser, S. Micali, and R. Rivest. "A Digital Signature Scheme Secure Against Adaptive Chosen-Message Attacks". *SIAM Journal of Computing* Vol. 17, No 2, (April 1988), pp. 281-308.

16. Goldwasser S., and R. Ostrovsky "Invariant Signatures and Non-Interactive Zero-Knowledge Proofs are Equivalent" CRYPTO 92.

17. L.C. Guillou and J.J. Quisquater. "A Practical Zero-Knowledge Protocol Fitter to Security Microprocessor Minimizing Both Transmission and Memory". EUROCRYPT 88.

18. M. Naor. "Bit Commitment Using Pseudo-Randomness". Crypto-89.

19. M. Naor and M. Yung. "Universal One-Way Hash Functions and their Cryptographic Applications". STOC 89.

20. T. Okamoto. "Provably Secure and Practical Identification Schemes and Corresponding Signature Schemes" CRYPTO 92.

21. D. Pointcheval and J. Stern. "Security Proofs for Signature Schemes". Eurocrypt 96.

22. D. Pointcheval and J. Stern. "Provably Secure Blind Signature Schemes". Asiacrypt 96.

23. R.L. Rivest, A. Shamir, and L. Adleman. "A Method for Obtaining Digital Signatures and Public Key Cryptosystems". Comm. ACM, Vol 21, No 2, 1978.

24. J. Rompel. "One-way Functions are Necessary and Sufficient for Secure Signatures". STOC 90.

25. C.P. Schnorr. "Efficient Identification and Signatures for Smart Cards". CRYPTO 89.

26. A. C. Yao. "How to Generate and Exchange Secrets". Proc. of 27th FOCS, 1986, pp. 162-167.

Digital Signcryption or How to Achieve Cost(Signature & Encryption) << Cost(Signature) + Cost(Encryption) *

Yuliang Zheng

Monash University, McMahons Road, Frankston, Melbourne, VIC 3199, Australia
Email: yuliang@mars.fcit.monash.edu.au

Abstract. Secure and authenticated message delivery/storage is one of the major aims of computer and communication security research. The current standard method to achieve this aim is "(digital) signature followed by encryption". In this paper, we address a question on the cost of secure and authenticated message delivery/storage, namely, *whether it is possible to transport/store messages of varying length in a secure and authenticated way with an expense less than that required by "signature followed by encryption"*. This question seems to have never been addressed in the literature since the invention of public key cryptography. We then present a positive answer to the question. In particular, we discover a new cryptographic primitive termed as "signcryption" which *simultaneously* fulfills both the functions of digital signature and public key encryption in a logically single step, and with a cost *significantly* lower than that required by "signature followed by encryption". For typical security parameters for high level security applications (size of public moduli = 1536 bits), signcryption costs 50% (31%, respectively) less in computation time and 85% (91%, respectively) less in message expansion than does "signature followed by encryption" based on the discrete logarithm problem (factorization problem, respectively).

Keywords

Authentication, Digital Signature, Encryption, Key Distribution, Secure Message Delivery/Storage, Public Key Cryptography, Security, Signcryption.

1 Introduction

To avoid forgery and ensure confidentiality of the contents of a letter, for centuries it has been a common practice for the originator of the letter to sign his/her name on it and then seal it in an envelope, before handing it over to a deliverer.

---

* Patent pending (PO3234/96, filed on October 25, 1996). The full version of this paper can be obtained from http://www-pscit.fcit.monash.edu.au/~yuliang/

Public key cryptography discovered nearly two decades ago [7] has revolutionized the way for people to conduct secure and authenticated communications. It is now possible for people who have never met before to communicate with one the computational cost by counting the number of dominant operations involved. Typically these operations include private key encryption and decryption, hashing, modulo addition, multiplication, division (inversion), and more importantly, exponentiation. In addition to computational cost, digital signature and encryption based on public key cryptography also require extra bits to be appended to a message. We call these extra "redundant" bits the *communication overhead* involved. We say that a message delivery method is superior to another if (the aggregated value of) the computational cost and communication overhead required by the former is less than that by the latter.

The first part of Table 2 indicates the computational cost and communication overhead of "Schnorr signature-then-ElGamal encryption" against that of "DSS-then-ElGamal encryption" and "RSA signature-then-RSA encryption". Note that, although not shown in the table, other combinations such as "Schnorr signature-then-RSA encryption" and "RSA signature-then-ElGamal encryption" may also be used in practice. As discussed in [21], with the current state of the art, computing discrete logarithm on $GF(p)$ and factoring a composite n of the same size are equally difficult. This simplifies our comparison of the efficiency of a cryptographic scheme based on RSA against that based on discrete logarithm, as we can assume that the moduli n and p are of the same size.

We close this section by examining the increasingly disproportionate cost for secure and authenticated message delivery in the currently standard signature-then-encryption approach, with an example text of 10000 bits (which corresponds roughly to a 15-line email message). For current and low security level applications, when RSA is used, the computational cost is centered around the execution of four (4) exponentiations modulo 512-bit integers, and the communication overhead is 1024 bits. When Schnorr signature and ElGamal encryption are used, the computational cost consists mainly of six (6) exponentiations modulo 512-bit integers, and the communication overhead is about 750 bits.

However, if the contents of the text are highly sensitive, or a text of the same length will be transmitted in 2010, then very large moduli, say of 5120 bits, might have to be employed. In such a situation, if RSA is used, four (4) exponentiations modulo (very large!) 5120-bit integers have to be invested in computation [2], and the communication overhead is 10240 bits, which is now longer than the original 10000-bit text ! If, instead, Schnorr signature and ElGamal encryption is used, then the computational cost is six (6) exponentiations modulo (again very large!) 5120-bit integers, and the communication overhead of about 5560 bits is more than half of the length of the original message. From this example, one can see that in the signature-then-encryption approach, the cost, especially communication overhead, for secure and authenticated message delivery, is becoming disproportionately large for future, or current but high-level security, applications. This observation serves as further justification on the necessity of inventing a new and more economical method for secure and authenticated message delivery.

---

[2] The number of bit operations required by exponentiation modulo an integer is a cubic function of the size of the modulo.

3 Digital Signcryption — A More Economical Approach

Over the past two decades since public key cryptography was invented, signature-then-encryption has been a standard method for one to deliver a secure and authenticated message of arbitrary length, and no one seems to have ever questioned whether it is absolutely necessary for one to use the sum of the cost for signature and the cost for encryption to achieve both contents confidentiality and origin authenticity.

Having posed a question that is of fundamental importance both from a theoretical and a practical point of view, we now proceed to tackle it. We will show how the question can be answered positively by the use of a new cryptographic primitive called "signcryption" whose definition follows.

Intuitively, a digital *signcryption* scheme is a cryptographic method that fulfills both the functions of secure encryption and digital signature, but *with a cost smaller than that required by signature-then-encryption*. Using the (informal) terminology in cryptography, it consists of a pair of (polynomial time) algorithms (S, U), where S is called the *signcryption algorithm*, while U the *unsigncryption algorithm*. S in general is probabilistic, but U is most likely to be deterministic. (S, U) satisfy the following conditions:

1. *Unique unsigncryptability* — Given a message m, the algorithm S *signcrypts* m and outputs a *signcrypted text* c. On input c, the algorithm U *unsigncrypts* c and recovers the original message un-ambiguously.
2. *Security* — (S, U) fulfill, simultaneously, the properties of a secure encryption scheme and those of a secure digital signature scheme. These properties mainly include: confidentiality of message contents, unforgeability, and non-repudiation.
3. *Efficiency* — The computational cost, which includes the computational time involved both in signcryption and unsigncryption, and the communication overhead or added redundant bits, of the scheme is *smaller* than that required by the best currently known signature-then-encryption scheme with comparable parameters.

A direct consequence of having to satisfy both the second and third requirements is that "signcryption $\neq$ signature-then-encryption". These two requirements also justify our decision to introduce the new word *signcryption* which clearly indicates the ability for the new approach to achieve both the functions of digital signature and secure encryption in a logically single operation.

The rest of this section is devoted to seeking for concrete implementations of signcryption. We first identify two (types of) efficient ElGamal-based signature schemes. Then we show how to use a common property of these schemes to construct signcryption schemes.

3.1 Shortening ElGamal-Based Signatures

ElGamal digital signature scheme [9] involves two parameters public to all users: (1) p — a large prime, and (2) g — an integer in $[1, \ldots, p-1]$ with order

$p-1$ modulo p. User Alice's secret key is an integer x_a chosen randomly from $[1,\ldots,p-1]$ with $x_a \nmid (p-1)$ (i.e., x_a does not divide $p-1$), and her public key is $y_a = g^{x_a} \bmod p$.

Alice's signature on a message m is composed of two numbers r and s:

$$r = g^x \bmod p$$
$$s = (hash(m) - x_a \cdot r)/x \bmod (p-1)$$

where $hash$ is a one-way hash function, and x is chosen independently at random from $[1,\ldots,p-1]$ with $x \nmid (p-1)$ every time a message is to be signed by Alice. Given (m,r,s), one can verify whether (r,s) is Alice's signature on m by checking whether $g^{hash(m)} = y_a^r \cdot r^s \bmod p$ is satisfied.

Since its publication in 1985, ElGamal signature has received extensive scrutiny by the research community. In addition, it has been generalized and adapted to numerous different forms (see for instance [23, 4, 18, 20] and especially [11] where an exhaustive survey of some 13000 ElGamal based signatures has been carried out.) Two notable variants of ElGamal signature are Schnorr signature [23] and DSS or Digital Signature Standard [18]. With DSS, g is an integer in $[1,\ldots,p-1]$ with order q modulo p, where q is a large prime factor of $p-1$. Alice's signature on a message m is composed of two numbers r and s which are defined as

$$r = (g^x \bmod p) \bmod q$$
$$s = (hash(m) + x_a \cdot r)/x \bmod q$$

where x is a random number chosen from $[1,\ldots,q-1]$. Given (m,r,s), one accepts (r,s) as Alice's valid signature on m if $(g^{hash(m)/s} \cdot y_a^{r/s} \bmod p) \bmod q = r$ is satisfied.

For most ElGamal based schemes, the size of the signature (r,s) on a message is $2|p|$, $|q|+|p|$ or $2|q|$, where p is a large prime and q is a prime factor of $p-1$. The size of an ElGamal based signature, however, can be reduced by using a modified "seventh generalization" method discussed in [11]. In particular, we can change the calculations of r and s as follows:

1. Calculation of r — Set $r = hash(k,m)$, where $k = g^x \bmod q$ ($k = g^x \bmod (p-1)$ if the original r is calculated modulo $(p-1)$), x is a random number from $[1,\ldots,q]$ (or from $[1,\ldots,p-1]$ with $x \nmid (p-1)$), and $hash$ is a one-way hash function such as Secure Hash Standard or SHS [19].
2. Calculation of s — For an *efficient* ElGamal based signature scheme, the calculation of (the original) s from x_a, x, r and optionally, $hash(m)$ involves only simple arithmetic operations, including modulo addition, subtraction, multiplication and division. Here we assume that x_a is the secret key of Alice the message originator. Her matching public key is $y_a = g^{x_a} \bmod p$. We can modify the calculation of s in the following way:
   (a) If $hash(m)$ is involved in the original s, we replace $hash(m)$ with a number 1, but leave r intact. The other way may also be used, namely we change r to 1 and then replace $hash(m)$ with r.

(b) If s has the form of $s = (\cdots)/x$, then changing it to $s = x/(\cdots)$ does not add additional computational cost to signature generation, but may reduce the cost for signature verification.

To verify whether (r, s) is Alice's signature on m, we recover $k = g^x \bmod p$ from r, s, g, p and y_a and then check whether $hash(k, m)$ is identical to r.

To illustrate how to shorten ElGamal based signatures, now we consider DSS. It should be stressed that many other ElGamal based signature schemes, in particular those defined on a sub-group of order q (see for example [11, 20]), can be shortened in the same way and are all equally good candidates for signcryption. Table 1 shows two shortened versions of DSS, which are denoted by SDSS1 and SDSS2 respectively. Here are a few remarks on the table: (1) the first letter "S" in the name of a scheme stands for "shortened", (2) the parameters p, q and g are the same as those for DSS, (3) x is a random number from $[1, \ldots, q]$, x_a is Alice's secret key and $y_a = g^{x_a} \bmod p$ is her matching public key, (4) $|t|$ denotes the size or length (in bits) of t, (5) the schemes have the same signature size of $|hash(\cdot)| + |q|$, (6) SDSS1 is slightly more efficient than SDSS2 in signature generation, as the latter involves an extra modulo multiplication.

Recently Pointcheval and Stern [22] have proven that Schnorr signature is unforgeable by any adaptive attacker who is allowed to query Alice's signature generation algorithm with messages of his choice [10], in a model where the one-way hash function used in the signature scheme is assumed to behave like a random function (the random oracle model). The core idea behind the unforgeability proof by Pointcheval and Stern is based on an observation that the signature has been converted from a 3-move zero-knowledge protocol (for proof of knowledge) with respect to a honest verifier. With such a signature scheme, unforgeability against a non-adaptive attacker who is not allowed to possess valid message-signature pairs follows from the soundness of the original protocol. Furthermore, as the protocol is zero-knowledge with respect to a honest verifier, the signature scheme converted from it can be efficiently simulated in the random oracle model. This implies that an adaptive attacker is not more powerful than a non-adaptive attacker in the random oracle model.

Turning our attention to SDSS1 and SDSS2, both can be viewed as being converted from a 3-move zero-knowledge protocol (for proof of knowledge) with respect to a honest verifier. Thus Pointcheval and Stern's technique is applicable also to SDSS1 and SDSS2. Summarizing the above discussions, both SDSS1 and SDSS2 are unforgeable by adaptive attackers, under the assumptions that discrete logarithm is hard and that the one-way hash function behaves like a random function.

3.2 Implementing Signcryption with Shortened Signature

An interesting characteristic of a shortened ElGamal based signature scheme obtained in the method described above is that although $g^x \bmod p$ is not explicitly contained in a signature (r, s), it can be recovered from r, s and other public parameters. This motivates us to construct a signcryption from a shortened signature scheme.

Shortened schemes	Signature (r, s) on a message m	Recovery of $k = g^x \bmod p$	Length of signature
SDSS1	$r = hash(g^x \bmod p, m)$ $s = x/(r + x_a) \bmod q$	$k = (y_a \cdot g^r)^s \bmod p$	$|hash(\cdot)| + |q|$
SDSS2	$r = hash(g^x \bmod p, m)$ $s = x/(1 + x_a \cdot r) \bmod q$	$k = (g \cdot y_a^r)^s \bmod p$	$|hash(\cdot)| + |q|$

p: a large prime (public to all),
q: a large prime factor of $p - 1$ (public to all),
g: a (random) integer in $[1, \ldots, p-1]$ with order q modulo p (public to all),
$hash$: a one-way hash function (public to all),
x_a: Alice's secret key,
y_a: Alice's public key ($y_a = g^{x_a} \bmod p$).

Table 1. Examples of Shortened and Efficient Signature Schemes

We exemplify our construction method using the two shortened signatures in Table 1. The same construction method is applicable to other shortened signature schemes based on ElGamal. As a side note, Schnorr's signature scheme, without being further shortened, can be used to construct a signcryption scheme which is slightly more advantageous in computation than other signcryption schemes from the view point of a message originator.

In describing our method, we will use E and D to denote the encryption and decryption algorithms of a private key cipher such as DES [17] and SPEED [25]. Encrypting a message m with a key k, typically in the cipher block chaining or CBC mode, is indicated by $E_k(m)$, while decrypting a ciphertext c with k is denoted by $D_k(c)$. In addition we use $KH_k(m)$ to denote hashing a message m with a key-ed hash algorithm KH under a key k. An important property of a key-ed hash function is that, just like a one-way hash function, it is computationally infeasible to find a pair of messages that are hashed to the same value (or collide with each other). This implies a weaker property that is sufficient for signcryption: given a message m_1, it is computationally intractable to find another message m_2 that collides with m_1. In [2] two methods for constructing a cryptographically strong key-ed hash algorithm from a one-way hash algorithm have been demonstrated. For most practical applications, it suffices to define $KH_k(m) = hash(k, m)$, where $hash$ is a one-way hash algorithm.

Assume that Alice also has chosen a secret key x_a from $[1, \ldots, q]$, and made public her matching public key $y_a = g^{x_a} \bmod p$. Similarly, Bob's secret key is x_b and his matching public key is $y_b = g^{x_b} \bmod p$.

The signcryption and unsigncryption algorithms constructed from a shortened signature are remarkably simple. For Alice to signcrypt a message m for Bob, she carries out the following:

Signcryption by Alice the Sender

1. Pick x randomly from $[1, \ldots, q]$, and let $k = y_b^x \bmod p$. Split k into k_1 and k_2 of appropriate length. (Note: one-way hashing, or even simple folding, may be applied to k prior splitting, if k_1 or k_2 is too long to fit in E or KH, or one wishes k_1 and k_2 to be dependent on all bits in k.)
2. $r = KH_{k_2}(m)$.
3. $s = x/(r + x_a) \bmod q$ if SDSS1 is used, or
   $s = x/(1 + x_a \cdot r) \bmod q$ if SDSS2 is used instead.
4. $c = E_{k_1}(m)$.
5. Send to Bob the signcrypted text (c, r, s).

The unsigncryption algorithm works by taking advantages of the property that $g^x \bmod p$ can be recovered from r, s, g, p and y_a by Bob. On receiving (c, r, s) from Alice, Bob unsigncrypts it as follows:

Unsigncryption by Bob the Recipient

1. Recover k from r, s, g, p, y_a and x_b:
   $k = (y_a \cdot g^r)^{s \cdot x_b} \bmod p$ if SDSS1 is used, or
   $k = (g \cdot y_a^r)^{s \cdot x_b} \bmod p$ if SDSS2 is used.
2. Split k into k_1 and k_2.
3. $m = D_{k_1}(c)$.
4. accept m as a valid message originated from Alice only if $KH_{k_2}(m)$ is identical to r.

In the following, the two examples of signcryption schemes will be denoted by SCS1 and SCS2 respectively. For the purpose of a detailed comparison, the cost of these signcryption schemes has been analyzed and listed, along with other signature-then encryption schemes, in Table 2.

Finally two remarks follow: (1) signcryption schemes can also be derived from shortened signature schemes based on the discrete logarithm problem on elliptic curves [13]. (2) the functions, especially non-repudiation and unforgeability, of signcryption may not be fully implemented by the use of a shared key between Alice and Bob, such as $g^{x_a \cdot x_b} \bmod p$ or a key obtained via a Key Pre-distribution Scheme [16], unless tamper-resistant devices and/or trusted third parties are involved.

3.3 Working with Signature-Only and Encryption-Only Modes

Not all messages require both confidentiality and integrity. Some messages may need to be signed only, while others may need to be encrypted only. For the two digital signcryption schemes SCS1 and SCS2, when a message is sent in clear, they degenerate to signature schemes with verifiability by the recipient only. As will be argued in Section 6, limiting verifiability to the recipient only still preserves non-repudiation, and may represent an advantage for some applications where the mere fact that a message is originated from Alice needs to be kept secret. Furthermore, if Alice uses g instead of Bob's public key y_b in the calculation of k, the schemes becomes corresponding shortened ElGamal based signature schemes with universal verifiability.

To work with the encryption-only mode, one may simply switch to the El-Gamal encryption, or any other public key encryption scheme.

Various schemes	Computational cost	Communication overhead (in bits)
signature-then-encryption based on RSA	EXP=2, HASH=1, ENC=1 [EXP=2, HASH=1, DEC=1]	$\|n_a\| + \|n_b\|$
signature-then-encryption based on DSS + ElGamal encryption	EXP=3, MUL=1, DIV=1 ADD=1, HASH=1, ENC=1 [EXP=3, MUL=1, DIV=2 ADD=0, HASH=1, DEC=1]	$2\|q\| + \|p\|$
signature-then-encryption based on Schnorr signature + ElGamal encryption	EXP=3, MUL=1, DIV=0 ADD=1, HASH=1, ENC=1 [EXP=3, MUL=1, DIV=0 ADD=0, HASH=1, DEC=1]	$\|KH.(\cdot)\| + \|q\| + \|p\|$
signcryption SCS1	EXP=1, MUL=0, DIV=1 ADD=1, HASH=1, ENC=1 [EXP=2, MUL=2, DIV=0 ADD=0, HASH=1, DEC=1]	$\|KH.(\cdot)\| + \|q\|$
signcryption SCS2	EXP=1, MUL=1, DIV=1 ADD=1, HASH=1, ENC=1 [EXP=2, MUL=2, DIV=0 ADD=0, HASH=1, DEC=1]	$\|KH.(\cdot)\| + \|q\|$

where
EXP = the number of modulo exponentiations,
MUL = the number of modulo multiplications,
DIV = the number of modulo division (inversion),
ADD = the number of modulo addition or subtraction,
HASH = the number of one-way or key-ed hash operations,
ENC = the number of encryptions using a private key cipher,
DEC = the number of decryptions using a private key cipher,
Parameters in the brackets indicate the number of operations involved in "decryption-then-verification" or "unsigncryption".

Table 2. Cost of Signature-Then-Encryption v.s. Cost of Signcryption

4 Cost of Signcryption v.s. Cost of Signature-Then-Encryption

The most significant advantage of signcryption over signature-then-encryption lies in the dramatic reduction of computational cost and communication overhead which can be symbolized by the following inequality:

$$\text{Cost(signcryption)} < \text{Cost(signature)} + \text{Cost(encryption)}.$$

With SCS1 and SCS2, this advantage is shown in Tables 3 and 4.

Note that when comparing with RSA based signature-then-encryption, we have assumed that a relatively small public exponent e is employed for encryption or signature verification, although cautions should be taken in light of recent progress in cryptanalysis against RSA with an small exponent (see for example [6]). Therefore the main computational cost for RSA based signature-then-encryption is in decryption or signature generation which generally involves a modulo exponentiation with a *full size* exponent d. We have further assumed that the Chinese Remainder Theorem is used, so that the computational expense for RSA decryption can be reduced, theoretically, to a quarter of the expense with a full size exponent.

security parameters $|p|, |q|, |KH.(\cdot)|(= |hash(\cdot)|)$	saving in comp. cost	saving in comm. overhead
768, 152, 80	50%	76.8%
1024, 160, 80	50%	81.0%
2048, 192, 96	50%	87.7%
4096, 256, 128	50%	91.0%
8192, 320, 160	50%	94.0%
10240, 320, 160	50%	96.0%

$$\text{saving in comp. cost} = \frac{3 \text{ modulo exponentiations}}{6 \text{ modulo exponentiations}} = 50\%$$

$$\text{saving in comm. cost} = \frac{|hash(\cdot)| + |q| + |p| - (|KH.(\cdot)| + |q|)}{|hash(\cdot)| + |q| + |p|}$$

Table 3. Saving of Signcryption over Signature-Then-Encryption Using Schnorr Signature and ElGamal Encryption

4.1 How the Parameters are Chosen

Advances in fast computers help an attacker in increasing his capability to break a cryptosystem. To compensate this, larger security parameters, including $|n_a|$, $|n_b|$, $|p|$, $|q|$ and $|KH.(\cdot)|$ must be used in the future. From an analysis by Odlyzko [21] on the hardness of discrete logarithm, one can see that unless there is an algorithmic breakthrough in solving the factorization or discrete logarithm

175

security parameters $|p|(=|n_a|=|n_b|),|q|,|KH.(\cdot)|$	advantage in comp. cost	advantage in comm. overhead
768, 152, 80	0%	84.9%
1024, 160, 80	6.25%	88.3%
2048, 192, 96	43.8%	93.0%
4096, 256, 128	62.0%	95.0%
8192, 320, 160	77.0%	97.0%
10240, 320, 160	81.0%	98.0%

$$\text{advantage in comp. cost} = \frac{0.375(|n_a|+|n_b|)-4.5|q|}{0.375(|n_a|+|n_b|)}$$
$$\text{advantage in comm. cost} = \frac{|n_a|+|n_b|-(|KH.(\cdot)|+|q|)}{|n_a|+|n_b|}$$

Table 4. Advantage of Signcryption over RSA based Signature-Then-Encryption with *Small Public Exponents*

problem, $|q|$ and $|KH.(\cdot)|$ can be increased at a smaller pace than can $|n_a|$, $|n_b|$ and $|p|$. Thus, as shown in Tables 3 and 4, the saving or advantage in computational cost and communication overhead by signcryption will be more significant in the future when larger parameters must be used.

The selection of security parameters $|p|$, $|q|$, $|n_a|$ and $|n_a|$ in Tables 3 and 4, has been partially based on recommendations made in [21]. The parameter values in the tables, however, are indicative only, and can be determined flexibly in practice. We also note that choosing $|KH.(\cdot)| \approx |q|/2$ is due to the fact that using Shank's baby-step-giant-step or Pollard's rho method, the complexity of computing discrete logarithms in a sub-group of order q is $O(\sqrt{q})$ (see [14]). Hence choosing $|KH.(\cdot)| \approx |q|/2$ will minimize the communication overhead of the signcryption schemes SCS1 and SCS2. Alternatively, one may decide to choose $KH.(\cdot) \in [1,\ldots,q]$ which can be achieved by setting $|KH.(\cdot)| = |q| - 1$. This will not affect the computational advantage of the signcryption schemes, but slightly increase their communication overhead.

5 Applications of Signcryption

As discussed in the introduction, a major motivation of this work is to search for a more economical method for secure and authenticated transactions/message delivery. If digital signcryptions are applied in this area, the resulting benefits are potentially significant: for every single secure and authenticated electronic transaction, we may save 50% in computational cost and 85% in communication overhead.

The proposed signcryption schemes are compact and particularly suitable for smart card based applications. We envisage that they will find innovative applications in many areas including digital cash payment systems, EDI and personal heath cards. Of particular importance is the fact that signcryption may be used to design more efficient digital cash transaction protocols that are often required to provide with both the functionality of digital signature and encryption.

In the full paper we also show how to adapt a signcryption scheme into one for broadcast communication which involves multiple recipients. Such an adapted scheme shares a comparable computational cost with a broadcast scheme proposed in RFC1421. The communication overhead required by the scheme based on signcryption, however, is multiple times lower than that required the scheme in RFC1421.

Another surprising property of the proposed signcryption schemes is that it enables us to carry out fast, secure, unforgeable and non-repudiatable key transport *in a single block whose size is smaller than* $|p|$. In particular, using either of the two signcryption schemes, we can transport highly secure and authenticated keys in a single ATM cell (48 byte payload + 5 byte header). A possible combination of parameters is $|p| \geq 512$, $|q| = 160$, and $|KH(\cdot)| = 80$, which would allow the transport of an unforgeable and non-repudiatable key of up to 144 bits. Advantages of such a key transport scheme over interactive key exchange protocols such as those proposed in [8] are obvious, both in terms of computational efficiency and compactness of messages. Compared with previous attempts for secure, but un-authenticated, key transport based on RSA (see for example [1, 12]), our key transport scheme has a further advantage in that it offers both unforgeability and non-repudiation. In a similar way, a multi-recipient signcryption scheme can be used as a very economical method for generating conference keys among a group of users.

6 Unforgeability, Non-repudiation and Confidentiality of Signcryption

Like any cryptosystem, security of signcryption in general has to address two aspects: (1) to protect what, and (2) against whom. With the first aspect, we wish to prevent the contents of a signcrypted message from being disclosed to a third party other than Alice, the sender, and Bob, the recipient. At the same time, we also wish to prevent Alice, the sender, from being masquerade by other parties, including Bob. With the second aspect, we consider the most powerful attackers one would be able to imagine in practice, namely adaptive attackers who are allowed to have access to Alice's signcryption algorithm and Bob's unsigncryption algorithm.

We say that a signcryption scheme is secure if the following conditions are satisfied:

1. Unforgeability — it is computationally infeasible for an adaptive attacker (who may be a dishonest Bob) to masquerade Alice in creating a signcrypted text.

2. Non-repudiation — it is computationally feasible for a third party to settle a dispute between Alice and Bob in an event where Alice denies the fact that she is the originator of a signcrypted text with Bob as its recipient.

3. Confidentiality — it is computationally infeasible for an adaptive attacker (who may be any party other than Alice and Bob) to gain any partial information on the contents of a signcrypted text.

A detailed description of the proofs/arguments of the security of the signcryption schemes SCS1 and SCS2 can be found in the full paper. Here are the key ideas used in the proofs/arguments:

1. Unforgeability — this can be done using the technique of Pointcheval and Stern [22].
2. Non-repudiation — A dispute between Alice and Bob can be settled by a trusted third party (say a judge), by the use of a zero-knowledge proof protocol between the judge and Bob. In particular, they can use a very simple 4-move zero-knowledge interactive proof protocol proposed by Chaum in [5].
3. Confidentiality — We achieve our goal by reduction: we will reduce the confidentiality of another encryption scheme called C_{kh}, whose confidentiality is relatively well-understood, to the confidentiality of a signcryption scheme (say SCS1). With the encryption scheme C_{kh}, the ciphertext of a message m is defined as ($u = g^x \bmod p$, $c = E_{k_1}(m)$, $r = KH_{k_2}(m)$) where k_1 and k_2 are defined in the same way as in SCS1. C_{kh} is a slightly modified version of a scheme that has received special attention in [24, 3] (see also earlier work [26].)

   Now assume that there is an attacker for SCS1. Call this attacker A_{SCS1}. We show how A_{SCS1} can be translated into one for C_{kh}, called $A_{C_{kh}}$. Note that for a message m, the input to A_{SCS1} includes q, p, g, $y_a = g^{x_a} \bmod p$, $y_b = g^{x_b} \bmod p$, $u = g^x \bmod p$, $c = E_{k_1}(m)$, $r = KH_{k_2}(m)$. With the attacker $A_{C_{kh}}$ for C_{kh}, however, its input includes: q, p, g, $y_b = g^{x_b} \bmod p$, $u = g^x \bmod p$, $c = E_{k_1}(m)$, and $r = KH_{k_2}(m)$. One immediately identifies that two numbers that correspond to y_a and s which are needed by A_{SCS1} as part of its input are currently missing from the input to $A_{C_{kh}}$. Thus, in order for $A_{C_{kh}}$ to "call" the attacker A_{SCS1} "as a sub-routine", $A_{C_{kh}}$ has to create two numbers corresponding to y_a and s in the input to A_{SCS1}. Call these two yet-to-be-created numbers y_a' and s'. y_a' and s' have to have the right form so that $A_{C_{kh}}$ can "fool" A_{SCS1}. It turns out that such y_a' and s' can be easily created by $A_{C_{kh}}$ as follows: (1) pick a random number s' from $[1, \ldots, q]$. (2) let $y_a' = u^{1/s'} \cdot g^{-r} \bmod p$.

A final note on signcryption follows. Unlike signature-then-encryption, the verifiability of a signcryption is in normal situations limited to Bob the recipient, as his secret key is required for unsigncryption. At the first sight, the limited verifiability of a signcryption, namely the direct verifiability by the sender only (and indirect verifiability by a judge with the cooperation of Bob), may be seen as a drawback of signcryption. Here we argue that the limited direct verifiability will not pose any problem in practice and hence should not be an obstacle to practical applications of signcryption. In the real life, a message sent to Bob in a secure and authenticated way is meant to be readable by Bob only. Thus if there is no dispute between Alice and Bob, direct verifiability by Bob only is precisely what the two users want. In other words, in normal situations where no disputes between Alice and Bob occur, the full power of universal verifiability provided by digital signature is never needed. (For a similar reason, traditionally one uses signature-then-encryption, rather than encryption-then-signature !) In a situation where repudiation does occur, interactions between Bob and a judge would follow. This is very similar to a dispute on repudiation in the real world, say between a complainant (Bob) and a defendant (Alice), where the process for a judge to resolve the dispute requires in general interactions between the judge and the complainant, and furthermore between the judge and an expert in hand-written signature identification, as the former may rely on advice from the latter in correctly deciding the origin of a message.

7 Conclusion

We have introduced a new cryptographic primitive called signcryption for secure and authenticated message delivery, which fulfills all the functions of digital signature and encryption, but with a far smaller cost than that required by the current standard signature-then-encryption methods. Security of the signcryption schemes has been proven, and extensions of the schemes to multiple recipients has been carried out. We believe that the new primitive will open up a number of avenues for future research into more efficient security solutions.

The signcryption schemes proposed in this paper have been based on ElGamal signature and encryption. We have not been successful in searching for a signcryption scheme employing RSA or other public key cryptosystems. Therefore it remains a challenging open problem to design signcryption schemes based factorization or other computationally hard problems.

References

1. Basturk, E., Bellare, M., Chow, C.-S., Guerin, R.: Secure transport protocols for high-speed networks. IBM Research Report Report RC 19981 IBM T. J. Watson Research Center Yorktown Heights, NY 10598 1994.
2. Bellare, M., Canetti, R., Krawczyk, H.: Keying hash functions for message authentication. In Advances in Cryptology - CRYPTO'96 (Berlin, New York, Tokyo, 1996) vol. 1109 of Lecture Notes in Computer Science Springer-Verlag pp. 1–15.
3. Bellare, M., Rogaway, P.: Random oracles are practical: A paradigm for designing efficient protocols. In Proceedings of the First ACM Conference on Computer and Communications Security (New York, November 1993) The Association for Computing Machinery pp. 62–73.
4. Brickell, E., McCurley, K.: Interactive identification and digital signatures. AT&T Technical Journal (1991) 73–86.
5. Chaum, D.: Zero-knowledge undeniable signatures. In Advances in Cryptology - EUROCRYPT'90 (Berlin, New York, Tokyo, 1990) vol. 473 of Lecture Notes in Computer Science Springer-Verlag pp. 458–464.
6. Coppersmith, D., Franklin, M., Patarin, J., Reiter, M.: Low-exponent RSA with related messages. In Advances in Cryptology - EUROCRYPT'96 (Berlin, 1996) vol. 1070 of Lecture Notes in Computer Science Springer-Verlag pp. 1–9.

7. Diffie, W., Hellman, M.: New directions in cryptography. IEEE Transactions on Information Theory **IT-22** (1976) 472–492.
8. Diffie, W., Oorschot, P. V., Wiener, M.: Authentication and authenticated key exchange. Designs, Codes and Cryptography **2** (1992) 107–125.
9. ElGamal, T.: A public key cryptosystem and a signature scheme based on discrete logarithms. IEEE Transactions on Information Theory **IT-31** (1985) 469–472.
10. Goldwasser, S., Micali, S., Rivest, R.: A digital signature scheme secure against adaptively chosen message attacks. SIAM J. on Computing **17** (1988) 281–308.
11. Horster, P., Michels, M., Petersen, H.: Meta-ElGamal signature schemes. In Proceedings of the second ACM Conference on Computer and Communications Security (New York, November 1994) ACM pp. 96–107.
12. Johnson, D., Matyas, S.: Asymmetric encryption: Evolution and enhancements. CryptoBytes **2** (1996) 1–6.
13. Koblitz, N.: Elliptic curve cryptosystems. Mathematics of Computation **48** (1987) 203–209.
14. Lenstra, A. K., Lenstra, H. W.: Algorithms in Number Theory vol. A of Handbook in Theoretical Computer Science. Elsevier and the MIT Press 1990.
15. Linn, J.: Privacy enhancement for internet electronic mail: Part I: Message encryption and authentication procedures. Request for Comments RFC 1421 IAB IRTF PSRG, IETF PEM WG 1993.
16. Matsumoto, T., Imai, H.: On the key predistribution systems: A practical solution to the key distribution problem. In Advances in Cryptology - CRYPTO'87 (Berlin, New York, Tokyo, 1987) vol. 239 of Lecture Notes in Computer Science Springer-Verlag pp. 185–193.
17. National Bureau of Standards: Data encryption standard. FIPS PUB 46 U.S. Department of Commerce January 1977.
18. National Institute of Standards and Technology: Digital signature standard (DSS). FIPS PUB 186 U.S. Department of Commerce May 1994.
19. National Institute of Standards and Technology: Secure hash standard. FIPS PUB 180-1 U.S. Department of Commerce April 1995.
20. Nyberg, K., Rueppel, R.: Message recovery for signature schemes based on the discrete logarithm problem. Designs, Codes and Cryptography **7** (1996) 61–81.
21. Odlyzko, A.: The future of integer factorization. CryptoBytes **1** (1995) 5–12.
22. Pointcheval, D., Stern, J.: Security proofs for signature schemes. In Advances in Cryptology - EUROCRYPT'96 (Berlin, New York, Tokyo, 1996) vol. 1070 of Lecture Notes in Computer Science Springer-Verlag pp. 387–398.
23. Schnorr, C. P.: Efficient identification and signatures for smart cards. In Advances in Cryptology - CRYPTO'89 (Berlin, New York, Tokyo, 1990) vol. 435 of Lecture Notes in Computer Science Springer-Verlag pp. 239–251.
24. Zheng, Y.: Improved public key cryptosystems secure against chosen ciphertext attacks. Technical Report 94-1 University of Wollongong Australia January 1994.
25. Zheng, Y.: The SPEED cipher. In Proceedings of Financial Cryptography'97 (Berlin, New York, Tokyo, 1997) Lecture Notes in Computer Science Springer-Verlag.
26. Zheng, Y., Seberry, J.: Immunizing public key cryptosystems against chosen ciphertext attacks. IEEE Journal on Selected Areas in Communications **11** (1993) 715–724.

How to Sign Digital Streams

Rosario Gennaro and Pankaj Rohatgi

I.B.M. T.J.Watson Research Center
P.O.Box 704, Yorktown Heights, NY 10598, U.S.A.
Email: rosario,rohatgi@watson.ibm.com

Abstract. We present a new efficient paradigm for signing digital streams. The problem of signing digital streams to prove their authenticity is substantially different from the problem of signing regular messages. Traditional signature schemes are message oriented and require the receiver to process the entire message before being able to authenticate its signature. However, a stream is a potentially very long (or infinite) sequence of bits that the sender sends to the receiver and the receiver is required to consumes the received bits at more or less the input rate and without excessive delay. Therefore it is infeasible for the receiver to obtain the entire stream before authenticating and consuming it. Examples of streams include digitized video and audio files, data feeds and applets. We present two solutions to the problem of authenticating digital streams. The first one is for the case of a finite stream which is entirely known to the sender (say a movie). We use this constraint to devise an extremely efficient solution. The second case is for a (potentially infinite) stream which is not known in advance to the sender (for example a live broadcast). We present proofs of security of our constructions. Our techniques also have applications in other areas, for example, efficient authentication of long files when communication is at a cost and signature based filtering at a proxy server.

1 Introduction

Digital Signatures (see [5, 17]) are the cryptographic answer to the problem of information authenticity. When a recipient receives digitally signed information and she is able to verify the digital signature then she can be certain that the information she received is exactly the same as what the sender (identified by his public key) has signed. Moreover, this guarantee is *non-repudiable*, i.e., the entity identified by the public key cannot later deny having signed the information. Thus, the recipient can hold the signer responsible for the content she receives.[1]

However current digital signature technology was designed to ensure message authentication and its straightforward application does not yield a satisfactory

---

[1] This distinguishes digital signatures from *message authentication codes* (MAC) which allow the receiver to have confidence on the identity of the sender, but not to prove to someone else this fact, i.e. MAC's are repudiable.

solution when applied to information resources which are not message-like. In this paper we discuss one such type of resource: *streams*. We point out shortcomings in several approaches (some of them used in practice) to tackle the problem of signing streams and then present our solution which does not have such shortcomings.

1.1 Streams Defined

A stream is a potentially very long (infinite) sequence of bits that the sender sends to the receiver. The stream is usually sent at a rate which is negotiated between the sender and receiver or there may be a demand-response protocol in which the receiver repeatedly sends requests for additional (finite) amount of data. The main features of streams which distinguish them from messages is that the receiver must consume the data it receives at more or less the input rate, i.e., it can't buffer large amounts of unconsumed data. In fact in many applications the receiver stores relatively very small amounts of the stream. In some cases the sender itself may not store the entire sequence, i.e., it may not store the information it has already sent out and it may not know anything about the stream much beyond what it has sent out.

There are many examples of digital streams. Common examples include *digitized video* and *audio* which is now routinely transported over the Internet and also to television viewers via various means, e.g., via direct broadcast satellites and very shortly via cable, wireless cable, telephone lines etc. This includes both pre-recorded and stored audio/video programming as well as live feeds. Apart from audio/video, there are also *data feeds* (e.g., news feeds, stock market quotes etc) which are best modeled as a stream. The Internet and the emerging interactive TV industry also provides another example of an information resource which is best modeled as a stream, i.e., *applets*. Most non-trivial applets are actually very large programs which are organized into several modules. The consumer's machine first downloads and executes the startup module and as the program proceeds, additional modules are downloaded and executed. Also, modules which are no longer in use may be discarded by the consumer machine. This structure of applets is forced by two factors. Firstly the amount of storage available on the consumer machine may be limited (e.g., in the emerging interactive TV industry set-top boxes have to be cheap and therefore resource limited) . Secondly (in the case of the Internet), the bandwidth available to download code may be limited and applets must be designed to start executing as soon as possible. Also it is quite likely that some of the more sophisticated applets may have data-rich components generated on the fly by the applet server. Therefore applets fit very nicely into the demand/response streams paradigm.

Given the above description, it is clear that message oriented signature schemes cannot be directly used to sign streams since the receiver cannot be expected to receive the entire stream before verifying the signature. If a stream is infinitely long (e.g., the 24-hours news channel), then it is impossible for the receiver to receive the entire stream and even if a stream is finite but long the receiver would have to violate the constraint that the stream needs to be consumed at roughly the input rate and without delay.

1.2 Previous Solutions and their Shortcomings

Up to the authors' knowledge there has been no proposed specific solution to the problem of signing digital streams in the crypto literature. One can envision several possible solutions, some of them are actually proposed to be used in practice.

One type of solution splits the stream in blocks. The sender signs each individual block and the receiver loads an entire block and verifies its signature before consuming it. This solution also works if the stream is infinite. However this solution forces the sender to generate a signature for each block of the stream and the receiver to verify a signature for each block. With today's signature schemes either one or both of these operations can be very expensive computationally. Which in turns means that the operations of signing and verifying can create a bottleneck to the transmission rate of the stream.

Another type of solution works for only finite streams. In this case, once again the stream is split into blocks. Instead of signing each block, the sender creates a table listing cryptographic hashes of each of the blocks and signs this table. When the receiver asks for the authenticated stream, the sender first sends the signed table followed by the stream. The receiver first receives and stores this table and verifies the signature on it. If the signature matches then the receiver has the authenticated cryptographic hash of each of blocks in the stream and thus each block can be verified when it arrives. The problem with this solution is that it requires the storage and maintenance of a potentially very large table on the receiver's end. In many realistic scenarios the receiver buffer is very limited compared to the size of the stream, (e.g., in MPEG a typical movie may be 20 GBytes whereas the receiver buffer is only required to be around 250Kbytes). Therefore the hash table can itself become fairly large (e.g., 50000 entries in this case or 800Kbytes for the MD5 hash function) and it may not be possible to store the hash table itself. Also, the hash table itself needs to be transmitted first and if it is too large then there will be a significant delay before the first piece of the stream is received and consumed. To address the problem of large tables one can also come up with a hybrid scheme in which the stream is split in consecutive pieces and each piece is preceded by a small signed table of contents.[2]

---

[2] This is the case now (Java Developer Kit 1.1) for large signed java applets which are distributed as a collection of Java archives (JAR) where each archive has a signed table of hashes of contents and the archives are loaded in the order given in the HTML page in which the applet is embedded.

The above solution can be further modified by using an authentication tree: the blocks are placed as the leaves of a binary tree and each internal node takes as a value the hash of its children (see [13].) This way the sender needs to sign and send only the root of this tree. However in order to authenticate each following block the sender has to send the whole authentication path (i.e. the nodes on the path from the root to the block, plus their siblings) to the receiver. This means that if the stream has k blocks, the authentication information associated with each block will be $O(\log k)$.

As we will see briefly our solution eliminates all these shortcomings. The basic idea works for both infinite and finite streams, only one expensive digital signature is ever computed, there are no big tables to store, and the size of the authentication information associated with each block does not depend on the size of the stream.

NON-REPUDIATION. Notice that if the receiver were only interested in establishing the identity of the sender, a solution based on MAC would suffice. Indeed once the sender and receiver share a secret key, the stream could be authenticated block by block using a MAC computation on it. Since MAC's are usually faster than signatures to compute and verify, this solution would not incur the computational cost associated with the similar signature-based solution described above.

However a MAC-based approach would not enjoy the non-repudiation property. We stress that we require such property for our solution. Also in order for this property to be meaningful in the context of streams we need to require that *each* prefix of the stream to be non-repudiable. That is, if the stream is $B = B_1, B_2, \ldots$ where each B_i is a block, we require that each prefix $B_i = B_1 \ldots B_i$ is non-repudiable. This rules out a solution in which the sender just attaches a MAC to each block and then signs the whole stream at the end.

This is to prevent the sender from interrupting the transmission of the stream before the non-repudiability property is achieved. Also it is a guarantee for the receiver. Consider indeed the following scenario: the receiver notices that the applets she is downloading are producing damages to her machine. She interrupts the transfer in order to limit the damage, but at the same time she still wants some proof to bring to court that the substream downloaded so far did indeed come from the sender.

1.3 Our solution in a nutshell

Our solution makes some reasonable/practical assumptions about the nature of the streams being authenticated.

First of all we assume that it is possible for the sender to embed authentication information in the stream. This is usually the case, see Section 7 to see how to do this in most real-world situations like MPEG video/audio. We also assume that the receiver has a "small" buffer in which it can first authenticate the received bits before consuming them. Finally we assume that the receiver has processing power or hardware that can compute a small number of fast cryptographic checksums faster than the incoming stream rate while still being able to play the stream in real-time.

The basic idea of our solution is to divide the stream into blocks and embed some authentication information in the stream itself. The authentication information of the ith block will be used to authenticate the $(i+1)^{st}$ block. This way the signer needs to sign just the first block and then the properties of this single signature will "propagate" to the rest of the stream through the authentication information. Of course the key problem is to perform the authentication of the internal blocks fast. We distinguish two cases.

In the first scenario the stream is finite and is known in its entirety to the signer in advance. This is not a very limiting requirement since it covers most of the Internet applications (digital movies, digital sounds, applets). In this case we will show that a *single* hash computation will suffice to authenticate the internal blocks. The idea is to embed in the current block a hash of the following block (which in turns includes the hash of the following one and so on...)

The second case is for (potentially infinite) streams which are not known in advance to the signer (for example live feeds, like sports event broadcasting and chat rooms). In this case our solution is less optimal as it requires several hash computations to authenticate a block (although depending on the embedding mechanism these hash computations can be amortized over the length of the block). The size of the embedded authentication information is also an issue in this case. The idea here is to use fast 1-time signature schemes (introduced in [11, 12]) to authenticate the internal blocks. So block i will contain a 1-time public key and also the 1-time signature of itself with respect to the key contained in block $i-1$. This signature authenticates not only the stream block but also the 1-time key attached to it.

1.4   Related Work

Some of the ideas involved in the solution for unknown streams have appeared previously, although in different contexts and with different usage.

Mixing "regular" signatures with 1-time signatures, for the purpose of improving efficiency is discussed in [7]. However in that paper the focus is in making the signing operation of a message M efficient by dividing it in two parts. An off-line part in which the signer signs a 1-time public key with his long-lived secret key even before the messages M is known. Then when M has to be sent the signer computes a 1-time signature of M with the authenticated 1-time public key and sends out M tagged with the 1-time public key and the two signatures. Notice that the receiver must compute two signature verifications: one on the long-lived key and one on the 1-time key. In our scheme we need to make both signing and verification extremely fast, and indeed in our case each block (except for the first) is signed (and hence verified) only once with a 1-time key.

We also use the idea to of using old keys in order to authenticate new keys. This has appeared in several places but always for long-lived keys. Examples include [1, 15, 18] where this technique is used to build provably secure signature schemes. We stress that the results in [1, 15, 18] are mostly of theoretical interest and do not yield practical schemes. Our on-line solution somehow mixes these two ideas in a novel way, by using the chaining technique with 1-time keys, embedding the keys inside the stream flow so that old keys can authenticate at the same time *both* the new keys and the current stream block.

The chaining technique can also be seen as a weak construction of *accumulators* as introduced in [2]. An accumulator for k blocks $B_1, \ldots, B_k$ is a single value ACC that allows a signer to quickly authenticate any of the blocks in any particular order. Accumulators based on the RSA assumption were proposed in [2]. In our case we have a much faster construction based on collision–free hash functions, since we exploit the property that the blocks must be authenticated in a specific order.

2 Preliminaries

In the following we denote with n the security parameter. We say that a function $\epsilon(n)$ is *negligible* if for all c, there exists an n_0 such that, for all $n > n_0$, $\epsilon(n) < 1/n^c$.

COLLISION-RESISTANT HASH FUNCTIONS. Let H be a function that map arbitrarily long binary strings into elements of a fixed domain D. We say that H is a *collision-resistant hash function* if any polynomial time algorithm who is given as input the values $H(x_i)$ on several adaptively chosen values x_i, finds a collision, i.e. a pair (x, y) such that $x \neq y$ and $H(x) = H(y)$, only with negligible probability. MD5 [16] and SHA-1 [14] are conjectured collision-resistant hash functions.

SIGNATURE SCHEMES. A *signature scheme* is a triplet (G, S, V) of probabilistic polynomial-time algorithms satisfying the following properties:

- G is the *key generator* algorithm. On input 1^n it outputs a pair $(SK, PK) \in \{0, 1\}^{2n}$. SK is called the secret (signing) key and PK is called the public (verification) key.
- S is the *signing* algorithm. On input a message M and the secret key SK, it outputs a signature σ.
- V is the *verification* algorithm. For every $(PK, SK) = G(1^n)$ and $\sigma = S(SK, M)$, it holds that $V(PK, \sigma, M) = 1$.

In [9] security for signature schemes is defined in several variants. The strongest variant is called "existential unforgeability against adaptively chosen message attack". That is, we require that no efficient algorithm will be able to produce a valid signed message, even after seeing several signed messages of its choice.

STREAM SIGNATURES. We define a *stream* to be a (possibly infinite) sequence of *blocks* $B = B_1, B_2, \ldots$ where each $B_i \in \{0,1\}^c$ for some constant[3] c.
We distinguish two cases. In the first case we assume that the stream is finite and known to the sender in advance. We call this the *off-line* case. Conversely in the *on-line* case the signer must process one (or a few) block at the time with no knowledge of the future part of the stream.

Definition 1. An *off-line stream signature scheme* is a triplet (G, S, V) of probabilistic polynomial-time algorithms satisfying the following properties:
- G is the *key generator* algorithm. On input 1^n it outputs a pair $(SK, PK) \in \{0,1\}^{2n}$. SK is called the secret (signing) key and PK is called the public (verification) key.
- S is the *signing* algorithm. On input a finite stream $B = B_1, \ldots, B_k$ and the secret key SK algorithm S outputs a new stream $B' = B'_1, \ldots, B'_k$ where $B'_i = (B_i, A_i)$.
- V is the *verification* algorithm. For every $(PK, SK) = G(1^n)$ and $B' = S(SK, B)$, it holds that $V(PK, B'_1, \ldots, B'_i) = 1$ for $1 \le i \le k$.

Notice that we modeled the off-line property by the fact that the signing algorithm is given the whole stream in advance. Yet the verifier is required to authenticate *each* prefix of the scheme without needing to see the rest of the stream. As it will become clear in the following our algorithms will not require the off-line verifier to store the whole past stream either.

Definition 2. An *on-line stream signature scheme* is a triplet (G, S, V) of probabilistic polynomial-time algorithms satisfying the following properties:
- G is the *key generator* algorithm. On input 1^n it outputs a pair $(SK, PK) \in \{0,1\}^{2n}$. SK is called the secret (signing) key and PK is called the public (verification) key.
- S is the *signing* algorithm. Given a (possibly infinite) stream $B = B_1, \ldots,$ algorithm S with input the secret key SK process each block one at the time, i.e.,

$$S(SK, B_1, \ldots, B_i) = B'_i = (B_i, A_i)$$

---

[3] The assumption that the blocks have all the same size is not really necessary. We just make it for clarity of presentation.

– V is the *verification* algorithm. For every $(PK, SK) = G(1^n)$ and $B'_1, B'_2, \ldots$ such that $B'_i = S(SK, B_1, \ldots, B_i)$ for all i, it holds that $V(PK, B'_1, \ldots, B'_i) = 1$ for all i.

Notice that in the on-line definition we have the signer process each block "on the fly" so knowledge of future blocks is not needed. In this case also the definition seems to requires knowledge of all past blocks for both signer and verifier, however this does not have to be the case (indeed in our solution some past blocks may be discarded).

The above definitions say nothing about security. In order to define security for stream signing we use the same notions of security introduced in [9]. That is, we require that no efficient algorithm will be able to produce a valid signed stream, even after seeing several signed streams. However notice that given our definition of signed streams, a prefix of a valid signed stream is itself a valid signed stream. So the forger can present a "different" signed stream by just taking a prefix of the ones seen before. However this hardly constitutes forgery, so we rule it out in the definition. With $B^{(1)} \subseteq B^{(2)}$ we denote the fact that $B^{(1)}$ is a prefix of $B^{(2)}$.

Definition 3. We say that an off-line (resp. on-line) stream signature scheme (G, S, V) is *secure* if any probabilistic polynomial time algorithm F, who is given as input the public key PK and adaptively chosen signed streams $B'^{(j)}$, outputs a new previously unseen valid signed stream $B' \not\subseteq B'^{(j)}$ $\forall j$ only with negligible probability.

For signed streams we slightly abuse the notation: when we write $B'^{(1)} \not\subseteq B'^{(2)}$ we mean that not only $B'^{(1)}$ is not a prefix of $B'^{(2)}$ but also the underlying "basic" unsigned streams are in the same relationship, i.e. $B^{(1)} \not\subseteq B^{(2)}$.

This is the definition of existential unforgeability against adaptive chosen message attack, the strongest of the notions presented in [9]. Following [9] weaker variants can be defined.

3   The Off-Line Solution

In this case we assume that the sender knows the entire stream in advance. (e.g., music/movie broadcast). Assume for simplicity that the stream is such that it is possible to reserve 20 bytes of extra authentication information in a block of size c.

The stream is logically divided into blocks of size c. The receiver has a buffer of size c. The receiver first receives the signature on the 20 byte hash (e.g., SHA-1) of the first block. After verification of the signature the receiver knows what the hash of the first block should be and then starts receiving the full stream and starts computing its hash block by block. When the receiver receives the first block, it checks its hash against what the signature was verified upon. If it matches, it plays the block otherwise it rejects it and stops playing the stream. How are other blocks authenticated? The key point is that the first block contains the 20 byte hash of the second block, the second block contains the 20 byte hash of the third block and so on... Thus, after the first signature check, there are just hashes to be checked for every subsequent block.

In more detail, let (G, S, V) be a regular signature scheme. The sender has a pair of secret-public key $(SK, PK) = G(1^n)$ of such signature scheme. Also let H be a collision-resistant cryptographic hash function. If the original stream is

$$B = B_1, B_2, \ldots, B_k$$

and the resulting signed stream is

$$B' = B'_0, B'_1, B'_2, \ldots, B'_k$$

the processing is done *backwards* on the original stream as follows:

$$B'_k = < B_k, 00 \ldots 0 >$$

$$B'_i = < B_i, H(B'_{i+1}) > \quad \text{for } i = 1, \ldots, k-1$$

$$B'_0 = < H(B'_1), S(SK, H(B'_1)) >$$

Notice that on the sender side, computing the signature and embedding the hashes requires a single *backwards* pass on the stream, hence the restriction that the stream is fully known in advance.

The receiver verifies the signed stream as follows: on receiving $B'_0 = < B, A_0 >$ she checks that

$$V(PK, A_0, B) = 1$$

then on receiving $B'_i = < B_i, A_i >$ (for $i \geq 1$) the receiver accepts B_i if

$$H(B'_i) = A_{i-1}$$

Thus the receiver has to compute a single public-key operation at the beginning, and then only one hash evaluation per block. Notice that no big table is needed in memory.

4 The On-Line Solution

In this case the sender does not know the entire stream in advance (e.g, live broadcast). In this scenario it is important that also the operation of signing (and not just verification) be fast, since the sender himself is bound to produce an authenticated stream at a potentially high rate.

ONE-TIME SIGNATURES. In the following we will use a special kind of signature scheme introduced in [11, 12]. These are signatures which are much faster to compute and verify than regular signatures since they are based on one-way functions and do not require a trapdoor function. Conjectured known one-way functions (as DES or SHA-1) are much more efficient then the known conjectured trapdoor functions as RSA. However these schemes cannot be used to sign an arbitrary number of messages but only a prefixed number of them (usually one). Several other 1-time schemes have been proposed [7, 3, 4]; in Section 6 we discuss possible instantiations for our purpose.

In this case also the stream is split into blocks. Initially the sender sends a signed public key for a 1-time signature scheme. Then he sends the first block along with a 1-time signature on its hash based on the 1-time public key sent in the previous block. The first block also contains a new 1-time public key to be used to verify the signature on the 2nd block and this structure is repeated in all the blocks.

More in detail: let us denote with (G, S, V) a regular signature scheme and with (g, s, v) a 1-time signature scheme. With H we still denote a collision-resistant hash function. The sender has long-lived keys $(SK, PK) = G(1^n)$. Let

$$\mathcal{B} = B_1, B_2, \ldots$$

be the original stream (notice that in this case we are not assuming the stream to be finite) and

$$\mathcal{B}' = B_0', B_1', B_2', \ldots$$

the signed stream constructed as follows. For each $i \geq 1$ let us denote with $(sk_i, pk_i) = g(1^n)$ the output of an independent run of algorithm g. Then

$$B_0' = < pk_0, S(SK, pk_0) >$$

(public keys of 1-time signature schemes are usually short so they need not to be hashed before signing)

$$B_i' = < B_i, pk_i, s(sk_{i-1}, H(B_i, pk_i)) > \text{ for } i \geq 1$$

Notice that apart from a regular signature on the first block, all the following signatures are 1-time ones, thus much faster to compute (including the key generation, which however does not have to be done on the fly.)

The receiver verifies the signed stream as follows. On receiving $B_0' =< pk_0, A_0 >$ she checks that

$$V(PK, A_0, pk_0) = 1$$

and then on receiving $B_i' =< B_i, pk_{i+1}, A_i >$ she checks that

$$v(pk_{i-1}, A_i, H(B_i, pk_i)) = 1$$

whenever one of these checks fails, the receiver stops playing the stream. Thus the receiver has to compute a single public-key operation at the beginning, and then only one 1-time signature verification per block.

5 Proofs of Security

We were able to prove the security of our stream signature schemes according to the definitions presented in Section 2, provided that the underlying components used to build the schemes are secure on their own. The proofs of the following theorems appear in the full version of the paper [8] due to space limitations.

THE OFF-LINE CASE. Let us denote with $(\mathcal{G}_{off}, \mathcal{S}_{off}, \mathcal{V}_{off})$ the off-line stream signature scheme described in Section 3. With (G, S, V) let us denote the "regular" signature scheme and with H the hash function used in the construction. The following holds.

Theorem 4. *If (G, S, V) is a secure signature scheme and H is a collision-resistant hash function then the resulting stream signature scheme $(\mathcal{G}_{off}, \mathcal{S}_{off}, \mathcal{V}_{off})$ is secure.*

THE ON-LINE CASE. Let us denote with $(\mathcal{G}_{on}, \mathcal{S}_{on}, \mathcal{V}_{on})$ the on-line stream signature scheme described in Section 4. With (G, S, V) let us denote the "regular" signature scheme, with (g, s, v) the one-time signature scheme and with H the hash function used in the construction. The following holds.

Theorem 5. *If (G, S, V) and (g, s, v) are secure signature schemes and H is a collision-resistant hash function then the resulting stream signature scheme $(\mathcal{G}_{on}, \mathcal{S}_{on}, \mathcal{V}_{on})$ is secure.*

Remark 1: In the bodies and proofs of the above theorems we meant security as "existential unforgeability against adaptively chosen message attack". However the theorems hold for any notion of security defined in [9], that is the stream signature scheme inherits the same kind of security of the underlying signature scheme(s) provided that the hash function is collision-resistant.

Remark 2: The statements of the above theorems are valid not only in asymptotic terms, but have also a concrete interpretation which ultimately is reflected in the key lengths used in the various components in order to achieve the desired level of security of the full construction. It is not hard to see, by a close analysis of the proofs, that the results are pretty tight. That is, a forger for the stream signing scheme can be transformed into an attacker for one of the components (the hash function, the regular signature scheme and, a little less optimally, the 1-time signature scheme) which runs in about the same time, asks the same number of queries and has almost the same success probability. This is turns means that there is no major degradation in the level of security of the compound scheme and thus the basic components can be run with keys of ordinary length.

6 Implementation Issues

6.1 The Choice of the One-Time Signature Scheme

Several one-time schemes have been presented in the literature, see for example [11, 12, 7, 3, 4]. The main parameters of these schemes are signature length and verification time. In the solutions we know, these parameters impose conflicting requirements, i.e. if one wants a scheme with short signatures, verification time goes up, while schemes with longer signatures can have a much shorter verification time. In our on-line solution we would like to keep both parameters down. Indeed the verification should be fast enough to allow the receiver to consume the stream blocks at the same input rate she receives them. At the same time, since the signatures are embedded in the stream, it's important to keep them small so that they will not reduce the throughput rate of the original stream.

We present a scheme which obtain a reasonable compromise. In the final paper we will present several more schemes. The scheme is based on a 1-way function F in a domain D. It also uses a collision resistant hash function H. The scheme allows signing of a single m-bit message. It is based on a combinations of ideas from [11, 19]. Here are the details of the scheme.

Key Generation. Choose $m + \log m$ elements in D, let them be $a_1, \ldots, a_{m+\log m}$. This is the secret key. The publick key is

$$pk = H(F(a_1), \ldots, F(a_{m+\log m}))$$

Signing Algorithm. Let M be the message to be signed. Append to M the binary representation of the number of zero's in M's binary representation. Call M' the resulting binary string. The signature of M is $s_1, \ldots, s_{m+\log m}$ where $s_i = a_i$ if the ith bit of M' is 1 otherwise $s_i = F(a_i)$.

Verification Algorithm. Check if

$$H(t_1, \ldots, t_{m+\log m}) = pk$$

where $t_i = s_i$ if i^{th} bit of M' is 0 otherwise $t_i = F(s_i)$.

Security. Intuitively this scheme is secure since it is not possible to change a 0 into a 1 in the binary representation of the message M without having to invert the function F. It is possible to change a 1 into a 0, but that will increase the number of 0's in the binary representation of M causing a bit to flip from 0 to 1 in the last $\log m$ bits of M', and so forcing the attacker to invert F anyway.

Parameters. This scheme has signature length $|D|(m + \log m)$ where $|D|$ is the number of bits required to represent elements of D. The receiver has to compute 1 hash computation of H plus on the average $\frac{m+\log m}{2}$ computations of F.

In practice we assume that F maps 64-bit long strings into 64-bit long strings. Since collision resistance is not required from F we believe this parameter is sufficient. Conjectured good F's can be easily constructed from efficient block ciphers like DES or from fast hash function like MD5 or SHA-1. [4] Similarly H can be instantiated to MD5 or SHA-1. In general we may assume m to be 128 or 160 if the message to be signed if first hashed using MD5 or SHA-1.

The SHA-1 implementation has then signatures which are 1344 bytes long. The receiver has to compute F around 84 times on the average. With MD5 the numbers become 1080 bytes and 68 respectively. When used in our off-line scheme one also has to add 16 bytes for the embedding of the public key in the stream.

Remark: Comparing the RSA signature scheme with verification exponent 3 with the above schemes, one could wonder if the verification algorithm is really more efficient (2 multiplications verses 84 hash computations). Typical estimates today are that an RSA verification is comparable to 100 hash computations. However we remind the reader that we are trying to improve *both* signature generation and verification as this scheme is used in the on-line case and as such both operations have to be performed on-line and thus efficiently. The improvement in signature generation is much more substantial.

---

[4] As a cautionary remark to prevent attacks where the attacker builds a large table of evaluations of F, in practice F could be made different for each signed stream (or for each large portion of the signed stream) by defining $F(x)$ to be $G(Salt||X)$ where G is a one-way 128 bit to 64 bit function, and the $Salt$ is generated at random by the signer once for each stream or large pieces thereof.

6.2 Non-Repudiation

In case of a legal dispute over a content of a signed stream the receiver must bring to court some evidence. If the receiver saves the whole stream, then there is no problem. However in some cases, for example because of memory limitations, the receiver may be forced to discard the stream data after having consumed it. In these cases what should she save to protect herself in case of a legal dispute?

In the off-line solution, assuming the last block of the signed stream always has a special reserved value for the hash-chaining field, (say all 0's) she needs to save only the first signed block. Indeed this proves that she received something from the sender. Now we could conceivably move the burden of proof to the sender to reconstruct the whole stream that matches that first block and ends with the last block which has the reserved value for the hash-chaining field.

Similarly in the on-line solution, at a minimum the receiver needs to save the first signed block and all 1-time signatures and have the sender reconstruct the stream. However in practice, this may still be too much to save. In the final paper we will discuss various modifications to the on-line scheme that can be used in practice to substantially reduce the data that the receiver needs to store.

6.3 Hybrid Schemes

In the on-line scheme, the length of the embedded authentication information is of concern as it could cut into the throughput of the stream. In order to reduce it hybrid schemes can be considered. In this case we assume that some asynchrony between the sender and receiver is acceptable.

Suppose the sender can process a group (say 20) of stream blocks at a time before sending them. With a pipelined process this would only add an initial delay before the stream gets transmitted. The sender will sign with a one-time key only 1 block out of 20. The 20 blocks in between these two signed blocks will be authenticated using the off-line scheme. This way the long 1-time signatures and the the verification time can be amortized over the 20 blocks.

A useful feature of our proposed 1-time signature scheme is that it allows the verification of (the hash of) a message bit by bit. This allows us to actually "spread out" the signature bits and the verification time among the 20 blocks. Indeed if we assume that the receiver is allowed to play at most 20 blocks of unauthenticated information before stopping if tampering is detected we can do the following. We can distribute the signature bits among the 20 blocks and verify the hash of the first block bit by bit as the signature bits arrive. This maintains the stream rate stable since we do not have long signatures sent in a single block and verification now takes 3-4 hash computations per block, on *every* block.

It is also possible to remove the constraint on playing 20 blocks of unauthenticated information before tampering is detected. This requires a simple modification to our on-line scheme. Instead of embedding in block B_i its own 1-time signature, we embed the signature of the next block B_{i+1}. This means that in the on-line case blocks have to be processed two at a time now. When this modification is applied to the hybrid scheme, the signature bits in the current 20 blocks are used to authenticate the following 20 blocks so unauthenticated information is never played. However this means that now the sender has to process 40 blocks at a time in the hybrid scheme.

7 Applications

MPEG VIDEO AND AUDIO. In the case of MPEG video and audio, there are several methods for embedding authentication data. Firstly, the Video Elementary stream has a USER-DATA section where arbitrary user defined information can be placed. Secondly, the MPEG system layer allows for an elementary data stream to be multiplexed synchronously with the packetized audio and video streams. One such elementary stream could carry the authentication information. Thirdly, techniques borrowed from digital watermarking can be used to embed information in the audio and video itself at the cost of slight quality degradation. In the case of MPEG video since each frame is fairly large, (hundreds of kilobits) and the receiver is required to have a buffer of at least 1.8Mbits, both the off-line as well as the on-line solutions can be deployed without compromising picture quality. In the case of audio however, in the extreme case the bit rate could be very low (e.g., 32Kbits/s) and each audio frame could be small (approx. 1000 bytes) and the receiver's audio buffer may be tiny (< 2 Kbytes). In such extreme cases the on-line method, which requires around 1000 bytes of authentication information per block cannot be used without seriously cutting into audio quality. For these extreme cases, the best on-line strategy would be either to send the authentication information via a separate but multiplexed MPEG data stream. For regular MPEG audio, if the receiver has a reasonably sized buffer (say 32K) then by having a large audio block (say 20K) our on-line scheme would a server-introduced delay of approximately 5-6 seconds and a 5% quality degradation. If the receiver buffer is small but not tiny (say 3 K) a hybrid scheme would work: as an example of a scheme that can be built one could use groups of 33 hash—chained blocks of length 1000 bytes each; this would typically result in a 5% degradation and a server initial delay in the 20 second range.

JAVA. In the original version of java (JDK 1.0), for an applet coming from the network, first the startup class was loaded and then additional classes were (down) loaded by the class loader in a lazy fashion as and when the running applet first attempted to access them. Since our ideas apply not only to streams which are a linear sequence of blocks but in general to trees as well (where one block can invoke any of its children), based on our model, one way to sign java applets would be to sign the startup class and each downloaded class would have embedded in it the hashes of the additional classes that it downloads directly. However for code signing, Javasoft has adopted the multiple signature and hash table based approach in JDK1.1, where each applet is composed of one or several Java archives, each of which contains a signed table of hashes (the manifest) of its components. It is our belief that once java applets become really large and complex the shortcomings of this approach will become apparent: (1) the large size of the hash table in relation to the classes actually invoked during a run. This table has to fully extracted and authenticated before any class gets authenticated; (2) the computational cost of signing each of the manifests if an applet is composed of several archives; (3) accommodating classes or data resources which are generated on the fly by the application server based on a client request.

These could be addressed by using some of our techniques. Also the problem of how to sign audio/video streams will have to be considered in the future evolution of Java, since putting the hash of a large audio/video file in the manifest would not be acceptable.

BROADCAST APPLICATIONS. Our schemes (both the off-line and the on-line one) can be easily modified to fit in a broadcast scenario. Assume that the stream is being sent to a broadcast channel with multiple receivers who dynamically join or leave the channel. In this case a receiver who joins when the transmission is already started will not be able to authenticate the stream since she missed the first block that contained the signature. Both schemes however can be modified so that every once in a while apart from the regular chaining information, there will also be a regular digital signature on a block embedded in the stream. Receivers who are already verifying the stream via the chaining mechanism can ignore this signature whereas receivers tuned in at various time will rely on the first such signature they encounter to start their authentication chain. A different method to authenticate broadcasted streams, with weaker non-repudiation properties than ours, was proposed in [10].

LONG FILES WHEN COMMUNICATION IS AT COST. Our solution can be used also to authenticate long files in a way to reduce communication cost in case of tampering. Suppose that a receiver is downloading a long file from the Web. There is no "stream requirement" to consume the file as it is downloaded, so the receiver could easily receive the whole file and then check a signature at the end. However if the file has been tampered with, the user will be able to detect this fact only at the end. Since communication is at a cost (time spent online, bandwidth wasted etc) this is not a satisfactory solution. Using our solution the receiver can interrupt the transmission as soon as tampering is detected thus saving precious communication resources.

SIGNATURE BASED CONTENT-FILTERING AT PROXIES. Recently there has been interest in using digital signatures as a possible way to filter content admitted in by proxy servers through firewalls. Essentially when there is a firewall and one wishes to connect to an external server, then this connection can only be done via a proxy server. In essence one establishes a connection to a proxy and the proxy establishes a separate connection to the external server (if that is permitted). The proxy then simulates a connection between the internal machine and the external machine by copying data between the two connections. There has been some interest in modifying proxies so that they would only allow signed data to flow from the external server to the internal machine. However, since the proxy is only copying data as it arrives from the external connection into the internal connection and it cannot store all the incoming data before transferring it, the proxy cannot use a regular signature scheme for solving this problem. However, it is easy to see that in the proxy's view the data is a stream. Hence if there could be some standardized way to embed authentication data in such streams then techniques from this paper would prove useful in solving this problem.

8 Acknowledgments

We would like to thank Hugo Krawczyk for his advice, guidance and encouragement for this work. We would also like to thank Ran Canetti and Mike Wiener for helpful discussions.

References

1. M. Bellare, S. Micali. How to Sign Given any Trapdoor Permutation. *J. of the ACM*, 39(1):214–233, 1992.
2. J. Benaloh, M. de Mare. One-Way Accumulators: A Decentralized Alternative to Digital Signatures. *Advances in Cryptology–EUROCRYPT'93*. LNCS, vol.765, pp.274–285, Springer–Verlag, 1994.
3. D. Bleichenbacher, U. Maurer. Optimal Tree-Based One-time Digital Signature Schemes. *STACS'96*, LNCS, Vol. 1046, pp.363–374, Springer-Verlag.

4. D. Bleichenbacher, U. Maurer. On the efficiency of one-time digital signatures. *Advances in Cryptology–ASYACRYPT'96*, to appear.

5. W. Diffie, M. Hellman. New Directions in Cryptography. *IEEE Transactions on Information Theory*, IT-22(6):74–84, 1976.

6. T. ElGamal. A Public-Key Cryptosystem and a Signature Scheme based on Discrete Logarithms. *IEEE Transactions on Information Theory*, IT-31(4):469–472, 1985.

7. S. Even, O. Goldreich, S. Micali. On–Line/Off–Line Digital Signatures. *J. of Cryptology*, 9(1):35–67, 1996.

8. R. Gennaro, P. Rohatgi. How to Sign Digital Streams.
   Final version available from
   `http://www.research.ibm.com/security/papers1997.html`

9. S. Goldwasser, S. Micali, R. Rivest. A Digital Signature Scheme Secure Against Adaptive Chosen Message Attack. *SIAM J. Comp.* 17(2):281–308, 1988.

10. G. Itkis. Asymmetric MACs. Rump talk at Crypto'96.

11. L. Lamport. Constructing Digital Signatures from a One-Way Function. *Technical Report SRI Intl.* CSL 98, 1979.

12. R. Merkle. A Digital Signature based on a Conventional Encryption Function. *Advances in Cryptology–Crypto '87.* LNCS, vol.293, pp. 369–378, Springer–Verlag, 1988.

13. R. Merkle. A Certified Digital Signature. *Advances in Cryptology–Crypto '89.* LNCS, vol.435, pp. 218–238, Springer–Verlag, 1990.

14. National Institute of Standard and Technology. Secure Hash Standard. NIST FIPS Pub 180-1, 1995.

15. M. Naor, M. Yung. Universal One-Way Hash Functions and their Cryptographic Applications. *Proceedings of STOC 1989*, pp.33–43.

16. R. Rivest. The MD5 Message Digest Algorithm. Internet Request for Comments. April 1992.

17. R. Rivest, A. Shamir, L. Adleman. A Method for Obtaining Digital Signatures and Public Key Cryptosystems. *Comm. of the ACM*, 21(2):120–126, 1978.

18. J. Rompel. One-Way Functions are Necessary and Sufficient for Secure Signatures. *Proceedings of STOC 1990*, pp.387–394.

19. Winternitz. Personal communication to R. Merkle.

Merkle-Hellman Revisited: A Cryptanalysis of the Qu-Vanstone Cryptosystem Based on Group Factorizations

Phong Nguyen
Phong.Nguyen@ens.fr

Jacques Stern
Jacques.Stern@ens.fr

École Normale Supérieure
Laboratoire d'Informatique
45, rue d'Ulm
F – 75230 Paris Cedex 05

Abstract. Cryptosystems based on the knapsack problem were among the first public key systems to be invented and for a while were considered quite promising. Basically all knapsack cryptosystems that have been proposed so far have been broken, mainly by means of lattice reduction techniques. However, a few knapsack-like cryptosystems have withstood cryptanalysis, among which the Chor-Rivest scheme [2] even if this is debatable (see [16]), and the Qu-Vanstone scheme proposed at the Dagstuhl'93 workshop [13] and published in [14]. The Qu-Vanstone scheme is a public key scheme based on group factorizations in the additive group of integers modulo n that generalizes Merkle-Hellman cryptosystems. In this paper, we present a novel use of lattice reduction, which is of independent interest, exploiting in a systematic manner the notion of an orthogonal lattice. Using the new technique, we successfully attack the Qu-Vanstone cryptosystem. Namely, we show how to recover the private key from the public key. The attack is based on a careful study of the so-called Merkle-Hellman transformation.

1 Introduction

The knapsack problem is as follows : given a set $\{a_1, a_2, \ldots, a_n\}$ of positive integers and a sum $s = \sum_{i=1}^{n} x_i a_i$, where each $x_i \in \{0, 1\}$, recover the x_i. It is well known that this problem is NP-complete, and accordingly it is considered to be quite hard in the worst case. However some knapsacks are very easy to solve : if the set $S = \{a_1, a_2, \ldots, a_n\}$ of positive integers is a *superincreasing sequence*, e.g.

$$\forall i \geq 2 \quad a_i > \sum_{j=1}^{i-1} a_j,$$

then the corresponding knapsack can easily be solved in linear time. Most of the public key schemes based on knapsacks are of the following form :

The Public Key: a set of positive integers $\{a_1, a_2, \ldots, a_n\}$.

The Private Key: a method to transform the presumed hard public knap-snack into an easy knapsack.

The Message Space: all $0 - 1$ vectors of length n.

Encryption: a message $M = (x_1, x_2, \ldots, x_n)$ is enciphered into $C = \sum_{i=1}^{n} x_i a_i$.

In 1978, Merkle and Hellman [10] devised a method to convert superincreasing sequences into what they believed were hard knapsacks. If $S = \{a_1, a_2, \ldots, a_n\}$ is a superincreasing sequence and $a = \sum_{i=1}^{n} a_i$, select two coprime integers m and w such that $m > a$. The *Merkle-Hellman transformation* associated with the pair (m, w) is the function f that maps any $x \in \{0, 1, \ldots, m - 1\}$ to the least positive residue of wx modulo m. This function is a permutation, and its reciprocal f^{-1} maps any $y \in \{0, 1, \ldots, m - 1\}$ to the least positive residue of $w^{-1}y$ modulo m, where w^{-1} is an inverse of w modulo m. Merkle and Hellman applied such a transformation f to form a new knapsack $\bar{S} = \{b_1, b_2, \ldots, b_n\}$ where $b_i = f(a_i)$. To decrypt a ciphertext $c = \sum_{i=1}^{n} x_i b_i$, one computes $f^{-1}(c)$. Since

$$f^{-1}(c) \equiv \sum_{i=1}^{n} x_i b_i w^{-1} \equiv x_i a_i \pmod{m},$$

with $\sum_{i=1}^{n} x_i a_i \le a < m$, we have $f^{-1}(c) = \sum_{i=1}^{n} x_i a_i$. By solving the easy knap-sack $S = \{a_1, a_2, \ldots, a_n\}$, one recovers the x_i. Applying a sequence of Merkle-Hellman transformations is not equivalent to a single application, and hence, should enhance the security of the system. Unfortunately, these systems were both shown to be insecure (see [17, 1]). Despite the failure of Merkle-Hellman cryptosystems, researchers continued to search for knapsack-like cryptosystems because such systems are very easy to implement and can attain very high encryption/decryption rates. But most of the proposed knapsack-like cryptosystems have been broken (for a survey, see [12]), either by specific attacks or by the so-called low-density attacks.

The *density* of a knapsack $S = \{a_1, a_2, \ldots, a_n\}$ is defined to be $d = \frac{n}{N}$ where $N = \max_{1 \le i \le n} \log_2 a_i$. When the density is small (namely, less than 0.94...), one can prove the knapsack problem can be solved using lattice reduction with high probability (see [4]). Such attacks are called low-density attacks. The attack has recently been improved by [16], but is still uneffective against high-density knap-sacks. The few knapsack cryptosystems that have so far withstood all attacks use knapsacks of high density. In [13], Qu and Vanstone showed that Merkle-Hellman knapsack cryptosystems could be viewed as special cases of knapsack-like cryptosystems arising from subset factorizations in finite groups. They proposed a generalization of these knapsack cryptosystems by constructing a supposedly hard factorization of finite group, using Merkle-Hellman-like transformations and superincreasing sequences. This hard factorization problem can be restated as a knapsack problem of density higher than 3. We will attack the Qu-Vanstone system by showing how to recover the hidden easy factorization (the private key) from the presumed hard factorization (the public key), in a reasonable time.

2   The orthogonal lattice

We will use the word *lattice* for any integer lattice, that is any additive subgroup of $\mathbf{Z}^n$. Background on lattices can be found in [5, 3]. We denote vectors by bold-face lowercase letters. Let Λ be a lattice in $\mathbf{Z}^n$.

Let $\mathbf{b}_1, \ldots, \mathbf{b}_d$ be vectors of Λ. These d vectors form a *basis* of Λ if they are linearly independent over $\mathbf{Z}$, and if any element of Λ can be expressed as a linear combination of the $\mathbf{b}_i$'s with integral coefficients. There exists at least one basis of Λ. The bases of Λ all have the same cardinality, called the *dimension* of Λ. We say that Λ is a *sublattice* of a lattice Ω in $\mathbf{Z}^n$ if Ω contains Λ and if both have the same dimension. All bases of Λ span the same $\mathbf{Q}$-vector subspace of $\mathbf{Q}^n$, which we denote by E_Λ. The dimension of E_Λ over $\mathbf{Q}$ is equal to the dimension of Λ. Define the lattice $\overline{\Lambda} = E_\Lambda \cap \mathbf{Z}^n$. Λ is a sublattice of $\overline{\Lambda}$. We say that Λ is a *complete lattice* if $\Lambda = \overline{\Lambda}$. In particular, $\overline{\Lambda}$ is a complete lattice.

Let $(\mathbf{x}, \mathbf{y}) \mapsto \mathbf{x}.\mathbf{y}$ be the usual euclidian inner product, and $\|.\|$ be its corresponding norm. Let $F = (E_\Lambda)^\perp$ be the orthogonal vector subspace with respect to this inner product. We define the *orthogonal lattice* to be $\Lambda^\perp = F \cap \mathbf{Z}^n$. Thus, $\Lambda^\perp$ is a complete lattice in $\mathbf{Z}^n$, with dimension $n - d$ if d is the dimension of Λ. This implies that $(\Lambda^\perp)^\perp$ is equal to $\overline{\Lambda}$. Let $\mathcal{B} = (\mathbf{b}_1, \ldots, \mathbf{b}_d)$ be a basis of Λ. Decompose each $\mathbf{b}_j$ over the canonical basis of $\mathbf{Z}^n$ as :

$$\mathbf{b}_j = \begin{pmatrix} b_{1,j} \\ b_{2,j} \\ \vdots \\ b_{n,j} \end{pmatrix}$$

Define the $n \times d$ integral matrix $B = (b_{i,j})_{1 \le i \le n, 1 \le j \le d}$. The lattice Λ is spanned by the columns of B : we say that Λ is *spanned* by B. Let $Q = {}^t B B$ be the $d \times d$ symmetric Gram matrix. The determinant of Q is a positive integer independent of $\mathcal{B}$. The *determinant* of Λ is defined as $\det(\Lambda) = \sqrt{\det(B)}$.

Theorem 1. *Let Λ be a complete lattice in $\mathbf{Z}^n$. Then $\det(\Lambda^\perp) = \det(\Lambda)$.*

Proof. We have $\Lambda = E_\Lambda \cap \mathbf{Z}^n$ and $\Lambda^\perp = E_\Lambda^\perp \cap \mathbf{Z}^n$. We know from [9] that :

$$\det(\mathbf{Z}^n) = \frac{\det(E_\Lambda \cap \mathbf{Z}^n)}{\det((E_\Lambda)^\perp \cap (\mathbf{Z}^n)^*)},$$

where $(\mathbf{Z}^n)^*$ denotes the polar lattice of $\mathbf{Z}^n$. But $\det(\mathbf{Z}^n) = 1$ and $(\mathbf{Z}^n)^* = \mathbf{Z}^n$, therefore $\det(\Lambda^\perp) = \det(\Lambda)$.   □

Corollary 2. *Let Λ be a lattice in $\mathbf{Z}^n$. Then $\det((\Lambda^\perp)^\perp) = \det(\Lambda^\perp) = \det(\overline{\Lambda})$.*

In 1982, Lenstra, Lenstra and Lovász introduced the famous LLL-algorithm [8], a polynomial time algorithm that computes a so-called LLL-reduced basis of any given lattice. For definitions and proofs regarding LLL-reduced bases, we refer to [8, 3]. In this paper, we only need the following properties of LLL-reduced bases :

Theorem 3. *Let* $(\mathbf{b}_1, \ldots, \mathbf{b}_d)$ *be an LLL-reduced basis of a lattice* Λ *in* $\mathbf{Z}^n$. *Then :*

1. $\det(\Lambda) \leq \prod_{i=1}^{d} \|\mathbf{b}_i\| \leq 2^{d(d-1)/4} \det(\Lambda)$.

2. *For any linearly independent vectors* $\mathbf{x}_1, \ldots, \mathbf{x}_t \in \Lambda$, *and* $1 \leq j \leq t$:

$$\|\mathbf{b}_j\| \leq 2^{(d-1)/2} \max(\|\mathbf{x}_1\|, \ldots, \|\mathbf{x}_t\|).$$

We now describe a basic algorithm to compute an LLL-reduced basis of an orthogonal lattice. Let $\mathcal{B} = (\mathbf{b}_1, \ldots, \mathbf{b}_d)$ be a basis of Λ, and $B = (b_{i,j})$ be its corresponding $n \times d$ matrix. Let c be a positive integer constant. Define Ω to be the lattice in $\mathbf{Z}^{n+d}$ spanned by the following $(n + d) \times n$ matrix :

$$B^{\perp} = \begin{pmatrix} c \times b_{1,1} & c \times b_{2,1} & \cdots & c \times b_{n,1} \\ c \times b_{1,2} & c \times b_{2,2} & \cdots & c \times b_{n,2} \\ \vdots & \vdots & \ddots & \vdots \\ c \times b_{1,d} & c \times b_{2,d} & \cdots & c \times b_{n,d} \\ 1 & 0 & \cdots & 0 \\ 0 & 1 & \ddots & \vdots \\ \vdots & \vdots & \ddots & 0 \\ 0 & 0 & \cdots & 1 \end{pmatrix}$$

A similar matrix is used in [7]. The matrix $B^{\perp}$ is divided in two blocks : the upper $d \times n$ block is $c\,{}^t B$ and the lower $n \times n$ block is the identity matrix. Let $p_{\uparrow}$ and $p_{\downarrow}$ be the two projections that map any vector of $\mathbf{Z}^{n+d}$ to respectively the vector of $\mathbf{Z}^d$ made of its first d coordinates, and the vector of $\mathbf{Z}^n$ of its last n coordinates, all with respect to the canonical basis. Let $\mathbf{x}$ be a vector of Ω and denote $\mathbf{y} = p_{\downarrow}(\mathbf{x})$. Then

$$p_{\uparrow}(\mathbf{x}) = c \begin{pmatrix} \mathbf{y}.\mathbf{b}_1 \\ \vdots \\ \mathbf{y}.\mathbf{b}_d \end{pmatrix}.$$

Hence, $\mathbf{y} \in \Lambda^{\perp}$ if and only if $p_{\uparrow}(\mathbf{x}) = 0$. Furthermore, if $\|\mathbf{x}\| \leq c$, then $p_{\uparrow}(\mathbf{x}) = 0$.

Theorem 4. *Let* $(\mathbf{x}_1, \mathbf{x}_2, \ldots, \mathbf{x}_n)$ *be an LLL-reduced basis of* Ω. *If*

$$c > 2^{(n-1)/2+(n-d)(n-d-1)/4} \det(\overline{\Lambda}),$$

then $(p_{\downarrow}(\mathbf{x}_1), p_{\downarrow}(\mathbf{x}_2), \ldots, p_{\downarrow}(\mathbf{x}_{n-d}))$ *is an LLL-reduced basis of* $\Lambda^{\perp}$.

Using Hadamard's inequality, we derive the following algorithm :

Algorithm 5. *Given a basis* $(\mathbf{b}_1, \mathbf{b}_2, \ldots, \mathbf{b}_d)$ *of a lattice* Λ *in* $\mathbf{Z}^n$, *this algorithm computes an LLL-reduced basis of* $\Lambda^{\perp}$.

1. *Select* $c = \lceil 2^{(n-1)/2+(n-d)(n-d-1)/4} \prod_{j=1}^{d} \|\mathbf{b}_j\| \rceil$.

2. *Compute the $(n + d) \times n$ integral matrix $B^\perp$ from c and the $n \times d$ matrix $B = (b_{i,j})$ corresponding to $\mathbf{b}_1, \mathbf{b}_2, \ldots, \mathbf{b}_d$.*

3. *Compute an LLL-reduced basis $(\mathbf{x}_1, \mathbf{x}_2, \ldots, \mathbf{x}_n)$ of the lattice spanned by $B^\perp$.*

4. *Output $(p_\downarrow(\mathbf{x}_1), p_\downarrow(\mathbf{x}_2), \ldots, p_\downarrow(\mathbf{x}_{n-d}))$.*

One can prove that this is a deterministic polynomial time algorithm with respect to the space dimension n, the lattice dimension d and any upper bound of the bit-length of the $\|\mathbf{b}_j\|$'s. In practice, one does not need to select such a large constant c because the theoretical bounds of the LLL algorithm (theorem 3) are quite pessimistic. We will use this algorithm throughout the attack.

3 The cryptanalysis of the Qu-Vanstone scheme

3.1 High level description of the Qu-Vanstone scheme

Since the Qu-Vanstone scheme is quite complicated, we give a simplified exposure. Additional information can be found in appendix and in [13, 14].

Let n be a positive integer of the form $n = d_1 d_2 d_3 d_4 d_5$, where $2^{s-1} \le d_\ell < 2^s$ (for $\ell = 1, 2, 3, 4$), $d_5 \le 16$, and s is some fixed even positive integer. Let $G = \mathbf{Z}_n$ be the additive group of integers modulo n. Qu and Vanstone found a way to build an efficient subset factorization in G. Namely, with help of 4 superincreasing sequences and 4 Merkle-Hellman transformations, they construct s blocks $C_i = \{c[i, j] : 0 \le j \le 15\}$ of 16 integers of G where $1 \le i \le s$ such that : for any $g \in G$ of form $g = \sum_{i=1}^{s} c[i, j_i] \pmod{n}$, one can quickly recover the j_i's from g and a trapdoor. This construction is intricate, a detailed description can be found in the appendix.

Qu and Vanstone further use k additional Merkle-Hellman-like transformations and s permutations to hide the subset factorization. The process consists of k iterations, starting with $m^{(0)} = n$ and $c^{(0)}[i, j] = c[i, j]$, $1 \le i \le s$, $0 \le j \le 15$. Consider the eth iteration $(1 \le e \le k)$:

- select s positive integers $a_1^{(e-1)}, \ldots, a_s^{(e-1)}$ such that $0 \le a_i^{(e-1)} < m^{(e-1)}$, and define $\bar{c}^{(e)}[i, j] = c^{(e-1)}[i, j] + a_i^{(e-1)} \pmod{m^{(e-1)}}$.

- select $m^{(e)}$ strictly greater than $\sum_{j=1}^{s} \max_{0 \le j \le 15} \bar{c}^{(e)}[i, j]$. Choosing $w^{(e)}$ coprime to $m^{(e)}$, define $c^{(e)}[i, j] = w^{(e)} \bar{c}^{(e)}[i, j] \pmod{m^{(e)}}$.

These Merkle-Hellman-like transformations differ from the original Merkle-Hellman transformations by the use of a modular addition which is performed before the modular multiplication. Now that the $c^{(k)}[i, j]$'s are defined, select s permutations $\pi_1, \ldots, \pi_s$ acting on $\{0, 1, \ldots, 15\}$. Let

$$d[i, j] = c[i, \pi_i^{-1}(j)] - c[i, \pi_i^{-1}(0)] \pmod{m^{(k)}}.$$

Notice that $d[i, 0] = 0$. Let $C = \sum_{i=1}^{s} c^{(k)}[i, \pi_i^{-1}(0)] \pmod{m^{(k)}}$.

The public key consists of the s blocks $D_i = \{d[i,j] \ : 0 \leq j \leq 15\}$, that is $15s$ non-zero positive integers. The private key is : C, π_i's, $w^{(e)}$'s, $m^{(e)}$'s, $\sum_{i=1}^{s} a_i^{(e)}$'s and the trapdoor corresponding to the subset factorization.

The message space is $\mathbf{Z}_{16^s}$, and the numbers in each D_i are roughly around $s^k n$. Qu and Vanstone suggested $s \geq 32$ and $k \geq 3$. In the smallest case, the message space is $\mathbf{Z}_{128}$, and the maximum element in the public key is less than 2^{151}. The keys are quite large but encryption/decryption rates are high.

To encode a message m, write m in its base 16 expansion as $m = p_1 + 16p_2 + \ldots + 16^{s-1}p_s$, where $0 \leq p_i \leq 15$. Then the ciphertext associated with m is

$$c = d[1, p_1] + d[2, p_2] + \ldots + d[s, p_s].$$

To decrypt c, compute $c^{(k)} = c + C \pmod{m^{(k)}}$. Then invert Merkle-Hellman-like transformations by applying the following process with $e = k, k-1, \ldots, 1$:

$$
\begin{aligned}
\overline{c}^{(e)} &= (w^{(e)})^{-1} c^{(e)} \pmod{m^{(e)}}, \\
c^{(e-1)} &= \overline{c}^{(e)} - \sum_{i=1}^{s} a_i^{(e-1)} \pmod{m^{(e-1)}}.
\end{aligned}
$$

At this point, $c^{(0)} = \sum_{i=1}^{s} c[i, j_i] \pmod{n}$ where each $j_i = \pi_i^{-1}(p_i)$ is still unknown. From the subset factorization trapdoor, the j_i's can be recovered. This technical step is described in the appendix. From the j_i's, one computes $p_i = \pi_i(j_i)$ and the message $m = p_1 + 16p_2 + \ldots + 16^{s-1}p_s$.

This public key system has features similar to the original knapsack scheme. The security rests on the Merkle-Hellman-like transformations that hide the 4 superincreasing sequences and the coset structure. The knapsack based on the blocks D_i has density higher than 3, so it looks immune to the usual low-density attacks. Qu and Vanstone discuss several attacks on this system in their paper [13]. We now describe our attack which mainly consists of two steps : we first attack Merkle-Hellman-like transformations by reducing several orthogonal lattices, then we compute successive orthogonal lattices to reveal the secret key. The first step is quite general but the second step is based on the particular structure of the hidden subset factorization. We advise to read the further description of the Qu-Vanstone scheme given in appendix in order to fully understand the second step.

3.2 Peeling off Merkle-Hellman transformations

Let N be an integer and $\mathbf{c}^{(0)}, \mathbf{c}^{(1)}, \ldots, \mathbf{c}^{(k)}$ be vectors of $\mathbf{Z}^N$ such that :

$$
\begin{aligned}
\|\mathbf{c}^{(e)}\| &\leq \sqrt{N} m^{(e)}, \ 0 \leq e \leq k & (1) \\
\mathbf{c}^{(e)} &\equiv w^{(e)} \mathbf{c}^{(e-1)} \pmod{m^{(e)}}, \ 1 \leq e \leq k & (2)
\end{aligned}
$$

Note that in equation (2), we mean component-wise operations and that we only assume congruences, not necessarily equalities. Under these hypotheses (1) and (2), we will see that $\mathbf{c}^{(0)}$ and $\mathbf{c}^{(k)}$ almost share the same orthogonal lattice.

Heuristic 6. *Let Λ be the lattice spanned by $\mathbf{c}^{(k)}$. Let $(\mathbf{e}_1,\dots,\mathbf{e}_{N-1})$ be an LLL-reduced basis of $\Lambda^\perp$. If we denote by Γ the lattice spanned by $(\mathbf{e}_1,\dots,\mathbf{e}_{N-k-1})$, then $\mathbf{c}^{(0)} \in \Gamma^\perp$.*

This heuristic confines $\mathbf{c}^{(0)}$ in a low-dimensional lattice that we can determine just by knowing $\mathbf{c}^{(k)}$. When $m^{(0)} \ll m^{(1)} \ll \dots \ll m^{(k)}$, this heuristic works well in practice. Namely, if we define

$$m = \min\left\{ \frac{m^{(1)}}{m^{(0)}}, \frac{m^{(2)}}{m^{(1)}}, \dots, \frac{m^{(k)}}{m^{(k-1)}} \right\},$$

experiments show that the heuristic is verified as soon as $m \geq 8$ and $N \geq 50$. We are unable to prove this heuristic, but we can offer some explanations.

Lemma 7. *Let $\mathbf{x}$ be a vector of $\mathbf{Z}^N$ such that $\mathbf{x}\perp\mathbf{c}^{(k)}$ and $\|\mathbf{x}\| < m/\sqrt{N}$. Then $\mathbf{x}$ is orthogonal to $\mathbf{c}^{(k-1)}, \mathbf{c}^{(k-2)}, \dots, \mathbf{c}^{(0)}$.*

Proof. We have $\mathbf{x}.\mathbf{c}^{(k-1)} \equiv 0 \pmod{m^{(k)}}$ since $\mathbf{c}^{(k)} \equiv w^{(k)}\mathbf{c}^{(k-1)} \pmod{m^{(k)}}$ and $\mathbf{x}.\mathbf{c}^{(k)} = 0$. If we assume that $\mathbf{x}$ is not orthogonal to $\mathbf{c}^{(k-1)}$, then $|\mathbf{x}.\mathbf{c}^{(k-1)}| \geq m^{(k)}$. Therefore, by Cauchy-Schwarz and inequality (1) :

$$m^{(k)} \leq \|\mathbf{x}\|.\|\mathbf{c}^{(k-1)}\| \leq \|\mathbf{x}\| m^{(k-1)}\sqrt{N}.$$

This contradicts the fact that $\|\mathbf{x}\| \leq m/\sqrt{N}$. Thus $\mathbf{x}\perp\mathbf{c}^{(k-1)}$. Iterating this process, we find that $\mathbf{x}$ is orthogonal to $\mathbf{c}^{(k-2)},\dots,\mathbf{c}^{(0)}$. □

This means that if $\mathbf{x} \in \Lambda^\perp$ is short enough, then $\mathbf{x}$ is orthogonal to $\mathbf{c}^{(0)}$. Now we will see that there exist $N - k - 1$ independent vectors of $\Lambda^\perp$ that are short, and hopefully short enough.

Lemma 8. *Let Ω be the lattice spanned by*

$$\left(\mathbf{c}^{(0)}, \lfloor\frac{w^{(1)}\mathbf{c}^{(0)}}{m^{(1)}}\rfloor, \lfloor\frac{w^{(2)}\mathbf{c}^{(1)}}{m^{(2)}}\rfloor, \dots, \lfloor\frac{w^{(k)}\mathbf{c}^{(k-1)}}{m^{(k)}}\rfloor \right).$$

Then $(\mathbf{c}^{(0)}, \mathbf{c}^{(1)}, \dots, \mathbf{c}^{(k)})$ is a sublattice of Ω, and

$$\det(\Omega) \leq \|\mathbf{c}^{(0)}\| N^{k/2} < m^{(0)} N^{(k+1)/2}.$$

Proof. We have $\mathbf{c}^{(e)} = w^{(e)}\mathbf{c}^{(e-1)} - m^{(e)}\lfloor\frac{w^{(e)}\mathbf{c}^{(e-1)}}{m^{(e)}}\rfloor$ for $1 \leq e \leq k$. Therefore $(\mathbf{c}^{(0)}, \mathbf{c}^{(1)}, \dots, \mathbf{c}^{(k)})$ is a sublattice of Ω. Furthermore :

$$\det(\Omega) = \|\mathbf{c}^{(0)} \wedge \lfloor\frac{w^{(1)}\mathbf{c}^{(0)}}{m^{(1)}}\rfloor \wedge \dots \wedge \lfloor\frac{w^{(k)}\mathbf{c}^{(k-1)}}{m^{(k)}}\rfloor\|$$

$$= \|\mathbf{c}^{(0)} \wedge (\frac{w^{(1)}\mathbf{c}^{(0)}}{m^{(1)}} - \lfloor\frac{w^{(1)}\mathbf{c}^{(0)}}{m^{(1)}}\rfloor) \wedge \dots \wedge (\frac{w^{(k)}\mathbf{c}^{(k-1)}}{m^{(k)}} - \lfloor\frac{w^{(k)}\mathbf{c}^{(k-1)}}{m^{(k)}}\rfloor)\|.$$

Since $\|\frac{w^{(e)}\mathbf{c}^{(e-1)}}{m^{(e)}} - \lfloor\frac{w^{(e)}\mathbf{c}^{(e-1)}}{m^{(e)}}\rfloor\| \leq \sqrt{N}$, this proves that :

$$\det(\Omega) \leq \|\mathbf{c}^{(0)}\| N^{k/2} < m^{(0)} N^{(k+1)/2}.$$

□

Since $\det(\Omega^\perp) = \det(\overline{\Omega})$, we can thus hope that there exists a basis of $\Omega^\perp$ whose vectors have norm less than $m' = (m^{(0)} N^{(k+1)/2})^{1/(N-k-1)}$. But these $N-k-1$ vectors also belong to $\Lambda^\perp$, so the first $N-k-1$ vectors of any LLL-reduced basis of $\Lambda^\perp$ are likely to have norm less than m'. Since m' is very small (smaller than $m/\sqrt{N}$ most of the time), it is not surprising that by lemma 7 $\mathbf{e}_1, \mathbf{e}_2, \ldots, \mathbf{e}_{N-k-1}$ are orthogonal to $\mathbf{c}^{(0)}$, which implies that $\mathbf{c}^{(0)} \in \Gamma^\perp$.

Although the Qu-Vanstone scheme uses Merkle-Hellman-like transformations instead of Merkle-Hellman transformations, there are particular vectors related to the scheme that satisfy conditions (1) and (2).

Let $N = 15s$. We index the coordinates of any vector of $\mathbf{Z}^N$ by $\gamma(i,j) = 15(i-1)+j-1$, where $1 \le i \le s, 1 \le j \le 15$. From the public key, we construct the vector $\mathbf{c}^{(k)}$ whose $\gamma(i,j)$ entry is $d[i,j]$. For $e = k-1, k-2, \ldots, 0$, let $\mathbf{c}^{(e)}$ be the (unknown) vector whose $\gamma(i,j)$ entry is $\bar{c}^{(e)}[i, \pi_i^{-1}(j)] - \bar{c}^{(e)}[i, \pi_i^{-1}(0)]$.

Lemma 9. The vectors $\mathbf{c}^{(0)}, \mathbf{c}^{(1)}, \ldots, \mathbf{c}^{(k)}$ satisfy conditions (1) and (2).

Proof. Since the coordinates of each $\mathbf{c}^{(e)}$ are less than $m^{(e)}$ in absolute value, we have (1). Write the Merkle-Hellman-like equations defining $d[i,j]$'s starting with $c^{(k)}[i,j]$'s, ending with $c^{(0)}[i,j]$'s. Collecting additions and multiplications that use the same modulus, $a_i^{(e)}$'s disappear by subtraction, proving (2). $\qquad \square$

From the description of the scheme, we know that $m \approx s \ge 32$, therefore heuristic 6 is likely to be satisfied. Hence, applying algorithm 5 twice, we can construct $k+1$ vectors $\mathbf{e}_1, \ldots, \mathbf{e}_{k+1}$ of $\mathbf{Z}^N$ such that there exist $\lambda_1, \ldots, \lambda_{k+1} \in \mathbf{Z}$ satisfying

$$\mathbf{c}^{(0)} = \lambda_1 \mathbf{e}_1 + \lambda_2 \mathbf{e}_2 + \ldots + \lambda_{k+1} \mathbf{e}_{k_1}.$$

In the second step of the attack, we determine these unknown integers λ_j. The knowledge of $\mathbf{c}^{(0)}$ then reveals the trapdoor and the rest of the secret key : this is sketched in the appendix because it is based on the structure of the subset factorization. We emphasize that the difficult part of the attack is to determine $\mathbf{c}^{(0)}$, not to obtain the secret key from $\mathbf{c}^{(0)}$ which is rather easy.

3.3 Breaking the kernel of the system

We say that $C_i = \{c[i,j] : 0 \le j \le 15\}$ is a *weak block* if $f(i)$ is of form $(4, i')$. For the definition of f, we refer to the description of the scheme in appendix. Clearly, half of the s blocks C_i are weak blocks. We call these blocks weak due to the following :

Lemma 10. Let $C_i = \{c[i,j] : 0 \le j \le 15\}$ be a weak block.

1. For $j \in \{0, 4, 8, 12\}$, we have

$$c[i, j+1] + c[i, j+2] \equiv c[i, j] + c[i, j+3] \pmod{d_1 d_2 d_3 d_4}.$$

2. There exist distinct $j_1(i)$, $j_2(i)$ and $j_3(i)$ computable from π_i such that

$$\hat{c}[i, j_1(i)] + \hat{c}[i, j_2(i)] - \hat{c}[i, j_3(i)] \equiv 0 \pmod{d_1 d_2 d_3 d_4},$$

where $\hat{c}[i,j]$ denotes the $\gamma(i,j)$ entry of $\mathbf{c}^{(0)}$.

Proof. From the definition of the $c[i,j]'s$, if $\lfloor j/4 \rfloor = \lfloor j'/4 \rfloor$ then

$$c[i,j] - c[i,j'] \equiv a^{v+2u}[f(i),w] - a^{v+2u}[f(i),w'] \pmod{n},$$

where $j = w + 4v + 8u$ and $j' = w' + 4v + 8u$. Since $f(i)$ is of form $(4,i')$, we obtain 1 by definition of $a^t[4,i',w]$. To prove 2, write $\pi_i^{-1}(0)$ as $j + \ell$ where $j = \lfloor \pi_i^{-1}(0)/4 \rfloor$. Apply 1 to find distinct j_1^*, j_2^*, j_3^* such that

$$c[i,j_1^*] + c[i,j_2^*] \equiv c[i,j_3^*] + c[i,j+\ell] \pmod{d_1 d_2 d_3 d_4}.$$

Conclude with $j_1(i) = \pi_i(j_1^*)$, $j_2 = \pi_i(j_2^*)$ and $j_3 = \pi_i(j_3^*)$. □

To simplify the exposition of the attack, we now assume that f and the π_i's are known to the attacker. We will show how to adapt the attack to the general case at the end of the section. We define a transformation ϕ that maps any $\mathbf{x}$ of $\mathbf{Z}^N$ to

$$\phi(x) = \begin{pmatrix} x[i_1,j_1(i_1)] + x[i_1,j_2(i_1)] - x[i_1,j_3(i_1)] \\ x[i_2,j_1(i_2)] + x[i_2,j_2(i_2)] - x[i_2,j_3(i_2)] \\ \vdots \\ x[i_{s/2},j_1(i_{s/2})] + x[i_{s/2},j_2(i_{s/2})] - x[i_{s/2},j_3(i_{s/2})] \end{pmatrix},$$

where $C_{i_1}, C_{i_2}, \ldots, C_{i_{s/2}}$ denote the $s/2$ weak blocks, and $x[i,j]$ denotes the $\gamma(i,j)$ entry of $\mathbf{x}$. One sees that ϕ is linear, which implies that

$$\phi(\mathbf{c}^{(0)}) = \lambda_1 \phi(\mathbf{e}_1) + \ldots + \lambda_{k+1} \phi(\mathbf{e}_{k+1}).$$

By lemma 10, the vector $\frac{\phi(\mathbf{c}^{(0)})}{d_1 d_2 d_3 d_4}$ has integral entries, so it must belong to $\overline{\Omega}$ where Ω denotes the lattice spanned by $\phi(\mathbf{e}_1), \phi(\mathbf{e}_2), \ldots, \phi(\mathbf{e}_{k+1})$, which we can determine. But this vector is unusually short : indeed, each coordinate of $\frac{\phi(\mathbf{c}^{(0)})}{d_1 d_2 d_3 d_4}$ is less than $3(d_5 - 1) \leq 45$ in absolute value, which makes a norm less than $45\sqrt{s/2}$ (note that this is a very pessimistic bound). Therefore it must have small coordinates with respect to any LLL-reduced basis because an LLL-reduced basis is almost orthogonal :

Lemma 11. *Let $(\mathbf{b}_1, \mathbf{b}_2, \ldots, \mathbf{b}_d)$ be an LLL-reduced basis of a lattice Λ. If $\mathbf{x} = \sum_{j=1}^d x_j \mathbf{b}_j$ where $x_j \in \mathbf{R}$ then, for $1 \leq j \leq d$,*

$$|x_j| \cdot \|\mathbf{b}_j\| \leq \|\mathbf{x}\| \sqrt{2^{j-1} \frac{(9/2)^{d-j} + 6}{7}}.$$

(this statement can be found in an unpublished draft [11] by P. Montgomery)

Proof. Denote by $(\mathbf{b}_1^*, \ldots, \mathbf{b}_d^*)$ the corresponding orthogonal Gram-Schmidt $\mathbf{Q}$-basis. Decompose $\mathbf{x}$ as $\mathbf{x} = \sum_{j=1}^d x_j^* \mathbf{b}_j^*$. By orthogonality, one finds that

$$x_j = x_j^* - \sum_{i=j+1}^d x_i \frac{\mathbf{b}_i . \mathbf{b}_j^*}{\|\mathbf{b}_j\|^2}.$$

It follows by induction on $d - j$ that :

$$|x_j| \le |x_j^*| + \frac{1}{3} \sum_{i=j+1}^{d} (3/2)^{i-j} |x_i^*|.$$

Since $(\mathbf{b}_1, \ldots, \mathbf{b}_d)$ is an LLL-reduced basis, if $j \le i$ then $\|\mathbf{b}_j\| \le 2^{(i-1)/2} \|\mathbf{b}_i^*\|$. From this and Cauchy-Schwarz, we obtain

$$|x_j|^2 \|\mathbf{b}_j\|^2 \le 2^{j-1} (\sum_{i=j}^{d} |x_i^*|^2 \|\mathbf{b}_i^*\|^2)(1 + \frac{1}{3^2}((9/2) + \cdots + (9/2)^{d-j}),$$

and the result follows. □

Hence, we compute an LLL-reduced basis $(\mathbf{b}_1, \ldots, \mathbf{b}_d)$ of $\overline{\Omega} = (\Omega^{\perp})^{\perp}$ by applying algorithm 5 twice. The unknown vector $\frac{\phi(\mathbf{c}^{(0)})}{d_1 d_2 d_3 d_4}$ has integral coordinates x_j with respect to $(\mathbf{b}_1, \ldots, \mathbf{b}_d)$. We make an exhaustive search on the x_j's within the bounds given by lemma 11. Since we are in low dimension ($d \le k + 1$), these bounds are very small, making exhaustive search possible.

Assume that one wants to check whether $\mathbf{x} = \sum_{j=1}^{d} x_j \mathbf{b}_j$ is the expected $\frac{\phi(\mathbf{c}^{(0)})}{d_1 d_2 d_3 d_4}$. Decompose each $\mathbf{b}_j$ as a linear combination of $\phi(\mathbf{e}_\ell)$'s with rational coefficients. Derive an integral linear dependence relation of form

$$\mu \mathbf{x} = \mu_1 \phi(\mathbf{e}_1) + \cdots + \mu_{k+1} \phi(\mathbf{e}_{k+1}),$$

where $\mu > 0$ and $gcd(\mu, \mu_1, \ldots, \mu_{k+1}) = 1$. Since

$$d_1 d_2 d_3 d_4 \frac{\phi(\mathbf{c}^{(0)})}{d_1 d_2 d_3 d_4} = \lambda_1 \phi(\mathbf{e}_1) + \cdots + \lambda_{k+1} \phi(\mathbf{e}_{k+1}),$$

it is likely that $d_1 d_2 d_3 d_4 = \mu$ and $\lambda_j = \mu_j$ if $\mathbf{x}$ is the expected vector. Since $d_1 d_2 d_3 d_4$ has bit-length $4s$, we can quickly check whether μ is consistent. Furthermore, we obtain $d_1 d_2 d_3 d_4$ and the λ_j's, which gives $\mathbf{c}^{(0)}$. But we can easily check whether this is a consistent $\mathbf{c}^{(0)}$, because $\mathbf{c}^{(0)}$ reveals the trapdoor corresponding to the subset factorization (see the appendix). Hence, the exhaustive search is really feasible and provides the secret key.

Now if we do not know the permutations π_i's and the bijection f, we construct the linear transformation ϕ by choosing randomly 2 distinct integers i_1, i_2 between 1 and s : for each of these 2 integers, we select randomly 3 distinct j_1, j_2, j_3 between 1 and 15 such that $j_1 < j_2$. The probability that both i_1 and i_2 correspond to weak blocks is $1/4$. For each of these 2 integers, we have to test at most $15 \times \frac{14 \times 13}{2} = 1365$ triplets (j_1, j_2, j_3) to find one that satisfies lemma 10. This means that we have to check at most $4 \times 1365^2 = 7452900$ choices of ϕ. But such a check can be done very quickly : if ϕ is correct, then $\overline{\Omega}$ has a very small vector (at least as short as $\frac{\phi(\mathbf{c}^{(0)})}{d_1 d_2 d_3 d_4}$), and otherwise, there is no reason that such a situation happens. Since computing $\overline{\Omega}$ can be done in less than a second (involved lattices have very small dimension), we can check all choices of ϕ in a reasonable time (namely, less than one week with 10 workstations). Once a suitable ϕ has been found, we perform an exhaustive search on $\frac{\phi(c^{(0)})}{d_1 d_2 d_3 d_4}$ as before. If one wants to improve success probabilities, one can increase the number of components of ϕ by adding new integers i, once a suitable ϕ with two components has been found. Each additional integer i costs at most 1365 tests and we can determine them successively, therefore we can easily determine the $s/2$ weak blocks, which reveals f. Then we apply the previous strategy in order to obtain the rest of the secret key.

3.4 Experiments

The attack has been successfully implemented using blockwise Korkine-Zolotarev lattice reductions [15] instead of LLL reductions to improve the reduced basis for heuristic 6. We used the package previously developped by A. Joux [6] in our lab. Timings are given for a 50Mhz Sparc 4, with parameters $s = 32$ and $k = 3$. It takes about 9 hours to obtain the $k + 1$-dimensional lattice from the 32 blocks of 16 integers that form the public key. In our implementation, we assumed that the permutations π_i's and the bijection f were known, which gave the secret key almost immediately : both the computation of $\overline{\Omega}$ and the exhaustive search of $\frac{\phi(c^{(0)})}{d_1 d_2 d_3 d_4}$ are performed in a few minutes. In practice, the vector $\frac{\phi(c^{(0)})}{d_1 d_2 d_3 d_4}$ happens to be a very small linear combination of the LLL-reduced basis vectors (coefficients less than 10 in absolute value). In the case where we do not know the permutations π_i's and the bijection f, initial experiments confirm the above discussion.

4 Conclusion

We introduced the basic notion of an orthogonal lattice. This concept first leads to an efficient attack against both Merkle-Hellman and Merkle-Hellman-like transformations. This attack differs from Shamir's and Brickell's attacks against original Merkle-Hellman cryptosystems. It points out that one should be cautious with the cryptographic use of Merkle-Hellman transformations. The notion of an orthogonal lattice also enables us to exploit weaknesses in the subset factorization (the trapdoor). These two applications of lattice reduction form an attack against the Qu-Vanstone scheme that works for any choice of the parameters. The attack has been successfully implemented and reveals the secret key from the public key in a reasonable time.

Acknowledgements

We would like to thank Anne-Marie Bergé for the proof of theorem 1. We also thank Jean-Marc Couveignes for his help.

A Appendix

In this appendix, we describe the subset factorization used in the Qu-Vanstone scheme and we provide the missing proofs of sections 2 and 3.

A.1 Further description of the Qu-Vanstone scheme

A.1.1 Construction of the s blocks C_i

Recall that n is a positive integer of the form $n = d_1 d_2 d_3 d_4 d_5$, where $2^{s-1} \leq d_\ell < 2^s$ (for $\ell = 1, 2, 3, 4$), $d_5 \leq 16$, and s is some fixed even positive integer. In the additive group $G = \mathbf{Z}_n$, we distinguish the subgroups G_1, G_2, G_3 and G_4 where G_ℓ is generated by $d_1 d_2 \ldots d_\ell$.

For each d_ℓ, $1 \leq \ell \leq 4$, select a superincreasing sequence $h[\ell, 1], \ldots, h[\ell, s]$ such that $\sum_{i=1}^{s} h[\ell, i] < d_\ell$. Choose integers q_1, q_2, q_3 and q_4 such that q_ℓ and d_ℓ are coprime. Apply a Merkle-Hellman transformation to get $\bar{h}[\ell, i] = h[\ell, i] q_\ell \pmod{d_\ell}$ where $0 < \bar{h}[\ell, i] < d_\ell$.

Select a permutation ξ_1 on $\{1, 2, \ldots, s\}$. For $1 \leq i \leq s$, select two positive integers $x[1, i, 0]$, $x[1, i, 1] < d_2 d_3 d_4 d_5$ and define two elements in distinct cosets of G_1 in G by :

$$
\begin{aligned}
a[1, i, 0] &= x[1, i, 0] d_1, \\
a[1, i, 1] &= \bar{h}[1, \xi_1(i)] + x[1, i, 1] d_1 \pmod{n}.
\end{aligned}
$$

Select a permutation ξ_2 on $\{1, 2, \ldots, s\}$. For $1 \leq i \leq s$ and $u = 0, 1$, select two positive integers $x^u[2, i, 0], x^u[2, i, 1] < d_3 d_4 d_5$ and define two elements in distinct cosets of G_2 in G_1 by :

$$
\begin{aligned}
a^u[2, i, 0] &= x^u[2, i, 0] d_1 d_2, \\
a^u[2, i, 1] &= \bar{h}[2, \xi_2(i)] d_1 + x^u[2, i, 1] d_1 d_2 \pmod{n}.
\end{aligned}
$$

Select a bijection g_1 from $\{1, 2, \ldots, s/2\}$ to $\{s/2 + 1, s/2 + 2, \ldots, s\}$. For $t, l = 0, 1, 2, 3$ and $i = 1, 2, \ldots, s/2$, select a positive integer $x^t[3, i, l] < d_4 d_5$. Define four elements in distinct cosets of G_3 in G_2 by :

$$
\begin{aligned}
a^t[3, i, 0] &= x^t[3, i, 0] d_1 d_2 d_3, \\
a^t[3, i, 1] &= \bar{h}[3, i] d_1 d_2 + x^t[3, i, 1] d_1 d_2 d_3 \pmod{n}, \\
a^t[3, i, 2] &= \bar{h}[3, g_1(i)] d_1 d_2 + x^t[3, i, 2] d_1 d_2 d_3 \pmod{n}, \\
a^t[3, i, 3] &= (\bar{h}[3, i] + \bar{h}[3, g_1(i)]) d_1 d_2 + x^t[3, i, 3] d_1 d_2 d_3 \pmod{n}.
\end{aligned}
$$

Select a bijection g_2 from $\{1, 2, \ldots, s/2\}$ to $\{s/2 + 1, s/2 + 2, \ldots, s\}$. For $t, l = 0, 1, 2, 3$ and $i = 1, 2, \ldots, s/2$, select a positive integer $x^t[4, i, l] < d_5$. Define four elements in distinct cosets of G_4 in G_3 by :

$$
\begin{aligned}
a^t[4, i, 0] &= x^t[4, i, 0] d_1 d_2 d_3 d_4, \\
a^t[4, i, 1] &= \bar{h}[4, i] d_1 d_2 d_3 + x^t[4, i, 1] d_1 d_2 d_3 d_4 \pmod{n}, \\
a^t[4, i, 2] &= \bar{h}[4, g_2(i)] d_1 d_2 d_3 + x^t[4, i, 2] d_1 d_2 d_3 d_4 \pmod{n}, \\
a^t[4, i, 3] &= (\bar{h}[4, i] + \bar{h}[4, g_2(i)]) d_1 d_2 d_3 + x^t[4, i, 3] d_1 d_2 d_3 d_4 \pmod{n}.
\end{aligned}
$$

Let f be a bijection from $\{1, 2, \ldots, s\}$ to $\{3, 4\} \times \{1, 2, \ldots, s/2\}$. For $1 \leq i \leq s$ and $0 \leq j \leq 15$, define $c[i, j] = a[1, i, u] + a^u[2, i, v] + a^{v+2u}[f(i), w]$ $(\bmod\, n)$, where j is uniquely decomposed as $j = w + 4v + 8u$ with $0 \leq u \leq 1$, $0 \leq v \leq 1$ and $0 \leq w \leq 3$. Qu and Vanstone proved in [13] that the s blocks $C_i = \{c[i, j] : 0 \leq j \leq 15\}$ form a direct sum in G. The trapdoor consists of the d_ℓ's, $h[\ell, i]$'s, $a[1, i, l]$'s, $a^u[2, i, l]$'s, $a^t[3, i, l]$'s, $a^t[4, i, l]$'s, the bijections f, g_1, g_2; the permutations ξ_1, ξ_2.

A.1.2   Factoring with the trapdoor

We now describe how, given any $g \in G$ of form $g = \sum_{i=1}^{s} c[i, j_i]$ $(\bmod\, n)$, one can quickly recover j_i's just by knowing g and the trapdoor.

Let $g_1 = g = \sum_{i=1}^{s} c[i, j_i]$ $(\bmod\, n)$. Recall that in G, we have :

$$c[i, j_i] = a[1, i, u] + a^u[2, i, v] + a^{v+2u}[f(i), w] \ (\bmod\, n), \ \ j_i = w + 4v + 8u.$$

We recover the values of u, v, w for each j_i value by solving 4 sub-knapsack problems based on appropriate superincreasing sequence :

Step 1. Compute $S_1 = q_1^{-1} g_1$ $(\bmod\, d_1)$, and solve the superincreasing knapsack $\sum_{i=1}^{s} u_i h[1, \xi_1(i)] = S_1$, $u_i \in \{0, 1\}$. Compute $g_2 = g_1 - \sum_{i=1}^{s} a[1, i, u_i]$.

Step 2. Compute $S_2 = q_2^{-1} \frac{g_2}{d_1}$ $(\bmod\, d_2)$, and solve the superincreasing knapsack $\sum_{i=1}^{s} v_i h[2, \xi_2(i)] = S_2$, $v_i \in \{0, 1\}$. Compute $g_3 = g_2 - \sum_{i=1}^{s} a^{u_i}[2, i, v_i]$.

Step 3. Compute $S_3 = q_3^{-1} \frac{g_3}{d_1 d_2}$ $(\bmod\, d_3)$, and solve the superincreasing knapsack $\sum_{i=1}^{s} x_i h[3, i] = S_3$, $x_i \in \{0, 1\}$. Compute

$$g_4 = g_3 - \sum_{f(i)=(3,1)}^{(3,s/2)} a^{v_i + 2u_i}[f(i), x_i + 2x_{g_1(i)}].$$

Step 4. Compute $S_4 = q_4^{-1} \frac{g_4}{d_1 d_2 d_3}$ $(\bmod\, d_4)$, and solve the superincreasing knapsack $\sum_{i=1}^{s} y_i h[4, i] = S_4$, where $y_i \in \{0, 1\}$. For $i = 1, 2, \ldots, s/2$ define $w_i = x_i + 2x_{g_1(i)}$ and $w_{i+s/2} = y_i + 2y_{g_2(i)}$.

Finally, we recover $j_i = w_i + 4v_i + 8u_i$ for $i = 1, 2, \ldots, s$.

A.2   Proof of theorem 4

Assume $c > 2^{(n-1)/2 + (n-d)(n-d-1)/4} \det(\overline{\Lambda})$ and let $(x_1, x_2, \ldots, x_n)$ be an LLL-reduced basis of Ω. Let $(b_1, b_2, \ldots, b_{n-d})$ be an LLL-reduced basis of $\Lambda^{\perp}$. Define $y_1, y_2, \ldots, y_{n-d}$ in Ω by $p_\uparrow(y_j) = 0$ and $p_\downarrow(y_j) = b_j$. These $n - d$ vectors are linearly independent, therefore by theorem 3 (2), for $1 \leq j \leq n - d$:

$$\|x_j\| \leq 2^{(n-1)/2} \max(\|y_1\|, \ldots, \|y_{n-d}\|)$$
$$\leq 2^{(n-1)/2} \max(\|b_1\|, \ldots, \|b_{n-d}\|).$$

But theorem 3 (1) ensures us that $\|\mathbf{b}_j\| \leq 2^{(n-d)(n-d-1)/4} \det(\Lambda^\perp)$. Thus :

$$\|\mathbf{x}_j\| \leq 2^{(n-1)/2} 2^{(n-d)(n-d-1)/4} \det(\Lambda^\perp) < c.$$

This implies that $p_\uparrow(\mathbf{x}_j) = 0$ and $p_\downarrow(\mathbf{x}_j) \in \Lambda^\perp$ for $1 \leq j \leq n - d$. Therefore $p_\downarrow(\mathbf{x}_1), \ldots, p_\downarrow(\mathbf{x}_{n-d})$ are linearly independent and they form a $\mathbf{Q}$-basis of $E_\Lambda^\perp$.

Now, let $\mathbf{y} \in \Lambda^\perp$. There exist $\lambda_1, \lambda_2, \ldots, \lambda_{n-d} \in \mathbf{Q}$ such that :

$$\mathbf{y} = \lambda_1 p_\downarrow(\mathbf{x}_1) + \lambda_2 p_\downarrow(\mathbf{x}_2) + \cdots + \lambda_{n-d} p_\downarrow(\mathbf{x}_{n-d}).$$

Defining $\mathbf{x} \in \Omega$ by $p_\uparrow(\mathbf{x}) = 0$ and $p_\downarrow(\mathbf{x}) = \mathbf{y}$, we have :

$$\mathbf{x} = \lambda_1 \mathbf{x}_1 + \lambda_2 \mathbf{x}_2 + \cdots + \lambda_{n-d} \mathbf{x}_{n-d}.$$

But there also exist $\mu_1, \mu_2, \cdots, \mu_n \in \mathbf{Z}$ such that $\mathbf{x} = \mu_1 \mathbf{x}_1 + \mu_2 \mathbf{x}_2 + \cdots + \mu_n \mathbf{x}_n$. Therefore :

$$(\mu_1 - \lambda_1)\mathbf{x}_1 + \cdots + (\mu_{n-d} - \lambda_{n-d})\mathbf{x}_{n-d} + \mu_{n-d+1}\mathbf{x}_{n-d+1} + \cdots + \mu_n \mathbf{x}_n = 0.$$

Since $\mathbf{x}_1, \ldots, \mathbf{x}_n$ are linearly independent, we deduce that $\lambda_j = \mu_j \in \mathbf{Z}$. Hence $(p_\downarrow(\mathbf{x}_1), \ldots, p_\downarrow(\mathbf{x}_{n-d}))$ is a $\mathbf{Z}$-basis of the lattice $\Lambda^\perp$.

Furthermore, for $1 \leq i \leq n - d$ and $1 \leq j \leq n - d$, $\|p_\downarrow(\mathbf{x}_j)\| = \|\mathbf{x}_j\|$ and $p_\downarrow(\mathbf{x}_i).p_\downarrow(\mathbf{x}_j) = \mathbf{x}_i.\mathbf{x}_j$. Since $(\mathbf{x}_1, \ldots, \mathbf{x}_n)$ is an LLL-reduced basis, this proves that $(p_\downarrow(\mathbf{x}_1), \ldots, p_\downarrow(\mathbf{x}_{n-d}))$ is an LLL-reduced basis too.

A.3 Recovering the secret key from $\mathbf{c}^{(0)}$

Notice that $\hat{c}[i,j] \equiv c[i, \pi_i^{-1}(j)] - c[i, \pi_i^{-1}(0)] \pmod{n}$. Since we know $d_1 d_2 d_3 d_4$, we recover $n = d_1 d_2 d_3 d_4 d_5$ from the size of each $\hat{c}[i,j]$. But the form of each $c[i, \pi_i^{-1}(j)]$ is very particular :

$$c[i,j] = a[1,i,u] + a^u[2,i,v] + a^{v+2u}[f(i),w] \pmod{n}.$$

By enumerating all possible cases, one notices that the knowledge of $c[i, \pi_i^{-1}(j)] - c[i, \pi_i^{-1}(0)] \pmod{n}$ reveals d_1, $d_1 d_2$, $d_1 d_2 d_3$ by particular gcd's, hence the π_i's by looking at the order in each block of 15 integers. By subtractions, we then obtain the $\bar{h}[1,i]$'s, $\bar{h}[2,i]$'s, $\bar{h}[3,i]$'s and the $\bar{h}[4,i]$'s. Since we now know the d_ℓ's, we derive the q_ℓ's and the $h[\ell,i]$'s. This reveals the $a[1,i,l]$'s, $a^u[2,i,l]$'s, $a^t[3,i,l]$'s, $a^t[4,i,l]$'s, the bijections g_1, g_2 and the permutations ξ_1, ξ_2 (looking at the order of superincreasing sequences). We now know the complete trapdoor. The coordinates λ_j's are actually closely related to the $w^{(e)}$'s and the $m^{(e)}$'s : one can derive equivalent $w^{(e)}$'s and $m^{(e)}$'s so that $\mathbf{c}^{(k)}$ is obtained by Merkle-Hellman-like transformations from $\mathbf{c}^{(0)}$. Since we now know the $c[i,j]$'s from the trapdoor, we also find out equivalent $\sum_{i=1}^s a_i^{(e)}$'s. Hence, we recovered the complete secret key.

References

[1] E. Brickell. Are most low density polynomial knapsacks solvable in polynomial time ? In *Proc. 14th Southeastern Conference on Combinatorics, Graph Theory, and Computing*, 1983.

[2] B. Chor and R.L. Rivest. A knapsack-type public key cryptosystem based on arithmetic in finite fields. *IEEE Trans. Inform. Theory*, 34, 1988.

[3] H. Cohen. *A course in computational algebraic number theory*. Springer-Verlag, Berlin, 1993.

[4] M.J. Coster, A. Joux, B.A. LaMacchia, A.M. Odlyzko, C.-P. Schnorr, and J. Stern. Improved low-density subset sum algorithms. *Comput. Complexity*, 2:111–128, 1992.

[5] P. M. Gruber and C. G. Lekkerkerker. *Geometry of numbers*. North-Holland, Amsterdam, 1969.

[6] A. Joux. *La réduction des réseaux en cryptographie*. PhD thesis, École Polytechnique, 1993.

[7] A. Joux and J. Stern. Lattice reduction: a toolbox for the cryptanalyst. (to appear in J. of Cryptology).

[8] A. K. Lenstra, H. W. Lenstra, and L. Lovász. Factoring polynomials with rational coefficients. *Math. Ann.*, 261:515–534, 1982.

[9] J. Martinet. *Les réseaux parfaits des espaces euclidiens (perfect lattices in euclidean spaces)*. Editions Masson, 1996.

[10] R. Merkle and M. Hellman. Hiding information and signatures in trapdoor knapsacks. *IEEE Trans. Inform. Theory*, IT-24:525–530, September 1978.

[11] P. L. Montgomery. Square roots of products of algebraic numbers. Draft of June, 1995.

[12] A. M. Odlyzko. The rise and fall of knapsack cryptosystems. In *Cryptology and Computational Number Theory*, volume 42 of *Proceedings of Symposia in Applied Mathematics*, pages 75–88. A.M.S., 1990.

[13] M. Qu and S. A. Vanstone. New public-key cryptosystem based on the subset factorizations in $\mathbf{Z}_n$. (to appear).

[14] M. Qu and S. A. Vanstone. The knapsack problem in cryptography. In *Finite Fields: Theory, Applications, and Algorithms*, volume 168 of *Contemporary Mathematics*, pages 291–308. A.M.S., 1994.

[15] C.-P. Schnorr. A hierarchy of polynomial lattice basis reduction algorithms. *Theoretical Computer Science*, 53:201–224, 1987.

[16] C.P. Schnorr and H.H. Hörner. Attacking the Chor-Rivest cryptosystem by improved lattice reduction. In *Advances in Cryptology : Proceedings of Eurocrypt' 95*, volume 921 of *LNCS*, pages 1–12. Springer-Verlag, 1995.

[17] A. Shamir. A polynomial time algorithm for breaking the basic Merkle-Hellman cryptosystem. In *Proceedings of the 23rd Annual Symposium on the Foundations of Computer Science (IEEE)*, pages 145–152, 1982.

Failure of the McEliece Public-Key Cryptosystem Under Message-Resend and Related-Message Attack

Thomas A. Berson

Anagram Laboratories
P.O. Box 791
Palo Alto, CA 94301 USA
berson@anagram.com

Abstract: The McEliece public-key cryptosystem fails to protect any message which is sent to a recipient more than once using different random error vectors. In general, it fails to protect any messages sent to a recipient which have a known linear relation to one another. Under these conditions, which are easily detectable, the cryptosystem is subject to a devastating attack which reveals plaintext with a work factor which is 10^{15} times better than the best general attack.

Keywords: McEliece, public-key cryptosystem, randomization, error-correcting codes, error vectors, message-resend attack, related-message attack, protocol failure, cryptanalysis.

1   Introduction

The McEliece public-key cryptosystem was proposed nearly 20 years ago [14]. The system is simple to explain and is very fast in execution. It is based on an NP-hard problem in coding theory, and features the ability of a hidden error-correcting code to recover plaintext from ciphertexts which the sender intentionally garbles with random errors. Although it has received much attention from the cryptologic community, the system remains unbroken to this day.

Despite these advantages, the McEliece public-key cryptosystem it is not widely used. Perhaps this is because it has a large public key and a low information rate. But changes in technology and economics, for example the plummeting cost of storage, keep it on the list of candidates for some applications.

In this paper we analyze and exploit the failure of the McEliece public-key cryptosystem to protect plaintext when any message is sent to a recipient more than once using different random error vectors. Our *message-resend* attack succeeds in βk^3 time, where β is a small constant, and k is the message size of the underlying code. We then generalize our attack to a *related-message* attack, which recovers any messages sent to a recipient when a linear relation between the messages is known, again in βk^3 time.

2 The McEliece Public-Key Cryptosystem

Without loss of generality we will describe the McEliece public-key cryptosystem system using the code and parameter sizes proposed originally by McEliece.

The private key consists of three matrices:

- a generator matrix for a ($n = 50$, $k = 524$, $t = 50$) Goppa code $G \in F_2^{524 \times 1024}$ (Goppa codes are a large class of error-correcting codes which have efficient decoding algorithms);

- an invertible scrambler matrix $S \in F_2^{524 \times 524}$, and;

- a permutation $P \in F_2^{1024 \times 1024}$.

The public key is the matrix product SGP. Note that S and P disguise G as a general linear code.

Now suppose a message $m \in F_2^{524}$ is to be sent. The parameters of the Goppa code (an irreducible polynomial $g(x) \in F_2[X]$ of degree 50 and an ordering of $F_{2^{10}}$) allow for the fast error correction of up to 50 errors. So a random error vector $e \in F_2^{1024}$ is chosen where the Hamming weight $wt(e) = 50$, and the cryptogram

$$c = mSGP + e$$

is sent.

The intended recipient then computes

$$cP^{-1} = mSG + eP^{-1}.$$

Since P is a permutation, $wt(eP^{-1}) = 50$. So decoding the Goppa code recovers mS, from which, finally, the intended recipient recovers $m = (mS)S^{-1}$.

Remarks

A great many workers, starting with Adams and Meijer [1,2], Hin [9], and Jorissen [10], have explored the relationship between the parameters of the underlying code, the security of the cryptosystem, and the data rate. For a description of this line of research see van Tilburg [17]. Optimizations have been suggested where $n = 1024$, k ranges from 524 to 654, and t ranges from 37 to 50. Our attack is not blunted by such adjustments.

Other workers have explored replacing the Goppa code with other types of error-correcting code. For example, Gabidulin et al. [5] tried using maximum-rank-distance codes. These schemes were shown to be insecure by Gibson [6,7]. In any event, such code replacements would not prevent our attack, which does not depend on the structure of the code.

3   Cryptanalytic Background

McEliece stated that the most promising line of attack on his public-key cryptosystem consists of decoding an arbitrary linear code containing correctable errors. Therefore, the security of the cryptosystem seems to be based on solving the corresponding the BHDD[1] problem.

The obvious [14,1] attack is this: if a cryptanalyst could guess 524 coordinates of c that are not garbled by e, then the restriction to those 524 columns of the cryptogram and the public key

$$\bar{c} = m\overline{SGP}$$

relates m to $\bar{c}$ by a known $\overline{SGP} \in F_2^{524 \times 524}$. If $\overline{SGP}$ is invertible, then m can be recovered.

Notice that this is a per-message attack; the secret key of the system remains unknown to the cryptanalyst.

What is the work factor for this attack? The cryptanalyst must correctly guess 524 ungarbled columns out of the possible 974 = 1024-50. So we can calculate that it will require

$$\frac{\binom{1024}{524}}{\binom{974}{524}} \approx 1.37 \times 10^{16} \text{ guesses to succeed.}$$

So the work factor is

$$w = \alpha \cdot 1.37 \times 10^{16},$$

where α is the cost of inverting a 524-square matrix, roughly 524^3.

Notice that the relatively low-weight error vector is crucial to the success of the Goppa decoding algorithm, and that it also impacts the work necessary for the cryptanalyst.

Remarks

The attack described above can be, and has been, improved slightly by taking partial information into account. See Lee and Brickell [12], Li, Deng and Wang [13], and van Tilburg [16].

---

[1] BHDD (Binary Hamming Distance Decoding) is the name given to the problem of decoding an arbitrary binary word to the nearest codeword in an arbitrary linear code under the restriction that the "arbitrary" binary word be at distance at most $(d-1)/2$ from a codeword. Berlekamp, McEliece and van Tilborg [4] showed that BHDD is NP-hard.

There was some excitement and confusion about the cryptanalysis of the McEliece public-key cryptosystem a few years ago. Korzhik and Turkin, announced that they had broken the cryptosystem. They gave a "demonstration" of their "attack" at Eurocrypt '91 [11]. However, the demonstration was only a toy: in place of the Goppa code it used a BCH code of dimension 36 in $F_2^{63} = GF(2)^{63}$, with minimum distance 11, and an error vector of weight 5. Even with this simplification, their attack achieved only a five-fold speedup over exhaustion. The details have never appeared. More generally, Korzhik and Turkin claimed to have found a polynomial time algorithm for BHDD, which is known to be NP-hard. But the published description and analysis of their algorithm are not precise, and its correct functioning within the claimed time bound has never been confirmed. In summary, their attack on the McEliece public-key cryptosystem is not believed to be effective.

4 Failure Under Message-Resend Conditions

Suppose now that, through some accident, or as a result of action in the part of the cryptanalyst, both

$$c_1 = mSGP + e_1$$

and
$$e_1 \neq e_2$$
$$c_2 = mSGP + e_2$$

are sent. We call this a *message-resend* condition. In this case it is easy for the cryptanalyst to recover m from the system of c_i. (We will examine only the case where the number of different cryptograms of the same message, which we call the *resend depth*, is 2. The attack is even easier at greater resend depths.)

Notice that $c_1 + c_2 = e_1 + e_2$ (mod 2).

The cryptanalyst can easily detect a message-resend condition by observing the Hamming weight of the sum of any two cryptograms. When the underlying messages are different, the expected weight of the sum is about 512. When the underlying messages are identical, the weight of the sum cannot exceed 100. Heiman [8] showed that a message-resend condition can be detected; we will show how to exploit it.

4.1 Method of Attack

We will compute two sets from $(c_1 + c_2)$. The set L_0 will be the locations where $(c_1 + c_2)$ contains zeroes. The set L_1 will be the locations where $(c_1 + c_2)$ contains ones.

Let

$$L_0 = \left\{ l \in \{1,2,\cdots,1024\} : c(l)_1 + c_2(l) = e_1(l) + e_2(l) = 0 \right\}$$

and
$$L_1 = \left\{ l \in \{1,2,\cdots,1024\} : c_1(l) + c_2(l) = e_1(l) + e_2(l) = 1 \right\}.$$

We aim to take advantage of the fact (and to quantify the claim) that

- $l \in L_0 \Rightarrow$ most probably neither $c_1(l)$ nor $c_2(l)$ is garbled by an error vector, while

- $l \in L_1 \Rightarrow$ certainly precisely one of $c_1(l)$ or $c_2(l)$ is garbled by an error vector.

Every $l \in L_0$ means that either $e_1(l) = 0 = e_2(l)$ or $e_1(l) = 1 = e_2(l)$. Assuming the error vectors e_1 and e_2 are chosen independently, then for any l

$$\Pr\big(e_1(l) = 1 = e_2(l)\big) = \left(\frac{50}{1024}\right)^2 \approx 0.0024 .$$

In other words, most $l \in L_0$ signify $e_1(l) = 0 = e_2(l)$. Thus the cryptanalyst should try to guess 524 ungarbled columns from those indexed by L_0.

How good is this strategy? Let p_i be the probability that precisely i coordinates are simultaneously garbled by e_1 and e_2. Then

$$p_i = \Pr\big(\big|\{l : e_1(l) = 1\} \cap \{l : e_2(l) = 1\}\big| = i\big) = \frac{\binom{50}{i}\binom{974}{50-i}}{\binom{1024}{50}}$$

since, say, e_2 must choose i error locations from those 50 of e_1 and the remaining $50-i$ from those ungarbled by e_1, this out of a total of $\binom{1024}{50}$ possible error vectors.

Therefore the expected cardinality of L_1 is

$$E\big(|L_1|\big) = \sum_{i=0}^{50} (100 - 2i) p_i \approx 95.1$$

since every l for which $e_1(l) = 1 = e_2(l)$ reduces $|L_1|$ by two.

For example, suppose $|L_1| = 94$. Then $|L_0| = 930$, of which only 3 are garbled. We see that the probability of guessing 524 ungarbled columns from those indexed by L_0 is

$$\frac{\binom{927}{524}}{\binom{930}{524}} \approx 0.0828$$

so the cryptanalyst expects to succeed in this case with only 12 guesses, at a cost of 12α.

When $|L_1| = 96$ only about 5 guesses are required!

These results are a factor of 10^{15} better than the exhaustive attack analyzed in Section 3.

Note that this attack does not recover the private key. We do not claim to have broken the McEliece public-key cryptosystem. But we have shown how a cryptanalyst may recover the plaintext of a resent message with very little work.

5   Failure Under Related-Message Conditions

We will now generalize the message-resend attack. Suppose that there are two cryptograms

$$c_1 = m_1 SGP + e_1$$
and
$$m_1 \neq m_2, \ e_1 \neq e_2$$
$$c_2 = m_2 SGP + e_2$$

and that the cryptanalyst knows a linear relation, for example $m_1 + m_2$, between the messages. We call this a *related-message* condition. In this case the cryptanalyst may recover the m_i from the set of c_i by doing one encoding and by then following the attack method of Section 4.1. Here are the details.

Combining the two cryptograms we get

$$c_1 + c_2 = m_1 SGP + m_2 SGP + e_1 + e_2.$$

Notice that $m_1 SPG + m_2 SGP = (m_1 + m_2)SGP$, a value the cryptanalyst may calculate in a related-message condition from the known relationship and the public key.

The cryptanalyst solves

$$c_1 + c_2 + (m_1 + m_2)SGP = e_1 + e_2$$

and proceeds with the attack as in Section 4.1, using $(c_1 + c_2 + (m_1 + m_2)SGP)$ in place of $(c_1 + c_2)$.

Remark

The message-resend attack is that special case of the related-message attack where $m_1 + m_2 = 0$.

6   Conclusions

The McEliece public-key cryptosystem fails to protect any message which is sent to a recipient more than once using different random error vectors.

The McEliece public-key cryptosystem fails to protect messages sent to a recipient which are have a known linear relation to one another.

Our attack is a general attack on the class of public-key cryptosystems which use an error-correcting code and the introduction of random errors by the sender.

Our attack under these conditions is a factor of 10^{15} better than the best attack under general conditions.

Users of the McEliece public-key cryptosystem, and of cryptosystems with similar structure, should guard against sending related messages. One countermeasure which comes to mind is to introduce an element of local randomness into any message before it is encrypted. But note that the obvious $c = (m\|r)SGP$ falls quickly to a synthesized related-message attack. A scheme is required which spreads randomness through the plaintext in some complicated fashion. Bellare and Rogaway's OAEP [3] *et seq.* are instructive. Of course, any such scheme extracts a penalty in data rate.

Cryptosystems which are based on the use of linear codes but without per-message error vectors, for example Neiderreiter [15], are not directly threatened by our attack. However, prudence dictates that all such systems now be reexamined for vulnerability to message-resend or related-message attack.

Acknowledgments

This work was begun while the author was a visitor at the Isaac Newton Institute for Mathematical Sciences in the University of Cambridge.

The author is grateful to Susan Langford, to Kevin McCurley, and to Matt Robshaw for their clarifying discussions.

References

1. C. ADAMS AND H. MEIJER, "Security-related comments regarding McEliece's public-key cryptosystem", *Advances in Cryptology—Crypto '87 (LNCS 293)*, 224-228, 1988.

2. C. ADAMS AND H. MEIJER, "Security-related comments regarding McEliece's public-key cryptosystem", *IEEE Transactions on Information Theory*, 35 (1989), 454-455.

3. M. BELLARE AND P. ROGAWAY, "Optimal asymmetric encryption", *Advances in Cryptology – EUROCRYPT '94 (LNCS 950)*, 232-249, 1994.

4. E.R. BERLEKAMP, R.J. MCELIECE, AND H.C.A. VAN TILBORG, "On the inherent intractability of certain coding problems", *IEEE Transactions on Information Theory*, 24 (1978), 384-386.

5. E.M. GABIDULIN, A.V. PARAMONOV, AND O.V. TRETJAKOV, "Ideals over a non-commutative ring and their application in cryptology", *Advances in Cryptology—EUROCRYPT '91 (LNCS 547), 482-489*, 1991.

6. J.K. GIBSON, "Severely denting the Gabidulin version of the McEliece public key cryptosystem", *Designs, Codes and Cryptography*, 6 (1995), 37-45.

7. J.K. GIBSON, "The security of the Gabidulin public key cryptosystem", *Advances in Cryptology—EUROCRYPT '96 (LNCS 1070)*, 212-223, 1996.

8. R. HEIMAN, "On the security of cryptosystems based on linear error-correcting codes", M.Sc. Thesis, Feinburg Graduate School, Weitzmann Institute of Science, Rehovot, August, 1987.

9. P.J.M. HIN, "Channel-error-correcting privacy cryptosystems", M.Sc. Thesis, Delft University of Technology, Delft, 1986.

10. F. JORISSEN, "A security evaluation of the public-key cipher system proposed by R.J. McEliece, used as a combined scheme", Technical report, Katholieke Universiteit Leuven, Dept. Elektrotechniek, January, 1986.

11. V.I. KORZHIK AND A.I. TURKIN, "Cryptanalysis of McEliece's public-key cryptosystem", *Advances in Cryptology—EUROCRYPT '91 (LNCS 547)*, 68-70, 1991.

12. P.J. LEE AND E.F. BRICKELL, "An observation on the security of McEliece's public-key cryptosystem", *Advances in Cryptology—EUROCRYPT '88 (LNCS 330)*, 275-280, 1988.

13. Y.X. LI, R.H. DENG, AND X.M. WANG, "On the equivalence of McEliece's and Neiderreiter's public-key cryptosystem", *IEEE Transactions on Information Theory*, 40 (1994), 271-273.

14. R.J. MCELIECE, "A public-key cryptosystem based on algebraic coding theory", DSN Progress Report 42-44, Jet Propulsion Laboratory, Pasadena, 1978.

15. H. NEIDERREITER, "Knapsack-type cryptosystems and algebraic coding theory", *Problems of Control and Information Theory*, 15 (1986), 159-166.

16. J. VAN TILBURG, "On the McEliece public-key cryptosystem", *Advances in Cryptology—Crypto '88 (LNCS 403)*, 119-131, 1990.

17. J. VAN TILBURG, "Security analysis of a class of cryptosystems based on linear error-correcting codes", Ph.D. Thesis, Technische Universiteit Eindhoven, Eindhoven, November, 1994.

A Multiplicative Attack Using LLL Algorithm on RSA Signatures with Redundancy

Jean-François Misarsky

France Télécom - Branche Développement
Centre National d'Etudes des Télécommunications
42, rue des Coutures, B.P. 6243
14066 Caen Cedex, FRANCE
jeanfrancois.misarsky@cnet.francetelecom.fr

Abstract. We show that some RSA signature schemes using fixed or modular redundancy and dispersion of redundancy bits are insecure. Our attack is based on the multiplicative property of RSA signature function and extends old results of De Jonge and Chaum [DJC] as well as recent results of Girault and Misarsky [GM]. Our method uses the lattice basis reduction [LLL] and algorithms of László Babai [B]. Our attack is valid when the length of redundancy is roughly less than half the length of the public modulus. We successfully apply our attack to a scheme proposed for discussion inside ISO. Afterwards, we also describe possible adaptations of our method to attack schemes using mask or different modular redundancies. We explain limits of our attack and how to defeat it.

Keywords. Multiplicative attack, LLL algorithm, redundancy, RSA.

1 Introduction

Let n be a RSA modulus [RSA], e the public exponent, and d the secret exponent. We can define $P(x) = x^e \pmod{n}$ the public function and $S(x) = x^d \pmod{n}$ the secret one. The multiplicative property of RSA, i.e. the fact that $S(xy) = S(x)S(y) \pmod{n}$, leads to potential weaknesses, especially when used for signatures. We will make an extensive use of this property in our attack.

When a forger wants the signature of a message m, he generates two messages x and y that satisfy $m = xy \pmod{n}$. If he obtains the signatures of x and y, as exponentiation preserves the multiplicative structure of the input, he simply computes the signature of m as the product of $S(x)$ and $S(y)$, $S(m) = S(x) S(y) \pmod{n}$. This is a chosen-message attack.

Two standard ways exist to eliminate this potential weakness. One is to sign a hashed value of the message rather than the message itself. The other is to add some redundancy to the message to be signed. These different signature schemes are sometimes called, respectively, schemes with appendix and schemes with message recovery ([MOV], pp.428-432).

Only the redundancy solution is concerned by this paper. It is of particular interest when the message is short, because it prevents from specifying and implementing a hash-function (a rather delicate cryptographic challenge), and it allows to construct very compact signed messages, since messages can be recovered from the signatures themselves (and hence need not any longer be transmitted or stored). Let R be the invertible redundancy function. The signature of a message m is $\Sigma(m) = S[R(m)]$ and the signer only sends $\Sigma(m)$ of the receiver. The latter applies P to $\Sigma(m)$, and verifies that the result complies with the redundancy rule, i.e. is an element of the image set of R. Then he recovers m by discarding the redundancy, i.e. by applying R^{-1} to this result.

At Crypto'85 conference, De Jonge and Chaum [DJC] showed that simple redundancy does not avoid all the chosen-message attacks. In their paper, they show that it is not sufficient to append trailing '0' bits to the right or the left of the message. They study the case when redundancy is an affine function of m, i.e. the signature $\Sigma(m)$ to m is computed as $\Sigma(m) = S(\omega m + a)$. Their attack is based on Euclid's algorithm and is valid for any message m for:

- $a = 0$, and any value of ω such that the amount of redundancy is less than half the length of the public modulus n.
- $\omega = 1$, a small value of a, and when the amount of redundancy is less than one third of the length of the public modulus n.

Girault and Misarsky [GM] recently extended these results. Their attack uses an affine variant of Euclid's algorithm due to Okamoto and Shiraishi [OS]. It is valid for any constant ω, any constant a, any message m provided that the amount of redundancy is less than half the length of the public modulus n. Moreover, they study the case when modular redundancy is used, i.e. when the amount of redundancy is obtained by appending to m the remainder of m modulo some fixed value. In this case, the signature is still subject to a chosen-message attack when redundancy is less than half the length of the public modulus, minus the length of remainder. They give three solutions that prevent their attack; one of them consists in dispersing the message in different parts and another one in using two different modular redundancies.

We show in this paper that a multiplicative attack is feasible on signature scheme that uses dispersion of redundancy bits and fixed or modular redundancy. We precisely explain our attack in this case. But our attack is also valid on more simple schemes or schemes with mask or different modular redundancies.

Our method makes use of the lattice basis reduction, which has not been used in multiplicative attacks yet. But, lattice reduction has already been applied successfully in cryptoanalysis: against Merkle-Hellman public key cryptosystem [S], against Okamoto's cryptosystems [VGT1], against RSA cryptosystem with small exponent [H], or against RSA encryption with small exponents and random padding [C], for instance.

We successfully apply our method on ISO 9796 Part 3, Working Draft, December 1996 [ISO2], a scheme using dispersion of redundancy bits and modular redundancy. Afterwards, we explain limits of our attack and how to defeat it.

Throughout this paper, we call bitlength (or length in short) of an integer the number of bits of its binary representation. We denote by $|m|$ the bitlength of m.

2 Our Results

We describe a method using lattice basis reduction that finds solutions x and y of the equation $R(m)R(x) = R(y) \pmod{n}$ where:

- R is a redundancy function
- m is a message of which we want to forge a signature

If signatures of x and y can be obtained, i.e. respectively $\Sigma(x) = S(R(x)) \pmod{n}$ and $\Sigma(y) = S(R(y)) \pmod{n}$, then the signature of m can be easily forged:

$$\Sigma(m) = \frac{\Sigma(y)}{\Sigma(x)} \pmod{n}$$

In the sequel, we denote by:

$\omega_1, \omega_2, \dots$: miscellaneous multiplicative redundancies constants

a : fixed redundancy constant

m : a message

k_1 : the number of parts of m

m_i : the ith part of m. The message m is split up into k_1 parts which have not necessary the same length:

$$m = \boxed{\quad m_1 \quad | \quad m_2 \quad | \ \dots\dots \ | \quad m_i \quad | \ \dots\dots \ | \ m_{k_1} \ }$$

$\varphi(m)$: modular redundancy of the message m i.e. the remainder of m modulo a fixed value

k_2 : the number of parts of $\varphi(m)$

$\varphi(m)_j$: the jth part of $\varphi(m)$. The modular redundancy is split up into k_2 parts which have not necessary the same length:

$$\varphi(m) = \boxed{\quad \varphi(m)_1 \quad | \quad \varphi(m)_2 \quad | \ \dots\dots \ | \quad \varphi(m)_j \quad | \ \dots\dots \ | \ \varphi(m)_{k_2} \ }$$

n : RSA modulus

m_r : redundancy modulus ($\varphi(m) = m \pmod{m_r}$)

The redundancy function R can take several forms, with increasing complexity:

i) $R(m) = \omega m + a$

ii) $R(m) = \omega_1 m + \omega_2 \varphi(m) + a$

iii) $R(m) = \sum_{i=1}^{k_1} m_i \omega_i + a$

iv) $R(m) = \sum_{i=1}^{k_1} m_i \omega_i + \sum_{j=1}^{k_2} \varphi(m)_j \omega_{j+k_1} + a$

The case iv) generalizes the others and we only study it in the sequel.

Example: when all ω_i are powers of two in the case iv), one could have:

$$R(m) = \boxed{10111\dots | m_1 | \varphi(m)_1 | ..1001.. | \varphi(m)_i | m_i | \varphi(m)_{i+1} | \dots | ..1011..}$$

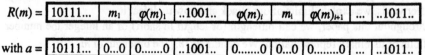

$$\text{with } a = \boxed{10111\dots | 0\dots0 | 0\dots\dots0 | ..1001.. | 0\dots\dots0 | 0\dots0 | 0\dots\dots0 | \dots | ..1011..}$$

Remark: we call *the number of bits of redundancy* the length of *n* minus the length of *m*. Note that the number of bits of modular redundancy is included in the number of bits of redundancy.

Main result:

If a signature scheme uses this kind of redundancy function:

$$R(m) = \sum_{i=1}^{k_1} m_i \omega_i + \sum_{j=1}^{k_2} \varphi(m)_j \omega_{j+k_1} + a$$

then our attack is valid when the number of bits of redundancy is roughly less than half the length of the public modulus n, minus the number of bits of modular redundancy (when the latter is present):

$$|\,redundancy\,| < \frac{1}{2}|n| - |m_r|$$

Another version of our attack, requiring more computation and memory, is valid when the number of bits of redundancy is roughly less than half the length of the public modulus n.

3 System: Definition and Solution

Solving $R(m)R(x) = R(y) \pmod{n}$ is equivalent to finding the different parts of $R(x)$ and $R(y)$, i.e. respectively $(x_i)_{1 \le i \le k_1}, (\varphi(x)_j)_{1 \le j \le k_2}$, and $(y_i)_{1 \le i \le k_1}, (\varphi(y)_j)_{1 \le j \le k_2}$.

Let $(X_i)_{1 \le i \le k}$ be the different parts of $R(x)$ to find, i.e. all or only part of $(x_i)_{1 \le i \le k_1}$ and $(\varphi(x)_j)_{1 \le j \le k_2}$. Let $(Y_i)_{1 \le i \le k}$ be the different parts of $R(y)$ to find, i.e. all or only part of $(y_i)_{1 \le i \le k_1}$ and $(\varphi(y)_j)_{1 \le j \le k_2}$. The modular redundancy, the fact that $x = \varphi(x) \pmod{m_r}$ and $y = \varphi(y) \pmod{m_r}$, implies two equations:

(i) $a_1 X_1 + a_2 X_2 + ... + a_{k-1} X_{k-1} = X_k + h_1 \pmod{m_r}$

(ii) $b_1 Y_1 + b_2 Y_2 + ... + b_{k-1} Y_{k-1} = Y_k + h_2 \pmod{m_r}$

with $(a_i)_{1 \le i \le k-1}, (b_i)_{1 \le i \le k-1}, h_1$ and h_2 fixed integers.

Note that:

- h_1 and h_2 are present only when some parts of $R(x)$ and $R(y)$ are fixed, i.e. one or several $x_i, \varphi(x)_j, y_i$ or $\varphi(y)_j$ are fixed.
- our method requires the coefficients of X_k and Y_k to be equal to one. It is easily obtained by a division modulo m_r. We have deliberately omitted to describe this step.

$R(m)R(x) = R(y) \pmod{n}$ also implies an other equation:

(iii) $c_1 X_1 + c_2 X_2 + ... + c_k X_k + d_1 Y_1 + d_2 Y_2 + ... + d_{k-1} Y_{k-1} = Y_k + h_3 \pmod{n}$

with $(c_i)_{1 \le i \le k}, (d_i)_{1 \le i \le k-1}$ and h_3 fixed integers.

Let (*SI*) be the system:

$$(SI) \begin{cases} a_1X_1+...+a_{k-1}X_{k-1} & = X_k+h_1 \pmod{m_r} & (i) \\ b_1Y_1+...+b_{k-1}Y_{k-1} & = Y_k+h_2 \pmod{m_r} & (ii) \\ c_1X_1+...+c_kX_k+d_1Y_1+...+d_{k-1}Y_{k-1} & = Y_k+h_3 \pmod{n} & (iii) \end{cases}$$

(*SI*) is a system with constraints on value of $(X_i)_{1 \le i \le k}$ and $(Y_i)_{1 \le i \le k}$.
We have for $1 \le i \le k$:

$$0 \le X_i < 2^{\text{Length of the part } X_i \text{ in bits}}$$
$$0 \le Y_i < 2^{\text{Length of the part } Y_i \text{ in bits}}$$

When modular redundancy is not used in the signature scheme, (*i*) and (*ii*) are useless. Only (*iii*) is necessary.
In the first part of our study, we define a lattice where all points give a solution to this system without second member, $h_1 = h_2 = h_3 = 0$, and without constraints on values of $(X_i)_{1 \le i \le k}$ and $(Y_i)_{1 \le i \le k}$. Next, we define a method to find a solution to (*SI*) without constraints on values of $(X_i)_{1 \le i \le k}$ and $(Y_i)_{1 \le i \le k}$ by using this lattice. After, we explain how to obtain solutions to the system (*SI*) with additional constraints on values of $(X_i)_{1 \le i \le k}$ and $(Y_i)_{1 \le i \le k}$. Finally we study the efficiency of our method.

3.1 First Step: Determination of the Lattice

We define an integer lattice L such that any element of this lattice is solution to (*S*). (*S*) is the system (*SI*) without second member and without constraints on values of $(X_i)_{1 \le i \le k}$ and $(Y_i)_{1 \le i \le k}$.

$$(S) \begin{cases} a_1X_1+...+a_{k-1}X_{k-1} & = X_k \pmod{m_r} \\ b_1Y_1+...+b_{k-1}Y_{k-1} & = Y_k \pmod{m_r} \\ c_1X_1+...+c_kX_k+d_1Y_1+...+d_{k-1}Y_{k-1} & = Y_k \pmod{n} \end{cases}$$

Hence, we define a lattice L of dimension $2k$ such that any vector $v = (v_1, v_2, ..., v_{2k-1}, v_{2k})$ verifies:

(*a*) $\qquad\qquad\qquad\qquad a_1v_1 + a_2v_2 +...+ a_{k-1}v_{k-1} = v_k \pmod{m_r}$
(*b*) $\qquad\qquad\qquad\qquad b_1v_{k+1} + b_2v_{k+2} +...+ b_{k-1}v_{2k-1} = v_{2k} \pmod{m_r}$
(*c*) $\quad c_1v_1 + c_2v_2 +...+ c_kv_k + d_1v_{k+1} + d_2v_{k+2} +...+ d_{k-1}v_{2k-1} = v_{2k} \pmod{n}$

Let M be the matrix of lattice L. Columns vectors of M are a basis of L, and for any element v of L, there is a column vector α with integer components such that:

$$M\alpha = v$$

Now, we construct this matrix M.
We denote by M_1 an identity matrix of dimension $2k$ where the row k is replaced by:

$$(a_1, a_2,...,a_{k-1}, m_r, 0, 0,...,0)$$

Then, for any vector α with integer components, $v = M_1 \alpha$ is a vector with components satisfying (*a*). Remark that $v_i = \alpha_i$ with $1 \le i \le 2k$, $i \ne k$ and:

(d) $$v_k = a_1v_1 + a_2v_2 + ... + a_{k-1}v_{k-1} + \alpha_k m_r$$

We gather equations (b) and (c) together with the Chinese Remainder Theorem. It is possible because n is the public modulus of RSA and is prime with m_r (otherwise we have a factor of n !).

We denote by Chinese the Chinese remainder function:

Chinese(a (mod m), b (mod n)), with m and n relatively primes, returns x such that:

$$\begin{cases} x = a \pmod{m} \\ x = b \pmod{n} \end{cases}$$

Let $(f_i)_{1 \le i \le 2k}$ such that:

$f_i = $ Chinese(0 (mod m_r), c_i (mod n))   when $1 \le i \le k$

$f_i = $ Chinese(b_{i-k} (mod m_r), d_{i-k} (mod n))   when $k + 1 \le i \le 2k - 1$

We obtain:

(f) $f_1v_1 + f_2v_2 + ... + f_{k-1}v_{k-1} + f_kv_k + f_{k+1}v_{k+1} + f_{k+2}v_{k+2} + ... + f_{2k-1}v_{2k-1} = v_{2k} \pmod{m_r n}$

But α_k is different from v_k. We use (d) to replace v_k in (f). We finally obtain an equation (e), equivalent to (b) and (c), that has this form:

(e) $e_1v_1 + e_2v_2 + ... + e_{k-1}v_{k-1} + e_k\alpha_k + e_{k+1}v_{k+1} + e_{k+2}v_{k+2} + ... + e_{2k-1}v_{2k-1} = v_{2k} \pmod{m_r n}$

with: $e_i = f_i + a_if_k$ when $1 \le i \le k-1$

$e_k = f_k m_r$

$e_i = f_i$ when $k + 1 \le i \le 2k-1$

Finally, the matrix M is the matrix M_1 where the latest row is replaced by the vector:

$(e_1, e_2, ... , e_{k-1}, e_k, e_{k+1}, ... , e_{2k-1}, m_r, n)$

We have:

$$M = \begin{pmatrix} 1 & 0 & \cdots & \cdots & \cdots & \cdots & \cdots & \cdots & 0 \\ 0 & \ddots & \ddots & & & & & & \vdots \\ \vdots & \ddots & \ddots & \ddots & & & & & \vdots \\ 0 & \cdots & 0 & 1 & \ddots & & & & \vdots \\ a_1 & \cdots & \cdots & a_{k-1} & m_r & 0 & & & \vdots \\ 0 & \cdots & \cdots & \cdots & 0 & 1 & \ddots & & \vdots \\ \vdots & & & & & \ddots & \ddots & 0 & \vdots \\ 0 & \cdots & \cdots & \cdots & \cdots & 0 & 1 & 0 \\ e_1 & e_2 & \cdots & \cdots & \cdots & \cdots & e_{2k-2} & e_{2k-1} & m_r & n \end{pmatrix}$$

A solution to the system (S) is obtained by multiplying matrix M by an integer vector α. The result v gives a solution to (S): $v_1,...., v_k$ will be $X_1,...., X_k$ and $v_{k+1},...., v_{2k}$ will be $Y_1,...., Y_k$. The reciprocal can be easily demonstrated and consequently we have:

Proposition 3.1.1:
A vector is in L if and only if it is a solution to (S).

3.2 Second step: System with a Second Member

Let (S') be the system (S) with a second member. (S') is the initial system (SI) but without constraints on values of solutions. The same lattice L is used to solve (S').

Proposition 3.2.1:
Let $v = (v_1, v_2, ...,v_{2k-1}, v_{2k})$ be a vector of L.
Let $P = (0, 0, ... ,0, p_k, 0, ..., 0, p_{2k})$ with

$$p_k = h_1$$
$$p_{2k} = Chinese(h_2 \ (mod \ m_r), \ h_3 \ (mod \ n)) + Chinese(0 \ (mod \ m_r), \ c_k \ (mod \ n)).h_1$$

Then $\beta = v - P$ gives a solution to (S').
$\beta_1, ..., \beta_k$ will be $X_1,...., X_k$ and $\beta_{k+1},...., \beta_{2k}$ will be $Y_1,...., Y_k$

Proof:

$$\beta_i = v_i \qquad when \ i \in \{1, 2, 3, ..., k-1, k+1, k+2, ..., 2k-1\}$$
$$\beta_k = v_k - p_k$$
$$= a_1v_1 + a_2v_2 + ...+ a_{k-1}v_{k-1} + \alpha_k \ m_r - h_1$$
$$= a_1\beta_1 + a_2\beta_2 + ...+ a_{k-1}\beta_{k-1} + \alpha_k \ m_r - h_1$$
$$\beta_{2k} = v_{2k} - p_{2k}$$
$$= e_1v_1 + ... + e_{k-1}v_{k-1} + e_k\alpha_k + e_{k+1}v_{k+1} + e_{k+2}v_{k+2} + ... + e_{2k-1}v_{2k-1} + \alpha_{2k} \ m_r \ n - p_{2k}$$

And we have:

$$\beta_{2k} \ (mod \ m_r) = b_1v_{k+1} + b_2v_{k+2} + ...+ b_{k-1}v_{2k-1} - h_2$$
$$= b_1\beta_{k+1} + b_2\beta_{k+2} + ...+ b_{k-1}\beta_{2k-1} - h_2$$
$$\beta_{2k} \ (mod \ n) = c_1v_1 + c_2v_2 + ... + c_kv_k + d_1v_{k+1} + ... + d_{k-1}v_{2k-1} - h_3 - c_kh_1$$
$$= c_1\beta_1 + c_2\beta_2 + ... + c_k(v_k - h_1) + d_1\beta_{k+1} + ... + d_{k-1}\beta_{2k-1} - h_3$$

As $v_k - h_1 = v_k - p_k = \beta_k$, we have:

$$\beta_{2k} \ (mod \ n) = c_1\beta_1 + c_2\beta_2 + ... + c_k\beta_k + d_1\beta_{k+1} + ... + d_{k-1}\beta_{2k-1} - h_3$$

Thus, β gives a solution to (S'). ∎

3.3 Third step: Additional Constraints

We always consider the system (S'), but we take into account the initial constraints on values of $(X_i)_{1 \le i \le k}$ and $(Y_i)_{1 \le i \le k}$. Hence, we solve (SI).

First case: same bounds

Let B be a positive integer. Find X_i and Y_i such that $0 \le X_i \le B$ and $0 \le Y_i \le B$ for any i such that $1 \le i \le k$.

Proposition 3.3.1:
Let HC be a ball of radius B/2, relative to the norm sup, centred on $Q = P + (B/2, B/2, ..., B/2)$, where the point P is defined in the proposition 3.2.1. Let v be a vector of L inside HC, and $\beta = v - P$.
Then β gives a solution to (S') and satisfies additional constraints.

Proof:

Proposition 3.2.1 shows that β gives a solution to (S').

v inside HC implies $0 \le v_i - p_i \le B$, i.e. $0 \le \beta_i \le B$, for any $1 \le i \le 2k$. ∎

Second case: distinct bounds

Let $(B_i)_{1 \le i \le 2k}$ be a family of positive integers. Find X_i and Y_i such that: $0 \le X_i \le B_i$ and $0 \le Y_i \le B_{k+i}$ for any i such that $1 \le i \le k$.

We apply a method of expansion-contraction to the lattice L to obtain another lattice L', see [VGT1] and [VGT2] for more details. We denote by M' the matrix of lattice L'.

Define B as $B^{2k} = \prod_{i=1}^{2k} B_i$. Let $(\lambda_i)_{1 \le i \le 2k}$ such that $\lambda_i = \dfrac{B}{B_i}$. Then the product $\prod_{i=1}^{2k} \lambda_i$ is equal to 1. M' is the matrix M where each row i, $1 \le i \le 2k$, is multiplied by λ_i:

$$M' = \begin{pmatrix} \lambda_1 & 0 & \cdots & \cdots & \cdots & \cdots & \cdots & \cdots & 0 \\ 0 & \ddots & \ddots & & & & & & \vdots \\ \vdots & \ddots & \ddots & \ddots & & & & & \vdots \\ 0 & \cdots & 0 & \lambda_{k-1} & 0 & & & & \vdots \\ \lambda_k a_1 & \cdots & \cdots & \lambda_k a_{k-1} & \lambda_k m_r & 0 & & & \vdots \\ 0 & \cdots & \cdots & \cdots & 0 & \lambda_{k+1} & 0 & & \vdots \\ \vdots & & & & & \ddots & \ddots & \ddots & \vdots \\ 0 & \cdots & \cdots & \cdots & \cdots & & 0 & \lambda_{2k-1} & 0 \\ \lambda_{2k} e_1 & \lambda_{2k} e_2 & \cdots & \cdots & \cdots & \cdots & \lambda_{2k} e_{2k-2} & \lambda_{2k} e_{2k-1} & \lambda_{2k} m_r n \end{pmatrix}$$

Remark that $\det(M) = \det(M') = m_r^2 n$.

Proposition 3.3.2:

Let $P' = (0, 0, \dots, 0, p_k \lambda_k, 0, \dots, 0, p_{2k} \lambda_{2k})$, where P is defined in the proposition 3.2.1. Let HC be a ball, relative to the norm sup, of radius $B/2$ centred on $Q = P' + (B/2, B/2, \dots, B/2)$. Let v' be an element of L' inside HC and $\beta' = v' - P'$. Then $\beta = (\lambda_1^{-1} \beta_1', \dots, \lambda_{2k}^{-1} \beta_{2k}')$ gives a solution to (S') and satisfies additional constraints.

Proof:

$$\begin{aligned} \beta = (\lambda_1^{-1} \beta_1', \dots, \lambda_{2k}^{-1} \beta_{2k}') &= (\lambda_1^{-1}(v_1' - p_1'), \dots, \lambda_{2k}^{-1}(v_{2k}' - p_{2k}')) \\ &= (\lambda_1^{-1} v_1' - p_1, \dots, \lambda_{2k}^{-1} v_{2k}' - p_{2k}) \end{aligned}$$

Let $v = (\lambda_1^{-1} v_1', \dots, \lambda_{2k}^{-1} v_{2k}')$ and $\beta = v - P$.

Then $v \in L$ and proposition 3.2.1 shows that β gives a solution to (S').

v' inside HC implies $0 \le v_i' - p_i' \le B$, i.e. $0 \le \beta_i \le B_i$, for any $1 \le i \le 2k$. ∎

Remark: the first case is a particular case of the second. In the sequel, we will consider always the lattice L' and its matrix M'.

3.4 How to Generate a Solution?

Proposition 3.3.2 shows that a point of lattice L' inside HC gives a solution to the system (SI). To find one, take a point x inside HC and find a close lattice point inside HC.

First, apply the LLL algorithm [LLL] to the matrix M'. A reduced basis of L' is obtained. Next, apply one of two algorithms of László Babai, *Rounding Off* or *Nearest Plane*, described in [B] to find a solution.

Let u be the nearest lattice point of x and d the dimension of lattice L'.

• **ROUNDING OFF**: this algorithm finds a lattice point v' such that:
$$|x - v'| \le C_d |x - u| \text{ with } C_d = 1 + 2d(9/2)^{d/2}$$

• **NEAREST PLANE**: this algorithm finds a lattice point v' such that:
$$|x - v'| \le C_d |x - u| \text{ with } C_d = 2^{d/2}$$

Remark that, if the dimension d of lattice increases, then the probability that one of these algorithms finds a lattice point inside HC decreases.

3.5 Efficiency: Heuristic Approach

Heuristically, if the ratio of the HC volume to the lattice determinant is greater than 1, then there is at least one lattice point in HC.

The *Nearest Plane* algorithm certainly finds this point when the dimension d of the lattice is not too large. When d increases, the term $C_d = 2^{d/2}$ increases too, and the probability to obtain a point inside HC decreases.

We study the general case where the redundant version of m is:

$$R(m) = \sum_{i=1}^{k_1} m_i \omega_i + \sum_{j=1}^{k_2} \varphi(m)_j \omega_{j+k_1} + a$$

This is the most complicated case, and the solutions to the others can be derived from the following analysis.

We denote by:

t : the bitlength of n

b_i: the length of the part m_i of m

b : such that $\sum_{i=1}^{k_1} b_i = b$

c_j: the length of the part $\varphi(m)_j$ of $\varphi(m)$

c : such that $\sum_{j=1}^{k_2} c_j = c$

First method: modular redundancies are fixed

Modular redundancies $\varphi(x)$ and $\varphi(y)$ are fixed. Finding two messages x and y such that $R(m)R(x) = R(y) \pmod{n}$ is equivalent to solve (SI) with $k = k_1$.

Lattice dimension   : $d = 2k_1$

Lattice determinant   : $\det(L) = \det(L') = m_r^2 n < (2^c)^2 2^t$

HC volume $\qquad : \left(2^{\frac{2\sum_i^d h_i}{d}}\right)^d = 2^{2b}$

Heuristically, there is one point in HC if:

$$\frac{2^{2b}}{2^{2c+t}} > 1$$

$$2b > 2c+t$$

$$b > \frac{t}{2}+c$$

$$t-b < \frac{t}{2}-c$$

$$\boxed{\;|redundancy| < \frac{1}{2}|n| - |m_r|\;}$$

Second method: modular redundancies are not fixed

Finding two messages x and y such that $R(m)R(x) = R(y)$ (mod n) is equivalent to solve (SI) with $k = k_1 + k_2$. But, there is a disadvantage when modular redundancies are not fixed. The dimension of lattice increases and therefore the probability to find a lattice point in HC with Babai's algorithms decreases.

Lattice dimension $\quad : d = 2(k_1 + k_2)$

Lattice determinant $\quad : \det(L)= m_r^2 n < (2^c)^2 2^t$

HC volume $\qquad : \left(2^{\frac{2\left[\sum_i h_i + \sum_j c_j\right]}{d}}\right)^d = 2^{2(b+c)}$

Heuristically, there is one point in HC if:

$$\frac{2^{2(b+c)}}{2^{2c+t}} > 1$$

$$2(b+c) > 2c+t$$

$$b > \frac{t}{2}$$

$$t-b < \frac{t}{2}$$

$$\boxed{\;|redundancy| < \frac{1}{2}|n|\;}$$

4 Application

We applied our attack on a project of digital signature schemes giving message recovery ISO/IEC 9796-3, Working Draft, December 1996 [ISO2]. It is supposed to avoid the known attacks against RSA [GQLS]. This part of ISO/IEC 9796 specifies a digital signature scheme for messages of limited length, so that the message is completely recovered from the signature. It uses a check-function to save bits and computations. This check-function is a modular redundancy, it is the remainder of the message to be signed modulo $2^{79}+1$. The modular redundancy takes the form:

$$R(m) = \sum_{i=1}^{k_1} m_i \omega_i + \sum_{j=1}^{k_2} \varphi(m)_j \omega_{j+k_1} + a$$

with all $(\omega_i)_{1 \leq i \leq k_1 + k_2}$ powers of two. We experiment our attacks on this scheme with a public modulus n of 640 bits of length. In this case the project defines an intermediate string IS:

Structure of the intermediate string IS (640-80 = 560 bits)

Header	Padding Field	Data field	Trailer
Three bits	640 - k_m - 87 bits	k_m bits	Four bits
Set at 010	640 - k_m - 88 bits set to 0 followed by one bit set to 1	Message m	Set to 0110

The structure of the valid message (640 bits) is:

Binary pattern (check-code in bold)

12 + **4** + 28 + **4** + 28 + **4** + ...+ 28 + **4** + 28 + **4** +16 = 640 bits

We applied the first method, i.e. we fixed the check-code. We found several solutions by using the *Rounding Off* algorithm. We give an example of solution:

Public modulus:

```
n =   ffffffff  78f6c555  06c59785  e871211e
      e120b0b5  dd644aa7  96d82413  a47b2457
      3f1be574  5b5cd995  0f6b389b  52350d4e
      01e90009  669a8720  bf265a28  65994190
      a661dea3  c7828e2e  7ca1b196  51adc2d5
```

Message m:

```
m =   fedcba98  76543210  fedcba98  76543210
      fedcba98  76543210  fedcba98  76543210
      fedcba98  76543210  fedcba98  76543210
      fedcba98  76543210  fedcba98  76543210
```

Check-code:

```
c =   0f6e  4af3  a0b1  3571  358b
```

Valid message of *m*:

$Sr(m)$ =

4bb0bbbb	bbbfafed	cba69876	543e210f
edc4ba98	765a4321	0fefdcba	98736543
210afedc	ba908765	432b10fe	dcb1a987
65433210	fed5cba9	87675432	10f1edcb
a9837654	32150fed	cba89876	543b2106

Message *x*:

x =

fedcba0e	2fff215a	00200b1f	17a18638
3212ac94	21061f58	0619a4f0	f912910d
bd3220e3	f4b8064c	89f15211	880c5445
6127d8c9	1a336791	5b962f17	a8386210

Message *y*:

y =

fedcba14	7597b137	39d20f85	33b07f20
cd1335d1	308be96c	14b053d1	4230e40f
02b2f14a	39f709a6	e6a0ede5	ae1f6313
50f4eaf1	1a2f2381	064c2f0f	f3ffa210

Valid message of *x*:

$Sr(x)$ =

4bb0bbbb	bbbfafed	cba60e2f	ff2e15a0
02040b1f	17aa1863	832f12ac	94231061
f58a0619	a4f00f91	291b0dbd	32210e3f
4b83064c	89f51521	18870c54	4561127d
8c931a33	679515b9	62f817a8	386b2106

Valid message of *y*:

$Sr(y)$ =

4bb0bbbb	bbbfafed	cba61475	97be1373
9d240f85	33ba07f2	0cdf1335	d13308be
96ca14b0	53d01423	0e4b0f02	b2f114a3
9f7309a6	e6a50ede	5ae71f63	13510f4e
af131a2f	23851064	c2f80ff3	ffab2106

We obtained this result within 30 minutes on a Pentium 166MHz by using GP/PARI CALCULATOR Version 1.39 (ftp: megrez.math.u-bordeaux.fr/pub/pari). It is the time necessary to apply LLL algorithm to the initial matrix. After, we can easily obtain different messages *x* and *y* in a few seconds by using *Rounding Off* or *Nearest Plane* algorithm on different points inside *HC*.

5 Extensions

We have described an attack on a signature scheme using one modular redundancy. But it is possible to increase the number of modular redundancies. If the different moduli are relatively prime, they can be gathered into one equation with the Chinese Remainder theorem and solved with the first method. If these moduli are not relatively prime, we use the second method, then the probability to find a solution is lower because the dimension is high.

We denote by mask a k_2-bit fixed string. Our attack also succeeds on a scheme that uses a modular redundancy and a mask, i.e. you apply the function exclusive OR between modular redundancy and the mask. In this case we use the first method.

6 How to defeat this forgery

If you want to use fixed or modular redundancy, it is recommended to have the same amount of redundancy as the number of bits of message m, and to have a big dispersion of redundancy bits. It is not sure that you cannot apply our attack but the probability of success will be small.

Another way to avoid this attack is to split the message and define bits of redundancy as parity bits (such as those determined by Hamming codes) of its different parts. ISO 9796 [ISO1] is another possible solution, but it doubles the length of the bit pattern you sign. Our attack cannot apply to the latter schemes because the redundancy depends on different bits of message m and we cannot adjust our attack to this case.

7 Conclusion

This paper describes two attacks to forge a signature of a message m when the bits of redundancy are dispersed and/or when a modular redundancy is used. The first one is valid when the length of redundancy is less than half the length of public modulus, minus the length of modular redundancy. The second attack is valid when the length of redundancy is less than half the length of public modulus, but the probability to find a forgery is smaller (because the lattice dimension grows); however, we have noticed that the *Nearest Plane* and *Rounding Off* algorithms [B] generally give better results than expected.

Afterwards, we have briefly described possible adaptations of our method to attack schemes using mask or different modular redundancies. Hence, we have shown the weakness of many attractive redundancy functions for the purpose of RSA digital signatures.

Finally, we advise to use, for RSA signature scheme with fixed or modular redundancy, the same length of redundancy that the length of the message and to disperse message bits in the valid message. But the best solution remains to use ISO 9796 [ISO1] or the parity bits scheme briefly described above, because they apparently cannot be attacked by our techniques.

Acknowledgments

We would like to thank Marc Girault for encouraging this research and for helpful comments on this paper. We are grateful to Brigitte Vallée for help on the lattice theory and for pointing out corrections to the initial draft. We also thank Louis Guillou for stimulating this research, Jacques Traoré and the referees for their useful comments.

References

[B]   L. Babai, "On Lovász' lattice reduction and the nearest lattice point problem", Combinatorica 6, pp. 1-14.

[C]   Don Coppersmith, "Finding a Small Root of Univariate Modular Equation", Proceedings of Eurocrypt '96, Lecture Note in Computer Science, vol. 1070, pp. 155-165.

[DJC]   W. De Jonge, D. Chaum, "Attacks on some RSA Signatures", Advances in Cryptology, Crypto '85 Proceedings, Lecture Notes In Computer Science, vol. 218, Springer-Verlag, Berlin, 1986, pp. 18-27.

[GQLS]   L.C. Guillou, J.J. Quisquater, P. Landrock, C. Shaer, "Precautions taken against various potential attacks in ISO/IEC DIS 9796, Digital signature scheme giving message recovery", Eurocrypt '90 Proceedings, Lecture Notes in Computer Science, vol.473, Springer-Verlag, pp 465-473.

[GM]   M. Girault, J.F. Misarsky, "Selective Forgery of RSA Signatures Using Redundancy", Advances in Cryptology - Eurocrypt '97, Lecture Notes in Computer Science, vol. 1233, Springer-Verlag, pp 495-507.

[H]   J. Hastad, "Solving simultaneous modular equations of low degree", SIAM J. Comput. vol.17, No.2, April 1988.

[ISO1]   ISO/IEC 9796, December 1991, "Digital signature scheme giving message recovery".

[ISO2]   ISO/IEC 9796-3, Working Draft, December 1996, "Digital signature schemes giving message recovery; Part 3: Mechanisms using a check-function".

[LLL]   A. K. Lenstra, H. W. Lenstra, L. Lovász, "Factoring Polynomials with Rational Coefficients", Mathematische Annalen, vol. 261, n. 4, 1982, pp. 515-534.

[OS]   T. Okamoto and A. Shiraishi, "A fast signature scheme based on quadratic inequalities", Proc. of the 1985 Sympsium on Security and Privacy, Apr.1985, Oakland, CA.

[MOV]   A.J. Menezes, P.C. Van Oorschot, S.A. Vanstone, "Handbook of Applied Cryptography", CRC Press.

[RSA]   R.L. Rivest, A. Shamir and L. Adleman, "A method for obtaining digital signatures and public-key cryptosystems", CACM, Vol. 21, n°2, Feb. 1978, pp. 120-126.

[S]   A. Shamir, "A polynomial-time algorithm for breaking the basic Merkle-Hellman cryptosystem", Proceedings of the 23rd IEEE Symposium on Foundations of Computer Science, pp 145-152. IEEE, 1982.

[VGT1]   B. Vallée, M. Girault, P. Toffin, "How to break Okamoto's cryptosystems by reducing lattice bases", Proceedings of Eurocrypt'87, Lecture notes in Computer Science.

[VGT2]   B. Vallée, M. Girault, P. Toffin, "How to guess L-th roots modulo n by reducing lattice bases", Proc. of Conference of ISSAC-88 and AAECC-6, Jul. 88.

On the Security of the KMOV Public Key Cryptosystem

D. Bleichenbacher

Bell Laboratories
700 Mountain Ave.
Murray Hill, NJ 07974
E-mail: bleichen@research.bell-labs.com

Abstract. This paper analyzes the KMOV public key cryptosystem, which is an elliptic curve based analogue to RSA. It was believed that this cryptosystem is more secure against attacks without factoring such as the Håstad-attack in broadcast application. Some new attacks on KMOV are presented in this paper that show the converse. In particular, it is shown that some attacks on RSA which work only when a small public exponent e is used can be extended to KMOV, but with no restriction on e. The implication of these attacks on related cryptosystems are also discussed.

1 Introduction

In 1985, Koblitz and Miller independently proposed new public key cryptosystems based on elliptic curves [9, 16]. These cryptosystems rely on the difficulty to solve the discrete logarithm problem for elliptic curves. Other cryptosystems based on the same problem have been proposed thereafter. We refer to [15] for more information. A more recent overview is [1].

Koyama, Maurer, Okamoto and Vanstone proposed another kind of elliptic curve based cryptosystems [11]. Their schemes are based on the difficulty of factoring large numbers and are similar to RSA and the Rabin scheme. The most practical of these schemes (Type 1) is generally called the KMOV public key cryptosystem, according to the first letters of the author's names. This scheme was the base for a few similar cryptosystems. Demytko proposed a scheme, which uses only one coordinate of a point over an elliptic curve to represent messages and ciphertexts [5]. Koyama proposed a scheme that is based on singular cubic curves [10]. Another closely related cryptosystem proposed by Koyama and Kuwakado in [14].

It is believed that breaking these systems as well as RSA completely is as difficult as factoring. However, there exist a few attacks on RSA which do not require to factor the modulus. Such attacks are sometimes possible when the ciphertexts and some additional information is known, i.e. (i) when some parts of the plaintext is known, (ii) the encryption of the same or related plaintexts is sent to different users (e.g. in a broadcast application) or (iii) when the encryptions of two related plaintexts are sent to the same user.

A few authors have shown that such attacks can be extended to elliptic curve cryptosystems [14, 12, 20, 8]. These attacks are based on division polynomials whose degree e^2 grows quadraticly with the public parameter e. Because of these results and the more complex structure of KMOV, it is sometimes believed that KMOV is more resistant against this kind of attacks.

In this paper, we present new attacks on KMOV which do not depend on e. In particular, it is shown that the plaintext can be found with high probability (but not always) in each of the following situations:

(i) **partially known plaintext:** The ciphertext and one half of the plaintext is known.

(ii) **broadcast application:** 3 encryptions of the same message or 6 encryptions of linearly related messages are known. All messages are encrypted with distinct public keys.

(iii) **related messages for the same user:** The encryptions of two (linearly) related messages are known. Here, both messages are encrypted with the same public key.

2 Definition of Elliptic curves

This section gives a summary of basic facts about elliptic curves over the field $\mathbb{Z}/(p)$. Let a, b be two integers, such that $4a^3 + 27b^2 \not\equiv 0 \pmod{p}$. By $E_{a,b}(p)$ we denote the group whose elements are given by $\{(x_1, y_1) \in (\mathbb{Z}/(p))^2 : y_1^2 \equiv x_1^3 + ax_1 + b \pmod{p}\} \cup \{\mathcal{O}\}$. By $\mathcal{O}$ we denote the point at infinity, which will also be the neutral element of $E_{a,b}(p)$. The inverse of a point (x_1, y_1) is $(x_1, -y_1)$. The sum $(x_3, y_3) = (x_1, y_1) + (x_2, y_2)$ of two points that are not the inverse of each other can be computed by

$$\lambda \equiv \begin{cases} \dfrac{3x_1^2 + a}{2y_1} & \pmod{p} \text{ if } x_1 \equiv x_2 \pmod{p} \\ \dfrac{y_1 - y_2}{x_1 - x_2} & \pmod{p} \text{ if } x_1 \not\equiv x_2 \pmod{p} \end{cases}$$

$$x_3 \equiv \lambda^2 - x_1 - x_2 \pmod{p}$$
$$y_3 \equiv \lambda(x_1 - x_3) - y_1 \pmod{p}$$

A multiplication of a point P by an integer t will be denoted by $t \cdot P$

$$t \cdot P = \overbrace{P + \cdots + P}^{t}.$$

Let p, q be two distinct primes and $n = pq$. Then $E_{a,b}(n)$ will be defined by

$$E_{a,b}(n) = E_{a,b}(p) \times E_{a,b}(q)$$

If a point $(x_1, y_1) \in (\mathbb{Z}/(n))^2$ satisfies $y_1^2 \equiv x_1^3 + ax_1 + b \pmod{n}$ then we will associate (x_1, y_1) with the point $((x_1 \bmod p, y_1 \bmod p), (x_1 \bmod q, y_1 \bmod q)) \in E_{a,b}(p) \times E_{a,b}(q)$. Two points represented like this can be added by using the same arithmetic operation as in the definition, however, computed over $\mathbb{Z}/(n)$. The points $(\mathcal{O}, P)$ and $(P, \mathcal{O}) \in E_{a,b}(n)$ can not be represented like this. Finding such a point is, however, very unlikely and would lead to a factorization of n.

3 Description of KMOV

Koyama, Maurer, Okamoto and Vanstone proposed three cryptosystems based on elliptic curves in [11]. We describe their Type 1 scheme here. We will not consider the Type 0 scheme, which seems less practical than the Type 1 scheme, because the order of a general elliptic curve must be computed and the Type 2 scheme, which is a Rabin-type generalization.

The private key of the Type 1 scheme consists of two large primes $p \equiv q \equiv 2 \pmod 3$. The public key consists of the product $n = pq$ and an integer e that is relatively prime to $(p+1)(q+1)$. A message is a pair (m_x, m_y) where $m_x, m_y \in \mathbb{Z}/(n)$. It is encrypted by computing $(c_x, c_y) \equiv e \cdot (m_x, m_y) \pmod n$ over the elliptic curve $E_{0,b}(n)$ where $b \equiv m_y^2 - m_x^3 \pmod n$ is determined by the message. Both values c_x and c_y are sent to the receiver. The receiver can determine the curve over which the message was encrypted (even though this computation is in fact not necessary) from the ciphertext since (c_x, c_y) and (m_x, m_y) are points on the same curve and therefore

$$b \equiv c_y^2 - c_x^3 \equiv m_y^2 - m_x^3 \pmod n. \tag{1}$$

Then, he can decrypt the message by computing $(m_x, m_y) \equiv d \cdot (c_x, c_y) \pmod n$ where $d \equiv e^{-1} \pmod{(p+1)(q+1)}$. This follows from the fact that the order of the curve $E_{0,b}(n)$ divides $(p+1)(q+1)$.

4 The value of a modular polynomial equation of small degree

All RSA-based cryptosystems are based on the difficulty of solving polynomial equations over $\mathbb{Z}/(n)$. No method for solving a univariate equation $f(m) \equiv 0 \pmod n$ for m where $f(x)$ is a polynomial of degree > 1 is known without factoring n. However, if some additional information on a solution m is known then this situation may change. Such situations are described in this section.

Coppersmith describes an algorithm that finds a root of a univariate polynomial if this root is small enough [3]. In particular, he proved the following result.

Theorem 1 (Coppersmith). *Let $f(x)$ be a monic integer polynomial of degree k and N a positive integer of unknown factorization. In time polynomial in $\log N$ and k, we can find all integer solutions m to $f(m) \equiv 0 \pmod N$ with $|m| < N^{1/k}$.*

Coppersmith also discusses the application of his method to general multivariate polynomials [3, Section 3], but we will only describe the implication to polynomials in two variables here. If $f(x, y)$ is a polynomial of total degree k then he showed, that it is often possible to find efficiently a solution m_x, m_y to $f(m_x, m_y) \equiv 0 \pmod N$ if

$$\max(|m_x|, |m_y|) < N^{1/2k - \epsilon}$$

for some $\epsilon > 0$. Even though there is no guarantee that a solution will be found, our experiments have almost always been successful, because the lattice reduction algorithm had found the small lattice vector related to the solution.

Another method that will be used in this paper was discovered by Copper-smith, Franklin, Patarin and Reiter [4]. The authors observed that an unknown value m can often be found when two polynomial equations of small degree $f(m) \equiv g(m) \equiv 0 \pmod{n}$ are known, since $(x - m)$ must divide the gcd of $f(x)$ and $g(x)$ and since $\gcd(f(x), g(x))$ is very likely a linear polynomial. Their attack is practical if the degrees of the polynomials are smaller than about 2^{32}.

Description of an improved algorithm. In this paper, we will study some attacks that are based on the more general problem where only $f(x)$ is a polynomial of small degree and where $g(x)$ is a rational function of large degree that can be computed in a small number of arithmetic steps. For example, if one coordinate is fixed then the encryption function in KMOV defines such a rational function $g(x)$, that can be computed in a small number of operations (i.e. $O(\log(e))$) even though the degree of the function may be large (i.e. e^2).

Even though, the small polynomial equation $f(m) \equiv 0 \pmod{n}$ can generally not be solved, $f(x)$ can be regarded as an implicit representation of m. Using this representation it is possible to perform arithmetic operations on m in almost the same way as arithmetic operations with algebraic numbers are performed. For example, it will be possible to compute an encryption on a message m given implicitly by a small polynomial equation $f(m) \equiv 0 \pmod{n}$.

The first step of our algorithm is a square free factorization on $f(x)$. Thus allows us to assume that $f(x)$ is in fact square free over $\mathbb{Z}[x]/(n)$. In the following, we will perform arithmetic operations in the quotient ring $R = \mathbb{Z}[x]/(n, f(x))$. x will be an implicit representation of m. More generally, any polynomial $h(x) \in R$ will represent $h(m)$ and hence we can define a ring homomorphism $\phi : R \to \mathbb{Z}/(n)$ given by

$$\phi : h(x) \mapsto h(m).$$

Note that ϕ is initially not known explicitly since the solution m is unknown. Note also that ϕ is well defined since $f(m) \equiv 0 \pmod{n}$, i.e. ϕ does not depend on representatives of an equivalence class in R as

$$\phi(h(x) + h'(x)f(x)) \equiv h(m) + h'(m)f(m) \equiv h(m) \equiv \phi(h(x)) \pmod{n}.$$

Let t be the degree of $f(x)$. We will now show that arithmetic operations with m known implicitly can be performed efficiently by representing all intermediary results $r \in \mathbb{Z}/(n)$ with polynomials $h(x) \in R$ of degree smaller than $\deg(f(x)) = t$ such that $h(m) \equiv r \pmod{n}$.

Given two polynomials $h(x), h'(x) \in R$ of degree smaller than t. Then polynomials representing the sum and product of $h(m)$ and $h'(m)$ can be found by adding respectively multiplying $h(x)$ and $h'(x)$ together and finally reducing the result modulo $f(x)$. A polynomial $r(x)$ representing the inverse of $h(m)$ can be found by using the extended Euclidean algorithm, i.e. by finding two polynomials $r(x)$ and $s(x)$ such that $r(x)h(x) + s(x)f(x) = \gcd(h(x), f(x))$. The inverse of

$h(x)$ is $r(x)$ if the gcd is 1. Otherwise, if $\gcd(h(x), f(x)) \neq 1$ then we have either found a nontrivial factor of n or $f(x)$. It is also possible to test equality, since $h(m) \equiv h'(m) \pmod{n}$ implies $\deg(\gcd(h(x) - h'(x), f(x))) \geq 1$. Either we have $\deg(\gcd(h(x) - h'(x), f(x))) = t$ and thus $h(m) \equiv h'(m) \pmod{n}$ or we have $\deg(\gcd(h(x) - h'(x), f(x))) = 1$ and $h(m) \not\equiv h'(m) \pmod{n}$ or we have found a nontrivial factor of $f(x)$. Hence we have shown that we can compute efficiently a polynomial $g'(x)$ of degree smaller than t such that

$$g(m) \equiv g'(m) \pmod{n}$$

or find a nontrivial factor of either n or $f(x)$.

A factor of n would mean that the secret key is found. If a factor of $f(x)$ is found then we can rerun the algorithm with $g(x)$ and each of the new factors of $f(x)$. Since the degree of $f(x)$ is t we will compute $g(x)$ in at most $2t$ rings $R_i = \mathbf{Z}[x]/(n, f_i(x))$ where $f_i(x)$ are factors of $f(x)$.

On the other hand if we find $g'(x)$ then we compute the gcd of $g'(x)$ and $f(x)$. From $g'(m) \equiv f(m) \equiv 0 \pmod{n}$ follows that $(x - m)$ is a divisor of the gcd. Thus when this gcd is in fact a linear polynomial then we can find m. We can now describe the algorithm as follows.

Algorithm 2. Given an RSA-modulus n with unknown factorization, a polynomial $f(x)$ of small degree and a rational function given by a short straight-line program (i.e. a short sequence of arithmetic operations to compute $g(x)$ from the set $\{x\} \cup \mathbf{Z}/(n)$). Then this algorithm tries to find a solution m to $f(m) \equiv g(m) \equiv 0 \pmod{n}$.

Step 1: Use square free factorization (e.g. [2, Algorithm 3.4.2]) to find

$$f(x) = \prod_{i=1}^{t} f_i(x)^i$$

where $f_i(x)$ are square free polynomials. If no factorization is found here (i.e. $f(x)$ is square free) then continue with step 2. Otherwise call this algorithm recursively with $n, f_i(x)$ and $g(x)$ for all $1 \leq i \leq t$ and return the union of solutions found.

Step 2: Let $R = \mathbf{Z}[x]/(n, f(x))$ and compute $g(x)$ over R.

If in any step a nontrivial factor of n is found then print this factor and terminate the algorithm.

If in any step a nontrivial factor $f'(x)$ of $f(x)$ is found then call this algorithm recursively with $n, f'(x), g(x)$ and with $n, f(x)/f'(x), g(x)$ and return the union of the solutions of this two calls.

Step 3: If no exception in Step 2 occurs then we get $g'(x) = g(x)$ over R where $g'(x)$ is a polynomials whose degree is smaller than $\deg(f(x))$. Now compute $r(x) = \gcd(f(x), g'(x))$.

If $r(x)$ is a constant then return 'no solution has been found'. If $r(x)$ is a linear polynomial then try to solve $r(m) \equiv 0 \pmod{n}$. This either finds a solution m or a nontrivial factor of n.

If $r(x)$ is a polynomial of degree larger than 1 then return that the algorithm is unable to solve $r(m) \equiv 0 \pmod{n}$.

Remark. In the situation, where we use this algorithm m will be the unique solution over $\mathbb{Z}/(n)$. This is not a sufficient condition since the algorithm can not find m when there is more than one solution to $g(m) = 0$ over $\mathbb{Z}[x]/(n, f(x))$. Therefore, our algorithm may sometimes fail. Fortunately, some of the attacks presented later in this paper allow a more rigorous analysis.

5 Partially known plaintext attack

In this section, we consider the security of the cryptosystem under the assumption that some part of the plaintext is known. We ask for the largest fraction of plaintext that can be recovered from the ciphertext when the rest of the plaintext is known. Hereby we assume an ideal situation for the attacker, i.e. we assume that the known bits are consecutive or simply those that help most.

Coppersmith has shown that $1/k$ of the bits of the plaintext can be recovered if a univariate equation of degree k over the plaintext is known. (See Theorem 1) This shows that up to $1/e$ unknown plaintext bits can be recovered from an RSA-encryption when the rest of the plaintext is known.

The attacks on KMOV in this section are based on the fact that the ciphertext (c_x, c_y) and the plaintext (m_x, m_y) are points on the same curve, i.e. that we can derive b from the ciphertext such that

$$m_x^3 + b \equiv m_y^2 \pmod{n}. \tag{2}$$

When the plaintext is partially known then we can eventually solve this equation. The multivariate version of Coppersmith's algorithm [3, Section 3]) can be applied when about $1/6$ of the bits of the plaintext are unknown.

Here, we present another method that can tolerate up to $1/2$ of unknown plaintext but that is less flexible since either m_x or m_y must be completely known.

Theorem 3. *Let n, e be a public key for KMOV and $C = (c_x, c_y)$ be the encryption of a message $M = (m_x, m_y)$. Then M can be computed efficiently given n, e, C and either m_x or m_y.*

Proof. Since (c_x, c_y) and (m_x, m_y) are points on the same elliptic curve we have

$$c_x^3 - c_y^2 \equiv m_x^3 - m_y^2 \pmod{n}. \tag{3}$$

When either m_x or m_y are known then Equation (3) becomes a univariate equation of degree 2 or 3 in the missing plaintext. Hence, we can apply the algorithm described in Section 4. Since there is no guarantee that the algorithm works in general we have to analyze this special case, which fortunately is simple enough to be analyzed rigorously.

Assume that m_x is known and m_y is unknown. Then we compute $e \cdot (m_x, y)$ over $\mathbb{Z}[y]/(y^2 - m_x^3 - b, n)$. It can be shown by induction over k and using the definition of the addition on elliptic curves that $k \cdot (m_x, y) \equiv (r_k, s_k y)$ (mod n) for two integers r_k, s_k. Thus we will finally get an equation $\phi(s_e y) \equiv c_y$ (mod n), which is solvable when $s_e \not\equiv 0$ (mod n). If, however, $s_e \equiv 0$ (mod n) then C is a point of order 2 and it follows from $M \equiv d \cdot C$ (mod n) that $C = M$. Hence, M is always computable.

Now assume that m_y is known and m_x is unknown. Then we will compute $e \cdot (x, m_y)$ over $\mathbb{Z}[x]/(x^3 + b - m_y^2, n)$. As before it can be shown by induction over k that $k \cdot (x, m_y) \equiv (r_k x, s_k)$ (mod n) for some integers r_k, s_k. Thus we finally have to solve the equation $\phi(r_e x) \equiv c_x$ (mod n), which is possible when $r_e \not\equiv 0$ (mod n). Again we have to treat the case $r_e \equiv 0$ (mod n) specially. It can be observed that $c_x \equiv 0$ (mod n) and $a \equiv 0$ (mod n) implies $2 \cdot C \equiv -C$ (mod n). Therefore C is a point of order 3 and hence M can be found easily. $\square$

Example. Let $n = 493$ and $e = 7$ be the public key of KMOV. Assume that we know the ciphertext $C = (214, 358)$ and $m_y \equiv 229$ (mod n), which is one half of the plaintext. First, we derive b from the ciphertext C and have

$$m_x^3 + b - m_y^2 \equiv m_x^3 + 297 \equiv 0 \quad (\text{mod } 493).$$

Now we encrypt the point $P = (x, 229)$ over $\mathbb{Z}[x]/(x^3 + 297, 493)$ and get

$$C \equiv (12\,x, 358) \quad (\text{mod } 493).$$

Therefore we have $m_x \equiv 214 \cdot 12^{-1} \equiv 100 \quad (\text{mod } 493)$.

6 Attacks in broadcast applications

In this section, we consider the situation of a broadcast application where a message is encrypted with different public keys and sent to the corresponding users. An attacker who intercepts some of these messages can sometimes combine the information he gained in such a way that he can learn the encrypted message. In particular, we will consider the following two situations:

1. All ciphertexts c_i are the encryption of the same message m.
2. The ciphertexts c_i are the encryption of linearly related messages

$$m_i \equiv \alpha_i m + \beta_i \quad (\text{mod } n_i)$$

for some m where α_i and β_i are known constants.

We will review the security of RSA in broadcast applications before describing the attack against KMOV, since it is sometimes overlooked that Coppersmith has improved Håstad's result [6].

A simple method can be used when at least e RSA encryptions of the same message m encrypted with the same public exponent e are known, i.e. when the ciphertext c_i are known such that

$$c_i \equiv m^e \quad (\text{mod } n_i) \text{ for } 1 \leq i \leq e.$$

From these equations we can derive C such that

$$C \equiv m^e \pmod{\prod_{i=1}^{e} n_i}.$$

Since $m^e < \prod_{i=1}^{e} n_i$ it follows that m can be found from C by computing the e-th root of C over $\mathbf{Z}$.

This simple method is no longer possible when we have k messages that are encrypted are not equal but linearly related. In this case, we will generally have a polynomial equation we can derive k equations

$$f_i(m) \equiv 0 \pmod{n_i} \text{ for } 1 \le i \le k.$$

We multiply the polynomials $f_i(m)$ by the inverse of their leading coefficient and possibly by a power of m such that the resulting polynomials are all monic polynomials of equal degree. Then we use the Chinese Remainder Theorem to derive an equation

$$F(m) \equiv 0 \pmod{N} \text{ where } N = \prod_{i=1}^{k} n_i. \tag{4}$$

The polynomial F is monic and thus we can use Coppersmith's algorithm if $|m| < N^{1/\deg(F)}$ to find m. This attack has apparently been described by Shimizu in [19].

A small improvement of this method is possible when the degrees of the polynomials $f_i(m)$ are different. Instead of multiplying them by a power of m it might be possible to compute powers of the polynomials itself. Since $f_i(m) \equiv 0 \pmod{n_i}$ implies $f_i(m)^{t_i} \equiv 0 \pmod{n_i^{t_i}}$ we can thus gain an equation $F(m) \equiv 0 \pmod{N}$ for a larger N. Given for example two RSA encryptions with $e_1 = 5$ and $e_2 = 3$ and a Rabin encryption of linearly related messages $m_i \equiv \alpha_i m + \beta_i \pmod{n_i}$ we can derive the following equations

$$(\alpha_1 m + \beta_1)^5 - c_1 \equiv 0 \pmod{n_1}$$
$$(\alpha_2 m + \beta_2)^3 - c_2 \equiv 0 \pmod{n_2}$$
$$(\alpha_3 m + \beta_3)^2 - c_3 \equiv 0 \pmod{n_3}$$

From these equations we compute

$$m\left((m + \beta_1 \alpha_1^{-1})^5 - c_1 \alpha^{-5}\right) \equiv 0 \pmod{n_1}$$
$$\left((m + \beta_2 \alpha_2^{-1})^3 - c_1 \alpha^{-3}\right)^2 \equiv 0 \pmod{n_2^2}$$
$$\left((m + \beta_3 \alpha_3^{-1})^2 - c_1 \alpha^{-2}\right)^3 \equiv 0 \pmod{n_3^3}.$$

All these equations are defined by monic polynomials of degree 6. Thus we can use the Chinese Remainder Theorem to get an equation

$$F(m) \equiv 0 \pmod{n_1 n_2^2 n_3^3}$$

where F is monic of degree 6 and $|m| < (n_1 n_2^2 n_3^3)^{1/6}$.

When KMOV is used then a message (m_x, m_y) can often be found when only 3 encryptions of the same message but with 3 different public keys are known. In particular, we have the following theorem.

Theorem 4. *Let* $t \geq 1$, n_1, n_2, n_3 *be the moduli of 3 different KMOV keys,* $n = \max(n_1, n_2, n_3)$ *and* $\hat{n} = \min(n_1, n_2, n_3)$. *Given the 3 ciphertexts of a randomly chosen message* $M = (m_x, m_y) \in \{0, \ldots, \hat{n} - 1\}^2$ *encrypted with these 3 keys then* M *can be found in time* $O(t^2 \log(n)^3)$ *with probability* $1 - 1/t$.

Proof. Because of Equation (1) we can derive b_i from the ciphertext such that

$$-b_i \equiv m_x^3 - m_y^2 \pmod{n_i} \text{ for } i \in \{1, 2, 3\}.$$

Thus we can find

$$b \equiv m_x^3 - m_y^2 \pmod{n_1 n_2 n_3}$$

for some $-\hat{n}^2 \leq b < n_1 n_2 n_3 - \hat{n}^2$. Moreover, since $m_x^3 - m_y^2$ must lie in the same interval it follows $b = m_x^3 - m_y^2$. We expect that m_y^2 is much smaller than m_x^3 and that therefore $m_x \approx b^{1/3}$, and we will show that m_x can be found with high probability by using this approximation.

Let $m_0 = \lceil b^{1/3} \rceil$. Then it is possible to find M in time $O(t^2 \log(n)^3)$ when $m_0 \leq m_x \leq m_0 + (4/3)t^2$. Indeed, we test for every $m_0 \leq m_x' \leq m_0 + (4/3)t^2$ whether the integer $m_x'^3 - b$ is a square. If this is the case then we let $m_y' = (m_x'^3 - b)^{1/2}$ and check whether the encryption of (m_x', m_y') with one of the public keys is equal to the corresponding ciphertext. This can be done in time $O(\log(n)^3)$ for every m_x'.

Thus it remains to show that the probability for $m_0 \leq m_x \leq m_0 + (4/3)t^2$ is at least $1 - 1/t$ for a randomly chosen message. Assume that $m_x \geq \hat{n}/t$ and let $\gamma = (4/3)t^2$. Then we have $m_y/t \leq m_x$ and thus $m_y^2 \leq (3/4)\gamma m_x^2 \leq \gamma((3/4)m_x^2 + (\gamma - (3/2)m_x)^2) = m_x^3 - (m_x - \gamma)^3$. Hence it follows $m_x^3 \geq m_x^3 - m_y^2 \geq (m_x - \gamma)^3$ and therefore that $m_0 \leq m_x \leq m_0 + (4/3)t^2$. Hence, when the attack fails we have $m_x < \hat{n}/t$ and the probability of this event is $1/t$. $\qquad \square$

When the messages are not equal but linearly related then we can derive equations of the form $(\gamma_i m_y + \delta_i)^2 - (\alpha_i m_x + \beta_i)^3 \equiv b_i \pmod{n_i}$. Such equations can be combined with the Chinese Remainder Theorem to one equation of degree 3 in two unknowns m_x and m_y, i.e. from k messages we find f such that

$$f(m_x, m_y) \equiv 0 \pmod{\prod_{i=1}^{k} n_i}.$$

Applying Coppersmith's result we can hope for a solution if

$$\max(|m_x|, |m_y|) < \left(\prod_{i=1}^{k} n_i\right)^{(1/6 - \epsilon)}$$

for some $\epsilon > 0$. This implies that the ciphertexts of 6 related messages might give enough information to recover the plaintext. This theoretical result should, however, be compared to our experimental results in section 9.

7 Attacks based on related messages for the same user

In this section, we discuss the situation where two related messages are both encrypted with the same public key. This situation has been analyzed by Coppersmith et al. for RSA [4]. They have shown that the ciphertext can be found from two encryptions with exponent e if it is computationally feasible to compute the gcd of two polynomials of degree e. They conclude that the attack is possible if the size of the public exponent e is smaller than about 32 bits.

The attack on KMOV presented here does not depend on the public parameter e. It is therefore not possible to prevent this attack by choosing e large. Let (m_x, m_y) and $(\tilde{m}_x, \tilde{m}_y)$ be two plaintexts that are related by known linear relations

$$\tilde{m}_x \equiv \alpha m_x + \gamma \tag{5}$$

$$\tilde{m}_y \equiv \beta m_y + \delta. \tag{6}$$

Assume that we know the encryption of these two messages, which is given by

$$(c_x, c_y) \equiv e \cdot (m_x, m_y) \pmod{n} \tag{7}$$

$$(\tilde{c}_x, \tilde{c}_y) \equiv e \cdot (\tilde{m}_x, \tilde{m}_y) \pmod{n} \tag{8}$$

From the ciphertext we can derive the curves $E_{0,b}(n)$ and $E_{0,\tilde{b}}(n)$ on which the points (m_x, m_y) and $(\tilde{m}_x, \tilde{m}_y)$ must lie. Thus we have

$$m_x^3 + b - m_y^2 \equiv 0 \pmod{n}$$

$$(\alpha m_x + \gamma)^3 + \tilde{b} - (\beta m_y + \delta)^2 \equiv 0 \pmod{n}$$

These two equations allow us to express m_y as a polynomial w in m_x. If we set

$$w(x) = \frac{(\alpha x + \gamma)^3 - \beta^2 x^3 - \delta^2 + \tilde{b} - \beta^2 b}{2\beta\delta}$$

then $w(m_x) \equiv m_y \pmod{n}$. Now let $f(x) = x^3 - w(x)^2 + b$, which is a polynomial of degree 6. From (1) follows $f(m_x) \equiv 0 \pmod{n}$. Next, we compute $e \cdot (x, w(x)) \equiv (h(x), j(x)) \pmod{n}$ over $\mathbb{Z}[x]/(n, f(x))$. Since we know the result of this encryption explicitly we have the equations

$$h(m_x) \equiv c_x \pmod{n} \tag{9}$$

$$j(m_x) \equiv c_y \pmod{n}. \tag{10}$$

Finally, we compute $\gcd(f(x), h(x) - c_x)$ and hope to find a linear polynomial of the form $\lambda(x - m_x)$, which allows us to find m_x.

Remark. The same attack would work even when the relation between (m_x, m_y) and $(\tilde{m}_x, \tilde{m}_y)$ is not linear but given by a polynomial relation whose degree is small.

When only one relation is known (e.g. between m_x and $\tilde{m}_x$ but not between m_y and $\tilde{m}_y$) then it is still possible to recover the plaintext when e is small. In that case, we have to compute the gcd between two polynomials of degree e^2. And this seems possible if e is smaller than about 2^{16} [7]. If this method is successful it finds m_x, afterwards m_y can be found using the method of Section 5. Thus, it is sometimes possible to recover the plaintext of two related messages even if one of the two text blocks (m_x, m_y) is chosen randomly for every message.

8 Implication on related cryptosystems

Demytko's cryptosystem [5] uses, contrary to KMOV, only one coordinate to represent messages. This difference seems to be crucial, as the attacks presented in this paper can not be applied to Demytko's cryptosystem. Other proposed cryptosystems only vary the type of curve that is used for the encryption and use like KMOV both coordinates of a point to represent messages [10, 13]. Our attacks work in almost the same way against these cryptosystems too. For example Koyama uses in [10] singular cubic curves of the form

$$y^2 + axy \equiv x^3 \pmod{n}, \tag{11}$$

where the plaintext is a pair (m_x, m_y) and a is chosen such that the Equation (11) with $x = m_x$ and $y = m_y$ is satisfied. Koyama claimed that this cryptosystem is provably as secure as RSA, but faster than RSA. However, this claim hold only one day. Shamir presented at Eurocrypt'95 an attack which showed that one half of the plaintext can be found when the other half is known. Because of (11) the plaintext can also be recovered when at most $1/6$ of m_x and m_y is unknown. When 6 linearly related messages are known then a Håstad attack is possible. The claim in [10] that the scheme is as secure as RSA in broadcast application is therefore not justified. The author wrongly assumes in the proof of section 5.2 of [10] that his elliptic curve cryptosystem cannot be weaker than RSA since he shows only that a successful attack on RSA would imply a successful attack on his scheme. The inversion of this implication is missing. Finally, we can perform a similar attack on Koyama's scheme as on KMOV when the ciphertexts of two related messages encrypted with the same public key are known. Again, this contradicts the conclusions the author draws from Theorem 4 as our attack shows that there is no reason to assume that Koyama's scheme is as secure as RSA.

9 Experimental results

This paper contains a few algorithms that have not been proven to work in all cases. We have therefore implemented all attacks in order to check their effectiveness. The attacks based on lattice basis reduction (i.e. based on [3]) are very computation expensive. This is specially the case when lattices of large dimension are involved. Therefore, we could not verify experimentally that all theoretical bounds are reachable. Fortunately, more knowledge can often help to reduce the dimension of the lattices.

A Håstad attack against KMOV with a 512-bit modulus and 6 linearly related messages seems to be computationally infeasible. But if 8 linearly related messages are known then a lattice basis of size 17 and 2400-digit integers has to be reduced. The package LiDIA, which implements a very sophisticated lattice basis reduction algorithm proposed by Schnorr and Euchner [18], can reduce such a lattice basis in about 2 weeks on an Ultra Sparc. The same attack with 9 linearly related messages can be done in about 15 minutes by reducing a lattice basis of size 6.

We observed that the attacks in Section 6 succeed almost always and we have not observed a failure in any of the other attacks.

10 Countermeasures

One possibility to avoid the attacks in this paper is to randomize some parts of the plaintext before the encryption. We propose that somewhat more than 1/5 of the bits in both coordinates of a point should be chosen randomly. This avoids the attacks presented in Section 5. Moreover, e should not be chosen too small, since a small e would give yet other small modular equations over the plaintext that can be combined with $m_x^3 + b \equiv m_y^2 \pmod{n}$ for even more effective attacks. Since the degree of the equations resulting from division polynomials (see e.g. [12]) is e^2 we suggest to choose e at least 16 bits long. These propositions require, of course, a careful analysis.

11 Conclusions

The attacks in this paper show that it is very dangerous when a cryptosystem leaks a modular relation of small degree on the message. Furthermore these attacks are an example for the fact that a more complex looking cryptosystem not necessarily is more secure than a simple looking one. The comparison of RSA and Koyama's scheme shows that a security analysis that considers only complete messages (i.e. showing that ability to decrypt all messages in one system implies that messages encrypted with the other system can also be decrypted) should not be used alone for comparing the security of two cryptosystems.

Acknowledgements

I'm grateful to Marc Joye for many comments on the paper. I would also like to thank Arjen Lenstra, Dan Boneh and Shai Halevi for answering questions about lattice basis reductions.

References

1. J. Borst. Public key cryptosystems using elliptic curves. Master's thesis, Eindhoven University of Technology, Feb. 1997.

2. H. Cohen. *A Course in Computational Algebraic Number Theory*. Number 138 in Graduate Texts in Mathematics. Springer Verlag, 1993.
3. D. Coppersmith. Finding a small root of a univariate modular equation. In *Advances in Cryptology - EUROCRYPT '96*, volume 1070 of *Lecture Notes in Computer Science*, pages 155–165. Springer Verlag, 1996.
4. D. Coppersmith, M. Franklin, J. Patarin, and M. Reiter. Low exponent RSA with related messages. In *Advances in Cryptology - EUROCRYPT '96*, volume 1070 of *Lecture Notes in Computer Science*, pages 1–9. Springer Verlag, 1996.
5. N. Demytko. A new elliptic curve based analogue of RSA. In T. Helleseth, editor, *Advances in Cryptology — EUROCRYPT '93*, volume 765 of *Lecture notes in computer science*, pages 40–49. Springer-Verlag, 1994.
6. J. Håstad. Solving simultaneous modular equations of low degree. *SIAM J. Computing*, 17(2):336–341, Apr. 1988.
7. M. Joye and J.-J. Quisquater. Overview and security analysis of RSA-type cryptosystems against various attacks. In *Proc. of DIMACS workshop on network threats*, Nov. 1996.
8. M. Joye and J.-J. Quisquater. Protocol failure for RSA-like functions using Lucas sequences and elliptic curves over a ring. In M. Lomas, editor, *Security Protocols*, volume 1189 of *Lecture Notes in Computer Science*, pages 93–100. Springer Verlag, 1997.
9. N. Koblitz. Elliptic curve cryptosystems. *Mathematics of Computation*, 48(177):203–209, 1987.
10. K. Koyama. Fast RSA-type schemes based on singular cubic curves $y^2 + axy = x^3$ (mod n). In *Advances in Cryptology - EUROCRYPT '95*, volume 921 of *Lecture Notes in Computer Science*, pages 329–340. Springer, 1995.
11. K. Koyama, U. Maurer, T. Okamoto, and S. Vanstone. New public-key schemes based on elliptic curves over the ring Z_n. In J. Feigenbaum, editor, *Advances in Cryptology - CRYPTO '91*, volume 576, pages 252–266. Springer Verlag, 1992. Lecture Notes in Computer Science.
12. K. Kurosawa, K. Okada, and S. Tsujii. Low exponent attack against elliptic curve RSA. In *Advances in Cryptology - ASIACRYPT 94*, volume 917, pages 376–383. Springer Verlag, 1995.
13. H. Kuwakado and K. Koyama. Efficient cryptosystems over elliptic curves based on a product of form-free primes. *IEICE Transactions on fundamentals of electronics, communications and computer sciences*, E77-A(8):1309–1318, Aug. 1994.
14. H. Kuwakado and K. Koyama. Security of RSA-type cryptosystems over elliptic curves against Håstad attack. *Electronic Letters*, 30(22):1843–1844, Oct. 1994.
15. A. Menezes, editor. *Elliptic Curve Public Key Cryptosystems*. Kluwer Academic Publishers, 1993.
16. V. S. Miller. Use of elliptic curves in cryptography. In H. C. Williams, editor, *Advances in Cryptology - CRYPTO '85*, volume 218 of *Lecture Notes in Computer Science*, pages 417–426. Springer, 1986.
17. L. Rivest, A. Shamir, and L. Adleman. A method for obtaining digital signatures and public-key cryptosystems. *Communications of the ACM*, 21(2):120–126, 1978.
18. C. P. Schnorr and M. Euchner. Lattice basis reduction: Improved practical algorithms and solving subset sum problems. In L. Budach, editor, *Proceedings of Fundamentals of Computation Theory (FCT '91)*, volume 529 of *Lecture Notes in Computer Science*, pages 68–85. Springer Verlag, Sept. 1991.
19. H. Shimizu. On the improvement of the Håstad bound. In *1996 IEICE Fall Conference*, volume A-162, 1996. (In Japanese).

20. T. Takagi and S. Naito. The multi-variable modular polynomial and its applications to cryptography. In *7th International Symposium on Algorithm and Computation, ISAAC'96*, volume 1178 of *Lecture Notes in Computer Science*, pages 386–396. Springer Verlag, 1996.

A Key Recovery Attack on Discrete Log-based Schemes Using a Prime Order Subgroup*

Chae Hoon Lim[1] and Pil Joong Lee[2]

[1] Information and Communications Research Center, Future Systems, Inc., 372-2,
Yang Jae-Dong, Seo Cho-Gu, Seoul, 137-130, KOREA
E-mail: chlim@future.co.kr
[2] Dept. of Electronic and Electrical Engineering, Pohang University of Science and
Technology (POSTECH), Pohang, 790-784, KOREA
E-mail: pjl@postech.ac.kr

Abstract. Consider the well-known oracle attack: somehow one gets a
certain computation result as a function of a secret key from the secret
key owner and tries to extract some information on the secret key. This
attacking scenario is well understood in the cryptographic community.
However, there are many protocols based on the discrete logarithm problem that turn out to leak many of the secret key bits from this oracle
attack, unless suitable checkings are carried out. In this paper we present
a key recovery attack on various discrete log-based schemes working in
a prime order subgroup. Our attack may reveal part of, or the whole
secret key in most Diffie-Hellman-type key exchange protocols and some
applications of ElGamal encryption and signature schemes.

1 Introduction

Many cryptographic protocols have been developed based on the discrete logarithm problem. The main objective of developers is to design a protocol that is
as difficult to break as the underlying discrete logarithm problem under some
reasonable assumptions. On the other hand, the goal of attackers is to find a way
to extract the secret key involved or to pretend to be a legitimate user without
knowing the secret key. Though provable security guarantees that there is no
efficient attack on the protocol, it should be carefully interpreted for practical
security; the most important would be to use secure parameters and follow the
assumed conditions or requirements as closely as possible. As an illustrative example, we refer to two recent papers on ElGamal-type signature schemes; one
regarding security proof by Pointcheval and Stern [34] and the other regarding
signature forgery by Bleichenbacher [4] (see also Stern [39] for further discussions
on their apparent contradiction).

The purpose of this paper is to point out the insecurity of various discrete
log-based schemes using a prime order subgroup. More specifically, we present a

---

* This work was done while the first author was in POSTECH Information Research
Laboratories, POSTECH, Pohang, Korea and the second author was in NEC Research Institute, Princeton, NJ, during his sabbatical leave.

key recovery attack on these protocols, which can find all or part of the secret key bits. Our attack is closely related to the choice of parameters and the checking of protocol variables. Thus, as is usual, our attack, once identified, can be easily prevented by adding suitable checking steps or by using 'secure' parameters. Here 'secure' means that the parameters are secure against our attack. And this also implies that the usual parameters commonly used in the literature are not secure against our attack. The presented attack demonstrates the importance of checking protocol variables in designing discrete log-based schemes.

Pohlig-Hellman Decomposition and Pollard's Methods : The discrete logarithm problem over Z_p^* can be broken down into a number of small such sub-problems defined over small order subgroups of Z_p^* (Pohlig-Hellman decomposition [33]). Then these sub-problems can be solved using Pollard's rho and lambda methods [35] and the resulting partial logarithms can be combined using the Chinese Remainder Theorem (CRT) to give the pursued discrete logarithm.

For simplicity, suppose that $p-1 = \prod_{i=1}^{n} q_i$ (q_i prime). Let α be a generator of Z_p^*. Given $y = \alpha^x \bmod p$, we can reduce the problem of finding $x \bmod p-1$ to the following sub-problems: find $x_i = x \bmod q_i$ from $y_i = y^{(p-1)/q_i} = \alpha_i^{x_i} \bmod p$ for each i, where $\alpha_i = \alpha^{(p-1)/q_i}$ (an element of order q_i). Each such sub-problem can be solved using Pollard's rho method (see [40] for linear speedup with multiple processors). Once x_i's are found for all i, they can be combined using the CRT to yield the logarithm $x \bmod p-1$. Pollard's rho method can compute a logarithm in a subgroup of prime order q in time $O(\sqrt{q})$, while Pollard's lambda method can compute a logarithm that is known to lie within some restricted interval of width w in time $O(\sqrt{w})$. Thus, both methods have similar square-root running time for a given size of an unknown exponent. In particular, the lambda method is very useful for computing a logarithm in a prime order subgroup when part of the logarithm is known. (For details, see van Oorschot and Wiener [41].)

The Attacking Scenario : In this paper we pay our attention to DL-based schemes using a prime order subgroup. Thus, as is usual, we assume that a prime p is chosen at random such that $p-1$ has a large prime factor q. Let g be an element of order q and $ord(\beta)$ denote the order of $\beta \bmod p$. Then, for a given y such that $y = g^x \bmod p$, it is completely infeasible under current technology to find x using Pollard's rho method, if we take for example $|q| = 160$, since it requires about 2^{80} operations. Our observation is that if we could obtain $z = \gamma^x \bmod p$ somehow by attacking a protocol, where $\gamma = \prod \beta_i$ (a product of distinct smooth order elements mod p), then we could find x modulo $ord(\gamma)$ using the Pohlig-Hellman decomposition. Here we assume that $(p-1)/q$ may have many small prime factors, which is usually the case for a randomly chosen prime p. And finally the remaining part of x could be found from the public key y using Pollard's lambda method. A special case of the attack is to find $x \bmod ord(\beta)$, given $z = f(\beta^x \bmod p)$ for any function f. In this case, one can find $j = x \bmod ord(\beta)$ by checking that $z = f(\beta^j \bmod p)$ for $j = 0, 1, \cdots, ord(\beta) - 1$, where we assume that the range of f is large enough compared to β, so that the probability of collisions occurring under f is negligible, which is usually the case in most instances of our attack.

The main problem in the above attacking scenario is how to obtain a Pohlig-Hellman decomposition for the secret key. This should be impossible in well-designed protocols. However, we could find many DL-based schemes susceptible to the above attack in the literature. Most Diffie-Hellman type key exchange protocols are vulnerable to the above attack. Other examples include shared decryption of ElGamal encryption, shared verification of ElGamal signatures and undeniable signatures. Our attack was possible in all these schemes, since the involved parties do not check relevant protocol variables. Though there are several papers pointing out the importance of checking public parameters and protocol variables (e.g., see [4,42,1,41,2]) in DH key exchange and digital signature schemes, no literature addresses such an explicit attack revealing the involved secret. Our attack may find the whole secret key in many cases.

Related Work : Previous work most relevant to our attack is the middleperson attack on the original Diffie-Hellman key exchange protocol [16] (see [41,2]). Two parties A and B agree on a prime p and a generator α of $\mathbf{Z}_p^*$, exchange random exponentials, $r_A = \alpha^{k_A} \bmod p$ and $r_B = \alpha^{k_B} \bmod p$, and then compute a shared secret $K = r_B^{k_A} = r_A^{k_B} = \alpha^{k_A k_B} \bmod p$. Suppose that $p - 1 = qw$ with w smooth. An attacker may replace r_A and r_B with $r_A^q \bmod p$ and $r_B^q \bmod p$ respectively. Then the shared key becomes $K = (\alpha^q)^{k_A k_B} \bmod p$, which can also be computed by the attacker since he can find $k_i \bmod w$ from r_i. This attack can be easily prevented by authenticating the random exchange, as in the STS [17] and SKEME [21] protocols.[1]

The above attack motivates the use of a prime order subgroup, which also substantially increases the efficiency in computation and parameter generation (see [41,2] for further discussions). Thus most DL-based schemes have been designed using a prime order subgroup since its first invention by Schnorr [38]. However, this paper will show potential weaknesses in such a setting. Our attack on key exchange protocols is quite similar to the above attack, except that our target protocols use a prime order subgroup and that our objective is to find the long-term secret key of the involved party (usually by the other legitimate party). Our attack can be applied to any protocol involving a DH shared secret.

The rest of this paper is organized as follows: We present in Sec.2 a key recovery attack on DH-type key exchange protocols and in Sec.3 a similar attack on other DL-based schemes such as ElGamal encryption and signatures. Sec.4 deals with the generation of secure primes and public key certificates as possible countermeasures to minimize security loss by our attack. And we conclude in Sec.5.

---

[1] It is very important to authenticate the exchanged random messages themselves, rather than the shared secret computed from them. For example, the modified STS protocol by Boyd and Mao [5] may be vulnerable to the middleperson attack, since it only authenticates the hashed version of the shared secret.

2 Extracting Secret Keys in Key Exchange Protocols

One of the well-known design principles for public key protocols states that a message received should not be assumed to have a particular form unless it can be checked [1]. In particular, it is very dangerous to apply one's secret to a number received from the other. However, this principle is hard to apply to Diffie-Hellman-type key exchange protocols. This has given rise to a lot of attacks or weaknesses under a variety of attacking scenarios. Most attacks aim at finding a session key (e.g., see [10,43]) or causing authentication failure (e.g., see [27]). In this section we present a key recovery attack that can be applied to many DH-type key exchange protocols published in the literature[2] unless proper precautions are taken additionally.

2.1 Basic Diffie-Hellman Key Exchange

We first consider the case where a user A successfully obtained a certificate on the public key $y_A = \beta g^{x_A} \bmod p$ with β of small order mod p. This is possible unless a certification authority checks that $y_A^q = 1 \bmod p$ before issuing a certificate for y_A. The CA usually requires that each user prove knowledge of a secret key corresponding to the public key to be certified, since otherwise there exist some protocols that can be attacked with a faked public key (e.g., see [27]). However, even in this case it may still be possible to register a public key of the form $y_A = \beta g^{x_A} \bmod p$ if it is not checked that $y_A^q = 1 \bmod p$.

For example, suppose that for registration the CA requires a user's digital signature on the certificate message which contains all necessary information for certification, including the public key, as defined by X.509. In this case it is easy for A to generate a valid signature corresponding to the public key $y_A = \beta g^{x_A} \bmod p$ when $ord(\beta)$ is small. For example, suppose that Schnorr's signature scheme [38] is used for this purpose. Given message m, A can find $r' \in (0, ord(\beta)]$ such that $r' = h(\beta^{r'} g^k \bmod p, m) \bmod ord(\beta)$ in about $ord(\beta)$ steps, where $k \in_R \mathbf{Z}_q$ and h denotes a secure hash function. Thus A can generate a signature $\{r, s\}$ on m by computing $r = h(\beta^{r'} g^k \bmod p, m)$ and $s = k - xr \bmod q$. It is easy to see that the resulting $\{r, s\}$ is a valid signature on m with the public key $y_A = \beta g^{x_A} \bmod p$. On the other hand, suppose that Schnorr's identification scheme is used instead. Then A can pass the protocol with probability $1/ord(\beta)$ on average, irrespective of the size of a challenge by B (A similar observation has been made before by Burmester [9]). Therefore, it is essential that the CA should first check that $y_A^q = 1 \bmod p$.

---

[2] Diffie-Hellman-type key exchange protocols can be divided into two broad classes. The first is to exchange random exponentials and then authenticate the exchange using a separate authentication mechanism. The STS and SKEME protocols belong to this class. Such protocols seem to be the most robust against various attacks, including our one. Most other protocols involve the fixed secret/public key pair for key exchange and (possibly) authentication. Our attack can be applied to most of such protocols.

We now present a key recovery attack under the assumption that an attacking user i has a public key $y_i = \beta g^{x_i} \bmod p$. This attack will demonstrate the importance of the checking step in the certification process. We first consider the zero-message DH key exchange with fixed keys (e.g., used in [22])[3] : Two users A and B share a session key K by computing $K = h(y_B^{x_A} \bmod p, d) = h(y_A^{x_B} \bmod p, d)$, where d is time/date information. In this protocol, suppose that user B with public key $y_B = g^{x_B} \bmod p$ uses a session key computed by $K = h(y_A^{x_B} \bmod p, d)$ to send a message m to user A with public key $y_A = \beta g^{x_A} \bmod p$. Then, when receiving $\{c = E_K(m), d\}$ from B, A can extract $|ord(\beta)|$ bits of B's secret key by an exhaustive search. For this, A computes $K_j = h(y_B^{x_A} \cdot \beta^j \bmod p, d)$ for $0 \le j < ord(\beta)$, decrypts c with each K_j and checks that the result is a meaningful message. If a meaningful message is found for some j, then A has found $j = x_B \bmod ord(\beta)$.

User A may repeat this attack by updating his public key with β of a different order and combine the resulting partial secrets using the Chinese Remainder Theorem. This will give about t bits of x_B if t is the bit-length of small prime factors of $p - 1$ that can be used for this attack. Now the remaining $(|q| - t)$ bits of x_B can be found in about $2^{(|q|-t)/2}$ steps using Shanks' method or Pollard's lambda method (see [41] for further discussions). Note that if $ord(\beta)$ is small, say of 20 bits, A has little difficulty in reading the ciphertext directed to him. Also note that this attack can be applied to any protocol if the protocol reveals an equation involving the fixed DH key $g^{x_A x_B} \bmod p$.

As another example, let us consider the following non-interactive, symmetric protocol (modified from Protocol 2 in [27]).

1. A computes $r_A = g^{k_A} \bmod p$ with $k_A \in_R \mathbf{Z}_q$, $K_{A1} = y_B^{k_A} \bmod p$, $e_A = h(K_{A1}, r_A, A, B)$ and $s_A = k_A - x_A e_A \bmod q$. A then sends $\{r_A, s_A\}$ to B.
2. B computes $r_B = g^{k_B} \bmod p$ with $k_B \in_R \mathbf{Z}_q$, $K_{B1} = y_A^{k_B} \bmod p$, $e_B = h(K_{B1}, r_B, B, A)$ and $s_B = k_B - x_B e_B \bmod q$. B then sends $\{r_B, s_B\}$ to A.
3. A computes $K_{A2} = r_B^{x_A} \bmod p$ and $e_B = h(K_{A2}, r_B, B, A)$, and checks that $g^{s_B} y_B^{e_B} = r_B \bmod p$. If the check succeeds, A computes the session key $K_A = h(K_{A1} K_{A2} \bmod p)$. Otherwise, A stops the protocol with failure.
4. B computes $K_{B2} = r_A^{x_B} \bmod p$ and $e_A = h(K_{B2}, r_A, A, B)$, and checks that $g^{s_A} y_A^{e_A} = r_A \bmod p$. If the check succeeds, B computes the session key $K_B = h(K_{B1} K_{B2} \bmod p)$. Otherwise, B stops the protocol with failure.

Suppose that A has a valid public key $y_A = \beta g^{x_A} \bmod p$. Then A can find $j = k_B \bmod ord(\beta)$ by checking that $g^{s_B} y_B^{e_B} = r_B \bmod p$ with $e_B = h(r_B^{x_A} \beta^j \bmod p, r_B, B, A)$ for all possible values of j. This gives $|ord(\beta)|$ bits of information on the secret key x_B, since $x_B = (s_B + k_B)e_B^{-1} \bmod q$, and thus the effective secret bits of x_B is reduced to $(|q| - |ord(\beta)|)$ bits. The reason why our attack can

---

[3] The SKIP protocol [3] being widely implemented in the industry for IP layer security also employs this scheme to get a long-term shared secret, which is used as a key-encrypting key. However, the SKIP documentation recommends to use a safe prime p, i.e., a prime p such that $(p-1)/2$ is also prime. Thus our attack on this protocol only discloses one bit of the secret, the parity bit.

apply to this protocol is that the same random secret k_i is used for authentication and session key computation. This shows that a robust protocol should avoid using the same secret (even if it is a one-time random number) for two different purposes [1]. In this respect the approach taken in the STS [17] and SKEME [21] seems to be a better way to design key exchange protocols.

2.2 Authenticated Key Exchange

The attack presented above can be easily prevented by a proper precaution in the certificate issuing process. We now extend our attack to the case where each user has a correct public key. As an example, we consider the following key exchange protocol, which is an authenticated version of the MTI (Matsumoto-Takashima-Imai) protocol [26]. This protocol, with slight changes, is widely studied in the literature (e.g., see [27,20]) and is also being standardized in ISO/IEC JTC1/SC27 [44].

1. A randomly picks $k_A \in \mathbf{Z}_q$, computes $r_A = g^{k_A} \bmod p$ and sends r_A to B.
2. B randomly picks $k_B \in \mathbf{Z}_q$, computes $r_B = g^{k_B} \bmod p$, $K_B = y_A^{k_B} r_A^{x_B} \bmod p$ and $e_B = h(K_B, r_B, r_A, B, A)$, and sends $\{r_B, e_B\}$ to A.
3. A computes $K_A = y_B^{k_A} r_B^{x_A} \bmod p$ and $e'_B = h(K_A, r_B, r_A, B, A)$, and checks that $e_B = e'_B$. If $e_B \neq e'_B$, then A stops the protocol with failure. (Optional) Otherwise, A computes $e_A = h(K_A, r_A, r_B, A, B)$ and sends e_A to B.
4. (Optional) B computes $e'_A = h(K_B, r_A, r_B, A, B)$ and checks that $e_A = e'_A$. If $e_A \neq e'_A$, then B stops the protocol with failure.

The session key K can be derived from the shared secret, for example, as $K = h(K_A) = h(K_B)$. The critical point relevant to our attack is the key authentication based on the shared secret $K_A = K_B$. That is, B applies his secret key to the number received from A and returns e_B as a function of the (assumed) session key K_B. Suppose that A sends $r_A = \beta g^{k_A} \bmod p$ in step 1. Then an honest user B will compute $K_B = y_A^{k_B} r_A^{x_B} = r_B^{x_A} y_B^{k_A} \beta^{x_B} \bmod p$ and return e_B computed with this K_B. Once receiving $\{r_B, e_B\}$, A may abort the protocol if a response is required. Since A can compute the first exponential in K_B, $y_A^{k_B} = r_B^{x_A} \bmod p$, it can find $j = x_B \bmod ord(\beta)$ in $O(2^{|ord(\beta)|})$ steps by checking the equality $e_B = h(K_B, r_B, r_A, B, A)$ with $K_B = r_B^{x_A} y_B^{k_A} \beta^j$ for all possible values of j (i.e., $j = 0, 1, \cdots, ord(\beta) - 1$).

The above attack may be repeated using different smooth order elements for which it is feasible to do the exhaustive search. Thus if $p - 1$ has several prime factors of small size (say, less than 40 bits), then it would be possible to find the whole secret in reasonable time. We note that the attack can be mounted against any authenticated key exchange protocol as long as authentication is performed using the shared secret (note that such authentication is possible only if each user's secret key is involved in the computation of the shared secret). This implies that almost all key exchange protocols providing explicit authentication without using a separate authentication channel (e.g., as in STS [17] or SKEME [21]) may be vulnerable to our attack.

Our attack can also be applied to key exchange protocols with implicit authentication, since the agreed upon session key will be used anyway in later communications. For example, suppose that user A mounted the attack in the original MTI protocol, where each user exchanges random exponential r_i and computes the session key as above. Now, if user B first uses the resulting session key for message authentication (or key authentication), A obtains a known equation involving the session key computed by B. Then the situation, in view of our attack, is the same as in the above authenticated protocol. On the other hand, if B sends a ciphertext for an unknown message, then A can find the intended partial secret by decrypting the ciphertext with all possible values of the session key that B is supposed to compute and then finding a meaningful message. Note that usual known-key attacks assume knowledge of the whole shared secret from which the session key is derived (e.g., see [43,10]), but in our attack it is sufficient to obtain any function of the shared secret (even a ciphertext suffices).

We next show that some key exchange protocols using a signature scheme for authentication may also be vulnerable to our attack. For example, consider the following protocol (developed from Protocol 4 in [23]):

1. A picks a random integer $k_A \in \mathbf{Z}_q$, computes $r_A = g^{k_A} \bmod p$ and sends r_A to B.
2. B picks a random integer $k_B \in \mathbf{Z}_q$, computes $r_B = g^{k_B} \bmod p$. B also computes $K_B = r_A^{k_B} \bmod p$, $e_B = h(K_B, r_B, r_A, B, A)$ and $s_B = k_B - x_B e_B \bmod q$, and sends $\{r_B, s_B\}$ to A.
3. A computes $K_A = r_B^{k_A} \bmod p$, $e_B = h(K_A, r_B, r_A, B, A)$ and checks that $g^{s_B} y_B^{e_B} = r_B \bmod p$. If the check fails, then A stops the protocol with failure. (Optional) Otherwise, A computes $e_A = h(K_A, r_A, r_B, A, B)$ and $s_A = k_A - x_A e_A \bmod q$, and sends s_A to B.
4. (Optional) B computes $e_A = h(K_B, r_A, r_B, A, B)$ and checks that $g^{s_A} y_A^{e_A} = r_A \bmod p$. If it does not hold, then B stops the protocol with failure.

This protocol uses a digital signature on the shared secret $K_A = K_B = g^{k_A k_B} \bmod p$ (for a honest run) to authenticate each other. However, the same random number is used for authentication and session key computation (as in the last example in Sec.2.1). This fact can be exploited by A to extract partial information on the secret key x_B. As before, A sends $r_A = \beta g^{k_A} \bmod p$ and does the exhaustive search for $k_B \bmod ord(\beta)$ using the verification equation $g^{s_B} y_B^{h(K_A, r_B, r_A, B, A)} = r_B \bmod p$ with $K_A = r_B^{k_A} \beta^{k_B} \bmod p$. This reduces the effective secret bits of x_B to $(|q| - |ord(\beta)|)$ bits. Note, however, that repetition of the attack with β of a different order does not help to find further bits of the secret in this case, since a different k_B is used each time and $ord(\beta)$ does not divide q.

The attack described in this section can be easily prevented by checking that $r_i^q = 1 \bmod p$ for each random exponential exchanged before raising it to the secret key. However, this considerably increases the computational load. A better solution would be to choose a prime p such that $(p-1)/2q$ has prime factors at least larger than q (see Sec.4). Such a p only leaks the parity bit of the secret key by our attack. Note that no key exchange protocol can protect the parity bit of the involved secret if the order of the received number is not checked as explained above, since there always exist an element of order 2 (i.e., p-1). This is also true for the following one-way key exchange protocol useful for email applications: A computes $r_A = g^{k_A} \mod p$ with random $k_A \in \mathbf{Z}_q$ and the session key $K = h(y_B^{k_A} \mod p, r_A, d)$, encrypts a message m as $c = E_K(m)$ and sends $\{r_A, d, c\}$ to B, where d is a timestamp. B can then compute $K = h(r_A^{x_B} \mod p, r_A, d)$ and decrypt c. In this protocol A may send $r_A = -g^{k_A} \mod p$. If B does not respond or claims a garbage mail, then A knows that x_B is odd. This attack may be repeated t times, revealing the last t bits of x_B, if $2^t | p - 1$.

3 Extracting Secret Keys in Other DL-based Schemes

There are many other discrete logarithm-based protocols which may be susceptible to our attack. In this section we present several such examples which we found in the literature. They include threshold cryptosystems based on ElGamal encryption [15], anonymous channels used in electronic voting schemes [30,36] and undeniable signatures [12,8,29].

3.1 Shared Decryption of ElGamal Encryption

ElGamal encryption of message m for user A consists of $\{c_1, c_2\}$, where $c_1 = g^k \mod p$ with $k \in_R \mathbf{Z}_q$ and $c_2 = my_A^k \mod p$ [18]. The receiver A can decrypt the ciphertext $\{c_1, c_2\}$ by computing $m = c_2 c_1^{-x_A} \mod p$. In some group-oriented applications we may need to encrypt the message in such a way that only an authorized subset of receivers can decrypt the ciphertext. This can be done using ElGamal encryption and Shamir's secret sharing scheme [37].

As an example, we consider a prime field implementation of the threshold cryptosystem proposed by Desmedt and Frankel [15]. Let $\mathbf{G}$ be a group of n members and $y_G = g^{x_G} \mod p$ be a public key of the group. We want to encrypt a message m so that any subset of t or more members in $\mathbf{G}$ can read the message. For this, in the system setup phase a trusted authority picks a random polynomial f of degree $t - 1$ in $\mathbf{Z}_q$ such that $f(0) = x_G$, i.e., $f(z) = a_{t-1} z^{t-1} + \cdots + a_1 z + x_G$ with $a_j \in_R \mathbf{Z}_q$, computes secret shares $x_{Gi} = f(i) \mod q$ for $i = 1, 2, \cdots, n$ and securely sends x_{Gi} to each member i of $\mathbf{G}$. (See [24] for a more flexible scheme not requiring such pre-distribution of secret shares.) Now, suppose that a ciphertext $\{c_1, c_2\}$, where $c_1 = g^k \mod p$ and $c_2 = my_G^k \mod p$, is received and that a subset $\mathbf{H}$ of t members in $\mathbf{G}$ agreed to decrypt the ciphertext. Then each member $j \in \mathbf{H}$ computes $w_j = c_1^{-b_j x_{Gj}} \mod p$, where $b_j = \prod_{i \in \mathbf{H}, i \neq j} \frac{-i}{j-i} \mod q$, and sends w_j to a combiner (e.g., one designated member). The combiner then computes $w = \prod_{j \in \mathbf{H}} w_j \mod p$, which should be $c_1^{-x_G} \mod p$ if all members involved worked correctly. Therefore, the message m can be recovered by $m = c_2 w \mod p$.

It is easy to see that our attack can be successful for the above scheme, if each shareholder does not check that $c_1^q = 1 \mod p$. In this case, our attack can extract much more secret key bits at a time. Let $\gamma = \prod \beta_i$ (a product of smooth order elements). The attacker sends a ciphertext $\{c_1, c_2\}$ such that $c_1 = \gamma g^k$ mod p and $c_2 = my_G^k$ mod p. Since $w_j^q = (\gamma^q)^{-b_j x_{Gj} \bmod q}$ mod p, once obtaining w_j, he can easily compute the logarithm $(-b_j x_{Gj} \bmod q) \bmod ord(\gamma)$ using a Pohlig-Hellman decomposition. The remaining part of $-b_j x_{Gj}$ mod q can be found from the value $y_j = g^{x_{Gj}} = w_j^{(-b_j k)^{-1}}$ mod p. This reveals the secret share of a shareholder j. If w_j's are transmitted through a secure channel, the attacker need to collude with the combiner.

Note the efficiency of the above attack. Unlike in key exchange protocols, where the attacker can only obtain a function of the shared secret (e.g., a hash value), in the above scheme the attacker has direct access to the shared secret itself (i.e., a value exponentiated with the secret key). This allows the attacker to get a Pohlig-Hellman decomposition for the secret key. Since now Pollard's ρ-method can be used to solve the decomposed problems, it would be quite feasible to use a β of order about 80 bits. Thus, for a random prime p such that $q|p-1$ and $|q| = 160$, the attack could reveal the whole secret key in most cases.

Anonymous channels proposed by Park et al.[30] uses a special case of the threshold cryptosystem described above, i.e., the case of $t = n$. The anonymous channel is primarily used to protect the secrecy of votes in electronic voting schemes. Later Pfitzmann [32] developed successful attacks on these channels. To defeat such attacks, Sako and Kilian [36] used a prime order subgroup in their election scheme, instead of the full multiplicative group $\mathbf{Z}_p^*$ originally used in [30]. However, in this case our attack can be applied again. To see this, we briefly describe the modified version in [36].

Each MIX M_i ($1 \leq i \leq n$) has a secret key $x_i \in_R \mathbf{Z}_q$ and publishes its public key $y_i = g^{x_i}$ mod p. Let $w_j = \prod_{i=j+1}^{n} y_i$ mod p for $j < n$ and $w_n = 1$. For each ciphertext $C_0 = \{c_{0,1}, c_{0,2}\} = \{g^k, mw_0^k\}$, each MIX M_i for $i = 1, 2, \cdots, n-1$ transforms the C_{i-1} posted by M_{i-1} into $C_i = \{c_{i-1,1}g^{r_i}, c_{i-1,2}w_i^{r_i}c_{i-1,1}^{-x_i}\}$ with random $r_i \in \mathbf{Z}_q$ and posts the C_i on the public board in alphabetical order (all computations are done in mod p). In [36] this is done in two phases: in the first phase M_i posts $z_i = c_{i-1,1}^{x_i}$ mod p and in the second phase it posts $\{c_{i-1,1}g^{r_i}, c_{i-1,2}w_i^{r_i}/z_i\}$. In each phase M_i also proves the correctness of computation. Now the final MIX M_n can recover m by computing $c_{n-1,2}c_{n-1,1}^{-x_n}$ mod p.

As is clear, this protocol is vulnerable to our attack. A voter V can submit a faked vote $C_0 = \{\gamma c_{0,1} \bmod p, c_{0,2}\}$ and then find x_i mod $ord(\gamma)$ from z_i as before. Therefore, it is essential that each MIX M_i should verify that $c_{i-1,1}^q = 1$ mod p before beginning its processing.

3.2 Undeniable Signatures

Our attack can also be applied to some digital signature applications. The most obvious case is to produce an undeniable signature on message $m \in \mathbf{Z}_q$ as m^x mod p without checking that $m^q = 1$ mod p, where x is the signer's secret key. We could find several other examples in the literature.

As a first example, we consider the validator issuing protocol by Chaum and Pedersen (see Sec.4 in [13]). The purpose of this protocol is that a center Z issues a validator to a 'wallet with observer' (consisting of a computer C and a tamper-proof module T embedded inside C). The validator is an unlinkable certificate for the public key $y_T = g^{x_T}$ of T. In some steps of the protocol the computer C blinds y_T as $m = y_T^k \bmod p$ with random k and sends it to the center Z, who then returns $z_0 = m^{x_Z} \bmod p$. Obviously, if C sends $m = \gamma y_T^k \bmod p$ with $\gamma = \prod \beta_i$, it can find $x_Z \bmod ord(\gamma)$ from $z_0^q = (\gamma^q)^{x_Z} \bmod p$ using the Pohlig-Hellman method. Note that C can still obtain the desired signature by computing $z_0 \gamma^{-x_Z} \bmod p$ after finding $x_Z \bmod ord(\gamma)$. The same attack can be applied to its privacy enhanced version [14] if the signer does not check that $m^q = 1 \bmod p$. The authors may omit this checking step in the thought that C can only obtain an undeniable signature for a random message, but this omission enables a fatal attack as shown above.

In Brands's electronic cash scheme using a wallet with observers [6] (see [7] for more details), each user computes $I = g_1^u \bmod p$ with $u \in_R \mathbf{Z}_q$ and sends it to the bank, which generates a signature $z = (Ig_2)^x \bmod p$ (g_1, g_2 generators of a subgroup of order q). In this case the user must prove to the bank that he knows u since I corresponds to the account number of the user (see also [11]). Thus our attack is not applicable here. However, as noted in Sec.2.1, it is essential to check $I^q = 1 \bmod p$ at the begining of the proof if a Schnorr-type identification scheme is used for this purpose (this is the case in [7]). Otherwise, the user can pass the proof with $I = \beta g_1^u \bmod p$ in success probability $1/ord(\beta)$. The successful pass will be fatal in this system: Not only the user can mount our attack to find partial information on the secret x, but also he can spend the same coin multiple times without being identified.

Another possibility for the attack exists in the confirmation protocol of undeniable signatures [12,8] and designated confirmer signatures [29]. For example, consider the convertible undeniable signature scheme by Boyar et al.[8].[4] In this scheme the signer S possesses two secret/public key pairs, $\{x \in_R \mathbf{Z}_q, y = g^x \bmod p\}$ and $\{z \in_R \mathbf{Z}_q, u = g^z \bmod p\}$. The signature on message m is a triple $\{t, r, s\}$, where $t = g^{k_1} \bmod p$, $r = g^{k_2} \bmod p$ and $s = k_2^{-1}(h(m)tzk_1 - xr) \bmod q$ ($k_1, k_2 \in_R \mathbf{Z}_q$). Thus the signature $\{t, r, s\}$ is valid iff $(t^{h(m)t})^z = y^r r^s \bmod p$. The confirmation protocol between S and V is as follows:

1. S and V computes $w = t^{h(m)t} \bmod p$ and $v = y^r r^s \bmod p$ from the signature $\{t, r, s\}$.
2. V computes a challenge $ch = w^a g^b \bmod p$ with $a, b \in_R \mathbf{Z}_q$ and sends ch to S.
3. S computes $h_1 = ch \cdot g^c \bmod p$ with $c \in_R \mathbf{Z}_q$ and $h_2 = h_1^z \bmod p$, and sends $\{h_1, h_2\}$ to V.
4. V reveals a and b to S.

---

[4] This scheme is taken as an example only to illustrate our attack, not to show that it is insecure. Michels et al. [28] have already shown that this scheme could be broken when used as a totally convertible signature scheme. We would like to thank one of anonymous referees for pointing out this attack.

5. S checks that $ch = w^a g^b \bmod p$. if it holds true, then S reveals c to V. Otherwise, S stops the protocol.

6. V checks that $h_1 = w^a g^{b+c} \bmod p$ and $h_2 = v^a u^{b+c} \bmod p$.

This protocol is complete, sound and proven zero-knowledge. However, suppose that the verifier V sends, as a challenge in step 2, any value of order q multiplied by small order elements, say $ch = \gamma g^b \bmod p$. Then the received value h_2 in step 3 satisfies $h_2^q = (\gamma^q)^z \bmod p$, from which the verifier can find z $\bmod\, ord(\gamma)$. This shows that the confirmation protocol cannot be zero-knowledge against a dishonest verifier, unless the prover checks that $ch^q = 1 \bmod p$ in step 3. In a variant by Pedersen [31], S computes h_1, h_2 as $h_1 = (ch)^c \bmod p$ with $c \in_R \mathbf{Z}_q$ and $h_2 = h_1^z \bmod p$. This variant is also vulnerable to our attack, since one can still obtain the equation $h_2^q = (h_1^q)^z \bmod p$ by sending $ch = \gamma g^b \bmod p$ (here note that $ord(h_1^q) = ord(\gamma)$).

The above attack suggests that a prime p should be chosen as $p = 2q + 1$ (q prime) if an undeniable signature is computed as $m^x \bmod p$ and if the above protocol is to be used for confirmation, as in Chaum's undeniable signature [12].[5] On the other hand, Jakobsson and Yung [19] proposed an oblivious decision proof protocol for proving validity/invalidity of undeniable signatures, where the prover does not have to know whether the signature in question is valid or not. Thus the protocol does not necessarily require the message m to be in $\mathbf{Z}_q$. They choose the system parameters for use in Chaum's scheme as: $p = ql + 1$ (p, q prime, l integer), g an element of order q and $\{x \in_R \mathbf{Z}_q, y = g^x \bmod p\}$ as the secret/public key pair of the signer. However, careful examination shows that their oblivious protocol with such parameters is also vulnerable to our attack. This protocol cannot be repaired by simply adding checking steps as noted above. To avoid our attack, we have to choose p as $p = 2q + 1$ (q prime). Or we may use the full multiplicative group $\mathbf{Z}_p^*$ as the underlying group.

We note that the signer or the confirmer(s) must verify the validity of a signature before executing the confirmation protocol if distinct protocols are used for confirmation and disavowal.[6] Otherwise, the verifier may change the first component t of the signature as $t' = \gamma t \bmod p$ and requests S to confirm its validity by sending $\{t', r, s\}$. Then V will be able to obtain $\gamma^z \bmod p$ at the end of the protocol, since the check in step 5 will succeed if V sends $ch = (w')^a g^b$ $\bmod p$ with $w' = (t')^{t' h(m)} \bmod p$ in step 2. If the secret z is distributed into a set of designated confirmers [31,29], an authorized subset of confirmers should run a shared verification protocol to validate the signature. Then there may exist

---

[5] Note that it is infeasible to generate m as an element of order q for any meaningful message or its hash value if q is chosen small compared to p. Thus there is no way to detect our attack unless p is chosen as $p = 2q + 1$. Alternatively, we may choose g as a generator of $\mathbf{Z}_p^*$. Then our attack will be useless; the public key itself already reveals the smooth part of the secret key.

[6] There is another weakness when the validity of a signature is not verified before confirmation, as described in Appendix A in [19]: If a verifier who does not have a valid signature for m is allowed to participate in the confirmation protocol, then the verifier can easily obtain a valid signature for m.

another possibility for the combiner to mount a similar attack to find secret shares of confirmers as in the shared decryption of ElGamal ciphertexts (see Sec.3.1), unless each confirmer checks that $w^q = 1 \bmod p$.

4 Generating and Registering Public Parameters

The presented attack shows that it is essential in discrete log-based schemes using a prime order subgroup to check the order of received numbers before applying the secret key. However, such explicit checking requires additional exponentiation. As a better countermeasure, we recommend to use a secure prime, i.e., a prime p such that $(p-1)/2q$ is also prime or each prime factor of $(p-1)/2q$ is larger than q. (see also [25] for a method of generating primes which can substantially reduce the modular reduction time and storage usage). Such a prime can be generated much faster than, and seems as strong against any known attack as, a safe prime (i.e., a prime of the form $p = 2q + 1$).

To generate a prime p such that $p = 2qp_1 + 1$, we first choose a random prime p_1 of length $|p| - |q| - 1$ and then find a desired prime p by testing $p = 2qp_1 + 1$ for primality with random primes q's of size $|q|$. Thus we need to generate a number of q's to find a p (e.g., about 710 for a 1024-bit prime p, considering the density of primes, $1/\ln x$). However, this does not require much time, since the size of q is quite small compared to that of p (usually $|q| = 160$ and $|p| = 768$ or 1024).

It is much cheaper to generate a prime p such that $p = 2qp_1p_2 \cdots p_n + 1$, where p_i's are primes almost equal to q. We first determine the number n from the inequality $l = |p_i| \approx (|p| - |q| - 1)/n \geq |q|$. Then we generate a pool of primes for p_i's. Suppose that the pool contains m primes of size l. Then we have $\binom{m}{n}$ candidates for p. Considering the density of primes, we can make this number large enough to guarantee that there are enough possibilities for a prime p to be found with this prime pool. For example, for a 1024-bit prime p and a 160-bit prime q we may choose $l \approx 173$, $n = 5$ and $m = 15$. This choice of parameters gives about 3000 candidates for p. Thus testing this many candidates will produce a prime p with very high probability.

There are other advantages in our proposed prime generating methods. They give a complete factorization of $p - 1$, which may be useful if we need to find a primitive element mod p.[7] Furthermore, since $p - 1$ contains many prime factors of similar size, we may use different subgroups of prime order for different applications (e.g., one for ordinary signatures, one for undeniable signatures and one for key exchange, etc.). This will prevent any potential weakness from the misuse of key parameters. Of course, using the same prime p in different primitives may not be desirable in view of security against most discrete logarithm algorithms, such as the index calculus method and the number field sieve. However, this gives us efficiency in storage and communication.

---

[7] Note that there is no known algorithm to find a primitive element mod p without knowing the factorization of $p - 1$. The simplest is to check the order of elements successively (say, $2, 3, 5, \cdots$) using the prime factors of $p - 1$.

The attacks presented in this paper and in Menezes et al.[27] show that the CA must check for knowledge of a secret key corresponding to the public key before issuing a certificate. In particular, the CA and each user must first check the order of the received base g and public key y. If such an interactive proof is hard to carry out in some environments, we propose to use the following certification procedure, which seems to preclude any known weakness even without such a proof. The basic idea is to allow the CA to also contribute to the randomness of a user's secret key. On receiving a user's part of a public key $y' = g^{x'} \bmod p$, the CA computes the actual public key y as $y = (y')^a g^b \bmod p$, where $a = \frac{p-1}{q} c$ and $b, c \in_R \mathbf{Z}_q$. The CA now generates a certificate for y and sends $\{a \bmod q, b\}$, along with the certificate, to the user. On receiving $\{a, b\}$, the user checks that $a \neq 0 \bmod q$, computes the actual secret key x as $x = x'a + b \bmod q$ and finally checks that the certified public key y is equal to his own computation $g^x \bmod p$.

The exponent a makes vanish any component of y' which does not belong to the subgroup of order q. This prevents users from registering an improper form of public keys. The multiplicative factor $g^b \bmod p$ makes it impossible for a user to register a public key as a power of other user's public key. This prevents the attack presented in [27]. Furthermore, users' secret keys can be made more pseudorandom by the CA's contribution if the CA is equipped with a true random generator or uses a cryptographically strong pseudorandom generator (in this case we assume that the CA sends $\{a \bmod q, b\}$ through a secure channel). This may be advantageous since most users may not be so careful in choosing their secrets.

5 Conclusion

We have demonstrated that many discrete logarithm protocols may be insecure against a key recovery attack unless suitable checking steps are added. The presented attack may reveal part of, in many cases the whole, secret key of a victim in a reasonable time for many DL-based schemes published in the literature. The attack exploits small order subgroups in $\mathbf{Z}_p^*$ to compute part of the secret key in a protocol working in a subgroup of prime order q. This is possible since in most schemes a prime p is chosen at random such that $q|p - 1$. Therefore, our attack can be easily prevented if relevant protocol variables are properly checked, that is, if each party checks that received numbers belong to the underlying subgroup of prime order. However, such explicit checkings substantially decrease efficiency. Thus a better alternative would be to minimize possible leakage of secret key bits by using a secure prime, a prime p such that all prime factors of $(p - 1)/2q$ are larger than q. Such a prime only leaks one bit of the secret by our attack.

References

1. R.Anderson and R.Needham, Robustness principles for public key protocols, In *Advances in Cryptology - CRYPTO'95*, LNCS 963, Springer-Verlag, 1995, pp.236-247.

262

2. R.Anderson and S.Vaudenay, Minding your p's and q's, In *Advances in Cryptology - ASIACRYPT'96*, LNCS 1163, Springer-Verlag, 1996, pp.15-25.
3. A.Aziz, T.Markson and H.Prafullchandra, Simple key-management for Internet protocols (SKIP), *draft-ietf-ipsec-skip-07.txt*, Aug. 1996. (see also the SKIP home page *http : //skip.incog.com/* for more information.)
4. D.Bleichenbacher, Generating ElGamal signatures without knowing the secret, In *Advances in Cryptology - EUROCRYPT'96*, LNCS 1070, Springer-Verlag, 1996, pp.10-18.
5. C.Boyd and W.Mao, Design and analysis of key exchange protocols via secure channel identification, In *Advances in Cryptology - ASIACRYPT'94*, LNCS 917, Springer-Verlag, 1995, pp.171-181.
6. S.Brands, Untraceable off-line cash in wallet with observers, In *Advances in Cryptology - CRYPTO'93*, LNCS 773, Springer-Verlag, 1994, pp.302-318.
7. S.Brands, An efficient off-line electronic cash system based on the representation problem, *Technical Report CS-R9323*, CWI, Amsterdam, 1993.
8. J.Boyar, D.Chaum, I.Damgard and T.Pedersen, Convertible undeniable signatures, In *Advances in Cryptology - CRYPTO'90*, LNCS 537, Springer-Verlag, 1991, pp.189-205.
9. M.Burmester, A remark on the efficiency of identification schemes, In *Advances in Cryptology - EUROCRYPT'90*, LNCS 473, Springer-Verlag, 1991, pp.493-495.
10. M.Burmester, On the risk of opening distributed keys, In *Advances in Cryptology - CRYPTO'94*, LNCS 839, Springer-Verlag, 1994, pp.308-317.
11. A.Chan, Y.Frankel and Y.Tsiounis, Mis-representation of identities in E-cash schemes and how to prevent it, In *Advances in Cryptology - ASIACRYPT'96*, LNCS 1163, Springer-Verlag, 1996, pp.276-285.
12. D.Chaum, Zero-knowledge undeniable signatures, In *Advances in Cryptology - EUROCRYPT'90*, LNCS 473, Springer-Verlag, 1991, pp.458-464.
13. D.Chaum and T.Pedersen, Wallet databases with observers, In *Advances in Cryptology - CRYPTO'92*, LNCS 740, Springer-Verlag, 1993, pp.89-105.
14. R.Cramer and T.Pedersen, Improved privacy in wallets with observers, In *Advances in Cryptology - EUROCRYPT'93*, LNCS 765, Springer-Verlag, 1994, pp.329-343.
15. Y.Desmedt and Y.Frankel, Threshold cryptosystems, In *Advances in Cryptology - CRYPTO'89*, LNCS 435, Springer-Verlag, 1990, pp.307-315.
16. W.Diffie and M.E.Hellman, New directions in cryptography, *IEEE Trans. Info. Theory*, 22(6), 1976, pp.644-654.
17. W.Diffie, P.van Oorschot and M.Wiener, Authentication and authenticated key exchange, *Designs, Codes and Cryptography*, 2, 1992, pp.107-125.
18. T.ElGamal, A public key cryptosystem and a signature scheme based on discrete logarithms, *IEEE Trans. Inform. Theory*, IT-31, 1985, pp.469-472.
19. M.Jakobsson and M.Yung, Proving without knowing: on oblivious, agnostic and blindfolded provers, In *Advances in Cryptology - CRYPTO'96*, LNCS 1109, Springer-Verlag, 1996, pp.186-200.
20. M.Just and S.Vaudenay, Authenticated multi-party key agreement, In *Advances in Cryptology - ASIACRYPT'96*, LNCS 1163, Springer-Verlag, 1996, pp.36-49.
21. H.Krawczyk, SKEME: A versatile secure key exchange mechanisms for Internet, In *Proc. of 1996 Symp. on Network and Distributed Systems Security*.
22. A.K.Lenstra, P.Winkler and Y.Yacobi, A key escrow system with warrant bounds, In *Advances in Cryptology - CRYPTO'95*, LNCS 963, Springer-Verlag, 1995, pp.197-207.
23. C.H.Lim and P.J.Lee, Several practical protocols for authentication and key exchange, *Information Processing Letters*, 53, 1995, pp.91-96.

24. C.H.Lim and P.J.Lee, Directed signatures and application to threshold cryptosystems, In *Pre-Proc. of 1996 Cambridge Workshop on Security Protocols*, The Isaac Newton Institute, Cambridge, April 1996.

25. C.H.Lim and P.J.Lee, Generating efficient primes for discrete log cryptosystems, submitted for publication (also presented at ASIACRYPT'96 Rump Session).

26. T.Matsumoto, Y.Takashima and H.Imai, On seeking smart public-key distribution systems, *The Transactions of the IEICE of Japan*, E69, 1986, pp.99-106.

27. A.J.Menezes, M.Qu and S.A.Vanstone, Some new key agreement protocols providing implicit authentication, In *Proc. SAC'95*, Carleton Univ., Ottawa, Ontario, May 1995, pp.22-32.

28. M.Michels, H.Petersen and P.Horster, Breaking and repairing a convertible undeniable signature scheme, *Proc. of 3rd ACM Conference on Computer and Communications Security*, Mar. 1996.

29. T.Okamoto, Designated confirmer signatures and public-key encryption are equivalent, In *Advances in Cryptology - CRYPTO'94*, LNCS 839, Springer-Verlag, 1995, pp.61-74.

30. C.S.Park, K.Itoh and K.Kurosawa, Efficient anonymous channel and all/nothing election scheme, In *Advances in Cryptology - EUROCRYPT'93*, LNCS 765, Springer-Verlag, 1994, pp.248-259.

31. T.Pedersen, Distributed provers with applications to undeniable signatures, In *Advances in Cryptology - EUROCRYPT'91*, LNCS 547, Springer-Verlag, 1991, pp.221-242.

32. B.Pfitzmann, Breaking an efficient anonymous channel, In *Advances in Cryptology - EUROCRYPT'94*, LNCS 950, Springer-Verlag, 1995, pp.332-340.

33. S.C.Pohlig and M.E.Hellman, An improved algorithm for computing logarithms over $GF(p)$ and its cryptographic significance, *IEEE Trans. Inform. Theory*, IT-24 (1), 1978, pp.106-110.

34. D.Pointcheval and J.Stern, Security proofs for signature schemes, In *Advances in Cryptology - EUROCRYPT'96*, LNCS 1070, Springer-Verlag, 1996, pp.387-398.

35. J.M.Pollard, Monte Carlo methods for index computation (mod p), *Math. Comp.*, 32(143), 1978, pp.918-924.

36. K.Sako and J.Kilian, Receipt-free mix-type voting scheme, In *Advances in Cryptology - EUROCRYPT'95*, LNCS 921, Springer-Verlag, 1995, pp.pp.393-403.

37. A.Shamir, How to share a secret, *Commun. ACM*, 22, 1979, pp.612-613.

38. C.P.Schnorr, Efficient identification and signatures for smart cards, In *Advances in Cryptology - CRYPTO'89*, LNCS 435, Springer-Verlag, 1990, pp.235-251.

39. J.Stern, The validation of cryptographic algorithms, In *Advances in Cryptology - ASIACRYPT'96*, LNCS 1163, Springer-Verlag, 1996, pp.301-310.

40. P.C.van Oorschot and M.J.Wiener, Parallel collision search with applications to hash functions and discrete logarithms, In *Proc. 2nd ACM Conference on Computer and Communications Security*, Fairfax, Virginia, Nov. 1994, pp.210-218.

41. P.C.van Oorschot and M.J.Wiener, On Diffie-Hellman key agreement with short exponents, In *Advances in Cryptology - EUROCRYPT'96*, LNCS 1070, Springer-Verlag, 1996, pp.332-343.

42. S.Vaudenay, Hidden collisions on DSS, In *Advances in Cryptology - CRYPTO'96*, LNCS 1109, Springer-Verlag, 1996, pp.83-88.

43. Y.Yacobi, A key distribution paradox, In *Advances in Cryptology - CRYPTO'90*, LNCS 537, Springer-Verlag, 1991, pp.268-273.

44. ISO/IEC JTC1/SC27, Information technology - Security techniques - Key management - Part 3: Mechanisms using asymmetric techniques.

The Prevalence of Kleptographic Attacks on Discrete-Log Based Cryptosystems

Adam Young*   and   Moti Yung**

Abstract. The notion of a Secretly Embedded Trapdoor with Universal Protection (SETUP) and its variations on attacking black-box cryptosystems has been recently introduced. The basic definitions, issues, and examples of various setup attacks (called Kleptographic attacks) have also been presented. The goal of this work is to describe a methodological way of attacking cryptosystems which exploits certain relations between cryptosystem instances which exist within cryptosystems. We call such relations "kleptograms". The identified kleptogram is used as the base for searching for a setup.

In particular, we employ as a discrete log based kleptogram a basic setup that was presented for the Diffie-Hellman key exchange. We show how it can be embedded in a large number of systems: the ElGamal encryption algorithm, the ElGamal signature algorithm, DSA, the Schnorr signature algorithm, and the Menezes-Vanstone PKCS. These embeddings can be extended directly to the MTI two-pass protocol, the Girault key agreement protocol, and many other cryptographic systems. These attacks demonstrate a systematic way to mount kleptographic attacks. They also show the vulnerability of systems based on the difficulty of computing discrete logs.

The setup attack on DSA exhibits a large bandwidth channel capable of leaking information which hardware black-box implementations (e.g., the Capstone chip) can use. We also show how to employ such channels for what we call "device marking".

Finally, note that it has been perceived that the DSA signature scheme was originally designed to be robust against its abuse as a public-key channel– to distinguish it from RSA signatures (where the signing function is actually a decryption function). In this paper we refute this "perceived advantage" and show how the DSA system (in hardware or software) can be easily modified to securely leak private keys and secure messages between two cooperating parties.

Key words: DSA signature, ElGamal encryption, ElGamal signature, Menezes-Vanstone PKCS, Schnorr signature algorithm, setup, Discrete-Log, Diffie-Hellman, subliminal channels, protocol abuse, kleptography, leakage-bandwidth, randomness, pseudorandomness, cryptographic system implementations.

* Dept. of Computer Science, Columbia University Email: ayoung@cs.columbia.edu.
** CertCo New York, NY, USA. Email: moti@certco.com, moti@cs.columbia.edu

1 Introduction

Recently, it has been shown that Black-Box cryptosystems can be designed so as to conform to public specifications and be polynomially indistinguishable from the known public specifications, and at the same time securely and subliminally leak secret key information to the implementor (either through keys at key generation or during run-time).

Young and Yung laid the foundation for these attacks, defined the basic notions, and demonstrated them [YY96, YY97]. These attacks imply that Black-Box systems (whose internals cannot be scrutinized) should not be automatically trusted (e.g., trust should be based on cryptosystems coming from a trustworthy source and not from the technology of tamper-resistant black-boxes, say).

Typically a cryptosystem produces a ciphertext for a given message or a signature for a given message. However, a cryptosystem with a setup produces a ciphertext/signature for a given message that also contains an internal ciphertext for a totally different message. We call such an output of a cryptosystem (with an inner ciphertext) a *kleptogram*. Kleptograms are undetectable in poly-time by the user, they are strong encryptions, and they coexist in the same ciphertext bits as normal public key encryptions.

In this paper a methodology for finding setup attacks is given. The methodology has two steps:

1. First we find a relation within a cryptographic function between its application and another inner encryption (this relation is called a kleptogram).
2. Then, given a cryptosystem and its workings, we identify how the kleptogram of the underlying function is embeddable in the system (and what leakage level is possible), which gives us a setup.

One of the setup attacks we present which is perhaps the main result of this work, is a (1,2)-leakage bandwidth setup for DSA. That is, we present a setup mechanism for DSA that is capable of leaking the user's private key through two (wlog) consecutive digital signatures. We then extend the attack to allow the user to send 160 subliminal bits of his choosing *in addition* to the private key. Furthermore, the user is free to re-key at any time and the attack will still work. The kleptographic attack therefore effectively leaks 80 key bits and 80 chosen bits per signature. This contrasts with the channel described by Simmons which leaks approximately 14 chosen bits per signature [Sim93]. Also, in the context of tamper-proof devices we show how the SETUP can be employed for "device marking", where the mark is added subliminally to the signature.

The above setup can be used to easily turn DSA into an effective public key (key exchange or message exchange) system. This spoofing, motivated by the potential of protocol abuse via kleptographic methods, shows that the claim that DSA is inherently different from RSA in this respect (the RSA signing function can obviously be used as a decryption function) is a myth! We refer the reader to the NIST response on DSA which alluded to this fact [SB92].

In this paper we show that these kinds of setup attacks are possible in other discrete log based cryptosystems such as the ElGamal encryption algorithm, ElGamal and Schnorr digital signatures, and various authenticated key exchange algorithms. Also, systems based on Elliptic Curves (Menezes-Vanstone) can be attacked. The attack methodology based on the discrete log kleptogram is therefore widely applicable to discrete log based cryptosystems.

2 Definitions and Background

A setup (Secretly Embedded Trapdoor with Universal Protection) is a mechanism that allows the secure leakage of private information within the output of a cryptosystem. The notion of a setup is due to Young and Yung [YY96]. The definitions of weak, regular, and strong setups and (m,n)-leakage are from [YY97].

Definition 1. Assume that C is a black-box cryptosystem with a publicly known specification. A (regular) SETUP mechanism is an algorithmic modification made to C to get C' such that:

1. The input of C' agrees with the public specifications of the input of C.
2. C' computes efficiently using the attacker's public encryption function E (and possibly other functions as well), contained within C'.
3. The attacker's private decryption function D is not contained within C' and is known only by the attacker.
4. The output of C' agrees with the public specifications of the output of C. At the same time, it contains published bits (of the user's secret key) which are easily derivable by the attacker (the output can be generated during key-generation or during system operation like message sending).
5. Furthermore, the output of C and C' are polynomially indistinguishable (as in [GM84]) to everyone except the attacker.
6. After the discovery of the specifics of the setup algorithm and after discovering its presence in the implementation (e.g. reverse-engineering of hardware tamper-proof device), users (except the attacker) cannot determine past (or future) keys.

Definition 2. A *weak setup* is a regular setup except that the output of C and C' are polynomially indistinguishable to everyone except the attacker and the owner of the device who is in control (knowledge) of his or her own private key (i.e., requirement 5 above is changed).

Definition 3. A *strong setup* is a regular setup, but in addition we assume that the users are able to hold and fully reverse-engineer the device after its past usage and before its future usage. They are able to analyze the actual implementation of C' and deploy the device. However, the users still cannot steal previously generated/future generated keys, and if the setup is not always applied to future keys, then setup-free keys and setup keys remain polynomially indistinguishable.

Another important notion is that of (m,n)-leakage bandwidth. A setup that has (m,n)-leakage bandwidth leaks m secret messages over the course of n messages that are output by the cryptographic device (or n of its executions).

Let us define a relation between a hidden encrypted value and another encryption/ signature.

Definition 4. A *kleptogram* is an encryption of a value (hidden value) that is displayed within the bits of an encryption/signature of a plaintext value (outer value).

Note that we say that a kleptogram is an encryption of a value, not a plaintext message. It is often the case in kleptography that the device is not free to choose this value. The device may calculate this hidden value, and then use it (for the 'randomness') in a subsequent computation, thus compromising that computation.

Related Work

The notion of a subliminal channel is due to Gus Simmons [Sim85, Sim93]. SETUPs exploiting such channels and a SETUP in the RSA public keys (as well as in other systems) were given in [YY96]. In [YY97] a SETUP was presented for Diffie-Hellman that does not make use of explicit subliminal channels, but rather exploits a number of executions of the system to "create a channel". This paper makes extensive use of this SETUP, which is a basic relation based on an underlying cryptographic problem. Due to its extensive applications as a relation between encrypted values, we call it the "discrete log kleptogram". The following section describes this relation in detail.

2.1 Discrete Log Kleptogram

Suppose that the only information that we are allowed to display is $g^c \bmod p$ for some $c < p - 1$. The following is a way to leak a value c_2, over the single message $m_1 = g^{c_1} \bmod p$, such that the subsequent message $m_2 = g^{c_2} \bmod p$ is compromised. In this attack we assume that the device is free to choose the exponents used. Let the attacker's ElGamal private key be X, and let the corresponding public key be Y. Let W be a fixed odd integer, and let H be a cryptographically strong pseudorandom function with a hidden seed. WLOG, assume that H outputs values less than $\phi(p)$. The following algorithm is based on the operation of a Diffie-Hellman device that is used two times in succession. Let a and b be fixed constants.

1. For the first usage, $c_1 \in Z_{p-1}$ is chosen uniformly at random
2. The device outputs $m_1 = g^{c_1} \bmod p$.
3. c_1 is stored in non-volatile memory for the next time the device is used.
4. For the second usage, $t \in \{0, 1\}$ is chosen uniformly at random.

5. $z = g^{c_1 - Wt} Y^{-ac_1 - b} \bmod p$.
6. $c_2 = H(z)$
7. The device outputs $m_2 = g^{c_2} \bmod p$.

The attacker need only passively tap the communications line, and obtain m_1 and m_2, in order to calculate c_2. We call z a *hidden field element*. The value for c_2 is found as follows.

1. $r = m_1{}^a g^b \bmod p$
2. $z_1 = m_1 / r^X \bmod p$
3. if $m_2 = g^{H(z_1)} \bmod p$ then output $H(z_1)$
4. $z_2 = z_1 / g^W$
5. if $m_2 = g^{H(z_2)} \bmod p$ then output $H(z_2)$

The value c_2 can be used by the attacker to determine the key from the second DH key exchange. Note that only the attacker can perform these computations since only the attacker knows X. In order for c_2 to be able to take on any value less $p - 1$ we assume that $g_1 = g^{-Xb-W}$, $g_2 = g^{-Xb}$, and $g_3 = g^{1-aX}$ are generators mod p. This secure disclosure of an encrypted value inside the 'encryption' of another value will be exploited to attack a number of cryptosystems. As a side remark, note that technically nothing is encrypted in a DH key exchange. However, we may regard the resulting shared key as having been conceptually encrypted by both parties.

3 Setups in ElGamal Systems

Indeed, we are now ready to take the second step of our methodology, whereby we apply the above discrete log kleptogram mechanism to discrete log based cryptosystems. We start with applications to ElGamal, and then explain the more complicated setup attack on DSA in the next section.

3.1 Strong Setup in the ElGamal Encryption Scheme

In ElGamal [ElG85], the first part of the ciphertext of a message is $a = g^k \bmod p$. Note that a from the ciphertext (a, b) displays an exponentiation mod p. We can use this to implement a strong setup in ElGamal. In fact, the discrete log kleptogram *is* the strong setup for ElGamal encryption. It is straightforward to implement a (1,2)-leakage bandwidth scheme by leaking a hidden field element in the a of the first encryption (a, b), and using the hash of this element as the k for the second encryption. Note however that we are leaking messages m instead of private keys in this case. There is no known way to efficiently recover k from an ElGamal encryption, even if the user's private key x is known. So, it can be shown that this is a strong setup.

Theorem 1 *ElGamal encryption has a strong setup version.*

Typically, when public key cryptography is needed to encrypt bulk data, hybrid cryptosystems are used. Thus, in this mode of usage, the setup can leak keys. It can leak the randomly generated symmetric keys used to encrypt the data. We can implement a (1,1)-leakage scheme in an ElGamal based hybrid system as follows. We use the discrete log attack to setup up the $g^k \bmod p$ portion of the ciphertext. We then choose the one way hash of the hidden field element as all or part of the symmetric encryption key. It is therefore imperative to verify the source code of hybrid systems based on ElGamal.

3.2 Regular Setup in ElGamal Signature Scheme

In [YY96] an attack on the ElGamal signature scheme was proposed. This attack is novel in that it allows Alice to securely give her private key to Bob through signatures alone. However, the presence of the setup can be readily detected without knowledge of the attacker's ElGamal public key, and hence constitutes a weak setup.

The problem is that a user can always recover the choices of k of his own device using his own private key [Sim85]. Given the (wlog consecutive) ith and (i+1)th signatures, the user can compute $Y = (k_{i+1}^{-1})^{1/k_i} \bmod p$. After a few such coincidences, the user will conclude that Y is in fact the public key of the attacker. Note that the attack would still be very effective in hiding the key exchange from a warden (overseeing the communication) as in the original scenario of Gus Simmons. We can modify the attack to be a regular setup by using the fixed private parameters a, b, and W in conjunction with Y in the usual way, rather than simply setting $k_{i+1}^{-1} = Y^{k_i}$.

This setup attack can also be carried out using the discrete log attack. For concreteness, we will simply point out that the first ElGamal signature value $g^k \bmod p$ is a exponentiation mod p. Thus we leak a hidden field element as before, and this mechanism is a (1,2)-leakage setup. This kleptographic attack constitutes a regular setup. The fact that the value k_2 can't be compromised was shown in [YY97]. This assumes that the DH problem is hard. There is also the issue of detectability by the signer who knows k_1 and k_2. It can be shown that if H is a pseudorandom function whose seed is kept private by the implementor and hidden in the black-box device, the signer can't even detect the presence of the setup. The private values a and b are thus used as an extra precaution. This is not a strong setup, since a user knowing his own private key can recover the choices of k and detect the presence of the setup mechanism, given the seed, a, and b. The existence of a strong setup for the ElGamal digital signature scheme is left as an open problem.

Theorem 2 *The ElGamal digital signature algorithm has a regular setup version.*

4   SETUPing and Spoofing DSA

It has been assumed that the DSA [DSS91] system was designed as a signature system that is hard to "abuse". Namely, that it was designed so that it would not be used directly as public-key system, a key exchange system, or any system providing for confidential information exchange (see [SB92]). Therefore, it was quite interesting that a low bandwidth (14-bit) subliminal channel was found in it [Sim93]. Here we show a much larger leakage potential for black-box implementations of the DSA; we note that such implementations exist (i.e., the Capstone technology).

We will now briefly review DSA. q is a 160 bit prime which divides $p - 1$. p is a prime that is at least 512 bits and at most 1024 bits in length. g is a qth root of 1 mod p. All three of these parameters are public. Alice's private key is x, where $x < q$. Alice's public key is y, where $y = g^x$ mod p. Let H denote the Secure Hash Algorithm. To compute the signature (r, s) of a message m, Alice does the following.

1. chooses a value k at random such that $k < q$.
2. computes $r = (g^k \ mod \ p) \ mod \ q$.
3. computes $s = k^{-1}(H(m) + xr) \ mod \ q$.
4. outputs the signature (r, s).

To verify that the signature is valid, Bob checks to make sure that r is equal to $(g^{s^{-1}H(m)}y^{s^{-1}r} \ mod \ p) \ mod \ q$.

It is clear from the discrete log kleptogram that we need only find a modular exponentiation $(mod \ p)$ that is displayed in DSA to find a setup. Note that a modular exponentiation $mod \ q$ won't suffice since q is too small. We could setup DSA keys, since $y = g^x$ mod p, but this indicates that the user must generate new keys (or a new signature) to be an effective setup attack for the attacker. The existence of the modular exponentiation that constitutes the (1,2)-leakage setup for signatures is indeed a bit more subtle. But, it turns out that the quantity $g^{s^{-1}H(m)}y^{s^{-1}r}$ mod p is in fact simply g^k mod p, and can thus be used as a kleptogram. It follows that over the course of two (wlog) consecutive signatures, a DSA device can securely leak the second choice of k. The attacker, given the value for k, can readily recover the user's private key x. This setup attack is rather odd since the kleptogram is not overtly displayed. Instead, it is recovered during the signature verification process.

The presence of the mechanism cannot be detected in a tamper-resistant black-box implementation (i.e., Capstone which is a key escrow technology which also employs the DSA system), for the same reasons as in the ElGamal digital signature setup. Note that this kleptographic attack assumes that the device was implemented with a priori knowledge of the values for g, p, and q that the user will use. One can envision a scenario in which the NIST invites several corporations to agree on a choice of parameters using the NIST designed prime number generation method reiterated in [Schneier]. With this setup attack, "trapdoor primes" (originally suspected in DSA) are not needed, any primes will do.

Theorem 3 *DSA has a regular setup version.*

If we trust that the device is indeed tamper-proof, we can leak a private key over two signatures while at the same time letting the user choose his own values for p, q, and g. This can be done by including the attacker's private key X (say, 511 bits in length) within the device. The attack is clearly not a setup attack (it includes the attacker's private key!), but leaks keys at a very high bandwidth, and is very flexible. This attack obviously relies heavily on the tamper-proof nature of the device in question.

4.1   Generalized Information Leakage in DSA

We have shown how the private key x can be leaked over two DSA signatures. Clearly, we can then use the Simmons channel and leak a message mod q in a third signature by setting k equal to that message. But, we can do better. In this section we will show how to leak a message mod q of our choosing over two signatures, in addition to the private key x. But, before doing so we will point out two weaknesses in the Simmons channel. Suppose Alice wants to send Bob the subliminal message "160 bit long string." on two separate occasions. Since the string is 160 bits long, it will occupy the entire value k (assuming we send ASCII text). Hence, the two values for r, where $r = (g^k \bmod p) \bmod q$ will be identical, and are easily noticed by the warden. Thus, we are forced to break down the message in order to introduce randomness. But then we need more bandwidth to send the message. Also note that without introducing randomness, the warden can mount guessed plaintext attacks. The warden simply guesses that Alice will send the string k = "160 bit long string." and then verifies that $r = (g^k \bmod p) \bmod q$.

The method we will now describe accomplishes this generalized (1,2)-leakage and avoids these drawbacks. The method for accomplishing this is subtle, and uses a "feed-back" like algorithm. Suppose that Alice is in prison and wants to send a 160 bit message $m < q$ to Bob, who is on the outside. To do so, Alice takes the first message M_1 to be signed, and computes g_1 based on M_1 where g_1 is an element in Z_p with order q. To compute g_1, Alice finds the smallest value $w > 0$ such that $H^w(M_1) \bmod p$ generates Z_p. Alice then sets $g_1 = H^w(M_1)^{(p-1)/q} \bmod p$. Alice computes the signatures (r_1, s_1) and (r_2, s_2) using her private key x as follows:

1. $k_1 = ((g_1{}^x \bmod p) \bmod q)m \bmod q$
2. $r_1 = (g^{k_1} \bmod p) \bmod q$
3. $s_1 = k_1{}^{-1}(H(M_1) + xr_1) \bmod q$
4. Calculate k_2 using the regular setup in DSA
5. $r_2 = (g^{k_2} \bmod p) \bmod q$
6. $s_2 = k_2{}^{-1}(H(M_2) + xr_2) \bmod q$

Upon receiving the two signatures, Bob can recover m as follows:

1. Bob recovers x using the regular setup in DSA

2. $k_1 = s_1^{-1}(H(M_1) + xr_1) \bmod q$

3. Bob computes g_1 using M_1 in the same way that Alice did

4. $m = k_1((g_1^x \bmod p) \bmod q)^{-1} \bmod q$

In the above algorithm, Alice sends the subliminal message m and a kleptogram in the first signature. The kleptogram is then used to securely compromise the second signature. Bob then recovers k_2, and thus x from the second signature. Bob then takes x and goes back to the first signature, and recovers k_1. Using k_1 Bob recovers the message m. So, Bob takes x from the second signature and "feeds it back" into the first signature to reveal the subliminal message.

Note that x is a shared secret between Alice and Bob. Thus $(g_1^x \bmod p) \bmod q$ is a shared secret between Alice and Bob. It is this secret that is used to blind the subliminal message m. The attack is therefore not subject to guessed plaintext attacks, since the warden must guess $(g_1^x \bmod p) \bmod q$, in addition to m. Also, since there is a one-to-one mapping between the shared secrets $(g_1^x \bmod p) \bmod q$ and the messages M_1 being signed, and since there is no need to sign the same message twice, Alice can send the same subliminal message twice and the values for r will be different.

This whole attack has the drawback that the device will always choose the same k for a given message M being signed. So, when designing black-box devices like Capstone, we might want to randomize some of the upper order bits of m so that it will *look like* the device is really choosing k randomly (joke). So, not only is it possible to leak DSA private keys over two DSA signatures, but it is also possible for devices to leak 160 bits of the devices own choosing at the same time. This channel is ideal for leaking symmetric keys chosen by the user.

4.2 Rogue Use of DSA Easily Implies a "Public Key Cryptosystem"

Recall that $g^k \bmod p$ can be recovered from (r, s) by computing the expression $g^{s^{-1}H(M)}y^{s^{-1}r} \bmod p$. So, Alice can send a DSA signed message to Bob that is effectively public key encrypted as follows. Alice chooses k randomly and raises Bob's DSA public key y to this k, thereby yielding a secret Diffie-Hellman key $z \bmod p$. Alice then encrypts this message using z in a symmetric cipher. Alice signs the resulting ciphertext file using her DSA private key and k. Using (r, s), Bob can recover $g^k \bmod p$. By raising $g^k \bmod p$ to his private key, Bob recovers z. The only information that is sent is the encrypted file (which is 'plaintext') and the signature (r, s). Note that z will be different each time a message is sent, because k will be different.

In [NR94] it was shown "How to Securely Integrate the DSA to Key Distribution" by sending the pair (r, s) with $H(M) = 1$. Here we are not fixing $H(M)$ which would have been quite noticeable. We are in fact *spoofing* normal DSA signed messages to send (effectively) public key encrypted signed messages at the same time. Thus we have shown that:

Theorem 4 *A DSA message/signature pair $(m, (r, s))$ signed by Alice and sent to Bob can be abused to be a pair consisting of a probabilistic public key encrypted message m' encrypted for Bob and signed by Alice.*

Note that Alice and Bob can establish the secret key z, thus:

Theorem 5 *A DSA message signature pair $(m, (r, s))$ signed by Alice and sent to Bob can be abused to be a key exchange message establishing a secret key between Alice and Bob.*

Let us recall that one of the criticisms of the DSA was that DSA does not provide for secret key distribution. In response, [SB92] stated *"The DSA does not provide for secret key distribution because DSA is not intended for secret key distribution"*. Yet we have shown that DSA can be used essentially and quite directly as a PKCS.

4.3 Device Marking

Note that the DSA setup requires that only 160 bits of the hidden field element (which is at least 512 bits) be used for k in the subsequent signature. The remaining 352 bits can be used to compromise other algorithms in a black-box implementation. Furthermore, note that since the user's private key x can be securely derived, the device can also leak information securely using k. This makes for the following rather inviting facility.

Each device could have a unique 26-bit serial number. The device could take 160 bits of the hidden field element, and use 134 of them as the lower order bits of the subsequent k. The other 26 bits can be XORed with the serial number. The result can be used to form the upper order bits of the subsequent k. Note that the attacker must now try 2^{26} possibilities to derive the correct k. However, once the user's private key is recovered, along with the 26 bit pad, the attacker knows exactly which device was used to compute the signature. This allows the device to securely and subliminally mark signatures that it outputs. This marking is essentially the signature of the device embedded within the signature of the user. If users primarily use their own devices to sign their own documents, then this mechanism can both help detect forgeries if x becomes known to an adversary and can be used to, for example, find thieves (who steal the device itself).

5   Regular Setup in the Schnorr Digital Signature Scheme

The following is a quick overview of Schnorr [Sc91]. Let p and q be primes such that q divides $p - 1$. Let g be a number such that $g^q = 1 \; mod \; p$. Let $s < q$ be the randomly chosen private key. The public key is $v = g^{-s} \; mod \; p$.

1. Alice picks $r < q$ randomly and computes $x = g^r \; mod \; p$
2. Alice computes $e = H(m, x)$ and sets $y = r + se \; mod \; q$
3. the signature of m is (e, y)

Here H is a one-way hash function. To verify the signature, Bob computes $z = g^y v^e \; mod \; p$ and then makes sure that $e = H(m, z)$. Note that $z = x = g^r \; mod \; p$. Thus, for a valid signature, z can be used to leak a hidden field element. In this respect, the (1,2)-leakage setup in Schnorr is very similar to the setup in DSA.

Theorem 6 *The Schnorr signature algorithm has a regular setup version.*

6 Setup Attacks on Elliptic Curve Cryptosystems

The discrete log setup extends directly to elliptic curve cryptosystems. Let E be an elliptic curve defined over F_q and let B be a publicly known point on E. The attacker chooses a random integer x of order of magnitude q, which he keeps private. The attacker includes in the cryptosystem the point $xB \in E$. B is analogous to g and xB is analogous to y in the discrete log attack. The attack proceeds in exactly the same way as described before, except that we calculate a pseudo-random point c on E as opposed to a pseudo-random hidden field element in F_p. This point is hashed in order to determine the subsequent value k to be used.

Note that a complication arises in trying to calculate the subsequent value k to be used in the cryptosystem. Let $\#E$ denote the number of points on E. We need a value k uniformly distributed in $[0..\#E - 1]$. Hasse's theorem asserts that $q + 1 - 2\sqrt{q} \leq \#E \leq q + 1 + 2\sqrt{q}$. Even if say, $q = \#E$, we could not simply set k to be the left coordinate of c since there could be many pairs (x, y) not on the curve with $x < q$. We cannot simply set k to be the right coordinate of c since there could be points (x, y) and (x', y) where $x \neq x'$ and $y < q$. It is well known that there is no convenient method known to deterministically generate points on E. Hence, finding a function that calculates an unbiased k given c is a difficult problem. One possible solution is to use a hash function as a random oracle to hash c to a value between 0 and $\#E - 1$. An elegant solution to the bias problem in elliptic curve setups is left as an open problem. We close this section by noting that the Menezes-Vanstone PKCS [MV93] can have a setup using this method.

Theorem 7 *Menezes-Vanstone PKCS has a regular setup version.*

7 SETUPs in key exchanges

We now describe how to employ the methodology in the authenticated key exchange protocols given in [Stin95]. The following is a review of MTI. Let g be a primitive element modulo the prime p. Each user U has an ID string, ID(U), a secret exponent a_u ($0 \leq a_u \leq p - 2$), and a corresponding public value $b_u = g^{a_u} \bmod p$. The TA has a signature scheme with a verification algorithm V_{TA} and a secret signing algorithm S_{TA}. Each user U will have a certificate $C(U) = (ID(U), b_u, S_{TA}(ID(U), b_u))$. To exchange keys, users U and V do the following.

1. U chooses r_u at random, $0 \leq r_u \leq p - 2$ and computes $s_u = g^{r_u} \bmod p$
2. U sends $(C(U), s_u)$ to V
3. V chooses r_v at random, $0 \leq r_v \leq p - 2$ and computes $s_v = g^{r_v} \bmod p$
4. V sends $(C(V), s_v)$ to U

5. U computes $K = s_v{}^{a_u} b_v{}^{r_u} \bmod p$, where b_v is obtained from $C(V)$. V computes $K = s_u{}^{a_v} b_u{}^{r_v} \bmod p$, where b_u is obtained from $C(U)$.

Note that in the first exchange, r_u and r_v for the second exchange can be leaked if the devices belonging to *both* U and V have a setup. The attacker then knows the K of the second round since $K = b_v{}^{r_u} b_u{}^{r_v} \bmod p$. Finding the setup in the Girault Key Agreement Protocol is left as an exercise for the reader.

8 Conclusion

We have demonstrated the prevalence of kleptographic attacks and the applicability of the kleptographic point of view. We presented a direct methodology for the systematic search for attacks based on kleptographic relations. The discrete log kleptogram was used in particular to implement setups in numerous systems, and influenced potential abuses of the DSA signature scheme.

References

[DSS91] Proposed Federal Information Processing Standard for Digital Signature Standard (DSS). In v. 56, n. 169 of *Federal Register*, pages 42980–42982, 1991.

[ElG85] T. ElGamal. A Public-Key Cryptosystem and a Signature Scheme Based on Discrete Logarithms. In *Advances in Cryptology—CRYPTO '84*, pages 10–18, 1985. Springer-Verlag.

[GM84] S. Goldwasser, S. Micali. Probabilistic Encryption. *J. Comp. Sys. Sci.* 28, pp 270-299, 1984.

[MV93] A. Menezes, S. Vanstone. Elliptic curve cryptosystems and their implementation. In *Journal of Cryptology*, volume 6, pages 209–224, 1993.

[NR94] K. Nyberg, R. Rueppel. Message Recovery for Signature Schemes Based on the Discrete Logarithm Problem. In *Advances in Cryptology—EUROCRYPT '94*, pages 182–193, 1994. Springer-Verlag.

[RSA78] R. Rivest, A. Shamir, L. Adleman. A method for obtaining Digital Signatures and Public-Key Cryptosystems. In *Communications of the ACM*, volume 21, n. 2, pages 120–126, 1978.

[SB92] M. Smid, D. Branstad. Response to Comments on the NIST Proposed Digital Signature Standard. In *Advances in Cryptology—CRYPTO '92*, pages 76–88, 1992. Springer-Verlag.

[Sc91] C. Schnorr. Efficient signature generation by smart cards. In *Journal of Cryptology*, volume 4, pages 161–174, 1991.

[Schneier] B. Schneier. Applied Cryptography, pages 309–310, 1994. John Wiley and Sons, Inc.

[Sim85] G. J. Simmons. The Subliminal Channel and Digital Signatures. In *Advances in Cryptology—EUROCRYPT '84*, pages 51–57, 1985. Springer-Verlag.

[Sim93] G. J. Simmons. Subliminal Communication Is Easy Using the DSA. In *Advances in Cryptology—EUROCRYPT '93*, 1993. Springer-Verlag.

[Stin95] D. R. Stinson. Cryptography: theory and applications, 1995, CRC Press.

[YY96] A. Young, M. Yung. The Dark Side of Black-Box Cryptography. In *Advances in Cryptology—CRYPTO '96*, pages 89–103, Springer-Verlag.

[YY97] A. Young, M. Yung. Kleptography: Using Cryptography against Cryptography. In *Advances in Cryptology—EUROCRYPT '97*, pages 62–74, 1997. Springer-Verlag.

"Pseudo-Random" Number Generation Within Cryptographic Algorithms: The DDS Case

Mihir Bellare[1], Shafi Goldwasser[2] and Daniele Micciancio[2]

[1] Dept. of Computer Science & Engineering, University of California at San Diego, 9500 Gilman Drive, La Jolla, California 92093, USA. E-Mail: mihir@cs.ucsd.edu. URL: http://www-cse.ucsd.edu/users/mihir.

[2] MIT Laboratory for Computer Science, 545 Technology Square, Cambridge, MA 02139, USA. E-Mail: {shafi,miccianc}@theory.lcs.mit.edu.

Abstract. The DSS signature algorithm requires the signer to generate a new random number with every signature. We show that if random numbers for DSS are generated using a linear congruential pseudorandom number generator (LCG) then the secret key can be quickly recovered after seeing a few signatures. This illustrates the high vulnerability of the DSS to weaknesses in the underlying random number generation process. It also confirms, that a sequence produced by LCG is not only predictable as has been known before, but should be used with extreme caution even within cryptographic applications that would appear to protect this sequence. The attack we present applies to truncated linear congruential generators as well, and can be extended to any pseudo random generator that can be described via modular linear equations.

1 Introduction

Randomness is a key ingredient for cryptography. Random bits are necessary not only for generating cryptographic keys, but are also often an integral part of steps of cryptographic algorithms. Examples are the DSS signature algorithm [16] which requires the choice of a new random number every time a new signature is generated, and CBC encryption, which requires the generation of a new random IV each time a new message is encrypted. (In fact, any secure, stateless encryption scheme must be probabilistic, requiring new randomness for each encryption [8].) In some cases, the random numbers chosen may have to be kept secret (as for DSS, where the leakage of one such random number compromises the secret key), whereas for other cases they can be made public (as in CBC encryption, where the IV may be sent in the clear).

In practice, the random bits will be generated by a pseudo random number generation process. For example, the DSS description [16] explicitly allows either using random or pseudo-random numbers. When this is done, the security of the scheme of course depends in a crucial way on the quality of the random bits produced by the generator. Thus, an evaluation of the overall security of a cryptographic algorithm should consider and take into account the choice of the pseudorandom generator.

It has been well accepted that a good notion of pseudorandomness for cryptographic purposes is *unpredictability* [18,20,3,7]: given an initial sequence produced by a pseudo-random number generator on an unknown seed, it is hard to predict with better probability than guessing at random, the next bit in the sequence output by the generator. Such generators can be constructed based on number-theoretic assumptions, but are computationally costly. Alternatively, one could build a generator out of DES which would be unpredictable assuming DES behaves like a pseudorandom function, but in some contexts this may be deemed costly too, or we might not want to make such a strong assumption. Since using a weaker generator does not necessarily mean the resulting cryptographic algorithm is insecure, in practice one usually uses some weak but fast generator.

The intent of our paper is to illustrate the extreme care with which one should choose a pseudo random number generator to use within a particular cryptographic algorithm. Specifically, we consider a concrete algorithm, the Digital Signature Standard [16], and a concrete pseudo random number generator, the linear congruential generator (LCG) or truncated linear congruential pseudo random generator. We then show that if a LCG or truncated LCG is used to produce the pseudo random choices called for in DSS, then DSS becomes completely breakable.

We remark that the Standard [16] recommends the use of a pseudo-random generator based on SHA-1 or DES. The attack we describe does not say anything about the use of DSS with such generators, but it does illustrates the high vulnerability of the DSS to the underlying random number generation process.

We remark that LCGs are known to be predictable if part of the pseudorandom sequence is made public (see section 1.2 for details). However in DSS none of the pseudo-random numbers used is ever revealed, and thus predictability does not imply insecurity here.

Let us now look at all this more closely.

1.1 Pseudorandom numbers in DSS

Recall that the DSS has public parameters p, q, g where p, q are primes, of 512 bits and 160 bits respectively, and g is a generator of an order q subgroup of Z_p^*. The signer has a public key $y = g^x$ where $x \in Z_q$. To sign a message $m \in Z_q$, the signer picks at random a number $k \in \{1, \ldots, q-1\}$ and computes a signature (r, s), where $r = (g^k \bmod p) \bmod q$ and $s = (xr + m)k^{-1} \bmod q$.

Here the "nonce" k is chosen at random, anew for each message. In practice, a sequence of nonces will be produced by a generator $\mathcal{G}$ which, given some initial seed k_0, produces a sequence of values $k_1, k_2, \ldots$; k_i will be the nonce for the i-th signature.

The adversary (cryptanalyst) sees the public key y, and triples (m_i, r_i, s_i) where (r_i, s_i) is a signature of m_i. Notice that the secrecy of the nonces is crucial. If ever a single nonce k_i is revealed to the adversary, then the latter can recover the secret key x, because $x = (s_i k_i - m_i)r_i^{-1} \bmod q$. However, the nonces appear to be very well protected, making it hard to exploit any such weakness. The cryptanalyst only sees $r_i = (g^{k_i} \bmod p) \bmod q$ from which he cannot recover k_i short of computing discrete logarithms, and in fact not even then, due to the second mod operation. So even if $\mathcal{G}$ is a predictable generator, meaning, say, that given k_1, k_2 we can find k_3, there is no a priori reason to think DSS is vulnerable with this generator, because how can the cryptanalyst ever get to know k_1, k_2 anyway?

This might encourage a user to think that even a weak (predictable) generator is OK for DSS. This view would be wrong. We indicate that in fact DSS is vulnerable, because without a sufficiently good pseudorandom number generation process, the "masking" of the nonces provided by the algorithm is not sufficient to protect the nonces, even though recovering them seems a priori to require solving the discrete logarithm problem. In fact we prove a quite general lemma showing why this masking is essentially ineffective for pretty much *any* pseudorandom generator, and show specifically how to recover the keys when the generator is an LCG or truncated LCG. Thus one should not succumb to the temptation of using a weak generator for DSS.

1.2 Linear congruential generators

Recall that linear congruential generators are pseudo-random number generators based on a linear recurrence $X_{n+1} = aX_n + b \bmod M$ where a, b and M are parameters initially chosen at random and then fixed, and the seed is the initial value X_0. The advantage of linear congruential generators is that they are fast, and it has been shown [11] that they have good statistical properties for appropriate choices of the parameters a, b, M.

On the other hand, their unpredictability properties are known to be quite weak. Clearly they are predictable in their simplest form: if the parameters a, b and M are known, given X_0 all the other X_n can be easily computed. Plumstead (Boyar) [17] shows that even if the parameters a, b, M are unknown the sequence of numbers produced by a linear congruential generator is still predictable given some of the X_i. Truncated LCG were suggested by Knuth [12] as a possible way to make a linear congruential generator secure. However these generators have also been shown to be predictable [5,9,19] as have more general congruential generators [4,13].

However, as indicated above, this predictability does not directly mean a cryptographic algorithm using the generator is breakable, since it is possible none of the bits of the random numbers used by the algorithm are ever made public. DSS is (was) a case in point.

1.3 Cryptanalysis of DSS with LCG

DSS WITH LCG. We consider what happens when the nonces in DSS are generated using an LCG with known parameters a, b, M and hidden seed k_0. The predictability of the generator does not a priori appear to be a problem, due to the masking provided by the algorithm as indicated above. However, given just three valid signatures, we show how to recover the secret key.

UNIQUENESS LEMMA. We begin with a general lemma which indicates why the above intuition that the DSS protects the nonces may be false. The lemma (called the Uniqueness Lemma) says that as long as the nonces are pseudorandomly generated then, even if we *ignore* the relations $r_i = (g^{k_i} \bmod p) \bmod q$, the DSS signature equations $s_i k_i - r_i x = m_i$ uniquely determine the secret key with high probability. This means the cryptanalyst can effectively ignore the masking that is supposed to protect the nonces. This is true for *any* pseudorandom generation process, even a cryptographically strong one, using an unpredictable generator. This lemma tells us we can concentrate on the signature equations.

SOLVING THE EQUATIONS. We begin the cryptanalysis of DSS with LCG by combining the DSS "signature equations" with the LCG generation equations to get a system of equations. (In the process we ignore the $r_i = (g^{k_i} \bmod p) \bmod q$ relations, invoking the Uniqueness Lemma to say that solving the signature equations suffices to find the secret key.) However, this system is not trivial to solve because it is a system of simultaneous *modular* equations in *different* moduli. Techniques like Gaussian elimination fail. Instead we turn to lattice reduction. We show how to use Babai's closest vector approximation algorithm to solve such a system. The main difficulty here is dealing with the fact that this algorithm only returns (not very good) approximations to the closest vector. We then extend this to the case of the truncated LCG.

1.4 Other results, discussion, and implications

We extend our techniques to provide a general algorithm for solving a system of simultaneous linear modular equations in different moduli. (Another way of doing this, when the number of equations is constant, is to reduce the problem to integer programming in constant dimension and apply the algorithms of [14,10]. Our alternative solution seems simpler and more direct.)

In many cryptographic algorithms, the random numbers used are processed in a way that the public information gives little information about the original numbers. This is the case for the nonces in DSS. In such a setting, it may be reasonable to think that weak random number generators can suffice: even predictable generators could be fine because not enough information about the random numbers is revealed to make predictability even come into play. We are indicating this may not always be true: the quality of random bits matters even when the only thing an adversary sees is the result of a one-way functions on these bits.

A common pseudo-random number generator that comes standard with various operating systems is a linear congruential generator with modulus 2^{32}. It is plausible that there are DSA implementations available where the k values are formed by concatenating 5 consecutive outputs from such a generator. Our attack easily extends to this case.

2 Preliminaries

2.1 The Digital Signature Standard

The Digital Signature Standard (DSS, see [16]) is an ElGamal-like [6] digital signature algorithm based on the hardness of computing the discrete logarithm in some finite fields.

THE SCHEME. The scheme uses the following parameters: a prime number p, a prime number q which divides $p - 1$ and an element $g \in Z_p^*$ of order q. (Chosen as $g = h^{(p-1)/q}$ where h is a generator of the cyclic group Z_p^*). These parameters may be common to all users of the signature scheme and we will consider them as fixed in the rest of the paper. The standard asks that $2^{159} < q < 2^{160}$ and $p > 2^{511}$. We let $G = \{ g^\alpha : \alpha \in Z_q \}$ denote the subgroup generated by g. Note it has prime order, and that the exponents are from a field, namely Z_q.

The secret key of a user is a random integer x in the range $\{0, \ldots, q - 1\}$, and the corresponding public key is $y = g^x \bmod p$. DSA (the Digital Signature Algorithm that underlies the standard) can be used to sign any message $m \in Z_q$, as follows. The signer generates a random number $k \in \{1, \ldots, q - 1\}$, which we call the *nonce*. It then computes the values $\lambda = g^k \bmod p$ and $r = \lambda \bmod q$. It sets $s = (xr + m) \cdot k^{-1} \bmod q$, where k^{-1} is the multiplicative inverse of k in the group Z_q^*. The signature of message m is the pair (r, s) and will be denoted by DSA(x, k, m). Note that a new, random nonce is chosen for each signature.

A purported signature (r, s) of message m can be verified, given the user's public key y, by computing the values $u_1 = m \cdot s^{-1} \bmod q$, $u_2 = r \cdot s^{-1} \bmod q$ and checking that $(g^{u_1} y^{u_2} \bmod p) \bmod q = r$. Notice that the values (r, s) output by DSA(x, k, m) satisfy the relation $sk - rx = m \pmod{q}$. We will make use of this relation in our attack on the DSS.

HASHING. The 160-bit "message" m above is not the actual text one wants to sign, but rather the hash of it, under a strong, collision resistant cryptographic hash function H. Specifically, if m is the actual text to be signed, the standard sets $H =$ SHA-1, the Secure Hash Algorithm of [15]. The hashing serves two purposes. The first is to enable one to sign messages of length longer than 160 bits. Second, it "randomizes" the message to prevent any possible attacks based on the algebraic structure of the scheme. Accordingly, following [2], we treat the hash function as a random oracle.

We stress that we are considering *attacks*. In this context, treating H as a random oracle only strengthens our results. If the scheme is breakable when H is a random oracle, we should definitely consider it insecure, because a random oracle is the "best" possible hash function!

Our attack on the DSS algorithm does not involve the hash function H other than to assume it random. Therefore we will assume that the messages are already integers in the range $\{0, \ldots, q - 1\}$ and that they are randomly distributed.

SECRECY OF THE NONCE. Recall that for every signature, the signer generates a new, random nonce k. An important feature (drawback!) of the DSS is that the security relies on the secrecy of the nonces. If any nonce k ever becomes revealed, at any time, even long after the signature (r, s) was generated, then given the nonce and the signature one can immediately recover the secret key x, via $x = (sk - m)r^{-1} \bmod q$. This is a key point in our attack.

2.2 Pseudo-Random Number Generators

Each time DSA is used to digitally sign a message m, a nonce k is needed. Ideally k should be a truly random number. In practice the nonces k are pseudo-random numbers produced by a pseudo-random number generator.

A pseudo-random number generator is a program $\mathcal{G}$ that on input a *seed* σ, generates a seemingly random sequence of numbers $\mathcal{G}(\sigma) = k_1, k_2, \ldots$.

The DSS algorithm can be used in conjunction with a pseudo-random number generator as follows. On input a secret key x, a seed σ to the generator, and a sequence of messages $m_1, \ldots, m_n$, run $\mathcal{G}(\sigma)$ to generate a sequence of pseudo-random numbers $k_1, \ldots, k_n$ and run DSA on input x, m_i, k_i for all $i = 1, \ldots, n$.

The pseudo-random number generators we consider in this paper are all variants of the linear congruential generator.

LINEAR CONGRUENTIAL GENERATORS. A linear congruential generator (LCG) is parameterized by a modulus M and two numbers $a, b \in Z_M$. The seed to $\mathcal{G}$ is just a number $\sigma = k_0 \in Z_M$. On input k_0, the generator produces a sequence of numbers, $\mathcal{G}(k_0) = k_1, k_2, \ldots$ defined by the linear recurrence $k_{i+1} = ak_i + b \bmod M$. The values k_i can be directly used by DSA as random nonces to sign the messages. (In which case they are treated modulo q. We assume that with high probability a k_i value will not be 0.)

TRUNCATED LINEAR CONGRUENTIAL GENERATORS. For security reasons it has been suggested that only some of the bits of the number produced by a linear congruential generator be used by applications, in our case the DSA algorithm. A truncated linear congruential generator does exactly this. Let's look at this more closely. A truncated linear congruential generator is parameterized by a modulus M, two numbers $a, b \in Z_M$ and two indices l, h such that $0 \leq l \leq h \leq \lg M$. The seed is a number $\sigma_0 \in Z_M$ and the generator produces numbers in the range $\{0, \ldots, 2^{h-l} - 1\}$. The generator computes a sequence σ_i according to the linear recurrence $\sigma_{i+1} = a\sigma_i + b \bmod M$. Then, each number σ_i is truncated by taking only bits $l, \ldots, h-1$ of the number, to get the number $k_i = ((\sigma_i - (\sigma_i \bmod 2^l)) \bmod 2^h)/2^l$ which is output by the generator.

3 The attack

We look at the security of the DSS when the nonces are generated using a LCG with parameters a, b, M. Later we will extend this to truncated LCGs.

3.1 Overview

Our attack on DSS exploits the relationship $sk - rx = m \bmod q$ holding for any digital signature $(r, s) = \mathrm{DSA}(x, k, m)$ produced by the DSA algorithm. The idea is this. Assume that we receive two messages m_1 and m_2 together with their digital signatures $(r_1, s_1) = \mathrm{DSA}(x, k_1, m_1)$ and $(r_2, s_2) = \mathrm{DSA}(x, k_2, m_2)$. We know that $s_1 k_1 - r_1 x = m_1 \bmod q$ and $s_2 k_2 - r_2 x = m_2 \bmod q$. The cryptanalyst knows $m_1, r_1, s_1, m_2, r_2, s_2$. He also knows the public parameters p, q, g of the DSS and the public key $y = g^x$ of the signer. What is hidden from him is the secret key x of the signer, and also the nonces k_1, k_2 which the signer used to produce the signatures.

At this point, the cryptanalyst is not expected to have any way of determining any of the unknowns short of computing discrete logarithms. However, now suppose we know that a linear congruential generator with parameters a, b, M has been used to produce the nonces. We assume the cryptanalyst knows the parameters a, b, M defining the LCG. (They were chosen at random, but then made public.) What is unknown to the cryptanalyst is the seed k_0 used by the signer to start the LCG. Now, we can combine the two signature equations above with the linear congruential equation $k_2 = ak_1 + b \bmod M$. These three equations together yield a system of three modular equations in three unknowns:

$$\begin{cases} s_1 k_1 - r_1 x = m_1 & \pmod q \\ s_2 k_2 - r_2 x = m_2 & \pmod q \\ -ak_1 + k_2 = b & \pmod M \end{cases} \tag{1}$$

Our approach is to try to solve these equations. Note it is a system of simultaneous modular linear equations in different moduli.

This approach at once raises two questions. One, of course, is how to solve such a system. But the other question may need to be addressed first. Namely, even if we solve it, how do we know the solutions we get are the desired ones? That is, there may be many different solutions, and finding a solution to the system (1) does not necessarily imply that we found the right one. (Meaning the one corresponding to the secret key x.)

This worry arises from a feature of this approach that we should highlight. We are not using all available information. We propose to ignore the fact that $r_i = (g^{k_i} \bmod p) \bmod q$. We will simply try to solve the equations, and see what we get. When we are ignoring what may seem a fundamental relation of the DSS signatures, it is not clear why solving the equations will bring us the right solutions: our system of equations might be under-determined.

We will answer this question in Section 3.2, showing that even disregarding the non-linear relationships $r_1 = (g^{k_1} \bmod p) \bmod q$ and $r_2 = (g^{k_2} \bmod p) \bmod q$, the solution to our equations is uniquely determined in most of the cases. Then we can turn to the problem of solving a system of modular linear equations.

If the moduli are the same, $M = q$, the equations can be easily solved by linear algebra. So, it is insecure to use q as the modulus in the LCG. However, if the modulus M is chosen (randomly and) independently from q, as we assume, one might still imagine that the equation $k_2 = ak_1 + b \bmod M$ does not help in finding the secret key because it is in a different modulus and cannot be easily combined with the other equations. In other words, we are faced with solving a system of simultaneous modular linear equations in different moduli. We address this via lattice reduction techniques in Section 3.3.

In later sections we extend the attack to truncated LCGs and also present a general method for solving systems of simultaneous linear modular equations in different modulii.

3.2 The Uniqueness Lemma

In this section we prove that when DSS is used with a pseudo-random number generator, a few signatures are usually enough for the linear equations $s_i k_i - r_i x = m_i$ to uniquely determine the secret key x, disregarding that it must be also that $r_i = g^{k_i} \bmod p \bmod q$. This answers the first question that we posed in section 3.1 and opens up the possibility of breaking DSS by solving a system of linear equations.

We stress this is true for any generator, not just LCG. The generator might be very strong (eg. cryptographically strong) or very weak, it does not matter. The number of signatures needed depends only on the length of the seed of the generator, growing linearly with this.

The statement we make is a probabilistic one: with high probability the system of equations obtained by using DSS with a linear congruential generator has a unique solution. The probability is taken over the choices of the messages to be signed only. (As discussed in Section 2, these are hashes of the real messages under some "strong" one-way hash function, and so considering them random is natural, especially from an attack point of view.) In other words, no matter how we had chosen x and σ, once they are fixed, if the messages m_i are randomly chosen the secret key x is uniquely determined with high probability.

Before stating the lemma we need some definitions. Fix a secret key $x \in Z_q$ of the DSS. Let $\mathcal{G}$ be some generator (not necessarily LCG) and let M be the total number of seeds that $\mathcal{G}$ can take. So we will think of a seed of $\mathcal{G}$ as being in Z_M. Now fix a seed $\sigma \in Z_M$ of the generator $\mathcal{G}$. Let $\mathcal{G}(\sigma) = k_1, k_2, \ldots, k_n$. Fix a message sequence $m_1, \ldots, m_n \in Z_q$ and let $(r_i, s_i) = \text{DSA}(x, k_i, m_i)$ be the signature of m_i using nonce k_i, for $i = 1, \ldots, n$. Let $x' \in Z_q$ and $\sigma' \in Z_M$, and let $\mathcal{G}(\sigma') = k_1', \ldots, k_n'$. We say (x', σ') is a *false solution* with respect to $x, \sigma, m_1, \ldots, m_n$ if $x \neq x'$ but $s_i k_i' - r_i x' = m_i \bmod q$ for all $i = 1, \ldots, n$. That is, the secret key is not the right one, but the equations work out anyway.

Lemma 1. *Fix a secret key $x \in Z_q$ of the DSS, and a seed $\sigma \in Z_M$ of the generator $\mathcal{G}$. Now, choose n messages $m_1, \ldots, m_n$ uniformly at random from Z_q. The probability, over the choices of the messages only, that there exists some (x', σ') which is a false solution with respect to $x, \sigma, m_1, \ldots, m_n$ is less than Mq^{1-n}. Moreover, the expected number of such false solutions is also less than Mq^{1-n}.*

Proof. Let $k_1, \ldots, k_n = \mathcal{G}(\sigma)$ be the output of the generator on seed σ. Since σ is fixed, so are $k_1, \ldots, k_n$. We will assume these are all in Z_q^*.

Fix $x' \in Z_q$ and $\sigma' \in Z_M$ such that $x \neq x'$, and let $k_1', \ldots, k_n' = \mathcal{G}(\sigma')$. For this fixed x', σ', and for a fixed i, we claim that the probability, over the choice of m_i, that $s_i k_i' - r_i x' = m_i$, is at most $1/q$. (Here $(r_i, s_i) = \text{DSA}(x, k_i, m_i)$, so that $r_i = g^{k_i}$ is a fixed quantity, while $s_i = (m_i + x r_i) k_i^{-1}$ is a random variable depending on the choice of m_i.) The reason this claim is not entirely obvious is that indeed s_i depends on m_i.

We first note that $s_i k_i' - r_i x' = m_i$ implies $k_i \neq k_i'$. To see this, note $m_i = s_i k_i - r_i x = m_i k_i' - r_i x'$. If $k_i = k_i'$ we would get $s_i k_i - r_i x = s_i k_i - r_i x'$ which yields $x = x'$ because $r_i \neq 0$. But we assumed $x \neq x'$.

Now, note that if $s_i k_i' - r_i x' = m_i$ then it must be that $(m_i + x r_i) k_i^{-1} k_i' - r_i x' = m_i$, or $m_i(1 - k_i^{-1} k_i') = r_i x k_i^{-1} k_i' - r_i x'$. But $k_i \neq k_i'$ implies $1 - k_i^{-1} k_i' \neq 0$ whence

$$m_i = \frac{r_i(x k_i^{-1} k_i' - x')}{1 - k_i^{-1} k_i'} = \frac{r_i(x k_i' - x' k_i)}{k_i - k_i'}.$$

But the right hand side does not depend on the choice of m_i, because all the quantities there are fixed. (We use here that r_i does not depend on m_i, a property of DSS.) This means there is only one value m_i for which the above equation can be true. So if we pick m_i at random from Z_q, there is only a $1/q$ chance that the above equation can be true.

Now since the messages are chosen independently at random, the probability that $s_i k_i' - r_i x' = m_i$ for all $i = 1, \ldots, n$, is q^{-n}. Recall this is for fixed $x' \in Z_q$ ($x' \neq x$) and $\sigma' \in Z_M$. The probability that there exists x', σ' which is a false solution is thus, by the union bound, at most $(q-1)M \cdot q^{-n} < M q^{1-n}$. For the claim about the expected number of false solutions, use linearity of expectation instead of the union bound. ∎

Recall these results are true for any pseudo-random number generator $\mathcal{G}$. That is even if $\mathcal{G}$ is cryptographically strong, with high probability there will be only one secret key x and seed σ such that the equations $r_i x' + s_i k_i' = m_i$ are simultaneously satisfied. Clearly if $\mathcal{G}$ is cryptographically strong it will be hard to recover these x and σ from the signatures (r_i, s_i) and messages m_i only. But for the LCG it can be done.

3.3 Solving the equations

Lemma 1 shows that even if $M \neq q$, if M and q have the same size (i.e., $1/2 < M/q < 2$), the system of equations 1 will usually have only a few solutions. Therefore, if we can solve the system of equations we can also retrieve the secret key.

SOLVING VIA INTEGER PROGRAMMING. We remark that systems of linear equations in different moduli can be rewritten as integer programming problems by introducing a new variable for each equation. Since we have a constant number of equations, they can thus be solved using polynomial time algorithms for integer programming in constant dimensions as given in [14,10]. However these algorithms are relatively complex and slow. Instead, we we want to solve more directly and simply. We now present a simple lattice based algorithm that solves our system using a nearest lattice vector approximation algorithm as a subroutine.

THE NEAREST LATTICE VECTOR PROBLEM. Let $B = \{b_1, \ldots, b_n\}$ be a finite set of vectors in $\mathbf{R}^n$. The lattice generated by B is the set of all integer combinations of the vectors in B and is denoted by $L(B)$. Given B and a vector $x \in R^n$ not in $L(B)$, the nearest lattice vector problem asks for a lattice vectors $w \in L(B)$ such that $\|w - x\| = \min_{v \in L(B)} \|v - x\|$. In [1], Babai gave a simple polynomial time algorithm to find an approximate solution to the nearest lattice vector problem: given the basis B and the target vector x, Babai's algorithm returns a lattice vector w such that $\|w - x\| \leq c \cdot \min_{v \in L(B)} \|v - x\|$, where $c = 2^{n/2}$ is an approximation factor depending only on the dimension of the lattice.

THE LATTICE. In order to solve the system of equations 1, we set up the following lattice. Let $x' = q/2$, $k'_1 = k'_2 = M/2$ and define also $\gamma_x = \min\{x', q - x'\}$, $\gamma_{k_1} = \min\{k'_1, M - k'_1\}$, $\gamma_{k_2} = \min\{k'_2, M - k'_2\}$. Consider the lattice L generated by the columns of the matrix

$$B = \begin{bmatrix} -r_1 & s_1 & 0 & q & 0 & 0 \\ -r_2 & 0 & s_2 & 0 & q & 0 \\ 0 & -a & 1 & 0 & 0 & M \\ \gamma_x^{-1} & 0 & 0 & 0 & 0 & 0 \\ 0 & \gamma_{k1}^{-1} & 0 & 0 & 0 & 0 \\ 0 & 0 & \gamma_{k2}^{-1} & 0 & 0 & 0 \end{bmatrix}.$$

Notice that multiplying the first three columns of the matrix by x, k_1, k_2 and subtracting the appropriate multiples of the remaining columns to perform modular reduction, we obtain the lattice vector

$$X = (m_1, m_2, m_3, x/\gamma_x, k_1/\gamma_{k1}, k_2/\gamma_{k2})^T$$

from which we can easily recover the secret key x.

Running Babai's nearest lattice vector algorithm on lattice $L(B)$ and target vector $Y = (m_1, m_2, m_3, x'/\gamma_x, k'_1/\gamma_{k1}, k'_2/\gamma_{k2})^T$ we obtain a lattice vector W such that $\|Y - W\| < 8\|Y - X\|$. Now, if $|x - x'| < \gamma_x/14$, $|k_1 - k'_1| < \gamma_{k1}/14$ and $|k_2 - k'_2| < \gamma_{k2}/14$, then $\|Y - W\| < 1$ and since the first three entries of W are integers they must coincide with the corresponding entries in Y and we have $W = (m_1, m_2, m_3, x''/\gamma_x, k''_1/\gamma_{k1}, k''_2/\gamma_{k2})^T$ for some x'', k''_1, k''_2 satisfying the equations 1. Moreover the following two inequalities are satisfied

$$x'' = (x'' - x') + x' \geq -\gamma_x + \gamma_x = 0$$
$$x'' = (x'' - x') + x' < \gamma_x + (q - \gamma_x) = q.$$

Inequalities $0 \leq k''_1, k''_2 < M$ can be proved analogously.

If the vector W does not have the desired form, that means that our initial guess (x', k'_1, k'_2) was not a good enough. If this is the case we simply repeat all the above steps with a different value for x', k'_1, k'_2. One can check that if we let x' range in the set $\{q/2 \pm (1 - (1 - 1/8)^j)q/2 \mid j = 0, \ldots, 8\lg q/2\}$, and k_1, k_2 in the set $\{M/2 \pm (1 - (1 - 1/8)^j)M/2 \mid j = 0, \ldots, 8\lg M/2\}$, there will be some x', k'_1, k'_2 such that $|x - x'| < \gamma_x/14$, $|k_1 - k'_1| < \gamma_{k1}/14$ and $|k_2 - k'_2| < \gamma_{k2}/14$.

The number of possible x', k'_1, k'_2 to start with, to be sure of finding a solution to the system is polynomial in $\lg q$ and $\lg M$, so we can try all of them in polynomial time.

Once we have found a solution x', k'_1, k'_2 to the equations 1, we can check that we actually found the secret key x by computing $g^{x'} \bmod p$ and comparing it with the public key y. If $g^{x'} \bmod p \neq y$, then $x \neq x'$ and we did not found the solution that we wanted. In this case we can use the method just described to find a solution to the equations 1 in the range $0 \leq x < x'$ or $x' < x < q$. Since by Lemma 1 the total number of x such that system 1 has solution is less then

2 on the average, with high probability we will find the right x after one or two steps.

This completes the description of the attack to DSS when used with linear congruential generators.

4 Solving Simultaneous Modular Equations

The technique described in section 3.3 can be generalized to work on arbitrary systems of linear equations in different moduli. These kind of systems arises in the cryptanalysis of DSS when used with more sophisticated pseudo-random number generators, such as truncated linear congruential generators.

In this section we state the problem of solving a system of linear equations in different moduli in its full generality and give an algorithm to find a solution to such a systems. When the number of equations and variables is fixed, the running time of the algorithm is polynomial in the logarithms of all numbers involved in the description of the equations.

We consider the problem of finding "small" solutions to a system of modular linear equations in different moduli. More precisely, let $U_1, \ldots, U_n$ be positive integers and let V_U be the set of vectors $\{x \in Z^n \mid \forall i. |x_i| < U_i\}$. Let also $A = \{a_{i,j}\}$ be an $m \times n$ integer matrix, and b and M be two vectors in Z^m. We want to find an integer vector $x \in V_U$ such that $A \cdot x = b \pmod{M}$, i.e., $|x_i| < U_i$ for all $i = 1, \ldots, n$ and the following modular equations are simultaneously satisfied

$$a_{1,1}x_1 + \ldots + a_{1,n}x_n = b_1 \pmod{M_1}$$
$$\vdots$$
$$a_{m,1}x_1 + \ldots + a_{m,n}x_n = b_m \pmod{M_m}.$$

We first assume that the above system has a solution x and that a good approximation to this solution is known and devise a method to find the exact solution.

Definition 2. Let x and y be two vectors in V_U. We say that vector y c-approximates x iff for all $i = 1, \ldots, n$ we have $|x_i - y_i| < (U_i - |y_i|)/(c\sqrt{n})$.

Lemma 3. Let c be a constant greater than $2^{(m+n)/2}$. There exists a polynomial time algorithm that on input $U_1, \ldots, U_n, A, b, M$ as above and a c-approximation y to a solution $x \in V_U$ to $A \cdot x = b \pmod{M}$, finds a (possibly different) solution $w \in V_U$ to $A \cdot x = b \pmod{M}$.

Proof. Let $\Gamma = \{\gamma_{i,j}\}$ be the $n \times n$ diagonal matrix defined by $\gamma_{i,i} = 1/(U_i - |y_i|)$ and let M be the $m \times m$ diagonal matrix whose diagonal entries are $M_1, \ldots, M_m$. Consider the lattice generated by the columns of the matrix $L = \begin{bmatrix} A & M \\ \Gamma & 0 \end{bmatrix}$ and define the vectors

$$X = \begin{bmatrix} b \\ \Gamma \cdot x \end{bmatrix} \quad Y = \begin{bmatrix} b \\ \Gamma \cdot y \end{bmatrix}$$

Notice that X is a lattice vector and

$$\|X - Y\| = \sqrt{\sum_{i=1}^{n}(x_i - y_i)^2 \gamma_{i,i}^2} \leq \sqrt{\sum_{i=1}^{n}\frac{1}{c^2 n}} = \frac{1}{c}.$$

Running Babai's nearest lattice vector algorithm [1] on lattice L and target vector Y we obtain a lattice vector W such that $\|W - Y\| < c\|X - Y\| \leq 1$. Since the first m elements of W and Y are integers, they must be the same. So, the vector W is equal to $\begin{bmatrix} b \\ \Gamma \cdot w \end{bmatrix}$ for some integer vector w satisfying $A \cdot w = b$ (mod M). It remains to be proved that $w \in V_U$. Now for all $i = 1, \ldots, n$ we have

$$(w_i - y_i)^2 \gamma_{i,i}^2 \leq \sum_i (w_i - y_i)^2 \gamma_{i,i}^2 = \|W - Y\|^2 < 1,$$

so that $|w_i - y_i| < 1/\gamma_{i,i} = U_i - |y_i|$ and by triangular inequality

$$|w_i| \leq |y_i| + |w_i - y_i| < |y_i| + U_i - |y_i| = U_i.$$

This proves $w \in V_U$.

We have shown how to solve a system of modular linear equations, given a good approximation to a solution. We now prove that for any fixed n and m there exists a set of vectors $D \subset V_U$ of size polynomial in the $\lg U_i$ such that for any $x \in V_U$ there exists a vector $y \in D$ such that y is a good approximation of x. This gives a polynomial time algorithm to solve modular linear systems in a fixed number of variables and equations.

Lemma 4. *Let $\delta > (1 + c\sqrt{n})/2$ and let D be the set $D_1 \times D_2 \times \cdots \times D_n$ where*

$$D_i = \{\pm(1 - (1 - 1/\delta)^j)U_i \mid j = 0, \ldots, \delta \lg U_i\}.$$

Then for any $x \in V_U$ there exists a vector $y \in D$ such that y is a c-approximation of x.

Proof. Clearly it is sufficient to show that for all i and for all $x \in \{-U_i + 1, \ldots, U_i - 1\}$, there is some y in D_i such that $|x - y| < (U_i - |y|)/(c\sqrt{n})$. Since the set D_i is symmetric with respect to the origin, we can assume without loss of generality that $x \geq 0$. Now, notice that the sequence $y_j = (1 - (1 - 1/\delta)^j)U_i$ is increasing. Moreover $y_0 = 0$ and $y_{\delta \lg U_i} > U_i - 1$. Therefore there exists a $j \in \{0, \ldots, \delta \lg U_i - 1\}$ such that $y_j \leq x < y_{j+1}$. Now, let $x' = y_j + U(1 - 1/\delta)^j/(c\sqrt{n})$. If $x < x'$ then we have $c\sqrt{n}(x - y_j) < U_i(1 - 1/\delta)^j = U_i - y_j$ and $|x - y_j| \leq (U_i - |y_j|)/(c\sqrt{n})$. Otherwise $x \geq x'$ and we have

$$\begin{aligned} c\sqrt{n}(y_{j+1} - x) &< c\sqrt{n}y_{j+1} - c\sqrt{n}y_j - U_i(1 - 1/\delta)^j \\ &= U_i(1 - 1/\delta)^{j+1}(c\sqrt{n}/(\delta - 1) - 1) \\ &< U_i(1 - 1/\delta)^{j+1} = U_i - y_{j+1} \end{aligned}$$

and $|x - y_{j+1}| \leq (U_i - |y_{j+1}|)/(c\sqrt{n})$.

Theorem 5. *There is an algorithm which on input m modular equations in n variables and n positive integers $U_1, \ldots, U_n$, finds a solution $x_1, \ldots, x_n$ to the* *equations such that $|x_i| < U_i$ for all $i = 1, \ldots, n$ and for any fixed n and m the running time of the algorithm is polynomial in the sizes of the numbers.*

In the above theorem the interval in which the variables x_i ranges need not be centered around the origin, as if we want $L_i < x_i < U_i$ we can simply substitute $x_i - (U_i + L_i)/2$ for x_i and obtain an equivalent linear system to be solved in the interval $|x_i| < (U_i - L_i)/2$.

Corollary 6. *There is an algorithm which on input m modular equations in n variables and positive integers $L_1, U_1, \ldots, L_n, U_n$, finds a solution $x_1, \ldots, x_n$ to the equations such that $L_i < x_i < U_i$ for all $i = 1, \ldots, n$ and for any fixed n and m the running time of the algorithm is polynomial in the sizes of the numbers.*

5   Other Pseudo-Random Number Generators

In section 3.1 we presented an attack to DSS that involves the solution of a system of three modular equations in different moduli. The attack easily extends to any pseudo-random number generator expressible by modular linear equations. As an example we consider truncated linear congruential generator and generators where a long nonce is obtained by concatenating shorter random numbers.

5.1   Truncated Linear Congruential Generators

We recall that a truncated linear congruential generator computes a sequence of values σ_i starting from a seed σ_0 according a modular linear recurrence relation, and for each i outputs a number k_i obtained by taking the bits of σ_i between positions l and h. The computation of the σ_i can be easily be expressed by the equation

$$\sigma_i = a\sigma_{i-1} + b \pmod{M} \qquad (0 \le \sigma_i < M) \tag{2}$$

where a, b and M are the parameters of the generator. Now let's look at how to express the truncation operation. If $l = 0$, then we can simply write

$$k_i = \sigma_i \pmod{2^h} \qquad (0 \le k_i < 2^h). \tag{3}$$

If $l \ne 0$ we need two equations. First we extract the l-lowest order bits of σ_i via

$$d_i = \sigma_i \pmod{2^l} \qquad (0 \le d_i < 2^l). \tag{4}$$

Then we use d_i to zero the l-lowest order bits of σ_i and extract the relevant bits of σ_i via

$$2^l k_i = \sigma_i - d_i \pmod{2^h} \qquad (0 \le k_i < 2^{h-l}). \tag{5}$$

Notice that $\sigma_i - d_i$ is always an integer multiple of 2^l, so equation (5) has solution despite of the fact that 2^l has not an inverse modulo 2^h.

So, the entire process of computing $k_1, k_2, \ldots, k_n$ from a seed σ_0 can be expressed by modular linear equations (2),(4) and (5) for $i = 1, \ldots, n$.

Consider now the use of DSA with a truncated linear congruential generator $\mathcal{G}$ of parameters M, a, b, l, h. For concreteness, we assume that half of the bits are truncated, i.e., $h - l = (\lg M)/2$. Since we want to use the numbers output by the generator as nonces in the DSA algorithm, we also assume that $h - l = \lg q$. Consider the system of equations

$$
\begin{cases}
s_1 k_1 - r_1 x = m_1 & \pmod{q} \\
s_2 k_2 - r_2 x = m_2 & \pmod{q} \\
s_3 k_3 - r_3 x = m_3 & \pmod{q} \\
s_4 k_4 - r_4 x = m_4 & \pmod{q}
\end{cases}
$$

together with equations (2),(4) and (5) for $i = 1, \ldots, 4$.

By Corollary 6, we can find in polynomial time a solution to the above equations such that $0 \le x, k_i < q$, $0 \le \sigma_i < M$ and $0 \le d_i < 2^l$. By Lemma 1 we know that with probability $1 - M/q^3 > 1 - 2q^2/q^3 = 1 - 2/q$ the equations have no false solution. Therefore, with high probability, the solution x we found is the DSA secret key.

5.2 Linear Congruential Generators with Concatenation

If the numbers σ_i output by the pseudo-random generator are too short, the nonces k_i can be obtained by concatenating several σ_i together.

For example, a common pseudo-random number generator that comes standard with various operating systems is a linear congruential generator with modulus 2^{32}. The 160 bit number k required to sign with DSA can be obtained by concatenating 5 consecutive outputs from such a generator.

Our attack immediately applies to these schemes. Let σ_i be the sequence of random numbers defined by a linear congruential generator modulo $M = 2^{32}$.

The concatenation operation is easily expressed as a linear equation:

$$
k_j = \sigma_{\alpha j} + M\sigma_{\alpha j + 1} + \cdots + M^{\alpha - 1}\sigma_{\alpha j + \alpha - 1} \pmod{M^\alpha} \qquad (0 \le k_i < M^\alpha) \quad (6)
$$

where $\alpha = \lceil \lg q / \lg M \rceil = 5$.

This time just two signature equations are enough to guarantee uniqueness of the solution with probability $1 - M/q \approx 1 - 2^{-128}$, and the secret key x can be easily found solving the system of modular equations

$$
\begin{cases}
s_1 k_1 - r_1 x = m_1 & \pmod{q} \\
s_2 k_2 - r_2 x = m_2 & \pmod{q}
\end{cases}
$$

together with equations (2) and (6) for $i = 1, \ldots, 2\alpha - 1$ and $j = 1, 2$.

The attack easily generalizes to generators involving any combination of truncation and concatenation operations.

Acknowledgments

The first author is supported in part by a 1996 Packard Foundation Fellowship in Science and Engineering, and NSF CAREER award CCR-9624439. The second and third authors are supported in part by DARPA contract DABT63-96-C-0018.

References

1. L. Babai. On Lovász' lattice reduction and the nearest lattice point problem. *Combinatorica*, 6(1):1–13, 1986.

2. M. Bellare and P. Rogaway. Random oracles are practical: A paradigm for designing efficient protocols. *Proceedings of the First Annual Conference on Computer and Communications Security*, ACM, 1993.

3. M. Blum and S. Micali. How to generate cryptographically strong sequences of pseudo-random bits. *SIAM J. Computing*, 13(4):850–863, November 1984.

4. Joan Boyar. Inferring sequences produced by pseudo-random number generators. *Journal of the ACM*, 36(1):129–141, January 1989.

5. A. M. Frieze, R. Kannan, and J. C. Lagarias. Linear congruential generators do not produce random sequences. In *Proc. 25th IEEE Symp. on Foundations of Comp. Science*, pages 480–484, Singer Island, 1984. IEEE.

6. Taher El Gamal. A public key cryptosystem and a signature scheme based on discrete logarithms. In G. R. Blakley and D. C. Chaum, editors, *Proc. CRYPTO 84*, pages 10–18. Springer, 1985. Lecture Notes in Computer Science No. 196.

7. O. Goldreich, S. Goldwasser, and S. Micali. How to construct random functions. In *Proc. 25th IEEE Symp. on Foundations of Comp. Science*, pages 464–479, Singer Island, 1984. IEEE.

8. S. Goldwasser and S. Micali. Probabilistic Encryption. *Journal of Computer and System Sciences* 28:270–299, April 1984.

9. J. Hastad and A. Shamir. The cryptographic security of truncated linearly related variables. In *Proc. 17th ACM Symp. on Theory of Computing*, pages 356–362, Providence, 1985. ACM.

10. R. Kannan. Minkowski's convex body theorem and integer programming. *Mathematics of operations research*, 12(3):415–440, 1987.

11. Donald E. Knuth. *Seminumerical Algorithms*, volume 2 of *The Art of Computer Programming*. Addison-Wesley, 1969. Second edition, 1981.

12. Donald E. Knuth. Deciphering a linear congruential encryption. *IEEE Transactions on Information Theory*, IT-31(1):49–52, January 1985.

13. H. Krawczyk. How to predict congruential generators. In G. Brassard, editor, *Proc. CRYPTO 89*, pages 138–153. Springer, 1990. Lecture Notes in Computer Science No. 435.

14. H.W. Lenstra. Integer programming with a fixed number of variables. *Mathematics of operations research*, 8(4):538–548, 1983.

15. National Institute of Standards and Technology (NIST). *FIPS Publication 180: Secure Hash Standard (SHS)*, May 11, 1993.

16. National Institute of Standards and Technology (NIST). *FIPS Publication 186: Digital Signature Standard*, May 19, 1994.

17. J. Plumstead (Boyar). Inferring a sequence generated by a linear congruence. In *Proc. 23rd IEEE Symp. on Foundations of Comp. Science*, pages 153–159, Chicago, 1982. IEEE.

18. Adi Shamir. The generation of cryptographically strong pseudo-random sequences. In Allen Gersho, editor, *Advances in Cryptology: A Report on CRYPTO 81*, pages 1–1. U.C. Santa Barbara Dept. of Elec. and Computer Eng., 1982. Tech Report 82-04.

19. J. Stern. Secret linear congruential generators are not cryptographically secure. In *Proc. 28th IEEE Symp. on Foundations of Comp. Science*, pages 421–426, Los Angeles, 1987. IEEE.

20. A. C. Yao. Theory and application of trapdoor functions. In *Proc. 23rd IEEE Symp. on Foundations of Comp. Science*, pages 80–91, Chicago, 1982. IEEE.

Unconditional Security Against Memory-Bounded Adversaries

Christian Cachin*   Ueli Maurer

Department of Computer Science
ETH Zürich
CH-8092 Zürich, Switzerland
cachin@acm.org   maurer@inf.ethz.ch

Abstract. We propose a private-key cryptosystem and a protocol for key agreement by public discussion that are unconditionally secure based on the sole assumption that an adversary's memory capacity is limited. No assumption about her computing power is made. The scenario assumes that a random bit string of length slightly larger than the adversary's memory capacity can be received by all parties. The random bit string can for instance be broadcast by a satellite or over an optical network, or transmitted over an insecure channel between the communicating parties. The proposed schemes require very high bandwidth but can nevertheless be practical.

1 Introduction

One of the most important properties of a cryptographic system is a proof of its security under reasonable and general assumptions. However, every design involves a trade-off between the strength of the security and further important qualities of a cryptosystem, such as efficiency and practicality.

The security of all currently used cryptosystems is based on the difficulty of an underlying computational problem, such as factoring large numbers or computing discrete logarithms in the case of many public-key systems. Security proofs for these systems show that the ability of the adversary to defeat the cryptosystem with significant probability contradicts the assumed difficulty of the problem [24]. Although the hardness of these problems is unquestioned at the moment, it can be dangerous to base the security of the global information economy on a very small number of mathematical problems. Recent advances in quantum computing show that precisely these two problems, factoring and discrete logarithm, could be solved efficiently if quantum computers could be built [27].

An alternative to proofs in the computational security model is offered by the stronger notion of information-theoretic or *unconditional* security where no limits on an adversary's computational power are assumed. The first information-theoretic definition of perfect secrecy by Shannon [26] led immediately to his famous impracticality theorem, which states, roughly, that the shared secret key in any perfectly secure cryptosystem must be at least as long as the plaintext

---

* Current address: MIT Laboratory for Computer Science, 545 Technology Square, Cambridge MA 02139, USA.

to be encrypted. Vernam's one-time pad is the prime example of a perfectly secure but impractical system. Unconditional security was therefore considered too expensive for a long time.

However, recent developments show how Shannon's model can be modified [16] to make practical provably secure cryptosystems possible. The first modification is to relax the requirement that perfect security means complete independence between the plaintext and the adversary's knowledge and to allow an arbitrarily small correlation. The second, crucial modification removes the assumption that the adversary receives exactly the same information as the legitimate users. The following two primitives are perhaps the most realistic mechanisms proposed so far for limiting the information available to the adversary.

Quantum Channel: Quantum cryptography was developed mainly by Bennett and Brassard during the 1980's [3]. It makes use of photons, i.e. polarized light pulses of very low intensity, that are transmitted over a fiber-optical channel. In the basic quantum key agreement protocol, this allows two parties to generate a secret key by communicating about the received values. The unconditional secrecy of the key is guaranteed by the uncertainty principle of quantum mechanics. Current implementations of quantum key distribution span distances of 20–30 kilometers.

Noisy Channel: In this model proposed by Maurer, the output of a random source is transmitted to the participants over partially independent noisy channels that insert errors with certain probabilities [19]. Two parties can then generate a secret key from their received values by public discussion. The secrecy of the key is based on the information differences between the channel outputs and on the assumption that no channel is completely error-free. This system is practical because it works also in the realistic case where the adversary receives the random source via a much better channel than the legitimate users. The power of a noisy channel was also demonstrated by Crépeau and Kilian who showed that unconditionally secure bit commitment and oblivious transfer can be based on this primitive [11,10].

In this paper, we show how to realize unconditionally secure encryption based on a third assumption: a limit on the memory size of the adversary. This means that an enemy can use unlimited computing power to compute any probabilistic function of some huge amount of public data, which is infeasible to store. As long as the function's output size does not exceed the number of available storage bits, we can prove that the proposed private-key system and public key agreement protocol are information-theoretically secure from this sole assumption.

The public data is the output of a random source that is broadcast at very high rate. The legitimate users Alice and Bob randomly select a small subset of the broadcast each and store these values. (How this selection is performed will be described below.) Because of the random selection process, the average fraction of the information of an adversary Eve about the selected subset is roughly the same as her fraction of information about the complete broadcast. By applying privacy amplification [2], Alice and Bob can then eliminate Eve's partial knowledge about the selected subset. (The random source does not have to be independent from the users, e.g. Alice could produce the random data herself and transmit it to Bob over a public channel.)

We describe how two different cryptographic tasks can be implemented using this mechanism, depending on how Alice and Bob select the random subset. First, if they share a short secret key initially that can be used to select identical subsets, the system realizes *private-key encryption*. Second, even if Alice and Bob do not share any secret information at the beginning, they can perform a *key agreement protocol* by public discussion: They select and store independently a random subset of the broadcast data. After some predetermined interval they publicly exchange the indices of their selected positions and determine the positions contained in both subsets. Privacy amplification is applied to the part of the broadcast they have in common.

Our model seems realistic because current communication and high-speed networking technologies allow broadcasting at rates of multiple gigabits per second. Storage systems that are hundreds of terabytes large, on the other hand, require a major investment by a potential adversary. Although this is within reach of government budgets, for example, the method is attractive for the following three reasons: First, the security can be based only on the assumption about the adversary's memory capacity. Second, storage costs scale linearly and can therefore be estimated accurately. Third, the system offers 'proactive' security in the sense that a future increase in storage capacity cannot break the secrecy of messages encrypted earlier.

A precursor of this system is the Rip van Winkle cipher proposed by Massey and Ingemarsson [17,16]. This private-key system is provably computationally secure but totally impractical because a legitimate receiver must wait even longer for receiving a message than it takes an adversary to decrypt it.

Related to our work is a paper by Maurer [18] that describes a system based on a large public randomizer which cannot be read entirely within feasible time. Maurer's paper contains also the idea of realizing provably secure encryption based only on assumptions about an enemy's available memory. Such a system for key agreement was described by Mitchell [20], but without security proof. Our analysis provides the first proof that unconditional security can be achieved against memory-bounded adversaries. (Recently, Aumann and Rabin [22] proved a conjecture of Maurer's paper [18] with the same effect.)

We borrow some methods from the work of Zuckerman and others on so-called extractors of uniform randomness from weak random sources [29]. Extractors are tools developed for running randomized algorithms with non-perfect randomness instead of uniform random bits. Nisan [21] presents a highly readable introduction to extractors and a survey of their applications.

The paper is organized as follows. After reviewing some information-theoretic concepts in Section 2, we introduce the building blocks of our system in Section 3. Our main result concerning Eve's information about the randomly selected subset is given in Section 4. Sections 5 and 6 describe how to realize private-key encryption and public key agreement, respectively. The paper concludes with a discussion of the underlying assumptions and future perspectives.

2 Preliminaries

We assume that the reader is familiar with the notion of entropy and the basic concepts of Shannon's information theory [9]. We repeat some fundamental definitions in this section and introduce the notation. All logarithms in this paper are to the base 2. The cardinality of a set S is denoted by $|S|$.

A random variable X induces a probability distribution P_X over an alphabet $\mathcal{X}$. Random variables are denoted by capital letters. If not stated otherwise, the alphabet of a random variable is denoted by the corresponding script letter. A sequence $X_1, \ldots, X_n$ of random variables with the same alphabet is denoted by X^n.

The *(Shannon) entropy* of a random variable X with probability distribution P_X and alphabet $\mathcal{X}$ is defined as

$$H(X) = -\sum_{x \in \mathcal{X}} P_X(x) \log P_X(x).$$

The *binary entropy function* is $h(p) = -p \log p - (1-p) \log(1-p)$. The *conditional entropy* of X conditioned on a random variable Y is

$$H(X|Y) = \sum_{y \in \mathcal{Y}} P_Y(y) H(X|Y = y)$$

where $H(X|Y = y)$ denotes the entropy of the conditional probability distribution $P_{X|Y=y}$. The *mutual information* of X and Y is the reduction of the uncertainty of X when Y is learned:

$$I(X;Y) = H(X) - H(X|Y).$$

The *variational distance* between two probability distributions P_X and P_Y over the same alphabet $\mathcal{X}$ is

$$\|P_X - P_Y\|_v = \max_{\mathcal{X}_0 \subseteq \mathcal{X}} \left| \sum_{x \in \mathcal{X}_0} P_X(x) - P_Y(x) \right| = \frac{1}{2} \sum_{x \in \mathcal{X}} \left| P_X(x) - P_Y(x) \right|.$$

$\|P_X - P_Y\|_v \leq \epsilon$ implies that X behaves like Y except with probability at most ϵ, i.e., any property of X is shared by Y with probability at least $1 - \epsilon$.

The *Rényi entropy of order* α of a random variable X with alphabet $\mathcal{X}$ is

$$H_\alpha(X) = \frac{1}{1-\alpha} \log \sum_{x \in \mathcal{X}} P_X(x)^\alpha$$

for $\alpha \geq 0$ and $\alpha \neq 1$ [23]. Because the limiting case of Rényi entropy for $\alpha \to 1$ is Shannon entropy, we can extend the definition to $H_1(X) = H(X)$. In the other limiting case $\alpha \to \infty$, we obtain the *min-entropy*, defined as

$$H_\infty(X) = -\log \max_{x \in \mathcal{X}} P_X(x).$$

For a fixed random variable X, Rényi entropy is a continuous positive decreasing function of α. For $0 < \alpha < \beta$, we have $H_\alpha(X) \geq H_\beta(X)$, with equality if and only if X is the uniform distribution over $\mathcal{X}$ or over a subset of $\mathcal{X}$. In particular, $\log |\mathcal{X}| \geq H_\alpha(X) \geq 0$ for $\alpha \geq 0$ and $H(X) \geq H_\alpha(X)$ for $\alpha > 1$.

3   Pairwise Independence and Entropy Smoothing

This section contains a short review of entropy smoothing with universal hashing. We start by repeating the construction of a sequence of pairwise independent random variables using universal hash functions.

Universal hash functions were introduced by Carter and Wegman [8,28] and have found many applications in theoretical computer science [15]. A *2-universal hash function* is a set $\mathcal{G}$ of functions $\mathcal{X} \to \mathcal{Y}$ if, for all distinct $x_1, x_2 \in \mathcal{X}$, there are at most $|\mathcal{G}|/|\mathcal{Y}|$ functions g in $\mathcal{G}$ such that $g(x_1) = g(x_2)$.

A *strongly 2-universal hash function* is a set $\mathcal{G}$ of functions $\mathcal{X} \to \mathcal{Y}$ if, for all distinct $x_1, x_2 \in \mathcal{X}$ and all (not necessarily distinct) $y_1, y_2 \in \mathcal{Y}$, exactly $|\mathcal{G}|/|\mathcal{Y}|^2$ functions from $\mathcal{G}$ take x_1 to y_1 and x_2 to y_2.

A strongly 2-universal hash function can be used to generate a sequence of pairwise independent random variables in the following way: Select $G \in \mathcal{G}$ uniformly at random and apply it to any fixed sequence $x_1, \ldots, x_l$ of distinct values in $\mathcal{X}$. Let $Y_j = G(x_j)$ for $j = 1, \ldots, l$. It can easily be verified that $Y_1, \ldots, Y_l$ are pairwise independent and uniformly distributed random variables over $\mathcal{Y}$. The advantage of this technique, compared to selecting n independent samples of Y, is that it requires only $2 \log |\mathcal{Y}|$ instead of $n \log |\mathcal{Y}|$ random bits.

An often-used example for a strongly 2-universal hash function from $GF(2^n)$ to $GF(2^m)$ is the set

$$\mathcal{G} = \left\{ g(x) = \mathrm{msb}_m(a_1 x + a_0) \mid a_0, a_1 \in GF(2^n) \right\}$$

where $\mathrm{msb}_m(x)$ denotes the m most significant bits of x and $a_1 x + a_0$ is computed in $GF(2^n)$. This construction has the nice property that when $\mathcal{G}$ is used to generate a sequence of pairwise independent random variables, all values in the sequence are distinct if and only if $a_1 \neq 0$. We will assume that $a_1 \neq 0$ whenever the pairwise independence construction is used and refer to the resulting distribution as "uniform and pairwise independent" although repeating values are excluded.

The strongly 2-universal family $\mathcal{G}$ is 2-universal even if a_0 is always set to 0. Thus, a member of the 2-universal family can be specified with only n bits.

2-universal hash functions are also the main technique to concentrate the randomness inherent in a probability distribution by a result known in different contexts as Entropy Smoothing Theorem, Leftover Hash Lemma [14], or Privacy Amplification Theorem [2].

In cryptography, privacy amplification is used to extract a short secret key from shared information about which an adversary has partial knowledge. Assume Alice and Bob share a random variable W, while an eavesdropper Eve knows a correlated random variable V that summarizes her knowledge about W. The details of the distribution P_{WV}, and thus of Eve's information V about W, are unknown to Alice and Bob, except that they assume a lower bound on the Rényi entropy of order 2 of $P_{W|V=v}$ for the particular value v that Eve observes.

Using a public channel, which is susceptible to eavesdropping but immune to tampering, Alice and Bob wish to agree on a function g such that Eve knows nearly nothing about $g(W)$. The following theorem by Bennett, Brassard, Crépeau, and Maurer [2] shows that if Alice and Bob choose g at random from a universal hash function $\mathcal{G} : \mathcal{W} \to \mathcal{Y}$ for suitable $\mathcal{Y}$, then Eve's information about $Y = g(W)$ is negligible.

Theorem 1 ([2]). *Let X be a random variable over the alphabet $\mathcal{X}$ with Rényi entropy $H_2(X)$, let G be the random variable corresponding to the random choice (with uniform distribution) of a member of a 2-universal hash function $\mathcal{G} : \mathcal{X} \to \mathcal{Y}$, and let $Y = G(X)$. Then*

$$H(Y|G) \geq \log |\mathcal{Y}| - \frac{2^{\log |\mathcal{Y}| - H_2(X)}}{\ln 2}. \tag{1}$$

To apply the theorem in the described scenario, replace P_X by the conditional probability distribution $P_{W|V=v}$. Cachin has recently extended the theorem to Rényi entropy of order α for any $\alpha > 1$ [6].

4 Extracting a Secret Key from a Randomly Selected Subset

We are now going to show how and why Alice and Bob can exploit the fact that an adversary Eve cannot store the complete output of a public random source that is broadcast to the participants. The security proof consists of three steps. In the first step, we use the fact that Eve's storage capacity is limited to establish a lower bound on the min-entropy of Eve about the broadcast bits. The second step shows that Eve's min-entropy about a randomly selected subset of the broadcast bits is large with high probability. In the third step, we apply privacy amplification to the selected subset to obtain the secret key.

Suppose the output of a uniformly distributed binary source R is broadcast over an error-free channel and can be received by all participants. The source can be independent from the participants or it can be operated by one of the legitimate users, e.g. Alice can generate R and transmit it over an authenticated public channel to Bob. More generally, any source that is trusted to output random bits and has a sufficient capacity can be used. The channel must have high capacity, which could be realized, for example, using satellite technology for digital TV broadcasting or all-optical networks. The channel is used n times in succession and the broadcast bits are denoted by $R^n = R_1, \ldots, R_n$. We assume that Eve has a total of $m < n$ storage bits available and therefore cannot record the complete broadcast, leaving her only with partial knowledge about R^n.

During the broadcast, Eve may compute an arbitrary function of R^n with unlimited computing power and can also use additional private random bits. We model the output of the function to be stored in her m bits of memory by the random variable Z with alphabet $\mathcal{Z}$, subject to $\log |\mathcal{Z}| \leq m$.

Because R^n is uniformly distributed, its Rényi entropy of any order $\alpha \geq 0$ and its Shannon entropy satisfy $H_\alpha(R^n) = H(R^n) = H_\infty(R^n) = n$. The following lemma shows that the min-entropy of R^n given Z, which corresponds to Eve's knowledge about R^n, is at least $n - m$ for all but a negligible fraction of the values of Z. More precisely, the lemma implies that for any $r > 0$, the particular value z that Z takes on satisfies $H_\infty(R^n|Z = z) \geq n - m - r$, except with probability at most 2^{-r}.

Lemma 2. *Let X be a random variable with alphabet $\mathcal{X}$, let Z be an arbitrary random variable with alphabet $\mathcal{Z}$, and let $r > 0$. Then with probability at least $1 - 2^{-r}$, Z takes on a value z for which*

$$H_\infty(X|Z = z) \geq H_\infty(X) - \log|\mathcal{Z}| - r.$$

Proof. Let $p_0 = 2^{-r}/|\mathcal{Z}|$. Thus, $\sum_{z:P_Z(z)<p_0} P_Z(z) < 2^{-r}$. It follows for all z with $P_Z(z) \geq p_0$

$$
\begin{aligned}
H_\infty(X|Z = z) &= -\log \max_{x \in \mathcal{X}} P_{X|Z=z}(x) \\
&= -\log \max_{x \in \mathcal{X}} \frac{P_X(x)P_{Z|X=x}(z)}{P_Z(z)} \\
&\geq -\log \max_{x \in \mathcal{X}} \frac{P_X(x)}{p_0} \\
&= H_\infty(X) - r - \log|\mathcal{Z}|
\end{aligned}
$$

which proves the lemma. $\qquad\square$

For the rest of this section, we denote Eve's knowledge of R^n, given the particular value $Z = z$ she observed, by the random variable $X^n = X_1, \ldots, X_n$ with alphabet $\mathcal{X}^n = \{0,1\}^n$. The distribution of X^n is arbitrary and only subject to $H_\infty(X^n) \geq n - m - r$ by Lemma 2.

The strategy of the legitimate users Alice and Bob is to select the values at l positions

$$\mathbf{S} = [S_1, \ldots, S_l] \quad \text{with} \quad S_1, \ldots, S_l \in \{1, \ldots, n\}$$

randomly from the broadcast symbols X^n. $\mathbf{S}$ is a vector-valued random variable taking on values $\mathbf{s} \in \{1, \ldots, n\}^l$ and the list of selected positions $X_{S_1}, \ldots, X_{S_l}$ is denoted by $X^{\mathbf{S}}$. Because this selection is performed with uniform distribution according to the pairwise independence construction of a sequence of l values from $\{1, \ldots, n\}$ as described in Section 3, the resulting $S_1, \ldots, S_l$ are all distinct and $\mathbf{S}$ can also be viewed as a set of l values. In addition, $\mathbf{S}$ can be determined efficiently from $2 \log n$ bits.

We assume that the value of $\mathbf{S}$ is known whenever the random variable $X^{\mathbf{S}}$ is used. In the private-key system described later, Eve is thus supposed to obtain $\mathbf{S}$ from an oracle *after* the public random string is broadcast.

How much does Eve know about the bits selected by Alice and Bob? Intuitively, one would expect that the fraction of bits in $X^{\mathbf{S}}$ known to Eve corresponds to the fraction of bits in X^n that Eve knows (here a bit is not to be understood as a binary digit, but in the information-theoretic sense). This is indeed the case, as was observed before by Zuckerman and others in the context of weak random sources [29,21]. It is easy to show that the fraction of Eve's Shannon information corresponds to the expected value [5, Theorem 5.10]. However, privacy amplification can only be applied if a lower bound on the Rényi entropy of order 2 of X^S is known, which follows from the stronger bound on the min-entropy by Lemma 3.

The cited proof for Shannon information works only because Shannon entropy has the intuitive property that side information can only reduce the average uncertainty. This is not the case for expected conditional Rényi entropy of order $\alpha > 1$ and is the main obstacle for extending the proof to Rényi entropy. However, the following stronger result by Zuckerman [29] shows that also the fraction of Eve's min-entropy in the selected positions is, with high probability, close to the corresponding fraction of the total min-entropy. Because the min-entropy of a random variable is a lower bound for its Rényi entropy for any $\alpha > 0$, the lemma is sufficient for applying privacy amplification to the selected subset.

Lemma 3. *Let X^n be a random variable with alphabet $\{0,1\}^n$ and min-entropy $H_\infty(X^n) \geq \delta n$ (where $\frac{1}{n} \leq \delta \leq 0.9453$), let $\mathbf{S} = [S_1, \ldots, S_l]$ be chosen pairwise independently as described in Section 3, let $\rho \in [0, \frac{1}{3}]$ be such that $h(\rho) + \rho \log \frac{1}{\delta} + \frac{1}{n} = \delta$, and let $\epsilon = \sqrt{4/(\rho l) + 2^{\rho n \log \delta}}$. Then, for every value $\mathbf{s}$ of $\mathbf{S}$ there exists a random variable $A^l(\mathbf{s})$ with alphabet $\{0,1\}^l$ and min-entropy $H_\infty(A^l(\mathbf{s})) \geq \rho l/2$ such that with probability at least $1 - \epsilon$ (over the choice of $\mathbf{S}$), $P_{X^{\mathbf{s}}}$ is ϵ-close to $P_{A^l(\mathbf{S})}$ in variational distance, i.e.*

$$\forall \mathbf{s} : \exists A^l(\mathbf{s}) : \mathrm{P}\big[\|P_{X^{\mathbf{s}}} - P_{A^l(\mathbf{S})}\|_v \leq \epsilon \big] \geq 1 - \epsilon.$$

Remark. For fixed, large n, the value of ρ resulting from the choice in the lemma increases monotonically with δ and for δ smaller than about 0.9453 there always exists a unique $\rho \in [0, \frac{1}{3}]$ satisfying $h(\rho) + \rho \log \frac{1}{\delta} = \delta$, as can be verified easily.

Proof. The statement of the lemma is slightly different from Zuckerman's asymptotic result [29, Lemma 9] with respect to ρ (that we use in place of α) and ϵ, but follows also from the original proof. We describe here only the differences that lead to our formulation of the lemma.

It is straightforward to verify that $\binom{n}{i-1} = \frac{i}{n-i+1}\binom{n}{i}$ and therefore $\binom{n}{i-1} < \frac{1}{2}\binom{n}{i}$ for $i \leq n/3$. This implies $\binom{n}{i-j} < 2^{-j}\binom{n}{i}$ for $i \leq n/3$ and $0 \leq j \leq i$, from which the bound

$$\sum_{i=0}^{k} \binom{n}{i} < \binom{n}{k} \sum_{i=0}^{k} 2^{-i} < 2\binom{n}{k} \tag{2}$$

for any $k \leq n/3$ follows immediately. The approximation of $\binom{n}{i}$ by the binary entropy function [9], $\frac{1}{n+1} 2^{nh(\frac{i}{n})} \leq \binom{n}{i} \leq 2^{nh(\frac{i}{n})}$, implies

$$2 \cdot 2^{nh(\rho)} \geq 2\binom{n}{\lfloor \rho n \rfloor} > \sum_{i=0}^{\lfloor \rho n \rfloor} \binom{n}{i},$$

where the second inequality follows from (2) for $\rho \le \frac{1}{3}$. Thus choosing ρ as described in the statement of the lemma guarantees that

$$2^{-\delta n} \cdot \sum_{i=0}^{\lfloor \rho n \rfloor} \binom{n}{i} < 2^{-\delta n} \cdot 2^{nh(\rho)+1} = 2^{-\rho n \log \frac{1}{3}}$$

as required in the proof of Lemma 12 in [29]. The choice of ϵ is the value resulting at the end of the proof of Lemma 10 in [29]. □

We are now ready to state the main result of this section. First, we summarize the scenario and the choice of the parameters.

Let R^n be a random n-bit string with uniform distribution that is broadcast to Alice and Bob who want to generate a secret key and to the adversary Eve who has a total of $m < n$ bits of memory available. Let the random variable Z denote Eve's knowledge about R^n, let $\varepsilon_1, \varepsilon_2 > 0$ be arbitrary error probabilities, and let $\Delta > 0$ be a parameter that denotes the amount of information that may leak to Eve. Let the parameters

1. $\delta = \min\{0.9453, \frac{1}{n}(n - m - \log\frac{1}{\varepsilon_1})\}$;
2. ρ such that $h(\rho) + \rho \log\frac{1}{3} + \frac{1}{n} = \delta$;
3. $l = \lfloor (\rho\varepsilon_2^2 - \rho 2^{-\rho n \log\frac{1}{3} - 2})^{-1} \rfloor$;
4. $r = \lfloor \log\Delta + \rho l/2 - 1 \rfloor$.

Alice and Bob select $\mathbf{S} = [S_1, \ldots, S_l]$ randomly from $\{1, \ldots, n\}$ with the pairwise independence construction as described in Section 3 and store the bits $R^{\mathbf{S}} = R_{S_1}, \ldots, R_{S_l}$. Then they select a function $G \in \mathcal{G}$ uniformly at random from a 2-universal hash function $\mathcal{G}$ from l-bit strings to r-bit strings and compute $K = G(R^{\mathbf{S}})$ as their secret key. The random experiment consists of the choices of R^n, Z, $\mathbf{S}$, and G. As mentioned before, the theorem is proved under the (weaker) assumption that $\mathbf{S}$ is known to Eve, although this may not even be the case.

Theorem 4. *In the described scenario, there exists a security event $\mathcal{E}$ that has probability at least $1 - \varepsilon_1 - \varepsilon_2$ such that Eve's information about K, given G, given her particular knowledge $Z = z$ about R^n, given $\mathbf{S} = \mathbf{s}$, and given $\mathcal{E}$, is at most Δ. Formally,*

$$\exists \mathcal{E} : P[\mathcal{E}] \ge 1 - \varepsilon_1 - \varepsilon_2 \quad and \quad I(K; G|Z = z, \mathbf{S} = \mathbf{s}, \mathcal{E}) \le \Delta.$$

Proof. Applying Lemma 2 with error probability ε_1 shows that

$$H_\infty(R^n|Z = z) \ge n - m - \log\frac{1}{\varepsilon_1},$$

leading to the value of δ. Lemma 3 shows that $\mathbf{S}$ takes on a value $\mathbf{s}$ such that there is a distribution $P_{A^l(\mathbf{s})}$ within $\varepsilon_2/2$ of $P_{R^\mathbf{s}|Z=z}$ with probability $1 - \varepsilon_2/2$. Privacy amplification can be applied because

$$H_2(A^l(\mathbf{s})) \ge H_\infty(A^l(\mathbf{s})) \ge \rho l/2.$$

The choice of r guarantees $H(K|G, Z = z, \mathbf{S} = \mathbf{s}) \geq r - \Delta$ by Theorem 1 because $2^{r - H_2(A^l(\mathbf{s}))}/\ln 2 \leq \Delta$.

Failure of the uniformity bound, which is equivalent to the event $\overline{\mathcal{E}}$, consists of the union of the following three events. First, the bound of Lemma 2 can fail with probability at most ε_1. Second, $A^l(\mathbf{s})$ may deviate from the random variable with distribution $P_{R^\bullet|Z=z}$ with probability at most $\varepsilon_2/2$ and third, an $\mathbf{s}$ such that the distance $\|P_{X^\bullet} - P_{A^l(\mathbf{s})}\|_v$ is outside of the allowed range occurs with probability at most $\varepsilon_2/2$ in Lemma 3. Applying the union bound, we see that $P[\mathcal{E}] \geq 1 - \varepsilon_1 - \varepsilon_2$ and $H(K|G, Z = z, \mathbf{S} = \mathbf{s}, \mathcal{E}) \geq r - \Delta$. The theorem now follows from the definition of mutual information upon noting that $H(K|Z = z, \mathbf{S} = \mathbf{s}, \mathcal{E}) \leq r$. $\square$

In a realistic cryptographic application of Theorem 4, the choice of the parameters is somewhat simplified because m is typically very large and because choosing a reasonable safety margin implies $n \gg m$. In this case, the parameters are $\delta = 0.9453$ and $\rho = \frac{1}{3}$, and l depends almost only on ε_2 and is close to $3/\varepsilon_2^2$. Thus, the storage required by Alice and Bob and the size of the resulting secret key are inverse proportional to the square of the desired error probability.

5   A Private-Key System

We now describe an example of a practical private-key encryption system that offers virtually the same security as the one-time pad. Assume Alice and Bob share a secret key K_0 and have both access to the broadcast public random source R^n. In addition, they are connected by an authenticated public channel on which Eve can read but not modify messages. For the pairwise independent selection of $\mathbf{S}$, the size of K_0 must be $2 \log n$ bit. However, no initial communication between the partners is needed because the interval to observe R and other parameters like l, r, ε_1, and ε_2 are fixed. The authenticated public channel is needed to exchange the description of the hash function G, which is used to extract the secret value K from $R^\mathbf{S}$.

In a straightforward implementation, Alice and Bob need $l(\log n + 1)$ bit of storage to hold $\mathbf{S} = [S_1, \ldots, S_l]$ and the values of $R^\mathbf{S}$. Because R^n is broadcast at high speed, the positions to observe must be precomputed and be recalled in increasing order. The legitimate users must only be able to synchronize on the broadcast channel and to read one bit from time to time. An adversary, however, needs equipment with high bandwidth from the channel interface through to mass storage in order to store a substantial part of R^n.

The following considerations demonstrate that this system is on the verge of being practical. The broadcast channel could be realized by a satellite. Typically, current communications satellites have a capacity of 1–10 Gbit/s [25]. Commercial satellite communications services offer broadcast data rates up to 0.8 Gbit/s at consumer electronics prices. Far more capacity is offered by fiber optical networks [12]. The test bed of the All-Optical Networking Consortium, for example, has a capacity of 1 Tbit/s and has been demonstrated at 130 Gbit/s (which was only limited by the number of sources available). On the other hand, tape libraries with capacities in the PByte range (1 PetaByte $= 2^{50}$ or about 10^{15} bytes) are a major investment [13].

As an example for the private-key system, consider a 16 Gbit/s satellite channel that is used for one day, making $n = 1.5 \times 10^{15}$. The size of the secret key K_0 is only 102 bit. Assume the adversary can store 100 TByte ($m = 8.8 \times 10^{14}$). Using $\Delta = 10^{-20}$ and error probabilities $\varepsilon_1 = 10^{-20}$ and $\varepsilon_2 = 10^{-4}$, we see that $\delta = 0.41$, $\rho = 0.060$, $l = 1.7 \times 10^9$, and $r = 5.0 \times 10^7$, that is, about 6.0 MByte of virtually secret information K can be extracted. The legitimate users need only 10 GByte of storage each to hold the indices and the selected bits. For privacy amplification, one of them has to announce the randomly chosen universal hash function, which takes about 197 MByte. An adversary knows not more than 10^{-20} bit of K except with probability about 10^{-4}. K can be used directly for encryption with a one-time pad, for example.

The memory requirements of Alice and Bob can be reduced if fast computation enables an implicit representation of the indices $\mathbf{S}$. This seems feasible because only simple operations are needed for the pairwise independence selection method. Assuming for example that the l values can be computed in one minute, only the positions to be observed within the next minute must be stored. With the figures of the preceding example, this reduces the storage requirements to only 7 MByte for the current block of indices plus a total of 197 MByte for $R^{\mathbf{S}}$. If the computation of the indices takes longer, observation of the random broadcast could also be halted until the indices are available.

The system can be used repeatedly with only one initial key K_0, because a small part of the secret key K obtained in the first round can be used safely as the secret key for the subsequent round and so forth. In addition, some part of K can be employed to relax the authenticity requirement for the public channel using unconditionally secure message authentication techniques [28].

6 Key Agreement by Public Discussion

Our methods can also be used to establish a secret key between two users not sharing secret information who have access to the random broadcast and are linked by a public channel. Communication on the public channel is assumed to be authenticated, i.e. the adversary can read but not modify messages. This system offers public key agreement with virtually the same security as the one-time pad under the sole assumption that the adversary's memory capacity is limited. (The public communication channel is different from the public broadcast channel whose only purpose is to distribute a large number of random bits.)

To agree on a secret key, Alice and Bob independently select and store a subset of the broadcast random bits R^n. After a predetermined amount of time, they announce the chosen set of positions on the public channel. The secret key can then be extracted from the values of R^n at the common positions using privacy amplification. To keep the communication and storage requirements for Alice and Bob at a reasonable level, it is crucial that they use a memory-efficient description of the index set. Fortunately, the pairwise independent selection method achieves this.

Both Alice and Bob select a sequence of q uniform and pairwise independent indices $T_1, \ldots, T_q$ and $U_1, \ldots, U_q$, respectively, from $\{1, \ldots, n\}$ as described in Section 3. (The values of q and the other parameters $n, l, r, \varepsilon_1, \varepsilon_2$ are fixed and also known to Eve.) Alice stores the values of R^n at the indices in $\mathbf{T} = [T_1, \ldots, T_q]$, denoted by $R^{\mathbf{T}}$, and Bob stores $R^{\mathbf{U}}$ for his indices $\mathbf{U} = [U_1, \ldots, U_q]$. We assume that they use a memory-efficient, implicit representation of the index set as described earlier for the private-key system, with on-line recomputation of the indices when necessary. In this way, Alice and Bob need approximately $\log q$ bits of memory each.

Because of the pairwise independent selection, both index sets can be determined from $2 \log n$ bits each. The descriptions of $\mathbf{T}$ and $\mathbf{U}$ exchanged on the public channel are therefore short. In order to apply Theorem 4 to the set $\{S_1, \ldots, S_l\} = \{T_1, \ldots, T_q\} \cap \{U_1, \ldots, U_q\}$ of common positions, we need the following lemma to make sure that also $S_1, \ldots, S_l$ have a uniform and pairwise independent distribution. It is easy to see that the expected number of common indices is $l = q^2/n$.

Lemma 5. *Let $T_1, \ldots, T_q$ and $U_1, \ldots, U_q$ be independent sequences of uniform and pairwise independent random variables, respectively, with alphabet $\{1, \ldots, n\}$ and distribution as described in Section 3, and let $S_1, \ldots, S_q$ be the sequence $T_1, \ldots, T_q$ restricted to those values occurring in $U_1, \ldots, U_q$, i.e. $S_j = T_j$ if there is an index h such that $U_h = T_j$ and $S_j = \omega$ otherwise. Then, the sequence $S_1, \ldots, S_q$ restricted to those positions different from ω is pairwise independent.*

Proof. Because the pairwise independence construction of Section 3 is used, no values in the $\mathbf{U}$ and $\mathbf{T}$ sequences are repeated. This implies

$$P[T_i = x_1 \wedge T_j = x_2] = \frac{1}{n(n-1)}$$

for all $i, j \in \{1, \ldots, q\}$ and all $x_1, x_2 \in \{1, \ldots, n\}$ with $x_1 \neq x_2$. The sequence $S_1, \ldots, S_l$ satisfies

$$P[S_i = x_1 \wedge S_j = x_2 | S_i \neq \omega \wedge S_j \neq \omega] = \frac{P[S_i = x_1 \wedge S_j = x_2]}{P[S_i \neq \omega \wedge S_j \neq \omega]} \quad (3)$$

for any $i, j \in \{1, \ldots, q\}$ and $x_1, x_2 \in \{1, \ldots, n\}$. Considering only those positions of the sequence $S_1, \ldots, S_q$ with values different from ω, we see that for any $i, j \in \{1, \ldots, q\}$ and all $x_1, x_2 \in \{1, \ldots, n\}$ such that $x_1 \neq x_2$ and $S_i \neq \omega$ and $S_j \neq \omega$,

$$P[S_i = x_1 \wedge S_j = x_2] = P[T_i = x_1 \wedge \exists h_1 : U_{h_1} = x_1 \wedge T_j = x_2 \wedge \exists h_2 : U_{h_2} = x_2]$$
$$= P[T_i = x_1 \wedge T_j = x_2] \cdot P[\exists h_1, h_2 : U_{h_1} = x_1 \wedge U_{h_2} = x_2]$$
$$= \frac{1}{n(n-1)} \cdot \frac{q(q-1)}{n(n-1)}.$$

Furthermore, for all $i, j \in \{1, \ldots, q\}$, we have

$$P[S_i \neq \omega \wedge S_j \neq \omega] = P[\exists h_1, h_2 : U_{h_1} = T_i \wedge U_{h_2} = T_j] = \frac{q(q-1)}{n(n-1)}$$

because every pair of distinct x_1, x_2 occurs with the same probability in the sequence $U_1, \ldots, U_q$. Thus, the probability in (3) is equal to $\frac{1}{n(n-1)}$ for any $i, j \in \{1, \ldots, q\}$ and all $x_1 \neq x_2$, and the lemma follows. □

To illustrate a concrete example of the system, assume Alice and Bob both have access to a 40 Gbit/s broadcast channel. We need more network capacity for public key agreement than for private-key encryption to achieve a similar error probability. The channel is used for 2×10^5 seconds (about two days), thus $n = 8.6 \times 10^{15}$. Eve is allowed to store 1/2 PByte or $m = 4.5 \times 10^{15}$ bit. With $\Delta = 10^{-20}$ and error probabilities $\varepsilon_1 = 10^{-20}$ and $\varepsilon_2 = 10^{-3}$, the parameters are $\delta = 0.476$, $\rho = 0.077$, $l = 1.3 \times 10^7$, and $r = 5.0 \times 10^5$. In order to have l common indices on the average, Alice and Bob must store $q = \sqrt{ln} = 3.3 \times 10^{11}$ bit or about 39 GByte each (assuming the index sequences $\mathbf{T}$ and $\mathbf{U}$ are represented implicitly). The public communication between Alice and Bob consists of $2 \log n = 106$ bit in each direction for the selected indices plus 1.5 MByte in one direction for privacy amplification. Except with probability about 10^{-3}, Eve knows less than 10^{-20} bit about the 61 KByte secret key that Alice and Bob obtain.

Because l is on the order of the inverse squared error probability ε_2, the probabilities in the example are relatively large to keep the storage requirements of Alice and Bob at a reasonable level. Generating a shorter key does not help to reduce the storage space, which depends primarily on ε_2. It is an interesting open question whether Lemma 3 can be improved in order to reduce the influence on the error probability.

The large size of the hash function that has to be communicated for privacy amplification can be reduced by using "almost universal" hash functions based on almost k-wise independent random variables that can be constructed efficiently [1]. Such functions $g : \mathcal{X} \to \mathcal{Y}$ can be described with about $5 \log |\mathcal{Y}|$ instead of $\log |\mathcal{X}|$ bits.

7 Discussion

Our results show that unconditional security can be based on assumptions about the adversary's available memory. In essence, such a system exploits the capacity gap between fast communication and mass storage technology. We discuss a few implications of this fact.

First of all, generating random bits at a sufficiently high rate may be more expensive than merely transmitting them. However, a large investment in a random source can be amortized by the potentially high number of participants that can use the source simultaneously.

A drawback of our system is that the security margin is linear in the sense that memory costs are directly proportional to the offered storage capacity, at least up to technological advances. In most computationally secure encryption systems, the complexity of a brute-force attack grows exponentially in the length of the keys.

Our system is provably secure taking into account the current storage capacity of an adversary because the only possible attack is to store the broadcast data when it is sent. In contrast, most computationally secure systems can be broken retroactively, once better algorithms are discovered or faster processing becomes possible.

We have used the broadcast channel as an error-free black-box communication primitive in our system, although the legitimate users do not need its full functionality: They need not receive the complete broadcast, but only a small part of it. It is conceivable that a receiving device could be much simpler and less expensive if it can only synchronize and read a small, but arbitrary part of the traffic. Such receivers could also allow for a greater capacity of the channel.

The described protocols offer no resilience to errors on the broadcast channel. To take into account such errors, Alice and Bob can perform information reconciliation [4] on the selected subset. Methods for bounding the effect of this additional information provided to Eve are known [7].

The system rests on the gap between two technologies—fast communication and mass storage. Impressive future developments can be expected in both fields. We only mention the big potential of all-optical networks on one side and the recent developments in holographic and molecular storage on the other side.

Acknowledgment

It is a pleasure to thank Jan Camenisch, Markus Stadler, and Stefan Wolf for interesting discussions on this subject and the anonymous referees for helpful remarks.

References

1. N. Alon, O. Goldreich, J. Håstad, and R. Peralta, "Simple constructions of almost k-wise independent random variables," *Random Structures and Algorithms*, vol. 3, no. 3, pp. 289–304, 1992. Preliminary version presented at 31st FOCS (1990).
2. C. H. Bennett, G. Brassard, C. Crépeau, and U. M. Maurer, "Generalized privacy amplification," *IEEE Transactions on Information Theory*, vol. 41, pp. 1915–1923, Nov. 1995.
3. G. Brassard and C. Crépeau, "25 years of quantum cryptography," *SIGACT News*, vol. 27, no. 3, pp. 13–24, 1996.
4. G. Brassard and L. Salvail, "Secret-key reconciliation by public discussion," in *Advances in Cryptology — EUROCRYPT '93* (T. Helleseth, ed.), vol. 765 of *Lecture Notes in Computer Science*, pp. 410–423, Springer-Verlag, 1994.
5. C. Cachin, *Entropy Measures and Unconditional Security in Cryptography*. Ph.D. dissertation No. 12187, ETH Zürich, 1997.
6. C. Cachin, "Smooth entropy and Rényi entropy," in *Advances in Cryptology — EUROCRYPT '97* (W. Fumy, ed.), vol. 1233 of *Lecture Notes in Computer Science*, pp. 193–208, Springer-Verlag, 1997.
7. C. Cachin and U. Maurer, "Linking information reconciliation and privacy amplification," *Journal of Cryptology*, vol. 10, no. 2, pp. 97–110, 1997.
8. J. L. Carter and M. N. Wegman, "Universal classes of hash functions," *Journal of Computer and System Sciences*, vol. 18, pp. 143–154, 1979.

306

9. T. M. Cover and J. A. Thomas, *Elements of Information Theory*. Wiley, 1991.
10. C. Crépeau, "Efficient cryptographic protocols based on noisy channels," in *Advances in Cryptology — EUROCRYPT '97* (W. Fumy, ed.), vol. 1233 of *Lecture Notes in Computer Science*, pp. 306–317, Springer-Verlag, 1997.
11. C. Crépeau and J. Kilian, "Achieving oblivious transfer using weakened security assumptions," in *Proc. 29th IEEE Symposium on Foundations of Computer Science (FOCS)*, 1989.
12. R. Cruz, G. Hill, A. Kellner, R. Ramaswami, G. Sasaki, and Y. Yamabashi, Eds., "Special issue on optical networks," *IEEE Journal on Selected Areas in Communications*, vol. 14, pp. 761–1052, June 1996.
13. *Proc. 14th IEEE Symposium on Mass Storage Systems*, IEEE Computer Society Press, 1995.
14. M. Luby, *Pseudorandomness and Cryptographic Applications*. Princeton University Press, 1996.
15. M. Luby and A. Wigderson, "Pairwise independence and derandomization," Tech. Rep. 95-035, International Computer Science Institute (ICSI), Berkeley, 1995.
16. J. L. Massey, "Contemporary cryptography: An introduction," in *Contemporary Cryptology: The Science of Information Integrity* (G. J. Simmons, ed.), ch. 1, pp. 1–39, IEEE Press, 1991.
17. J. L. Massey and I. Ingemarsson, "The Rip van Winkle cipher: A simple and provably computationally secure cipher with a finite key," in *Proc. 1985 IEEE International Symposium on Information Theory*, p. 146, 1985.
18. U. M. Maurer, "Conditionally-perfect secrecy and a provably-secure randomized cipher," *Journal of Cryptology*, vol. 5, pp. 53–66, 1992.
19. U. M. Maurer, "Secret key agreement by public discussion from common information," *IEEE Transactions on Information Theory*, vol. 39, pp. 733–742, 1993.
20. C. J. Mitchell, "A storage complexity based analogue of Maurer key establishment using public channels," in *Cryptography and Coding: 5th IMA Conference, Cirencester, UK* (C. Boyd, ed.), vol. 1025 of *Lecture Notes in Computer Science*, pp. 84–93, Springer, 1995.
21. N. Nisan, "Extracting randomness: How and why — a survey," in *Proc. 11th Annual IEEE Conference on Computational Complexity*, 1996.
22. M. Rabin. Personal Communication, 1997.
23. A. Rényi, "On measures of entropy and information," in *Proc. 4th Berkeley Symposium on Mathematical Statistics and Probability*, vol. 1, (Berkeley), pp. 547–561, Univ. of Calif. Press, 1961.
24. R. L. Rivest, "Cryptography," in *Handbook of Theoretical Computer Science* (J. van Leeuwen, ed.), ch. 13, pp. 717–755, Elsevier, 1990.
25. L. P. Seidman, "Satellites for wideband access," *IEEE Communications Magazine*, pp. 108–111, Oct. 1996.
26. C. E. Shannon, "Communication theory of secrecy systems," *Bell System Technical Journal*, vol. 28, pp. 656–715, Oct. 1949.
27. P. W. Shor, "Algorithms for quantum computation: Discrete log and factoring," in *Proc. 35th IEEE Symposium on Foundations of Computer Science (FOCS)*, pp. 124–134, 1994.
28. M. N. Wegman and J. L. Carter, "New hash functions and their use in authentication and set equality," *Journal of Computer and System Sciences*, vol. 22, pp. 265–279, 1981.
29. D. Zuckerman, "Simulating BPP using a general weak random source," *Algorithmica*, vol. 16, pp. 367–391, 1996. Preliminary version presented at 32nd FOCS (1991).

Privacy Amplification
Secure Against Active Adversaries

Ueli Maurer   Stefan Wolf

Department of Computer Science
Swiss Federal Institute of Technology (ETH Zürich)
CH-8092 Zürich, Switzerland
E-mail addresses: {maurer,wolf}@inf.ethz.ch

Abstract. Privacy amplification allows two parties Alice and Bob knowing a partially secret string S to extract, by communication over a public channel, a shorter, highly secret string S'. Bennett, Brassard, Crépeau, and Maurer showed that the length of S' can be almost equal to the conditional Rényi entropy of S given an opponent Eve's knowledge. All previous results on privacy amplification assumed that Eve has access to the public channel but is passive or, equivalently, that messages inserted by Eve can be detected by Alice and Bob. In this paper we consider privacy amplification secure even against active opponents. First it is analyzed under what conditions information-theoretically secure authentication is possible even though the common key is only partially secret. This result is used to prove that privacy amplification can be secure against an active opponent and that the size of S' can be almost equal to Eve's min-entropy about S minus $2n/3$ if S is an n-bit string. Moreover, it is shown that for sufficiently large n privacy amplification is possible when Eve's min-entropy about S exceeds only $n/2$ rather than $2n/3$.

Keywords: Privacy amplification, Secret-key agreement, Unconditional secrecy, Authentication codes, Information theory, Extractors.

1 Introduction and Preliminaries

Privacy amplification introduced by Bennett *et. al.* [2] is a technique for transforming a string that is only partially secret into a highly secret (but generally shorter) string. More precisely, two parties Alice and Bob who share a string S about which an opponent Eve has partial information agree, by communication over an insecure channel, on a string S' such that Eve's information about S' is negligible, i.e., such that $H(S'|U = u) \geq \log|S'| - \varepsilon$ holds with very high probability for some small $\varepsilon > 0$, where the random variable U summarizes Eve's complete knowledge about S', and where u is the particular value known to Eve. (All the logarithms in this paper are to the base 2, unless otherwise stated.) Privacy amplification is an important sub-protocol in many information-theoretic protocols such as protocols in quantum cryptography and secret-key agreement by public discussion [8].

Before we formalize the main problem considered in this paper, we give some definitions and state previous results on privacy amplification.

1.1 Entropy Measures

We recall the definitions of some entropy measures we need in this paper. We assume that the reader is familiar with the basic information-theoretic concepts. For a good introduction, we refer to [4]. Let R be a discrete random variable with range $\mathcal{R}$. Then the *(Shannon) entropy* $H(R)$ is defined as

$$H(R) := - \sum_{r \in \mathcal{R}} P_R(r) \cdot \log(P_R(r)) .$$

The *Rényi entropy* $H_2(R)$ is defined as

$$H_2(R) := - \log \left(\sum_{r \in \mathcal{R}} P_R^2(r) \right) .$$

Finally, the *min-entropy* $H_\infty(R)$ is

$$H_\infty(R) := - \log \max_{r \in \mathcal{R}} (P_R(r)) .$$

It is not difficult to see that for any random variable R the entropy measures H, H_2, and H_∞ satisfy

$$H(R) \geq H_2(R) \geq H_\infty(R) \geq H_2(R)/2 .$$

Equality of the first three expressions holds if and only if R is uniformly distributed over some set, in which case this value is the logarithm of the cardinality of this set.

1.2 Universal and Strongly Universal Hashing

In the technique presented in this paper, hashing is used for two different purposes: universal hashing for privacy amplification and strongly universal hashing for authentication.

Definition 1. A class $\mathcal{F}$ of functions $\mathcal{A} \to \mathcal{B}$ is called *universal$_2$* (or simply *universal*) if, for any x_1, x_2 in $\mathcal{A}$ with $x_1 \neq x_2$, the probability that $f(x_1) = f(x_2)$ is at most $1/|\mathcal{B}|$ when f is chosen from $\mathcal{F}$ according to the uniform distribution.

The following is a well-known example of such a class of hash functions $\{0,1\}^n \to \{0,1\}^r$ containing 2^n distinct functions. Let $b \in GF(2^n)$, and interpret $x \in \mathcal{A} = \{0,1\}^n$ also as an element of $GF(2^n)$. Consider the function f_b assigning to the argument x the first r bits of the element $b \cdot x$ of $GF(2^n)$. The set of these functions f_b for $b \in GF(2^n)$ is a universal class of functions for $1 \leq r \leq n$.

Definition 2. Let $\varepsilon > 0$. A class $\mathcal{H}$ of (hash) functions $\mathcal{A} \to \mathcal{B}$ is called ε-*almost-strongly-universal$_2$* (or ε-ASU$_2$ for short) if the following two conditions are satisfied:

1. For every $a \in \mathcal{A}$ and $b \in \mathcal{B}$, the number of functions $h \in \mathcal{H}$ with $h(a) = b$ is $|\mathcal{H}|/|\mathcal{B}|$.

2. For every distinct $a_1, a_2 \in \mathcal{A}$ and for every $b_1, b_2 \in \mathcal{B}$, the number of hash functions $h \in \mathcal{H}$ for which both $h(a_1) = b_1$ and $h(a_2) = b_2$ hold is at most $\varepsilon \cdot |\mathcal{H}|/|\mathcal{B}|$.

An $(1/|\mathcal{B}|)$-ASU$_2$ class is also called *strongly-universal*$_2$ (or SU$_2$).

Some constructions of ε-ASU$_2$ classes are described in [12], and lower bounds on the size of such classes are proved. An SU$_2$ class of functions mapping n-bit strings to n-bit strings can be constructed similarly to the universal class described above: the class $\mathcal{H} = \{h_{ab} : (a,b) \in (GF(2^n))^2\}$, where $h_{ab}(x) := a \cdot x + b$, is an SU$_2$ class of hash functions $\{0,1\}^n \rightarrow \{0,1\}^n$ with 2^{2n} elements. It is shown in [12] that ε-ASU$_2$ classes can be obtained which are close to strongly-universal, but substantially smaller.

1.3 Privacy Amplification by Authenticated Public Discussion

Bennett *et. al.* [1] analyzed the privacy amplification technique of [2] under the assumption that the two parties Alice and Bob are connected by an authentic (but otherwise insecure) channel, or equivalently, that the opponent is not able to insert or modify messages without being detected. The idea of this technique is to take a hash value of the string S as the highly secret key. More precisely, Alice chooses a hash function h at random from a universal class and sends this function to Bob. Then they both compute $S' := h(S)$.

It was shown that the amount of almost secret key that can be extracted is at least equal to the conditional Rényi entropy H_2 of S, given Eve's knowledge $U = u$. This fact is an immediate consequence of the following result of [1] which states that if a random variable X is used as the argument of universal hashing, where the output Y is an r-bit string, and r is equal to $H_2(X)$ minus a security parameter, then the resulting string Y has almost maximal Shannon entropy r, given the hash function (which is chosen uniformly from the universal class).

Theorem 3. [1] *Let X be a random variable with probability distribution P_X and Rényi entropy $H_2(X)$, and let G be the random variable corresponding to the random choice (with uniform distribution) of a member of a universal class of hash functions mapping $\mathcal{X}$ to r-bit strings, and let $Y = G(X)$. Then*

$$r \geq H(Y|G) \geq H_2(Y|G) \geq r - \frac{2^{r-H_2(X)}}{\ln 2} .$$

Of course the theorem also holds when all the probabilities are conditioned on a particular event (e.g., $U = u$).

1.4 Privacy Amplification by NOT Authenticated Public Discussion

In this paper we consider the generalized problem of privacy amplification when dropping the condition that the channel connecting the two parties Alice and Bob be authentic, i.e., privacy amplification secure even against active adversaries who are able to insert or modify messages.

Privacy amplification is often used as the final phase of unconditional secret-key agreement. In [6], it was investigated under what conditions secret-key agreement by not authenticated public discussion is possible when the parties Alice, Bob, and Eve have access to random variables X, Y, and Z, respectively (the "initialization phase"). Several impossibility results were shown, whereas a positive result was derived in [6] only for the special case where the information that the parties obtain consists of many independent repetitions of a random experiment. Privacy amplification, which was not treated in [6], corresponds to the situation where $X = Y$, and where the random experiment is not repeated.

We make precise what we mean by a protocol for privacy amplification by communication over a non-authentic insecure channel. Assume that two parties Alice and Bob both know a random variable S, for example an n-bit string, and that the adversary Eve has some information about S. Let again the random variable U summarize Eve's entire information about the random variable S. In the following, all the results are stated for some particular value $u \in \mathcal{U}$ (where $\mathcal{U}$ is the range of the random variable U), i.e., for a fixed event $U = u$, and hence all the probabilities are conditioned on $U = u$. The type of the opponent's information about S is not necessarily precisely specified, i.e., $P_{S|U=u}$ is not assumed to be known. However, the amount of information is limited in some way, for example in terms of the conditional min-entropy.

Formally, a protocol for privacy amplification consists of two phases. During the first phase (the communication phase), Alice and Bob exchange messages $C_1, C_2, \ldots$ over some channel (where Alice sends the messages $C_1, C_3, \ldots$, and Bob sends $C_2, C_4, \ldots$). Each of these messages can depend on the sender's knowledge when sending the message and some random bits. In the second phase, both parties decide whether they accept or reject the outcome of the protocol. In case of acceptance, Alice and Bob compute strings S'_A and S'_B, respectively. (Note that it is not required that Alice and Bob are synchronized in the sense that they both either reject or accept. This would be impossible to achieve in the presence of an active adversary, who could for instance delete all messages from Alice to Bob after Alice has accepted.) Definition 4 defines security of such a protocol.

Definition 4. A protocol is called an $(n, l, n', \varepsilon, \delta)$-*protocol for privacy amplification over an insecure and non-authentic channel* if it is a protocol for privacy amplification with the following properties. If there exists a random variable S with $|\mathcal{S}| \leq 2^n$ (i.e., we can assume that $\mathcal{S} \subset \{0,1\}^n$) that is known to Alice and Bob, and such that given Eve's entire knowledge $U = u$ about S, the conditional min-entropy of S is at least l, i.e.,

$$H_\infty(S|U = u) \geq l ,$$

then the protocol satisfies the following conditions. In the case of a *passive* (only wire-tapping) adversary, Alice and Bob always accept at the end of the protocol and obtain a common n'-bit string S' ($= S'_A = S'_B$) such that Eve's knowledge about S' is virtually 0 or, more precisely,

$$H(S'|SC) = 0 ,$$

and

$$H(S'|C, U = u) \geq n' - \varepsilon , \tag{1}$$

where C summarizes the entire communication $(C_1, C_2, \ldots)$ between Alice and Bob. In the case of an *active* adversary, with probability at least $1 - \delta$ one of the following conditions must be satisfied: either the adversary's presence is detected by at least one of the two parties (who hence rejects), or Alice and Bob both accept and successfully agree on a common string S' ($= S'_A = S'_B$) satisfying (1).

This definition can be generalized to different ways of limiting Eve's knowledge about S, for example in terms of the Rényi entropy instead of the min-entropy.

1.5 Outline

The rest of this paper is organized as follows. In Section 2 we investigate the general problem of information-theoretically secure message authentication under the (weakened) condition that two parties share a partially (rather than completely) secret key. In Section 3 we show a first result concerning privacy amplification. It states that privacy amplification (by communication over a non-authentic channel) is possible if Eve's min-entropy about S exceeds two thirds of the length n of the string, and the maximal length of the generated highly secret string is roughly $H_\infty(S|U = u) - 2n/3$. In Section 4 it is demonstrated that this result is not optimal: it is sufficient that Eve's min-entropy about S is greater than *half* of the length of the string (where the length of the extracted highly secret string is a constant fraction of $H_\infty(S|U = u) - n/2$) if the string is sufficiently long. Section 5 provides evidence that some of the results of Sections 2, 3, and 4 are optimal, and Section 6 states some open problems.

2 Unconditionally-Secure Authentication with a Partially Secret Key

All previous results on unconditionally-secure authentication require a key that is completely secret, i.e., the opponent's a priori probability distribution of the key is uniform. In this section we consider authentication where the opponent is allowed to have some partial information about the key.

There exists a variety of constructive results as well as impossibility results on information-theoretically secure authentication (see for example [11], [7], or [12]). The following two types of attacks are possible. In an *impersonation attack*, the opponent tries to generate a (correctly authenticated) message, and in a *substitution attack*, the adversary observes a correctly authenticated message and tries to replace it by a different correctly authenticated message. The success probabilities are denoted by p_{imp} and p_{sub}, respectively. (General lower bounds on these probabilities are given in [7].)

One possibility for realizing information-theoretically secure authentication is by using strongly-universal (or almost-strongly-universal) classes of hash functions (see for example [12]). The secret key then determines a hash function of the class, and the message is authenticated by appending its hash value. The authentication code corresponding to an ε-ASU$_2$ class of hash function satisfies

$$p_{imp} = 1/|\mathcal{B}| \qquad \text{and} \qquad p_{sub} \leq \varepsilon .$$

There are also different ways to realize authentication codes than with strongly universal hashing. One example is given in [5], where a construction is described with a smaller amount of secret key, but which requires more communication.

Let us now investigate the general scenario in which the key is not entirely secret, i.e., where the opponent Eve has a certain amount of information about the key. We first prove a bound on the information that is gained by Eve when observing a correctly authenticated message. The following lemma states that the min-entropy of the key, given Eve's information $U = u$, decreases by more than the length of the authenticator only with exponentially small probability. (A related result for different entropy measures is proved in [3].) For simplicity, the condition $U = u$ is omitted in the lemma and the proof. Of course the analogous result holds also when all the probabilities are conditioned on $U = u$.

Lemma 5. *Let S, X, and Y be arbitrary discrete random variables (with ranges $\mathcal{S}$, $\mathcal{X}$, and $\mathcal{Y}$, respectively) such that S and X are independent (i.e., $P_{SX} = P_S \cdot P_X$). Then for all real numbers $\ell > 0$*

$$H_\infty(S) - H_\infty(S|X = x, Y = y) \leq \log|\mathcal{Y}| + \ell$$

holds with probability greater than $1 - 2^{-\ell}$ or, more precisely,

$$P_{XY}\left[\{(x,y) \in \mathcal{X} \times \mathcal{Y} : H_\infty(S) - H_\infty(S|X = x, Y = y) > \log|\mathcal{Y}| + \ell\}\right] < 2^{-\ell} .$$

Proof. Let $p_0 := 2^{-\ell}/|\mathcal{Y}|$. Then we have for all $x \in \mathcal{X}$

$$P_{Y|X=x}[\{y : P_{Y|X=x} < p_0\}] < 2^{-\ell} ,$$

and hence

$$P_{XY}\left[\{(x,y) \in \mathcal{X} \times \mathcal{Y} : P_{Y|X=x}(y) < p_0\}\right] < 2^{-\ell} .$$

This inequality implies that

$$P_{S|XY}(s,x,y) = \frac{P_{SXY}(s,x,y)}{P_{XY}(x,y)} = \frac{P_S(s) \cdot P_X(x) \cdot P_{Y|SX}(y,s,x)}{P_X(x) \cdot P_{Y|X}(y,x)}$$

$$\leq \frac{P_S(s)}{P_{Y|X}(y,x)} \leq \frac{P_S(s)}{p_0} = P_S(s) \cdot |\mathcal{Y}| \cdot 2^\ell ,$$

holds with probability greater than $1 - 2^{-\ell}$ (over values x and y). The statement of the lemma follows by maximizing over all $s \in S$, and by taking negative logarithms. □

We will show in Section 5 that the bounds of this lemma (and hence also those of the following theorem) are almost tight.

We can now prove a result concerning authentication with a partially secret key which states that information-theoretically secure authentication is possible under the sole condition that no conditional probability of a certain key, given Eve's information, exceeds a bound which is roughly $1/\sqrt{|S|}$.

Theorem 6. *Assume that two parties Alice and Bob have access to a random variable S, which is a binary string of length n (n even), and that S is used as the key in the authentication scheme based on strongly-universal hashing described in Section 2. Assume further that an adversary Eve knows a random variable U, jointly distributed with S according to some probability distribution, and that Eve has no further information about S. Let*

$$H_\infty(S|U = u) \geq \left(\frac{1}{2} + t\right) \cdot n$$

for a particular realization u of U, and let $\mathcal{D}$ be the event that Eve can either insert a message (successful impersonation attack) or modify a message sent by Alice or Bob (successful substitution attack) without being detected. Then for every strategy, the conditional probability of $\mathcal{D}$, given $U = u$, can be upper bounded as follows:

$$P(\mathcal{D}|U = u) \leq 2^{-(tn/2-1)} \tag{2}$$

holds under the condition that the correctly authenticated message observed by Eve is independent of S, given $U = u$.

Remark. Note that in Theorem 6 it need not be assumed that the message observed by Eve be independent of S (but independent of S *given $U = u$*). For example (2) holds also when the message is selected by Eve herself.

Proof. First we prove an upper bound on the success probability p_{imp} of the impersonation attack. For every possible message $x \in GF(2^{n/2})$ and for every authenticator $y \in GF(2^{n/2})$ there exist exactly $2^{n/2}$ possible keys such that the authentication is correct. The probability of such a set of keys, given $U = u$, is upper bounded by

$$2^{n/2} \cdot 2^{-H_\infty(S|U=u)} \leq 2^{n/2} \cdot 2^{-(1/2+t)n} = 2^{-tn} ,$$

and hence

$$p_{imp} \leq 2^{-tn} .$$

In a substitution attack the adversary sees a message-authentication pair $(X, Y) \in GF(2^{n/2})^2$, where X is independent of S given $U = u$. According to

Lemma 5 (applied to distributions conditioned on $U = u$), we have for every $r > 0$ that

$$H_\infty(S|X = x, Y = y, U = u) \geq \frac{n}{2} + tn - \log|\mathcal{Y}| - rn = (t - r)n$$

holds with probability greater than $1 - 2^{-rn}$. A successful substitution attack immediately yields the key $S = (A, B)$ because the equations $Y = AX + B$ (from the observed message) and $Y' = AX' + B$ (from the modified message) uniquely determine the key (and can efficiently be solved). Hence the success probability of such an attack can be upper bounded as follows:

$$p_{sub} < 2^{-rn} \cdot 1 + (1 - 2^{-rn}) \cdot 2^{-(t-r)n} < 2^{-rn} + 2^{-(t-r)n} . \tag{3}$$

The reason for this is that with probability greater than $1 - 2^{-rn}$, the maximal probability of a particular key is at most $2^{-(t-r)n}$. Inequality (3) is true for every $r > 0$; the choice $r = t/2$ gives

$$p_{sub} \leq 2 \cdot 2^{-tn/2} = 2^{-(tn/2-1)} .$$

The probability $P(\mathcal{D}|U = u)$ is equal to the maximum of p_{imp} and p_{sub}, given $U = u$. This concludes the proof. $\qquad\square$

3 Privacy Amplification with Universal Hashing

The results in this and the next section are of the following type: If the min-entropy of a partially secret string S of length n, given the opponent's knowledge, is greater than a certain fraction of n, then Alice and Bob can, by communication over a non-authentic and insecure channel, agree on a common string about which Eve has virtually no information. The maximal length of the resulting highly secret string depends on Eve's knowledge about S and the security conditions. The idea is to use the partially secret string in a first step to authenticate a message containing the description of a function from a suitable class of hash functions. In the second step, this hash function is used for privacy amplification, and the string is used again as the input to this hash function. There are two possibilities to proceed: one can divide the string into two parts and use the first part for authentication and the remaining part as the argument for the final privacy amplification. The second possibility is to use the whole string for both authentication and as argument for privacy amplification. The disadvantage of the second possibility is that the authenticator gives Eve information about the argument of the hashing. A drawback of the first method is that Eve's information about S could be about either string (in fact about both, see below). However, the following lemma implies a tight bound on Eve's information about substrings.

Lemma 7. Let $S = (S_1, S_2, \ldots, S_n)$ be a random variable consisting of n binary random variables. For any k-tuple $\underline{i} = (i_1, i_2 \ldots, i_k)$, where $1 \leq i_1 < i_2 < \ldots < i_k \leq n$, let $S_{\underline{i}}$ be the string $(S_{i_1}, S_{i_2}, \ldots, S_{i_k})$. Then

$$H_\infty(S_{\underline{i}}) \geq H_\infty(S) - (n - k) .$$

Proof. A string $(s_{i_1}, s_{i_2}, \ldots, s_{i_k})$ corresponds to exactly 2^{n-k} strings $(s_1, \ldots, s_n)$. Hence the maximal probability of such a k-bit string is at most 2^{n-k} times the maximal probability of a string in S, i.e., $H_\infty(S) - H_\infty(S_{\underline{i}}) \leq n - k$. ☐

Remark. Note that when the string S is split into two parts S_l and S_r, then the bounds of Lemma 7 applied to S_l and S_r are tight simultaneously. For example let $s = (s_l, s_r)$ be a particular n-bit string, and let s_l and s_r be the first and second half of s. Define (for some $v \leq n/2 - 1$) $P_S((s_l, \bar{s})) = P_S((\bar{s}, s_r)) := 2^{v-n}$ for all $n/2$-bit strings $\bar{s}$ (and a uniform distribution for the remaining n-bit strings), i.e., $H_\infty(S) = n - v$. Then

$$H_\infty(S_l) = H_\infty(S_r) = n/2 - v = H_\infty(S) - n/2 .$$

Intuitively speaking but counter to intuition, Eve's information about S in terms of min-entropy appears entirely in both substrings S_l and S_r.

The following theorem states that if Eve's knowledge (in terms of H_∞) is less than one third of the length of the entire string (this is an intuitive, but somewhat imprecise description of $H_\infty(S|U = u) > 2n/3$), then privacy amplification by not authenticated public discussion is possible using two thirds of the string to authenticate a hash function from a universal class, and the remaining third as the input to the hash function. We can assume that the length of the string is divisible by 3 (otherwise Alice and Bob discard one or two bits).

Theorem 8. *For every n (multiple of 3) and for all positive numbers $t \leq 1/3$ and r such that $(t - r)n$ is a positive integer, there exists a*

$$\left(n, (2/3 + t)n, (t - r)n, 2^{-rn}/\ln 2, 2^{-(tn/2-1)}\right) \text{ - protocol}$$

for privacy amplification over an insecure and non-authentic channel.

Proof. Let $n = 3k$, and let S be the random variable known to Alice and Bob where $S \subset \{0,1\}^n$. Let further U be the opponent Eve's information about S, and let finally

$$H_\infty(S|U = u) \geq \left(\frac{2}{3} + t\right) \cdot n$$

for a particular $u \in \mathcal{U}$. We denote by S_1 the string consisting of the first $2k = 2n/3$ bits of S (more precisely, S_1 is interpreted as a pair (A, B) of elements of $GF(2^k)$), and by S_2 the remaining k bits (i.e., $S_2 \in GF(2^k)$). The idea of the protocol is to use S_1 for authenticating an element of the universal class of hash functions described in Section 1, and S_2 as the input to this function. According to Lemma 7 applied to conditional distributions (with respect to $U = u$),

$$H_\infty(S_1|U = u) \geq \left(\frac{1}{3} + t\right) \cdot n$$

and
$$H_\infty(S_2|U=u) \geq tn .$$

Alice randomly chooses an element X of $GF(2^k)$, which she sends to Bob together with the authenticator $Y = AX + B$ (see Section 2). According to Theorem 6, the probability $P(\mathcal{D}|U=u)$ of undetected modification is bounded by

$$P(\mathcal{D}|U=u) \leq 2^{-(tn/2-1)} .$$

Let the hash function be specified by X (see Section 1). The argument S_2 of the hash function satisfies

$$H_2(S_2|U=u) \geq H_\infty(S_2|U=u) \geq tn .$$

Let S' be the first $(t-r)n$ bits of $S_2 \cdot X$ (where the product is taken in $GF(2^k)$). Then

$$H(S'|XY, U=u) \geq (t-r)n - \frac{2^{-rn}}{\ln 2}$$

follows from Theorem 3. □

It is not difficult to verify that the use of authentication codes based on the ε-ASU$_2$ classes of hash functions explicitly given in [12] do not lead to a better result than stated in Theorem 8. For a more detailed discussion of the optimality of our results, see Section 5.

4 Privacy Amplification with Extractors

It appears that the condition in Theorem 8 on Eve's min-entropy about S can be weakened if the description of the hash function is shorter. Extractors are a method for extracting all or part of the min-entropy of a random source into an almost uniformly distributed string by requiring only a small amount of truly random bits. By using extractors instead of universal hashing for privacy amplification, we show that privacy amplification can be secure against an active opponent, provided his min-entropy about S exceeds half of the length of the string. The length of the resulting secret string can be a constant fraction of Eve's min-entropy about S minus $n/2$.

In [10], extractors are defined as follows (for an introduction to the theory of extractors, see for example [9] or [10]).

Definition 9. [10] A function $E : \{0,1\}^n \times \{0,1\}^w \to \{0,1\}^{n'}$ is called a (δ, ε')-extractor if for any distribution P on $\{0,1\}^n$ with min-entropy $H_\infty(P) \geq \delta n$, the distance of the distribution of $[V, E(X,V)]$ to the uniform distribution of $\{0,1\}^{w+n'}$ is at most ε' when choosing X according to P and V according to the uniform distribution in $\{0,1\}^w$. The distance between two distributions P and P' on a set $\mathcal{X}$ is defined as

$$d(P, P') := \frac{1}{2} \sum_{x \in \mathcal{X}} |P(x) - P'(x)| .$$

Various possible constructions of extractors have been described. The following theorem of [10] states that it is possible to extract a constant fraction of the min-entropy of a given source where the number of required random bits is polynomial in the logarithm of the length of the string and in $\log(1/\varepsilon')$.

Theorem 10. [10] *For any parameters $\delta = \delta(n)$ and $\varepsilon' = \varepsilon'(n)$ with $1/n \leq \delta \leq 1/2$ and $2^{-\delta n} \leq \varepsilon' \leq 1/n$, there exists an efficiently computable (δ, ε')-extractor $E : \{0,1\}^n \times \{0,1\}^w \to \{0,1\}^{n'}$, where $w = O(\log(1/\varepsilon') \cdot (\log n)^2 \cdot (\log(1/\delta))/\delta)$ and $n' = \Omega(\delta^2 n / \log(1/\delta))$.*

We also need the following lemma, which states that a random variable whose distribution is close to uniform (in terms of the distance d) has a Shannon entropy close to maximal.

Lemma 11. *Let Z be a random variable with range $\mathcal{Z} \subset \{0,1\}^k$. Then*

$$H(Z) \geq k \cdot (1 - d(U_k, P_Z) - 2^{-k}) \,,$$

where U_k stands for the uniform distribution over $\{0,1\}^k$.

Proof. Let $d := d(U_k, Z)$. We can assume that $d < 1 - 2^{-k}$ because otherwise the inequality is trivially satisfied. The distribution P_Z of Z can be thought of as obtained from the uniform distribution U_k by increasing some of the probabilities (by total amount d) and decreasing some others (by the same total amount). The function

$$\frac{\mathrm{d}}{\mathrm{d}p}\left(-p\log p\right) = -\frac{\ln p + 1}{\ln 2}$$

is monotonically decreasing, hence increasing [decreasing] a *smaller* probability increases [decreases] the entropy more than modifying a greater probability by the same amount. Hence a distribution with distance d from U_k with minimal entropy can be obtained by adding d to one of the probabilities, and by reducing as many probabilities as possible to 0, leaving the other probabilities unchanged. One of the probabilities of the new distribution equals $2^{-k}+d$, $\lfloor 2^k d \rfloor$ probabilities are equal to 0, one probability equals $2^{-k}(2^k d - \lfloor 2^k d \rfloor)$ (if this is not 0), and $\lfloor 2^k (1 - d) \rfloor - 1$ probabilities are unchanged and hence equal to 2^{-k}. Thus the entropy of the new random variable Z can be bounded from below by

$$H(Z) \geq 2^{-k}(2^k d - \lfloor 2^k d \rfloor) \cdot k + (\lfloor 2^k (1 - d) \rfloor - 1) \cdot 2^{-k} \cdot k = k \cdot (1 - d - 2^{-k}) \,.$$

$\square$

For certain values of d equality can hold in the above inequality. In particular $H(Z) \geq k \cdot (1 - d(U_k, P_Z))$ is false in general: $H(Z) = 0$ is possible when $d(U_k, P_Z) = 1 - 2^{-k} < 1$.

Theorem 12 below states that if Eve's min-entropy about S is greater than half of the length of S, then a constant (where this constant is not explicitly specified) fraction of the difference of this entropy and half of the length of S (plus a security parameter) can be extracted by privacy amplification using public discussion over a non-authentic channel, provided that S is sufficiently long. In contrast to the proof of Theorem 8, the entire string S is used twice here: once for authentication, and once as the input of the extractor.

Theorem 12. *Let t and r be positive numbers such that $r < t < 1/2$. There exists a constant c with the following property. Let $\varepsilon'(n)$ be a function such that*

$$\lim_{n \to \infty} n \cdot \sqrt{\varepsilon'(n)} = 0 \tag{4}$$

and

$$\varepsilon'(n) = 2^{-o(n)/(\log n)^2} \tag{5}$$

(i.e., $\log(1/\varepsilon'(n)) \cdot (\log n)^2/n \to 0$ for $n \to \infty$). Then there exists a bound n_0 such that for all $n \geq n_0$ there exists an $n' \geq c(t-r)n$ and an

$$\left(n, (1/2+t)n, n', n' \cdot \left(2\sqrt{\varepsilon'(n)} + 2^{-rn} + 2^{-n'}\right), 2^{-(tn/2-1)}\right) \text{- protocol}$$

for privacy amplification over an insecure and non-authentic channel.

Remark. The function $\varepsilon'(n)$ is directly related to the tolerable amount of information that Eve obtains about the key S' as a function of the length n of the string S. Possible functions $\varepsilon'(n)$ satisfying both (4) and (5) are $\varepsilon'(n) = 2^{-n/(\log n)^3}$, $\varepsilon'(n) = 2^{-n^\alpha}$ for any $0 < \alpha < 1$, or $\varepsilon'(n) = 1/(n^2(\log n)^2)$. The choice of a more restrictive $\varepsilon'(n)$ with respect to Eve's knowledge increases the bound n_0.

Proof. The number w of random bits required as the second part of the input for a (δ, ε')-extractor according to Theorem 10, where $\delta = t - r$ is constant, is

$$w = O(\log(1/\varepsilon') \cdot (\log n)^2) , \tag{6}$$

and the length n' of the output is $\Omega((t-r)n)$, i.e., $n' \geq c(t-r)n$ for some constant c.

Because of (5) and (6) there exists an n_0 (depending on $\varepsilon'(n)$) such that $n \geq n_0$ implies $w \leq n/2$. Let $n \geq n_0$ (and we can assume that n is even). The message sent from Alice to Bob is a random element $X \in GF(2^{n/2})$ (of which the first w bits are used as the second input V to the extractor) and is authenticated by $Y = S_1 \cdot X + S_2 \in GF(2^{n/2})$, i.e., the authentication scheme based on strongly universal hashing (see Section 2) is used with S as partially secret key, and where S_1 and S_2 are the first and second half of S, interpreted as elements of $GF(2^{n/2})$. According to Theorem 6, Eve's probability of undetected modification satisfies

$$P(\mathcal{D}|U = u) \leq 2^{-(tn/2-1)} .$$

Lemma 5 implies that

$$H_\infty(S|X = x, Y = y, U = u) \geq (t-r)n \tag{7}$$

holds with probability greater than $1 - 2^{-rn}$. We can assume that $(t-r)n$ is an integer. If (7) holds, the extractor's output satisfies

$$d([V, E(S, V)], U_{w+n'}) \leq \varepsilon'(n) . \tag{8}$$

Here and below the random variable S is meant to be distributed according to $P_{S|U=u}$, i.e., Eve's point of view is taken. It is easy to see that the distance in (8) is the expected value of the distances $d(E(S,V),U_{n'})$, where V is chosen at random from $\{0,1\}^w$. We conclude that for every K,

$$P_V\left[d(E(S,V),U_{n'}) \leq K \cdot \varepsilon'(n)\right] \geq 1 - \frac{1}{K}, \tag{9}$$

where V is uniformly distributed in $\{0,1\}^w$. From (7) and (9), with the special choice $K = \sqrt{1/\varepsilon'(n)}$, we obtain that

$$d(E(S,V),U_{n'}) \leq \sqrt{\varepsilon'(n)}$$

holds with probability at least $1 - \sqrt{\varepsilon'(n)} - 2^{-rn}$. With Lemma 11, this leads to

$$H(S'|XY,U=u) \geq (1 - \sqrt{\varepsilon'(n)} - 2^{-rn}) \cdot (1 - \sqrt{\varepsilon'(n)} - 2^{-n'}) \cdot n'$$
$$\geq n' - n' \cdot (2\sqrt{\varepsilon'(n)} + 2^{-rn} + 2^{-n'}).$$

$\square$

5 Optimality Considerations

This section provides evidence that the result of Section 2 (and the condition on Eve's knowledge in Theorem 12) is optimal: if Eve's min-entropy about S is less than half of the length of the string, then no non-trivial upper bound on the probability of undetected modification $P(\mathcal{D}|U=u)$ can be shown. This fact also implies that no better result than that of Section 3 can hold if one splits the string into two parts, one of which is used for authentication and the other for privacy amplification.

We will show that the bound given in Lemma 5 is tight (Lemma 13) and that this implies that when using the authentication code based on an ε-ASU$_2$ class of hash functions, a substantially better result than Theorem 6 cannot be derived. The omission of the additional random variable X in Lemma 13 is for simplicity. It is obvious that the same tightness result also holds in the situation of Lemma 5.

Lemma 13. *For every integer $k > 0$ and for every number $\ell \geq 0$ there exist random variables S and Y (with ranges $\mathcal{S}$ and $\mathcal{Y}$) such that $|\mathcal{Y}| = k$, and such that*

$$H_\infty(S|Y=y) = H_\infty(S) - \log|\mathcal{Y}| - \ell \tag{10}$$

holds with probability

$$\left(1 - \frac{1}{|\mathcal{Y}|}\right) \cdot 2^{-\ell}, \tag{11}$$

and even with probability 1 in the case $\ell = 0$.

Proof. Let $R := 2^\ell k$ and $S := \{s_1, s_2, \ldots, s_R\}$ with $P_S(s_i) = 1/R$ for $1 \le i \le R$. Let further

$$P_{Y|S}(y_i, s_i) = 1$$

for $1 \le i \le k - 1$, and

$$P_{Y|S}(y_k, s_i) = 1$$

for $i \ge k$. Then we have $H_\infty(S) = \log k + \ell = \log |\mathcal{Y}| + \ell$, and $H_\infty(S|Y = y_i) = 0$ for $1 \le i \le |\mathcal{Y}| - 1$. Hence (10) holds with the probability given in (11). In the case $\ell = 0$, let $Y = S$, and the result follows immediately. $\quad\square$

Let us now assume that an ε-ASU$_2$ class $\mathcal{H}$ of hash functions mapping $\mathcal{X}$ to $\mathcal{Y}$ (where $|\mathcal{Y}| \le |\mathcal{X}|$ and $1/\varepsilon \le |\mathcal{Y}|$) is used for authentication. We show that when Eve observes a correctly authenticated message, then the min-entropy of the key must be reduced by at least half of the key size to obtain a lower bound for the min-entropy of the correct authenticator of a different message. This implies the optimality of our results in the earlier sections when using this authentication method.

According to Lemma 13 we must assume that the min-entropy of the key, given Eve's information, is decreased by at least $\log |\mathcal{Y}|$ when Eve observes a correctly authenticated message x. On the other hand, given an arbitrary additional message-authenticator pair (x', y') (with $x' \ne x$), it is possible that $\varepsilon \cdot |\mathcal{H}|/|\mathcal{Y}|$ keys are compatible with both pairs. Hence the conditional min-entropy of the correct authenticator for a given message x' (this min-entropy is directly linked with the substitution attack success probability) can, in the worst case, be smaller than the min-entropy of the key by $\log |\mathcal{H}| - \log |\mathcal{Y}| - \log(1/\varepsilon)$. Both reductions of the initial min-entropy together are, in the worst case, of size $R := \log |\mathcal{H}| - \log(1/\varepsilon)$. Because of

$$|\mathcal{H}| \ge \frac{|\mathcal{Y}| - 1}{\varepsilon} \ge \frac{1}{\varepsilon^2}(1 - \varepsilon)$$

(see Theorem 4.2 in [12]), we have

$$\log\left(\frac{1}{\varepsilon}\right) \le \frac{1}{2} \cdot \log |\mathcal{H}| - \frac{1}{2} \cdot \log(1 - \varepsilon) \le \frac{1}{2} \cdot \log |\mathcal{H}| + \frac{\varepsilon}{2\ln 2}$$

and

$$R \ge \frac{1}{2} \cdot \log |\mathcal{H}| - \frac{\varepsilon}{2\ln 2} \ . \tag{12}$$

The lower bound in (12) is almost $(\log |\mathcal{H}|)/2$. Hence these worst-case estimates suggest that the result of Theorem 6 and the condition in Theorem 12 are optimal.

6 Open Problems

It is conceivable that stronger results than those of Theorems 6, 8, and 12 can be shown under certain additional conditions on Eve's information. We state as an open problem to find such conditions, as well as the question whether the results of the previous sections can be improved by using different authentication protocols (e.g., [5]), or even a completely different type of protocol for privacy amplification by not authenticated public discussion. Finally, are there different kinds of scenarios, besides the situations of independent repetitions of a random experiment [6] and of privacy amplification, for which a positive result can be proved for secret-key agreement by not authenticated public discussion?

Acknowledgments

We would like to thank Christian Cachin for interesting discussions on the subject of this paper.

References

1. C. H. Bennett, G. Brassard, C. Crépeau, and U. M. Maurer, Generalized privacy amplification, *IEEE Transactions on Information Theory*, Vol. 41, Nr. 6, 1995.
2. C. H. Bennett, G. Brassard, and J.-M. Robert, Privacy amplification by public discussion, *SIAM Journal on Computing*, Vol. 17, pp. 210-229, 1988.
3. C. Cachin, Smooth entropy and Rényi entropy, *Advances in Cryptology - EUROCRYPT '97*, Lecture Notes in Computer Science, Vol. 1233, pp. 193-208, Springer-Verlag, 1997.
4. T. M. Cover and J. A. Thomas, *Elements of information theory*, Wiley Series in Telecommunications, 1992.
5. P. Gemmell and M. Naor, Codes for interactive authentication, *Advances in Cryptology - CRYPTO '93*, Lecture Notes in Computer Science, Vol. 773, pp. 355-367, Springer-Verlag, 1993.
6. U. Maurer, Information-theoretically secure secret-key agreement by NOT authenticated public discussion, *Advances in Cryptology - EUROCRYPT '97*, Lecture Notes in Computer Science, Vol. 1233, pp. 209-225, Springer-Verlag, 1997.
7. U. M. Maurer, A unified and generalized treatment of authentication theory, *Proceedings 13th Symp. on Theoretical Aspects of Computer Science - STACS '96*, Lecture Notes in Computer Science, Vol. 1046, pp. 387-398, Springer-Verlag, 1996.
8. U. M. Maurer, Secret key agreement by public discussion from common information, *IEEE Transactions on Information Theory*, Vol. 39, No. 3, pp. 733-742, 1993.
9. N. Nisan, Extracting randomness: how and why - a survey, preprint, 1996.
10. N. Nisan and D. Zuckerman, Randomness is linear in space, *Journal of Computer and System Sciences*, Vol. 52, No. 1, pp. 43-52, 1996.
11. G. J. Simmons, A survey of information authentication, *Proc. of the IEEE*, Vol. 76, pp. 603-620, 1988.
12. D. R. Stinson, Universal hashing and authentication codes, *Advances in Cryptology - CRYPTO '91*, Lecture Notes in Computer Science, Vol. 576, pp. 74-85, Springer-Verlag, 1992.

Visual Authentication and Identification*

Moni Naor** and Benny Pinkas***

Dept. of Applied Mathematics and Computer Science, Weizmann Institute of Science,
Rehovot 76100, Israel.

Abstract. The problems of authentication and identification have received wide interest in cryptographic research. However, there has been no satisfactory solution for the problem of authentication by a *human* recipient who does not use any trusted computational device, which arises for example in the context of smartcard–human interaction, in particular in the context of electronic wallets. The problem of identification is ubiquitous in communication over insecure networks.

This paper introduces *visual authentication* and *visual identification* methods, which are authentication and identification methods for human users based on visual cryptography. These methods are very natural and easy to use, and can be implemented using very common "low tech" technology. The methods we suggest are efficient in the sense that a single transparency can be used for several authentications or for several identifications. The security of these methods is rigorously analyzed.

Keywords: authentication, identification, visual cryptography.

1 Introduction

Authentication and identification are among the main issues addressed in Cryptography. In an authentication protocol an *informant* tries to transmit some message to a *recipient*, while an *adversary* controls the communication channel by which the informant and the recipient communicate and might change the messages transmitted through that channel. At the end of the protocol the recipient outputs what he considers to be the message sent to him by the informant. If the adversary does not alter the communication, then this output should be equal to the original message. If however the adversary does change the communication, the recipient should detect this with high probability and report that the communication has been tampered. In an identification protocol, a *user* has to prove his identity to a *verifier*. Any adversary trying to pose as the user should not be able (except with small probability) to convince the verifier that he is communicating with the user.

Authentication and identification protocols have been studied extensively in various setups and under different assumptions on the power of the different parties. This paper concentrates on a scenario in which the recipient in the authentication protocol or the user in the identification protocol is human and as such cannot perform complicated computations or store large amounts of data. We do

---

* A full version of this paper is available in [9].
** Incumbent of the Morris and Rose Goldman Career Development Chair. Research supported by BSF Grant 32-00032. E-mail: naor@wisdom.weizmann.ac.il.
*** E-mail: bennyp@wisdom.weizmann.ac.il.

not require this human to use any secure computational device except his or her natural capabilities. This case is interesting since a system is as secure as its weakest component, and yet we do not know of any rigorous treatment of the human factor in cryptographic protocols. Here we analyze cryptographic systems in which the human part can be isolated and examined: Authentication by a human recipient is a cryptographic system in which a human has to solve a decision problem – whether to accept or reject the received message. Identification of a human user is a protocol in which an adversary should not be able to replicate the role of the human user, even if this user does not use any computational device. Another motivation to investigate these problems is to construct functional cryptographic protocols in which the human party does not need to use any device except natural human capabilities. The implementation of such protocols may be cheaper since there is need for less hardware.

Although humans cannot perform computations which are easily carried out by computers, the human visual perception can easily perform tasks which may be considered as "complicated computations". The systems we present utilize the visual capabilities of the human user. In our systems the human party and the other party share some secret information, and the human receives, stores and uses this information as an image on a transparency. The systems we suggest are based on the idea of *visual cryptography*, which was introduced in [10]. We describe the basic concepts of visual cryptography in subsection 1.3.

All the systems we suggest are rigorously analyzed. The security of the systems does not depend on any computational assumptions. Instead it is reduced to assumptions regarding human visual capabilities, which can be verified by empirical tests. We therefore present a new framework for proving the security of systems which involve human participants.

1.1 Motivation

The motivation for human identification is clear to anyone who has used a password. Such a system should enable the user to prove his identity to a remote computer, and yet should not enable an adversary who controls the communication to identify himself as the original user. There are systems which perform secure human identification using hand held computing devices or through biometric approaches. Compared to such systems our visual identification system is very "low tech". It does not require special hardware and can actually be independently implemented by anyone who wishes to use it, thus freeing security from being dependent on external hardware suppliers.

Authentication by a human recipient is intended to aid users who receive messages from a remote party through an insecure channel[1]. We will refer to the different parties as follows: the human recipient is Harry (Human), the informant is Sally (since in some applications the informant is a Smartcard), and the adversary is Peggy (in some applications the adversary is the Point of sale). One

---

[1] It can also be used to authenticate messages that human users send to remote parties, if a second round of communication is used. In this round the remote party answers with an authenticated message which contains the message it received, and the human should acknowledge the correctness of this message using a password.

application can be a user using an a terminal and a network which are insecure to connect to his remote computer. Another application might be the authentication of messages received by facsimile. A major application answers a well known threat to electronic payments: to authenticate the messages sent from an electronic wallet (most commonly a smartcard) to its owner.

It should be stressed that a straightforward application of visual cryptography to perform authentication is insecure, as is any straightforward application of a one-time-pad for authentication. In the scheme we suggest Harry is equipped with a (small) transparency. When Harry places the transparency over an image sent to him by Sally, the combination of both images will be the message that is sent to Harry.

The idea of supplying Harry with a transparency to help him in the authentication or to allow him to identify himself might seem strange. However, this procedure has some clear advantages: A transparency is much cheaper than a computing device and the systems we propose use transparencies which can be small enough to be carried in a wallet. Moreover, the production of the transparencies is very simple and so users can build their own authentication or identification systems without having to base their security on external hardware manufactures. The authentication and identification processes are very simple, the user just has to place the transparency on a screen or a printed message and view the result[2], he does not have to key numbers into a computer or consult a codebook. The visual authentication methods we suggest have the additional advantage of being applicable to any kind of visual image, not just for textual messages. The security of the authentication and identification methods does not depend on any computational assumptions and an upper bound on the (small) probability of failure can be computed.

1.2 Previous Work

Human–computer cryptographic interaction has been previously studied in both contexts we examine, authentication and identification. The problem of authentication was previously investigated mostly in the context of electronic payment systems [1, 2, 4] but no satisfactory solution was given for *standard* smartcards. All the suggested solutions require a secure channel between the user (who is the recipient) and his secure hand held computer (the informant). These methods are also only applicable for textual (or even just numerical) messages.

The second problem, human identification which does not require external devices, is very important in the context of access control since it frees the human user from carrying auxiliary computing devices for the identification process. This problem was addressed in [8, 7] but the methods suggested there are not proved to be secure for performing several identifications. Another solution is for the user to carry a list of one-time passwords, such as in [5, 11], but our system offers a much larger "density" for the information that the user carries. That is, it enables a much larger number of identifications for a certain amount of "storage" required from the user. This property enables the user to perform secure identifications with several verifiers, as we describe in subsection 5.2.

---

[2] The problem of correct alignment between the two images can be solved by providing a solid frame into which the transparency is entered, which fixes it in the right place.

1.3 Visual Cryptography

Visual cryptography was introduced by Naor and Shamir in [10]. It is a perfectly secure encryption mechanism, and the decryption process is performed by the human visual system. The ciphertext is a printed page and the key is a printed transparency of the same size. When the two are stacked together and carefully aligned the plaintext is revealed. Knowing just one of these two shares does not reveal any new information about the plaintext. This encryption scheme can be also considered as a 2-out-of-2 secret sharing scheme (the two shares being the ciphertext and the key), and it can be generalized to a k out of n secret sharing scheme. More information on visual cryptography can be found in the full version of this paper [9] or in [12].

In this paper we will only use the basic 2-out-of-2 visual secret sharing of [10]. In this scheme the plaintext is treated as an image, a collection of pixels. Each pixel in the plaintext is represented by a square of 2×2 real pixels (that is, real dots that are printed on a sheet of paper or on a transparency), these are called subpixles. Each plaintext pixel is divided into two shares such that in each share exactly two of the subpixels are black and the other two are transparent. Suppose that in the first share the two upper subpixels are black. If in the other share the two *lower* subpixels are black, then stacking the two shares together composes an image in which all four subpixles are black. If on the other hand the two *upper* subpixels in the second share are black (as in the first share) then stacking the two shares together yields an image in which only two subpixels are black. The former possibility is used to encrypt a black pixel, whereas the latter one is used to encrypt a white pixel[3]. There are six ways to place two black subpixels in the 2×2 square. For each pixel, one of these options will be chosen randomly for the first share. The second share will be the same as the first one if the pixel is white, or it will contain the complementary subpixels if the pixel is black. Note that since each single share is random, a single share does not add any information to the a-priori information that is known about the shared secret.

A straightforward implementation of visual cryptography for authentication is insecure. For a secure authentication Peggy must have some ambiguity regarding the contents of the share that Harry holds even after knowing the message sent by Sally, as in the case of standard authentication [3].

1.4 Organization of the Paper

In the next section we define the model of the authentication process we investigate, and the exact power of the different parties. Section 3 describes general methods for visual authentication, including efficient methods for performing several authentications using a single transparency. Section 4 defines and section 5 describes methods for secure visual identification of a human user. Section 6 concludes and suggests some open problems.

---

[3] Note that a white pixel is represented by a square which is not completely white but rather half white. This causes a reduction in the contrast of the image but the image is still easily readable by the human eye.

2 Model and Definitions for Visual Authentication

First we define the *visual authentication scenario*, and based on it we define what is a *visual authentication protocol* which is performed in this scenario. Together they constitute a *visual authentication system*. We then define the security requirements that a visual authentication system should have.

Definition 1 (visual authentication scenario). There are three entities in the visual authentication scenario: H (Harry), P (Peggy) and S (Sally). H is human and has human visual capabilities. For each protocol the capabilities that are required from H must be stated. These capabilities must include the ability to identify an image resulting from the composition of two shares of a 2-out-of-2 visual secret sharing. Other capabilities might be the ability to verify that a certain area is black, the ability to check whether two images are similar, etc. There is a security parameter n, such that the storage capacities and computing power of S and P are polynomial in n.

In the initialization phase S produces a random string r and creates a transparency T_r and some auxiliary information A_r as a function of r. Their size is polynomial in the security parameter n. S sends T_r and A_r to H through an off-line private initialization channel to which P has no access (this is the only time this private channel is used). S also sends to H a set of instructions that H should perform in the protocol (e.g. checking at a certain point whether a certain area in the image is black, comparing two areas, etc.). These instructions are public and might get known to P, but she is unable to change them.

Following the initialization phase all the communication is done through a channel controlled by P, who might change the communicated messages.

It is hard to rigorously analyze processes which involve humans since there is no easy mathematical model of human behavior. In order to prove the security of such protocols the human part in the protocol should be explicitly defined, thus isolating the capabilities required from the human participant. The security of the protocol must be reduced to the assumption that a "normal" person has these capabilities. This assumption can then be verified through empirical tests. Although we restrict P's power to be polynomial in the security parameter we do not make use of this limitation, the schemes we suggest are secure against an adversary with unbounded computing and memory capabilities.

Definition 2 (visual authentication protocol). S wishes to communicate to H an information piece m, the content of which is known to P.

- S sends a message c to H, which is a function of m and r.
- P might change c before H receives it[4].

---

[4] In our applications a message c is an image. Therefore it might be possible for P to change it so that it will not be in the form of a black and white image. For instance, m' might contain blinking pixels or, if the resolution is good enough, grey pixels. However, we assume that H either detects such messages as illegal, or assigns each pixel a value of either black or white.

- Upon receiving a message c' H outputs either FAIL or $\langle \text{ACCEPT}, m' \rangle$ as a function of c' and of T_r and A_r. When he outputs ACCEPT he also outputs m', what he considers to be the information sent to him by S.

Next we define the security requirements from visual authentication systems. The first definition ensures that the adversary cannot convince the human recipient to receive *any* message different from the original message. The second definition only ensures that for any a-priori determined message m' the adversary cannot convince the recipient that the received message was m'.

Definition 3 (security). Assume that H has the capabilities required from him for the protocol, that he acts according to the instructions given in the protocol, and that the visual authentication system has the property that when P is faithful then H always outputs $\langle \text{ACCEPT}, m \rangle$. We call the system

- $(1 - p)$-authentic if for any message m communicated from S to H, the probability that H outputs $\langle \text{ACCEPT}, m' \rangle$ is at most p (m' should of course be different from m).
- $(1 - p)$-single-transformation-secure ($(1 - p)$-sts) if for any message m communicated from S to H and any $m' \neq m$ (which was determined a-priori) the probability that H outputs $\langle \text{ACCEPT}, m' \rangle$ is at most p.

A $(1 - p)$-sts visual authentication system is obviously less secure than a $(1 - p)$-authentic system, but it suffices for many applications and in particular for smartcard payment systems: we can demand that the customer receives the amount of money that his smartcard has to pay (m') directly from the point of sale, and if it does not equal the communicated message then the customer rejects.

In our model the adversary P can change the message sent from S to H at its will. However a legal share of a visual secret sharing scheme should contain exactly two black subpixels in every 2×2 square representing a pixel. There are two types of changes which can be made by P:

1. She can change the position of the two black subpixels in the squares in the image. This change cannot be noticed by the recipient H.
2. She can put more than or less than two black subpixels in a square. This produces an illegal share. However, this deviation will probably go unnoticed by H unless it is done in too many pixels[5]. We will further discuss and quantify this issue in the following section.

We do assume that the image that the human user views does not change after he has placed his transparency. This can be easily achieved if the image is first printed and then used by H (however, this requires the use of a printer which might be

---

[5] It is not easy to detect such pixels since there is no clear separation between different squares. H can detect these pixels more easily if he is supplied with two "chess board" transparencies: one with the pixels (i, j) with odd $i + j$ blackened, and the other with the even pixels blackened. He will be instructed to put each of these transparencies on the displayed image before putting his "secret" transparency. The first transparency isolates the pixels in the "even" locations and makes it easier to detect illegal pixels in these locations. The second transparency has the same effect for the "odd" pixels.

too expensive for some applications, e.g. for vending machines). We also assume that the contents of H's transparency remain secret. For example, this requires that there is no hidden camera behind H's back that reads the contents of the transparency (a solution against peeping eyes is suggested in [6]).

The definitions we gave define one-time systems. That is, they do not guarantee the security of the system if it is used to authenticate more than a single message. When we will suggest systems for several authentications we will explicitly define them as n-times secure, i.e. good for securely authenticating n messages.

There are two types of measures for complexity. Physical measures include the size of the information that the user has to carry, the storage and computation requirements from S, and the length of the communication. The second type includes the complexity of the operations that the human user has to perform in the authentication process.

In all the systems we propose the physical requirements are linear in the size of the message and logarithmic in the fault probability p (note also that the communication channel between current smartcards and a host computer runs at 9600 bps, and this throughput is enough for the methods we suggest). The complexity of the operations that the human user has to perform cannot be measured in "number of basic operations" as is done with machine computations. For each scheme we explicitly define what capabilities the human participant should have in order for the scheme to be secure. In some cases these capabilities are quantified (e.g. the human participant notices if the displayed image is different from a "legal" image in more than t pixels), and the other complexity measures are connected to the parameters of this quantification. The assumptions made about human capabilities can be verified through experiments. When these assumptions are verified the protocol is completely proved to be secure.

3 Authentication Schemes

This section describes visual authentication methods which are applicable for any kind of visual data: numerical, textual or graphical. The first three methods are one-time methods that can be used for only a single authentication. We then describe an efficient many-times method which can be used for several authentications. It is also possible to define visual methods which are good only for authenticating textual or numerical messages. Such methods use the fact that such messages are composed of characters which are elements from a small alphabet (i.e. digits or letters). We do not describe these methods since they are of much less interest than methods for general visual messages.

3.1 Method 1 — Content Areas and Black Areas

Initialization: The user H receives a transparency which is a share of a 2-out-of-2 visual secret sharing scheme. It is divided into two areas, one of them (which was chosen at random) is denoted as the *content* area, and the other is denoted as the *black* area.

Authenticated communication: S sends to H a message which is a share of a 2-out-of-2 visual secret sharing scheme. The image which is the combination of the transparency and this share has the message m in the content area and a black area which is completely black (see fig. 1). If the black area is not totally black then H should regard this message as a fraud attempt.

It is easy to prove that the adversary P has success probability at most $1/2$ if the two following assumptions on H's capabilities holds: (a) For any two semantically different messages m and m', H can notice if the share he receives from S has $|m \triangle m'|$ or more pixels in which the number of black subpixels is not two (this assumption seems reasonable since if $|m \triangle m'|$ is too small then the two messages are not semantically different). (b) H is capable of noticing any white subpixel in the black area (since this areas is completely black).

The first assumption prevents P from changing the message using only changes of type 2. The second assumption prevents it from doing any changes of type 1 to the black area. Therefore she must decide which is the content area, and her probability of success is at most $1/2$.

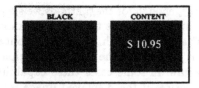

Fig. 1. The result of the composition of the user's transparency and the communicated image, for the "content areas black areas" method.

To reduce P's probability of success we can use k areas: There are $2^k - 1$ possibilities to partition k areas into black areas and content areas such that there is at least one content area. One of these partitions is selected at random and H is told in advance which areas are content areas. The image he observes should have the same message in all the content areas and all the other areas should be black. If P wishes to change the displayed message she must decide exactly which are the content areas, and her probability of success is at most $\frac{1}{2^k-1}$. This is more efficient than repeating the basic scheme to achieve this probability, which would have required k (possibly concurrent) repetitions, using $2k$ areas.

Theorem 4. *There is a $\left(1 - \frac{1}{2^k-1}\right)$-authentic visual authentication scheme which uses a transparency with k areas such that each is large enough to accommodate the transmitted message. The method assumes H has the capability to detect a white pixel in a black region, to distinguish for every two semantically different messages m and m' between the case that there are more than $|m \triangle m'|$ pixels with more than or less than two black subpixels in the message he receives and the case that there are none, and to compare up to k areas in order to check whether they all contain the same message.*

There is a variation of this method which is slightly less efficient but does not require the user to check the image he receives for illegal pixels before placing his transparency on it. We describe it in the full version of the paper.

3.2 Method 2 — Position on the Screen

Initialization: Assume the image is composed of $r \times c$ pixels. A "bounding box" of size $r' \times c'$ pixels is drawn with a thin line at a random location on the transparency that is given to H.

Authenticated communication: The combination of the transparency and the communicated share should have the message displayed inside the bounding box, in white on a black background which covers all pixels inside and outside the bounding box. Figure 2 illustrates a transparency with a marked bounding box and a composed image with the message in the bounding box.

(a)   (b)

Fig. 2. (a) The user's transparency with the bounding box. (b) The composed image.

It should be shown that for any message $m' \neq m$ the adversary P has small success probability in changing m to m'. The task of P is to reverse the pixels of $m_d = m \triangle m' = (m \cap \overline{m'}) \cup (\overline{m} \cap m')$ for the image located inside the bounding box. We should prove that her chances in achieving this are small.

It is easy to prove security if we assume H to be very sharp-eyed and to notice if the displayed image is different from m' by even a single pixel: Let $m_d^{i,j}$ be the set of pixels which correspond to the set m_d in the bounding box located at coordinates (i, j). If P does not flip exactly the pixels in $m_d^{i,j}$, she fails. For any two different locations (i, j) and (i', j') it holds that $m_d^{i,j} \triangle m_d^{i',j'} \neq \emptyset$. There are $(r - r')(c - c')$ equally likely different locations and therefore P's probability of success is at most $\frac{1}{(r-r')(c-c')}$.

A more relaxed assumption on the capabilities of the user is that he can detect differences of t pixels or more between the displayed message and the image with m' in the correct bounding box. If the difference is at least this big then P fails. The following theorem is proved in the full version of the paper (the proof can be applied to other metrics, as is described in the full paper).

Theorem 5. *Let r and c be the number of rows and columns of the image. Let r' and c' be these values regarding the bounding box. Let m be the message communicated by S and let m' be a semantically different message. Assume that the human recipient H has the following capabilities: any image with hamming distance greater than t from m' is not captured by H as being m', and H notices if more than t' pixels in the image displayed to him have more than or less than two black subpixels. Then the authentication system we described is a $\left(1 - \frac{4(t+t')}{(r-r')(c-c')}\right)$-single-transformation-secure visual authentication system.*

3.3 Method 3 — Black and Grey

The security of the following method is exponential in the hamming distance between the original message and the message that P wishes to display to H. The drawback of this method is that it reduces the contrast of the displayed image.

We previously used the 2-out-of-2 visual secret sharing method in which all four subpixels of a black pixel are black, whereas a white pixel has two black subpixels. We can also define a *grey* pixel as a pixel with three black subpixles. Let the two shares of a pixel be denoted as s_1 and s_2. Given a share s_1 of a black pixel it is easy to construct another share s_1' such that together with s_2 it composes a grey pixel. However, given a share s_1 of a grey pixel the probability of constructing a share s_1' that together with s_2 composes a black pixel is at most 1/4. When the message m is written in black on a grey background it is therefore hard for the adversary to change a background pixel into a message pixel. Similarly, when the message is written in grey on a black background it is hard for the adversary to "erase" a pixel of the message and change it to a background pixel. The scheme we suggest displays the message in two areas. In one area it is displayed in black on grey and in the other area in grey on black. The user is instructed to verify that the messages on both areas are equal. The following theorem is easily proved using the Chernoff bound.

Theorem 6. *Let t' be an upper bound on the number of pixels of the share sent by S, in which the number of black subpixles is different from two, that still goes unnoticed by the user. For any message m', define $t_{m'}$ as the maximum hamming distance of a displayed message from m' such that a user may accept the displayed message as m'. Let t be an upper bound on $t_{m'}$ over all messages m'. If the message is displayed in the scheme suggested here and the hamming distance between any two semantically different messages m and m' is at least $2 \cdot (t' + \frac{4}{3}(1 + \varepsilon)t)$, then this is a $(1 - p)$-authentic visual authentication system, where $p = 2e^{-2\frac{\varepsilon^2}{1+\varepsilon}t}$.*

3.4 Many-Times Methods

The three authentication methods we suggested in the previous subsections were all secure for only a single authentication. It is obviously preferable to have methods which are secure for several authentications. A straightforward construction of a many-times scheme is to take any of the previous one-time schemes and store several independent copies of it in different areas of a single transparency. The number of copies in a single transparency depends on the security parameters which define the size of the area that is used by each copy, and on the size of the transparency. This construction is not too bad since the methods we suggested are relatively efficient in the transparency space they use, especially the "black on grey" method of subsection 3.3 which has exponential security. However, we would like to do better than this, since in practice there is great importance for the size of the transparency (which should be minimized) and for the number of possible secure authentications (which should be maximized). Next we define many-times security and demonstrate how to construct an efficient many-times authentication scheme from the "position on the screen" scheme.

Definition 7 (n-times security). A visual authentication system is n-times $(1-p)$-single-transformation-secure (n-times $(1-p)$-sts) if the following is true for any n messages $\langle m_1, \ldots, m_n \rangle$. For any message m_i $(1 \leq i \leq n)$ communicated from S to H, and any message m' different from m_i, the probability that H outputs $\langle$ACCEPT,$m'\rangle$ is at most p. If P is faithful then H should always output $\langle$ACCEPT,$m\rangle$.

The many-times authentication scheme we suggest uses the following parameters. The messages to be authenticated are of size $r' \times c'$ pixels, and r_0 and c_0 are the security parameters. The size of the transparency is $r \times c$, where $r = r_0 + n_r r'$ and $c = c_0 + n_c c'$. The transparency is used for $n = n_r n_c$ authentications.

Initialization: A random starting point (i_0, j_0) is chosen s.t. $1 \leq i_0 \leq r_0$ $1 \leq j_0 \leq c_0$. A grid of n areas, each composed of $r'c'$ pixels, is drawn with a thin line on the transparency starting from location (i_0, j_0). The ith area is defined as the area in the intersection of row $\lceil i/n_c \rceil$ and column $(i \bmod n_c) + 1$. Figure 3 illustrates the configuration of the transparency in this scheme.

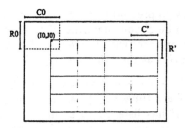

Fig. 3. The user's transparency in the many-times visual authentication scheme.

i-th authentication: S sends her share of the message m_i (written in white over a black background) in the ith area of the grid, and in all the other pixels of the share that she sends there are exactly two black subpixels in two random locations (in the 2×2 square). The human recipient H verifies that the message he sees when he puts his transparency is in the ith area.

Theorem 8. *Assume that if the hamming distance between the displayed image and an image m' is greater than t then the human recipient H does not perceive the displayed image as m'. Also assume that the user notices if in more than t' pixels of the communicated image the number of black subpixels is not two. Then a transparency of size $(r_0 + n_r r') \times (c_0 + n_c c')$ pixels can be used to get an $n_r n_c$-times $(1-p)$-single-transformation-secure visual authentication system, where each message is of size $r' \times c'$ pixels, and where $p = \frac{4(t+t')}{r_0 c_0}$.*

4 Model and Definitions for Visual Identification

The scenario of visual identification is identical to the visual authentication scenario of definition 1. However the goal of the identification protocol is different, to allow the *human user H* to prove his identity to the *verifier S* without consulting any computational device. The objective of the adversary P is to convince the verifier that she (P) is actually the human user. There is no point in constructing visual identification protocols which enable only a single secure identification since this can be achieved by supplying the user with a simple password. We will therefore consider only many-times identification protocols, i.e. protocols in which a single transparency can be used for many identifications. The protocol is a *challenge-response* type protocol in which the verifier sends a challenge to the user, who should answer it based on the secret information he holds.

Definition 9 (visual identification protocol). We define the protocol for the i-th identification of H to S:

- S sends a challenge c_i to H, which is a function of the secret data r.
- Upon receiving c_i the human user H computes a response a_i as a function of c_i and his secret information T_r and A_r, and sends it back to S.
- S decides whether the other party is H based on the messages c_i and a_i, and the secret data r. She then answers either ACCEPT or REJECT.

The adversary P might try to pretend to be H. In this case she might even try to question H by claiming to be S and requiring H to prove his identity. Then she initiates the identification protocol with the verifier S and sends a response which she hopes would convince S that the other party is H.

Definition 10 (ℓ-times $(1-p)$-secure visual identification protocol). A visual identification protocol is ℓ-times $(1-p)$-secure if the following two conditions hold after the adversary P has listened to at most ℓ_1 identifications that were answered by H and has pretended to be the verifier in at most ℓ_2 identifications of H, subject to the constraint $\ell_1 + \ell_2 \leq \ell$.

- S always accepts when H answers according to the protocol.
- If an adversary P receives the message c_i sent from S and answers it with a message b_i which is a function of c_i and any previous ℓ communications (where ℓ_1 of them were initiated by S and ℓ_2 by P, and all were answered by H), then S accepts with probability at most p.

A stronger definition is security against coalitions of k corrupt verifiers. That is, there are many verifiers and the user might need to prove his identity to any one of them. No coalition of at most k verifiers should be able to pretend to be the user in a conversation with a verifier which is not a member of the coalition. The visual identification scenario against coalitions of size k is identical to the single verifier visual identification scenario, except for the creation and distribution of the random data r and its derivatives: a central trusted authority generates r, sends each verifier S_i some secret data r_i which is a function of r and of i, and as before sends H the transparency T_r and the auxiliary information A_r. The visual identification protocol against coalitions of size k is as in the single verifier case except for S_i basing her operation on the data r_i and not on r. The definition of security is identical for the former security definition, but security is required even when the coalition members use all the secret information r_i they have and the information they gathered while tapping to or initiating at most ℓ identifications of the user.

5 Visual Identification Methods

The methods we suggest for visual identification do not use any visual secret sharing scheme since there is no need to construct an image to be viewed by H. Instead H has to prove to the verifier S that he knows some property of the transparency. We use colored transparencies, or more concretely ten different colors which we assume to be easily discernible from each other: black, white, green, blue, red, yellow, purple, brown, pink and orange. A different set of colors can be used and the security depends on the number of colors in the set.

A very attractive property of our methods is that they are very "low tech" in comparison to current secure identification methods that require the user to consult a hand held computing device, to connect a smartcard into a special port in the remote computer, or even to use biometric identification devices. Visual identification methods enable everyone with access to a color printer (or even to a black and white printer) to build a secure identification scheme which can be used for example to permit access to certain areas or to identify parties for communication. Furthermore, since the world-wide-web introduces a universal graphic interface a visual identification can be performed when a user connects from a remote host, and use a web browser to display the image that is sent from the verifier to the user. In this case no special software should be installed on the remote computer for the purpose of identification.

The visual authentication methods we suggest demand very little of the verifier. Therefore the roles of the verifier and prover can be reversed, i.e. the verifier is human and he verifies the identity of a computer with which he communicates. The human can then demand a remote computer to prove its identity to him before he sends it some confidential information (e.g. his credit card number).

5.1 A Secure Visual Identification Scheme for a Single Verifier

Here the basic unit we consider in the transparency is not a pixel but rather a *square*, which is a collection of a few pixels (for example, a square of 4×4 pixels). At the initialization phase the user H receives a transparency which is divided into many squares, and each square is randomly colored with one of the ten possible colors. The order of the colors is kept secret and is known only to H and to the verifier S (S either knows the order explicitly, or alternatively the order can be determined by the output of a pseudorandom number generator and S should only store its seed).

Let N be the number of squares in the transparency, and let d be the number of squares which are queried about in a single identification. The identification protocol goes as follows: S chooses d random squares. She sends H an image which is completely black, except for the locations of the d squares which are white. The user H puts his transparency over this image and sends back to S the colors in the locations of the white squares, by some predefined order (to make the system easier to use H can send his response using a point-and-click interface). The verifier S accepts only if H's answer is correct for all the d squares.

It is clear that H can always identify himself successfully. The best strategy for P is to query the user ℓ times and learn the color of $d\ell$ squares. P does not have any information about the colors of the other squares. Her probability of success is expected[6] to be $(\frac{1}{10} + \frac{9d\ell}{10N})^d$. A transparency with N squares can therefore be definitely used for $\ell = \frac{N}{9d}$ identifications and the security is still greater than $1 - 5^{-d}$. This result is summed up in the following theorem:

Theorem 11. *A transparency with* N *squares colored with* 10 *colors can be used for an* ℓ-*times* $(1 - (\frac{1}{10} + \frac{9d\ell}{10N})^d)$-*secure visual identification scheme, such that in each identification the user should send to the verifier the colors of d squares.*

5.2 A Visual Identification Scheme Secure Against Coalitions of Verifiers

In this scheme the secret information r_i that each verifier S_i receives contains the colors of a random subset of $(1 - q)N$ squares in the transparency that the user holds (where $0 < q < 1$). The identification protocol is identical to the previous identification protocol except for the verifier questioning the user about the colors of random squares from the set of squares whose colors the verifier knows. The "density" of the visual identification scheme, i.e. the large number of squares which can be stored in a single transparency, enables this scheme to be secure against relatively large coalitions.

Theorem 12. *When* $\ell \leq \frac{Nq^k}{2d}$ *a transparency with* N *squares colored with* 10 *colors can be used for an* ℓ-*times* $1 - (1 - \frac{9}{20}(1 - \frac{d}{(1-q)N})^\ell)^d$-*secure against k-verifiers, visual identification scheme, in which the user has to send the values of d colors in each identification.*

6 Conclusions and Open Questions

We have suggested methods for visual authentication and identification, and have given rigorous analysis of their security. All methods are secure regardless of the computational capabilities of the adversary. We also demonstrated a secure many-times visual identification method which is very "low tech" and can be implemented with almost no investment.

Comparing the one-time visual authentication methods, the advantage of the first method ("black area content area") is that its security depends on relatively easy requirements from the human user. Its disadvantage is the loss in area which implies that the security may not be as small as we would like. The advantage of the "position on the screen" method is that the error probability is proportional to the number of pixels and not to the redundancy in area. Its disadvantages are that the probability might not be small enough, and more capabilities are required of the human user. The advantage of the "black and grey" method is that the probability of non-detection is exponentially small in the distance between semantically different messages. Its disadvantages are the loss in contrast, and the additional capabilities required of the user. In comparison to the one-time methods

---

[6] This follows since P knows the colors of at most $d\ell$ squares. S chooses squares randomly and the expected success probability of P is $\sum_{i=0}^{d} \binom{d}{i} (d\ell/N)^i (1 - d\ell/N)^{(d-i)} 10^{-(d-i)} = (\frac{1}{10} + \frac{9d\ell}{10N})^d$.

the many-times authentication method has the advantage of substantially reducing the amount of transparency area that is needed per authentication in order to achieve a certain security level.

There are many open questions left. It should be interesting to find an authentication method whose security is exponential in the *size* of the message, or a method which does not reduce the contrast and whose security is exponential in the hamming difference between the messages. Another open problem is to devise more efficient methods which are secure only against polynomial adversaries. An important issue is to check which human capabilities can be easily verified and to base the security of the visual methods on these capabilities (in particular a better measure than hamming distance can be used to define similarity between images). It should also be interesting to design a method that enables a human informant to authenticate a message it sends, *without* requiring two-way interaction. A related problem is to devise a one-way function which is easily computable by humans.

7 Acknowledgments

We thank Omer Reingold for his careful reading and valuable suggestions, and the anonymous referees for their helpful comments.

References

1. Abadi M., Burrows M., Kaufman C. and Lampson B., Authentication and delegation with smart-cards, *Sci. of Comp. Prog.*, 21 (2), Oct. 1993, 93-113.
2. Boly J., Bosselaers A., Cramer R., Michelsen R., Mjolsnes S., Muller F., Pedersen T., Pfitzmann B., de Rooij P., Schoenmakers B., Schunter M., Vallee L. and Waidner M., The esprit project cafe – high security digital payment system, in *Computer Security – ESORICS 94*, Springer-Verlag LNCS Vol. 875, 1994.
3. Gilbert E., MacWilliams F. and Sloane N., Codes which detect deception, *Bell Sys. Tech. J.*, Vol. 53, No. 3, 1974, 405–424.
4. Gobioff H., Smith S., Tygar J. D. and Yee B., Smartcards in hostile environments, in *Proc. of The 2nd USENIX Workshop on Elec. Commerce*, Nov. 1996, 23-28.
5. Haller N. M., The S/KEY one-time password system, in *Internet Soc. Symp. on Network and Dist. Sys. Sec.*, 1994.
6. Kobara K. and Imai H., Limiting the visible space visual secret sharing schemes and their application to human identification, in *Asiacrypt '96*, Springer-Verlag LNCS Vol. 1163, 1996, 185–195.
7. Matsumoto T., Human-computer cryptography: an attempt, in *ACM Conf. on Comp. and Comm. Sec.*, ACM Press, March 1996, 68-75.
8. Matsumoto T. and Imai H., Human identification through insecure channel, in *Eurocrypt '91*, Springer-Verlag LNCS Vol. 547, 1991, 409–421.
9. Naor M. and Pinkas B., Visual authentication and identification, in *Theory of Cryptography Library*, http://theory.lcs.mit.edu/~tcryptol.
10. Naor M. and Shamir A., Visual cryptography, in *Eurocrypt '94*, Springer-Verlag LNCS Vol. 950, 1995, 1–12.
11. Rubin A. D., Independent one-time passwords, *Computing Systems*, The USENIX Association, Vol. 9, No. 1996, 15–27.
12. Stinson D. R., An introduction to visual cryptography, presented at *Public Key Solutions '97*. Available at http://bibd.unl.edu/~stinson/VCS-PKS.ps.

Quantum Information Processing:
The Good, the Bad and the Ugly

Gilles BRASSARD *

Université de Montréal, Département d'informatique et de recherche opérationnelle
C. P. 6128, Succursale Centre-Ville, Montréal (Québec), CANADA H3C 3J7
email: brassard@iro.umontreal.ca

Abstract. Quantum mechanics has the potential to play a major role in the future of cryptology. On the one hand, it could bring to its knees most of the current trends in contemporary cryptography. On the other hand, it offers an alternative for the protection of privacy whose security cannot be matched by classical means.

> *God only knows*
> *God makes his plan*
> *The information's unavailable*
> *To the mortal man*
>
> — *Paul Simon*

1 Good or Bad? It's a Matter of Perspective

Following pioneering work by Paul Benioff [2], the idea that quantum mechanics could be harnessed to the cause of computational speed was planted by Richard Feynman [15, 16] in the early eighties and championed by David Deutsch [13] shortly afterwards. At least in principle, a quantum computer working on a few thousand quantum bits of memory can quickly perform an amount of computation greater than possible with a classical computer the size of the Earth running for the lifetime of the universe. Nevertheless, quantum computing was but a fringe pursuit for more than a decade because (1) building a quantum computer seemed totally out of reach from current and foreseeable technology, and (2) nobody knew of a *practical* computational problem that quantum computers could solve faster than classical computers.

This all changed in 1994 when Peter Shor made his momentous discovery: quantum computers can factor large numbers and extract discrete logarithms in expected polynomial time [21]. Even better—or worse, depending on the perspective—the time needed to factor an RSA integer [20] is in the same order as the time needed to use that same integer as modulus for a *single* RSA

---

* Research supported in part by NSERC, FCAR and the Canada Council.

encryption. In other words, it takes no more time to break RSA on a quantum computer (up to a multiplicative constant) than to use it legitimately on a classical computer [8]! Of course this has no practical consequences as long as quantum computers remain the stuff of dreams [18], but Shor's breakthrough gave a remarkable boost to the quest for understanding better the feasibility of quantum computation. In just a few years, this has lead to encouraging advances in the experimental manipulation of quantum information [19]. Although large-scale quantum computations—such as the factorization of a two-hundred digit number—are still rather speculative, nontrivial quantum computations involving perhaps as many as 10 or 20 quantum bits are on the horizon. Perhaps the most exciting recent discovery in quantum information theory is that of quantum error correction [19], which makes it theoretically possible to compute reliably with unreliable components.

Even if large-scale quantum computers—or perhaps special-purpose quantum factoring devices—become a reality, this would not doom all of classical cryptography. (Of course, "classical" is used here to mean non-quantum, and it includes secret-key and public-key cryptography on the same footing, just as "classical physics" lumps together Newton's mechanics with Einstein's relativity.) For one thing, quantum computing does not weaken information-theoretic secure schemes such as the one-time pad. Actually, it makes such schemes all the more important since they could remain the only safe alternative for classical cryptography. Not even all of public-key cryptography is threatened by quantum computing: it has been argued [3] that strong one-way functions that can be computed efficiently with classical computers but cannot be inverted efficiently even with a quantum computer may well exist. This would suffice to achieve computationally secure cryptographic pseudorandom generation, bit commitment schemes and zero-knowledge protocols for all of **NP**. Even though most public-key cryptosystems currently in use are based on the presumed difficulty of either factoring large numbers or extracting discrete logarithms, which would not survive widespread use of laptop quantum computers, alternative quantum-resistant public-key systems are not ruled out to the best of our current knowledge.

Shor's algorithms are not directly relevant to the security of secret-key cryptosystems such as the DES, *provided users do not establish their secret session key with public-key techniques* such as the Diffie–Hellman key distribution scheme [14]. However, a more recently discovered quantum algorithm, due to Lov K. Grover [17], could have an impact on the security of secret-key systems by significantly speeding up exhaustive search. For example, given a single matching pair of plaintext/ciphertext, single-key DES encipherment can be broken after a mere 185 million expected quantum encipherments of the known plaintext when the solution is unique [6, 10]. This number is admittedly large, yet it is hundreds of millions of times smaller than the $2^{55} \approx 3.6 \times 10^{16}$ expected number of DES encipherments required by classical exhaustive search. It is still an open question whether or not a quantum computer could break double-key encipherment of classical cryptosystems faster than a classical computer that uses the meet-in-the-middle attack.

A new application of Grover's algorithm was recently discovered in collaboration with Peter Høyer and Alain Tapp [12]: there is a quantum algorithm that finds collisions in arbitrary two-to-one functions after only $O(\sqrt[3]{N})$ expected evaluations of the function, where N is the cardinality of the domain. This should be compared with the best possible classical algorithm, which requires $O(\sqrt{N})$ expected evaluations when the function is provided as a black box. This has obvious consequences for the cryptanalysis of hash functions, unconditionally concealing bit commitment schemes and signature schemes based on claw-free pairs of functions.

More thorough, yet elementary, introductions to quantum computation can be found in [7, 1, 9, 5].

2 The Other Side of the Coin

The previous section is not precisely good news for those of us who simultaneously believe in quantum mechanics and in the legitimate need for protecting privacy. Fortunately, quantum mechanics may provide the ultimate solution to secure communication. Quantum cryptography exploits the impossibility to measure quantum information reliably. (Remember the wise words of Paul Simon: "The information's unavailable to the mortal man".) When information is encoded with non-orthogonal quantum states, any attempt from an eavesdropper to access it necessarily entails a probability of spoiling it irreversibly, which can be detected by the legitimate users. Using protocols designed in collaboration with Charles H. Bennett [4], building on earlier work of Stephen Wiesner [22], this phenomenon can be exploited to implement a key distribution system that is provably secure even against an eavesdropper with unlimited computing power, indeed even if the eavesdropper is equipped with a quantum computer. This is achieved by the exchange of very tenuous signals that consist on the average of one-tenth of one photon per pulse. Several exciting experimental realizations have been successful so far, demonstrating the feasibility of quantum key distribution over tens of kilometres of ordinary optical fibre or hundreds of meters in free space (no wave guides), and even the possibility of quantum cryptographic networks capable of linking many users. Whether or not quantum cryptography can go beyond key distribution and the secure transmission of confidential information is an actively researched topic.

Rather than repeating material that I have written too many times already, I invite you to read my *Sigact News* survey on 25 years of quantum cryptography [11].

3 And the Ugly?

Use your imagination!

Acknowledgements

I am very grateful to Burt Kaliski Jr and the entire CRYPTO 97 Programme
Committee for inviting me to give this lecture. I also wish to acknowledge Charles
H. Bennett, Claude Crépeau, Christopher Fuchs and William Wootters, who
joined me on the occasion of the First Killam Workshop on Quantum Information
Theory: our grilled octopus eating brainstorming session lead to the wonderful
title of this talk. Chris provided the opening quote from *Slip Slidin' Away*.

References

1. Barenco, A., "Quantum physics and computers", *Contemporary Physics*, Vol. 38,
   1996, pp. 357–389.
2. Benioff, P., "Quantum mechanical Hamiltonian models of Turing machines",
   Journal of Statistical Physics, Vol. 29, no. 3, 1982, pp. 515–546.
3. Bennett, C. H., E. Bernstein, G. Brassard and U. Vazirani, "Strengths and weaknesses of quantum computing", *SIAM Journal on Computing*, to appear, 1997.
4. Bennett, C. H., F. Bessette, G. Brassard, L. Salvail and J. Smolin, "Experimental
   quantum cryptography", *Journal of Cryptology*, Vol. 5, no. 1, 1992, pp. 3–28.
5. Berthiaume, A., "Quantum computation", in *Complexity Theory Retrospective II*,
   L. A. Hemaspaandra and A. Selman, editors, Springer-Verlag, Berlin, to appear,
   1997.
6. Boyer, M., G. Brassard, P. Høyer and A. Tapp, "Tight bounds on quantum searching", *Proceedings of 4th Workshop on Physics and Computation*, Boston, November 1996, New England Complex Systems Institute, pp. 36–43. Available online in
   the *InterJournal* at URL http://interjournal.org. Improved version available
   from the authors.
7. Brassard, G., "A quantum jump in computer science", in *Computer Science
   Today*, Jan van Leeuwen (Editor), Lecture Notes in Computer Science, Vol. 1000,
   Springer–Verlag, 1995, pp. 1–14.
8. Brassard, G., "The impending demise of RSA?", *RSA Laboratories CryptoBytes*,
   Vol. 1, no. 1, 1995, pp. 1–4.
9. Brassard, G., "New trends in quantum computing", *Proceedings of 13th Annual
   Symposium on Theoretical Aspects of Computer Science*, February 1996, pp. 3–10.
10. Brassard, G., "Searching a quantum phone book", *Science*, Vol. 275, 31 January
    1997, pp. 627–628.
11. Brassard, G. and C. Crépeau, "Cryptology column — 25 years of quantum cryp-
    tography", *Sigact News*, Vol. 27, no. 3, 1996, pp. 13–24.
12. Brassard, G., P. Høyer and A. Tapp, "Cryptology column — Quantum crypt-
    analysis of hash and claw-free functions", *Sigact News*, Vol. 28, no. 2, June 1997,
    pp. 14–19.
13. Deutsch, D., "Quantum theory, the Church–Turing principle and the universal
    quantum computer", *Proceedings of the Royal Society*, London, Vol. A400, 1985,
    pp. 97–117.
14. Diffie, W. and M. E. Hellman, "New directions in cryptography", *IEEE Transac-
    tions on Information Theory*, Vol. IT–22, no. 6, 1976, pp. 644–654.
15. Feynman, R. P., "Simulating physics with computers", *International Journal of
    Theoretical Physics*, Vol. 21, nos. 6/7, 1982, pp. 467–488.

16. Feynman, R. P., "Quantum mechanical computers", *Optics News*, February 1985. Reprinted in *Foundations of Physics*, Vol. 16, no. 6, 1986, pp. 507–531.
17. Grover, L. K., "A fast quantum mechanical algorithm for database search", *Proceedings of the 28th Annual ACM Symposium on Theory of Computing*, 1996, pp. 212–219. Final version to appear in *Physical Review Letters* under title "Quantum mechanics helps in searching for a needle in a haystack".
18. Haroche, S. and J.–M. Raimond, "Quantum computing: Dream or nightmare?", *Physics Today*, August 1996, pp. 51–52.
19. Los Alamos National Laboratory, "Quantum physics e-print archive": for the latest papers on the implementation of quantum computation and on quantum error correction, as well as just about everything of interest for quantum computing, quantum cryptography and quantum information theory in general, surf the Web from URL http://xxx.lanl.gov/archive/quant-ph.
20. Rivest, R. L., A. Shamir and L. M. Adleman, "A method for obtaining digital signatures and public-key cryptosystems", *Communications of the ACM*, Vol. 21, no. 2, 1978, pp. 120–126.
21. Shor, P. W., "Algorithms for quantum computation: Discrete logarithms and factoring", *Proceedings of the 35th Annual IEEE Symposium on Foundations of Computer Science*, 1994, pp. 124–134. Final version to appear in *SIAM Journal on Computing* under title "Polynomial-time algorithms for prime factorization and discrete logarithms on a quantum computer".
22. Wiesner, S., "Conjugate coding", *Sigact News*, Vol. 15, no. 1, 1983, pp. 78–88; original manuscript written *circa* 1970.

Efficient Algorithms for Elliptic Curve Cryptosystems*

Jorge Guajardo
(guajardo@ece.wpi.edu)

Christof Paar
(christof@ece.wpi.edu)

ECE Department
Worcester Polytechnic Institute
Worcester, MA 01609, USA

Abstract. This contribution describes three algorithms for efficient implementations of elliptic curve cryptosystems. The first algorithm is an entirely new approach which accelerates the multiplications of points which is the core operation in elliptic curve public-key systems. The algorithm works in conjunction with the k-ary or sliding window method. The algorithm explores computational advantages by computing repeated point doublings directly through closed formulae rather than from individual point doublings. This approach reduces the number of inversions in the underlying finite field at the cost of extra multiplications. For many practical implementations, where field inversion is at least four times as costly as field multiplication, the new approach proofs to be faster than traditional point multiplication methods. The second algorithm deals with efficient inversion in composite Galois fields of the form $GF((2^n)^m)$. Based on an idea of Itoh and Tsujii, we optimize the algorithm for software implementation of elliptic curves. The algorithm reduced inversion in the composite field to inversion in the subfield $GF(2^n)$. The third algorithm describes the application of the Karatsuba-Ofman Algorithm to multiplication in $GF((2^n)^m)$. We provide a detailed complexity analysis of the algorithm for the case that subfield arithmetic is performed through table look-up. We apply all three algorithms to an implementation of an elliptic curve system over $GF((2^{16})^{11})$. We provide absolute performance measures for the field operations and for an entire point multiplication.

1 Introduction

Elliptic curve (EC) cryptosystems were first suggested by Miller [16] and Koblitz [8]. A main feature that makes EC attractive is the relatively short operand length relative to RSA and systems based on the discrete logarithm (DL) in finite fields. Cryptosystems which explore the DL problem over EC can be built with an operand length of 150–200 bits [15]. IEEE [11] and other standard bodies such as ANSI and ISO are in the process of standardizing EC cryptosystems. EC can provide various security services such as key exchange, privacy through

---

* This work was sponsored in part by GTE Corporation.

encryption, and sender authentication and message integrity through digital signatures. For these reasons it is expected that EC will become very popular for many information security applications in the near future. It is thus very attractive to provide algorithms which allow for efficient implementations of EC cryptosystems. Our contribution will deal with such algorithms.

Efficient algorithms for EC can be classified into high-level algorithms, which operate with the group operation, and into low-level algorithms, which deal with arithmetic in the underlying finite field. For efficient implementations it is obviously the best to optimize both types of algorithms. Our contribution will introduce three algorithms, one high-level algorithm for point multiplication and two low-level for finite field inversion and multiplication, respectively.

In Sect. 3 we introduce an entirely new approach for accelerating the multiplication of points on an EC. The approach works in conjunction with the k-ary and the sliding window methods. The method is applicable to EC over any field, but we provide worked-out formulae for EC over fields of characteristic two.

Although EC cryptosystems can be based on finite fields of any characteristic, practical systems have only been implemented over prime fields or Galois fields of characteristic two. Section 4 shows an algorithm for efficiently computing the inverse of an element in the composite Galois field $GF((2^n)^m) \cong GF(2^k)$. The algorithm is based on an idea by Itoh and Tsujii [6], but our approach is optimized for a standard basis representation and for binary field polynomials. The algorithm reduces inversion in the composite field to inversion in the subfield $GF(2^n)$. Unlike the inversion algorithms in [21, 22], the inversion algorithm is not based on Euclid's algorithm.

Section 5 provides a detailed treatment of the Karatsuba-Ofman algorithm (KOA) applied to field multiplication in $GF((2^n)^m)$. We provide a complexity analysis of the KOA for software implementations where arithmetic in the subfield $GF(2^n)$ is based on table look-up.

Section 6 shows the actual performance of all three algorithms in an implementation of an EC cryptosystem over $GF(2^{176}) \cong GF((2^{16})^{11})$. We provide absolute timing measurements for an entire EC multiplication as well as timings for individual operations.

2 Previous Work

As stated above, this contribution describes a new algorithm for point multiplication and optimized inversion and multiplication algorithms for arithmetic in composite Galois fields $GF((2^n)^m)$. In the sequel we will summarize some of the previous work in each of those areas.

The problem of multiplying a point P of an EC by a (large) integer n is analogous to exponentiation of an element in a multiplicative group to the nth power. The standard algorithm for this problem is the binary exponentiation method (or square-and-multiply algorithm) which is studied in detail in [7]. A generalization of the binary method is the k-ary method [2] which processes k exponent bits in one iteration. Further improvements of the k-ary method include the sliding window method [10]. In [21] a version of the k-ary method is applied to point multiplication for an EC. The method that we propose in Sect. 3 explores arithmetic advantages that occur if several point doublings are computed from closed expressions rather than from computing several individual doublings. This approach is unique to EC systems and, to our knowledge, has not been reported anywhere else. In [9] and [14] direct formulae to compute $Q = nP$ are introduced, where n is an integer. However, these formulae are based on the computation of division polynomials and they do not appear to be computationally efficient if used for the fast calculation of nP.

Software implementations of EC over composite Galois field $GF((2^n)^m)$ were first described in [5] for the field $GF((2^8)^{13})$. More recently, EC systems for the field $GF((2^{16})^{11})$ were described independently in [22] and [1]. In all cases, field multiplication is accomplished through table look-up in the subfield $GF(2^n)$. Neither reference explores advanced convolution algorithms, such as the Karatsuba-Ofman algorithm (KOA), as it is proposed here. The KOA has been studied for general polynomial multiplication in [3] and [7]. An application of the KOA to polynomials over finite fields in the context of computer hardware is described in [19]. None of the previous contributions provide a detailed complexity analysis of the algorithm as it is done in Sect. 5 of this work. Composite Galois fields are applied to a hardware architecture for public-key algorithms in [20].

Previously, there were two principle approaches to inversion in finite fields. One is based on the extended Euclidean algorithm, the other one is based on Fermat's Little Theorem. [21] optimizes a version of the Euclidean algorithm, named "almost inverse algorithm," for an EC implementation over $GF(2^{155})$. [22] applies the same algorithm for EC over the composite field $GF((2^{16})^{11})$. In [6] an efficient method for inversion in $GF(2^k)$ is discussed. In Sect. 6 of the reference the method is extended to composite Galois fields $GF((2^n)^m)$. The reference only considers a normal basis representation, whereas we will optimize the inversion algorithm for composite fields in standard basis representation and for binary field polynomials.

3 Point Multiplication with a Reduced Number of Inversions

3.1 Elliptic Curves over $GF(2^k)$

In this paper, we will only be concerned with non-supersingular elliptic curves. Thus, an elliptic curve E will be defined to be the set of points (x, y) with coordinates x and y lying in the Galois field $GF(2^k)$ and satisfying the cubic equation $y^2 + xy = x^3 + ax^2 + c$, where $a, c \in GF(2^k)$, $c \neq 0$, together with the point at infinity $\mathcal{O}$. The points (x, y) form an abelian group under "addition" where the group operation is defined as in [15]. In what follows, we will only be concerned with the doubling of a point P, $2P = (x_1, y_1)$. This is achieved by

$$x_1 = \left(x + \frac{y}{x}\right)^2 + \left(x + \frac{y}{x}\right) + a \tag{1}$$

$$y_1 = x^2 + \left(x + \frac{y}{x}\right)x_1 + x_1 \tag{2}$$

From (1) and (2) it can be seen that the doubling of two points in E will require one inverse, two multiplications, five additions, and two squarings in the underlying field $GF(2^k)$. Notice also that in most practical applications, inversion is by far the most expensive operation to perform. In the following we will develop a scheme which reduces the number of field inversions.

3.2 A New Approach to Point Doubling

Public-key schemes which use EC are based on the DL problem in the point group of the EC. Such schemes include analogies to the Diffie-Hellman key establishment protocol, the ElGamal encryption, and various digital signature schemes. The basic operation for the DL problem for EC is "multiplication" of a point $P \in E$ with an integer n, which is of the order of $\#E$. One way of performing this operation is analogous to the square and multiply algorithm for exponentiation [7] and it is known as "repeated double and add" [11]. A generalization of this method is the k-ary method [2, 10] which reduces the number of additions needed in the regular double and add. Theorem 1 was adapted from [13] and it describes the algorithm as it applies to EC.

Theorem 1 *Let $P \in E$ and $n = (e_t e_{t-1} \cdots e_1 e_0)_b$ be the radix representation of the multiplier n in base b where $b = 2^k$ for $k \geq 1$. Then, $Q = nP$ can be computed using the following algorithm.*

> **Algorithm** (Input: $P = (x, y)$; Output: $Q = nP$)
> 1. Precomputation
>   1.1 $P_0 \leftarrow \mathcal{O}$ (Point at infinity)
>   1.2 For $i = 1$ to $2^k - 1$
> $P_i = P_{i-1} + P$ (i.e., $P_i = i \cdot P$)
> 2. $Q \leftarrow \mathcal{O}$
> 3. For $i = t$ to 0
>   3.1 $Q \leftarrow 2^k Q$
>   3.2 $Q \leftarrow Q + P_{e_i}$
> 4. Return(Q)

Notice that Step 1.2 requires $2^k - 3$ additions and one doubling, Step 3.1 involves the doubling of point Q, k times, and Step 3.2 requires one point addition. The complexity of the k-ary method with t iterations is thus $kt + 1$ point doublings, $2^k - 3 + t$ point additions. Since point doublings are the most costly operations, it is extremely attractive to find ways of accelerating the doubling operation. In the following, we will introduce an entirely new approach to compute repeated point doublings over an EC.

Our new approach is based on the following principle. First, observe that the k-ary method relies on k *repeated* doublings. The new approach allows computation of $2^k P = (x_k, y_k)$ directly from $P = (x, y)$ without computing the intermediate points $2P, 2^2 P, \cdots, 2^{k-1} P$. Such direct formulae are obtained by inserting (1) and (2) into one another. For instance, for $4P = 2^2P = (x_2, y_2)$, we obtain

$$x_2 = \frac{\zeta^2 + (\delta\gamma)\zeta}{(\delta\gamma)^2} + a \tag{3}$$

$$y_2 = \frac{\zeta(\delta\gamma)x_2 + (\delta^2)^2}{(\delta\gamma)^2} + x_2, \tag{4}$$

where $\gamma = x^2$, $\eta = \gamma + y$, $\delta = \eta^2 + \eta x + a\gamma$, $\xi = \eta x + \gamma$, and $\zeta = \delta(\delta + \xi) + \gamma^2\gamma$. Notice that (3) and (4) imply that $2^2P = (x_2, y_2)$ can be computed with one inverse, nine multiplications, ten additions, and six squarings. The advantage of (3) and (4) is that they only require one inversion as opposed to the two inversions that two separate double operations would require for computing $4P$. The "price" that must be paid is $9 - 4 = 5$ extra multiplications if squarings and additions are ignored. For $k = 2$, the direct formulae (3) and (4) trade thus one inversion at the cost of 5 multiplications. It is easy to see that the formulae are an advantage in situations where multiplication is at least five times as costly as inversion. However, this "break-even point" decreases if the method is extended to the computation of 2^kP for $k > 2$ as described below.

We continued in a similar manner and found expressions for $2^3P = (x_3, y_3)$ and $2^4P = (x_4, y_4)$. Again, these expressions, shown in the appendix, only require one inversion as opposed to the three or four inversions that the regular double and add algorithm would require in each one of these cases. It is important to point out that the point P has to be an element with an order larger than 2^k. This last requirement ensures that $4P$, $8P$, or $16P$ will never equal $\mathcal{O}$. Notice that this is compliant with [11].

3.3 Comparison

For application in practice it is highly relevant to compare the complexity of our newly derived formulae with that of the double and add algorithm. If we note that our method reduces inversions at the cost of multiplications, the performance of the new method depends on the cost factor of one inversion relatively to the cost of one multiplication. For this purpose we introduce the notion of a "break-even point." Since it is possible to express the time that it takes to perform one inversion in terms of the equivalent number of multiplication times, we define the break even point as the number of multiplication times needed per inversion so that our formulae outperform the regular double and add algorithm. The results are summarized in Table 1.

3.4 Applications

Our new method for point doubling can directly be applied to all EC defined over fields of characteristic two, regardless of the specific field representation. For instance, the formulae in the appendix are applicable to a composite field representation $GF((2^n)^m)$ as well as to binary field representations $GF(2^l)$. In

Table 1. Complexity comparison: Individual doublings vs. direct computation of several doublings

Calculation	Method	Complexity				Break-Even Point
		Sq.	Add.	Mult.	Inv.	
4P	Direct Doublings	6	10	9	1	1 inv. > 5 mult.
	Individual Doublings	4	10	4	2	
8P	Direct Doublings	7	17	14	1	1 inv. > 4 mult.
	Individual Doublings	6	15	6	3	
16P	Direct Doublings	15	20	19	1	1 inv. > 3.7 mult.
	Individual Doublings	8	20	8	4	

addition, the choice of of the basis does not matter. Standard, dual, and normal basis representations are all possible. The latter observation is of special interest since efficient inversion methods for a normal basis representation appear not to be known, and inversion based on Fermat's Theorem using addition chains requires at least 7 multiplications if $l > 128$. We expect our method yields a considerable acceleration in such situations.

In situations where the ratio between an inversion time and a multiplication is three or smaller [22], our method may not give an advantage. However, we would like to point out that as shown in Sect. 6, the repeated point multiplication behaves better than predicted in our implementation, so that the break-even points might in practice be even lower than shown in Table 1.

4   Efficient Inversion in Composite Fields $GF((2^n)^m)$

Composite Galois fields for EC systems were explored earlier in [5, 1, 22]. The following notation will be used here. We consider arithmetic in an extension field of $GF(2^n)$. The extension degree is denoted by m, so that the field can be denoted by $GF((2^n)^m)$. This field is isomorphic to $GF(2^n)/(P(x))$, where $P(x) = x^m + \sum_{i=0}^{m-1} p_i x^i, p_i \in GF(2^n)$, is a monic irreducible polynomial of degree m over $GF(2^n)$. In the following, a residue class will be identified with the polynomial of least degree in this class. We consider a standard (or polynomial or canonical) basis representation of a field element A:

$$A(x) = a_{m-1}x^{m-1} + \cdots + a_1 x + a_0, \tag{5}$$

where $a_i \in GF(2^n)$. Note that it is possible to choose $P(x)$ with binary coefficients if $\gcd(n, m) = 1$ [12], a fact that will be explored for the inversion algorithm below.

As stated above, inversion is the most costly arithmetic operation in EC systems. In the following an inversion method based on Fermat's Little Theorem will be developed which is entirely different from the approach in [22, 21]. The basic property of the algorithm developed in this section is that inversion in $GF((2^n)^m)$ is reduced to inversion in the subfield $GF(2^n)$. It is important to point out that subfield inversion can be done extremely fast through table look-up provided n is moderate, say $n \leq 16$ [22]. We extend and optimize the idea in [6] for the case of a standard basis representation as suggested in [18] and we show a major computational advantage for the case that the field polynomial has only coefficients from $GF(2)$.

We want to determine the inverse of $A \in GF((2^n)^m)$, $A \neq 0$, and A is given as in (5). By applying Fermat's Theorem, we can readily obtain that $A^{2^{nm}-1} = AA^{2^{nm}-2} = 1 \bmod P(x)$, from which it follows that

$$A^{-1} = A^{2^{nm}-2}. \tag{6}$$

Equation (6) shows that the inverse of an element $A \in GF((2^n)^m)$ can be computed by raising it to the power of $2^{nm} - 2 = 2 + 2^2 + 2^3 + \cdots + 2^{nm-1}$ using addition chains [7]. However, by noticing that the inversion in the composite field $GF((2^n)^m)$ can be reduced to inversion in the ground field $GF(2^n)$, one obtains a better method to calculate the inverse of an element A.

Theorem 2 *[17] The multiplicative inverse of an element A of the composite field $GF((2^n)^m)$, $A \neq 0$, can be computed by*

$$A^{-1} = (A^r)^{-1}A^{r-1} \bmod P(x),$$

where $A^r \in GF(2^n)$ and $r = (2^{nm} - 1)/(2^n - 1)$.

A central observation is that A^r is an element of the subfield. Computing the inverse through Theorem 2 requires four steps: exponentiation in $GF((2^n)^m)$ (A^{r-1}), multiplication in $GF((2^n)^m)$ with $AA^{r-1} \in GF(2^n)$, inversion in $GF(2^n)$, and multiplication of $(A^r)^{-1}A^{r-1}$. Each of the steps will be analyze below.

4.1 Exponentiation in $GF((2^n)^m)$

The first step in the algorithm above is the computation of A^{r-1} where $A \in GF((2^n)^m)$. Notice that r can be expressed as a sum of powers as follows:

$$r - 1 = \frac{2^{nm} - 1}{2^n - 1} - 1 = 2^n + 2^{2n} + 2^{3n} + \cdots + 2^{(m-1)n}$$

A^{r-1} can now be computed using addition chains. The method requires

$$\lfloor \log_2(m - 1) \rfloor + H_w(m - 1) - 1 \tag{7}$$

general multiplications and at most $m - 1$ exponentiations to the power of 2^n [6], with both types of operations performed in $GF((2^n)^m)$ ($H_w()$ denotes the Hamming weight of the binary representation of the operand). Notice that the number of multiplications in $GF((2^n)^m)$ as given by (7) determines in essence the overall complexity of the inversion algorithms. Exponentiation is realized as explained below. Let B and C be elements of $GF((2^n)^m)$. We want to find $C(x) = B(x)^{2^n}$, where $B(x) = \sum_{i=0}^{m-1} b_i x^i$. This is done as follows [12]:

$$C(x) = \sum_{i=0}^{m-1} c_i x^i = \left(\sum_{i=0}^{m-1} b_i x^i \right)^{2^n} = \sum_{i=0}^{m-1} b_i x^{i2^n}, \quad b_i \in GF(2^n). \qquad (8)$$

Assuming $2^n > m-1$, there are $m-1$ powers of x which must be reduced modulo the field polynomial $P(x)$, namely the powers x^{i2^n}, $i = 1, 2, \ldots, m-1$. We use the following notation for the representation of these powers in the residue classes modulo $P(x)$:

$$x^{i2^n} = s_{0,i} + s_{1,i} x + \cdots + s_{m-1,i} x^{m-1} \bmod P(x), \quad i = 1, 2, \ldots, m-1.$$

Using the coefficients $s_{j,i}$, the exponentiations in (8) can be expressed in matrix form as

$$\begin{pmatrix} c_0 \\ c_1 \\ \vdots \\ c_{m-1} \end{pmatrix} = \begin{pmatrix} 1 & s_{0,1} & s_{0,2} & \cdots & s_{0,m-1} \\ 0 & s_{1,1} & s_{1,2} & \cdots & s_{1,m-1} \\ \vdots & \vdots & \vdots & \ddots & \vdots \\ 0 & s_{m-1,1} & s_{m-1,2} & \cdots & s_{m-1,m-1} \end{pmatrix} \begin{pmatrix} b_0 \\ b_1 \\ \vdots \\ b_{m-1} \end{pmatrix}. \qquad (9)$$

A main computational advantage occurs if $P(x)$ is chosen to have only binary coefficients, as suggested in [22, 1]. In this case, all powers $x^a \bmod P(x)$ belong to a subfield whose elements are represented by binary polynomials. In particular, all coefficients $s_{i,j}$ in (9) are binary, i.e., elements from $GF(2)$. Since both n and $P(x)$ are known ahead of time, and thus the entire exponentiation matrix in (9), one exponentiation is reduced to only $(m^2 - 3m + 2)/2$ additions in $GF(2^n)$ on average. Moreover, exponentiations of the form $B(x)^{2^{ln}}$, $l > 1$, which occur in the algorithm, can be computed with one *single* matrix multiplication which is analogous to (9).

4.2 Multiplication in $GF((2^n)^m)$, where the Product is an Element of $GF(2^n)$

The second step performs the operation

$$A^r = A^{r-1} A \bmod P(x),$$

where $A^r \in GF(2^n)$, and the two operands are elements in $GF((2^n)^m)$. We will show that this operation can be considerably less costly than general multiplication in $GF((2^n)^m)$ if $P(x)$ is chosen carefully. We consider the multiplication $H(x) = F(x)G(x) \bmod P(x)$ where $F, G \in GF((2^n)^m)$ and $H \in GF(2^n)$. First, we consider the pure polynomial multiplication of F and G:

$$H'(x) = F(x)G(x) = \left(\sum_{i=0}^{m-1} f_i x^i \right) \left(\sum_{i=0}^{m-1} g_i x^i \right) = \left(\sum_{i=0}^{2m-2} h'_i x^i \right).$$

We know that $H'(x) \equiv H(x) = h_0 \bmod P(x)$, i.e., that all but the zero coefficient of $H'(x)$ vanish after reduction modulo $P(x)$. Hence we only have to compute those coefficients h'_i, $i = 0, 1, \cdots, m - 2$, which influence h_0. For instance, for $m = 11$ and $P(x) = x^{11} + x^2 + 1$, it follows that $H(x) = h_0 = h'_0 + h'_{11} + h'_{20}$ which requires only 12 multiplications and 11 additions in $GF(2^n)$ as opposed to 121 multiplications and 100 additions for a general multiplication.

4.3 Inversion in $GF(2^n)$ and Multiplication of an Element from $GF(2^n)$ with an Element from $GF((2^n)^m)$

The third and fourth steps carry small complexities since both involve operations with elements of the subfield. First, we calculate the inverse of A^r with two table look-ups [22] since A^r is an element of the ground field. Finally, we compute $A^{-1} = (A^r)^{-1} A^{r-1}$ by multiplying $(A^r)^{-1}$, which is also an element of $GF(2^n)$, with A^{r-1}, an element of $GF((2^n)^m)$. This last operation requires m multiplications in $GF(2^n)$. Notice that there is no reduction modulo $P(x)$ since all arithmetic is done in $GF(2^n)$.

4.4 Inversion in $GF((2^{16})^{11})$

As a way of summarizing the previous sections, we consider the special case for which $n = 16$ and $m = 11$. In this case, we chose as field polynomial the trinomial $P(x) = x^{11} + x^2 + 1$. Using this polynomial, one finds A^{r-1} with 4 multiplications in $GF((2^{16})^{11})$ and 5 exponentiations to the 2^n, 2^{2n}, and 2^{4n} powers. $A^r = A^{r-1}A$ can be computed using 12 multiplications and 10 additions in $GF(2^{16})$, $(A^r)^{-1}$ requires one inversion in the subfield $GF(2^{16})$, and $(A^r)^{-1}A^{r-1}$ involves 11 subfield multiplications in $GF(2^{16})$. We would like to mention that the extended Euclidean algorithm in [22] allows an inversion in almost 3 multiplication times (assuming an optimized multiplication routine) for the same composite field. Although the algorithm is faster, we believe that our method can be of advantage in situations where *predictable* timings for the inversion are desired.

5 Fast Multiplication in Composite Galois Fields $GF((2^n)^m)$

With respect to complexity, field multiplication is the next costly operation in EC systems. Since the new point multiplication algorithm from Sect. 3 trades field inversions for field multiplications, it is especially attractive to provide efficient multiplication algorithms. In this section we apply the Karatsuba-Ofman Algorithm (KOA) to polynomials over Galois fields $GF(2^n)$ of degree $m-1$ which represent a field element in $GF((2^n)^m)$. First, we consider the general KOA as it is applied to two polynomials $A(x)$ and $B(x)$ with maximum degree $m-1$ over the field $\mathcal{F}$. We derive new formulae for the multiplicative and additive complexity for the cases $m = 2^t l$ and $m = 2^t l - 1$ based on the analysis by [19]. Finally, we define two new operations, table look-up (TLU) and exponent addition (EXPA), and derive their complexities for the two cases. These two operations are of central importance for an exact complexity analysis of a software implementation of the KOA in composite fields.

5.1 Complexity of the KOA for Polynomials of Degree $2^t l - 1$

In this section we generalize Theorem 1 from [19]. We consider the product of two polynomials $A(x)$ and $B(x)$ with maximum degree of $2^t l - 1$ over a field $\mathcal{F}$. In particular, we want to find $C(x) = A(x)B(x)$ such that $\deg(C(x)) \leq 2m - 2$ with $m = 2^t l$, t an integer. Since the algorithm can only run for t iterations (as many powers of 2 as there are in m), we obtain in the final step polynomials with maximum degree of $l - 1$ which are then multiplied using the school book method.

Theorem 3 *Consider two arbitrary polynomials in one variable of degree less than or equal to $m - 1$ where $m = 2^t l$, with coefficients in a field $\mathcal{F}$ of characteristic 2. Then, by using the Karatsuba-Ofman algorithm the polynomials can be multiplied with:*

$$\#MUL = l^2 \left(\frac{m}{l}\right)^{\log_2 3} = l^{2-\log_2 3} m^{\log_2 3}$$

$$\#ADD = (l-1)^2 \left(\frac{m}{l}\right)^{\log_2 3} + (8l - 2)\left(\frac{m}{l}\right)^{\log_2 3} - 8m + 2$$

Notice that for $l = 1$ the expressions in Theorem 3 reduce to those given in Theorem 1 of [19].

5.2 Complexity of the KOA for Polynomials of Degree $2^t l - 2$

In the previous section, we covered the case $m = 2^t l$. As stated earlier we often choose $\gcd(n, m) = 1$. Since it is often desired to have n even there is a need to consider the case for $m = 2^t l - 1$. In particular, we want to find $C(x) = A(x)B(x)$ such that $\deg(C(x)) \leq 2m - 2$ with $m = 2^t l - 1$, t an integer. Then, we can represent $A(x)$ and $B(x)$ by adding an extra term with coefficient $a_m = 0$. For convenience, we will introduce the parameter $r = m + 1$ and express $A(x)$ and $B(x)$ as follows:

$$A(x) = x^{\frac{r}{2}}(0x^{\frac{r}{2}-1} + a_{r-2}x^{\frac{r}{2}-2} + \cdots + a_{\frac{r}{2}}) + (a_{\frac{r}{2}-1}x^{\frac{r}{2}-1} + \cdots + a_0)$$
$$B(x) = x^{\frac{r}{2}}(0x^{\frac{r}{2}-1} + b_{r-2}x^{\frac{r}{2}-2} + \cdots + b_{\frac{r}{2}}) + (b_{\frac{r}{2}-1}x^{\frac{r}{2}-1} + \cdots + b_0) \quad (10)$$

Notice that now the polynomials $A(x)$ and $B(x)$ have an even number of coefficients ($r = m + 1 = 2^t l$), allowing us to split them in half and to apply the general KOA to (10) t times. This reduces this problem to the case for $m = 2^t l$, permitting us to apply the same equations. However, since we have one less coefficient the final multiplicative and additive complexities are reduced. Theorem 4 summarizes the results.

Theorem 4 *Consider two arbitrary polynomials in one variable of degree less than or equal to $m - 1$ where $m = 2^t l - 1$, with coefficients in a field $\mathcal{F}$ of characteristic 2. Then, by using the Karatsuba-Ofman algorithm the polynomials can be multiplied with:*

$$\#MUL = l^2 \left(\frac{m+1}{l}\right)^{\log_2 3} - 2l + 1$$

$$\#ADD \leq (l-1)^2 \left(\frac{m+1}{l}\right)^{\log_2 3} + (8l - 2) \left(\frac{m+1}{l}\right)^{\log_2 3} - 8(m+1) + 2$$

5.3 Complexity Analysis for Software Implementations

It was shown in [22, 1] that multiplication and inversion in $GF(2^n)$ can be done through table look-up. Since all non-zero elements of $GF(2^n)$ form a cyclic group we can express all elements $a_i \in GF(2^n)$ as a multiple of a primitive element α: $a_i = \alpha^i$. Then, we store all the pairs (a_i, i) in two tables, log and antilog, sorted by the first component (a_i) and second component (i), respectively. Thus, the product of two elements $a_j, a_k \in GF(2^n)$ can be obtained as follows:

$$a_j a_k = \text{antilog}(\log(a_j) + \log(a_k)) \pmod{2^n - 1} \tag{11}$$

Notice that (11) implies that two elements of the ground field $GF(2^n)$ can be multiplied using three table look-up operations and one addition modulo the order of the multiplicative group (exponent addition). It is important to point out that depending on the hardware platform (e.g., microprocessor, RISC, etc.) the relative speed for the two types of operations can differ dramatically. For instance, in our implementation where $n = 16$ it was found that access to the large look-up tables took 6 clock cycles on a DEC Alpha workstation, whereas element addition and exponent addition took about 2 clock cycles on average. Thus, in order to obtain valid performance predictions one needs exact counts of the number of operations. Based on these two new operations table look-up (TLU) and exponent addition (EXPA) we have derived new formulae and re-written Theorems 3 and 4 as follows. For more details on the derivation of the theorems and corollaries, refer to [4].

Corollary 1 *Consider two arbitrary polynomials in one variable of degree less than or equal to $m - 1$ where $m = 2^t l$, with coefficients in a field $\mathcal{F}$ of characteristic 2. Then, by using the Karatsuba-Ofman algorithm the polynomials can be multiplied with:*

$$\#ADD = (l-1)^2 \left(\frac{m}{l}\right)^{\log_2 3} + (8l - 2) \left(\frac{m}{l}\right)^{\log_2 3} - 8m + 2$$

$$\#TLU = l(l+2) \left(\frac{m}{l}\right)^{\log_2 3}$$

$$\#EXPA = l^2 \left(\frac{m}{l}\right)^{\log_2 3}$$

Corollary 2 *Consider two arbitrary polynomials in one variable of degree less than or equal to* $m - 1$ *where* $m = 2^t l - 1$, *with coefficients in a field* $\mathcal{F}$ *of characteristic 2. Then, by using the Karatsuba-Ofman algorithm the polynomials can be multiplied with:*

$$\#ADD \leq (l-1)^2 \left(\frac{m+1}{l}\right)^{\log_2 3} + (8l-2) \left(\frac{m+1}{l}\right)^{\log_2 3} - 8(m+1) + 2$$

$$\#TLU \leq \left(\frac{m+1}{l}\right)^{\log_2 3} l(l+2) - 2l - 1$$

$$\#EXPA = l^2 \left(\frac{m+1}{l}\right)^{\log_2 3} - 2l + 1$$

5.4 Multiplication in $GF((2^{16})^{11})$

We summarize this section by considering the complexity of a multiplication in $GF((2^n)^m)$ for $n = 16$ and $m = 11$. We can apply Corollary 2 and let $m = 11$, $l = 3$, and $t = 2$. From there we obtain that one needs 140 additions and 76 multiplications in $GF(2^{16})$ or equivalently, at most 140 additions, 124 table look-ups, and 76 exponent additions. On the other hand, when using the school book method for multiplication one would require 121 multiplications and 100 coefficient additions or, in terms of table look-ups, 100 coefficient additions, $121 + 22 = 143$ table look-ups, and 121 exponent additions. If one compares both complexities and ignores exponent and coefficient additions, one can readily see that the theoretical improvement in the timing for the multiplication operation when using the Karatsuba-Ofman algorithm would be about 12.5 percent.

6 Implementation and Timings

This section describes the application of the various algorithms to an actual EC system over the field $GF(2^{16})^{11}) \cong GF(2^{176})$. A DEC Alpha 3000, a 175 MHz RISC architecture, was used to perform all measurements. Table 2 shows the timings for several arithmetic operations in $GF((2^{16})^{11})$. It was found that by applying the KOA to field multiplication one obtains a 10% improvement over the school book method. Using the timings in Table 2, we computed estimates for the timings for repeated point doublings using individual doublings and direct doublings for the computation of $8P$ and $16P$. We compared these predictions with the actual timings as shown in Table 3. Interestingly, the actual timings are considerably better than expected. We attribute this observation to the reduced overhead in the software implementation (e.g., fewer function calls and variable initializations). Table 4 compares timings for point multiplication for different parameters k in the k-ary method and the improved k-ary method including the formulae for direct point doubling. It was found that the optimum value for the window size in the k-ary method was $k = 4$. We achieved a speed-up of almost 19% using the new formulae. Table 5 presents the timings for several algorithms

Table 2. Timings for various field operations in $GF((2^{16})^{11})$.

Type of Operation	Average Timing (μsec)
176 bit addition	1.19
176 bit squaring	4.23
176 x 176 bit multiplication	38.56
176 bit inverse	158.73

Table 3. Timing comparison: Individual doublings vs. direct computation of several doublings in $GF((2^{16})^{11})$.

Calculation	Method	Predicted Timing	Measured Timing	% Improvement Predicted	% Improvement Measured
8P	Direct Doublings	748.41 μsec	904.812 μsec	0.30	12.5
	Individual Doublings	750.78 μsec	1.035 msec		
16P	Direct Doublings	978.62 μsec	1.141 msec	2.24	17.85
	Individual Doublings	1.001 msec	1.389 msec		

used to compute nP. Notice that the last entry of Table 5 corresponds to the nP calculation where the point P is known ahead of time, thus it is possible to pre-compute a table of multiples of P. This implies that the formulae derived in Sect. 3 were not used in this algorithm.

Acknowledgements

We would like to thank Dan Beauregard for providing the EC implementation.

7 Appendix - Formulae for $8P$ and $16P$

In this section we present the improved formulae to find $8P = 2^3 P = (x_3, y_3)$ and $16P = 2^4 P = (x_4, y_4)$ where $P = (x, y)$ is an element of prime order belonging to the cyclic subgroup corresponding to the largest prime factor in the order of E.

1. Formulae for $8P = 2^3 P = (x_3, y_3)$

$$x_3 = \frac{\omega^2 + \omega\rho}{\rho^2} + a$$

$$y_3 = \frac{(v^2)^2 + \omega\rho x_3}{\rho^2} + x_3$$

Table 4. Comparison of average time required to perform the nP calculation using the regular k-ary method and the improved k-ary with four direct doublings

Method	Window Size k	Average Timing (in msec)
k-ary	3	87
	4	84
	5	88
k-ary with formulae	4	68

Table 5. Timings for elliptic curve operations

Operations	Method/Type of Operation	Average Timing
Multiply new elliptic curve point ($n \rightarrow 176bits$)	Double and Add Algorithm	95 msec
	k-ary method ($k = 4$)	84 msec
	k-ary method with formulae ($k = 4$)	68 msec
Multiply known elliptic curve point ($n \rightarrow 176bits$)	Brickell's Algorithm ($base = 2^4 = 16$)	20 msec

2. Formulae for $16P = 2^4 P = (x_4, y_4)$

$$x_4 = \frac{\theta^2 + \theta\mu\rho^2}{(\mu\rho^2)^2} + a$$

$$y_4 = \frac{(\mu^2)^2 + (\theta\mu\rho^2)x_4}{(\mu\rho^2)^2} + x_4$$

where $\gamma = x^2, \eta = \gamma + y, \delta = \eta^2 + \eta x + a\gamma, \xi = \eta x + \gamma, \zeta = \delta(\delta + \xi) + \gamma^2\gamma, \tau = \delta\gamma,$ $\upsilon = \zeta^2 + \tau\zeta + \tau^2 a, \rho = \upsilon\tau^2, \omega = \upsilon(\upsilon + \zeta\tau) + (\tau\delta^2)^2 + \rho, \mu = \omega^2 + \omega\rho + a\rho^2,$ and $\theta = \mu^2 + \mu(\omega\rho) + \mu\rho^2 + (\upsilon^2\rho)^2.$

References

1. D. Beauregard. Efficient algorithms for implementing elliptic curve public-key schemes. Master's thesis, ECE Dept., Worcester Polytechnic Institute, Worcester, MA, May 1996.
2. H. Cohen. *A Course in Computational Algebraic Number Theory*. Springer-Verlag, Berlin, 1993.
3. R.J. Fateman. Polynomial multiplication, powers and asymptotic analysis: Some comments. *SIAM J. Comput.*, 7(3):196–21, September 1974.
4. J. Guajardo. Efficient algorithms for elliptic curve cryptosystems. Master's thesis, ECE Dept., Worcester Polytechnic Institute, Worcester, MA, May 1997.

5. G. Harper, A. Menezes, and S. Vanstone. Public-key cryptosystems with very small key lengths. In *Advances in Cryptology — EUROCRYPT '92*, pages 163–173, May 1992.

6. T. Itoh and S. Tsujii. A fast algorithm for computing multiplicative inverses in $GF(2^m)$ using normal bases. *Information and Computation*, 78:171–177, 1988.

7. D.E. Knuth. *The Art of Computer Programming. Volume 2: Seminumerical Algorithms*. Addison-Wesley, Reading, Massachusetts, 2nd edition, 1981.

8. N. Koblitz. Elliptic curve cryptosystems. *Mathematics of Computation*, 48:203–209, 1987.

9. N. Koblitz. Constructing elliptic curve cryptosystems in characteristic 2. In *Advances in Cryptology — CRYPTO '90*, pages 156–167. Springer-Verlag, Berlin, 1991.

10. C. K. Koc. Analysis of sliding window techniques for exponentiation. *Computers and Mathematics with Applications*, 30(10):17–24, November 1995.

11. J. Koeller, A. Menezes, M. Qu, and S. Vanstone. Elliptic Curve Systems. Draft 8, IEEE P1363 Standard for RSA, Diffie-Hellman and Related Public-Key Cryptography, May 1996. working document.

12. R. Lidl and H. Niederreiter. *Finite Fields*, volume 20 of *Encyclopedia of Mathematics and its Applications*. Addison-Wesley, Reading, Massachusetts, 1983.

13. A. J. Menezes, P. C. van Oorschot, and S. A. Vanstone. *Handbook of Applied Cryptography*. CRC Press, Boca Raton, Florida, 1997.

14. A. J. Menezes, S. A. Vanstone, and R. J. Zuccherato. Counting points on elliptic curves over F_{2^m}. *Mathematics of Computation*, 60(201):407–420, January 1993.

15. A.J. Menezes. *Elliptic Curve Public Key Cryptosystems*. Kluwer Academic Publishers, 1993.

16. V. Miller. Uses of elliptic curves in cryptography. In *Advances in Cryptology — CRYPTO '85*, pages 417–426. Springer-Verlag, Berlin, 1986.

17. C. Paar. *Efficient VLSI Architectures for Bit-Parallel Computation in Galois Fields*. PhD thesis, (Engl. transl.), Institute for Experimental Mathematics, University of Essen, Essen, Germany, June 1994.

18. C. Paar. Some remarks on efficient inversion in finite fields. In *1995 IEEE International Symposium on Information Theory*, page 58, Whistler, B.C. Canada, September 17–22 1995.

19. C. Paar. A new architecture for a parallel finite field multiplier with low complexity based on composite fields. *IEEE Transactions on Computers*, 45(7):856–861, July 1996.

20. C. Paar and P. Soria-Rodriguez. Fast arithmetic architectures for public-key algorithms over galois fields $GF((2^n)^m)$. In *Advances in Cryptology — EUROCRYPT '97*, pages 363–378, 1997.

21. R. Schroeppel, H. Orman, S. O'Malley, and O. Spatscheck. Fast key exchange with elliptic curve systems. *Advances in Cryptology, Crypto 95*, pages 43–56, 1995.

22. E. De Win, A. Bosselaers, S. Vandenberghe, P. De Gersem, and J. Vandewalle. A fast software implementation for arithmetic operations in $GF(2^n)$. In *Asiacrypt '96*. Springer Lecture Notes in Computer Science, 1996.

An Improved Algorithm for Arithmetic on a Family of Elliptic Curves*

Jerome A. Solinas

National Security Agency, Ft. Meade, MD 20755, USA

Abstract. It has become increasingly common to implement discrete-logarithm based public-key protocols on elliptic curves over finite fields. The basic operation is *scalar multiplication:* taking a given integer multiple of a given point on the curve. The cost of the protocols depends on that of the elliptic scalar multiplication operation.

Koblitz introduced a family of curves which admit especially fast elliptic scalar multiplication. His algorithm was later modified by Meier and Staffelbach. We give an improved version of the algorithm which runs 50% faster than any previous version. It is based on a new kind of representation of an integer, analogous to certain kinds of binary expansions. We also outline further speedups using precomputation and storage.

Keywords: elliptic curves, exponentiation, public-key cryptography.

1 Introduction

It has become increasingly common to implement discrete-logarithm based public-key protocols on elliptic curves over finite fields. More precisely, one works with the points on the curve, which can be added and subtracted. If we add the point P to itself n times, we denote the result by nP. The operation of computing nP from P is called *scalar multiplication* by n. Elliptic public-key protocols are based on scalar multiplication, and the cost of executing such protocols depends mostly on the complexity of the scalar multiplication operation.

Scalar multiplication on an elliptic curve is analogous to exponentiation in the multiplicative group of integers modulo a fixed integer m. Various techniques have been developed [1] to speed modular exponentiation using memory and precomputations. Such methods, for the most part, carry over to elliptic scalar multiplication.

There are also efficiency improvements available in the elliptic case that have no analogue in modular exponentiation. There are three kinds of these:

1. One can choose the curve, and the base field over which it is defined, so as to optimize the efficiency of elliptic scalar multiplication. Thus, for example, one might choose the field of integers modulo a Mersenne prime, since modular

---

* This paper presents the results of cryptographic research conducted at NSA and does not necessarily represent the policies of the NSA or U.S. Government.

reduction is particularly efficient [2] in that case. This option is not available for, say, RSA systems, since the secret primes are chosen randomly in order to maintain the security of the system.

2. One can use the fact that subtraction of points on an elliptic curve is just as efficient as addition. (The analogous statement for integers (mod m) is false, since modular division is more expensive than modular multiplication.) The efficient methods for modular exponentiation all involve a sequence of squarings and multiplications that is based on the binary expansion of the exponent. The analogous procedure for elliptic scalar multiplication uses a sequence of doublings and additions of points. If we allow subtractions of points as well, we can replace [3] the binary expansion of the coefficient n by a more efficient *signed binary expansion* (i.e., an expansion in powers of two with coefficients 0 and ± 1).

3. One can use *complex multiplication*. Every elliptic curve over a finite field[2] comes equipped with a set of operations which can be viewed as multiplication by complex algebraic integers (as opposed to ordinary integers). These operations can be carried out efficiently for certain families of elliptic curves. In these cases, they can be utilized in various ways [5] to increase the efficiency of elliptic scalar multiplication.

It is the purpose of this paper to present a new technique for elliptic scalar multiplication. This new algorithm incorporates elements from all three of the above categories. The new method is 50% faster than any method previously known for operating on a non-supersingular elliptic curve.

2 Field and Elliptic Operations in $\mathbb{F}_{2^m}$

We begin with a brief survey of the various operations we will need in the field $\mathbb{F}_{2^m}$ and on elliptic curves over this field.

Squaring. We will assume that the field $\mathbb{F}_{2^m}$ is represented in terms of a *normal basis*: a basis over $\mathbb{F}_2$ of the form

$$\left\{ \theta, \theta^2, \theta^{2^2}, \ldots, \theta^{2^{m-1}} \right\} .$$

The advantage of this representation is that squaring a field element can be accomplished by a one-bit cyclic shift of the bit string representing the element. This property will be crucial in what follows. If m is not divisible by 8, then one can use Gaussian cyclotomic periods to construct easily [6] an efficient normal basis for $\mathbb{F}_{2^m}$. (Since our application will require m to be prime, we can always use the Gaussian method.)

Our emphasis in this paper will be the case in which the field arithmetic is be implemented in hardware. Although the algorithms that follow will be efficient

---

[2] We restrict our attention to elliptic curves that are not *supersingular*, since such curves are cryptographically weak. (See [4].)

in software as well, the full advantage of our method occurs in hardware, where the bit shifts (and therefore field squarings) are virtually free.

Addition and Multiplication. We may neglect the cost of additions in $\mathbb{F}_{2^m}$ since they involve only bitwise XORs. A multiplication (of distinct elements) takes about m times as long, just as in the case of integer arithmetic. The cost of an elliptic operation depends mostly on the number of field multiplications it uses.

Inversion. Multiplicative inversion in $\mathbb{F}_{2^m}$ can be performed in

$$L(m-1) + W(m-1) - 2$$

field multiplications using the method of [7]. Here $L(k)$ represents the length of the binary expansion of k, and $W(k)$ the number of ones in the expansion. This fact may be a consideration when choosing the degree m. (Alternatively, one can use the Euclidean algorithm [8], but one must first convert from the normal basis representation to the more familiar polynomial basis form, and then back again after the inversion.)

Elliptic Addition. The standard equation for an elliptic curve over $\mathbb{F}_{2^m}$ is the *Weierstrass equation*

$$E: \quad y^2 + xy = x^3 + ax^2 + b \tag{1}$$

where $b \neq 0$. Public key protocols based on this curve work on the group consisting of the points (x, y) on this curve, along with the group identity $\mathcal{O}$. (The element $\mathcal{O}$ is called the *point at infinity*, but it is most convenient to represent it[3] by $(0,0)$.) The following algorithm inputs the points $P_0 = (x_0, y_0)$ and $P_1 = (x_1, y_1)$ on E and returns their sum $P_2 = (x_2, y_2)$.

Algorithm 1. *(Elliptic Group Operation)*

> If $P_0 = \mathcal{O}$ then output $P_2 \leftarrow P_1$ and stop
> If $P_1 = \mathcal{O}$ then output $P_2 \leftarrow P_0$ and stop
> If $x_0 = x_1$
>> then
>>> if $y_0 + y_1 = x_1$ then output $\mathcal{O}$ and stop
>>> else
>>>> $\lambda \leftarrow x_1 + y_1/x_1$
>>>> $x_2 \leftarrow \lambda^2 + \lambda + a$
>>>> $y_2 \leftarrow x_1^2 + (\lambda + 1) x_2$
>> else
>>> $\lambda \leftarrow (y_0 + y_1)/(x_0 + x_1)$
>>> $x_2 \leftarrow \lambda^2 + \lambda + x_0 + x_1 + a$

---
[3] This does not cause confusion, because the origin is never on E.

$$y_2 \leftarrow (x_1 + x_2)\lambda + x_2 + y_1$$

Output $P_2 \leftarrow (x_2, y_2)$

To *subtract* the point $P = (x, y)$, one adds the point $-P = (x, x + y)$.

Except for the special cases involving $\mathcal{O}$, the above addition and subtraction operations each require 1 multiplicative inversion and 2 multiplications.[4] (As always, we disregard the cost of adding and squaring field elements.)

Elliptic Scalar Multiplication. The basic technique for elliptic scalar multiplication is the *addition-subtraction method.* This begins with the *nonadjacent form* (NAF) of the coefficient n: a signed binary expansion with the property that no two consecutive coefficients are nonzero. For example,

$$\text{NAF}(29) = \langle 1, 0, 0, -1, 0, 1 \rangle \tag{2}$$

since $29 = 32 - 4 + 1$.

Just as every positive integer has a unique binary expansion, it also has a unique NAF. Moreover, $\text{NAF}(n)$ has the fewest nonzero coefficients of any signed binary expansion of n (see [1]). There are several ways to construct the NAF of n from its binary expansion. We present the one that most resembles the new algorithm we will present in §3.

The idea is to divide repeatedly by 2. Recall that one can derive the binary expansion of an integer by dividing by 2, storing off the remainder (0 or 1), and repeating the process with the quotient. To derive a NAF, one allows remainders of 0 or ± 1. If the remainder is to be ± 1, one chooses whichever makes the quotient even.

Algorithm 2. *(NAF)*

```
Input n
Set k ← n
Set S ← ⟨⟩
While k > 0
        If k odd
                then set u ← 2 - (k (mod 4))
                else set u ← 0
        Set k ← k - u
        Prepend u to S
        Set k ← k/2
EndWhile
Output S
```

For example, to derive (2), one applies Alg. 2 with $n = 29$. The results are shown in Fig. 1.

---

[4] There does exist a faster algorithm for doubling a point, but we relegate it to the Appendix since it does not fit well with the best hardware implementations of normal bases.

Fig. 1. Computing a NAF.

k	u	S
29		$\langle \rangle$
28	1	
14		$\langle 1 \rangle$
14	0	
7		$\langle 0, 1 \rangle$
8	-1	
4		$\langle -1, 0, 1 \rangle$
4	0	
2		$\langle 0, -1, 0, 1 \rangle$
2	0	
1		$\langle 0, 0, -1, 0, 1 \rangle$
0	1	
0		$\langle 1, 0, 0, -1, 0, 1 \rangle$

Note that, although we have phrased the algorithm in terms of integer arithmetic, it can be implemented in terms of bit operations on the binary expansion of n. No arithmetic operations are needed beyond integer addition by 1.

In the derivation of the ordinary binary expansion, the sequence k is decreasing, but that is not true in general in Alg. 2. As a result, the NAF of a number may be longer than its binary expansion. Fortunately, it can be at most one bit longer, because

$$2^\ell < 3n < 2^{\ell+1}$$

where ℓ is the bit length of NAF(n). (See [3].)

We now implement elliptic scalar multiplication using the NAF. Given the NAF

$$n = \sum_{i=0}^{\ell-1} e_i 2^i \ ,$$

the elliptic scalar multiplication $Q = nP$ is performed as follows.

Algorithm 3. *(Addition-Subtraction Method)*

```
        Input  P
        Set Q ← P
        For i = ℓ − 2 downto 1 do
                Set Q ← 2Q
                If e_i = 1 then set Q ← Q + P
                If e_i = −1 then set Q ← Q − P
        Output Q
```

The cost of Alg. 3 depends on the bit length ℓ of $\text{NAF}(n)$, which we now estimate. It follows from the Hasse theorem that the order of an elliptic curve over $\mathbb{F}_{2^m}$ is

$$\#E(\mathbb{F}_{2^m}) = 2^m + O(2^{m/2}) \ . \tag{3}$$

Most public-key protocols on elliptic curves use a base point of prime order p. Since all of the curves (1) have even order, then p must be at most $2^{m-1} + O(2^{m/2})$. We can assume that $n < p$; thus[5] $\ell \leq m$.

It follows that Alg. 3 requires about m doubles at most. The number of additions is (about) the number of nonzero coefficients in $\text{NAF}(n)$. The average density of nonzero coefficients among NAF's is $1/3$ (see [3]). Therefore the average cost of Alg. 3 is $\sim m$ doubles and $\sim m/3$ additions, for a total of $\sim 4m/3$ elliptic operations. This is about one-eighth faster than the *binary method*, which uses the ordinary binary expansion in place of the NAF and therefore requires an average of $\sim m/2$ elliptic additions rather than $\sim m/3$.

3 Anamolous Binary Curves

Two extremely convenient families of curves [5] are the *anamolous binary curves* (or ABC's). These are the curves E_0 and E_1 defined over $\mathbb{F}_2$ by

$$E_a : y^2 + xy = x^3 + a\,x^2 + 1 \ .$$

We denote by $E_a(\mathbb{F}_{2^m})$ the group of $\mathbb{F}_{2^m}$-rational points on E_a. This is the group on which the public-key protocols are performed. The group should be chosen so that it is computationally difficult to compute discrete logarithms of its elements. Thus, for example, the order $\#E_a(\mathbb{F}_{2^m})$ should be divisible by a large prime (see [9]). Ideally, $\#E_a(\mathbb{F}_{2^m})$ should be a prime or the product of a prime and small integer. This can only happen when m is itself prime, for otherwise there are large divisors arising from subgroups $E_a(\mathbb{F}_{2^d})$ where d divides m.

Actually, the orders $\#E_a(\mathbb{F}_{2^m})$ are never prime, because they always contain the point $(0, 1)$, which is easily seen to have order 2. The best result to be hoped for, then, is that the order is twice a prime. This happens relatively frequently for E_1. The values of $m \leq 512$ for which $\#E_1(\mathbb{F}_{2^m})$ is twice a prime are

$$m = 3, 5, 7, 11, 17, 19, 23, 101, 107, 109, 113, 163, 283, 311, 331, 347, 359 \ .$$

The curves E_0 contain the points $(1, 0)$ and $(1, 1)$, which are easily seen to have order 4. The best result to be hoped for among the curves E_0, then, is that the order is 4 times a prime. The values of $m \leq 512$ for which this happens are

$$m = 5, 7, 13, 19, 23, 41, 83, 97, 103, 107, 131, 233, 239, 277, 283, 349, 409 \ .$$

---

[5] A further one-bit improvement on this bound is possible if we use the identity

$$n(x, y) = (p - n)(x, x + y)$$

whenver $n > p/2$. Moreover, if a has trace 0 over $\mathbb{F}_2$, we save yet another bit since the order of E must be divisible by 4.

Since the ABC's are defined over $\mathbb{F}_2$, they have the property that, if $P = (x, y)$ is a point on E_a, then so is the point (x^2, y^2). Moreover, one can verify from Alg. 1 that

$$(x^4, y^4) + 2(x, y) = (-1)^{1-a}(x^2, y^2) \tag{4}$$

for every (x, y) on E_a. This relation can be written more easily in terms of the Frobenius (squaring) map over $\mathbb{F}_2$:

$$\tau(x, y) = (x^2, y^2) \ .$$

Using this notation, (4) becomes

$$\tau(\tau P) + 2P = (-1)^{1-a}\tau P$$

for all $P \in E$. Symbolically, this can be written

$$(\tau^2 + 2)P = (-1)^{1-a}\tau P \ .$$

This means that the squaring map can be regarded as implementing multiplication by the complex number τ satisfying

$$\tau^2 + 2 = (-1)^{1-a}\tau \ .$$

Explicitly, this number is

$$\tau = \frac{(-1)^{1-a} + \sqrt{-7}}{2} \ .$$

By combining the squaring map with ordinary scalar multiplication, we can multiply points on E_a by any element of the ring $\mathbb{Z}[\tau]$. We say that E_a has *complex multiplication* by τ. (See [5].)

The reason why this property is useful for elliptic scalar multiplication is that multiplication by τ, being implemented by squaring, is essentially free when $\mathbb{F}_{2^m}$ is represented in terms of a normal basis. Thus it is worthwhile, when computing nP, to regard n as an element of $\mathbb{Z}[\tau]$ rather than as "just" an integer. More precisely, one replaces the (signed) binary expansion of the coefficient with the (signed) *τ-adic expansion*. That is, one represents n as a sum and difference of distinct powers of τ.

For example, with $a = 0$ we have

$$9 = \tau^5 - \tau^3 + 1 \ . \tag{5}$$

Thus, if $P = (x, y)$ is a point on E_0, then

$$9P = (x^{32}, y^{32}) - (x^8, y^8) + (x, y) \ .$$

The above example gives 9 as what we call a *τ-adic NAF*, since no two consecutive terms are nonzero. (Both [5] and [10] use signed τ-adic expansions, but neither kind has the nonadjacency property.) As we shall see, the use of the τ-adic NAF gives a significant reduction in the number of terms, just as the

NAF gives a significant improvement over the binary expansion in the case of integers.

The τ-adic NAF has a property analogous to the NAF for integers, namely that *every element of the ring* $\mathbf{Z}[\tau]$ *has a unique* τ-*adic NAF*. We shall prove the existence by providing the construction. (The proof of uniqueness is similar to that of the NAF for integers.)

We begin with the observation that $x + y\tau$ is divisible by τ if and only if x is even. One direction of this statement follows from the identity

$$(u + v\tau)\,\tau = -2v + (u + (-1)^{1-a}v)\,\tau \ ,$$

and the other from the fact that, if $x = 2v$, then

$$x + y\tau = (y + (-1)^{1-a}v - \tau v)\,\tau \ .$$

We now present the algorithm [11] for computing the τ-adic NAF. It is completely analogous to Alg. 2, but here we are dividing by τ rather than by 2. The ring $\mathbf{Z}[\tau]$ is Euclidean with norm function

$$N(x + y\tau) = x^2 + (-1)^{1-a}\,xy + 2y^2 \ .$$

Since τ has norm 2, the possible remainders upon division by τ are ± 1. Earlier algorithms chose the remainder that minimized the norm of the quotient; this is analogous to the basic division algorithm for generating the binary expansion of an integer. What we shall do instead is to choose the remainder that makes the quotient divisible by τ (i.e., having real part even). This is analogous to the computation of the NAF for integers.

Algorithm 4. *(τ-adic NAF)*

```
Input x₀, y₀
Set x ← x₀, y ← y₀
Set S ← ⟨⟩
While x ≠ 0 or y ≠ 0,
        If x odd,
                then set u ← 2 − (x − 2y (mod 4))
                else set u ← 0
        Set x ← x − u
        Prepend u to S
        Set (x, y) ← (y + (−1)ᵃ x/2, −x/2)
EndWhile
Output S
```

For example, to derive (5), one applies Alg. 4 with $a = 0$, $x = 9$, and $y = 0$. The results are shown in Fig. 2.

Note that the implementation of Alg. 4 involves nothing more complicated than integer addition. (This is slightly more than is required by Alg. 2, which only adds 1 to an integer.)

Fig. 2. Computing a generalized NAF.

x	y	u	S
9	0		$\langle\,\rangle$
8	0	1	
4	-4		$\langle 1 \rangle$
4	-4	0	
-2	-2		$\langle 0, 1 \rangle$
-2	-2	0	
-3	1		$\langle 0, 0, 1 \rangle$
-2	1	-1	
0	1		$\langle -1, 0, 0, 1 \rangle$
0	1	0	
1	0		$\langle 0, -1, 0, 0, 1 \rangle$
0	0	1	
0	0		$\langle 1, 0, -1, 0, 0, 1 \rangle$

An argument similar to the one [3] in the NAF case proves that the average density of nonzero terms among τ-adic NAF's is $1/3$. There is a drawback to this representation, however: the τ-adic NAF of an integer n is about twice as long as its ordinary NAF. This is because Alg. 4 begins with n, which is an element of $\mathbb{Z}[\tau]$ with norm n^2, and repeatedly divides by τ, which has norm 2.

The solution is to adopt the following modification from [10]. Recall that multiplication by τ is implemented by a one-bit circular shift of each of the m-long bit strings representing the coordinates of P. Multiplication by τ^m, then, involves m such shifts, returning each coordinate to its original state. In other words, $\tau^m P = P$ for all $P \in E_a(\mathbb{F}_{2^m})$. It follows that, if α and β are elements of $\mathbb{Z}[\tau]$ with $\alpha \equiv \beta \pmod{\tau^m - 1}$, then $\alpha P = \beta P$ for all P.

This means that, to multiply by n, one need not work with n itself, but rather the remainder obtained from dividing n by $\tau^m - 1$. Since $\mathbb{Z}[\tau]$ is Euclidean, this remainder will have norm smaller than that of $\tau^m - 1$. The norm of $\tau^m - 1$ is precisely the order of $E_a(\mathbb{F}_{2^m})$, and this is roughly 2^m by (3). Thus the τ-adic NAF of the remainder will have length $\sim m$, only half as long as the τ-adic NAF of n itself. Moreover, the average density is still only $1/3$. To see this, one must examine the distribution of the residues $\pmod{\tau^m - 1}$ of the integers; see [1].

To implement this improvement, one needs a division algorithm in $\mathbb{Z}[\tau]$. The following algorithm inputs the dividend $u + v\tau$ and divisor $r + s\tau$ and outputs a quotient $w + z\tau$ and remainder $x + y\tau$, the latter having smaller norm than the divisor.

Algorithm 5. *(Division in the Ring $\mathbb{Z}[\tau]$)*

      Input u, v, r, s
      Set $k \leftarrow ru + su + 2sv$,

$$\ell \leftarrow rv - su$$

Set $h \leftarrow r^2 + (-1)^{1-a} rs + 2s^2$

Set $w \leftarrow \lfloor k/h \rfloor,$
$z \leftarrow \lfloor \ell/h \rfloor$

Set $x \leftarrow u - rw + 2sz,$
$y \leftarrow v - sw - rz - sz$

Output w, z, x, y

To apply Alg. 5, one needs to express $\tau^m - 1$ as an expression of the form $r + s\tau$. This is done via Lucas sequences. Let $U_0 = 0$, $U_1 = 1$, and

$$U_k = (-1)^{1-a} U_{k-1} - 2 U_{k-2}$$

for $k \geq 2$. It is easy to prove that

$$\tau^m = U_m \tau - 2 U_{m-1} \ .$$

Thus we have the following procedure for computing nP in $E_a(\mathbb{F}_{2^m})$.

Algorithm 6. *(Scalar Multiplication on ABC's)*

   1. Divide n by $U_m \tau - (2 U_{m-1} + 1)$ via Alg. 5.
   2. Compute the τ-adic NAF

$$\langle e_\ell, e_{\ell-1}, \ldots, e_1, e_0 \rangle$$

   of the remainder via Alg. 4.

   3. Set $Q \leftarrow e_\ell P$
   4. For i from $\ell - 1$ downto 1 do
        Set $Q \leftarrow \tau Q (= \text{Shift}[Q])$
        If $e_i = 1$ then set $Q \leftarrow Q + P$
        If $e_i = -1$ then set $Q \leftarrow Q - P$
   5. Output Q.

Except for Step 1 (i.e. Alg. 5), the only arithmetic required by Alg. 6 is binary field arithmetic and integer addition. Alg. 5, on the other hand, requires several multiplications and divisions involving m-bit numbers. Thus it is less well suited to hardware, and more expensive in software, than the other steps.[6] The running time of Step 1, however, is negligible compared to the actual elliptic scalar multiplication (see [10]).

Since $\ell \approx m$, then Alg. 6 requires $\sim m/3$ additions and no doubles. This is at least 50% faster than any of the earlier versions, as is shown in Table 1.

---

[6] On the other hand, Alg. 5 can be replaced by simpler and more efficient algorithms that do much the same thing. For example, one might use a "double-and-add" method of "building up" to the integer n via its binary expansion, reducing when needed by suitable multiples of $\tau^m - 1$. Such reductions would involve additions rather than multiplications and divisions. Details are not available as of this writing, but it seems that an efficient implementation could be developed which would yield a τ-adic NAF of only a few bits over the output of Alg. 5.

Table 1. Comparison of Elliptic Scalar Multiplication Techniques.

Type of Curve	Method		Length of Expansion	Avg. Density	Avg. # of Elliptic Operations
General	Binary Method		m	1/2	$3m/2$
"	Addition-Subtraction	(1989)	m	1/3	$4m/3$
ABC	Koblitz, Balanced	(1991)	$2m$	3/8	$3m/4$
"	Meier-Staffelbach	(1992)	m	1/2	$m/2$
"	τ-adic NAF	(1997)	m	1/3	$m/3$

The "length" and "density" columns give the approximate length of the relevant representation of the number and the average density of nonzero terms. The density figure of 3/8 for Koblitz' "balanced" expansions is from experimental observation and may be only an approximation.

4 Precomputation and Memory Speedups

We can obtain still more dramatic savings by precomputing and storing some "small" τ-adic multiples of P. By this we mean the points αP for which $\alpha \in \mathbf{Z}[\tau]$ has a short τ-adic NAF. These precomputed values can then be used as needed when going through the τ-adic expansion of n. This is essentially a "τ-adic window method." We illustrate with a simple example: that of using windows of a fixed width w.

This method is very similar to the fixed-width version of the window method for ordinary NAF's of integers. Consider the following example. We let the width $w = 4$ and $n = 22310$. Then NAF(n) is given by

$$\langle 1, 0, -1, 0, -1, 0, 0, -1, 0, 0, 1, 0, 1, 0, -1, 0 \rangle \ . \tag{6}$$

We now rewrite (6) by allowing nonzero coefficients to take on the values ± 3, ± 5, ± 7, ± 9 as well as ± 1. (This choice reflects the fact that the odd numbers 1 through 9 are the ones with NAF of length 4 or less.) We go right to left, as in Fig. 3.

As a result, we have the expression

$$22310 = 2^{15} - 5 \cdot 2^{11} - 7 \cdot 2^5 + 3 \cdot 2 \ .$$

Therefore, we can multiply the point P by 22310 by precomputing $3P$, $5P$, $7P$, $9P$ and calculating

$$22310P = 2^{15}P - 2^{11}(5P) - 2^5(7P) + 2(3P)$$

via the suitable generalization of Alg. 3.

Fig. 3. Widening a NAF.

$$
\begin{array}{l}
\langle\, 1\,,\,0\,,\,\text{-}1\,,\,0\,,\,\text{-}1\,,\,0\,,\,0\,,\,\text{-}1\,,\,0\,,\,0\,,\,1\,,\boxed{0\,,\,1\,,\,0\,,\,\text{-}1}\,,\,0\,\rangle \\
\langle\, 1\,,\,0\,,\,\text{-}1\,,\,0\,,\,\text{-}1\,,\,0\,,\,0\,,\boxed{\text{-}1\,,\,0\,,\,0\,,\,1}\,,\,0\,,\,0\,,\,0\,,\,3\,,\,0\,\rangle \\
\langle\, 1\,,\boxed{0\,,\,\text{-}1\,,\,0\,,\,\text{-}1}\,,\,0\,,\,0\,,\,0\,,\,0\,,\,0\,,\,\text{-}7\,,\,0\,,\,0\,,\,0\,,\,3\,,\,0\,\rangle \\
\langle\, 1\,,\,0\,,\,0\,,\,0\,,\,\text{-}5\,,\,0\,,\,0\,,\,0\,,\,0\,,\,0\,,\,\text{-}7\,,\,0\,,\,0\,,\,0\,,\,3\,,\,0\,\rangle
\end{array}
$$

Applying the method for the general width-w case requires

$$C(w) = (2^w - (-1)^w)/3$$

values to be precomputed and stored. The resulting width-w NAF has the property that any w consecutive coefficients include at most one nonzero entry. The average density of nonzero coefficients among width-w NAF's is $(w + 1)^{-1}$.

The same width-w NAF calculations can be used in the τ-adic case. The example analogous to the above is multiplication by

$$\alpha = \tau^{15} - \tau^{13} - \tau^{11} - \tau^8 + \tau^5 + \tau^3 - \tau \ ,$$

since the τ-adic NAF of α is given by (6). To devise a width-w τ-adic NAF of α, we allow nonzero coefficients to take on the values $\pm\beta_3$, $\pm\beta_5$, $\pm\beta_7$, $\pm\beta_9$ as well as ±1, where β_k is the element of $\mathbf{Z}[\tau]$ whose τ-adic NAF is the same as the ordinary NAF of k. (Explicit values are given in Fig. 4.)

Fig. 4. Analogues of the Small Odd Integers.

$\mathrm{NAF}(3) = \langle 1, 0, -1 \rangle$	$\beta_3 = \tau^2 - 1$	$P_3 = (\tau^2 - 1)P$
$\mathrm{NAF}(5) = \langle 1, 0, 1 \rangle$	$\beta_5 = \tau^2 + 1$	$P_5 = (\tau^2 + 1)P$
$\mathrm{NAF}(7) = \langle 1, 0, 0, -1 \rangle$	$\beta_7 = \tau^3 - 1$	$P_7 = (\tau^3 - 1)P$
$\mathrm{NAF}(9) = \langle 1, 0, 0, 1 \rangle$	$\beta_9 = \tau^3 + 1$	$P_9 = (\tau^3 + 1)P$

The calculation shown in Fig. 3 shows that the width-4 τ-adic NAF of α is

$$\alpha = \tau^{15} - \beta_5 \cdot \tau^{11} - \beta_7 \cdot \tau^5 + \beta_3 \cdot \tau \ .$$

Thus one computes

$$\alpha P = \tau^{15}P - \tau^{11}P_5 - \tau^5 P_7 + \tau P_3$$

by precomputing and storing the points P_i given in Fig. 4.

To perform this procedure in general requires enough memory to store $C(w)$ points, including P itself. The precomputation requires $C(w) - 1$ elliptic additions, and no memory other than that used to store the $C(w)$ points. The main computation is the analogue of Alg. 6, performed on a length-m, width-w, τ-adic NAF with average density $(w+1)^{-1}$. The total work, then, is

$$\sim \frac{2^w}{3} + \frac{m}{w+1} \text{ elliptic additions.}$$

Table 2 gives the performance of this algorithm on the curve $E_1(\mathbb{F}_{2^{163}})$ for various widths. (Entries are rounded to the nearest integer.) The case $w = 2$ is the ordinary method of §3. By choosing $w = 4$ or 5, one saves roughly one-third the work. For larger w, the precomputation costs overshadow any savings on the real-time computation.

Table 2. Performance at Various Widths.

Width	Number of Elliptic Operations		
	Precomp-utation	Real Time (avg)	Total (avg)
2	0	52	52
3	2	39	41
4	4	31	35
5	9	26	35
6	20	23	43
7	42	19	61

It is remarkable that one can perform a general elliptic scalar multiplication on $E_1(\mathbb{F}_{2^{163}})$ using only about 35 multiplicative inversions and 70 field multiplications.

One could obtain still further speedups by using more general window methods. These would be straightforward adaptations of existing methods such as those found in [12]. On the other hand, such methods are less automatic than the above fixed-width-window technique, so that more complicated up-front calculations are needed.

Note added during review: the results of [10] have recently been generalized to curves defined over fields of 2^d elements for small d. For example, the curves with complex multiplication by $(\pm 1 + \sqrt{-15})/2$ are defined over $\mathbb{F}_{2^2}$. The results of this paper should also carry over to this more general situation.

References

1. D. Gordon, "A survey of fast exponentiation methods" *(to appear)*.
2. D. E. Knuth, *Seminumerical Algorithms*, Addison-Wesley, 1981, p. 272.

3. F. Morain and J. Olivos, "Speeding up the computations on an elliptic curve using addition-subtraction chains", *Inform. Theor. Appl.* **24** (1990), pp. 531–543.

4. A. Menezes, T. Okamoto and S. Vanstone, "Reducing elliptic curve logarithms to logarithms in a finite field", *Proc. 23rd Annual ACM Symp. on Theory of Computing* (1991), pp. 80–89.

5. N. Koblitz, "CM curves with good cryptographic properties", *Proc. Crypto '91*, Springer-Verlag, 1992, pp. 279–287.

6. D. W. Ash, I. F. Blake, and S. Vanstone, "Low complexity normal bases", *Discrete Applied Math.* **25** (1989), pp. 191–210.

7. T. Itoh, O Teechai, and S. Trojii, "A fast algorithm for computing multiplicative inverses in $GF(2^t)$", *J. Soc. Electron. Comm.* (Japan) **44** (1986), pp. 31–36.

8. E. Berlekamp, *Algebraic Coding Theory*, Aegean Park Press, 1984, pp. 36–44.

9. A. Menezes, P. van Oorschot, and S. Vanstone, *Handbook of Applied Cryptography*, CRC Press, 1997, pp. 107–109.

10. W. Meier and O. Staffelbach, "Efficient multiplication on certain non-supersingular elliptic curves", *Proc. Crypto '92*, Springer-Verlag, 1993, pp. 333–344.

11. R. Reiter and J. Solinas, "Fast elliptic arithmetic on special curves", *NSA/R21 Informal Tech. Report*, 1997.

12. K. Koyama and Y. Tsuruoka, "Speeding up elliptic cryptosystems by using a signed binary window method", *Proc. Crypto '92*, Springer-Verlag, 1993, pp. 345–357.

13. R. Schroeppel, H. Orman, S. O'Malley, and O. Spatscheck, "Fast key exchange with elliptic curve systems", *Proc. Crypto '95*, Springer-Verlag, 1995, pp. 43–56.

14. R. Schroeppel, H. Orman, S. O'Malley, and O. Spatscheck, "Fast key exchange with elliptic curve systems", *Univ. of Arizona Comp. Sci. Tech. Report* 95-03, 1995.

15. F. J. MacWilliams and N. J. A. Sloane, *The Theory of Error-Correcting Codes*, Elsevier, 1977, pp. 277–279.

A Elliptic Doubling with Normal Bases

The following technique[7] carries out the doubling

$$(x_2, y_2) = 2(x_1, y_1)$$

of a point (for which $x_1 \neq 0$) on the curve

$$y^2 + xy = x^3 + ax^2 + b$$

over $\mathbb{F}_{2^m}$, where the field is represented in terms of a normal basis. The usual algorithm requires 1 multiplicative inversion and 2 multiplications. The method given here replaces one of the general multiplications by a multiplication by a fixed constant (namely b). The operation of multiplying by a fixed constant is comparable in speed to field addition. Therefore the effective cost of this algorithm is 1 multiplicative inversion and 1 multiplication.

One begins by computing

$$x_2 = x_1^2 + \frac{b}{x_1^2} \ ,$$

---

[7] This method is of the kind alluded to in [13]. There it is credited to [14], which is not so easily available; hence its inclusion in this Appendix.

which is easily seen to equal the expression for x_2 appearing in Alg. 1. One then finds a root μ of the quadratic equation

$$\mu^2 + \mu = x_2 + a \ .$$

Since the field is being represented in terms of a normal basis, this process can be done without using anything more expensive than addition [15], so we can neglect its cost. The element μ will equal $\lambda + e$, where $e = 0$ or 1 and

$$\lambda = x_1 + \frac{y_1}{x_1} \ .$$

Therefore

$$\mu\, x_1 + x_1^2 + y_1 = e\, x_1 \ .$$

This equation allows us to find e and therefore λ. Notice that it is not necessary to perform the multiplication $\mu\, x_1$ in full, but rather to compute one coordinate of the product. (We can choose any coordinate where the corresponding coordinate of x_1 is 1.) Computing one coordinate of a product costs the same as an addition, so the derivation of λ is virtually cost-free. To complete the doubling, one computes

$$y_2 = x_1^2 + (\lambda + 1)\, x_2 \ .$$

Fast RSA-Type Cryptosystems
Using N-Adic Expansion

Tsuyoshi Takagi

NTT Software Laboratories
3-9-11, Midori-cho Musashino-shi, Tokyo 180, Japan
E-mail: ttakagi@slab.ntt.co.jp

Abstract. We propose two RSA-type cryptosystems using n-adic expansion, where n is the public key. These cryptosystems can have more than one block as a plaintext space, and the decrypting process is faster than any other multi-block RSA-type cryptosystem ever reported. Deciphering the entire plaintext of this system is as intractable as breaking the RSA cryptosystem or factoring. Even if a message is several times longer than a public key n, we can encrypt the message fast without repeatedly using the secret key cryptosystem.

1 Introduction

The RSA cryptosystem is one of the most practical public key cryptosystems and is used throughout the world [17]. Let n be a public key, which is the product of two appropriate primes, e be an encryption key, and d be a decryption key. The algorithms of encryption and decryption consist of the e-th and d-th power modulo n, respectively. We can make e small by considering the low exponent attacks [3] [4] [7]. The encryption process uses less computation and is fast. On the other hand, we must keep the decryption key d up to the same size as the public key n to preclude Wiener's attack [21]. Therefore, the cost of the decryption process is dominant for the RSA cryptosystem.

If a cryptosystem has more than one block of plaintexts, where each block is as large as the public-key n, we call it a multi-block cryptosystem. A lot of multi-block RSA-type cryptosystems have been proposed [5] [11] [12] [13] [15] [19]. Their advantage is that they allow us to encrypt data larger than the public-key at a time, and we can prove their security is equivalent to the original RSA cryptosystem or factoring. However, these algorithms are very slow and the attacks against the RSA cryptosystem are also applicable to them (See, for example, [8] [20].). We cannot find significant advantage over using the original RSA cryptosystem for each block.

In this paper, we propose two methods of constructing fast multi-block RSA-type cryptosystems. We express the plaintext as an n-adic expansion, where n is the public key. The features of this method are as follows. We can take an arbitrary number of blocks as a plaintext. To implement the proposed cryptosystems, we use only ordinary and elementary mathematical techniques i.e., the greatest common divisor, so the designer can easily make them. Deciphering the entire plaintext of the proposed cryptosystems is as hard as breaking the original RSA cryptosystem or factoring. Moreover, the decryption speed is much faster than any previously proposed multi-block RSA-type cryptosystems. Decryption time of the first block is dominant, because we calculate the modular multiplication of the encryption exponent and a greatest common divisor to decrypt blocks after the first one. Even if a message is several times longer than a public-key n, we can encrypt the message fast without repeatedly using the secret key cryptosystem.

Notation: $\mathbf{Z}$ is an integer ring. $\mathbf{Z}_n$ is a residue ring $\mathbf{Z}/n\mathbf{Z}$ and its complete residue class is $\{0, 1, 2, \ldots, n-1\}$. $\mathbf{Z}_n^\times$ is a reduced residue group modulo n. $\mathrm{LCM}(m_1, m_2)$ is the least common multiple of m_1 and m_2. $\mathrm{GCD}(m_1, m_2)$ is the greatest common divisor of m_1 and m_2. $_l C_m$ is permutation theory notation meaning the number of ways of choosing m from l.

2 The n-adic extension of RSA cryptosystem

In this section, we describe how to extend the RSA cryptosystem using n-adic expansion, and discuss its security and running time.

2.1 Algorithm

1. Generation of the keys: Generate two appropriate primes p, q, and let $n = pq$. Compute $L = \mathrm{LCM}\,(p-1, q-1)$, and find e, d which satisfies $ed \equiv 1 \pmod{L}$, $\mathrm{GCD}\,(e, L) = 1$ and $\mathrm{GCD}\,(e, n) = 1$. Then e, n are public keys, and d is the secret key.
2. Encryption: Let $M_0 \in \mathbf{Z}_n^\times$ and $M_1, \ldots, M_{k-1} \in \mathbf{Z}_n$ be the plaintext. We encrypt the plaintexts by the equation:

$$C \equiv (M_0 + nM_1 + \ldots + n^{k-1}M_{k-1})^e \pmod{n^k}. \tag{1}$$

3. Decryption: First, we decrypt the first block M_0 by the secret key d:

$$M_0 \equiv C^d \pmod{n}. \tag{2}$$

This is the same decryption process as in the original RSA. For the remaining blocks $M_1, M_2, \ldots, M_{k-1}$, we can decrypt by solving the linear equation modulo n.

2.2 Details of decryption

Assume that we have already decrypted M_0 by the decryption method of the original RSA cryptosystem, and we write down the process to find $M_1, M_2, \ldots, M_{k-1}$ as follows.

Consider that the encryption function (1) is the polynomial of the variables $X_0, X_1, \ldots, X_{k-1}$ such that

$$E(X_0, X_1, \ldots, X_{k-1}) = (X_0 + nX_1 + \ldots + n^{k-1}X_{k-1})^e.$$

Expand the polynomial $E(X_0, X_1, \ldots, X_{k-1})$ by the polynomial theorem:

$$\sum_{\substack{0 \leq s_0, s_1, \ldots, s_{k-1} \leq e \\ s_0 + s_1 + \cdots + s_{k-1} = e}} \frac{e!}{s_0! s_1! \ldots s_{k-1}!} X_0^{s_0} (nX_1)^{s_1} \ldots (n^{k-1} X_{k-1})^{s_{k-1}}.$$

And let

$$\Gamma_i := \{(s_0, s_1, \ldots, s_i) \mid s_1 + 2s_2 + \ldots + is_i = i,$$
$$s_0 + s_1 + \ldots + s_i = e, 0 \leq s_0, s_1, \ldots, s_i \leq e\},$$

where $(0 \leq i \leq k-1)$. Let $D_i(X_0, X_1, \ldots, X_i)$ be the coefficient of n^i $(0 \leq i \leq k-1)$. For $i = 0, 1, \ldots, k-1$, we can find $D_i(X_0, X_1, \ldots, X_i)$ by calculating

$$D_i(X_0, X_1, \ldots, X_i) = \sum_{(s_0, s_1, \ldots, s_i) \in \Gamma_i} \frac{e!}{s_0! s_1! \ldots s_i!} X_0^{s_0} X_1^{s_1} \ldots X_i^{s_i}. \qquad (3)$$

Here, we write them down with small i as follows:

$D_0(X_0) = M_0^e,$

$D_1(X_0, X_1) = eM_0^{e-1} M_1,$

$D_2(X_0, X_1, X_2) = {}_eC_2 M_0^{e-2} M_1^2 + eM_0^{e-1} M_2,$

$D_3(X_0, X_1, X_2, X_3) = {}_eC_3 M_0^{e-3} M_1^3 + 2{}_eC_2 M_0^{e-2} M_1 M_2 + eM_0^{e-1} M_3,$

$D_4(X_0, X_1, \ldots, X_4) = {}_eC_4 M_0^{e-4} M_1^4 + 3{}_eC_3 M_0^{e-3} M_1^2 M_2 + {}_eC_2 M_0^{e-2} M_2^2 + eM_0^{e-1} M_4,$

$D_5(X_0, X_1, \ldots, X_5) = {}_eC_5 M_0^{e-5} M_1^5 + 4{}_eC_4 M_0^{e-4} M_1^3 M_2 + 3{}_eC_3 M_0^{e-3} M_1 M_2^2$

$\qquad + 2{}_eC_2 M_0^{e-2} M_2 M_3 + 2{}_eC_2 M_0^{e-2} M_1 M_4 + eM_0^{e-1} M_5,$

$D_6(X_0, X_1, \ldots, X_6) = {}_eC_6 M_0^{e-6} M_1^6 + 5{}_eC_5 M_0^{e-5} M_1^4 M_2 + 4{}_eC_4 M_0^{e-4} M_1^3 M_3$

$\qquad + 3{}_eC_3 M_0^{e-3} M_1^2 M_4 + {}_eC_3 M_0^{e-3} M_2^3 + {}_eC_2 M_0^{e-2} M_3^2$

$\qquad + 2{}_eC_2 M_0^{e-2} M_2 M_4 + 2{}_eC_2 M_0^{e-2} M_1 M_5 + eM_0^{e-1} M_6,$

$\ldots$

$D_{k-1}(X_0, X_1, \ldots, X_{k-1}) = \{\text{polynomial of } M_0, M_1, \ldots, M_{k-1}\}.$

We show the algorithm of decryption. Note that the terms that include X_i do not appear in D_j $(j < i)$, and the only term that includes X_i in D_i is $eX_0^{e-1}X_i$ for $i = 0, 1, \ldots, k-1$. We define

$$D_i'(X_0, X_1, \ldots, X_{i-1}) = D_i(X_0, X_1, \ldots, X_i) - eX_0^{e-1}X_i.$$

Therefore, the terms $D_0, D_1, \ldots, D_{i-1}, D_i'$ are the polynomial of $X_0, X_1, \ldots, X_{i-1}$ $(0 \leq i \leq k-1)$.

From this relation, we can inductively decrypt M_i after decrypting $M_0, M_1, \ldots, M_{i-1}$ $(0 \leq i \leq k-1)$. Indeed, $M_1, M_2, \ldots, M_{k-1}$ are calculated as follows. At first, let $i = 1$. The relations $D_1'(X_0) = 0$ and $D_0(X_0) = X_0^e$ hold. So, the solution of the linear equation

$$eM_0^{e-1}x \equiv B_1 \pmod{n}, \quad B_1 = E_1/n, \qquad (4)$$

$$E_1 \equiv C - D_0(M_0) \pmod{n^2},$$

is M_1, because M_0 and e are in the reduced residue class modulo n such that $\mathbf{Z}_n^\times$. Provided that we decrypt $M_1, M_2, \ldots, M_{i-1}$, in the same manner we can uniquely decrypt M_i by solving the linear equation

$$eM_0^{e-1}x \equiv B_i \pmod{n}, \quad B_i = E_i/n^i, \tag{5}$$

$$E_i \equiv C - \sum_{j=0}^{i-1} n^j D_j(M_0, M_1, \ldots, M_j)$$

$$-n^i D_i'(M_0, M_1, \ldots, M_{i-1}) \pmod{n^{i+1}}.$$

Inductively, we can decrypt all plaintexts $M_1, M_2, \ldots, M_{k-1}$.

Here, we describe the decryption program written in the pidgin ALGOL in the following. For $x \in \mathbf{Z}$ and positive integer N, $[x]_N$ will denote the remainder of x modulo N, which is in $\{0, 1, \ldots, N-1\}$.

procedure **DECRYPTION**:
INPUT: $d, n, C(:= [(M_0 + nM_1 + \ldots + n^{k-1}M_{k-1})^e]_{n^k})$
OUTPUT: $M_0, M_1, \ldots, M_{k-1}$

(1) $C_0 := [C]_n$;
$M_0 := [C_0^d]_n$;

(2) $D_0 := [M_0^e]_{n^2}$;
$E_1 := [C - D_0]_{n^2}$;
$B_1 := E_1/n$ in $\mathbf{Z}$;
$A := [(eC_0)^{-1}M_0]_n$;
$M_1 := [AB_1]_n$;

(3) **FOR** $i = 2$ to $(k-1)$ do
begin
SUM := 0;
FOR $j = 0$ to $(i-1)$ do
begin
$D_j := [D_j(M_0, M_1, \ldots, M_j)]_{n^{i+1}}$;
SUM := $[\text{SUM} + n^j D_j]_{n^{i+1}}$
end
$D_i' := [D_i'(M_0, M_1, \ldots, M_{i-1})]_{n^{i+1}}$;
$E_i := [C - \text{SUM} - n^i D_i']_{n^{i+1}}$;
$B_i := E_i/n^i$ in $\mathbf{Z}$;
$M_i := [AB_i]_n$
end

2.3 Permutation

Let S be a finite set, and let $F(x)$ be a function from S to S. The function $F(x)$ is called a permutation function if every pair $x, y \in S$ that satisfies $F(x) = F(y)$ also satisfies $x = y$. If the encryption function $F(x)$ is not a permutation, we cannot uniquely decrypt a ciphertext. It is known that the encryption function of the RSA cryptosystem is a permutation, if and only if the relation $GCD(e, L) = 1$ holds with the same notation as in section 2.1. In the previous section, we showed that if the conditions $GCD(e, L) = 1$ and $GCD(e, n) = 1$ are satisfied, the proposed cryptosystem can be uniquely decrypted i.e., it is a one-to-one function.

Here, the encryption function of the proposed cryptosystem is defined from $Z_{n^k}^{\times}$ to $Z_{n^k}^{\times}$. We can prove this function is a permutation if and only if the conditions $GCD(e, L) = 1$ and $GCD(e, n) = 1$ hold.

Actually, the reduced residue group modulo n^k such that $Z_{n^k}^{\times}$ is decomposed into two products such that

$$Z_{n^k}^{\times} \cong Z_{p^k}^{\times} \times Z_{q^k}^{\times}. \tag{6}$$

Both groups are cyclic groups whose orders are $p^{k-1}(p-1)$ and $q^{k-1}(q-1)$, respectively. Therefore, the order of the group $Z_{n^k}^{\times}$ is $n^{k-1}(p-1)(q-1)$. All elements in $Z_{n^k}^{\times}$ are expressed by n-adic expansion such that

$$M = M_0 + nM_1 + \ldots + n^{k-1}M_{k-1},$$

where $M_0 \in Z_n^{\times}$ and $M_1, \ldots, M_{k-1} \in Z_n$. This is the reason that the plaintext must have the form in equation (1).

Let $E(x) \equiv x^e \pmod{n^k}$ be the encryption function of the proposed cryptosystem. Suppose $E(x) \equiv E(y) \pmod{n^k}$, and we get $(x/y)^e \equiv 1 \pmod{n^k}$. By Chinese remainder theorem, we reduce the equation into modulo p^k. Let g be a primitive root of modulo p^k, and let $x/y \equiv g^j \pmod{p^k}$ for some j. We get $g^{je} \equiv 1 \pmod{p^k}$, and

$$je \equiv 0 \pmod{p^{k-1}(p-1)}.$$

If $E(x)$ is a permutation, this equation must be solvable and all solutions are different, so $GCD(e, p) = 1$ and $GCD(e, (p-1)) = 1$ holds. Therefore, we have to choose e such that $GCD(e, n) = 1$ and $GCD(e, L) = 1$. The criteria in the key generation of the proposed cryptosystem are necessary.

2.4 Security

Theorem 1. *When plaintexts are uniformly distributed, finding the entire plaintext from the ciphertext for the RSA cryptosystem is as intractable as doing it for the proposed n-adic RSA-type cryptosystem.*

Proof. Using a black-box which can decipher the RSA cryptosystem, we can decipher the first block. Moreover, we can also decrypt blocks after the first one by using the decryption algorithm in section 2.2, so the entire plaintext is deciphered. Conversely, we are given ciphertext C, which is the result of encrypting a random M (mod n) by the RSA cryptosystem. Let C' be a random n-adic ciphertext, whose plaintext M' satisfies $M' \equiv M$ (mod n). All the bits of M' are uniquely distributed since M is random, and we can use the black box for the n-adic system to recover M'. Hence, we can decipher the plaintext M.

All the attacks against the RSA cryptosystem (See, for example, [14] [9].) are also applicable to the proposed system, because if we can decipher the first block M_0, then we can recover all the following blocks using relationships (4) or (5).

Here, we wonder whether the proposed cryptosystem has extra flaws in terms of using a non-square modulo n^k. The attacks that might break it are the message concealing [2] and the cycling attacks [22]. In the following two sections, we show these attacks never work against the proposed cryptosystem.

2.5 Message concealing

A function $F(x)$ is called unconcealed when $F(x) = x$ holds for all x. If a function of a cryptosystem is unconcealed, then we cannot encrypt any message by it. G. R. Blakley and I. Borosh showed that the encryption function of the RSA cryptosystem is unconcealed [2]. Let N be the number of residue classes x modulo n^k such that $x^e \equiv x$ (mod n^k). They proved

$$N = (1 + \text{GCD}(e - 1, p^{k-1}(p - 1)))(1 + \text{GCD}(e - 1, q^{k-1}(q - 1))).$$

If $\text{GCD}(e - 1, pq) > 1$ holds, then N becomes very large. We have to choose the system parameters p, q and e described in section 2.1 to preclude this failure. It must be noted that if e is selected smaller than p and q, then $\text{GCD}(e-1, pq) = 1$ holds.

Moreover, they also showed that if e is an odd integer larger than 2, then $N = 9$ if and only if

$$\text{GCD}(e - 1, \lambda) = 2, \quad \lambda = \text{LCM}((p - 1)p^{k-1}, (q - 1)q^{k-1}).$$

For example, the RSA cryptosystem has only 9 unconcealed messages if $\text{GCD}(e-1, L) = 2$. For small e, we have $\text{GCD}(e - 1, pq) = 1$, and N for the proposed cryptosystem is equal to that of the RSA cryptosystem.

2.6 Cycling attacks

It is known that the RSA cryptosystem is broken without factoring n when a ciphertext C has a period such that $C^{P(b)} \equiv 1$ (mod n), where $P(t)$ is a polynomial and $t = b$ is an integer. Actually, if the relation holds, the plaintext can be recovered by computing $M \equiv C^Q$ (mod n), where Q satisfies $eQ \equiv 1$ (mod P') and $P' = P(b)/\text{GCD}(e, P(b))$. Moreover, this analysis is true even if the modulo n is changed to n^k. To break the proposed n-adic RSA-type cryptosystem, an attacker would have to find the polynomial $P(t)$ and the value $t = b$, which have the relation $C^{P(b)} \equiv 1 \pmod{n^k}$. By decomposing of the group $\mathbf{Z}_{n^k}$ like (6), we reduce the relations to

$$P(t) \equiv 0 \pmod{p_i}, \quad P(t) \equiv 0 \pmod{q_i} \quad (i = 1, 2), \tag{7}$$

where $p_1 = p$; $p_2 = q$; and q_i is a large prime such that $q_i | p_i - 1 (i = 1, 2)$. H. C. Williams and B. Schmid [22] showed that the possibility of this polynomial satisfying equation (7) is very small, unless $P(t) = t \pm 1$ and $t = e^m$. Therefore, the designers must make m very large to preclude this attack. One method is to have $q_i - 1$ and $p_i - 1$ be divisible by large primes r_i and r_i' such that $r_i | p_i - 1$ and $r_i' | q_i - 1$; then $r_i | m$ and $r_i' | m$ hold for $i = 1, 2$ and m becomes very large. Since $p_i - 1 (i = 1, 2)$ must be divisible by a large prime to prevent the factoring algorithm called Pollard's $p - 1$ method, we do not need worry about the equations $e^m \equiv \pm 1 \pmod{p_i}$. Consequently, the proposed n-adic RSA-type cryptosystem is secure against this attack according to the same treatment as used for the original RSA cryptosystem.

2.7 Running time

Here, we discuss the running time of the proposed cryptosystem. In the encryption process, we have to compute the e-th power modulo $n^k (k \geq 2)$. As k increases, the running time becomes longer. However, it is possible to make the exponent of the encryption e small, since considering the low exponent attacks [3] [4] [7], the encryption cost is not so expensive.

Next, we consider the decryption process. The first block is decrypted by the same algorithm as in the RSA cryptosystem, and we should make the exponent d as large as the public modulus n to avoid Wiener's attack [21]. Therefore, the decryption of the first block is the most expensive task. After the first block, we have to generate linear equation (4) and maybe also (5), and solve it/them. The ciphertext $C_i (i \geq 1)$ is expressed by the polynomial of $M_i (i \geq 1)$ and the task of computing the polynomial is essentially to calculate M_0^e. Therefore, it costs the same as the encryption process to generate the linear equations. Solving a linear equation is fast, so the decryption time after the first block also becomes as fast as the encryption process. If we choose a very small e, this algorithm becomes very efficient. For example, let the number of blocks be two. We can generate the linear equation to compute equation (4), which are at most $2 \lfloor \log_2 e \rfloor$ multiplications modulo n^2 and one division of n^2, and to solve it, which are two multiplications modulo n and one inversion modulo n.

On the other hand, several multi-block RSA-type cryptosystems have been proposed [5] [11] [13]. Their decryption time is l times slower than the original RSA cryptosystem, where l is the number of blocks. Our proposed cryptosystem is much faster than these cryptosystems, as showed by the above analysis. [1]

---

[1] K. Koyama proposed a two-block cryptosystem having fast decryption by using singular cubic curves. But it only has two blocks [12].

2.8 Effectiveness

As we discussed in the previous sections, the proposed n-adic RSA-type cryptosystem has several effective features. The most significant points are being as hard as breaking the original RSA cryptosystem and providing fast decryption for messages longer than the public key n.

By the way, the RSA cryptosystem is slower than the secret-key cryptosystem, so the RSA cryptosystem is used to encrypt a session key of the secret-key cryptosystem to overcome this disadvantage. However, its theoretical security level must be estimated from the RSA cryptosystem and the secret-key. We do not have to use the secret-key cryptosystem, if the length of the data is shorter than a public-key n.

For a message that is several times longer than the public-key n, our proposed n-adic RSA-type cryptosystem is very efficient. We can encrypt such a message much faster.

Moreover, it is expected that the encryption speed of the RSA cryptosystem will reach 1 Mbits/second within a year or so [18]. The proposed method can contribute to the attainment of the fast encryption speed.

3 The n-adic extension of Rabin cryptosystem

In this section, we describe how to extend the Rabin cryptosystem using n-adic expansion. The discussion is similar to the extension of the RSA cryptosystem.

3.1 Algorithm

1. Generating keys: Generate two appropriate primes p, q, and let $n = pq$. Here, p and q are the secret keys, and n is the public key.
2. Encryption: Let $M_0 \in \mathbf{Z}_n^\times$ and $M_1, \ldots, M_{k-1} \in \mathbf{Z}_n$ be the plaintext. We encrypt the plaintext by

$$C \equiv (M_0 + nM_1 + \ldots + n^{k-1}M_{k-1})^2 \pmod{n^k}. \tag{8}$$

   And we send the ciphertext C.
3. Decryption: We solve the modular quadratic equation

$$x^2 \equiv C \pmod{n^k}. \tag{9}$$

   Then the solutions are just plaintext $M_0, M_1, \ldots, M_{k-1}$.

3.2 Details of decryption

First, we decrypt the first block M_0. We solve the quadratic equation $C \equiv M_0^2$ modulo primes p and q. Here, several algorithms to solve the quadratic equation modulo a prime p are known, and the fastest one can be computed in subquadratic polynomial time [10]. Next, we decrypt the first block of the plaintext M_0 by the Chinese remainder theorem. The degree of ambiguity is 4 for the decryption modulo n, because we have two solutions of each quadratic equation. And we can eliminate the ambiguousness by adding redundancy bits, and we can get the true plaintext.

Next, we discuss the decryption of the remaining blocks $M_1, M_2, \ldots, M_{k-1}$. The process is similar to the case in the RSA cryptosystem. For M_1, we have the linear equation modulo n^2,

$$M_0^2 + 2nM_0 x \equiv C \pmod{n^2}. \tag{10}$$

And this equation is solvable because $2M_0 \in (\mathbf{Z}/n\mathbf{Z})^\times$, and the solution is M_1. Here, assume that we already decrypt $M_0, M_1, \ldots, M_{i-1}$, and we can uniquely decrypt M_i by solving

$$2n^i M_0 x \equiv C - \sum_{\substack{0 \le l, m \le i-1}}^{l+m \le i} n^{l+m} M_l M_m \pmod{n^{i+1}}, \tag{11}$$

Therefore, we can decrypt all plaintext blocks $M_0, M_1, M_2, \ldots, M_{k-1}$.

We describe the decryption program written in the pidgin ALGOL in the following. For $x \in \mathbf{Z}$ and positive integer N, $[x]_N$ will denote the remainder of x modulo N, which is $\{0, 1, \ldots, N-1\}$.

procedure **DECRYPTION**:

INPUT: $p, q, n, C (:= [(M_0 + nM_1 + \ldots + n^{k-1}M_{k-1})^2]_{n^k})$

OUTPUT: $M_0, M_1, \ldots, M_{k-1}$

(1) $C_0 := [C]_n$;

       decrypt M_0 using p, q, C_0;

(2)   **FOR** $i = 1$ **to** $(k-1)$ **do**

        begin

        SUM := 0;

        FOR $l = 0$ **to** $(i-1)$ **do**

           FOR $m = 0$ **to** $(i-1)$ **do**

              WHILE $l + m \le i$ **do**

                begin

                $D := [n^{l+m} M_l M_m]_{n^{i+1}}$;

                SUM := $[\text{SUM} + D]_{n^{i+1}}$

              end

        $E_i := [C - \text{SUM}]_{n^{i+1}}$;

        $B_i := E_i/n^i$ in $\mathbf{Z}$;

        $M_i := [(2M_0)^{-1} B_i]_n$

    end

3.3 Security

Theorem 2. *Completely breaking the proposed n-adic Rabin-type cryptosystem is as intractable as factoring.*

Proof. Let p, q be primes, and let $n = pq$. The complexities of the following three algorithms only differ by polynomial time.

(I) to factor $n = pq$

(II) to find the solution of the quadratic equation modulo n

(III) to find the solution of the quadratic equation modulo n^k,

where k is an integer greater than 2. (I) and (II) are clearly equivalent because the security of the Rabin cryptosystem is the same as factoring [16]. (III) $\Rightarrow$ (II) is true by reducing the solution in (II) modulo n. (II) $\Rightarrow$ (III) is true because it is just the decryption process after the first block in the previous section, and the algorithm only takes polynomial time to generate and solve linear equations. Here, (III) is just the algorithm deciphering the proposed n-adic system.

The exponent of the Rabin cryptosystem is only 2, so the low exponent attacks are applicable to it [3] [4] [7]. However, we can preclude these attacks by padding a plaintext with random bits.

3.4 Running time and effectiveness

Here, we discuss the running time of the proposed cryptosystem. In the encryption process, we only compute the second power modulo $n^k (k \geq 2)$, which is very fast. For the decryption process, the first block is decrypted by the same decryption method as for the Rabin cryptosystem. The decryption of the first block is the most expensive task. After the first block, we have to generate the linear equation (10) and maybe also (11), and solve it/them. These are computed very fast, and the cost is very small compared with the cost of decrypting the first block. Therefore, the total cost of the decryption is essentially the cost of the first block.

On the other hand, several multi-block Rabin-type cryptosystems have been proposed [15] [19]. We have to solve a polynomial with more than two degrees over the finite field of a prime order. Solving polynomials of higher degree is more expensive than solving a quadratic polynomial, and makes the decryption process ambiguous and restricts the form of the secret primes. These cryptosystems have few advantages.

From the above analysis, our proposed cryptosystem is much faster than these cryptosystems, and easy to implement. Designers do not have to code a complicated algorithm and can use only ordinary mathematical tools such as the greatest common divisor.

As we discussed in section 2.8, for messages that are several times longer than the public-key n, our proposal n-adic Rabin cryptosystem is very efficient. We can encrypt a message with the running time of the first block.

4 Open problems and a partial solution

A plaintext of the proposed n-adic cryptosystem modulo n^k has the form $M \equiv M_0 + nM_1 + \ldots + n^{k-1}M_{k-1}$. Theorems 1 and 2 show that breaking the entire plaintext M is as hard as breaking the RSA cryptosystem or factoring. Here, we mention some problems concerning the security of each block $M_0, M_1, \ldots, M_{k-1}$.

If we have an algorithm that breaks the first block M_0, we can decipher the RSA or Rabin cryptosystem. However, it is an open problem whether you can find the blocks after the first one without deciphering the first block. One strategy for finding such an algorithm is to seek some algebraic relations between a ciphertext and blocks after the first one. Indeed, the most trivial relation is linear equation (4) or (5) whose solutions are the remaining blocks after the first one. But, we have to compute the value M_0^{e-1} to construct them, which is as hard as deciphering the RSA cryptosystem.

W. Alexi et al. showed that we can find the whole plaintext by using an algorithm that deciphers certain bits of the plaintext [1]. This also means that the proposed n-adic system can be broken by an algorithm that deciphers certain bits of the first block of the plaintext. It is an open problem whether there exists an algorithm that can decipher certain bits after first block of the plaintext.

Against the RSA cryptosystem, D. Coppersmith et al. showed that we can recover the original plaintext by algebraic calculation, if we send two ciphertexts whose plaintexts have a polynomial relationship [3]. It might be possible to recover the plaintext of the proposed n-adic system using a variation of this technique. It is an open problem whether you can recover the plaintext if there is a polynomial relationship between some blocks of one plaintext or between blocks of two plaintexts.

4.1 Security of the second block

Theorem 3. *Consider the n-adic RSA-type cryptosystem. Let $\mathcal{O}$ be an oracle which, given a ciphertext $C \equiv (M_0 + nM_1 + \ldots + n^{k-1}M_{k-1})^e \pmod{n^k}$, answers the second block of the plaintext M_1. The oracle $\mathcal{O}$ can be used to break the entire plaintext $(M_0, M_1, \ldots, M_{k-1})$.*

Proof. If we can decipher the first block M_0, then we can also do all the remaining blocks $M_2, \ldots, M_{k-1}$. Therefore, we can reduce the attack to the case of the two-block cryptosystem with modulo n^2. Let the plaintext $M = M_0 + nM_1$ $(0 \leq M_0, M_1 < n)$, and $C \equiv M^e \pmod{n^2}$ be the ciphertext. For $i = 0, 1, 2, \ldots, h$, expand

$$2^i M \equiv M_0^{(i)} + nM_1^{(i)} \pmod{n^2}, \quad 0 < M_0^{(i)}, M_1^{(i)} < n,$$

where $h = \lfloor \log_2 n \rfloor$. Here, $2^{ie}C \equiv (M_0^{(i)} + nM_1^{(i)})^e \pmod{n^2}$ holds, and we can get each second block $M_1^{(i)} = \mathcal{O}(2^{ie}C)$ by using the oracle $\mathcal{O}$. Here, note that $M_0^{(i)} < n/2$ if and only if $2M_0^{(i)} \pmod{n} = M_0^{(i+1)}$ for $i = 0, 1, 2, \ldots, h$. Hence $2M_1^{(i)} \pmod{n} = \mathcal{O}(2^{(i+1)e}C)$ if and only if $M_0^{(i)} < n/2$ for $i = 0, 1, 2, \ldots, h$. On the other hand, let $C_0 \equiv C \pmod{n}$, and we have $2^{ie}C_0 \equiv (2^i M_0)^e \equiv (M_0^{(i)})^e$ $\pmod{n}$ for $i = 0, 1, 2, \ldots, h$. This observation means that we can construct the half bit oracle $\mathcal{O}_H$, which computes $\mathcal{O}_H(2^{ie}C_0) = 0$ if $M_0^{(i)} < n/2$ and $\mathcal{O}_H(2^{ie}C_0) = 1$ if $M_0^{(i)} > n/2$. Indeed, define that

$$
\mathcal{O}_H(2^{ie}C_0) = \begin{cases} 0, & (2\mathcal{O}(2^{ie}C) \pmod{n}) = \mathcal{O}(2^{(i+1)e}C), \\ 1, & (2\mathcal{O}(2^{ie}C) \pmod{n}) \neq \mathcal{O}(2^{(i+1)e}C), \end{cases}
$$

for $i = 0, 1, 2, \ldots, h$. It is well-known this half bit oracle $\mathcal{O}_H$ recovers the plaintext M_0 such that $C_0 \equiv M_0^e \pmod{n}$[6]. Consequently, we can decipher the first block M_0.

5 Conclusion

Our proposed n-adic extensions of the RSA and Rabin cryptosystems perform decryption faster than any other multi-block RSA-type or Rabin-type cryptosystems ever reported. Deciphering the entire plaintext of this system is as intractable as breaking the original RSA cryptosystem or factoring. We also showed that the proposed n-adic RSA-type cryptosystem is a permutation function, and showed the criteria for message concealing and cycling attacks which are applicable to the RSA cryptosystem. Even if a message is several times longer than a public-key n, we can encrypt it fast without repeatedly using the secret-key cryptosystem.

Acknowledgments

I wish to thank S. Naito, M. Nishio, and members of the cryptology seminar in NTT Laboratories for their helpful discussions. I would also like to thank the anonymous referees for their valuable comments.

References

1. W. Alexi, B. Chor, O. Goldreich, C. P. Schnorr; "Rsa and Rabin functions: certain parts are as hard as the whole," SIAM Journal of Computing, 17, (1988), pp.194-209.
2. G. R. Blakley and I. Borosh, "Rivest-Shamir-Adelman public key cryptosystems do not always conceal messages," Comput. & Maths. with Appls., 5, (1979), pp.169-178.
3. D. Coppersmith, M. Franklin, J. Patarin and M. Reiter, "Low-exponent RSA with related messages," Advances in Cryptology – EUROCRYPT '96, LNCS 1070, (1996), pp.1-9.
4. D. Coppersmith, "Finding a small root of a univariate modular equation," Advances in Cryptology – EUROCRYPT '96, LNCS 1070, (1996), pp.155-165.
5. N. Demytko, "A new elliptic curves based analogue of RSA," Advances in Cryptology – EUROCRYPT '93, LNCS 765, (1994), pp.40-49.

6. S. Goldwasser, S. Micali, and P. Tong, "Why and how to establish a private code on a public network," Proc. of FOCS, (1982), pp.134-144.

7. J. Håstad, "Solving simultaneous modular equations of low degree," SIAM Journal of Computing, 17, (1988), pp.336-341.

8. B. S. Kaliski Jr., "A chosen message attack on Demytko's elliptic curve cryptosystem," Journal of Cryptology, 10, (1997), pp.71-72.

9. B. S. Kaliski Jr. and M. Robshaw, "Secure use of RSA," CRYPTOBYTES, 1 (3), (1995), pp.7-13.

10. E. Kaltofen and V. Shoup, "Subquadratic-time factoring of polynomials over finite fields", Proc. of STOC, (1995), pp.398-406.

11. K. Koyama, U. M. Maurer, T. Okamoto and S. A. Vanstone, "New public-key schemes based on elliptic curves over the ring Z_n," Advances in Cryptology – CRYPTO '91, LNCS 576, (1991), pp.252-266.

12. K. Koyama; "Fast RSA-type schemes based on singular cubic curves," Advances in Cryptology – EUROCRYPT '95, LNCS 921, (1995), pp.329-340.

13. J. H. Loxton, D. S. P. Khoo, G. J. Bird and J. Seberry, "A cubic RSA code equivalent to factorization," Journal of Cryptology, 5, (1992), pp.139-150.

14. A. J. Menezes, P. C. van Oorschot and S. A. Vanstone, "Handbook of applied cryptography," CRC Press, (1996).

15. B. Meyer and V. Müller, "A public key cryptosystem based on elliptic curves over Z/nZ equivalent to factoring," Advances in Cryptology – EUROCRYPT '96, LNCS 1070 (1996), pp.49-59.

16. M. O. Rabin, "Digitalized signatures and public-key functions as intractable as factorization", Technical Report No.212, MIT, Laboratory of Computer Science, Cambridge (1979), pp.1-16.

17. R. Rivest, A. Shamir and L. M. Adleman, "A method for obtaining digital signatures and public-key cryptosystems," Communications of the ACM, 21(2), (1978), pp.120-126.

18. RSA Laboratories, "Frequently asked questions about today's cryptography (Version 3.0)," http://www.rsa.com/rsalabs/, (1996).

19. J. Schwenk and J. Eisfeld, "Public key encryption and signature schemes based on polynomials over Z_n," Advances in Cryptology – EUROCRYPT '96, LNCS 1070, (1996), pp.60-71.

20. T. Takagi and S. Naito, "The multi-variable modular polynomial and its applications to cryptography," Proc. of ISAAC'96, LNCS 1178, (1996), pp.386-396.

21. M. J. Wiener, "Cryptanalysis of short RSA secret exponents," IEEE Transactions on Information Theory, IT-36, (1990), pp.553-558.

22. H. C. Williams and B. Schmid, "Some remarks concerning the M.I.T. public-key cryptosystem," BIT 19, (1979), pp.525-538.

A One Way Function Based on Ideal Arithmetic in Number Fields

Johannes Buchmann   Sachar Paulus

Department of Computer Science
TH Darmstadt
64283 Darmstadt
Germany

Abstract. We present a new one way function based on the difficulty of finding shortest vectors in lattices. This new function consists of exponentiation of an ideal in an order of a number field and multiplication by an algebraic number which can both be performed in polynomial time. The best known algorithm for inverting this function is exponential in the degree of the lattices involved.

Key words: one way function, number fields, shortest vectors in lattices.

1   Introduction

In the past 20 years several number theoretic problems have been identified on whose difficulty the security of cryptographic protocols can be based. Prominent examples are the factoring problem for integers [RSA78] and the discrete logarithm problem in the multiplicative group of a finite field [Odl85], in the class group of an order of a quadratic field [BW88], and on an elliptic curve curve over a finite field [Kob87]. There is, however, no guarantee that those problems remain difficult to solve in the future. On the contrary, as the experience with the factoring problem for integers shows unexpected breakthroughs are always possible. It is therefore important to design cryptographic schemes in such a way that the underlying mathematical problem can easily be replaced with another one. It is also important to search for mathematical problems on which secure one way functions can be based.

In this paper we show how to use ideal arithmetic in number fields to design a cryptographic one way function NF-EXP with the following properties:

- NF-EXP can be computed in polynomial time.
- Inverting NF-EXP is at least as hard as factoring integers.
- The only method currently known for inverting NF-EXP requires computing shortest vectors in lattices whose dimension is the degree of the number field. This currently requires exponential time in the degree of the number field.

This papers generalizes ideas which have been introduced in [BW90] for orders of quadratic number fields. However, for quadratic fields, or more generally for number fields of fixed degree NF-EXP can be inverted in subexponential time (see [Buc88]). The new idea in this paper is to introduce the problem of computing shortest vectors in lattices of growing dimension as the basis for cryptographic security. This is done by considering number fields of growing degree. The problem of computing shortest vectors in lattices is known to be very difficult. It does not play a role in any of the schemes based on factoring, finite fields, quadratic number fields, and curves over finite fields. We therefore believe that the new one way function has some potential in future cryptographic applications.

The paper is organized as follows: In a first section, we present the necessary concepts from algebraic number fields. Then, we explain NF-EXP and show how to compute it efficiently. In Section 3, we sketch how to invert NF-EXP and show its difficulty. Finally, we present an example.

2 Algebraic number fields

In this section we briefly introduce the reader to algebraic number fields, orders, fractional ideals, and the arithmetic in those structures. For more details see [BS96].

By an *algebraic number field* we mean an extension field of the field $\mathbb{Q}$ of rational numbers which, as a $\mathbb{Q}$-vector space, is finite dimensional. The dimension is called the *degree* of the number field. Let F be an algebraic number field of degree n. Then there is a monic irreducible polynomial $f \in \mathbb{Z}[X]$ such that F is isomorphic to the residue class field $\mathbb{Q}[x]/(f)$ where (f) denotes the ideal generated by f in $\mathbb{Q}[x]$. We call f a *generating polynomial* of F. In the sequel we will assume that the number field F is represented by a generating polynomial f and that $F = \mathbb{Q}[x]/(f)$.

The elements of F are residue classes, i.e. they are of the form $g + (f)$ with $g \in \mathbb{Q}[x]$. Any such residue class has a unique representative of degree less than n. We assume that the residue classes are represented by those representatives. They can be obtained from any representative by division with remainder by f. Addition, subtraction, and multiplication of elements of F can be effected by the addition, subtraction, and multiplication of the representing polynomials followed by a division with remainder by f. If $g + (f)$ is a non zero element of f then its inverse can be computed as follows. Using the extended Euclidean algorithm a polynomial h is computed with $gh \equiv 1 \bmod (f)$. Then $h + (f)$ is the inverse of $g + (f)$. Addition, subtraction, multiplication and inversion can obviously be performed in polynomial time.

Let $\alpha \in F$. Then the map $F \to F, \xi \mapsto \alpha\xi$ is an endomorphism of the $\mathbb{Q}$-vector space F. The *trace* of α is the trace of this endomorphism. It is denoted by $trace(\alpha)$. The *norm* of α is the the determinant of this endomorphism. It is denoted by $norm(\alpha)$.

Next we introduce orders and ideals thereof. An order $\mathcal{O}$ in F is a subring of F containing the element 1 which admits a $\mathbb{Z}$-basis $(\omega_1, \ldots, \omega_n)$. We represent $\mathcal{O}$ by such a $\mathbb{Z}$-basis, where the basis elements are represented as described above. The *discriminant* of $\mathcal{O}$ is $\Delta = \det((trace(\omega_i\omega_j)_{1 \leq i,j \leq n})$. The generating polynomial and the $\mathbb{Z}$-basis of $\mathcal{O}$ can always be chosen in such a way that $(\log|\Delta|)^{O(1)}$ bits are necessary to store them. We will assume that this is done.

An integral $\mathcal{O}$-ideal is an additive subgroup I of $\mathcal{O}$ which is an $\mathcal{O}$-module. This means that $\mathcal{O}I = \{\omega\alpha : \omega \in \mathcal{O}, \alpha \in I\} \subset I$. The *norm* of I is the index of I in $\mathcal{O}$. It is denoted by N(I). If I is an integral $\mathcal{O}$-ideal then any subset of F of the form $(1/d)I$, where d is a non zero integer, is called a *fractional $\mathcal{O}$-ideal*. Its norm is $\mathrm{N}(I)/d^n$. If I and J are fractional $\mathcal{O}$-ideals then their *product*

$$I \cdot J = \{ \sum_{(\alpha,\beta) \in S} \alpha\beta \mid S \subset I \times J \text{ finite } \}$$

is again a fractional $\mathcal{O}$-ideal. The set $\mathcal{I}$ of all fractional ideals of $\mathcal{O}$ forms a commutative semi-group. The *quotient* of I and J

$$I : J = \{\alpha \in F \mid \alpha J \subset I\}$$

is a fractional $\mathcal{O}$-ideal. A fractional $\mathcal{O}$-ideal I is called *invertible* if $I(\mathcal{O} : I) = \mathcal{O}$. The invertible $\mathcal{O}$-ideals form an Abelian group $\mathcal{J}$. The set $\mathcal{P} = \{\alpha\mathcal{O} \mid \alpha \in \mathcal{F}^*\}$ of *principal* ideals is a subgroup of this group. Two fractional invertible $\mathcal{O}$-ideals I and J are called *equivalent* if there is an $\alpha \in F$ with $J = \alpha I$. Equivalence of ideals is an equivalence relation which is compatible with ideal multiplication. The equivalence class of I is called the *ideal class* of I and is denoted by $[I]$. The set of ideal classes $Cl(\mathcal{O}) = \mathcal{J}/\mathcal{P}$ forms a finite Abelian group and is called the *class group* of $\mathcal{O}$.

We explain ideal arithmetic. A fractional $\mathcal{O}$-ideal I is represented by a $\mathbb{Z}$-basis $(\alpha_1, \ldots, \alpha_n)$ where $\alpha_i \in F$, $1 \leq i \leq n$. As we have explained above, those elements are represented by polynomials of degree less than n. We will now assume that such a polynomial $g(x) = g_1 x^{n-1} + \ldots + g_{n-1}x + g_n$ is represented by the coefficient vector $(g_1, \ldots, g_n)$. Then the ideal I can be viewed as a lattice in $\mathbb{Q}^n$. If I and J are fractional $\mathcal{O}$-ideals with bases $(\alpha_1, \ldots, \alpha_n)$ and $(\beta_1, \ldots, \beta_n)$, respectively, then $(\alpha_i\beta_j)_{1 \leq i,j \leq n}$ is a generating system of the product IJ as a lattice in $\mathbb{Q}^n$. Using Hermite normal form computation [Coh95, 65ff], a basis thereof can be found in polynomial time.

We explain module reduction. For this purpose we need yet another interpretation of a fractional $\mathcal{O}$-ideal as an n-dimensional lattice. Let $\rho^{(1)}, \ldots, \rho^{(n)}$ be the complex zeros of the generating polynomial f. For $\alpha = g + (f) \in F$ the numbers $\alpha^{(i)} = g(\rho^{(i)})$, $1 \leq i \leq n$ are the *conjugates* of α. There are positive integers r_1, r_2 with $r_1 + 2r_2 = n$ such that $\rho^{(i)} \in \mathbb{R}$ for $1 \leq i \leq r_1$, $\rho^{(i)} \notin \mathbb{R}$ for $r_1 < i \leq n$ and $\rho^{(i)} = \overline{\rho^{(i+r_2)}}$ for $r_1 < i \leq r_1 + r_2$. Define the map

$\varphi: \quad F \to \mathbb{R}^n$
$$\alpha \mapsto (\alpha^{(1)}, \ldots, a^{(r_1)}, Re\,\alpha^{(r_1+1)}, Im\,\alpha^{(r_1+1)}, \ldots, Re\,\alpha^{(r_1+r_2)}, Im\,\alpha^{(r_1+r_2)}).$$

If I is a fractional $\mathcal{O}$-ideal then $\varphi(I)$ is an n-dimensional lattice of determinant $2^{-r_2}N(I)\sqrt{|\Delta|}$. Let $c \in \mathbb{R}$, $c \geq 1$. An $\mathcal{O}$-ideal I is called c-reduced if the fractional $\mathcal{O}$-ideal $\mathcal{O} : I$ contains the element 1 and

$$\{\alpha \in \mathcal{O} : I \mid \forall i = 1, \ldots, n \ |\alpha^{(i)}| < 1/c\} = \{0\}.$$

If I is c-reduced then I is integral and $\mathrm{N}(I) \leq c^n \sqrt{|\Delta|}$. It follows that $(\log c + \log|\Delta|)^{O(1)}$ bits are sufficient to store I. A c-reduced ideal can be computed as follows. Determine $\mathcal{O} : I$. In $\mathcal{O} : I$ find a number α such that the length of the shortest non zero vector in the lattice $\varphi(\mathcal{O} : I)$ is at least as large as $\|\varphi(\alpha)\|/c$. Then αI is c-reduced. It follows that 1-reduced ideals can be computed using the shortest vector algorithm of [Kan87] in time $n^{O(n)}(\log|\Delta|)^{O(1)}$. Also, if we use LLL-reduction [Coh95, 83ff] then we obtain 2^n-reduced ideals in time $n^{O(1)}(\log|\Delta|)^{O(1)}$.

3 The one way function NF-EXP

Let F be an algebraic number field of degree n over $\mathbb{Q}$. Let $\mathcal{O}$ be an order in F and let I be an $\mathcal{O}$-ideal. Recall that $\mathcal{I}$ denotes the group of fractional $\mathcal{O}$-ideals. We define the following function parameterized by $\mathcal{O}$ and I:

$$\text{NF-EXP}: \quad F^* \times \mathbb{N} \to \mathcal{I}$$
$$(\alpha, k) \mapsto \alpha I^k.$$

There is a problem in using NF-EXP for cryptographic purposes: If we use a $\mathbb{Z}$-basis to represent the ideal αI^k then the number of bits needed to represent αI^k is $\Omega(k)$. Hence, we cannot even write down NF-EXP(α, k) in polynomial time. One solution is to restrict the set of arguments of NF-EXP such that the restricted function can be computed in polynomial time. This can be done in such a way that we get a compact representation of αI^k without affecting the security of NF-EXP.

We can for example choose (α, k) such that αI^k is 2^n-reduced. More precisely: We choose $k \in \{1, \ldots, |\Delta|\}$. Using fast exponentiation techniques we can compute a 2^n-reduced ideal J in $[I]^k$ and a number $\alpha \in F$ with $J = \alpha I^k$. The reduced ideal J is not uniquely determined. In fact, there are in general many such reduced ideals. Using techniques of [Buc88] and [Thi95] it is possible to choose α in such a way that J could be any of all 2^n-reduced ideals in its ideal class. Note that all those computations can be carried out in polynomial time since 2^n-reduction can be performed in polynomial time.

Using the same idea, one can obtain a compact representation of αI^k for general α and k as follows: Compute $\beta \in F$ such that $J = \beta\alpha I^k$ is 2^n-reduced and $\|\mathrm{Log}\beta\|$ is minimal, where

$$\mathrm{Log}(\alpha) = (\log|\alpha^{(1)}|, \ldots, \log|\alpha^{(r_1+r_2-1)}|).$$

This is again done by fast exponentiation. Then represent αI^k by J and a suitable approximation of $\mathrm{Log}\beta$. Using the results of [Thi95] it can be shown that this representation has polynomial length and can be computed in polynomial time.

In practice, we could for example choose a series of number fields of growing degree generated by a sparse polynomials of small discriminant, since the complexity of computing NF-EXP depends on the discriminant and on the number of non-zero coefficients of the generating polynomial. A careful analysis together with experiments in practice have to be done.

Using the arguments of [BBM+92] it follows that the one way function NF-EXP can be used to implement many cryptographic protocols, e.g. a Diffie-Hellman key exchange or a ElGamal signature scheme.

4 Inverting NF-EXP

It has been shown in [BW88] that there is a polynomial time reduction of the factorization problem for integers to inverting NF-EXP for imaginary quadratic orders.

Now we sketch an algorithm for inverting NF-EXP. We use the same notation as in the previous section. The problem of inverting NF-EXP will be called the NF-DL-problem: Given two ideals I and J of an Order $\mathcal{O}$ in a number field F find $\alpha \in F$ and $k \in \mathbb{N}$ such that $J = \alpha I^k$. Solving the NF-DL-problem is closely related to computing the unit group and the class group of the order $\mathcal{O}$. An algorithm computing these groups is described in [Buc88]. We will now explain how this algorithm can be modified to solve NF-DL. We will see that the running time of this algorithm which is the fastest we can currently obtain is exponential in the degree.

Suppose that I and J are fractional $\mathcal{O}$-ideals and that we wish to solve the NF-DL-problem

$$J = \alpha I^k.$$

Without loss of generality we assume that I and J are invertible $\mathcal{O}$-ideals. If they are not, then one can compute the so-called *order of multipliers* in polynomial time, where both ideals are invertible. First we determine an exponent k which is a solution of the equation

$$[J] = [I]^k.$$

For this purpose we choose a factor base FB which contains finitely many invertible prime ideals of $\mathcal{O}$ and also the ideals I and J. Then we determine exponent vectors $v = (v_P)_{P \in FB} \in \mathbb{Z}^{|FB|}$ with

$$\prod_{P \in FB} [P]^{v_P} = [\mathcal{O}].$$

Those vectors v are called *relations* on FB. The set of all relations on FB is an $|FB|$-dimensional $\mathbb{Z}$-lattice. If a generating system of this lattice is known, then the exponent k can be determined by means of linear algebra. This has been explained in [BD90]. The relations on FB are found by the method from [Buc88]. The basic idea is as follows. A random exponent vector $e \in \{1, \ldots, |\Delta|\}^{|FB|}$ is chosen. Then a reduced ideal K in the class $[\prod_{P \in FB} P^{e(P)}]$ is computed. If K can be factored over FB, i.e. if $K = \prod_{P \in FB} P^{f(P)}$ for some $f \in \mathbb{Z}^{|FB|}$ then

$v = e - f$ is a relation on FB. To guarantee that the decomposition is successful with a sufficient probability we use 1-reduction as in [Buc88]. This means that we have to compute shortest vectors in n-dimensional lattices. Using the best known algorithm [Kan87] for this problem this requires exponential time in n.

Once k is known we must find α. Note that $JI^{-k} = \alpha\mathcal{O}$. Since we know k this means that we have to find a generator of the principal ideal JI^{-k}. More generally let L be a principal ideal and suppose that we want to find a generator of L. This generator can again be found by means of linear algebra. With each relation v on FB that we have computed we have also obtained a number $\alpha = \alpha(v) \in F$ such that $\prod_{P \in FB} P^{v_P} = \alpha\mathcal{O}$. Now we choose again random vectors $e \in \{1, \ldots, |\Delta|\}^{|FB|}$. Then we compute $\gamma \in F$ and a reduced $\mathcal{O}$-ideal K such that $\gamma K = L \prod_{P \in FB} P^{e_P}$. If K can be factored over FB, i.e. if $K = \prod_{P \in FB} P^{f_P}$ for some $f \in \mathbb{Z}^{|FB|}$ then $L = \gamma \prod_{P \in FB} P^{e_P - f_P}$. Then $w = e - f$ is a relation on FB. Use linear algebra to write w as an integer linear combination on the generators of the relation lattice, i.e. $w = \sum_v w_v v$. Then $L = \gamma \prod_v \alpha(v)^{w_v}$. So we have found a generator for L.

A careful analysis of this algorithm together with experiments has to be done to evaluate the practical security of NF-EXP.

5   Two small examples

We present computations for the one way function NF-EXP in the order $\mathcal{O}$ of the equation $f(x) = x^4 + 989$. Elements of the number field $F = \mathbb{Q}[x]/(f(x))$ are multiplied as described above; this is realized using a multiplication table of size $4 \times 4 \times 4$, where there are 16 non-zero entries of which 10 equal to 1 and 6 equal to -989. A polynomial yielding such a *sparse* multiplication table is of special interest for practical purposes, since multiplication of number field elements is a basic operation in the evaluation of NF-EXP.

We start with the (prime) ideal I given by the matrix

$$\begin{pmatrix} 5 & 1 & 4 & 1 \\ 0 & 1 & 0 & 0 \\ 0 & 0 & 1 & 0 \\ 0 & 0 & 0 & 1 \end{pmatrix},$$

where the columns are the coefficient vectors representing the ideal basis elements. Now we choose $k = 200$; exponentiation of I by k yields a representation of I^k as a matrix with entries in the first row having more than 160 digits. Reducing this matrix using LLL with a large parameter (see [Coh95, 83ff]) induces an ideal J having a matrix representation

$$\begin{pmatrix} 1474 & 0 & 737 & 595 \\ 0 & 1474 & 0 & 1277 \\ 0 & 0 & 737 & 199 \\ 0 & 0 & 0 & 1 \end{pmatrix}$$

and a number α given by the coefficient vector

$$(9008155062745426317142256291951327,$$
$$-503841134277550505015598483379592542,$$
$$135374570150008211758174790992165237,$$
$$-3548184135796965661440536008930274)/1474$$

(we use a common denominator for all vector components) such that we have

$$\text{NF-EXP}_I(\alpha, k) = J.$$

The computation of J and α took 61 ms.

As claimed in Section 3, we cannot compute α and J from I and k in this way in general, since the size of the intermediate results and of α is exponential. We explain a fast exponentiation technique which computes only with numbers of polynomial length. It is called *square, multiply and reduce*. It proceeds as follows:

Let $(b_{l_k}, \ldots, b_0)_2$ be the binary expansion of k and denote $(J, \alpha) \leftarrow reduce(I)$ the computation of an equivalent, reduced ideal together with a number $\alpha \in F$ such that $I = \alpha J$.

Procedure Square, Multiply and Reduce:

1. IF $b_0 = 1$ THEN $E \leftarrow I$ ELSE $E \leftarrow \mathcal{O}$
2. $S \leftarrow I$
3. FOR $i \leftarrow 1$ TO l_k
3.1 $S^* \leftarrow S^2$, $(S, \beta_i) \leftarrow reduce(S^*)$
3.2 IF $b_i = 1$ THEN $E^* \leftarrow E \cdot S$, $(E, \alpha_i) \leftarrow reduce(E^*)$

This procedure yields a reduced ideal $J = E$ equivalent to I^k and a number α coded by the $\alpha_1, \ldots, \alpha_{l_k}, \beta_1, \ldots, \beta_{l_k}$. In our example, J is then represented by the matrix

$$\begin{pmatrix} 350 & 0 & 275 & 147 \\ 0 & 350 & 150 & 319 \\ 0 & 0 & 25 & 13 \\ 0 & 0 & 0 & 1 \end{pmatrix}$$

and the α_i and β_i are given by the following table. We mention only those β_i for which $b_i = 1$.

$$
\begin{array}{ll}
\alpha_1: & (-9, 1, 0, 0)/302 \\
\alpha_2: & (-32398, 3816, 122, -20)/1665 \\
\alpha_3: & (-8452, 71353, -2467, 163)/1185 \\
\beta_3: & (1, 0, 0, 0) \\
\alpha_4: & (-94051, -51202, -8704, 1592)/19489 \\
\alpha_5: & (3078261, -726226, -349243, 171480)/7458 \\
\alpha_6: & (-131260, -476600, -52456, 8698)/3205 \\
\beta_6: & (946075, 11920, -14675, 1180)/10954 \\
\alpha_7: & (-46517, 868, -661, 30)/10 \\
\beta_7; & (127134, 12057, -1464, -197)/175
\end{array}
$$

This computation took 51 ms.

A measure for the practical hardness of inverting NF-EXP is given by the time used to compute the class group of $\mathcal{O}$. The algorithm used can be modified to invert NF-EXP, as explained in the last section. Assuming some mathematical reasonable conjecture, the class group of $\mathcal{O}$ is generated by two ideals and is isomorphic to $\mathbb{Z}/2\mathbb{Z} \times \mathbb{Z}/36\mathbb{Z}$. The computation took 16 seconds.

Here another example: let $f(x) = x^6 - 11$. Let I be given by

$$
\begin{pmatrix}
19 & 0 & 0 & 0 & 0 & 3 \\
0 & 19 & 0 & 0 & 0 & 6 \\
0 & 0 & 19 & 0 & 0 & 12 \\
0 & 0 & 0 & 19 & 0 & 5 \\
0 & 0 & 0 & 0 & 19 & 10 \\
0 & 0 & 0 & 0 & 0 & 1
\end{pmatrix}.
$$

Choose $k = 1000$; the matrix representing I^k requires more than $14,000$ decimal digits, the reduction of this matrix took 97321 ms. Applying the *square, multiply and reduce* procedure, we get the following representation for I^k: an ideal J given by

$$
\begin{pmatrix}
4 & 0 & 2 & 2 & 0 & 1 \\
0 & 4 & 2 & 0 & 2 & 3 \\
0 & 0 & 2 & 0 & 0 & 1 \\
0 & 0 & 0 & 2 & 0 & 1 \\
0 & 0 & 0 & 0 & 2 & 1 \\
0 & 0 & 0 & 0 & 0 & 1
\end{pmatrix}
$$

and an number $\alpha \in \mathbb{Q}[x]/(f)$ given in the short representation explained above:

$$
\begin{aligned}
\alpha_1 &: \quad (-17, 99, 103, 16, 13, -12)/38 \\
\alpha_2 &: (-83, 73, -99, -113, -5, -13)/5 \\
\alpha_3 &: \quad\quad (2, 0, 0, 0, 0, 0) \\
\beta_3 &: \quad\quad (1, 0, 0, 0, 0, 0) \\
\alpha_4 &: (887, -183, 197, 27, -193, 37)/38 \\
\alpha_5 &: \quad (-34, 524, 206, -32, 26, 24)/43 \\
\beta_5 &: (59, 469, -171, -11, -51, -41)/2 \\
\alpha_6 &: (-312, -137, 88, -151, 70, 90)/2 \\
\beta_6 &: \quad\quad (1, 0, 0, 0, 0, 0) \\
\alpha_7 &: \quad\quad (1, 0, 0, 0, 0, 0) \\
\beta_7 &: \quad\quad (1, 0, 0, 0, 0, 0) \\
\alpha_8 &: \quad\quad (-7, 0, 3, 0, 1, 0)/3 \\
\beta_8 &: \quad\quad (1, 0, -1, 0, 1, 0) \\
\alpha_9 &: \quad\quad (-6, 0, 3, 0, 0, 0) \\
\beta_9 &: \quad\quad (1, 0, 0, 0, 0, 0)
\end{aligned}
$$

This computation took 124 ms. The class group was computed in 27,150 ms.

6 Conclusions

We have shown that the one way function NF-EXP can in principle be used to implement cryptographic primitives such as key exchange and digital signatures. We have also argued that the only known method for inverting NF-EXP requires computing shortest vectors in lattices whose dimension is the degree of the number field in which NF-EXP is implemented. This requires exponential time in the degree of the number field.

There are two important open problems:

- Can the algorithm for inverting NF-EXP be improved?
- Can NF-EXP be efficiently implemented?

As to the first question it is conceivable that an algorithm for solving NF-DL can be designed which uses c-reduction with $c > 1$. Certainly, $c = 2^n$, for which the reduction algorithm is polynomial, is not sufficient. But it may be possible to use some c which is subexponential in n. To implement such a c-reduction one can use the short vector algorithm described in [Sch87]. This algorithm is a candidate for a subexponential time reduction procedure.

To answer the second question much more research has to be done. We suspect that for example very efficient implementations are possible for families of number fields which are given by very sparse generating polynomials.

7 Acknowlegdement

We thank Stefan Neis for providing us with the computations in section 4. The algorithms have been implemented using the computer algebra C++ library LiDIA.

References

[BBM+92] I. Biehl, J. Buchmann, Bernd Meyer, Christian Thiel, and Christoph Thiel. Tools for proving zero knowledge. In *Proc. of EUROCRYPT'92*, Lecture Notes in Computer Science, pages 356–365. Springer, 1992.

[BD90] J. Buchmann and S. Düllmann. On the computation of discrete logarithms in class groups. In *Proc. of CRYPTO'90*, volume 537 of *Lecture Notes in Computer Science*, pages 134–139. Springer, 1990.

[BS96] E. Bach and J. Shallit. *Algorithmic number theory*. MIT Press, Cambridge, Massachusetts and London, England, 1996.

[Buc88] J. Buchmann. A subexponential algorithm for the determination of class groups and regulators of algebraic number fields. *Séminaire de théorie des nombres*, pages 28–41, 1988.

[BW88] J. Buchmann and H.C. Williams. A key-exchange system based on imaginary quadratic fields. *Journal of Cryptology*, 1:107–118, 1988.

[BW90] J. Buchmann and H.C. Williams. Quadratic fields and cryptography. *Number Theory and Cryptography, London Math. Soc. Lecture Note Series*, 154:9–26, 1990.

[Coh95] H. Cohen. *A course in computational algebraic number theory*. Springer, Heidelberg, 2nd edition edition, 1995.

[LiD97] The LiDIA Group. *LiDIA - A library for computational number theory*. Technische Hochschule Darmstadt, Germany.
http://www.informatik.th-darmstadt.de/TI/LiDIA.

[Kan87] R. Kannan. Minkowski's konvex body theorem and integer programming. *Mathematics of operations research*, 12, no. 5, 1987.

[Kob87] N. Koblitz. Elliptic curve cryptosystems. *Math. Comp.*, 48:203–209, 1987.

[Odl85] A.M. Odlyzko. Discrete logarithms in finite fields and their cryptographic significance. In *Proc. of EUROCRYPT'84*, volume 209 of *Lecture Notes in Computer Science*, pages 224–314. Springer, 1985.

[RSA78] R. L. Rivest, A. Shamir, and L. Adleman. A method for obtaining digital signatures and public key cryptosystems. *Communications of the ACM*, 21:120–126, 1978.

[Sch87] C.P. Schnorr. A hierarchy of polynomial time lattice basis reduction algorithms. *Theoretical Computer Science*, 53:201–224, 1987.

[Thi95] C. Thiel. *On some computational problems in algebraic number theory*. PhD thesis, Universität des Saarlandes, Saarbrücken, Germany, 1995.

Efficient Anonymous Multicast and Reception

(EXTENDED ABSTRACT)

Shlomi Dolev[1]*   Rafail Ostrovsky[2]

[1] Department of Mathematics and Computer Science, Ben-Gurion University of the Negev, Beer-Sheva 84105, Israel. Email: dolev@cs.bgu.ac.il.
[2] Bell Communications Research, 445 South St., MCC 1C-365B, Morristown, NJ 07960-6438, USA. Email: rafail@bellcore.com.

Abstract. In this work we examine the problem of efficient anonymous broadcast and reception in general communication networks. We show an algorithm which achieves anonymous communication with $O(1)$ amortized communication complexity on each link and low computational complexity. In contrast, all previous solutions require polynomial (in the size of the network and security parameter) amortized communication complexity.

1 Introduction

One of the primary objectives of an adversary is to locate and to destroy command-and-control centers – that is, sites that send commands and data to various stations/agents. Hence, one of the crucial ingredients in almost any network with command centers is to conceal and to confuse the adversary regarding which stations issue the commands. This paper shows how to use standard off-the-shelf cryptographic tools in a novel way in order to conceal the command-and-control centers, while still assuring easy communication between the centers and the recipients.

Specifically, we show efficient solutions that hide who is the sender and the receiver (or both) of the message/directive in a variety of threat models. The proposed solutions are efficient in terms of communication overhead (i.e., how much additional information must be transmitted in order to confuse the adversary) and in terms of computation efficiency (i.e., how much computation must be performed for concealment). Moreover, we establish rigorous guarantees about the proposed solutions.

1.1 The problem considered

Modern cryptographic techniques are extremely good in hiding all the *contents* of data, by means of encrypting the messages. However, hiding the contents of the message does not hide the fact that *some* message was sent from or received by a particular site. Thus, if some location (or network node) is sending and/or

---

* Part of this work was done while this author visited Bellcore with the support of DIMACS. Partially supported by the Israeli ministry of science and arts grant #6756196.

receiving a lot of messages, and if an adversary can monitor this fact, then even if an adversary does not understand what these messages are, just the fact that there are a lot of outgoing (or incoming) messages reveals that this site (or a network node) is sufficiently active to make it a likely target. The objective of this paper is to address this problem — that is, the problem of how to hide, in an efficient manner, which site (i.e. command-and-control center) transmits (or receives) a lot of data to (or from, respectively) other sites in the network. This question was addressed previously in the literature [6, 20] at the price of polynomial communication overhead for each bit of transmission per edge. We show an amortized solution which after a fixed pre-processing stage, can transmit an arbitrary polynomial-size message in an anonymous fashion using only $O(1)$ bits over each link (of a spanning tree) for every data bit transmission across a link.

1.2   General setting and threat model

We consider a network of processors/stations where each processor/station has a list of other stations with which it can communicate (we do not restrict here the means of communication, i.e. it could be computer networks, radio/satellite connections, etc.) Moreover, we do not restrict the topology of the network — our general methodology will work for an arbitrary network topology. One (or several) of the network nodes is a command-and-control center that wishes to send commands (i.e. messages) to other nodes in the network. To reiterate, the question we are addressing in this paper is how we can hide which site is broadcasting (or multicasting) data to (a subset of) other processors in the network. Before we explore this question further, we must specify what kind of attack we are defending against.

A simple attack to defend against is of a restricted adversary (called outside adversary) who is allowed only to monitor communication channels, but is *not* allowed to infiltrate/monitor the internal contents of any processor of the network. (As a side remark, such weak attack is very easy to defend against: all processors simply transmit either noise or encrypted messages on each communication channel – if noise is indistinguishable from encrypted traffic this completely hides a communication pattern.) Of course, a more realistic adversary, (and the one that we are considering in this paper) is the (internal) adversary that can monitor all the communication between stations and which *in addition* is also trying to infiltrate the internal nodes of the network.

That is, we consider the adversary that may mount a more sophisticated attack, where he manages to compromise the security of one or several *internal* nodes of the network, whereby he is now not only capable of monitoring the external traffic pattern but is also capable of examining every message and all the data which passes through (or stored at) this infiltrated node. Thus, we define an *internal k-listening adversary*, an adversary that can monitor all the communication lines between sites and *also* manages to monitor (the internal contents of) up to k sites of the network. (This, and similar definitions were considered before in the literature, see, for example [20, 5] and references therein). We remark, though, that in this paper we restrict out attention only to listening adversary, that only monitors traffic, but does not try to sabotage it, similar to

[10, 14], but with different objectives. Before we proceed, let us turn to a simple example, to better explain what are the issues that must be addressed.

1.3 A simple example

In this subsection, we examine a very simple special case, in order to illustrate the issues being considered and a solution to this special case. We stress, though, that we develop a general framework that works for the general case (e.g. the case of general communication graph, unknown receiver, etc.) as well.

Suppose we are dealing with a network having 9 nodes:

$$P_1 \longrightarrow P_2 \longrightarrow P_3 \longrightarrow P_4 \longrightarrow P_5 \longrightarrow P_6 \longrightarrow P_7 \longrightarrow P_8 \longrightarrow R$$

where R is the "receiver" node and one of the P_i is the command-and-control center which must broadcast commands to R. The other P_j's for $j \neq i$ are "decoys" which are used for transmission purposes from P_i to R and also are used to "hide" which particular P_i is the real command and control center. That is, in this simplified example, we only wish to hide from an adversary which of the P_i is the real command and control center which sends messages to R. Before we explain our solution, we examine several inefficient, but natural to consider simple strategies and then explain what are their drawbacks.

Communication-inefficient solution: One simple (but inefficient!) way to hide which P_i is the command-and-control center is for every P_i to broadcast an (encrypted) stream of messages to R. Thus, R receives 8 different streams of messages, ignores all the messages except those from the real command-and-control center, and decrypts that one. Every processor P_i forwards messages of all the smaller-numbered processors and in addition sends its own message. Clearly, an adversary who is monitoring all the communication channels and which can also monitor the internal memory of one of the P_i's (which is not the actual command-and-control center) does not know which P_j is broadcasting the actual message. **Drawback:** Notice that instead of one incoming message, R must receive 8 messages, thus the throughput of how much information the real command-and-control center can send to R is only $\frac{1}{8}$ of the total capacity! As the network becomes larger this solution becomes even more costly.

Computation-inefficient solution: In the previous example, the drawback was that the messages from decoy command-and-control nodes were taking up the bandwidth of the channel. In the following solution, we show how this difficulty can be avoided. In order to explain this solution, we shall use pseudo-random generators[3] [2, 12, 13]. We first pick 8 seeds $s_1, \ldots, s_8$ for the pseudo-random generator, and give to processor P_i seed s_i. Processor P_1 stretches its seed s_1 into long pseudo-random sequence, and sends, at each time step the next bit of its sequence to processor P_2. Processor P_2 takes the bit it got from

---

[3] Pseudo-random generator $G(s) = r_1, r_2, \ldots$ takes a small initial "seed" of truly random bits, and deterministically expands it into a long sequence of pseudo-random bits. There are many such commercially available pseudo-random generators, and any such "off-the-shelf" generator that is sufficiently secure and efficient will suffice.

processor P_1 and "xors" it with its own next bit from its pseudo-random sequence $G(s_2)$ and sends it to P_3 and so forth. The processor P_j which is the real command-and-control center additionally "xors" into each bit it sends out a bit of the actual message m_i. Processor R is given all the 8 seeds $s_1, \ldots s_8$, so it can take the incoming message, (which is the message from command-and-control center "xored" with 8 different pseudo-random sequences.) Hence, R can compute all the 8 pseudo-random sequences, subtract (i.e. xor) the incoming message with all the 8 pseudo-random sequences and get the original command-and-control message m. The advantage of this solution is that any P_j which is not a command-and-control center (and not R), clearly can not deduce which other processor is the real center. Moreover, the entire bandwidth of the channel between command-and-control processor and the receiver is used to send the messages from the center to the receiver. **Drawback:** The receiver must compute 8 different pseudo-random sequences in order to recover the actual message. As the network size grows, this becomes prohibitively expensive in terms of the computation that the receiver needs to perform in order to compute the actual message m.

Our solution for this simple example: Here, we present a solution that is *both computation-efficient and communication-efficient* and is secure against an adversary that can monitor all the communication lines and additionally can learn internal memory contents of any one of the intermediate processors. The seed distribution is as follows:

- Pick 9 random seeds for pseudo-random generator $s_0, s_1, \ldots, s_8$.
- Give to the real command-and-control processor seed s_0.
- Additionally, give to processor P_1 seed $\{s_1, s_2\}$; to processor P_2 two seeds $\{s_2, s_3\}$, to processor P_3 two seeds $\{s_3, s_4\}$, and so on. That is, we give to each processor P_i for $i > 1$ the seeds $\{s_i, s_{i+1}\}$.
- give to receiver, R, one seed s_0

Suppose processor P_4 is the real command-and-control center. Then the distribution of seeds is as follows:

$$\boxed{P_1(s_1, s_2)} \longrightarrow \boxed{P_2(s_2, s_3)} \longrightarrow \boxed{P_3(s_3, s_4)} \longrightarrow \boxed{P_4(s_4, s_5, s_0)} \longrightarrow \boxed{P_5(s_5, s_6)} \longrightarrow$$

$$\longrightarrow \boxed{P_6(s_6, s_7)} \longrightarrow \boxed{P_7(s_7, s_8)} \longrightarrow \boxed{P_8(s_8, s_1)} \longrightarrow \boxed{R(s_0)}$$

Now, the transmission of the message is performed in the same fashion as in the previous solution — that is, each processor receives a bit-stream from its predecessor, "xors" a single bit from each pseudo-random sequence that it has, and sends it to the next processor. The command-and-control center "xors" bits of the message into each bit that it sends out.

Notice, that adjacent processors "cancel" one of the pseudo-random sequences, by xoring it twice, but introduce a new sequence. For example, processor P_2 cancels s_2, but "introduces" s_3. Moreover, each processor must now only compute the output of at most three seeds. Yet, it can be easily verified that if the adversary monitors all the communication lines and in addition can learn seeds of any single processor P_i which is not a command and control center, then it can not gain any information as to which other P_i is the real command and control center, even after learning the two seeds that belong to processor P_i.

Of course, the simplified example that we presented works only provided that the adversary cannot monitor both the actual command-and-control center and can not monitor the memory contents of the receiver. (We note that these and other restrictions can be resolved – we address this further in the paper.) Moreover, it should be stressed that the restricted solution presented above does not work if the adversary is allowed to monitor more than one decoy processor. We should point out that in the rest of the paper we show how the above scheme can be extended to one that is robust against adversaries that can monitor up-to k stations, where in our solution every processor is required to compute the number of different pseudo-random sequences proportional to k only (in particular, at most $2k + 1$). Moreover, we also show how to generalize the method to arbitrary-topology networks/infrastructures. Additionally, we show how initial distribution of seeds can be done without revealing the command-and-control center and how the actual location of the command-and-control center can be hidden from the recipients of the messages as well. At last, we show how communication from stations back to the command-and-control center could be achieved without the stations knowing at which node of the network the center is located and how totally anonymous communication can be achieved.

1.4 Private-key solutions vs. Public-key solutions

The above simple solution is a private-key solution, that is, we assume that before the protocol begins, a set of seeds for pseudo-random function must be distributed in a private and anonymous manner. We note that we show how to distribute these seeds using a public-key solution, that is, a solution where we assume that all users/nodes only have corresponding public and private keys and do not share any information a-priori. Thus, our overall solution is a public-key solution, where before communication begins, we do not assume that users share any private data. As usual in many of such cryptographic setting, our overall efficiency comes from the fact that we *switch* from public-key to private key solution and then show how to (1) make an efficient private-key implement and (2) how to set up private keys in a pre-processing stage by using public keys in an anonymous and private manner.

1.5 Comparison with Previous Work

One of the first works (if not the first one) to consider the problem of hiding the communication pattern in the network is the work of Chaum [6] where he introduced the concept of a *mix*: A single processor in the network, called a mix, serves as a relay. A processor P that wants to send a message m to a processor Q encrypts m using Q's public key to obtain m'. Then P encrypts the pair (m', q) using the public key of the mix. The double encrypted message is sent to the mix. The mix decrypts the message (to get the pair (m', q)) and forwards m' to q. Further work in this direction appear in [15, 17, 18]. The single mix processor is not secure when this *single* processor is cooperating with the (outside) adversary; If the processor that serves as a mix is compromised, it can inform the adversary where the messages are forwarded to. Hence, as Chaum pointed out, a sequence of "mixes" must be employed at the price of additional communication and computation. Moreover, the single mix scheme operates under some statistic assumption on the pattern of communication. In case a single message is sent to the mix then an adversary that monitors the communication channels can observe the sender and the receiver of the particular message.

An extension of the mix scheme is presented by Rackoff and Simon [20] who embedded an n-element sorting network of depth polynomial in $\log(n)$ that mixes incoming messages and requires only polynomially many (in $\log(n)$) synchronous steps. In each such step every message is sent from one site of the network to another site of the network. Thus, the message delay may be proportional to $\log(n)$ times the diameter of the network. The statistic assumptions on the pattern of communication is somewhat relaxed in [20] by introducing dummy communication: Every processor sends a message simultaneously. However, the number of (real and dummy) messages arriving to each destination is available to the traffic analyzer. Rackoff and Simon also presented in [20] a scheme that copes with passive internal adversaries by the use of randomly chosen committees and multi-party computation (e.g., [11, 3, 7, 4, 5].)

More generally, secure multi-party computation can be used to hide the communication pattern in the network (see, for example, [11, 8, 19, 3, 7, 4, 5]) via secure function valuation. However, anonymous communication is a very restricted form of hiding participants' input and hence may benefit from less sophisticated and more efficient algorithms.

In this work we consider the problem of anonymous communication on a spanning tree of a general graph communication network. In a network of n processors our algorithm (after a pre-processing stage) sends $O(1)$ bits on each tree link in order to transmit a clear-text bit of data and each processor computes $O(k)$ pseudo-random bits for the transmission of a clear-text bit. Multiple anonymous transmission is possible by executing in parallel several instances of our algorithm. Each instance uses part of the bandwidth of the communication links. Our algorithm is secure for both *outside adversary* and *k-listening internal dynamic adversary*. (We remark, though, that we are only considering eavesdropping "listening" adversary, similar to [10, 14], and *do not* consider a Byzantine adversary which tries to actively disrupt the communication, as in [11].) Our algorithm starts with anonymous *seeds* distribution. These seeds are later used for the generation of pseudo-random sequences.

The rest of the paper is organized as follows. The problem statement appears in Section 2. The anonymous communication (our Xor-Tree Algorithm) which is the heart of our scheme appears in 3. Section 4 and 5 sketch the anonymous seeds transmission and the initialization and termination schemes, respectively. Extensions and concluding remarks appear in Section 6.

2 Problem Statement

A communication network is described by a communication graph $G = (V, E)$. The nodes, $V = \{1, \cdots, n\}$, represent the processors of the network. The edges of the graph represent bidirectional communication channels between the processors. Let us first define the assumptions and requirements used starting with the adversary models.

- An *outside adversary* is an adversary that can monitor all the communication links but *not* the contents of the processors memory.
- An *internal dynamic k-listening adversary* (*inside adversary*, in short) is an adversary that can choose to "bug" (i.e., listen to) the memory of up to *k* processors. The targeted processors are called *corrupted* or *compromised* processors. Corrupted processors reveal all the information they know to the adversary, however they still behave according to the protocol. The adversary does not have to choose the *k* faulty processors in advance. While the adversary corrupts less than *k* processors the adversary can choose the next processor to be corrupted using the information the adversary gained so far from the processors that are already corrupted.

The following assumptions are used in the first phase of our algorithm which is responsible for the *seeds distribution*. Each of the *n* processors has a public-key/private-key pair. The public key of a processor, P, is known to all the processors while the private key of P is known only to P.

The anonymity of the communicating parties can be categorized into four cases:

- *Anonymous to the non participating processors:* A processor P wishes to send a message to processor Q without revealing to the rest of the processors and to the inside and outside adversary the fact that P is communicating with Q.
- *Anonymous to the sender and the non participating processors:* P wishes to receive a message from Q without revealing its identity to any processor including Q as well as to an inside and outside adversary.
- *Anonymous to the receiver(s) and the non participating processors:* P wishes to send (or multicast) a message without revealing its identity to any processor as well as to an inside and an outside adversary.
- *Anonymous to the sender, to the receiver, and the non participating processors:* A processor P (and Q) wishes to communicate with some other processor, without knowing the identity of the processor, and without revealing its identity to any processor including Q, as well as to an inside and outside adversary. (This is similar to the "chat-room" world-wide-web applications, where two processors wish to communicate with one another totally anonymously, without revealing to each other or anybody else their identity.)

The *efficiency* of a solution is measured by the *communication overhead* which is the number of bits sent over each link in order to send a bit of clear-text data. The efficiency is also measured by the *computation overhead* which is the maximal number of *computation steps* performed by each processor in order to transfer a bit of clear-text data.

The algorithm is a combination of anonymous seeds transmission, initialization, communication and termination. In the anonymous seeds transmission phase, processors that would like to transmit, anonymously send seeds for a pseudo-random sequence generators to the rest of the processors. The anonymous seeds transmission phase also resolves conflicts of multiple requests for transmission by an anonymous back-off mechanism. Once the seeds are distributed the communication can be started. Careful communication initialization (and termination) procedure that hide the identity of the sender must be performed.

We first describe the core of our algorithm which is the communication phase. During the communication phase seeds are used for the production of pseudo-random sequences. The anonymous seeds distribution is presented following the description of the anonymous communication phase.

3 Anonymous Communication

3.1 Computation-inefficient $O(n)$ solution

The communication algorithm is designed for a spanning tree T of a general communication graph, where the relation *parent child* is naturally defined by the election of a root. We start with a simple but inefficient algorithm which requires $O(n)$ computation steps of a processor. (This algorithm is similar to the computation-inefficient solution presented in Section 1, but for the general-topology graph. We then show how to make it computation-efficient as well.) In this (computation-inefficient) solution the sender will chose a distinct seed for each processor. Then the sender can encrypt each bit of information using the seeds of all the processors including its own seeds. Each such seed is used for producing a pseudo-random sequence. The details of the algorithm appear in Figure 1. The symbol $\oplus$ is used to denote the binary xor operation.

Seeds Distribution — Assign (anonymously) distinct seed to each processor. In addition assign the receiver(s) with the sender seed and the sender with every assigned seed.

Upwards Communication :

    P is the sender — Let d_i be the i'th bit of data. Let $b_1, b_2, \cdots, b_l$ be the i'th bits received from the children (if any) of P, and let b' be the i'th bit of the pseudo random sequence obtained from the seed of P. The i'th bit P sends to its parent (if any) is $b_1 \oplus b_2 \oplus \cdots b_l \oplus b'_1 \oplus b'_2 \cdots \oplus b'_n \oplus d_i$, where $b'_1, b'_2, \cdots, b'_n$ are the i'th bits of the pseudo random sequence obtained from the seeds of all the processors.

    P is not the sender — The i'th bit that P communicants to its parent (if any) is $b_1 \oplus b_2 \oplus \cdots b_l \oplus b'$.

Downwards Communication — The root processor calculates an output as if it has a parent and sends the result to every of its children. Every processor which is not the root, sends to its children every bit received from its parent. The receiver(s) decrypts the downward communication by the use of the senders' seed.

Fig. 1. $O(n)$ Computation Steps Algorithm.

Note that the i'th bit produced by the root is a result of xoring twice every of the i'th bits of the pseudo-random sequences except the senders' sequence: once by the sender and then during the communication upwards. Each encrypted bit of data will be xored by the receiver(s) using the senders' seed to reveal the

403

clear-text. Note that the scheme is resilient to any number of colluding processors as long as the sender and the receiver(s) are non-faulty. This simple scheme requires a single node (the sender) to compute $O(n)$ pseudo-random bits for each bit of data. (We remark that in contrast, our Xor-Tree Algorithm, requires the computation of only $O(k)$ pseudo-random bits to cope with an outside adversary and an internal dynamic k-listening adversary.) The next Lemma state the communication and computation complexities of the algorithm presented in Figure 1.

Lemma 1. *The next two assertions hold for every bit of data to be transmitted over each edge of the spanning tree:*

- *The communication overhead of the algorithm is $O(1)$ per edge.*
- *The computation overhead of our algorithm is $O(n)$ pseudo-random bits to be computed by each processor per each bit of data.*

Proof. In each time unit two bits are sent in each link: one upwards and the other downwards. Since a bit of data is sent every time unit (possibly except the first and last h time units, where $h < n$ is the depth of the tree) the number of bits sent over a link to transmit a bit of data is $O(1)$. The second assertion follows from the fact that the sender computes the greatest number of pseudo random bits in every time unit, namely $O(n)$ pseudo-random bits in every time units. □

3.2 Towards our $O(k)$ solution: The choice of seeds

For the realization of the communication phase of our $O(k)$ solution we use $n(k+1)$ distinct seeds where k is less than $\lfloor n/2 - 1 \rfloor$. Each processor receives $2(k+1)$ seeds. To describe the seeds distribution decisions of the sender we use $k+1$ *levels* each consists of two *layers* of seeds.

The first level — Let $\mathcal{L}^1 = s_1^1, s_2^1, s_3^1, \cdots, s_n^1$ be the seeds that the sender (randomly) chooses for the first level. The sender uses the sequence of seeds $\mathcal{L}_1^1 = s_1^1, s_2^1, s_3^1, \cdots, s_n^1$ for the first layer of the first level and $\mathcal{L}_2^1 = s_2^1, s_3^1, \cdots, s_n^1, s_1^1$ for the second layer. Note that $\mathcal{L}_1^1 = \mathcal{L}^1$ and that $\mathcal{L}_2^1$ is obtained by rotating $\mathcal{L}^1$ once. P_i, $1 \leq i < n$, receives the seeds s_i^1 and s_{i+1}^1. P_n receives s_n^1 and s_1^1.

The l'th level — Similarly, for the l'th level $1 \leq l \leq k+1$ the sender (randomly) chooses n distinct seeds for this level $\mathcal{L}^l = s_1^l, s_2^l, \cdots, s_n^l$ to be the seeds of the l'th level and uses two sequences $\mathcal{L}_1^l = \mathcal{L}^l$ and $\mathcal{L}_2^l = s_{l+1}^l, s_{l+2}^l, \cdots, s_n^l, s_1^l, \cdots, s_l^l$; $\mathcal{L}_2^l$ is obtained by rotating $\mathcal{L}^l$ l times. P_i $1 \leq i \leq n-l$ receives the seeds s_i^l and s_{i+l}^l and P_j $n-l < j \leq n$ receives the seeds s_j^l and $s_{j-(n-l)}^l$.

Thus, at the end of this procedure every processor is assigned by $2k+2$ distinct seeds.

Fig. 2. The choice of seeds.

The seeds distribution procedure appears in Figure 2. An example for the choice of seeds for the processors appears in Figure 3.

Seeds of	P_1	P_2	P_3	P_4	P_5	P_6	P_7	P_8	P_9
Level 1	s_9	s_1	s_2	s_3	s_4	s_5	s_6	s_7	s_8
	s_1	s_2	s_3	s_4	s_5	s_6	s_7	s_8	s_9
Level 2	s'_8	s'_9	s'_1	s'_2	s'_3	s'_4	s'_5	s'_6	s'_7
	s'_1	s'_2	s'_3	s'_4	s'_5	s'_6	s'_7	s'_8	s'_9
Level 3	s''_7	s''_8	s''_9	s''_1	s''_2	s''_3	s''_4	s''_5	s''_6
	s''_1	s''_2	s''_3	s''_4	s''_5	s''_6	s''_7	s''_8	s''_9

Fig. 3. An example for the distribution of seeds, where $n = 9$ and $k = 2$.

The choice of seeds made by the sender has the following properties:

- Each seed is shared by exactly two processors.
- For every processor P, P shares a (distinct) seed with every of the $k+1$ processors that immediately follow P, (if there are at least such $k+1$ processors), or with the rest of the processors including P_n, otherwise.

3.3 The Xor-Tree Algorithm

Here, we present out main algorithm, the Xor-Tree Algorithm. The Xor-Tree Algorithm appears in Figure 4.

Seeds Distribution — Assign seeds to the processors as described in Figure 2. Assign the sender with one additional seed. Assign the receiver(s) with this additional seed of the sender.

Upwards Communication :

P **is the sender** — Let d_i be the i'th bit of data. Let $b_1, b_2, \cdots, b_l$ be the i'th bits received from the children (if any) of P, let $b'_1, b'_2, \cdots, b'_{2k+2}$ be the i'th bits of the pseudo-random sequences obtained from the seeds of P, and let b'_{2k+3} be the i'th bit of the pseudo-random sequence obtained from the additional seed of the sender. The i'th bit P sends to its parent (if any) is $b_1 \oplus b_2 \oplus \cdots b_l \oplus b'_1 \oplus b'_2 \cdots \oplus b'_{2k+2} \oplus b'_{2k+3} \oplus d_i$.

P **is not the sender** — The i'th bit that P communicants to its parent (if any) is $b_1 \oplus b_2 \oplus \cdots b_l \oplus b'_1 \oplus b'_2 \cdots \oplus b'_{2k+2}$.

Downwards Communication — The root processor calculates an output as if it has a parent and sends the result to every of its children. Every non root processor send to its children every bit received from its parent. The receiver(s) decrypts the downward communication, to obtain the clear-text, by the use of the senders' additional seed.

Fig. 4. The Xor-Tree Algorithm.

3.4 An abstract game

In this subsection we describe an abstract game that will serve us in analyzing and proving the correctness of the Xor-Tree Algorithm presented in the previous subsection.

Seeds Assignment — Assign seeds to the processors as described in Figure 2. Assign the sender with one additional seed.

Computation — Each processor, P, uses its seeds to compute pseudo-random sequences. At the i'th time unit the sender S computes the i'th bit of every of its pseudo-random sequences, xors these bits and the i'th bit of data and outputs the result. At the same time unit every other processor P computes the i'th bit of every of its pseudo-random sequences xors these bits and outputs the result.

Fig. 5. The Abstract Game.

The adversary get to see the outputs of all the players. The adversary can pick k out of the players and see their seeds. We claim, and later prove, that when the adversary does not pick the sender then every one of the remaining $(n - k)$ processors that are not picked by the adversary is equally likely to be the sender for any poly-bounded adversary[4].

We proceed by showing that the above assignment of seeds yields a *special seed* ds_P for each processor P. We choose ds_P out of the seeds assigned to each non-faulty processor P. We order the processors by their index in a cyclic fashion such that the processor that follows the i'th processor, $i \neq n$, is the processor with the index $i + 1$ and the processor that follows the n'th processor is the first processor. Then we assign a new index for each processor such that the sender has the index one, the processor that follows the sender has the index two and so on and so forth. These new indices are used for the interpretation of next, follows, prior and last in the description of the choice of special seeds that appears in Figure 6. Recall that with overwhelming probability every two processors share at most one seed.

Note that by our special seeds selection, described in Figure 6, the special seeds are not known to the k faulty processors.

Theorem 2. *In the abstract game any of the $(n - k)$ non-faulty processors is equally likely to be the sender for any poly-bounded internal k-listening adversary.*

Proof. We prove that the i'th bit produced by any non-faulty processors is equally likely to be 0 or 1 for any poly-bounded adversary. Let P be the first non-faulty processor that follows the sender (P is among the first $k+1$ processors that follow the sender). Let b be the special seed of the sender that is shared only with (the non-faulty processor) P. The i'th bit that the sender outputs is a

---

[4] If the adversary can predict who is the sender then we can use this adversary to break a pseudo-random generator.

The sender P_1 — Each of the $k + 1$ processors that immediately follows the sender shares exactly one seed with the sender. Since there are at most k colluding processors, one of these $k + 1$ processors must be non-faulty. Pick, P, the first such non-faulty processor. Assign ds_{P_1}, the special seed of the sender, to be the seed that the sender shares with P.

A processors P that is not among the $k + 1$ last processors — If P is not among the $k + 1$ last processors then P is assigned by $2k + 2$ seeds $k + 1$ seeds of these seeds are from the first layers of the $k + 1$ seed levels. These $k + 1$ seeds are *new* — they do not appear in any processor prior to P. Since there are at most k colluding processors, one of the next $k + 1$ processors is non-faulty. Let Q be the first such non-faulty processor and assign ds_P by the seed that P shares with Q. Repeat the procedure until you reach a non-faulty processor that is among the last $k + 1$ processors.

A processors Q that is among the $k + 1$ last processors — Note that Q does not introduce $k + 1$ new seeds since some of its seeds are assigned to the first processors (at least the one in the $k + 1$'th level). Fortunately, Q shares a single seed with every of the last processors. This fact allows us to continue the special seed selection procedure, by choosing the seed shared with the next non-faulty processor.

Fig. 6. Choice of special seeds.

result of a xor operation with the i'th bit of the pseudo-random sequence (among other pseudo-random sequences) obtained from b. Thus, it is equally likely to be 0 or 1 for any poly-bounded internal k-listening adversary. A similar argument hold for the output of P and in general the output of every non-faulty processor. The same argument holds if *any* of the $n - k$ non-faulty processors is the sender. Thus, for any polynomially-bounded k-internal and external adversary, the distribution of the output is indistinguishable of the identity of the sender. □

3.5 Reduction to the abstract game

In this subsection we prove that if there is an algorithm that reveals information on the identity of the sender in the tree then there exists an algorithm that reveals information on the identity of the sender in the abstract game. The above reduction together with Theorem 2 yields the proof of correctness for the Xor-Tree algorithm.

Assume that the adversary reveals information on the sender in a tree T of n processors. Then an abstract game of n nodes is mapped to the tree as follows:

Theorem 3. *In the Xor-Tree Algorithm any of the $(n - k)$ non-faulty processors is equally likely to be the sender for any poly-bounded internal k-listening adversary.*

Proof. If there exists an adversary $\mathcal{A}$ that reveals information on the identity of the sender in a tree T then there exists an abstract game with the same number of processors and the same seeds distribution, such that the application of the

---

1. Each processor of the abstract game is assigned to a node of the tree T.
2. The output of every processor to its parent is computed as follows: Let the *hight* of a processor P in T be the number of edges in the longest path $\mathcal{P}$ from P to a leaf such that $\mathcal{P}$ does not traverse the root. We start with the processors that are in hight 0 i.e. the leaves. The output of the processors that were assigned to the leaves of the tree is not changed i.e. it is identical to their output in the abstract game. Once we computed the output of processors in hight h we use these computed outputs to compute the outputs of processors in hight $h + 1$. Let Q be a processor in hight $h + 1$, let $b_1, b_2, \cdots, b_l$ be the i'th computed bits that are output by the children of Q, and let b_Q be the original i'th output bit of Q in the abstract game. The computed output of Q is $b_1 \oplus b_2 \oplus \cdots b_l \oplus b_Q$.

Fig. 7. The Reduction.

reduction in Figure 7 yields the communication pattern on T and reveals information on the sender identity in the abstract game. This contradicts Theorem 2 and thus contradicts the existence of $\mathcal{A}$. □

The next Lemma states the communication and computation overheads of the anonymous communication algorithm.

Lemma 4. *The next two assertions hold for every bit of data to be transmitted over each edge of the spanning tree:*

- *The communication overhead of the algorithm is $O(1)$ per edge.*
- *The computation overhead of our algorithm is $O(k)$ pseudo-random bits to be computed by each processor per each bit of data.*

Proof. In each time unit two bits are sent in each link: one upwards and the other downwards. Since a bit of data is sent every time unit (possibly except the first and last h time units, where $h < n$ is the depth of the tree) the number of bits send over each link to transmit a bit of data is $O(1)$. The second assertion follows from the fact that in each time unit each processor generates at most $2k + 3$ pseudo-random bits. □

4 Anonymous Seeds Transmission

The details of the anonymous seeds transmission are omitted from this extended abstract, here we only sketch the main ideas. Every processor has a public-key encryption, known to all other processors. A virtual ring defined by the Euler tour on the tree is used for the seeds transmission. Note that the indices of the processor used in this description are related to their location on the virtual ring. First all processors send messages to P_1 over the (virtual) ring. Those processors that wish to broadcast send a collection of seeds, and those processors that do not with to broadcast, send dummy messages of equal length. To do so in an anonymous fashion (so that P_1 does not know which message is from which processor), $k + 1$ of Chaum's *mixes* [6] are used, where $k + 1$ (real) processors just before P_1 in the Euler tour are used as mixes. Hence, P_1 can identify the number of non-dummy arriving messages but not their origin. In case more than one non-dummy message reaches P_1, a standard back-off algorithm is initiated by P_1. Once exactly one message (containing a collection of seeds) arrives to P_1 the seed distribution procedure described above (for sending a collection of seeds to P_1) is used to send the seeds to P_2 and so on. (At this point processors know that only one processor wishes to broadcast.) This procedure is repeated n times in order to allow the anonymous sender to transmit a collection of seeds to every processor. Notice that this process is quadratic in the size of the ring, the number of colluding processors k, and the length of the security parameter, (i.e., let g be a security parameter and k as before, then we send $O((gkn)^2)$ bits per edge.) Thus, as long the message size p to be broadcasted is greater than $O((gkn)^2)$ we achieve $O(1)$ overall amortized cost per edge, and otherwise we get $O((gkn)^2/p)$ amortized cost.

5 Initialization and Termination

When the seed distribution procedure is over, then the transmission of data may start. P_n broadcasts a signal on the tree that notifies the leaves that they can start transmitting data. The leaves start sending data in a way that ensures that every non-leaf processor receives the i'th bit from its children simultaneously. Thus, the delay in starting transmission of a particular leaf l is proportional to the difference between the longest path from a leaf to the root and the distance of l from the root. Each non leaf processor waits for receiving the i'th bit from each of its children, uses these bits and its seeds to compute its own i'th bit and sends the output to its parent. Note that buffers can be used in case the processors are not completely synchronized.

The sender can terminate the session by sending a termination message that is not encrypted by its additional seed. This message will be decrypted by the root that will broadcast it to the rest of the processors to notify the beginning of a new anonymous seeds transmission.

6 Extensions and Concluding Remarks

Our treatment so far considered the anonymous sender case, which is also anonymous to the non participating processors. A simple modification of the algorithm can support the anonymous receiver case: The receiver plays a role of a sender of the previous solution in order to communicate in an anonymous fashion an additional seed to the sender. Then the sender uses the same scheme for the anonymous sender case with the seed the sender got from the receiver.

To achieve anonymous communication in which both the sender and the receiver are anonymous, do the following: The two participants, P and Q, that would like to communicate (each) send anonymously distinct seeds to $P_1, \ldots, P_{k+1}$ (again, if there are several parties that are competing, a back-off mechanism is used). P_1 will encrypt and broadcast the two seeds it got, each seed encrypted (using distinct intervals of the pseudo-random expansions of the two seeds) by the other seed. Hence, each of the two processors will use its seed to reveal the seed of the other processor. The same procedure (after back-off protocol terminated in case of competing requests) continues for P_2, P_3, $P_4 \cdots P_{k+1}$. At this stage P has a set of $k + 1$ seeds that are used for encryption of messages sent to Q and Q has $k + 1$ seeds used for encryption messages send to P. Moreover, this is resilient against any coalition of size k and we are done.

Acknowledgment: We thank Oded Goldreich and Ron Rivest for helpful remarks.

References

1. M. Blum and S. Goldwasser, "An efficient probabilistic public-key encryption scheme which hides all partial information", *CRYPTO 84*.
2. M. Blum, and S. Micali "How to Generate Cryptographically Strong Sequences of Pseudo-Random Bits", FOCS 82 and *SIAM J. on Computing*, Vol 13, 1984, pp. 850–864.
3. M. Ben-or, S. Goldwasser, and A. Wigderson, "Completeness Theorems for Non-Cryptographic Fault-Tolerant Distributed Computation", *STOC 88*.
4. R. Canetti, U. Feige, O. Goldreich, and M. Naor, "Adaptively Secure Multi-Party Computation", *STOC 96*.
5. R. Canetti, E. Kushilevitz, R. Ostrovsky, and A. Rosén, "Randomness vs. Fault-Tolerance", *PODC 97*.
6. D. Chaum, "Untraceable Electronic Mail, Return Addresses, and Digital Pseudonyms", *Communication of the ACM*, vol. 24, no. 2 (1981), pp. 84-88.
7. D. Chaum, C. Crépeau, and I. Damgård, "Multiparty Unconditionally Secure Protocols", *STOC 88*.
8. D. Chaum, "The Dining Cryptographers Problem: Unconditional Sender and Recipient Untraceability", *Journal of Cryptology*, vol. 1 (1988), pp. 65-75.
9. D. Chaum, "Achieving Electronic Privacy", *Scientific American*, vol. 267, no. 2 (1992), pp. 96-101.
10. M. Franklin, Z. Galil and M. Yung "Eavesdropping Games: A Graph-Theoretic Approach to Privacy in Distributed Systems," *FOCS 93*.
11. O. Goldreich, S. Micali and A. Wigderson, "How To Play Any Mental Game", *STOC 87*.
12. J. Hastad, "Pseudo-Random Generators under Uniform Assumptions", *STOC 90*.
13. R. Impagliazzo, L. Levin, and M. Luby "Pseudo-Random Generation from One-Way Functions," *STOC 89*.
14. E. Kushilevitz, S. Micali, and R. Ostrovsky, "Reducibility and Completeness in Multi-Party Private Computations", *FOCS 94*.
15. A. Pfitzmann, "How to Implement ISDNs Without User Observability — Some Remarks", TR 14/85, Fakultat fur Informatik, Universitat Karlsruhe, 1985.
16. A. Pfitzmann, M. Waidner, "Network without User Observability," *Computer & Security* 6 (1987) 158-166.
17. A. Pfitzmann, B. Pfitzmann and M. Waidner, "ISDN-MIXes — Untraceable Communication with Very Small Bandwidth Overhead," *Proc. Kommunikation in verteilten Systemen* (1991), pp. 451-463.
18. P. F. Syverson, D. M. Goldsclag, M, G. Reed, "Anonymous Connections and Onion Routing" *Proc. of the Symposium on Security and Privacy* 1997.
19. M. Waidner and B. Pfitzmann, "The Dining Cryptographers in the Disco: Unconditional Sender and Recipient Untraceability with Computationally Secure Serviceability *Eurocrypt 89*.
20. C. Rackoff and D. Simon, "Cryptographic Defense Against Traffic Analysis", *STOC 93*

Efficient Group Signature Schemes
for Large Groups

(Extended Abstract)

Jan Camenisch

Department of Computer Science
Haldeneggsteig 4
ETH Zurich
8092 Zurich, Switzerland
camenisch@inf.ethz.ch

Markus Stadler

Ubilab
Union Bank of Switzerland
Bahnhofstr. 45
8021 Zurich, Switzerland
Markus.Stadler@ubs.com

Abstract. A group signature scheme allows members of a group to sign messages on the group's behalf such that the resulting signature does not reveal their identity. Only a designated group manager is able to identify the group member who issued a given signature. Previously proposed realizations of group signature schemes have the undesirable property that the length of the public key is linear in the size of the group. In this paper we propose the first group signature scheme whose public key and signatures have length independent of the number of group members and which can therefore also be used for large groups. Furthermore, the scheme allows the group manager to add new members to the group without modifying the public key. The realization is based on methods for proving the knowledge of signatures.

1 Introduction

A group signature scheme allows members of a group to sign messages on behalf of the group. Signatures can be verified with respect to a single group public key, but they do not reveal the identity of the signer. Furthermore, it is not possible to decide whether two signatures have been issued by the same group member. However, there exists a designated *group manager* who can, in case of a later dispute, *open* signatures, i.e., reveal the identity of the signer.

Group signatures could for instance be used by a company for authenticating price lists, press releases, or digital contracts. The customers need to know only a single company public key to verify signatures. The company can hide any internal organizational structures and responsibilities, but still can find out which employee (i.e., group member) has signed a particular document.

The concept of group signatures was introduced by Chaum and van Heyst [11] and they also proposed the first realizations. Improved solutions were later presented by Chen and Pedersen [12], Camenisch [7], and Petersen [22]. However, all previously proposed solutions have the following undesirable properties:

- the length of the group's public key and/or the size of a signature depends on the size of the group. This is very problematic for large groups.
- to add new group members, it is necessary to modify at least the public key.

In this paper we present the first efficient group signature schemes which overcome these problems[1]. The lengths of the public key and of the signatures

---

[1] The only previously proposed schemes with fixed size public keys [21,17] were broken.

are, as well as the computational effort for signing and verifying, *independent of the number of group members*. Furthermore, the public key *remains unchanged* if new members are added to the group. The schemes even conceal the size of the group.

For realizing such schemes we employ novel techniques of independent interest, such as efficient proofs of (or signatures of) knowledge of double discrete logarithms, of e-th roots of discrete logarithms, and of e-th roots of components of representations. Of particular interest is a method for proving the knowledge of a signature.

2 Group Signature Schemes

In this section we present the concept of a group signature scheme and explain the basic idea underlying our realizations.

2.1 The Concept of a Group Signature Scheme

A group signature scheme consists of the following four procedures:

Setup: a probabilistic interactive protocol between a designated group manager and the members of the group. Its result consists of the group's public key $\mathcal{Y}$, the individual secret keys x of the group members, and a secret administration key for the group manager.

Sign: a probabilistic algorithm which, on input a message m and a group member's secret key x, returns a signature s on m.

Verify: an algorithm which, on input a message m, a signature s, and the group's public key $\mathcal{Y}$, returns whether the signature is correct.

Open: on input a signature s and the group manager's secret administration key this algorithm returns the identity of the group member who issued the signature s together with a proof of this fact.

It is assumed that all communications between the group members and the group manager are secure. A group signature scheme must satisfy the following properties:

1. Only group members are able to correctly sign messages (unforgeability).
2. It is neither possible to find out which group member signed a message (anonymity) nor to decide whether two signatures have been issued by the same group member (unlinkability).
3. Group members can neither circumvent the opening of a signature nor sign on behalf of other group members; even the group manager cannot do so (security against framing attacks).

A consequence of the last property is that the group manager must not know the secret keys of the group members.

In an extended model it may be desirable to assign the different roles of the group manager, namely managing the membership list of the group and opening signatures, to different parties. Furthermore, these roles could be shared among several parties (i.e. among the group members) in order to increase the security against a cheating group manager.

With regard to the efficiency of a group signature scheme the following parameters are of particular interest:

- the size of the group public key $\mathcal{Y}$,
- the length of signatures,
- the efficiency of the algorithms Sign and Verify,
- and the efficiency of the protocols Setup and Open.

In all previously proposed schemes, the length of the public key is at least linear in the size of the group and therefore also the running time of the verification algorithm depends on the number of group members. In some schemes also the length of the signature and the running time of the signing algorithm depend on the group size. In Sections 4 and 6 we propose new group signature schemes which overcome these problems. Both solutions are based on the following idea.

2.2 Schemes with Fixed Size Public Key and Signatures

Using the techniques of Brassard et al. [6] or Boyar et al. [3] for proving the knowledge of a satisfying assignment of a boolean circuit, a group signature scheme with fixed length public key and signatures can be constructed as follows.

The group manager computes a key pair of an ordinary digital signature scheme, denoted (sig_M, ver_M), and a key pair of a probabilistic public-key encryption scheme, denoted $(encr_M, decr_M)$, and publishes the two public keys as the group public key. Alice can join the group in the following way: she chooses a random *secret key* x and computes a *membership key* $z = f(x)$, where f is a one-way function. She commits herself to z, e.g., by signing it, and then sends z to the group manager who returns to her the *membership certificate* $v = sig_M(z)$. Alice's group secret key consists of the triple (x, z, v).

To sign a message m on behalf of the group, Alice encrypts the pair (m, z) using the group manager's encryption key, i.e., $d = encr_M(r, (m, z))$, where r is a sufficiently large random string. She computes a non-interactive minimum-disclosure proof p that she knows values x', v', and r' satisfying the following equations:

$$d = encr_M(r', (m, f(x'))) \quad \text{and} \quad ver_M(v', f(x')) = \text{correct} .$$

The resulting signature on the message m consists of the pair (d, p) and can be verified by checking the proof p. To open this signature, the group manager decrypts the ciphertext d to obtain the membership key z which reveals Alice's identity. A proof of this fact consists of z, Alice's commitment to it, and a non-interactive proof that d encrypts (m, z).

It can easily be verified that all security properties hold:

1. Only group members who know a membership certificate can construct a valid proof p.
2. Because the proof p does not reveal information about x, z, or v, and because (m, z) is probabilistically encrypted, signatures are anonymous and unlinkable.

3. Group members cannot circumvent the opening of signatures because they prove that the value d contains their membership key.

Note that instead of encrypting the message m in d, one could instead make the proof p message-dependent (see Section 3).

The disadvantage of this solutions is that the general techniques for proving statements in minimum-disclosure make the resulting signatures very large and impractical. The rest of the paper describes techniques for the construction of more efficient scheme based on proofs (or signatures) about the knowledge of double discrete logarithms and about the knowledge of roots of logarithms.

3 Preliminaries and Techniques

After giving notational and number theoretic preliminaries, we present some well known techniques for proving knowledge of discrete logarithms and extend them to the building blocks for our group signature schemes.

3.1 Notations

The symbol $\|$ denotes the concatenation of two (binary) strings (or of binary representations of integers and group elements) and ' ' denotes the empty string. By $c[i]$ we denote the i-th rightmost bit of the string c. If A is a set, $a \in_R A$ means that a is chosen at random from A according to the uniform distribution. For an integer q, $\mathbb{Z}_q$ denotes the ring of integers modulo q and $\mathbb{Z}_q^*$ denotes the multiplicative group modulo q. Finally, we assume a collision resistant hash function $\mathcal{H} : \{0,1\}^* \to \{0,1\}^k$ ($k \approx 160$).

3.2 Number Theoretic Preliminaries

Let $G = \langle g \rangle$ be a cyclic group of order n, and let a be an element of $\mathbb{Z}_n^*$. The discrete logarithm of $y \in G$ to the base g is the smallest positive integer x satisfying $g^x = y$. Similarly, the double discrete logarithm of $y \in G$ to the bases g and a is the smallest positive integer x satisfying

$$g^{(a^x)} = y \, ,$$

if such an x exists. In the sequel, the parameters n, G, g, and a should be chosen such that computing discrete logarithms in G to the base g and in $\mathbb{Z}_n^*$ to the base a is infeasible.

An e-th root of the discrete logarithm of $y \in G$ to the base g is an integer x satisfying

$$g^{(x^e)} = y \, ,$$

if such an x exists. Note that if the factorization of n is unknown, for instance if n is an RSA modulus (see [24]), computing e-th roots in $\mathbb{Z}_n^*$ is assumed to be infeasible.

3.3 Signature of Knowledge of Discrete Logarithms

Throughout this paper we make use of "proof systems" that allow one party to convince other parties about its knowledge of certain values, such that no useful information is leaked. Various such systems have been proposed, for instance minimum-disclosure proofs [6] and zero-knowledge proofs of knowledge [15]. We will make use of constructions based on the Schnorr signature scheme [25] to prove knowledge. However, to avoid confusions with the notion of proofs of knowledge of [15] and to point out that these proofs also serve as signatures, we call them *signatures of knowledge*. All these signatures of knowledge can be proved secure in the random oracle model [2,15] and their interactive versions are zero-knowledge (given that several rounds with small challenges are used).

The first primitive we define is a signature of the knowledge of the discrete logarithm of y to the base g. It is basically a Schnorr signature [25] on a message m of the entity knowing the discrete logarithm of y.

Definition 1. A pair $(c, s) \in \{0, 1\}^k \times \mathbb{Z}_n^*$ satisfying $c = \mathcal{H}(m \| y \| g \| g^s y^c)$ is a signature of knowledge of the discrete logarithm of the element $y \in G$ to the base g on the message m. □

Such a signature can be computed if the secret key $x = \log_g(y)$ is known, by choosing r at random from $\mathbb{Z}_n^*$ and computing c and s according to

$$c := \mathcal{H}(m \| y \| g \| g^r) \quad \text{and} \quad s := r - cx \pmod{n}.$$

This technique for constructing a signature of the knowledge of a discrete logarithm can also be used to build signatures that involve more complex statements, such as the knowledge of a representation of y to the bases g and h, i.e., a pair (α, β) satisfying $y = g^\alpha h^\beta$ (see [4] for a discussion of the representation problem). Even signatures of knowledge of complex relationships among different representations are possible [5,8,16].

Before we define such signatures of knowledge let us explain our notation with the following example: a signature of knowledge, denoted

$$SKREP\left[(\alpha, \beta) : y = g^\alpha \wedge z = g^\beta h^\alpha\right](m),$$

is used for 'proving' the knowledge of the discrete logarithm of y to the base g and of a representation of z to the bases g and h, and in addition, that the h-part of this representation equals the discrete logarithm of y to the base g. This is equivalent to the knowledge of a pair (α, β) satisfying the equations on the right side of the colon. In the sequel, we use the convention that Greek letters denote the elements whose knowledge is proven and all other letters denote elements that are known to the verifier. We now generalize these types of signatures of knowledge.

Definition 2. A signature of the knowledge of representations of $y_1, \ldots, y_w$ with respect to the bases $g_1, \ldots, g_v$ on the message m is denoted as follows

$$SKREP\left[(\alpha_1, \ldots, \alpha_u) : \left(y_1 = \prod_{j=1}^{\ell_1} g_{b_{1j}}^{\alpha_{e_{1j}}}\right) \wedge \ldots \wedge \left(y_w = \prod_{j=1}^{\ell_w} g_{b_{wj}}^{\alpha_{e_{wj}}}\right)\right](m),$$

where the indices $e_{ij} \in \{1, \dots, u\}$ refer to the elements $\alpha_1, \dots, \alpha_u$ and the indices $b_{ij} \in \{1, \dots, v\}$ refer to the base elements $g_1, \dots, g_v$. The signature consists of an $(u + 1)$ tuple $(c, s_1, \dots, s_u) \in \{0, 1\}^k \times \mathbb{Z}_n^u$ satisfying the equation

$$c = \mathcal{H}\left(m\|y_1\|\dots\|y_w\|g_1\|\dots\|g_v\|\{\{e_{ij}, b_{ij}\}_{j=1}^{\ell_i}\}_{i=1}^{w}\|y_1^c\prod_{j=1}^{\ell_1} g_{b_{1j}}^{s_{e_{1j}}}\|\dots\|y_w^c\prod_{j=1}^{\ell_w} g_{b_{e_{wj}}}^{s_{e_{wj}}}\right) \quad \square$$

$SKREP$ can be computed in the same way as the simple signature of knowledge of a discrete logarithm if a u-tuple $(\alpha_1, \dots, \alpha_u)$ is known which satisfies the given equations. One first chooses $r_i \in_R \mathbb{Z}_n$ for $i = 1, \dots, u$, computes c as

$$c := \mathcal{H}\left(m\|y_1\|\dots\|\{\{e_{ij}, b_{ij}\}_{j=1}^{\ell_i}\}_{i=1}^{w}\|\prod_{j=1}^{\ell_1} g_{b_{1j}}^{r_{e_{1j}}}\|\dots\|\prod_{j=1}^{\ell_w} g_{b_{e_{wj}}}^{r_{e_{wj}}}\right),$$

and then sets $s_i := r_i - c\alpha_i \pmod{n}$ for $i = 1, \dots, u$.

Signatures of knowledge of representations are a powerful tool for constructing various cryptographic systems, but we will also employ signatures of the knowledge of double discrete logarithms (see [26]) and of roots of logarithms.

Definition 3. Let $\ell \leq k$ be a security parameter. An $(\ell + 1)$ tuple $(c, s_1, \dots, s_\ell) \in \{0, 1\}^k \times \mathbb{Z}^\ell$ satisfying the equation

$$c = \mathcal{H}(m\|y\|g\|a\|t_1\|\dots\|t_\ell) \quad \text{with} \quad t_i = \begin{cases} g^{(a^{s_i})} & \text{if } c[i] = 0 \\ y^{(a^{s_i})} & \text{otherwise} \end{cases}$$

is a signature the knowledge of a double discrete logarithm of y to the bases g and a, and is denoted $SKLOGLOG[\alpha : y = g^{(a^\alpha)}](m)$. $\square$

An $SKLOGLOG[\alpha : y = g^{(a^\alpha)}](m)$ can be computed only if the double discrete logarithm x of the group element y to the bases g and a is known. We assume that there is an upper bound λ on the length of x, i.e., $0 \leq x < 2^\lambda$ ($\lambda = |n|$ is an example, but for certain applications, smaller bounds can be used). Let $\epsilon > 1$ be a constant. One first computes the values

$$t_i^* := g^{(a^{r_i})}$$

for $i = 1, \dots, \ell$ with randomly chosen $r_i \in \{0 \dots, 2^{\epsilon\lambda} - 1\}$. Then, c is set to $\mathcal{H}(m\|y\|g\|a\|t_1^*\|\dots\|t_\ell^*)$, and finally,

$$s_i := \begin{cases} r_i & \text{if } c[i] = 0, \\ r_i - x & \text{otherwise.} \end{cases}$$

for $i = 1, \dots, \ell$. It can easily be verified that the resulting tuple $(c, s_1, \dots, s_\ell)$ satisfies the verification equation. Note that if the order of $a \in \mathbb{Z}_n^*$ is known, the computations of the s_i can be "reduced" modulo this order.

Definition 4. An $(\ell + 1)$ tuple $(c, s_1, \dots, s_\ell) \in \{0, 1\}^k \times \mathbb{Z}_n^{*\ell}$ satisfying the equation

$$c = \mathcal{H}(m \,\|\, y \,\|\, g \,\|\, e \,\|\, t_1 \,\|\, \dots \,\|\, t_\ell) \quad \text{with} \quad t_i = \begin{cases} g^{(s_i^e)} & \text{if } c[i] = 0 \\ y^{(s_i^e)} & \text{otherwise} \end{cases}$$

is a signature of the knowledge of an e-th root of the discrete logarithm of y to the base g, and is denoted $SKROOTLOG\,[\alpha : \; y = g^{\alpha^e}\,](m)$. □

Note that the values $s_1, \dots, s_\ell$ belong to $\mathbb{Z}_n^*$ and therefore must not be zero.

Such a signature can be computed if the e-th root x of the discrete logarithm of y to the base g is known. One first computes the values

$$t_i^* := g^{(r_i^e)}$$

for $i = 1, \dots, \ell$ with randomly chosen $r_i \in \mathbb{Z}_n^*$. Then, c is set to $\mathcal{H}(m \,\|\, y \,\|\, g \,\|\, e \,\|\, t_1^* \,\|\, \dots \,\|\, t_\ell^*)$, and finally,

$$s_i := \begin{cases} r_i & \text{if } c[i] = 0, \\ r_i/x \pmod{n} & \text{otherwise.} \end{cases}$$

for $i = 1, \dots, \ell$. It can easily be seen that the resulting tuple $(c, s_1, \dots, s_\ell)$ satisfies the verification equation.

4 The Basic Group Signature Scheme

In this section we propose a first realization of a group signature scheme based on the ideas presented in the end of Section 2. In this solution, the opening of signatures can even be realized in a simpler way.

4.1 System Setup

The group manager computes the following values:

- an RSA public key (n, e),
- a cyclic group $G = \langle g \rangle$ of order n in which computing discrete logarithms is infeasible (e.g. G could be a subgroup of $\mathbb{Z}_p^*$, for a prime p with $n|(p-1)$),
- an element $a \in \mathbb{Z}_n^*$ (a should be of large multiplicative order modulo both prime factors of n), and
- an upper bound λ on the length of the secret keys and a constant $\epsilon > 1$ (these parameters are required for the $SKLOGLOG$ signatures)

The group's public key is $\mathcal{Y} = (n, e, G, g, a, \lambda, \epsilon)$.

4.2 Generating Membership Keys and Certificates

When Alice is to join the group, she chooses her secret key $x \in_R \{0, \dots, 2^\lambda - 1\}$ and computes the value $y := a^x \pmod{n}$ and her membership key $z = g^y$. She commits herself to y, for instance by signing it. She then sends y and z to the group manager and proves to him that she knows the discrete logarithm of y to the base a (this can be done with techniques similar to those for signatures of knowledge of a discrete logarithm, with the difference that the group order is unknown). When the group manager is convinced that Alice knows this logarithm, he returns to her the membership certificate

$$v \equiv (y+1)^{1/e} \pmod{n} .$$

It seems infeasible to construct such a triple (x, y, v) without the help of the group manager: on one hand, if y is correctly formed then it is infeasible to compute the e-th root of $y + 1$ because the factorization of n is unknown. On the other hand, if $y + 1$ is computed as w^e for some value w then it is infeasible to compute the discrete logarithm of $w^e - 1$ to the base a. Furthermore, even if several group members pool their values, they still seem unable to construct a new such triple.

4.3 Signing Messages

To sign a message m, Alice computes the following values:

- $\tilde{g} := g^r$ for $r \in_R \mathbb{Z}_n^*$
- $\tilde{z} := \tilde{g}^y$
- $V_1 := SKLOGLOG\left[\alpha : \ \tilde{z} = \tilde{g}^{a^\alpha}\right](m)$
- $V_2 := SKROOTLOG\left[\beta : \ \tilde{z}\tilde{g} = \tilde{g}^{\beta^e}\right](m)$

The resulting signature on the message m consists of $(\tilde{g}, \tilde{z}, V_1, V_2)$ and can be verified by checking the correctness of the signatures of knowledge V_1 and V_2.

We now explain briefly why this signature really proves that Alice belongs to the group. On one hand, because of V_1 the value $\tilde{z}\tilde{g}$ must be of the form

$$\tilde{z}\tilde{g} = \tilde{g}^{a^\alpha + 1}$$

for an integer α Alice knows. On the other hand, V_2 proves that Alice knows an e-th root of $(a^\alpha + 1)$, which means that Alice knows the secret key and a membership certificate of her membership key.

4.4 Opening Signatures

Linking two signatures $(\tilde{g}, \tilde{z}, V_1, V_2)$ and $(\tilde{g}', \tilde{z}', V_1', V_2')$, i.e., deciding whether these signatures have been issued by the same group member or not, is only possible by deciding whether $\log_{\tilde{g}} \tilde{z} = \log_{\tilde{g}'} \tilde{z}'$. Generally, solving this problem is infeasible and therefore the signatures of the group members are anonymous and unlinkable. However, the group manager has an advantage: he knows the relatively few possible values of $\log_{\tilde{g}} \tilde{z}$, namely the discrete logarithms (to the base g) of the membership keys of the group members, and can therefore perform this test. Given only a signature $(\tilde{g}, \tilde{z}, V_1, V_2)$ for a message m, the group manager can find the group member who issued this signature by testing

$$\tilde{g}^{y_P} \stackrel{?}{=} \tilde{z}$$

for all group members P (where y_P denotes discrete logarithm of P's membership key z_P to the base g). A proof of this fact consists of the signer's membership key z_P, his commitment to this key, and a non-interactive proof of the equality of $\log_g z$ and $\log_{\tilde{g}} \tilde{z}$. Unfortunately, this method is impractical for very large groups. In Section 6 we present an extension that makes it possible to identify group members directly.

4.5 Security and Efficiency Considerations

The security of the basic group signature scheme presented in this section is based on the difficulty of the discrete logarithm problem and on the security of the Schnorr [25] and of the RSA [24] signature schemes. It is also based on the additional assumption that computing membership certificates of valid membership keys is infeasible if the factorization of the modulus n is unknown. With regard to the anonymity of group members, linking two signatures is as hard as deciding whether two discrete logarithms are equal (for instance, undeniable signatures [10] make also use of this assumption).

With the following values of the system parameters

$$k = 160, \ell = 64, \lambda = 170, \epsilon = 4/3, |n| = 600, \text{ and } e = 3,$$

a signature is less than 7 KBytes long and the operations for signing messages and for verifying signatures require the computation of approximately 140'000 modular multiplications with a 600 bit modulus (this corresponds to about 155 exponentiations with full 600 bit exponents).

5 Efficient *SKROOTLOG*

A disadvantage of the scheme presented in the previous section is that the signatures *SKROOTLOG* and *SKLOGLOG* are quite inefficient. In this section we show how an efficient *SKROOTLOG* can be realized when the exponent e is small.

5.1 A Simple Observation

If e is small then it is possible to efficiently convince somebody about one's knowledge of the e-th root of the discrete logarithm of $z = g^{x^e}$ to the base g by computing the following $e - 1$ values:

$$z_1 := g^x, \ z_2 := g^{x^2}, \ldots, \ z_{e-1} := g^{x^{e-1}}$$

and showing with a signature of knowledge

$$U := SKREP\left[\alpha : z_1 = g^\alpha \ \wedge \ z_2 = z_1^\alpha \ \wedge \ldots \wedge \ z = z_{e-1}^\alpha\right]$$

that the discrete logarithms 'between' two subsequent values in the list g, z_1, ..., z_{e-1}, z are all equal and known. Therefore the following equations

$$z = z_{e-1}^\alpha = z_{e-2}^{\alpha^2} = \ldots = z_1^{\alpha^{e-1}} = g^{\alpha^e}$$

must hold and the knowledge of an e-th root of z to the base g is assured. More generally, one could use any addition chain for the integer e, but we restrict ourselves to this simple case for the rest of the paper.

However, the problem of this approach is that the values $z_1, \ldots, z_{e-1}$ leak additional information. In the next section we show how these values can be randomized. This leads to an efficient *SKROOTLOG* presented in the next but one section.

5.2 Efficient Signatures Proving the Knowledge of Roots of Representations

From now on we assume that an element $h \in G$ is available whose discrete logarithm to the base g is unknown (for instance, h could be computed according to a suitable pseudo-random process with g as seed). The element h is now used to randomize (or blind) z and the z_i's of the previous section, i.e., v becomes $h^r z$ for some random r and one wants to 'prove' the knowledge of a pair (α, β) for which $v = h^\alpha g^{\beta^e}$ holds. Such a signature can be given efficiently by applying the method described above.

Similar techniques have already been used in [13,19] for the purpose of proving properties of bit commitments.

Definition 5. An efficient signature of the knowledge of the e-th root of the g-part of a representation of v to the bases h and g, denoted

$$E\text{-}SKROOTREP\left[(\alpha, \beta) : v = h^\alpha g^{\beta^e}\right](m) ,$$

consists of an $(e-1)$-tuple $(v_1, \ldots, v_{e-1}) \in G^{e-1}$ and of a signature of knowledge

$$U = SKREP\Big[(\gamma_1, \ldots, \gamma_e, \delta) : v_1 = h^{\gamma_1} g^\delta \ \wedge \ v_2 = h^{\gamma_2} v_1^\delta \ \wedge \ldots$$

$$\ldots \wedge \ v_{e-1} = h^{\gamma_{e-1}} v_{e-2}^\delta \ \wedge \ v = h^{\gamma_e} v_{e-1}^\delta\Big](m) .$$

The signature of knowledge can be verified by checking the correctness of U. □

The following equation explains why a verifier will be convinced of the prover's knowledge of (α, β):

$$v = h^{\gamma_e}\left(h^{\gamma_{e-1}}\left(\ldots h^{\gamma_2}(h^{\gamma_1} g^\delta)^\delta \ldots\right)^\delta\right)^\delta = h^{\gamma_e + \gamma_{e-1}\delta + \ldots + \gamma_2 \delta^{e-2} + \gamma_1 \delta^{e-1}} g^{\delta^e} =: h^\alpha g^{\beta^e}.$$

Such a signature can be computed if values r and x in $\mathbb{Z}_n$ are known for which $v = h^r g^{x^e}$: one first computes the values $v_i := h^{r_i} g^{x^i}$ for $i = 1, \ldots, e-1$ with randomly chosen $r_i \in \mathbb{Z}_n$. Then the signature of knowledge U is computed. Note that the elements v_i are truly random group elements and so do not leak any information.

5.3 Efficient Signatures proving the Knowledge of Roots of Logarithms

Based on the *E-SKROOTREP*'s, it is now easy to construct an efficient and secure *E-SKROOTLOG*, by showing that z itself is not blinded:

Definition 6. A efficient signature on the knowledge of the e-th root of the discrete logarithm of z to the base g, denoted

$$E\text{-}SKROOTLOG\left[\delta : z = g^{\delta^e}\right](m)$$

consists of the two signatures

$$E\text{-}SKROOTREP\left[(\alpha,\beta) : z = h^\alpha g^{\beta^e}\right](m) \quad \text{and} \quad SKREP[\gamma : z = g^\gamma](m)$$

where the discrete logarithm of h the base g must be unknown. $\qquad\Box$

Since one can know only one representation of z to the bases h and g, it follows that $\alpha \equiv 0 \pmod{n}$ and $\gamma \equiv \beta^e \pmod{n}$ and that the prover knows the e-th root of the discrete logarithm of z to the base g.

6 A More Efficient Variant

Because similar improvements as for *SKROOTLOG* signatures seem not be possible for *SKLOGLOG* signature, an evident solution to design a more efficient group signature scheme is to modify it in a way that allows to replace the *SKLOGLOG* by an *SKROOTLOG* signature. This is indeed possible if the membership key is computed using $y = x^e \pmod{n}$ instead of $y = h^x \pmod{n}$.

As an immediate consequence, the group manager must be prevented from learning the value y (otherwise he could compute an e-th root of y and sign on behalf of group members). This problem can be solved by sending the group manager only g^y and adapting the protocol for issuing the membership certificates accordingly. Furthermore, because the group manager no longer knows y, the method for opening signatures as described in Section 2.2 must be realized.

6.1 System Setup

The group manager computes the following values:

- an RSA modulus n and two public exponents e_1, $e_2 > 1$, such that e_2 is relatively prime to $\varphi(n)$,
- two integers f_1, $f_2 > 1$ whose e_1-th roots and e_2-th roots cannot be computed without knowing the factorization of n,
- a cyclic group $G = \langle g \rangle$ of order n in which computing discrete logarithms is infeasible,
- an element $h \in G$ whose discrete logarithm to the base g must not be known,
- his public key $y_R = h^\rho$ for a randomly chosen value $\rho \in \mathbb{Z}_n$.

The group's public key consists of $\mathcal{Y} = (n, e_1, e_2, f_1, f_2, G, g, h, y_R)$, whereas ρ and the factorization of n remain the group manager's secret key. Possible choices for the parameters e_1, e_2, f_1, and f_2 are discussed in section 6.5.

6.2 Membership Keys and Blind Issuing of Membership Certificates

To become a group member, Alice computes her membership key as follows:

- $y := x^{e_1} \pmod{n}$ for $x \in_R \mathbb{Z}_n^*$ (see also discussion in Section 6.5)
- $z := g^y$

A certificate in this scheme is of the form

$$v = (f_1 y + f_2)^{1/e_2} \pmod{n} .$$

To prevent the group manager from learning y, this certificate must be issued using the blind RSA-signature scheme of Chaum [9]. Additionally, Alice must send z to the group manager and convince him that the discrete logarithm of z to the base g is a valid membership key and is contained in the blinded certificate. More formally, Alice computes

- $\tilde{y} := r^{e_2}(f_1 y + f_2) \pmod{n}$ for $r \in_R \mathbb{Z}_n^*$
- $U := E\text{-}SKROOTLOG[\alpha : z = g^{\alpha^{e_1}}]$ (' ')
- $V := E\text{-}SKROOTLOG[\beta : g^{\tilde{y}} = (z^{f_1} g^{f_2})^{\beta^{e_2}}]$ (' ')

and sends $\tilde{y}, z, U$, and V to the group manager. If U and V are correct, the group manager sends Alice the blinded certificate

$$\tilde{v} = \tilde{y}^{1/e_2} \pmod{n},$$

which Alice unblinds and thereby obtains her membership certificate

$$v = \tilde{v}/r = (f_1 y + f_2)^{1/e_2} \pmod{n}.$$

Let us now explain what the signatures of knowledge U and V actually mean. The signature U shows that the element z is of the form $g^{\alpha^{e_1}}$ for some α Alice knows. The signature V assures that $\tilde{y} = \beta^{e_2}(f_1 \alpha^{e_1} + f_2) \pmod{n}$ holds for some β Alice knows, and therefore the group manager can conclude that $\tilde{y}$ is a correctly blinded membership key.

6.3 Signing Messages

To sign a message m on behalf of the group, Alice performs the following computations:

- $\tilde{z} := h^r g^y$ for $r \in_R \mathbb{Z}_n^*$
- $d := y_R^r$
- $V_1 := E\text{-}SKROOTREP[(\alpha, \beta) : \tilde{z} = h^\alpha g^{\beta^{e_1}}](m)$
- $V_2 := E\text{-}SKROOTREP[(\gamma, \delta) : \tilde{z}^{f_1} g^{f_2} = h^\gamma g^{\delta^{e_2}}](m)$
- $V_3 := SKREP[(\varepsilon, \zeta) : d = y_R^\varepsilon \wedge \tilde{z} = h^\varepsilon g^\zeta](m)$

The resulting signature on the message m consists of $(\tilde{z}, d, V_1, V_2, V_3)$ and is valid if the three signatures of knowledge V_1, V_2, and V_3 are correct.

The following explains briefly why such a signature convinces a verifier that the signer knows the secret key of a certified membership key. Consider the signature V_1: it 'proves' that the signer knows a representation (α, β^{e_1}) of $\tilde{z}$ to the bases h and g and that she knows the e_1-th root of the g-part of this representation, i.e., β. The signature V_2 'proves' the signer's knowledge of a representation (γ, δ^{e_2}) of $\tilde{z}^{f_1} g^{f_2}$ to the bases h and g and her knowledge of the e_2-th root of δ^{e_2}. As the signer can know *at most one* representation of $\tilde{z}^{f_1} g^{f_2}$ to the bases h and g it follows that

$$\gamma \equiv \alpha \pmod{n} \quad \text{and} \quad \delta^{e_2} \equiv f_1 \beta^{e_1} + f_2 \pmod{n}.$$

The fact that the signer knows an e_1-th root of β^{e_1} and an e_2-th root of δ^{e_2} means that she knows a membership certificate and the secret key of the corresponding membership key.

Finally, consider the element d and the signature V_3. The pair $(d, \tilde{z})$ is a (modified) ElGamal encryption [14] of g^y encrypted under the group manager's public key y_R and enables the group manager to open signatures. The signature V_3 guarantees that this encryption is formed correctly.

6.4 Opening Signatures

When the group manager wants to open a signature $(\tilde{z}, d, V_1, V_2, V_3)$ on the message m, he computes $\hat{z} := \tilde{z}/d^{1/\rho}$ which corresponds to the signer's membership key z. To prove that z is indeed encrypted in $\tilde{z}$ and d, the group manager computes

$$SKREP[\alpha : \tilde{z} = zd^{\alpha} \wedge h = y_R^{\alpha}]('\ '),$$

which he can do because $1/\alpha \pmod{n}$ corresponds to his administration key.

6.5 Security and Efficiency Considerations

The security of the group signature scheme presented in this section is based on the difficulty of the discrete logarithm problem and on the security of the RSA and Schnorr signature schemes. The security of the scheme relies also on the difficulty of computing certificates when the factorization of n is unknown. The latter depends on the choices for the values e_1, e_2, f_1, and f_2. For instance, the choice $e_1 = 2$, $e_2 = 2$, and $f_1 = 1$, related to the Ong-Schnorr-Shamir signature scheme [20], is not secure for any value of f_2 as is shown in [1,23]. Generally, it is regarded open problem to determine which types of polynomial congruences with composite moduli are hard to solve [18]. Furthermore, it is also important that when given several solutions of such a polynomial congruence it remains hard to compute other ones.

In order to make it harder to forge membership certificates, it is possible to modify the group signature scheme such that only solutions of the polynomial equation are accepted that meet additional requirements. For instance, by modifying the *E-SKROOTLOG* signature V_1, one can efficiently prove that the secret key x is smaller than $\sqrt{n}$ (the techniques are similar to those used for the *SKLOGLOG* signatures). As a challenge, we propose to use this approach with the following parameters:

$$e_1 = 5, e_2 = 3, f_1 = 1, \text{ and, } f_2 \text{ such that its 3rd root is hard to compute.}$$

For this choice and with $k = 160$ and $|n| = 600$, a signature is about 1.4 KByte long and the operations for signing for verifying signatures require the computation of approximately 18'000 modular multiplications with a 600 bit modulus (this corresponds to about 20 exponentiations with full 600 bit exponents).

7 Extensions

An obvious (and for the second scheme simple) extension would be to assign the different roles of the group manager to different entities, i.e., to a membership manager, who is responsible for adding new members to the group, and to a revocation manager, who is responsible for opening signatures. The functionality of these managers can also be shared among several entities. The realization is straightforward.

Acknowledgments

Many thanks to M. Abe, D. Bleichenbacher, C. Cachin, R. Cramer, U. Maurer, M. Michels, S. Wolf, and the anonymous referee for helpful comments and discussions, and to K. McCurley, D. Coppersmith, and J. Benaloh for pointing us to useful literature. The first author is supported by the Swiss Commission for Technology and Innovation (KTI) and by the Union Bank of Switzerland.

References

1. L. M. Adleman, D. R. Estes, and K. S. McCurley. Solving bivariate quadratic congruences in random polynomial time. *Mathematics of Computation*, 43(177):17–28, Jan. 1987.
2. M. Bellare and P. Rogaway. Random oracles are practical: A paradigm for designing efficient protocols. In *First ACM Conference on Computer and Communication Security*, pages 62–73. Association for Computing Machinery, 1993.
3. J. Boyar and R. Peralta. Short discreet proofs. In *Advances in Cryptology — EUROCRYPT '96*, volume 1070 of *Lecture Notes in Computer Science*, pages 131–142. Springer Verlag, 1996.
4. S. Brands. An efficient off-line electronic cash system based on the representation problem. Technical Report CS-R9323, CWI, Apr. 1993.
5. S. Brands. Rapid demonstration of linear relations connected by boolean operators. In *In Advances in Cryptology — EUROCRYPT '97*, volume 1233 of *Lecture Notes in Computer Science*, pages 318–333. Springer Verlag, 1997.
6. G. Brassard, D. Chaum, and C. Crépeau. Minimum disclosure proofs of knowledge. *Journal of Computer and System Sciences*, 37(2):156–189, Oct. 1988.
7. J. Camenisch. Efficient and generalized group signatures. In *In Advances in Cryptology — EUROCRYPT '97*, volume 1233 of *Lecture Notes in Computer Science*, pages 465–479. Springer Verlag, 1997.

8. J. Camenisch and M. Stadler. Proof systems for general statements about discrete logarithms. Technical Report TR 260, Institute for Theoretical Computer Science, ETH Zürich, Mar. 1997.

9. D. Chaum. Blind signature systems. In *Advances in Cryptology — CRYPTO '83*, page 153. Plenum Press, 1984.

10. D. Chaum and H. van Antwerpen. Undeniable signatures. In *Advances in Cryptology - CRYPTO '89*, volume 435 of *Lecture Notes in Computer Science*, pages 212–216. Springer Verlag, 1990.

11. D. Chaum and E. van Heyst. Group signatures. In *Advances in Cryptology — EUROCRYPT '91*, volume 547 of *Lecture Notes in Computer Science*, pages 257–265. Springer-Verlag, 1991.

12. L. Chen and T. P. Pedersen. New group signature schemes. In *Advances in Cryptology — EUROCRYPT '94*, volume 950 of *Lecture Notes in Computer Science*, pages 171–181. Springer-Verlag, 1995.

13. I. B. Damgård. Practical and provable secure release of a secret and exchange of signature. In *Advances in Cryptology — EUROCRYPT '93*, volume 765 of *Lecture Notes in Computer Science*, pages 200–217. Springer-Verlag, 1994.

14. T. ElGamal. A public key cryptosystem and a signature scheme based on discrete logarithms. In *Advances in Cryptology — CRYPTO '84*, volume 196 of *Lecture Notes in Computer Science*, pages 10–18. Springer Verlag, 1985.

15. U. Feige, A. Fiat, and A. Shamir. Zero-knowledge proofs of identity. *Journal of Cryptology*, 1:77–94, 1988.

16. E. Fujisaki and T. Okamoto. Witness hiding protocols to confirm modular polynomial relations. In *The 1997 Symposium on Cryptograpy and Information Security*, Fukuoka, Japan, Jan. 1997. The Institute of Electronics, Information and Communcation Engineers. SCSI97-33D.

17. S. J. Kim, S. J. Park, and D. H. Won. Convertible group signatures. In *Advances in Cryptology — ASIACRYPT '96*, volume 1163 of *Lecture Notes in Computer Science*, pages 311–321. Springer Verlag, 1996.

18. K. McCurley. Odds and ends from cryptology and computational number theory. In *Cryptology and computational number theory*, volume 42 of *Proceedings of Symposia in Applied Mathematics*, pages 145–166. American Mathematical Society, 1990.

19. T. Okamoto. Threshold key-recovery systems for RSA. In *Security Protocols Workshop*, Paris, 1997.

20. H. Ong, C. P. Schnorr, and A. Shamir. Efficient signature schemes based on polymonial equations. In *Advances in Cryptology — CRYPTO '84*, volume 196 of *Lecture Notes in Computer Science*, pages 37–46. Springer Verlag, 1984.

21. S. J. Park, I. S. Lee, and D. H. Won. A practical group signature. In *Proceedings of the 1995 Japan-Korea Workshop on Information Security and Cryptography*, pages 127–133, Jan. 1995.

22. H. Petersen. How to convert any digital signature scheme into a group signature scheme. In *Security Protocols Workshop*, Paris, 1997.

23. J. M. Pollard and C. P. Schnorr. An efficient solution of the congruence $x^2 + ky^2 = m \pmod{n}$. *IEEE Transactions on Information Theory*, 33(5):702–709, September 1987.

24. R. Rivest, A. Shamir, and L. Adleman. A method for obtaining digital signatures and public-key cryptosystems. *Communications of the ACM*, 21(2):120–126, Feb. 1978.

25. C. P. Schnorr. Efficient signature generation for smart cards. *Journal of Cryptology*, 4(3):239–252, 1991.

26. M. Stadler. Publicly verifiable secret sharing. In *Advances in Cryptology — EUROCRYPT '96*, volume 1070 of *Lecture Notes in Computer Science*, pages 191–199. Springer Verlag, 1996.

Efficient Generation of Shared RSA Keys

(Extended Abstract)

Dan Boneh[1]   Matthew Franklin[2]
dabo@bellcore.com   franklin@research.att.com

[1] Bellcore, 445 South St., Morristown, NJ, 07960, USA
[2] AT&T Labs, 180 Park Ave., Florham-Park, NJ, 07932, USA

Abstract. We describe efficient techniques for three (or more) parties to jointly generate an RSA key. At the end of the protocol an RSA modulus $N = pq$ is publicly known. None of the parties know the factorization of N. In addition a public encryption exponent is publicly known and each party holds a share of the private exponent that enables threshold decryption. Our protocols are efficient in computation and communication.

Keywords: RSA, Threshold Cryptography, Primality testing, Multiparty computation.

1 Introduction

We propose efficient protocols for three (or more) parties to jointly generate an RSA modulus $N = pq$ where p, q are prime. At the end of the computation the parties are convinced that N is indeed a product of two large primes. However, none of the parties know the factorization of N. We then show how the parties can proceed to compute a public exponent e and shares of the corresponding private exponent.

Our techniques require a number of steps including a new distributed primality test. The test enables two (or more) parties to test that a random integer N is a product of two large primes without revealing the primes themselves.

Several cryptographic protocols require an RSA modulus N for which none of the participants know the factorization. For examples see [11, 12, 14, 19, 20, 21]. Usually this is done by asking a dealer to generate N. Clearly, the dealer must be trusted not to reveal the factorization of N. Our results eliminate the need for a trusted dealer since the parties can generate the modulus N by themselves.

Threshold cryptography is a concrete example where shared generation of RSA keys is very useful. We give a brief motivating discussion and refer to [9] for a survey. A threshold RSA signature scheme involves k parties and enables any subset of t of them to generate an RSA signature of a given message. No subset of $t - 1$ parties can generate a signature. A complete solution to this problem was given in [8]. Unfortunately, the modulus N and the shares of the private key were assumed to be generated by a dealer. The dealer, or anyone who compromises the dealer, can forge signatures. Our results eliminate the need for a trusted dealer (as long as $t < \lceil k/2 \rceil$) since the k parties can generate N and the private shares themselves. Such results were previously known for the ElGamal public key system [22], but not for RSA.

We note that generic secure circuit evaluation techniques, e.g. [26, 17, 3, 6] can also be used to generate shared RSA keys. After all, a primality test can be represented as a boolean circuit. However, such general techniques are usually too inefficient.

Our protocols are useful even when only two parties are involved. However, some steps of the protocol require the parties to interact with a third "helper" party we call Henry. At the end of the protocol Henry learns nothing, but the value of N which is public. To simplify the exposition we first describe our results for the case of two parties with a third helper (Sections 2-6). In Section 7, we explain how our methods generalize to more parties. An overview of our techniques is given in Section 2, and the various stages of the protocol are given in Sections 3-6.

2 Overview

In this section we give a high level overview of the protocol. The parties are Alice and Bob, with a third helper party Henry (see Section 7 for a generalization to more parties). Alice and Bob wish to generate a shared RSA key. More precisely, they wish to generate an RSA modulus $N = pq$ and a public/private pair of exponents e, d. At the end of the computation N and e are public, and d is shared between Alice and Bob in a way which enables threshold decryption. Alice and Bob should be convinced that N is indeed a product of two primes, but neither of them know the factorization of N.

We assume a model of passive adversaries, i.e. all three parties follow the protocol as required. At the end of the protocol no party is able to factor N. We discuss the case of active adversaries at the end of the paper.

At a high level the protocol works as follows:

(1) **pick candidates:** The following two steps are repeated twice.
 (a) **secret choice:** Alice and Bob pick random n-bit integers p_a and p_b respectively, and keep them secret.
 (b) **trial division:** Using a private distributed computation Alice, Bob and Henry determine that $p_a + p_b$ is not divisible by small primes. If this step fails repeat step (a).
 Denote the secret values picked at the first iteration by p_a, p_b, and at the second iteration by q_a, q_b.
(2) **compute N:** Using a private distributed computation Alice, Bob and Henry compute

$$N = (p_a + p_b) \cdot (q_a + q_b)$$

Other than the value of N, this step reveals no further information about the secret values p_a, q_a, p_b, q_b.

(3) **primality test:** Alice and Bob (without Henry) engage in a private distributed computation to test that N is indeed the product of two primes. If the test fails, then the protocol is restarted from step 1.

(4) **key generation:** Alice and Bob engage in a private distributed computation to generate a public encryption exponent e and a shared secret decryption exponent d.

Notation Throughout the paper we adhere to the following notation: the RSA modulus is denoted by N and is a product of two n bit primes p, q. When $p = \sum p_i$ we denote by p_i the share in possession of party i. Similarly for q_i. When the p_i's themselves are shared among the parties we denote by $p_{i,j}$ the share of p_i that is sent to party j.

Performance issues Our protocol generates two random numbers and tests that $N = pq$ is a product of two primes. By the prime number theorem the probability that both p and q are prime is asymptotically $1/n^2$. Therefore, naively one has to perform n^2 probes on average until a suitable N is found. This is somewhat worse than the expected $2n$ probes needed in traditional generation of an RSA modulus (one first generates one prime using n probes and then a second prime using another n probes). This $n/2$ degradation in performance is usually unacceptable (typically $n = 512$).

Fortunately, thanks to trial division things aren't so bad. Our trial division tests each prime individually. Therefore, to analyze our protocol we must analyze the effectiveness of trial division. Suppose a random n-bit number p passes the trial division test where all primes less than B are tested. We take $B = c \cdot n$ for some constant c. How likely is p to be prime? Using a classic result due to Mertens, DeBruijn [7] shows that asymptotically

$$\Pr[p \text{ prime } \mid \text{ trial division up to } B] = e^{\gamma} \frac{\ln B}{\ln 2^n} \big(1 + o(1/n)\big) = 2.57 \frac{\ln B}{n} \big(1 + o(1/n)\big)$$

Hence, when $n = 512$ bits and $\ln B = 9$ (i.e. $B = 8103$) the probability that p is prime is approximately $1/22$. Consequently, traditional RSA modulus generation requires 44 probes while our protocol requires 484 probes. This eleven fold degradation in performance is unfortunate, but manageable.

Generation of shares In step (1) of the protocol each of Alice and Bob uniformly picked a random n bit integer p_a, p_b as its secret share. The prime p was taken to be the sum of these shares. Since the sum of uniform independent random variables over the integers is *not* uniformly distributed, p is picked from a distribution with slightly less entropy than uniform. We show that this is not a problem. For the generalization to k parties, each party i uniformly picks a random n bit integer p_i. Then $p = \sum_i p_i$ is at most an $n + \log k$ bit number. One can easily show that p is chosen from a distribution with at least n bits of entropy (since the n least significant bits of p are a uniformly chosen n bit string). Intuitively, these $\log k$ bits of "lost" entropy can not help an adversary, since they can be easily guessed (the number of parties k is small, certainly $k < n$). This is formally stated in the next lemma. We note that by allowing some communication between the parties it is possible to ensure that p is uniformly distributed among "most" n bit integers.

A second issue is the fact that the shares themselves leak some information about the factors of N. For instance, party i knows that $p > p_i$. We argue that this information does not help an adversary either.

The two issues raised above are dealt with in the following lemma. Due to space limitation we leave the proof for the full version. Let $\mathbf{Z}_n^{(2)}$ be the set of RSA moduli $N = pq$ that can be output by our protocol above when k parties are involved. We assume $k < \log N$.

Lemma 1. *Suppose there exists a polynomial time algorithm $\mathcal{A}$ that given a random $N \in \mathbf{Z}_n^{(2)}$ chosen from the distribution above and the shares $\{p_i\}$ of $k-1$ parties, factors N with probability at least $1/n^d$. Then there exists an expected polynomial time algorithm $\mathcal{B}$ that factors $1/n^{d+2}$ of the integers in $\mathbf{Z}_n^{(2)}$.*

3 Distributed primality test

We now consider the distributed primality test. We describe our protocol for the case of two parties, and discuss the case of $k > 2$ parties in Section 7.

In the case of two parties, Alice and Bob possess integers p_a, q_a and p_b, q_b respectively. Both parties know N, where $N = (p_a + p_b)(q_a + q_b)$. They wish to determine if N is the product of two primes. The primality test is a mix of the Solovay-Strassen [24] and the Rabin-Miller [23] primality tests. We assume that the secret values chosen by the parties satisfy $p_a = q_a = 3 \bmod 4$ and $p_b = q_b = 0 \bmod 4$. This can be agreed upon before hand and causes the resulting modulus N to be a Blum integer[3], since $p \equiv q \equiv 3 \bmod 4$. The test is as follows:

1. Alice and Bob agree on a random $g \in \mathbf{Z}_N^*$.
2. Alice computes the Jacobi symbol of g over N. If $\left(\frac{g}{N}\right) \neq 1$ the protocol is restarted at step (1).
3. Otherwise, Alice computes $v_a = g^{(N-p_a-q_a+1)/4} \bmod N$, and Bob computes $v_b = g^{(p_b+q_b)/4} \bmod N$. They exchange these values, and verify that

$$v_a = \pm v_b \pmod N$$

If the test fails then the parties declare that N is not a product of two primes. Otherwise they declare success.

Since $p_a = q_a = 3 \bmod 4$ and $p_b = q_b = 0 \bmod 4$ both exponents in the computation of v_a, v_b are integers after division by 4. The correctness and privacy of the protocol is proved in the next two lemmas.

---

[3] The primality test described in this section is best suited for Blum integers. For non-Blum moduli the test may leak a few bits of information depending on the power of two dividing $\mathrm{lcm}(p-1, q-1)$. For non-Blum integers $N = pq$ with $p = q = 1 \bmod 4$ these problems can be avoided by performing the test in a different group (i.e. not in $\mathbf{Z}_N^*$). We use a quadratic extension of $\mathbf{Z}_N$.

Lemma 2. *Let $N = pq$ be an integer with $p \equiv q \equiv 3$ mod 4. If N is a product of two distinct primes then success is declared in all invocations of the protocol. Otherwise, for all but an exponentially small fraction of N, the parties declare that N is not a product of two primes with probability at least $\frac{1}{2}$ (over the choice of g).*

Proof. Observe that the test in step (3) of the protocol checks that $g^{(N-p-q+1)/4} \equiv \pm 1$ mod N.

Suppose p and q are prime. In step (2) we verify that $\left(\frac{g}{N}\right) = 1$. This implies $\left(\frac{g}{p}\right) = \left(\frac{g}{q}\right)$. Also, since $\frac{q-1}{2}$ and $\frac{p-1}{2}$ are odd we have,

$$
\begin{cases}
g^{\phi(N)/4} = \left(g^{\frac{p-1}{2}}\right)^{\frac{q-1}{2}} \equiv \left(\frac{g}{p}\right)^{\frac{q-1}{2}} = \left(\frac{g}{p}\right) \quad (\text{mod } p) \\
g^{\phi(N)/4} = \left(g^{\frac{q-1}{2}}\right)^{\frac{p-1}{2}} \equiv \left(\frac{g}{q}\right)^{\frac{p-1}{2}} = \left(\frac{g}{q}\right) \quad (\text{mod } q)
\end{cases}
$$

Since $\left(\frac{g}{p}\right) = \left(\frac{g}{q}\right)$ it follows that $g^{\phi(N)/4} \equiv \pm 1$ mod N. Since $\phi(N) = N-p-q+1$ when p and q are prime, it follows that the test in step 3 always succeeds.

Suppose at least one of p, q is not prime. That is, $N = r_1^{d_1} \cdots r_s^{d_s}$ is a nontrivial factorization of N with $\sum d_i \geq 3$ and $s \geq 1$. Set $e = (N - p - q + 1)/4 = (p-1)(q-1)/4$ to be the exponent used in step (3). Note that e is odd since $p \equiv q \equiv 3$ mod 4. Define the following two subgroups of $\mathbf{Z}_N^*$:

$$
G = \{g \in \mathbf{Z}_N \text{ s.t. } \left(\frac{g}{n}\right) = 1\} \quad \text{and} \quad H = \{g \in G \text{ s.t. } g^e = \pm 1 \text{ mod } N\}
$$

To prove the lemma we show that $|H| \leq \frac{1}{2}|G|$. Since H is a subgroup of G it suffices to prove proper containment of H in G, i.e. prove the existence of $g \in G \setminus H$. There are four cases to consider.

Case 1. Suppose $s \geq 3$. Let a be a quadratic non-residue modulo r_3. Define $g \in \mathbf{Z}_N$ to be an element satisfying

$$
g \equiv 1 \text{ mod } r_1 \quad \text{and} \quad g \equiv -1 \text{ mod } r_2 \quad \text{and} \quad g \equiv \begin{cases} 1 \text{ mod } r_3 \text{ if } \left(\frac{-1}{r_2}\right) = 1 \\ a \text{ mod } r_3 \text{ if } \left(\frac{-1}{r_2}\right) = -1 \end{cases}
$$

and $g \equiv 1$ mod r_i for $i > 3$. Observe that $g \in G$. Since e is odd $g^e = g = 1$ mod r_1 and $g^e = g = -1$ mod r_2. Consequently, $g^e \neq \pm 1$ mod N i.e. $g \notin H$.

Case 2. Suppose $\gcd(p, q) > 1$. Then there exists an odd prime r such that r divides both p and q. Then r^2 divides N implying that r divides $\phi(N)$. It follows that in $\mathbf{Z}_N^*$ there exists an element g of order r. Since r is odd we have $\left(\frac{g}{N}\right) = \left(\frac{g^r}{N}\right) = \left(\frac{1}{N}\right) = 1$, i.e. $g \in G$. Since r divides both p and q we know that r does not divide $N - p - q + 1 = 4e$. Consequently $g^{4e} \neq 1$ mod N implying that $g^e \neq \pm 1$ mod N. Hence, $g \notin H$.

Case 3. The only way $N = pq$ does not fall into both cases above is if $p = r_1^{d_1}$ and $q = r_2^{d_2}$ where r_1, r_2 are distinct primes and at least one of d_1, d_2 is bigger than 1 (case 2 handles N that are a prime power $N = r^d$). By symmetry we may assume $d_1 > 1$. Since $\mathbf{Z}_p^*$ is a cyclic group of order $r_1^{d_1-1}(r_1 - 1)$ it contains an element of order $r_1^{d_1-1}$. It follows that $\mathbf{Z}_N^*$ also contains an element g of order $r_1^{d_1-1}$. As before, $\left(\frac{g}{N}\right) = 1$, i.e. $g \in G$. If $q \neq 1 \bmod r_1^{d_1-1}$ then $4e = N - p - q + 1$ is not divisible by $r_1^{d_1-1}$. Consequently, $g^{4e} \neq 1 \bmod N$, i.e. $g \notin H$.

Case 4. We are left with the case $N = pq$ with $p = r_1^{d_1}$, $q = r_2^{d_2}$, $d_1 > 1$ as above and $q \equiv 1 \bmod r_1^{d_1-1}$. In this case it may indeed happen that $H = G$. For example, $p = 3^n$ and $q = 2 \cdot 3^{n-1} + 1$ with n odd and q prime.
Observe that $r_1^{d_1-1} \geq \sqrt{p} \geq 2^{n/2}$. Consequently, since p and q are chosen independently the probability of $q \equiv 1 \bmod r_1^{d_1-1}$ is less than $1/2^{n/2}$. In addition, p has to be a prime power which happens with probability less than $n/2^{n/2}$. The probability that both events happen is less than $n/2^n$. Hence, this case occurs with exponentially small probability. $\quad\square$

Integers N that fall into Case 4 above incorrectly pass the test. This can be rectified by adding a fourth step to the protocol to filter out these integers. With this extra fourth step our protocol becomes a complete probabilistic test for proving that N is a product of two primes. Due to space limitation we only give a high level description in the next subsection.

Lemma 3. *Suppose p, q are prime. Then either party can simulate the transcript of the primality testing protocol. Consequently, neither party learns anything about the factors of N from this protocol.*

Proof Sketch. Since p, q are prime we know that $v_a = \pm v_b \bmod N$ where v_a, v_b are defined as in step (3) of the protocol. Consequently, given either of v_a or v_b, the simulator can compute the other one up to sign. If $v_a = v_b$ then $\left(\frac{g}{p}\right) = \left(\frac{g}{q}\right) = 1$, and if $v_a = -v_b$ then $\left(\frac{g}{p}\right) = \left(\frac{g}{q}\right) = -1$. That is, the sign determines whether g is a quadratic residue or not modulo N. If the simulator chooses the sign according to the flip of an unbiased coin, the resulting distribution is indistinguishable from the true distribution assuming the hardness of quadratic residuosity modulo a Blum integer. $\quad\square$

We note that step (2) of the protocol is crucial. Without it the condition of step (3) might fail (and reveal the factorization) even when p and q are prime. We also note that in practice the probability that a non RSA modulus passes even one iteration of this test is actually much less than a half.

3.1 A complete probabilistic primality test

Integers N that fall into Case 4 of Lemma 2 can be filtered out by adding an extra fourth step to our protocol. Due to lack of space we only give a high level description. There are two alternatives:

1. Let K be the group $K = \left(\mathbf{Z}_N[x]/(x^2+1)\right)^*/\mathbf{Z}_N^*$. When $N = pq$ is a product of two distinct primes K contains $(p+1)(q+1)$ elements (recall $p = q = 3 \bmod 4$). In this case all $g \in K$ satisfy $g^{(p+1)(q+1)} = 1$. One can show that when N falls into Case 4, at least half the elements in K do not satisfy the above condition. Hence, by picking a random $g \in K$ and jointly testing that $g^{(p+1)(q+1)} = 1$ Alice and Bob can eliminate all such N.

2. Alternatively, observe that N that fall into Case 4 satisfy $\gcd(N, p+q-1) > 1$. The parties can easily test this with the help of a third party, Henry. Alice picks a random r_a. Bob picks a random r_b. Using the protocol of the next section they compute $z = (r_a + r_b)(p_a + q_a + p_b + q_b - 1) \bmod N$ and test that $\gcd(z, N) > 1$. If so, then N is rejected. Unfortunately this test also eliminates a few valid RSA moduli, i.e. moduli $N = pq$ with p, q prime and $q = 1 \bmod p$.

4 Distributed computation of N

We now turn our attention to the computation of N. We describe our protocols for the case of two parties with a helper and discuss the case of $k > 2$ parties in Section 7.

In the case of two parties, Alice and Bob posses integers p_a, q_a and p_b, q_b respectively. They wish to compute the integer $N = (p_a + p_b)(q_a + q_b)$ such that at the end of the computation Alice has no information about p_b, q_b beyond what is revealed by the knowledge of N. The same should hold for Bob. To make the protocol secure in the information theoretic sense we require the help of a third "helper" party called Henry. Henry has no information about either p_i nor q_i (for $i = a, b$) and the same should hold at the end of the protocol. Clearly, Henry learns N (since N is public) but he learns nothing more.

BenOr, Goldwasser and Wigderson [3] (and similarly Chaum, Crépeau and Damgård [6]) describe an elegant protocol for private evaluation of general functions for three or more parties. Their full technique is an overkill for the simple function we have in mind. We adapt and optimize their protocol in several ways so as to minimize the amount of computation and communication between the parties. From here on, let $P > N$ be some prime. Unless otherwise stated, all arithmetic operations are done modulo P. The protocol works as follows:

Alice: Alice picks two random lines that intersect the y axis at p_a, q_a respectively. This is done by picking two integers $c_a, d_a \in \mathbf{Z}_P^*$ and using the lines $c_a x + p_a$ and $d_a x + q_a$. She evaluates each line at three points $x_a = 1, x_b = 2, x_h = 3$. Let $p_{a,i} = c_a x_i + p_a$ and $q_{a,i} = d_a x_i + q_a$ for $i = a, b, h$.

Next, Alice picks two random numbers $p_{b,a}, q_{b,a}$ and a random quadratic polynomial $r(x)$ such that $r(0) = 0$. Set $r_i = r(x_i)$ for $i = a, b, h$. She computes $N_a = (p_{a,a} + p_{b,a})(q_{a,a} + q_{b,a}) + r_a$.

Finally, she sends $p_{a,b}, q_{a,b}$ and $p_{b,a}, q_{b,a}$ and r_b to Bob. She sends $p_{a,h}, q_{a,h}, r_h$ and N_a to Henry.

Bob: Bob computes $c_b = (p_{b,a} - p_b)/x_a$ and $d_b = (q_{b,a} - q_b)/x_a$. Note that the two lines $c_b x + p_b$ and $d_b x + q_b$ intersect the y-axis at p_b, q_b respectively and evaluate to $p_{b,a}, q_{b,a}$ at x_a.

Next, Bob computes $p_{b,i} = c_b x_i + p_b$ and $q_{b,i} = d_b x_i + q_b$ for $i = b, h$. He computes $N_b = (p_{a,b} + p_{b,b})(q_{a,b} + q_{b,b}) + r_b$ and sends $p_{b,h}, q_{b,h}$ and N_b to Henry.

Henry: Henry computes $N_h = (p_{a,h} + p_{b,h})(q_{a,h} + q_{b,h}) + r_h$. He then interpolates the quadratic polynomial $\alpha(x)$ that passes through the points (x_a, N_a) ; (x_b, N_b) ; (x_h, N_h). We have $\alpha(0) = N$. Henry sends N to Alice and Bob.

To see that $\alpha(0) = N$ observe that the polynomial $\alpha(x)$ satisfies

$$\alpha(x) = ((c_a x + p_a) + (c_b x + p_b)) \cdot ((d_a x + q_a) + (d_b x + q_b)) + r(x)$$

Indeed $\alpha(x_i) = N_i$ for $i = a, b, h$.

Lemma 4. *Given N, Alice, Bob and Henry can each simulate the transcript of the protocol. Consequently, they learn nothing more than the value of N.*

Proof Sketch. This is clear for Alice and Bob. To simulate Henry's view the simulator picks $p_{a,h}, q_{a,h}, p_{b,h}, q_{b,h}, r_h$ at random and computes $N_h = (p_{a,h} + p_{b,h})(q_{a,h} + q_{b,h}) + r_h$. It then picks a random quadratic polynomial $\alpha(x)$ satisfying $\alpha(0) = N$ and $\alpha(x_h) = N_h$. It computes $N_a = \alpha(x_a)$ and $N_b = \alpha(x_b)$. These values are a perfect simulation of Henry's view. $\square$

The protocol's communication pattern is very simple: initially Alice sends one message to Bob and one to Henry. Then Bob sends a message to Henry. Finally, Henry publishes the value of N. Hence, during the protocol only three messages are sent. The protocol is also efficient in computation since only three multi-precision multiplications are performed.

The protocol differs from the BGW protocol in two ways. First, there is no need for a truncation step. Second, to minimize the number of messages we let Alice pick her shares $p_{b,a}$ and $q_{b,a}$ of Bob's secret. Bob then picks his polynomial to be consistent with Alice's choice.

5 Trial division

In this section, we consider the trial division step. We describe our protocol for the case of two parties with a helper and discuss the case of $k > 2$ parties in Section 7.

Let q be some random number. The first step in testing the primality of q is trial division, which tests if q is divisible by any small prime. In our case $q = q_a + q_b$ where Alice knows q_a and Bob knows q_b. Let $p_1, \ldots, p_j$ be the set of small primes to be considered. Together they wish to test that $q \neq 0 \bmod p_i$

for all i, $1 \leq i \leq j$, without revealing any other information about q_a, q_b. This is equivalent to testing that $q_a \bmod p_i \neq -q_b \bmod p_i$ for all i, $1 \leq i \leq j$. A number of simple protocols have been proposed for privately evaluating the equality predicate [15], including one with a third helper party, based on universal classes of hash functions [5, 25] (attributed to Noga Alon in [15]). Using this equality test, the trial division protocol is as follows:

Alice Pick random $c_i \in \mathbf{Z}_{p_i}$ and $d_i \in \mathbf{Z}_{p_i}^*$. Compute $u_i = c_i + d_i q_a \bmod p_i$, for all i, $1 \leq i \leq j$. Send $c_1, d_1, \ldots, c_j, d_j$ to Bob and $u_1, \ldots, u_j$ to Henry.

Bob Compute $v_i = c_i - d_i q_b \bmod p_i$ for all i, $1 \leq i \leq j$. Send $v_1, \ldots, v_j$ to Henry.

Henry Output "pass" if $u_i \neq v_i$ for all i, $1 \leq i \leq j$. Otherwise, output "fail".

Lemma 5. *The output of the protocol is "pass" if and only if $q \neq 0 \bmod p_i$ for all i, $1 \leq i \leq j$.*

Lemma 6. *When the output is "pass", each party can simulate its view of the transcript of the protocol. Consequently, when the output is "pass", the parties learn nothing about q other than the fact that $q \neq 0 \bmod p_i$ for all i, $1 \leq i \leq j$.*

6 Shared generation of public/private keys

In this section, we consider the step of key generation. We describe our protocol for the case of two parties and discuss the case of $k > 2$ parties in Section 7.

Suppose Alice and Bob have successfully computed $N = pq = (p_a + p_b)(q_a + q_b)$. They wish to compute shares of $d = e^{-1} \bmod \phi(N)$ for some agreed upon value of e. We have two approaches for computing shares of d. The first only works for small e (say $e < 1000$) but is very efficient requiring very little communication between the parties. The second works for any e and is still efficient, however it requires the help of Henry and takes more rounds of communication (but still constant).

6.1 Small public exponent

We begin by describing an efficient technique for generating shares of d when the public exponent e is small. For simplicity throughout the section we assume $e = 3$.

First, Alice and Bob compute $\phi(N) \bmod 3$, by exchanging $p_a + q_a \bmod 3$ and $p_b + q_b \bmod 3$. This reveals some little information (less than two bits) about $\phi(N)$; this information is of no use since it can be easily guessed. Observe that[4]:

$$\begin{cases} d = [\phi(N) + 1]/3 = \frac{1}{3}[N + 2 - (p_a + p_b + q_a + q_b)] & \text{if } \phi(N) = 2 \bmod 3 \\ d = [2\phi(N) + 1]/3 = \frac{2}{3}[N - (p_a + p_b + q_a + q_b)] + 1 & \text{if } \phi(N) = 1 \bmod 3 \end{cases}$$

---

[4] The case $\phi(N) = 0 \bmod 3$ is of no interest since in that case $e = 3$ can not be used as a public RSA exponent.

Consequently, knowing $\phi(N) \bmod 3$ enables Alice and Bob to locally compute shares of the decryption exponent d: If $\phi(N) \bmod 3 = 1$, then Alice sets her share to be $d_a = \lfloor \frac{N - 2p_a - 2q_a}{3} \rfloor + 1$ and Bob sets his share to be $d_b = \lceil \frac{N - 2p_b - 2q_b}{3} \rceil$. If $\phi(N) \bmod 3 = 2$, then $d_a = \lfloor \frac{N - p_a - q_a + 2}{3} \rfloor$ and $d_b = \lceil \frac{-p_b - q_b}{3} \rceil$. Either way $d = d_a + d_b \bmod \phi(N)$. This enables threshold decryption as described in [13], i.e., $c^d \equiv c^{d_a} c^{d_b} \bmod N$.

6.2 Arbitrary public exponent

Unlike the previous technique, our second method for generating shares of d works for arbitrary public exponent e and leaks no information. However, it requires the help of Henry.

Recall that the public modulus $N = (p_a + p_b)(q_a + q_b)$ satisfies $\phi(N) = (N - p_a - q_a + 1) - (p_b + q_b)$. We set $\phi_a = N - p_a - q_a + 1$ and $\phi_b = -p_b - q_b$. Then $\phi(N) = \phi_a + \phi_b$ is a sharing of $\phi(N)$ between Alice and Bob. The private exponent d is the inverse of $e \bmod \phi_a + \phi_b$. Unfortunately, traditional inversion algorithms, e.g. extended gcd, involve computations modulo $\phi_a + \phi_b$. When $\phi = \phi_a + \phi_b$ is shared among two users we do not know how to efficiently perform these computations. We therefore develop an inversion algorithm for computing $e^{-1} \bmod \phi$ that avoids any computation modulo ϕ.

When only a single user is involved the inversion algorithm works as follows: (1) Compute $\zeta = \phi^{-1} \bmod e$. (2) Set $T = -\zeta \cdot \phi + 1$. Observe that $T \equiv 0 \bmod e$. (3) Set $d = T/e$. Then $d = e^{-1} \bmod \phi$ since $d \cdot e \equiv 1 \bmod \phi$. Notice that the algorithm made no reductions modulo ϕ. Our inversion algorithm made use of the fact that $e^{-1} \bmod \phi$ can be immediately deduced from $\phi^{-1} \bmod e$.

We now show how the above inversion algorithm can be used to compute shares $d_a + d_b = e^{-1} \bmod \phi_a + \phi_b$. Clearly we may assume $\gcd(\phi(N), e) = 1$.

Step 1. Alice picks a random $r_a \in \mathbf{Z}_e$. Bob picks a random $r_b \in \mathbf{Z}_e$.

Step 2. Using the protocol of Section 4 compute $\Psi = (r_a + r_b) \cdot (\phi_a + \phi_b) \bmod e$. Since e is odd $(\gcd(\phi(N), e) = 1)$ all the required Lagrange coefficients indeed exist. At this point Ψ is known to both Alice and Bob. If Ψ is not invertible modulo e the protocol is restarted at Step 1.

Step 3. Alice sets $\zeta_a = r_a \Psi^{-1} \bmod e$. Bob sets $\zeta_b = r_b \Psi^{-1} \bmod e$.
Observe $\zeta_a + \zeta_b = (r_a + r_b)\Psi^{-1} = \phi^{-1} \bmod e$.

Step 4. Next they fix an arbitrary odd integer $P > 2N^2 e$, e.g. $P = 2N^2 e + 1$. They then regard the shares $0 \le \zeta_a, \zeta_b < e$ as elements of $\mathbf{Z}_P$. Using a modification of the BGW protocol of Section 4 they compute a sharing of

$$A + B = -(\zeta_a + \zeta_b)(\phi_a + \phi_b) + 1 \bmod P$$

such that Alice knows A and Bob knows B. Recall that in Section 4 Alice uses a random quadratic $r(x)$ such that $r(0) = 0$. Instead, Alice will choose a truly random quadratic $r(x)$. Then the final result computed by Henry is offset from the desired result by an additive factor of $r(0)$, where only Alice knows $r(0)$. If Henry gives his final result to Bob, then Alice and Bob have additive shares of the desired result. These shares could then be re-randomized if Alice adds, and Bob subtracts, an agreed-upon random value unknown to Henry.

Step 5. From here on we regard A and B as integers $0 \leq A, B < P$. Our objective is to ensure that over the integers

$$A + B = -(\zeta_a + \zeta_b)(\phi_a + \phi_b) + 1 \tag{1}$$

Observe that $0 \leq A + B \bmod P < P/N$ (since $\zeta_a + \zeta_b < 2e$ and $\phi(N) < N$). It follows that $A + B > P$ with probability more than $1 - \frac{1}{N}$ (the only way that $A+B < P$ is if both A and B are less than P/N). Therefore, if Alice sets $A \leftarrow A - P$ then equation (1) holds over the integers. In the very unlikely event (that occurs with probability $1/N$) that the relation doesn't hold over the integers, the wrong sharing of the private key will be generated. This will be detected when the parties do a trial decryption .

Step 6. At this point e divides $A + B$ since

$$A + B = (\zeta_a + \zeta_b)(\phi_a + \phi_b) + 1 = -(\phi_a + \phi_b)^{-1}(\phi_a + \phi_b) + 1 \equiv 0 \pmod{e}$$

Therefore $d = (A + B)/e$ since $de = A + B = k\phi(N) + 1 \equiv 1 \bmod \phi(N)$. Consequently, Alice sets $d_a = \lfloor A/e \rfloor$ and Bob sets $d_b = \lceil B/e \rceil$. Clearly $d = d_a + d_b$.

Notice that the value P we use in step 4 is quite large. As a result the shares d_a, d_b are of the order of N^2. In actual implementations there is no need for this to happen. The only reason P has to be this large is to ensure that step 5 succeeds with overwhelming probability. If one is willing to tolerate leakage of one bit in step 5 then the parties can use a much smaller P, e.g. $P = 2Ne + 1$. If the resulting A, B satisfy $A + B > P$ then the correct sharing of d is obtained. Otherwise, trial decryption will fail and the parties learn that $A + B < P$. In this case, Alice adds P back to her share A and step 6 is repeated again. The correct sharing of d is now obtained. This results in shares d_a, d_b of order N.

The computation of $\phi^{-1} \bmod e$ (steps 1-3 above) is based on a technique due to Beaver [2].

7 Generalizations to k parties

Our results thus far show how two parties can generate an RSA modulus $N = (p_a + p_b)(q_a + q_b)$ with the help of a third neutral party. In this section we discuss how these results generalize to the case of three or more parties. In this case, the k parties will be generating an RSA modulus $N = (p_1 + \ldots + p_k)(q_1 + \ldots + q_k)$, where each party i knows p_i, q_i. Afterwards, assuming that the parties follow the protocol as required, no coalition of $\lceil k/2 \rceil - 1$ parties can factor N.

The primality test from Section 3 generalizes easily to $k > 2$ parties. Assume that the secret values chosen by the parties satisfy $p_1 = q_1 = 3 \bmod 4$ while for all other parties $p_i = q_i = 0 \bmod 4$. Then party 1 computes $v_1 = g^{\frac{N - p_1 - q_1 + 1}{4}} \bmod$

N. Party i computes $v_i = g^{\frac{p_i + q_i}{4}}$, $2 \le i \le k$. They all publish their values and verify that $v_1 \equiv \pm v_2 v_3 \ldots v_k \bmod N$. The arguments for correctness and privacy are essentially the same. The resulting protocol is k-private.

To generalize the distributed computation of N of Section 4 to $k > 2$ parties use the BGW protocol with higher degree polynomials (rather than linear). The BGW protocol can be made private (i.e. no information about the p_i, q_i is leaked) even when $\lceil k/2 \rceil - 1$ parties collude.

Trial division (Section 5) with $k > 2$ parties can be done $\lceil k/2 \rceil - 1$ privately, but a different protocol must be used. We adapt an idea due to Beaver [2]. Let $q = q_1 + \ldots + q_k$ be an integer shared among k parties. Let p be a small prime. To test if p divides q each party picks a random $r_i \in \mathbf{Z}_p$. Using the BGW protocol they compute $qr = (\sum q_i)(\sum r_i) \bmod p$. If $qr \ne 0$ then p does not divide q. Furthermore, since r is unknown to any minority of parties, qr provides no other information about q. Note that if $qr = 0 \bmod p$ it could still be the case that p does not divide q. However, if the test is repeated twice for each small prime p, the probability that a good candidate is rejected is at most $1 - \prod_{p < B} (1 - \frac{1}{p^2}) < \frac{1}{2}$.

The first key generation protocol of Section 6 immediately generalizes to produce a k-out-of-k sharing of a private key d among $k > 2$ parties. The second protocol can produce a k-out-of-k sharing $d = \sum d_i$; however the computation is based on BGW and is therefore only $\lceil k/2 \rceil - 1$ private.

The more difficult case of t-out-of-k sharing of a private key among $k > 2$ parties is treated in the next subsection.

7.1 t-out-of-k sharing

Since the computation of N in Section 4 relies on the BGW protocol [3] we are a priori restricted to threshold t satisfying $t < \lceil k/2 \rceil$. A coalition of more than $k/2$ parties can already factor N. We show how to achieve any threshold $t < \lceil k/2 \rceil$.

To achieve t-out-of-k sharing of d, first share d using a k-out-of-k scheme as described above, i.e. each party computes a share d_i such that $d = \sum d_i \bmod \phi(N)$. Then each party i shares its share d_i with all other parties using a t-out-of-k scheme. We denote the share of d_i sent to party j by $d_{i,j}$. A coalition $\mathcal{C}$ of t parties can do threshold decryption using its shares of d and its shares $d_{i,j}$ for $i \notin \mathcal{C}$. Thus, we are left with the problem of showing how party i generates the $d_{i,j}$ given d_i. Secret sharing modulo $\phi(N)$ is not easy. An elegant solution was given in [8] where the authors show how a trusted dealer (who knows the factorization of N) can generate shares $d_{i,j}$ as required. We can show that when $N = (\sum p_i)(\sum q_i)$ where party i only knows p_i, q_i, there is no need for a trusted dealer. That is, the parties can engage in a multi-party protocol to compute the same shares $d_{i,j}$ that were generated by the dealer in [8]. Unfortunately, this requires multiple invocations of the BGW protocol described in Section 4.

Since we are mainly concerned with efficient solutions we describe an alternate approach which works well when the threshold t is small. When t is small

t-out-of-k sharing can be achieved through t-out-of-t sharing. Naively this can by done by giving each of the $\binom{k}{t}$ coalitions a t-out-of-t sharing of the secret. Other techniques [1] can reduce t-out-of-k sharing to t-out-of-t far more efficiently[5]. However, it is essential for these reductions that the instances of t-out-of-t sharing be independent. Because it is difficult to compute reduction modulo $\phi(N)$ efficiently without revealing $\phi(N)$, ordinary techniques for generating new sharing instances cannot be used.

We propose the following procedure for party i to generate many independent t-out-of-t sharings of d_i. To avoid unnecessary indices we refer to d_i as s. Pick $t - 1$ random integers $s_1, \ldots, s_{t-1} \in_R [-B, B]$ for some large B and compute $s_t = s - \sum_{j=1}^{t-1} s_j$ (where addition is over the integers). We show that $s_1, \ldots, s_t$ is a private t-out-of-t sharing of s for suitable choice of B. Note that this sharing scheme is at least as secure as the scheme where *every* share is a random elements in $[-B, B]$ and i publishes the difference between the secret and the sum of all the shares. When $s \in [1, b]$ the following lemma establishes that this scheme is sufficiently private when $B > tb^{2+\epsilon}$ for any fixed $\epsilon > 0$.

Lemma 7. *Let $s \in [1 \ldots b]$, and let $p_x = \text{prob}(s = x) = \frac{1}{b}$ for all $x \in [1, b]$. Let $(s_1, \ldots, s_t) \in_R [-B, B]^t$, and $\delta = \sum_{i=1}^{t} s_i - s$. For any coalition $\mathcal{C} \subset [1, t]$, let $p_{x,\mathcal{C}} = \text{prob}(s = x | \delta, \{s_i\}_{i \in \mathcal{C}})$. Then, for every coalition $\mathcal{C}$ and every $\epsilon > 0$, the distributions $\{p_x\}_x$ and $\{p_{x,\mathcal{C}}\}_x$ are statistically indistinguishable when $B > tb^{2+\epsilon}$.*

Due to space limitation we give the proof in the full version of the paper.

8   Summary and open problems

We presented techniques that allow two or more parties to generate an RSA modulus $N = pq$ such that all parties are convinced that N is indeed a product of two primes; however none of them can factor N. When only two parties are involved, interaction with a third helper party is needed to complete some steps of the protocol. Finally we show how the parties can generate shares of a private decryption exponent to allow threshold decryption.

Our protocols are practical, though there is some slowdown in comparison to single user generation of an RSA key. The main reason is that both primes p, q are generated at once. This increases the number of tries until a suitable N is found, as was discussed in Section 2. A possible approach for solving this

---

[5] For instance we show how to efficiently implement 2-out-of-k sharing from 2-out-of-2 sharing. Let d be a secret and $r = \lceil \log k \rceil$. Let $d = d_{1,0} + d_{1,1} = d_{2,0} + d_{2,1} = \cdots = d_{r,0} + d_{r,1}$ be r independent 2-out-of-2 sharings of the secret d. For an $i \in [0, k]$ let $i = i_r i_{r-1} \ldots i_0$ be the binary digits of i. Party i's share of the secret d is the set $\{d_{r,i_r}, d_{r-1,i_{r-1}}, \ldots, d_{0,i_0}\}$. Given two parties $i = i_r i_{r-1} \ldots i_0$ and $j = j_r j_{r-1} \ldots j_0$ there exists an s such that $i_s \neq j_s$. Then $d = d_{s,i_s} + d_{s,j_s}$, enabling the two parties to reconstruct the secret. Hence, we achieved 2-out-of-k sharing using only $\log k$ independent 2-out-of-2 sharings (as opposed to $\binom{k}{2}$ required by the naive solution).

is to generate N as $N = p_a p_b (q_a + q_b)$ where p_a, p_b are primes known to Alice, Bob respectively and q_a, q_b are random n bit integers. The number of probes until $q_a + q_b$ is found to be prime is just as in single user generation of N. Unfortunately, this approach doesn't scale well. To support k parties, N must be a product of $k+1$ primes. Also, one has to design a protocol for testing that such an N is indeed a product of *three* primes.

In the two party case our protocols require the use of a third helper party. The helper party is needed for the private computation of $N = (p_a + p_b)(q_a + q_b)$. Therefore, it is of some interest to develop efficient two party protocols for this specific function which do not make use of a third party. General two party computation protocols(e.g. [26]) are too inefficient.

Our protocols generate an RSA modulus which is the product of two large random primes. It would be useful to be able to generate moduli of some special form. For example, a modulus which is a product of "safe primes" (i.e., where both $\frac{p-1}{2}$ and $\frac{q-1}{2}$ are prime) has been considered for security purposes [18] as well as for technical reasons related to threshold cryptography [10, 16].

Throughout the paper we use a model in which parties honestly follow the protocol. The case of active adversaries that cheat during the protocol is of great interest as well. Since the RSA function is verifiable (the parties can simply check that they correctly decrypt encrypted messages) active adversaries are limited in the amount of damage they can cause. However, it may still be possible that one party can cheat during the protocol and consequently be able to factor the resulting N. Our techniques can be made to withstand some number of active adversaries though more parties must participate in the protocol. We leave the details for the full version of the paper.

Acknowledgments

We thank Yair Frankel and Don Beaver for several stimulating discussions on our results.

References

1. N. Alon, Z. Galil and M. Yung, "Dynamic-resharing verifiable secret sharing," ESA 1995.
2. D. Beaver, "Security, fault tolerance, and communication complexity in distributed systems," Ph.D. thesis, Harvard University, May 1990.
3. M. Ben-Or, S. Goldwasser, A. Wigderson, "Completeness theorems for non-cryptographic fault tolerant distributed computation", STOC 1988, pp. 1–10.
4. J. Benaloh (Cohen), "Secret sharing homomorphisms: keeping shares of a secret secret," Crypto '86, 251-260.
5. J. Carter and M. Wegman, "Universal classes of hash functions", *J. Comput. Syst. Sci.* 18 (1979), 143–154.
6. D. Chaum, C. Crépeau, and I. Damgård, "Multiparty unconditionally secure protocols," ACM STOC 1988, 11-19.

7. N. De Bruijn, "On the number of uncanceled elements in the sieve of Eratosthenes", Proc. Neder. Akad. Wetensch, vol. 53, 1950, pp. 803–812. Reviewed in LeVeque Reviews in Number Theory, Vol. 4, Section N-28, p. 221.

8. A. DeSantis, Y. Desmedt, Y. Frankel, M. Yung, "How to share a function securely", STOC 1994, pp. 522–533.

9. Y. Desmedt, "Threshold cryptography," European Transactions on Telecommunications and Related Technologies, Vol. 5, No. 4, July-August 1994, pp. 35–43.

10. Y. Desmedt and Y. Frankel, "Shared generation of authenticators and signatures", Crypto '91, 457–469.

11. U. Feige, A. Fiat, and A. Shamir, "Zero-knowledge proofs of identity," Journal of Cryptology 1 (1988), 77-94.

12. A. Fiat and A. Shamir, "How to prove yourself: Practical solutions to identification and signature problems," Crypto '86, 186-194.

13. Y. Frankel, "A practical protocol for large group oriented networks", Eurocrypt 89, pp. 56–61.

14. M. Franklin and S. Haber, "Joint encryption and message-efficient secure computation," Journal of Cryptology, 9 (1996), 217-232.

15. R. Fagin, M. Naor, P. Winkler, "Comparing information without leaking it", CACM, Vol 39, No. 5, May 1996, pp. 77–85.

16. R. Gennaro, S. Jarecki, H. Krawczyk, T. Rabin, "Robust and efficient sharing of RSA functions", Crypto 96, pp. 157–172.

17. O. Goldreich, S. Micali, A. Wigderson, "How to play any mental game", STOC 1987, pp. 218–229.

18. J. Gordon, "Strong primes are easy to find", Eurocrypt 84, pp. 216–223.

19. L. Guillou and J. Quisquater, "A practical zero-knowledge protocol fitted to security microprocessor minimizing both transmission and memory," Eurocrypt '88, 123-128.

20. K. Ohta and T. Okamoto, "A modification of the Fiat-Shamir scheme," Crpto '88, 232-243.

21. H. Ong and C. Schnorr, "Fast signature generation with a Fiat Shamir-like scheme," Eurocrypt '90, 432-440.

22. T. Pederson, "A threshold cryptosystem without a trusted party," Proceedings of Eurocrypt 91, pp. 522–526.

23. M. Rabin, "Probabilistic algorithm for testing primality", J. of Number Theory, vol. 12, pp. 128–138, 1980.

24. R. Solovay, V. Strassen, "A fast monte carlo test for primality", SIAM journal of computing, vol. 6, pp. 84–85, 1977.

25. M. Wegman and J. Carter, "New hash functions and their use in authentication and set equality", J. Comput. Syst. Sci. 22 (1981), 265–279.

26. A. Yao, "How to generate and exchange secrets", FOCS 1986, pp. 162–167.

Proactive RSA

Yair Frankel[1] *   Peter Gemmell[2] *   Philip D. MacKenzie[3] *   Moti Yung[4]

[1] CertCo, N.Y. NY, frankely@certco.com; on leave from Sandia National Labs
[2] Sandia National Labs, Albuquerque NM, psgemme@sandia.gov
[3] Boise State University, Boise ID, philmac@cs.idbsu.edu
[4] CertCo, N.Y. NY, moti@certco.com, moti@cs.columbia.edu.

Abstract. Distributed threshold protocols that incorporate proactive maintenance can tolerate a very strong "mobile adversary." This adversary may corrupt all participants throughout the lifetime of the system in a non-monotonic fashion (i.e., recoveries are possible) but the adversary is limited to the number of participants it can corrupt during any short time period. The proactive maintenance assures increased security and availability of the cryptographic primitive. We present a proactive RSA system in which a threshold of servers applies the RSA signature (or decryption) function in a distributed manner. Our protocol enables servers which hold the RSA key distributively to dynamically and co-operatively self-update; it is secure even when a linear number of the servers are corrupted during any time period; it efficiently maintains the security of the function; and it enables continuous function availability (correct efficient function application using the shared key is possible at any time). A major technical difficulty in "proactivizing" RSA was the fact that the servers have to update the "distributed representation" of an RSA key, while not learning the order of the group from which keys are drawn (in order not to compromise the RSA security). We give a distributed threshold RSA method which permits "proactivization".

1 Introduction

This work concerns algorithmic mechanisms to provide an increased level of security and availability to an RSA public-key system via distribution of the private key and active communication between shareholders. This improved level of security and availability counters a very strong "mobile adversary" who may corrupt **all** participants (servers, each with private memory) throughout the lifetime of the system but is not able to corrupt too many participants during any short period of time. The servers engage in a "proactive maintenance" of key shares that protects them against this mobile adversary who tries to learn the secret or disrupt their operation. *Proactive security* refers to security and availability

---

* This work was performed under U.S. Department of Energy contract number DE-AC04-94AL85000.

in the presence of a this mobile adversary which was first suggested by Ostro-vsky and Yung [OY91]. Proactive security is vital for dealing with the increasing number of threats (including viruses and hackers) to local and international net-work domains, and for securing long-lived cryptographic keys that cannot be replaced easily (e.g., basic cryptographic infrastructure functions). In addition to protection, proactive maintenance techniques provide flexible and dynamic key management. As companies change (through mergers, firing of executives, etc.) and governments change (through elections, appointments, etc.), trust re-lations, and thus shareholders, must change. Proactive maintenance techniques allow for easy enrollment and disenrollment of shareholders.

A number of very useful cryptographic mechanisms have been efficiently "proactivized", such as pseudorandomness and secret sharing [OY91, CH94, HJKY95]. More recently, [HJJKY96] developed proactive public-key schemes for keys with publicly known key-domain (essentially, those based on the Discrete Logarithm problem over groups of known order [DF89, GJKR]). Our result and [HJJKY96] both extend the notion of *threshold cryptosystems* by incorporating mechanisms to protect against a mobile adversary.

The previous proactivization techniques do not seem to be sufficient to con-struct an efficient proactively-secure RSA public-key system [RSA78]. One of the problems with distributing power to perform a keyed RSA has been how to distribute the shares without revealing $\phi(N)$ (knowledge of which implies breaking the key). We cannot store the secret distributively inside a distributed circuit state (inefficient techniques of circuit evaluation as was done proactively in [OY91]) since this is inherently inefficient (not allowing circuit computation embedding in the communication is a major distinction between inefficient and efficient protocols [FY93]). Previous efficient proactivization techniques require providing the order of the share domain to the shareholders, hence these re-sults are not useful for RSA. To resolve this problem our result generates shares over the integers. Even then using previous techniques would increase the size of the shares each time servers perform a share re-randomization to self-secure themselves, whereas we use small uniformly bounded ($O(\log N)$) share sizes.

Figure 1 depicts the development of the distribution of the RSA private key to enhance its security. It is presented in strict order of increasing security and availability. Note that our "proactive" solution is robust. That is, the RSA prim-itive is available and efficiently computable in the presence of adversarial share-holders. Figure 2 depicts the "proactivizing" of various cryptographic primitives.

Our contribution is a new way to distribute and maintain the RSA function (and its relatives) so that robust computation is possible at any point assuming a mobile adversary (it is also a new "robust threshold RSA" if one assumes a stationary adversary, but it permits "proactivization").

The primary techniques used in the result: We first employ a combinatorial reduction of r-out-of-r (verifiable) secret sharing (additive threshold scheme) to r-out-of-l (verifiable) secret sharing (for $r < l/(2 + \epsilon)$). This probabilistic con-struction allows for the verifiable distribution of shares of an RSA key by a key

442

[F89]	(l, l)- (additive) shared RSA
[DF91]	heuristic (t, l)-shared RSA scheme
[FD92, DDFY92]	provable (t, l)-shared RSA
[FGY96, GJKR96]	Efficiently Robust (i.e. verifiable) shared RSA
Our result	First "proactivized" (vs. Mobile Adversary) shared RSA

Fig. 1. History of Increased Security of Distributed RSA

[OY91]	mobile adv. and "proactive protocol" model introduced
[CH94]	proactive pseudorandomness
[HJKY95]	proactive secret sharing
[HJJKY96]	proactive public key with public key domain (Disc. Log based)
Our result	proactivized RSA

Fig. 2. Basic results on "proactivization"

generator and also allows the re-randomization of these shares by the servers; it also simplifies the domain over which sharing is done (when compared with [DDFY92]). This construction was originally designed for a specific verifiable secret sharing scheme which is based on the quadratic residue problem modulo Blum integers [AGY95]. (We extend the construction in [AGY95], by observing that their results will hold for more general sets of good and bad servers.) We use a simulatability argument (similar to one that was put forth in the static distribution of RSA [DDFY92]) to show that the distribution of shares is secure. We then employ the idea of witness-based cryptographic program checking [FGY96] which extends Blum's methodology of program result checking [Bl88] to a system where the checker itself is not trusted by the program. We then develop specific techniques that use the RSA properties (being an exponentiation cipher, and having certain algebraic structure) that complete the design.

We prove the security and robustness of the combined system throughout its lifetime, and thus show that RSA is efficiently "proactivizable". We design efficient protocols for arbitrary numbers $k' < k$ that are each a constant fraction of the number of shareholders. We show that k uncorrupted shareholders may compute RSA signatures efficiently in the presence of corrupted shareholders. We show that k' shareholders can not learn any information about the RSA key.

The protocol is geared towards practical adaptations where a small (constant) numbers of servers is expected. In this case the probabilistic construction of server assignment in [AGY95] we can replaced by a specific assignment based on combinatorial designs.

Organization: In Section 2 we present the model and our definitions of robustness (correctness) and security in the proactive RSA model. Section 3 describes the system and its protocols. Section 4 presents the proof of robustness, and Section 5 presents the proof of security.

2 Model and Definitions

Definition 1. Let h be the security parameter. Let key generator GE define a family of RSA functions to be $(e, d, N) \leftarrow GE(1^h)$ such that N is a composite number $N = P*Q$ where P, Q are prime numbers of $h/2$ bits each. The exponent e and modulus N are made public while $d \equiv e^{-1} \bmod \lambda(N)$ is kept private.[5]

The **RSA encryption function** is public, defined for each message $M \in Z_N$ as: $C = C(M) \equiv M^e \bmod N$. The **RSA decryption function** (also called signature function) is the inverse: $M = C^d \bmod N$. It can be performed by the owner of the private key d. The security of RSA is given in Definition 11.

Our system consists of l servers $\{s_1, \ldots, s_l\}$ and a function f_x for some key x. The system is synchronized and there are two types of time periods repeated in sequence: an *update* period and an *operational* period.

We say an adversary is (k', k, l)-**restricted** if it can corrupt at most $\min\{l - k, k'\}$ servers, and view the memory of at most k' servers (including those that it corrupts) during any time period. Our system is designed such that the following properties hold:

- During an operational period, any k uncorrupted servers can efficiently compute $f_x(\alpha)$, for any α, without revealing anything about f_x other than $f_x(\alpha)$.
- The function f_x is secure against any mobile adversary that is (k', k, l)-restricted. (We assume that the adversary is computationally bounded and therefore cannot break any of the underlying cryptographic primitives used.)

Hence our system is a robust function sharing system, in which the security of the function is maintained over many function applications (as opposed to secret sharing, which is a one-time reveal operation). Moreover, our system provides security against a mobile adversary that may have access to all of the servers throughout the lifetime of the system (albeit, no more than k' simultaneously in any period). Next, we discuss informally some of the issues and assumptions in our model.

The communication model: The communication model presented here is similar to [HJKY95]. The l servers communicate via an authenticated bulletin board [CF85] in a synchronized manner. The board is accessible by a Gateway (an efficient combining function which produces the correct final result) that can be assumed to be insecure. We assume that the adversary cannot jam communication. The board assumption models an underlying basic communication protocol (authenticated broadcasts) and allows us to disregard the low-level technical details.

---

[5] $\lambda(N) = \operatorname{lcm}(P - 1, Q - 1)$ is the smallest integer such that any element in Z_N^* raised by $\lambda(N)$ is the identity element. RSA is typically defined using $\phi(N)$, the number of elements in Z_N^*, but $\lambda(N)$ can be used instead. Knowing a value which is a multiple of $\lambda(N)$ implies breaking the system.

Time periods: Time is divided into *time periods* which are determined by the common global clock (e.g., a day, a week, etc.). There are two types of time periods repeated in sequence: an **update period** (odd times) and an **operational period** (even times). During the update period the servers engage in an interactive *update protocol*. At the end of an update period the servers hold new shares (which are used during the following operational period). We consider a server that is corrupted during an update phase as being corrupted during both its adjacent periods.

Definitions of the subprotocols discussed above will be given in Section 3. We next define what it means for the system to be robust.

Definition 2. (Robustness) Let h be the security parameter. Let key generator GE define a family of RSA functions (i.e., $(e, d, N) \leftarrow GE(1^h)$ be an RSA instance with security parameter h). A system $S(e, d, N)$ is a (k', k, l)-robust proactive RSA system if it contains probabilistic polynomial-time protocols for the following tasks: initial (typically centralized) *share distribution* (to l servers), *RSA function application*, *share renewal* (odd steps), *lost share detection* and *lost share recovery* (even steps); such that, for any probabilistic polynomial-time (k', k, l)-restricted adversary $\mathcal{A}$, for any polynomial-size history H (described below) and for any polynomial poly($\cdot$),

- (correctness) with probability greater than $1 - \frac{1}{\text{poly}(h)}$, for any operational round $2t$ (where, by definition, there are at least k uncorrupted servers in each), and for any $\alpha \in [0, N]$ (to be added to L), S can compute $\alpha^d \bmod N$ using the RSA function application protocol.
- (efficiency) the computational effort is polynomial.

Next we define what does it mean for the system to be secure.

An adversary can be either "stationary" or "mobile" and we consider the mobile case here. It can be "passive" or "arbitrary" (malicious) and we consider the later. Finally it can be oblivious (with a fixed corruption strategy) adaptive (where corruptions are based solely on previous function outputs, e.g. message/signature pairs) or communication-sensitive (i.e. fully adaptive, where corruptions may be dependent on ciphertexts exchanged, check shares, and partial function evaluations). We consider the adaptive adversary. We assume that the adversary collects everything from the public channel and stores all information gained in its view during its attack on the system. It is (k', k, l)-restricted, meaning that it can corrupt at most $\min\{l - k, k'\}$ servers, and view the memory of at most k' servers (including those that it corrupts) during any time period. The adversary collects *history* H consisting of (1) a list L of message/signature/time-period tuples ordered according to time, and (2) a sequence of corruptions and releases which remembers which server is under control at which time interval based on events in L. When the adversary no longer controls a server, it is "removed" by an underlying system management (that server is started from scratch and "rebooted" by the other servers). Given this adversarial behavior we define:

Definition 3. (Security) Let h be the security parameter. Let key generator GE define a family of RSA functions (i.e., $(e, d, N) \leftarrow GE(1^h)$ be an RSA instance with security parameter h). A system $S(e, d, N)$ is a (k', k, l)-robust secure proactive RSA system if for any probabilistic polynomial-time (k', k, l)-restricted adversary $\mathcal{A}$, for any polynomial size history H (described above) and for any polynomial poly$(\cdot)$: $\Pr[u^e \equiv w \bmod N : (e, d, N) \leftarrow GE(1^h); w \in_R \{0, 1\}^h; u \leftarrow A(1^h, w, \text{view}_{\mathcal{A},H}^{S(e,d,N)})] < \frac{1}{\text{poly}(h)}$.

3 The Protocol

3.1 Initialization Protocol

Family and Committee Assignments. We first distribute shares in multiple r-out-of-r secret sharing protocols. This technique is essentially from [AGY95]. The assignment of families and committees can be done by the dealer (but can also be done by the servers). Let $S = \{s_1, \ldots, s_l\}$ be the set of servers and $\mathcal{F} = \{F_1, \ldots, F_m\}$ be the set of families, where each $F_i = \{C_{i,1}, \ldots, C_{i,r}\}$ is a set of committees of servers. Each committee is of size c. Denote $I = \{1, \ldots, m\}$ and $J = \{1, \ldots, r\}$ the indices of families and committees, respectively. The parameters m, r, and c are chosen such that the result will be a (σ, τ)-terrific assignment, that is, one that obeys the following properties **for any** set of "bad" servers $B \subseteq S$ with $|B| \leq k' \leq l\sigma$ and any set of "good" servers $E \subseteq S$ with $|E| \geq k \geq l\tau$:

1. For all $i \in I$, there exists a $j \in J$ such that $B \cap C_{i,j} = \emptyset$. (For each family there is one committee with no bad servers which we call an *excellent* committee.)
2. For at least 90 percent of $i \in I$, for all j, $E \cap C_{i,j} \neq \emptyset$. (In 90 percent of the families, all committees have at least one good server. We call a family F_i with this property a *good* family.)

Given l, q, p, and security parameter $h \geq \max\{2l + 2, 100\}$, we will set $c = \lceil\{2\log h/\log(\frac{1-\sigma}{1-\tau})\}\rceil$, $r = (1 - \tau)^{-c}/h$, and $m = 10h$.

Lemma 4. *A randomly chosen assignment is (σ, τ)-terrific with overwhelming probability.*

We can set the probability of obtaining a non-(σ, τ)-terrific assignment to be smaller than that of breaking the RSA function given the security parameter.

Once the assignment is set, the servers run a protocol to generate and distribute public/private key pairs for secure communication amongst themselves. That is, these keys are used for secure (probabilistic) encryption ([GM84, L96]) which emulates a private channel between the sender of a message and the holder(s) of the private key. The protocol to perform the generation and distribution of keys is similar to the one in Section 3.3, except that the messages are authenticated with a renewable token (which is, e.g., given to the server by a trusted system administrator on booting up). As a result we can see that:

Lemma 5 [AGY95]. *The preceding protocol gives the servers in a (σ, τ)-terrific assignment a public/private key pair for each committee, and further: excellent committees have secure keys.*

Notation: For each $(i,j) \in I \times J$, $\text{ENC}_{i,j}(\alpha)$ will denote an encryption of α using the public key of $C_{i,j}$. For all $s \in S$, $\text{ENC}_s(\alpha)$ will denote a probabilistic encryption of α using the public key of server s. Remember, in our model the adversary is computationally bounded and thus it cannot get more than a negligible advantage in computing any function of α by seeing its encryption.

Distributing the secret Let us review the set-up protocol.

1. The dealer generates p, q, e, d, as in RSA: $N = pq$ and $ed \equiv 1 \mod \lambda(N)$.
2. The dealer generates[6] $g \in_R [2, N-2]$ and broadcasts
   [DISTRIBUTE.1, $N, e, g, g^d \mod N$].
3. For each $(i,j) \in I \times J \setminus \{r\}$, the dealer generates $a_{i,j}^0 \in_R [-2Nm^3cft^2, 2Nm^3cft^2]$. Then it sets $a_{i,r}^0 = d - \sum_{j \in J \setminus \{r\}} a_{i,j}^0$
4. For each $i \in I$ and $j \in J$, the dealer sets $\epsilon_{i,j} \equiv g^{a_{i,j}^0} \mod N$
5. The dealer broadcasts [DISTRIBUTE.2, $\{\epsilon_{i,j}\}_{i \in I, j \in J}$, $\{\text{ENC}_{i,j}(a_{i,j}^0)\}_{i \in I, j \in J}$].
6. Every server checks for all $i \in I$ that $\prod_{j \in J} g^{a_{i,j}^0} \equiv g^d \mod N$ and each server in $C_{i,j}$ checks that $\epsilon_{i,j} \equiv g^{a_{i,j}^0} \mod N$.
7. For each $(i,j) \in I \times J$, every server sets $b_{i,j}^0 = \epsilon_{i,j}$.

Rather than relying on a centralized dealer, we can employ a new result by [BF97]. It enables the generation of a distributed RSA key (N, e, d) (step 1). Given the distributed d we can distributively perform steps 2-7 (with many random generators).

3.2 Operational period (for round $2t$)

This is the protocol to be followed when the gateway obtains a message M to be signed in round $2t$. This protocol follows the one in [FGY96]. We use the fact that since $d = \sum_{j \in J} a_{i,j}^t$ then $M^d \equiv \prod_{j \in J} M^{a_{i,j}^t} \mod N$. We also need to verify correctness of the results using a witness.

1. The gateway broadcasts [SIGN.1, M].
2. For all $(i,j) \in I \times J$, each server $s \in C_{i,j}$ computes $r_{i,j} \equiv M^{a_{i,j}^{2t}} \mod N$ and broadcasts the message [SIGN.2, $s, i, j, M, r_{i,j}$].
3. For all $(i,j) \in I \times J$, each server $s' \in C_{i,j}$ checks each message [SIGN.2, s, i, j, $M, r_{i,j}$]. If M is not the same message broadcast by the gateway, then s'

---

[6] We assume here that the order of g is maximal (i.e., $\lambda(N)$). In practice, a trusted dealer will know the factorization of $P-1$ and $Q-1$ and then be able to generate such a g with overwhelming probability.

disregards the message, else if $r_{i,j} \not\equiv M^{a_{i,j}^{2t}} \bmod N$, then s' broadcasts the challenge[7] [SIGN.CHALLENGE, $s', i, j, a_{i,j}^{2t}$].

4. All servers verify all challenges (by checking if $b_{i,j}^{2t} \equiv g^{a_{i,j}^{2t}} \bmod N$) and inform the system management of any bad servers (i.e., those servers that sent a message with $r_{i,j} \not\equiv M^{a_{i,j}^{2t}} \bmod N$).

5. For some good family i, the gateway computes $\prod_{j \in J} r_{i,j} \equiv M^d \bmod N$. There is a vast majority of good committees that will give this value.

3.3 Update period (for round $2t + 1$)

So far the solution is for a stationary adversary; now we need to update and recover (needed against a mobile adversary). In the update period the public and private keys of the servers are updated, lost shares are detected and shares are updated. This is the self-maintenance portion of the proactive protocol.

Key Renewal: The public/private key pairs of each server are simply renewed as follows. Each shareholder chooses a new public/private key pair and broadcasts the new public key signed with the old private key. For each committee, the least lexicographically ordered member chooses a new committee public/private key pair and broadcasts the new public key. The server also sends the new private key, encrypted, to each other member of the committee.

Lost Share Detection:

1. Every server s sends out [LOSS.DETECT, $s, \{b_{i,j}^{2t}\}_{i \in I, j \in J}$].
2. Each server decides the correct shares by majority, and informs the system management.

Share Renewal/Lost Share Recovery: We have one protocol that handles share renewal and lost share recovery. (For efficiency, one could streamline this protocol, or separate the protocols.) Note that possibly ten percent of the families have committees that contained all bad servers who erased those committees' shares. Those families would not be able to reconstruct the secret d, and thus all the shares in those families are useless. In our protocol, these useless shares will be replaced by shares of shares from a good family. Actually, each family's shares will be replaced by shares of shares from a good family, and thus all shares will be renewed. To create the new shares for a family $F_{i'}$, every family sends shares of its shares to the committees in $F_{i'}$. $F_{i'}$ takes the shares of shares of some family (which it verifies to be valid) and creates its new shares by summing these shares of shares in each committee. This type of share recovery is unlike the share recovery protocols in previous proactive schemes. Now we give the protocol:

---

[7] If the server is uncomfortable providing $a_{i,j}^{2t}$ in this message, [FGY96] could be used to prove knowledge of $a_{i,j}^{2t}$ from $g^{a_{i,j}^{2t}}$ and $M^{a_{i,j}^{2t}}$. However, in our model, revealing $a_{i,j}^{2t}$ does not lessen the security.

1. Let f be a function that is super-polynomial in h. The shareholders will randomize their shares over intervals proportional to f, N, and other terms so as to maintain statistical uncertainty of the value of d. For all $(i, j, i') \in I \times J \times I$, each server s in $C_{i,j}$ does the following: s chooses $w_{s,i,j,i',j'} \in_R [-2Nm^3cft^2, 2Nm^3cft^2]$ for $j' \in J \setminus \{r\}$ and sets $c^{2t}_{s,i,j,i',j'} = w_{s,i,j,i',j'}$ for $j' \in J \setminus \{r\}$. Then s sets $c^{2t}_{s,i,j,i',r} = a^{2t}_{i,j} - \sum_{j' \in J \setminus \{r\}} c^{2t}_{s,i,j,i',j'}$. Then for all $j' \in J$, s computes $e_{s,i,j,i',j'} = \mathrm{ENC}_{i',j'}[c^{2t}_{s,i,j,i',j'}]$ and $\epsilon_{s,i,j,i',j'} = g^{c^{2t}_{s,i,j,i',j'}} \bmod N$.

2. For all $(i, j) \in I \times J$, each server s in $C_{i,j}$ broadcasts
   $$[\mathrm{RECOVER.1}, s, i, j, i', \{\epsilon_{s,i,j,i',j'}\}_{(i',j') \in I \times J}, \{e_{s,i,j,i',j'}\}_{(i',j') \in I \times J}].$$

3. Every server verifies, for all $(i, j, i') \in I \times J \times I$ and all $s \in C_{i,j}$, that $\prod_{j' \in J} \epsilon_{s,i,j,i',j'} = b^{2t}_{i,j} \bmod N$, and informs the system management if it doesn't hold for some s. From this point on, we only deal with messages from those s where it does hold.

4. For all $(i', j') \in I \times J$, if $s \in C_{i',j'}$, decrypt shares to $C_{i',j'}$ and verify $g^{c^{2t}_{s',i,j,i',j'}} \equiv \epsilon_{s',i,j,i',j'} \bmod N$ for all s'. For all $(i', j') \in I \times J \setminus \{r\}$, if $s \in C_{i',j'}$, also verify $|c^{2t}_{s',i,j,i',j'}| \leq 2Nm^3cft^2$ for all s'.

5. If server s finds that verification fails for a message from server s', s broadcasts
   $[\mathrm{RECOVER.ACCUSE}, s, i, j, i', j', s']$, to which s' responds by broadcasting
   $[\mathrm{RECOVER.DEFEND}, s', i, j, i', j', c^{2t}_{s',i,j,i',j'}]$.

6. All servers check all accusations and inform the system management of any bad servers (i.e., those that defended with an invalid value of $c^{2t}_{s',i,j,i',j'}$). Again, from this point on, we only deal with messages from the good servers.

7. If $s \in C_{i',j'}$, using the shares of the lexicographically first family F_i with shares that passed verification, using the lexicographically first servers in each committee in that family with shares that passed verification (call them $s_{i,j}$), compute $a^{2t+2}_{i',j'} = \sum_{j \in J} c^{2t}_{s_{i,j},i,j,i',j'}$, and $b^{2t+2}_{i',j'} \equiv \prod_{j \in J} \epsilon_{s_{i,j},i,j,i',j'} \bmod N$.

8. Everything is erased except $a^{2t+2}_{i,j}$ and $b^{2t+2}_{i,j}$ for all $(i, j) \in I \times J$.

4 Proof of Robustness

We will show that the proactive RSA system from Section 3 is robust as defined. It will be implied by two conditions: correctness (of the function representation), and verifiability (of correctness of evaluations), throughout. We then will show that the shares remain small throughout the protocol.

Theorem 6. *The proactive RSA system above is robust against any (k', k, l)-restricted adversary $\mathcal{A}$.*

Proof. We say the system is *correct at time* $2t$ when $d \equiv \sum_{j \in J} a^{2t}_{i,j} \bmod \lambda(N)$ for all good families F_i. (We note that the majority agreement on $b^{2t}_{i,j}$ implies that all good servers in a committee $C_{i,j}$ will either agree on one share $a_{i,j}^{2t}$ or agree they have none.)

At the time of function evaluation, the gateway G is given the outputs of all the committees in all the families, the opened shares at the committees that were challenged by having contradictory outputs and the public witnesses $\{b_{i,j}^{2t}\}_{(i,j)\in I \times J}$. We say that the system is *verifiable at time 2t* if G can pick the correct shares of all good committees. Verifiability implies that the gateway G can identify a good family and compute $M^d \equiv \prod_{j \in J} M^{a_{i,j}^{2t}} \bmod N$ for a good family F_i; this implies efficiency. Thus, correctness and verifiability imply robustness. We omit the inductive proof that the system maintains correctness and verifiability throughout.

Next we deal with boundedness of the shares throughout the protocol. The following lemmas show that the sizes of shares are bounded (by a polynomial in h) at good committees. The initial shares (sent by the trusted dealer) are in the range $[-r2Nm^3cf, r2Nm^3cf]$, and thus are of size at most $2h+\log(m^3cf)+\log r$. (Note that this could be verified by the servers in the distribution protocol. Also note that r is bounded by a polynomial in h.)

Lemma 7. *For any $t > 0$, and any good committee $C_{i',j'}$ with $j' \in J \setminus \{r\}$, $-2Nrm^3cft^2 \leq a_{i',j'}^{2t} \leq 2Nrm^3cft^2$.*

The proof follows from the verification in step 4 of the Share Renewal/Lost Share Recovery protocol. Robustness assures that modulo the (maximal) order of g, we maintain a correct representation of the function. Next we show that if the adversary can violate a certain bound on the representation size at some step, then that adversary knows a multiple of $\lambda(N)$, and thus has broken the RSA function:

Lemma 8. *For any $t > 0$, and any good family $F_{i'}$, if $\sum_{j' \in J} a_{i',j'}^{2t} \neq d$, then the adversary can break the underlying RSA function of the system.*

Thus we conclude that:

Corollary 9. *Assuming the system has not been broken (to be shown next), the size of shares is bounded by $h + \log(r^2m^3cft^2)$.*

The complexity of the scheme is mr times that of a single RSA.

Small scale implementations are much more efficient. For example 2-out-of-3 system can be done by 2-out-of-2 sharings for the pairs: (1,2), (1,3) and (2,3). A 3-out-of-4 can be done by two 3-out-of-3 sharings for the multisets: (1 and 2,3,4) and (1,2, 3 and 4). These designs can be tuned to achieve a desirable number of messages during update and signature generation based on performance. Also in practice, the combiner can first compute a result based on a single family and check for its correctness (by applying the inverse public function). In the typical case, this first result will be correct, thus saving a great deal of computation.

5 Proof of Security

We claim that:

Theorem 10. *The proactive RSA system above is secure against any (k', k, l)-restricted adversary $\mathcal{A}$.*

We prove the security of our system (in Subsection 5.2) by constructing a simulator (in Subsection 5.1) which reduces the security of the RSA function to the security of our scheme against the mobile adversary.

Definition 11. The RSA security assumption (with respect to a history): Let h be the security parameter. Let key generator GE define a family of RSA functions (i.e., $(e, d, N) \leftarrow GE(1^h)$ be an RSA instance with security parameter h). For any probabilistic polynomial-time adversary $\mathcal{A}$, given a history H containing polynomial-size list L of messages and their signatures, for any polynomial poly($\cdot$): $\Pr[u^e \equiv w \bmod N : (e, d, N) \leftarrow GE(1^h); w \in_R \{0, 1\}^h; u \leftarrow A(1^h, w, H)] < \frac{1}{\text{poly}(h)}$.

Remark: The definition above assumed history H, namely with respect to known cleartext-ciphertext pairs. When H is empty, this is the traditional definition of security. For secure uses of RSA (probabilistic encryption or signature schemes) the definition applies for properly chosen H's.

5.1 The Simulator

Here, for the sake of proof of security, we construct SIM to simulate the view of $\mathcal{A}$ with history H in the system we construct in Section 3. For simulation purposes, SIM will assume that all servers are controlled by $\mathcal{A}$ for the maximum time that the security requirements assume they are corrupted, i.e., if $\mathcal{A}$ controls server s sometime during round $2t + 1$, then it also controls s all during rounds $2t, 2t + 1, 2t + 2$, and if $\mathcal{A}$ controls server s sometime during round $2t$, then it also controls s all during round $2t$.

Important Notation: For each $i \in I$ and each round t, SIM sets j_i^t such that C_{i,j_i^t} is a committee which contains no servers with memory viewed by $\mathcal{A}$ during round t. (See property 1 from Section 3.1).

It is easiest to describe the simulation as follows. First assume $L = \langle (M_1, M_1^d), \ldots, (M_v, M_v^d) \rangle$ is the list of (message, signature) pairs obtained by $\mathcal{A}$ from a previous "execution of RSA". Now SIM creates the initial shares, of course, independent of secret key d. The difficulty is that consistency among all broadcasts is needed to satisfy the various test performed throughout the protocol. For all $i \in I$, for all $j \in J \setminus j_i^0$, SIM generates $a_{i,j}^{0,sim} \in_R [-2Nm^3cft^2, 2Nm^3cft^2]$ and computes $b_{i,j}^{0,sim} \equiv \epsilon_{i,j} \equiv g^{a_{i,j}^{0,sim}} \bmod N$. For any $C_{i,j_i^{2t}}$, SIM can not compute a valid value for $a_{i,j_i^{2t}}^{0,sim}$ (i.e., one that will make $\sum_{j \in J} a_{i,j}^{0,sim} = d$) or else it could compute d on its own. However, SIM needs to compute a valid value for $b_{i,j_i^0}^{0,sim}$, to pass the verification step. The simulator can do this as follows. For all $i \in I$, SIM generates $b_{i,j_i^0}^{0,sim} \equiv \epsilon_{i,j_i^0} = g^d / \prod_{j \in J, j \neq j_i^0} g_{a_{i,j}^{0,sim}}$. The rest of the protocol continues as specified.

Simulating the signing phase uses a similar approach. Again, we have the problem that SIM does not have the shares for $C_{i,j_i^{2t}}$. For each $i \in I$, for each $j \in J \setminus j_i$, each $s \in C_{i,j}$ computes $r_{i,j} \equiv M_v^{a_{i,j}^{2t,sim}} \mod N$. Now, For each $i \in I$, each $s \in C_{i,j_i^{2t}}$ computes (with the help of SIM) $r_{i,j_i^{2t}} \equiv M_v^d / \prod_{j \in J, j \neq j_i^{2t}} r_{i,j} \mod N$. The rest of the protocol continues as specified.

Share renewal and lost share recovery are more complex, but are based on the same basic idea. Key renewal is performed exactly as in the real protocol and is not discussed further.

Share Distribution Simulation

The Share Distribution Simulation is similar to the Share Distributions, except that firstly the pair (g, g^d) is produced by choosing $\rho \in_R [0, N]$ and letting $g = \rho^e$ (we can repeat the simulation with poly-log number of ρ's, one of which will surely give a maximal order element), and, secondly, that the shares are produced as follows.

- For each $i \in I$, for each $j \in J \setminus \{r\}$, SIM generates $a_{i,j}^{0,sim} \in_R [-2Nm^3cft^2, 2Nm^3cft^2]$. Then it sets $a_{i,r}^{0,sim} = -\sum_{j \in J \setminus \{r\}} a_{i,j}^{0,sim}$.
- For all $i \in I$, for all $j \in J \setminus j_i^0$, SIM computes $b_{i,j}^{0,sim} \equiv \epsilon_{i,j} \equiv g^{a_{i,j}^{0,sim}} \mod N$. For all $i \in I$, SIM generates $b_{i,j_i^0}^{0,sim} \equiv \epsilon_{i,j_i^0} \equiv g^d / \prod_{j \in J, j \neq j_i^0} g^{a_{i,j}^{0,sim}} \mod N$.

Signature Simulation

Simulating a signature of M during round $2t$ is done as in the Signature protocol with the following exception: Let $j_i = j_i^{2t}$.

- In step 2, for each $i \in I$, each $s \in C_{i,j_i}$ computes (with SIM's help) $r_{i,j_i} \equiv M^d / \prod_{j \in J, j \neq j_i} r_{i,j} \mod N$.

Lost Share Detection Simulation

The servers perform lost share detection exactly as in the real protocol.

Share Renewal/Lost Share Recovery Simulation

Simulating the Share Renewal/Lost Share Recover Protocol is done as in the original protocol except as follows. Assume it is the start of round $2t + 1$. Say $F_{i'}$ is the family whose shares must be recovered. For all $i \in I$, let $j_i = j_i^{2t}$ and $j_i' = j_i^{2t+2}$. (The following protocol assumes for each $i \in I \setminus \{i'\}$, $j_i, j_{i'} \neq r$. The other cases are similar.)

- In step 1, for all $(i, i') \in I \times I$, every $s \in C_{i,j_i}$ computes $\epsilon_{s,i,j_i,i',j_{i'}'} \equiv b_{i,j_i}^{2t,sim} / \prod_{j' \in J \setminus \{j_{i'}'\}} g^{c_{s,i,j_i,i',j'}^{2t,sim}} \mod N$.
- In step 5, for all $(i, i') \in I \times I$, no server $s \in C_{i',j_{i'}'}$ broadcasts an accusation of any server $s' \in C_{i,j_i}$.

5.2 Security proofs

First, we reduce the security to a simulator of certain properties, then we prove that the simulator constructed above indeed possesses these properties.

Main Security Lemma. The following is shown by reduction.

Lemma 12. *Let h be the security parameter. Let G be a family of RSA functions with security parameter h. Let $S(e, d, N)$ be a system that satisfies the robustness property of a proactive RSA system. If, for any probabilistic polynomial-time (k', k, l)-restricted adversary $\mathcal{A}$, and for any history H containing a polynomial-size list L of (message,signature) pairs and a polynomial-size sequence of corruptions and signature requests, there exists a probabilistic polynomial-time simulator $\mathrm{simu}(e, N, \mathcal{A}, H)$ such that $\mathrm{view}_{\mathcal{A}, H}^{\mathrm{simu}(e, N, \mathcal{A}, H)}$ is indistinguishable from $\mathrm{view}_{\mathcal{A}, H}^{\mathrm{real}(e, d, N)}$ then $S(e, d, N)$ is a (k', k, l)-secure robust proactive RSA system.*

The simulator and indistinguishability proofs. We reduce the problem to proving that the unencrypted parts of the real and simulated views are statistically indistinguishable. We use semantically secure probabilistic encryption [GM84] which the proof of the following relies on:

Lemma 13. *If the adversary $\mathcal{A}$ is restricted to probabilistic polynomial time and if the servers are using semantically secure (probabilistic) encryption (in which distinguishing between encryptions of two given messages is difficult), then:*

If views from executions of the real and simulated protocols assuming secure communication channels are statistically indistinguishable, then views from executions of the real and simulated protocols using semantically secure encryption are polynomial-time indistinguishable.

Proof omitted due to space limitation.

Lemma 14. *Assuming secure channels and $0 < \sigma < \tau < 1$, for any probabilistic $(l\sigma, l\tau, l)$-restricted adversary $\mathcal{A}$, $\mathrm{view}_{\mathcal{A}}^{\mathrm{simu}(e, N, \mathcal{A}, C)}$ is statistically indistinguishable from $\mathrm{view}_{\mathcal{A}}^{\mathrm{real}(e, d, N, C)}$.*

Proof. To simplify the proof we make the following assumptions (without loss of overall correctness):

1. we assume that $\mathcal{A}$'s random bits are fixed. We will show that, for every assignment of $\mathcal{A}$'s random bits, the two views are indistinguishable.

2. we assume, that for all i and even times $2t$, $\mathcal{A}$ sees all shares except $a_{i, j_i^{2t}}^{2t}$. j_i^{2t} is discussed in the simulation. We let $Bad_i^{2t} = J \setminus \{j_i^{2t}\}$ be the set of indices of family F_i's shares the adversary knows at round $2t$.

3. for all i, t, i', and j', we assume $\mathcal{A}$ sees all values of $c_{s, i, j, i', j'}^{2t}$ except for $\{c_{s, i, j_i^{2t}, i', j_{i'}^{2t+2}}^{2t}\}$

The view of $\mathcal{A}$ also consists of all other messages that are broadcast in each round. However, these will not include the encryptions, since we are assuming secure channels for those messages.

For random variables Y and Y' drawn from distributions $\mathcal{D}$ and $\mathcal{D}'$, respectively, we define $\textit{diff}(Y, Y') = \sum_{v \in \mathcal{D} \cup \mathcal{D}'} |\Pr[Y = v] - \Pr[Y' = v]|$.

To prove lemma 14, we need only show that

$$\textit{diff}(\text{view}_{\mathcal{A}(L)}^{\text{simu}(e,N,\mathcal{A},L)}, \text{view}_{\mathcal{A}(L)}^{\text{real}(e,d,N,L)})$$

is small.

Let $X = \prod_{i,j \neq j_i^0} a_{i,j}^0 \times \prod_{t,i,j,s \in C_{i,j}, i',j':(j,j') \neq (j_i^{2t}, j_{i'}^{2t+2})} c_{s,i,j,i',j'}^{2t}$, and let $X^{sim} = \prod_{i,j \neq j_i^0} a_{i,j}^{0,sim} \times \prod_{t,i,j,s \in C_{i,j}, i',j':(j,j') \neq (j_i^{2t}, j_{i'}^{2t+2})} c_{s,i,j,i',j'}^{2t,sim}$. Here multiplication denotes cross products.

Lemma 15. $\textit{diff}(\text{view}_{\mathcal{A}(L)}^{\text{simu}(e,N,\mathcal{A},L)}, \text{view}_{\mathcal{A}(L)}^{\text{real}(e,d,N,L)}) \leq \textit{diff}(X, X^{sim})$.

Proof omitted due to space limitation.

The following lemma completes the proof by showing that $\textit{diff}(X, X^{sim})$ is small (again the proof is omitted in this abstract).

Lemma 16.

$$\textit{diff}(X, X^{sim}) \leq \frac{dm^3 c}{Nm^3 cft^2}$$

This completes the sketch of the main steps of the proof of security of the system.

Acknowledgments We would like to thank Markus Jakobsson, Stas Jarecki, Hugo Krawczyk, and Tal Rabin for helpful discussions on proactive public key and on proactive RSA. We thank Kevin McCurley for many helpful discussions on number theory. Finally, thanks to Nancy Irwin for her implementation which demonstrated the workings of the protocols presented in this paper.

References

[AGY95] N. Alon, Z. Galil and M. Yung, *Dynamic-resharing Verifiable Secret Sharing*, European Symposium on Algorithms (ESA) 95, Springer-Verlag LNCS.

[B79] G.R. Blakley, *Safeguarding Cryptographic Keys*, AFIPS Con. Proc (v. 48), 1979, pp 313–317.

[Bl88] M. Blum, *Designing programs to check their work*, ICSI Technical report TR-88-009.

[B88] C. Boyd, *Digital Multisignatures*, IMA Conference on Cryptography and Coding, Claredon Press, 241–246, (Eds. H. Baker and F. Piper), 1986.

[BF97]   D. Boneh and M. Franklin, *Efficient Generation of Shared RSA Keys*, Crypto 97 (these proceedings).

[CH94]   R. Canetti and A. Herzberg, *Maintaining Security in the presence of transient faults*, Crypto '94, Springer-Verlag, 1994.

[CF85]   J. Cohen and M. Fischer, *A robust and verifiable cryptographically secure election scheme*, FOCS 85.

[DDFY92]   A. De Santis, Y. Desmedt, Y. Frankel and M. Yung, *How to Share a Function Securely*, ACM STOC 94. (First version May 92).

[DF89]   Y. Desmedt and Y. Frankel, *Threshold cryptosystems*, Crypto '89 Springer-Verlag, 1990.

[DF91]   Y. Desmedt and Y. Frankel, *Shared generation of authenticators and signatures*, Crypto 91, Springer-Verlag LNCS 576, 1992, pp. 307–315.

[F87]   P. Feldman, *A Practical Scheme for Non-Interactive Verifiable Secret Sharing*, Proc. of the 28th IEEE FOCS pp. 427-437, 1987

[F89]   Y. Frankel, *A practical protocol for large group oriented networks*, Eurocrypt '89, Springer-Verlarg LNCS 773, pp. 56-61.

[FD92]   Y. Frankel and Y. Desmedt. *Distributed reliable threshold multisignatures*, Tech. Report version TR–92–04–02, Dept. of EE & CS, Univ. of Wisconsin-Milwaukee, April 1992.

[FGY96]   Y. Frankel, P. Gemmell and M. Yung, *Witness-based Cryptographic Program Checking and Robust Function Sharing* Proc. of STOC 1996, pp. 499–508.

[FY93]   M. Franklin and M. Yung, *Secure and Efficient Digital Coin*, ICALP 93, Springer Verlag LNCS.

[GHY]   Z. Galil, S. Haber and M. Yung, *Minimum-Knowledge Interactive Proofs for Decision Problems*, SIAM Journal on Computing, vol. 18, n.4, pp. 711–739. (Previous version in FOCS 85).

[GJKR]   R. Gennaro, S. Jarecki, H. Krawczyk and T. Rabin, *Robust Threshold DSS Signatures*, Eurocrypt 96.

[GJKR96]   R. Gennaro, S. Jarecki, H. Krawczyk and T. Rabin, *Robust and Efficient Sharing of RSA*, Crypto 96.

[GM84]   S. Goldwasser and S. Micali, *Probabilistic Encryption*, J. Comp. Sys. Sci. 28, 1984, pp. 270-299.

[HJKY95]   A. Herzberg, S. Jarecki, H. Krawczyk and M. Yung, *How to Cope with Perpetual Leakage, or: Proactive Secret Sharing*, Crypto 95.

[HJJKY96]   A. Herzberg, M. Jakobsson, S. Jarecki, H. Krawczyk, and M. Yung, *Proactive public key and signature systems*, The 4-th ACM Symp. on Comp. and Comm. Security. April 1997.

[L96]   M. Luby, *Pseudorandomness and its Cryptographic Applications*, Princeton University Press, 1996.

[OY91]   R. Ostrovsky and M. Yung, *How to withstand mobile virus attacks*, ACM Symposium on Principles of Distributed Computing (PODC), 1991, pp. 51-61.

[RSA78]   R. Rivest, A. Shamir and L. Adleman, *A Method for Obtaining Digital Signature and Public Key Cryptosystems*, Comm. of the ACM, 21 (1978), pp. 120-126.

[S79]   A. Shamir. *How to share a secret*, Comm. of the ACM, 22 (1979), pp. 612-613.

Towards Realizing Random Oracles:
Hash Functions That Hide All Partial Information

Ran Canetti

IBM T.J. Watson Research Center. Email: *canetti@watson.ibm.com*

Abstract. The *random oracle model* is a very convenient setting for
designing cryptographic protocols. In this idealized model all parties
have access to a common, public random function, called a *random oracle*. Protocols in this model are often very simple and efficient; also
the analysis is often clearer. However, we do not have a general mechanism for transforming protocols that are secure in the random oracle
model into protocols that are secure in real life. In fact, we do not even
know how to meaningfully *specify* the properties required from such a
mechanism. Instead, it is a common practice to simply replace - often
without mathematical justification - the random oracle with a 'cryptographic hash function' (e.g., MD5 or SHA). Consequently, the resulting
protocols have no meaningful proofs of security.
We propose a research program aimed at rectifying this situation by
means of identifying, and subsequently realizing, the useful properties
of random oracles. As a first step, we introduce a new primitive that
realizes a specific aspect of random oracles. This primitive, called *oracle
hashing*, is a hash function that, like random oracles, 'hides all partial
information on its input'. A salient property of oracle hashing is that it is
probabilistic: different applications to the same input result in different
hash values. Still, we maintain the ability to *verify* whether a given hash
value was generated from a given input. We describe constructions of
oracle hashing, as well as applications where oracle hashing successfully
replaces random oracles.

1 Introduction

Existing collision resistant hash functions, such as MD5 [Ri] and SHA [SHA],
are very useful and popular cryptographic tools. In particular, these functions
(often nicknamed 'cryptographic hash functions') are used in a variety of settings
where far stronger properties than collision resistance are required.

Some of these properties are better understood and can be rigorously formulated (e.g., the use as pseudorandom functions [BCK1], or as message authentication functions [BCK2]). Often, however, these extra properties are not precisely
specified; even worse, it is often unclear whether the attributed properties can
at all be formalized in a meaningful way.

We very roughly sketch two salient such properties. One is 'total secrecy':
it is assumed that if h is a cryptographic hash function then $h(x)$ 'gives no
information on x'. The other is 'unpredictability': it is assumed to be infeasible
to 'find an x such that $x, h(x)$ have some desired property'. This is of course
only a sketch; each application requires different variants.

These uses of MD5, SHA, and other cryptographic hash functions are often justified by saying that 'using cryptographic hash functions is equivalent to using random oracles'. More specifically, the following random oracle paradigm is employed. Assume the security of some protocol (that makes use of a 'cryptographic hash function' h) needs to be proven. Then an idealized model is formulated, where there is a public and *random* function R such that everyone can query R on any value x and be answered with $R(x)$. Next modify the protocol so that each invocation of the hash function h is replaced with a query to R. Finally, it is suggested that if the modified construction (using R) is secure in this idealized model then the original construction (using h) is secure in 'real-life'. (We remark that here the random oracle can be viewed as an 'ideal hash function'. In particular, R satisfies both the 'total secrecy' and the 'unpredictability' properties sketched above, since $R(x)$ is a *random number* totally independent of x.)

However, the fact that a construction is secure in the random oracle model does not provide any concrete assurance in the security of this construction in 'real life'. In particular, there exist natural protocols that are secure if a random oracle is used, but are clearly insecure if the random oracle is replaced by *any deterministic function* (and in particular by any cryptographic hash function). The motivating scenario described below provides a good example. (In view of this criticism we stress that, with all its drawbacks, the random oracle model has proved instrumental in designing very useful protocols and applications, as well as new concepts, e.g. [FS, BDMP, M, BR1, BR2, BR3, PS]).

In this work we make a first step towards rigorously specifying some 'random-oracle-like' properties of hash functions. We concentrate on the 'total secrecy' property sketched above. That is, we propose a new primitive, called oracle hashing, that hides all partial information on its input, while maintaining the desired properties of a hash function.

The rest of the introduction is organized as follows. We first sketch a motivating scenario for the new primitive. Next we describe the new primitive, together with several constructions and applications.

A motivating scenario. Consider the following scenario. (It should be kept in mind that, while providing initial intuition for the properties desired from the new primitive, this scenario is of limited scope. In particular, some properties of the new primitive do not come to play here.) Alice intends to publish a puzzle in the local newspaper. She also wants to attach a short string c that will allow readers that solved the puzzle to verify that they have the right solution, but such that c will 'give away' *no partial information* about the solution, x, to readers who have not solved the puzzle themselves. In other words, Alice wants c to mimic an 'ideal scenario' where the readers can call the editor (as many times as they wish), suggest a solution and be answered only 'right' or 'wrong'.

A crypto-practitioner posed with this problem may say: "what's the big deal? c should be a cryptographic hash (e.g., MD5 or SHA) of the solution. It is easily verifiable, and since the hash is one-way c gives no information on the preimage."

Indeed, this ad-hoc solution may be good enough for some practical purposes. However, when trying to 'pin down', or formalize the requirements from a solution some serious difficulties are encountered. In particular, no known cryptographic primitive is adequate. For instance one-way functions are not good enough, since they only guarantee that the *entire* preimage cannot be computed given the function value. It is perfectly possible that a one-way function 'leaks' partial information, say half of the bits of its input.

Furthermore, *any deterministic function* (even ones that hide all the bits of the input, and even 'cryptographic hash functions') are inadequate here, since they are bound to disclose *some* information on the input: For any deterministic function f, $f(x)$ itself is some information on x. One way hash families [NY1] are inadequate for the same reason: they only guarantee that collisions are hard to find, and may leak partial information on the input.

Similarly, commitment schemes (even non-interactive ones) are inadequate since they require the committer to participate in the de-commit stage, whereas here the newspaper editor does not want to be involved in de-committals. (Also, de-committals by nature reveal the correct solution x, even if the suggested solution is wrong.)

In fact, it seems that the only known way to model such a primitive is via the random oracle model: Given access to a random oracle R, Alice can simply publish $c = R(x)$, where x is the solution to the puzzle. This way, given x it can be easily verified whether $c = R(x)$, and as long as the correct x is not found then $R(x)$, being a totally random string, gives no information on x.

We remark that $R(x)$ does in a way provide assistance in finding x since one can now exhaustively search the domain of solutions until a solution x such that $R(x) = c$ is found. This is, however, the same assistance provided by the newspaper in the 'ideal scenario' described above; thus it is a welcome property of a solution.

The new primitive: Oracle Hashing. The proposed primitive, oracle hashing, is designed to replace the random oracle R in the above scenario, as well as in several others. The idea behind this mechanism is quite simple. Traditionally, one thinks of a hash function as a *deterministic* construct, in the sense that two invocations on the same input will yield the same answer. We diverge from this concept, allowing the hash function, H, to be probabilistic in the sense that different invocations on the same input result in different outputs. That is, $H(x)$ is now a random variable depending on the random choices of H. It is this randomization that allows us to require that $H(x)$ will 'hide all partial information on x'.

Oracle hashing also diverges from the notion of (universal, or even one way) hash families [CW, NY1], since there it is usually the case that a *deterministic* function is randomly chosen 'beforehand', and then fixed for the duration of the application.

But now we may have lost the ability to verify hashes. So we require that there exists a verification algorithm, V, that correctly decides, given x and c, whether c a hash of x. (Using standard deterministic hashing, the verification procedure is simple: apply the hash function to x and check whether the result equals c. Here a different procedure may be required.)

This mechanism is somewhat reminiscent of signature schemes, where H takes the role of the signing algorithm and V takes the role of the signature verification algorithm. It is stressed, however, that here no secret keys are involved and both functions can be invoked by everyone. (Also, here additional properties will be required from the pair H, V.)

It remains to formulate the 'secrecy' requirement. This proves to be a non-trivial task. We want to capture the property that 'the hashed value gives no information on the input'. The natural concept that comes to mind is semantic security (originally used for encryption schemes [GM]): 'whatever can be computed given $H(x)$ can also be computed without it'. But semantic security is unachievable in our scenario, since given $H(x)$ and some value y one must be able to tell whether $x = y$. In particular, if the input x has only a small number of possible values (say 0 or 1) then it is easy to find x from $H(x)$ by running the verification algorithm on all possible inputs.

We thus introduce a new, weaker notion of secrecy, which we call oracle security. This notion essentially means that the only way in which $c = H(x)$ can be used to find information on x is by exhaustively trying different z's and checking if $V(z, c)$ accepts. Very roughly, this can be formulated as follows: Let I_x be the oracle that answers 1 to a query z iff $z = x$; Otherwise it answers 0. Then, "finding information on x given $H(x)$ is equivalent to finding information on x given only access to the oracle I_x". Thus, oracle security is valuable only if there is 'enough uncertainty' about the input, i.e. if no single input is too probable.

We present several equivalent formalizations of oracle security. Furthermore, as in the case of encryption, it is convenient to incorporate in the formalization the notion of 'a-priori information' on the secret value. However here (in contrast with the case of encryption) we don't know whether oracle security without a-priori information is equivalent to oracle security with a-priori information. We elaborate within.

On the constructions. We present a simple oracle hashing scheme based on number-theoretic primitives. Assume a large safe prime p is known. (p is safe if $q = (p-1)/2$ is a prime.) Then, given input x, choose a random element r in Z_p^* and let $H(x) = r^2, r^{2 \cdot h(x)}$, where the calculations are made modulo p, and h is any collision resistant hash function. Verification is straightforward (i.e., to verify whether a pair a, b is a hash of a known message x, check whether $a^{h(x)} \equiv b \pmod{p}$). Here the only requirement from the hash function h is collision resistance. The security of this construction is shown based on strong variants of the Diffie-Hellman assumption. (Different assumptions are needed to show different levels of security.) These assumptions may well be of independent interest.

The above scheme is somewhat costly, since it involves a modular exponentiation. We thus suggest simple constructions based on a cryptographic hash function h. (For instance, let $H(x, r) = r, h(r, h(x))$.) Here we of course make stronger assumptions on h than just collision resistance. We stress however that, in contrast to the 'random oracle heuristic', these are well-defined assumptions.

We remark that constructs similar to the ones described here are implicit in several previous works, sometimes for related purposes (e.g., [F, P, E]). None of these works, however, suggests any primitive similar to the one proposed here. Also, a similar idea is used in the BSD UNIX password file, where a random 'salt' is prepended to a password before encrypting it, and then stored together with the ciphertext.

Applications. A first, immediate application is for scenarios like the 'puzzle in the newspaper' scenario (i.e., whenever one wants to make public a verifiable hash that leaks no information on the hashed value.) Oracle hashing can also be used to replace the use of random oracles in known constructions. We demonstrate this on an encryption function introduced by Bellare and Rogaway [BR1]. This function was proven semantically secure only in the random oracle model described above. (It is suggested in [BR1], as a rule-of-thumb, to replace the random oracle with a cryptographic hash function.) We show that if one replaces oracle hashing for random oracles then the construction becomes secure without resorting to random oracles.

Another application is for content-concealing signatures: Assume that one wants to sign a message m and at the same time make sure that the signature itself hides all partial information on m (from parties who do not already know m). Then, given a message m one can simply sign $H(m)$ instead of signing m. See more details within.

Further research. This work can be viewed as a first step in a research program whose goal is to better understand the random oracle model. This model 'blends' in it several potentially unrelated desired characteristics hash functions, in a way that makes it hard to distinguish which property is being used at each application. Such properties are 'total secrecy' together with several quite different flavors of 'unpredictability'. As demonstrated here, some applications need only some properties and not others. Is it possible to identify additional such properties, and subsequently to realize them without resorting to random oracles?

2   Defining oracle hashing

The definition of oracle hashing consists of three requirements: Completeness and Correctness (that together comprise a validity requirement), together with Secrecy. The first two requirements are fairly standard. Formulating the Secrecy requirement, however, is non-trivial. We present several variants and briefly discuss their relations.

Completeness. This requirement is straightforward: *"Algorithm V will accept (except perhaps with negligible probability) pairs x, c where c is generated by applying H to x."*

Correctness. We would like to require that: *"It is infeasible to cheat V into accepting pairs x, c such that c was not generated by applying H to x."* Formalizing this requirement is somewhat tricky since the fact that H is probabilistic make the statement 'c was not generated as $H(x)$' ambiguous. In particular, this requirement takes different flavors depending on whether the producer of the hash is trusted to use H as specified (in which case one only needs to protect against non-malicious errors) or untrusted (in which case one need to protect against malicious efforts to generate ambiguous hashes). We get around these problems by making the stronger requirement that it is infeasible to find 'collisions', i.e. two different inputs x, y and a hash value c such that V accepts c as a legal hash of both x and y.

Secrecy (oracle security). We want to say that: *"Having $c = H(x)$ gives no information on x, besides the ability to exhaustively search the domain for*

x such that $c = H(x)$." This requirement takes different flavors, depending on which probability distributions on the inputs are considered, and on whether the attacker is assumed to have some a-priori information on x. We start with the case where no a-priori information on x is known. Here we present our chosen formalization, together with two other formalizations. We show that all three are equivalent.[1] We believe that comparing the different formalizations helps understanding the nature of oracle security. In particular, two of the formalizations are reminiscent of the two equivalent formalizations of the security of encryption functions (see [G]).

We first need the following definitions: Say that a function $f : N \rightarrow R$ is negligible if it approaches zero faster than any polynomial (when its input grows to infinity).

Definition 1. Let $\mathcal{X} = \{X_k\}_{k \in N}$ and $\mathcal{Y} = \{Y_k\}_{k \in N}$ be two ensembles of probability distributions. We say that $\mathcal{X}$ and $\mathcal{Y}$ are computationally indistinguishable (and write $\mathcal{X} \overset{\cdot}{\approx} \mathcal{Y}$) if for any polytime distinguisher D the difference $|\text{Prob}(D(x) = 1) - \text{Prob}(D(y) = 1)|$ is a negligible function of k, where x is drawn from X_k and y is drawn from Y_k.

Definition 2. A distribution ensemble $\mathcal{X} = \{X_k\}_{k \in N}$ is well-spread if for any polynomial $p(\cdot)$ and all large enough k, the largest probability of an element in X_k is smaller than $p(k)$ (i.e., $\max_a(X_k(a)) < p(k)$).
(In other words, the max-entropy of distributions in $\mathcal{X}$ must vanish super-logarithmically, see [CG]).

We proceed to the (basic) definition of oracle hashing. Consider a pair of algorithms H, V. Algorithm H, given a security parameter k and input x, chooses a random value in domain R_k and outputs a value c. Algorithm V, given k and input c, outputs a binary value. In the sequel the security parameter, k, is often implicit in our notation.

Definition 3. Say that H, V are an oracle hashing scheme if the following requirements hold.

1. **Completeness:** For all large enough k, for all input x and for $r \in_R R_k$ we have that $\text{Prob}(V(x, H(x, r)) \neq 1)$ is negligible (in k).[2]
2. **Correctness:** For any probabilistic polynomial time adversary $\mathcal{A}$, the probability that $\mathcal{A}$ outputs, on input k, a triplet c, x, y such that $x \neq y$ and $V(x, c) = V(y, c) = 1$ is negligible.[3]

---

[1] Here and for the rest of the discussion we assume non-uniform adversaries. I.e., an adversary is a family of circuits with polynomial size.

[2] $x \in_R D$ means that element x is independently and uniformly chosen from domain D.

[3] Note that in the case that such triplets c, x, y exist, a non-uniform adversary can have a fixed triplet 'wired in' its circuit for each value of k. Thus, it appears to make no sense to require that it is hard to find such triplets. We get around this problem by letting H, V be chosen a-priori from a family of functions, and requiring that any fixed triplet forms a collision only for a small fraction of the functions in the family. See [D].

3. Secrecy: For any poly time adversary $\mathcal{A}$ *with binary output*, and any well-spread distribution $\{X_k\}$:

$$\langle x, \mathcal{A}(H(x,r)) \rangle \overset{c}{\approx} \langle x, \mathcal{A}(H(y,r)) \rangle \tag{1}$$

where $r \in_R R_k$, and x, y are independently drawn from X_k.

Remarks: 1. The Secrecy requirement can be relaxed by taking into account only the uniform distribution on the inputs. We call this variant oracle hashing for random inputs.

2. It appears that limiting $\mathcal{A}$'s output to a binary value is essential for the Secrecy requirement to make sense. In particular, if $\mathcal{A}$ could have arbitrary length output then it could simply output its input, thus making distinguishing between the two sides of (1) easy.

We present two other formalizations of the secrecy requirement (i.e., of oracle security). A somewhat simplified sketch follows.

First is the formalization sketched in the Introduction. We call it Oracle simulatability: Let I_x be the oracle that answers 1 to a query z iff $z = x$; Otherwise it answers 0. Then, "For any algorithm C' that has access to hashes of x, there exists an algorithm C that has access only to I_x, such that for any distribution on the x's, and any predicate P, C' does not predicts $P(x)$ substantially better than C."

Second is a formalization reminiscent of security by indistinguishability of encryption functions. We call it Oracle indistinguishability: *For any distinguisher D there exists a set L of polynomially many inputs, such that for any $x, y \notin L$ we have that D distinguishes between hashes of x and hashes of y only with negligible probability."*

We preferred the formalization of Definition 3 since it naturally supports consideration of only specific distributions on the inputs, and since it extends easily to a reasonable definition for the case where a-priori information on the input is known (see Definition 6).

Theorem 4. *The following requirements are equivalent to the Secrecy requirement of Definition 3:*

3a. Oracle simulatability: *For any polytime adversary C' and any polynomial $p(\cdot)$ there exists a polytime adversary C, such that for any distribution ensemble $\{X_k\}$, for any polytime predicate $P(\cdot)$, and for all large enough k:*

$$\text{Prob}(C'(H(x,r)) = P(x)) - \text{Prob}(C^{I_x}() = P(x)) < \frac{1}{p(k)}$$

where $r \in_R R_k$, and x is drawn from X_k.

3b. Oracle indistinguishability: *For any polytime distinguisher D and any polynomial $p(\cdot)$ there exists a polynomial-size family $\{L_k\}$ of sets such that for all large enough k and for all $x, y \notin L_k$:*

$$\text{Prob}(D(H(x,r)) = 1) - \text{Prob}(D(H(y,r)) = 1) < \frac{1}{p(k)}$$

where $r \in_R R_k$.

Proof. See [C].

Oracle security with a-priori information. The secrecy requirement of Definition 3 assumes that no a-priori information on x is known. We formulate a definition requiring that the hashed value gives no *extra* information on the input x, even when some partial information is already known on x. This definition will be needed for the application described in Section 4.

A first attempt to incorporate a-priori information functions in oracle security may be: *"For any algorithm A and any a-priori information function f, we have that $\langle x, A(f(x), H(x, r)) \rangle$ and $\langle x, A(f(x), H(y, r)) \rangle$ are computationally indistinguishable, where r, x, y are chosen at random from their domains."* This requirement doesn't make sense, though, since $f(x)$ may 'leak' x in full (for instance it may be that $f(x) = x$), in which case A can use v to verify whether its second input is a hash of x.

But a-priori information functions f that leak all information on their inputs seem uninteresting here: why try to hide x from adversaries that already know it via $f(x)$? We therefore restrict our attention to functions f that do *not* give full information on x (i.e., functions where x can be computed from $f(x)$ only with negligible probability.) We call such functions uninvertible. Note that one-way functions are uninvertible; yet uninvertible functions are a much broader class than one-way functions. For instance, the null function $\forall x, f(x) = \emptyset$ is uninvertible but not one-way.[4] Furthermore, we allow uninvertible functions to be *probabilistic*, (i.e., the function value can be a random variable depending on internal random choices of f). See also the discussion in [GL].

Definition 5. A (probabilistic) function $f : \{0,1\}^* \rightarrow \{0,1\}^*$ is uninvertible with respect to distribution ensemble $\{X_k\}$ if for any probabilistic polynomial time algorithm A and for x taken from X_k, the probability $\text{Prob}(A(1^k, f(x)) = x)$ is negligible in k, where the probability is taken over the choices of f, A and x. (We let A have input 1^k to signify that it may run in time polynomial in k.) When no distribution is specified, uninvertibility with respect to the uniform distribution is implied.

Definition 6. Say that H, V are a strong oracle hash scheme if the Secrecy requirement of Definition 3 is replaced with:

3. Strong Secrecy (oracle security with a-priori information): For any algorithm A with *binary output*, for any well-spread distribution ensemble $\{X_k\}$, and and for any function f that is uninvertible for $\{X_k\}$:

$$\langle x, A(f(x), H(x, r)) \rangle \stackrel{c}{\approx} \langle x, A(f(x), H(y, r)) \rangle,$$

where $r \in_R R_k$, and x, y are independently drawn from X_k.

Remarks: 1. As in the case of Definition 3, the strong secrecy requirement can be relaxed by taking into account only the uniform distribution on the inputs. We call this variant strong oracle hashing for random inputs. In particular, this variant will suffice for the first application of Section 4.

---

[4] One-way functions require that it is infeasible to find *any* value in the preimage of $f(x)$.

3 Constructions

We describe some constructions of oracle hash. First we describe a construction based on a number theoretic assumption. Next we describe constructions based on cryptographic hash functions (such as MD5, SHA).

3.1 The r, r^x construction

The construction proceeds as follows. Let p be a large safe prime, i.e. $p = \alpha q + 1$ where α is a small integer (for simplicity we assume $\alpha = 2$). Let Q be the subgroup of order q in Z_p^* (i.e., Q is the group of squares modulo p). On input m and random input $r \in_R Q$, the oracle hash function H first computes $x = h(m)$ where h is a collision resistant hash function; next it outputs $H(m, r) = r, r^x$. (Here and in the sequel calculations are made modulo p.) The verification algorithm V is straightforward: given an input m and a hashed value $\langle a, b \rangle$, compute $x = h(m)$ and accept if $a^x = b$.

We analyze this construction based on three strong variants of the Diffie-Hellman assumption. The variants are used to show, respectively, that the construction satisfies oracle security with random inputs, oracle security, and oracle security with a-priori information. (These notions are defined in Section 2.)

Assumption 7 The Diffie-Hellman Indistinguishability Assumptions: Let k be a security parameter. Let $p = 2q + 1$ be a randomly chosen k-bit safe prime and let $g \in_R Q$ (where Q is the group of squares modulo p).

DHI Assumption I: Let $a, b, c \in_R Z_q^*$. Then, $\langle g^a, g^b, g^{ab} \rangle \stackrel{\circ}{\approx} \langle g^a, g^b, g^c \rangle$.

DHI Assumption II: For any well-spread distribution ensemble $\{X_q\}$ where the domain of X_q is Z_q^*, for a drawn from X_q and for $b, c \in_R Z_q^*$ we have $\langle g^a, g^b, g^{ab} \rangle \stackrel{\circ}{\approx} \langle g^a, g^b, g^c \rangle$.

DHI Assumption III: For any uninvertible function f and for $a, b, c \in_R Z_q^*$ we have $\langle f(a), g^b, g^{ab} \rangle \stackrel{\circ}{\approx} \langle f(a), g^b, g^c \rangle$.

Remarks: 1. It can be seen that Assumption III implies Assumption II, and that Assumption II implies Assumption I. We were unable to show implications in the other direction.

2. While these assumptions are considerably stronger than the standard Diffie-Hellman assumption (there it is only assumed that g^{ab} cannot be computed given p, g, g^a, g^b), they seem consistent with the current knowledge on the Diffie-Hellman problem. In particular, Assumption I appeared in the past, both explicitly and implicitly. It is not hard to see that it is *equivalent* to the semantic security of the El-Gamal encryption scheme [E]. Furthermore, the value exchanged via the DH key exchange is often assumed to be indistinguishable from random. An assumption equivalent to Assumption I is formulated in [B]. Also, this assumption underlies a new and efficient construction of pseudorandom functions [NR].

Although Assumptions II and III look quite strong, we were unable to contradict them. We propose the viability of these assumptions as an open question.

To gain assurance in the plausibility of these assumptions, we remark that it is a common practice to use Diffie-Hellman key exchange modulo a large prime of, say, 1024 bits, but to choose the secret exponents a and b as random numbers of only, say, 200 bits. It is then assumed that the resulting secret, g^{ab}, still has the full '100 bits of security'.[5] This practice implicitly relies on Assumption II (or, alternatively, III) for the case where the first 824 bits of a are fixed to 0.

3. Choosing a safe prime (and the restriction to the subgroup Q) is a standard procedure aimed at avoiding attacks based on the residuocity of a, b, c relative to small factors of $p - 1$. It also carries the advantage that any non-zero member of Q is a generator of Q.

4. Naor and Reingold show that if Assumption I is broken then it is possible to distinguish g^a, g^b, g^{ab} from g^a, g^b, g^c for any $a, b, c \in Z_q^*$ [NR].

For the analysis of the construction, we first consider a somewhat simplified version, where the collision resistant hash function h is omitted and the input is assumed to be taken from Z_q^*.

Theorem 8. *1. If DHI Assumption I holds then the function $H(x, r) = r, r^x$, together with its verification algorithm, are an oracle hashing scheme for random inputs.*

2. If DHI Assumption II holds then the function $H(x, r) = r, r^x$, together with its verification algorithm, are an oracle hashing scheme.

3. If DHI Assumption III holds then the function $H(x, r) = r, r^x$, together with its verification algorithm, are a strong oracle hashing scheme.

Proof. See [C].

The construction $H(m, r) = r, r^{h(m)}$. Strictly speaking, this construction does not satisfy our requirements since the functions h we have in mind are fixed, non-scalable constructs with no asymptotic behavior. Assume however, for sake of the following discussion, that h now describes a scalable collision resistant function where the probability of finding collisions is negligible in the security parameter. (In the next subsection we deal with the non-scalability of existing cryptographic hash functions in a more rigorous way.)

We examine compliance with Definition 3. Completeness still holds. Correctness is now based on the collision resistance of h. (I.e., if two inputs $m \neq m'$ and a hash value c are found such that $V(m, c) = V(m', c) = 1$, then $h(m) = h(m')$.) For the Secrecy requirement, note that as long as the input m is drawn from a well-spread distribution, the value $x = h(m)$ must also be well-spread (otherwise h-collisions may be found by straightforward sampling). Thus, as long as h is collision-resistant, Definition 3 is satisfied under DHI Assumption II; Definition 6 is satisfied under DHI Assumption III.

3.2 Constructions based on cryptographic hash functions

The construction described in the previous subsection is somewhat inefficient since it involves a modular exponentiation. In light of the efficiency of existing

---

[5] There are several ways to find discrete logarithms of $2k$ bit numbers in $O(2^k)$ steps, regardless of the size of the modulus. See details in [MOV].

cryptographic hash functions (such as MD5 and SHA), and of the general "diffusion and confusion" properties they seem to possess, it is natural to look for a construction based only on such functions. Here making additional new assumptions on these functions is unavoidable. However, in contrast with the 'random oracle heuristic' discussed in the introduction, these will be well defined assumptions.

We propose three simple constructions of oracle hashing, incorporating randomness in the input of the hash function. Each construction (or, mode of operation of the hash function) results in a different assumption on the underlying hash function. The assumption will simply be that using the hash function in the corresponding mode satisfies either Definition 3 or 6, respectively. We let further research and practical experience indicate which construction (if any) is preferable in terms of performance and security.

A first construction that comes to mind, given a cryptographic hash function h is $H(x, r) = r, h(r, x)$, where r is a random string of length β. (Setting $\beta = 128$ for MD5 and $\beta = 160$ for SHA seems appropriate.) Verification (and the Completeness property) are straightforward. Correctness follows directly from the collision resistance of h. The Secrecy requirement imposes the following requirement on h. Following the concrete (i.e., non-asymptotic) security approach of [BKR, BGR, BCK1] we say that:

Definition 9. A hash function h is (τ, δ)-secure with respect to the $H(x, r) = r, h(r, x)$ construction and some distribution Δ on $\{0, 1\}^*$ if for any adversary $\mathcal{A}$ and distinguisher D, each running in time τ, we have

$$|\text{Prob}(D(x, \mathcal{A}(r, h(r, x))) = 1) - \text{Prob}(D(x, \mathcal{A}(r, h(r, y))) = 1)| \leq \delta$$

where x, y are independently drawn from Δ and $r \in_R \{0, 1\}^\beta$.

(This assumption is obtained by simply plugging the construction in the Secrecy requirement of Definition 3.) A seemingly equivalent variant is $H(x, r) = r, h(r \oplus x)$, where $\oplus$ denotes bitwise exclusive or.

We remark that the "bit commitment scheme based on one way functions" described in [S], p. 87, is secure under the assumption that the one-way function in use satisfies Definition 9. In fact, this assumption seems *necessary* here.

Another possible construction is $H(x, r) = r, h(r, h(x))$. Completeness and correctness are as above. The resulting security assumption can be formulated analogously to the former one. Note that potentially this construction is 'more secure' than the former one, in the sense that if the latter construction fails then most probably the former one fails, but not necessarily vice-versa.

Yet another construction is based on the HMAC construction [BCK2]: let $H(x, r) = r, h(r_1, h(r_2, x))$, where $r_1 = r \oplus \text{opad}$, $r_2 = r \oplus \text{ipad}$, and opad and ipad are two fixed constants. Also here, Completeness and correctness are as above. This construction may be even 'more secure', again in the sense that if the HMAC construction fails then most probably so does the previous one, but not necessarily vice versa.

We remark that embedding the randomness in the IV may result in inferior constructions, since it may simplify violating Correctness. That is, let $h_r(x)$

denote the value of $h(x)$ when the IV is set to r. Consider the construction $H(x, r) = h_r(x)$. Now in order to violate Correctness it suffices to find r, r', x, x' such that $h_r(x) = h_{r'}(x')$. This is a much easier task; See [BB, MOV] for more details.

4 Applications

We describe two more applications, on top of the one described in the Introduction.

Avoiding random oracles. In a sequence of papers Bellare and Rogaway demonstrate how to construct, in the random oracle model, simple, efficient, and provably secure encryption and signature schemes, based on any trapdoor permutation (e.g., the RSA permutation) [BR1, BR2, BR3, BR4]. It is suggested as a 'rule-of-thumb' to replace, in practice, the random oracle with a cryptographic hash function. While the resulting constructions are very attractive and useful in practice, they lack rigorous proofs of security.

It is thus natural to attempt the following procedure with respect to these schemes: (a) replace the random oracle with oracle hashing, and (b) prove the security of the resulting schemes *without random oracles*. We do that to a simple encryption scheme described in [BR1].

The scheme proceeds as follows given a random oracle R, and a trapdoor permutation generator G that on input 1^k outputs a pair f, f^{-1} (where f is a one way permutation and f^{-1} is the inverse of f). The public encryption key is f and the private decryption key is f^{-1}. Given message m and a random input s, let the encryption be $E(m, s) = f(s), R(s) \oplus m$. Decryption is straightforward.

It is shown there that this scheme is semantically secure (in the random oracle model). There, semantic security means that for any two messages m_0, m_1, no polytime adversary $\mathcal{A}$ (with access to the encryption algorithm E and to R) can distinguish between encryptions of m_0 and encryptions of m_1 with more than negligible probability.

We show how to replace R with an oracle hashing scheme H. First however we need to make the following two technical assumptions on H. The first assumption is that the random input r appears explicitly in the output of $H(x, r)$. All the schemes described in this paper have this property. We call such schemes public randomness schemes and write $H(x, r) = r, \tilde{H}(x, r)$.

Let B_k denote the domain of hashes with security parameter k. The second assumption is that there is an 'easy to compute' encoding from B_k to $\{0, 1\}^{l(k)}$ for some 'reasonable' length function $l(k)$. The encoding should make sure that when a hash is chosen at random from B_k then the encoded value is distributed (close to) uniform in $\{0, 1\}^{l(k)}$. Again, the schemes described in this paper have this property: For the r, r^x scheme, one can use a standard encoding of Z_p^* in, say, $\{0, 1\}^{|p|-1}$. For the schemes based on cryptographic hash functions no encoding seems to be needed.

We suggest the following encryption scheme. Given message m and random input r, s compute:

$$E(m, r, s) = f(s), r, \tilde{H}(s, r) \oplus m \qquad (2)$$

Again, decryption is straightforward.

Proving semantic security of this construction, based on the fact that H is a strong oracle hash function for random inputs, is quite straightforward. In fact, we use only a considerably weaker secrecy property than the one in Definition 6, namely that $\langle f(x), h(x, r) \rangle \overset{s}{\approx} \langle f(x), h(y, r) \rangle$ where x, y, r are uniformly distributed in their domains.

Theorem 10. *The encryption scheme described in (2) is semantically secure, if H is a strong oracle hash function for random inputs with the additional technical properties described above.*

Proof (sketch): Assume an adversary $\mathcal{A}$ such that $\text{Prob}(\mathcal{A}(E(m_1, s)) = 1) - \text{Prob}(\mathcal{A}(E(m_0, s)) = 1) > \delta$ for some m_0, m_1 and δ. Let p_0 (resp., p_1) denote the probability that $\mathcal{A}$ outputs '1' if it is given $E(m_0, s)$ (resp., $E(m_1, s)$), and let p_* denote the probability that $\mathcal{A}$ outputs '1' given $E(m, s)$, where m is uniformly distributed in its domain. Clearly either $|p_* - p_0| \geq \delta/2$, or $|p_* - p_1| \geq \delta/2$. Assume that $|p_* - p_0| \geq \delta/2$.

Construct a distinguisher D between $\langle f(s), H(s, r) \rangle$ and $\langle f(s), H(s', r) \rangle$, where $s.s', r$ are randomly chosen. (Note that the function f is uninvertible.) Recall that here $H(s, r) = r, \tilde{H}(s, r)$, and that for uniformly chosen s, r the value $H(s, r)$ is uniform in $\{0, 1\}^l$ for some l. Given $f(s), r, \xi$ (where ξ is either $\tilde{H}(s, r)$ or $\tilde{H}(s', r)$), D will hand $\mathcal{A}$ the 'ciphertext' $f(s), r, \xi \oplus m_0$. Now, if $\mathcal{A}$ outputs 'm_0' then D outputs '$\xi = \tilde{H}(s, r)$'; otherwise it outputs '$\xi = \tilde{H}(s', r)$'.

Analyzing D is straightforward. (It distinguishes with probability $\delta/2$.) It should only be noted that if $\xi = \tilde{h}(s', r)$ then $\mathcal{A}$ is given an encryption of a uniformly chosen message. $\square$

Content-Concealing signatures. Assume that one wants to sign a document m in a way that if m is known then the signature can be verified as usual, and at the same time make sure that the signature itself hides all partial information on m from parties who do not already know m. We call a signature scheme that has this property content-concealing. Such signatures may become handy, for instance, when the document to be signed has been agreed by the parties in a private way, but the signature has to be broadcasted on a public channel where encryption is unavailable or costly. Another possible scenario is when the signer wants to publish beforehand a signature on a document (say, the quarterly earnings of IBM) but make the document public only at a later date.

As in the 'puzzle in the newspaper' problem, to crypto practitioners it may seem that this problem is already solved: Since cryptographic hash functions are assumed to 'hide all partial information on the input', and since the first step in any digital signature algorithm is to apply a cryptographic hash function to the document, then existing digital signatures are already content-concealing.

Also here, however, this is an illusion. No known (until now) cryptographic primitive solves this problem. Furthermore, also here there is a simple solution in the random oracle model: in the presence of a random oracle R one can simply sign $R(m)$ instead of signing m.

When formalizing the requirement that the signature 'hides all partial information on the input' and at the same time allows for verification, one ends up with the same notion of oracle security used for oracle hash. That is:

Definition 11. A signature scheme is (Strong) content-concealing if, in addition to being a signature scheme (as defined in, say, [GMR]), the signing algorithm satisfies the Secrecy requirement of Definition 3 (resp., 6).

Once content-concealing signatures are defined, a solution is straightforward: *To sign a message m, sign $c = H(m, r)$ (and attach c to the signature), where H, V are an oracle hash scheme and r is randomly chosen.* For verification, first verify the signature on c; next verify that c is a hash of m using the verification algorithm V.

Acknowledgments. First and special thanks are due to Hugo Krawczyk and Oded Goldreich, who spent considerable time trying to make sense of my rambling thoughts and early drafts. In particular, the idea to use the r, r^x construction is Hugo's, and some of the formalizations of oracle security are Oded's.

I thank Shafi Goldwasser for discussions on content-concealing signatures (and Juan Garay and Tal Rabin for suggesting the name). I also thank Rosario Gennaro, Moni Naor and Omer Reingold for very helpful discussions, and Dan Boneh for drawing my attention to [B].

References

[AS] N. Alon and J. Spencer, *The Probabilistic Method*, Wiley, 1992.

[BCK1] M. Bellare, R. Canetti and H. Krawczyk, "Pseudorandom functions revisited: The cascade construction and its concrete security", *37th FOCS*, 1996.

[BCK2] M. Bellare, R. Canetti and H. Krawczyk, "Keying hash functions for message authentication", *CRYPTO'96*, 1996.

[BGR] M. Bellare, R. Guérin and P. Rogaway, "XOR MACs: New methods for message authentication using finite pseudorandom functions," *CRYPTO'95*, 1995.

[BKR] M. Bellare, J. Kilian and P. Rogaway, "The security of cipher block chaining." *CRYPTO'94*, 1994.

[BR1] M. Bellare and P. Rogaway, "Random oracles are practical: a paradigm for designing efficient protocols", *1st ACM Conference on Computer and Communications Security*, 62-73, 1993.

[BR2] M. Bellare and P. Rogaway, "Optimal Asymmetric Encryption", *EUROCRYPT '94 (LNCS 950)*, 92-111, 1995.

[BR3] M. Bellare and P. Rogaway, "The exact security of digital signatures — How to sign with RSA and Rabin", *EUROCRYPT '96 (LNCS 1070)*, 1996.

[BR4] M. Bellare and P. Rogaway, 'Minimizing the use of random oracles in P1363 encryption schemes", Contribution on IEEE P1364. November 10, 1996.

[BDMP] M. Blum, A. De Santis, S. Micali and G. Persiano, "Non-interactive zero-knowledge", *SIAM Journal on Computing, 20(6):1084-1118*, December 1991.

[BB] B. den Boer and A. Bosselaers, "Collisions for the compression function of MD5", *EUROCRYPT'93*, 293-304, 1994.

[B] S. Brands, "An efficient off-line electronic cash system based on the representation problem", CWI TR CS-R9323, 1993.

[C] R. Canetti, "Towards realizing random oracles: Hash functions that hide all partial information", in Theory of Cryptology Library, No. 97-07. http://theory.lcs.mit.edu/tcryptol, 1997.

[CW] J. L. Carter and M. N. Wegman, " Universal classes of hash functions", *JCSS No. 18*, 143-154, 1979.

[CG] B. Chor and O. Goldreich, "Unbiased bits from sources of weak randomness and probabilistic communication complexity", *SIAM J. Comp., Vol. 17, No. 2*, 230-261, 1988.

[D] I.B. Damgård, "Collision free hash functions and public key signature schemes", *EUROCRYPT 87 (LNCS 304)*, pp. 203–216, 1988.

[DDN] D. Dolev, C. Dwork and M. Naor, "Non-malleable cryptography", *23rd STOC*, 1991.

[E] T. El-Gamal, *"Cryptography and logarithms over finite fields"*, Ph.D. Thesis, Stanford University, 1984.

[F] P. Feldman, "A practical scheme for non-interactive verifiable secret sharing", *28th FOCS*, 427-437, 1987.

[FS] A. Fiat and A. Shamir, "How to prove yourself: Practical solutions to identification and signature problems", *CRYPTO'86 (LNCS 263)*, 186-194, 1986.

[G] O. Goldreich, *"Foundations of Cryptography (Fragments of a book)"*, Weizmann Inst. of Science, 1995. (Avaliable at http://theory.lcs.mit.edu/~cryptol/)

[GM] Shafi Goldwasser and Silvio Micali, "Probabilistic encryption", *JCSS, Vol. 28, No 2*, 270-299, April 1984.

[GL] O. Goldreich and L. Levin, A Hard-Core Predicate to any One-Way Function, *21st STOC*, 1989, pp. 25-32.

[GMR] S. Goldwasser, S. Micali and R. Rivest, "A digital signature scheme secure against adaptive chosen-message attacks," *SIAM Journal of Computing*, 17(2):281–308, April 1988.

[MOV] A. J. Menezes, P. C. van Oorschot and S. A. Vanstone, "Handbook of applied cryptography", CRC Press, 1997.

[M] S. Micali, "CS proofs", *35th FOCS*, 436-453, 1994.

[NR] M. Naor and O. Reingold, "The Brain can Compute Pseudo-Random Functions, or Efficient Cryptographic Primitives Based on the Decisional Diffie-Hellman Assumption", manuscript.

[NY1] M. Naor and M. Yung, "Universal one-way hash functions and their cryptographic applications", *21st STOC*, 33-43, 1989.

[NY2] M. Naor and M. Yung, "Public key cryptosystems provably secure against chosen ciphertext attacks", *22nd STOC*, 427-437, 1990.

[P] T. P. Pedersen, "Distributed provers with applications to undeniable signatures", *EUROCRYPT'91*, 1991.

[PS] D. Pointcheval and J. Stern, "Security proofs for signature schemes", *Eurocrypt '96 (LNCS 1070)*, pp. 387-398, 1996.

[RS] C. Rackoff and D. Simon, "Non-interactive zero-knowledge proof of knowledge and chosen ciphertext attack", *CRYPTO'91, (LNCS 576)*, 1991.

[Ri] R. Rivest, "The MD5 message-digest algorithm," IETF Network Working Group, RFC 1321, April 1992.

[S] B. Schneier, *"Applied cryptography"*, 2nd edition, Wiley and sons, 1996.

[SHA] FIPS 180, "Secure Hash Standard", Federal Information Processing Standard (FIPS), Publication 180, National Institute of Standards and Technology, US Department of Commerce, Washington D.C., May 1993.

Collision-Resistant Hashing: Towards Making UOWHFs Practical

Mihir Bellare[1] and Phillip Rogaway[2]

[1] Dept. of Computer Science & Engineering, University of California at San Diego, 9500 Gilman Drive, La Jolla, California 92093, USA. E-Mail: mihir@cs.ucsd.edu. URL: http://www-cse.ucsd.edu/users/mihir.

[2] Dept. of Computer Science, Engineering II Bldg., University of California at Davis, Davis, CA 95616, USA. E-mail: rogaway@cs.ucdavis.edu. URL: http://wwwcsif.cs.ucdavis.edu/~rogaway.

Abstract. Recent attacks on the cryptographic hash functions MD4 and MD5 make it clear that (strong) collision-resistance is a hard-to-achieve goal. We look towards a weaker notion, the *universal one-way hash functions* (UOWHFs) of Naor and Yung, and investigate their practical potential. The goal is to build UOWHFs not based on number theoretic assumptions, but from the primitives underlying current cryptographic hash functions like MD5 and SHA-1. Pursuing this goal leads us to new questions. The main one is how to extend a compression function to a full-fledged hash function in this new setting. We show that the classic Merkle-Damgård method used in the standard setting fails for these weaker kinds of hash functions, and we present some new methods that work. Our main construction is the "XOR tree." We also consider the problem of input length-variability and present a general solution.

1 Introduction

A cryptographic hash function is a map F which takes a string of arbitrary length and maps it to a string of some fixed-length k. The property usually desired of these functions is *collision-resistance*: it should be "hard" to find distinct strings x and y such that $F(x) = F(y)$.

Cryptographic hash functions are much used, most importantly for digital signatures, and cheap constructions are highly desirable. But in recent years we have seen a spate of attacks bringing down our most popular constructions, MD4 and MD5 [8–11]. The conclusion is that the design of collision-resistant hash functions may be harder than we had thought.

What can we do? One approach is to design new hash functions. This is being done, with SHA-1 [17] and RIPEMD-160 [12] being new designs which are more conservative then their predecessors. In this paper we suggest a complementary approach: weaken the goal, and then make do with hash functions meeting this goal. Ask less of a hash function and it is less likely to disappoint!

Luckily, a suitable weaker notion already exists: *universal one-way hash functions* (UOWHF), as defined by Naor and Yung [16]. But existing constructions, based on general or algebraic assumptions [16,22,13], are not too efficient. We take a different approach. We integrate the notion with current hashing technology, looking to build UOWHFs out of MD5 and SHA-1 type primitives.

The main technical issue we investigate is how to extend the classic Merkle-Damgård paradigm [15,7] to the UOWHF setting. In other words, how to build "full-fledged" UOWHFs out of UOW compression functions. We address practical issues like key sizes and input-length variability. Our main construction, the "XOR tree," also turns out to have applications to reducing key sizes for existing subset-sum based constructions. To make for results more directly meaningful to practice we treat security "concretely," as opposed to asymptotically.

Unfortunately, the name UOWHFs does not reflect the property of the notion, which is a weak form of collision-resistance. We will call our non-asymptotic version *target collision-resistance* (TCR). We refer to the customary notion of collision resistance as *any collision-resistance* (ACR).

1.1 Background

A function $F : D \to \{0,1\}^k$ on some domain D is a *compression function* if D is the set of strings of some small length (eg., $D = \{0,1\}^{640}$ for the compression function of MD5). It is a (full-fledged) hash function if $D = \{0,1\}^*$, or at least $D = \{0,1\}^{\le \mu}$ for some large number μ. In either case, a *collision* for F is a pair $x, y \in D$ such that $x \ne y$ but $F(x) = F(y)$. Informally, F is said to be *any collision-resistant* (ACR) if it is computationally hard to find a collision.

THE MD METHOD. The Merkle-Damgård construction [15,7] turns an ACR compression function $F: \{0,1\}^{k+b} \to \{0,1\}^k$ into an ACR hash function MDF. Fix $C_0 \in \{0,1\}^k$. Given $M = M[1] \dots M[n]$, for $M[i] \in \{0,1\}^b$, compute $C_i = F(C_{i-1} \| M[i])$ and set $MDF(M) = C_n$. The property of this method is that if it is hard to find collisions in F, then it is hard to find collisions in MDF.

All of the popular hash functions (MD4, MD5, SHA-1, and RIPEMD-160) use the MD construction. Thus the crucial component of each algorithm is the underlying compression function, and we want it to be ACR. But the compression function of MD4 is not: following den Boer and Bosselaers [8], collisions were found by Dobbertin [10]. Then collisions were found for the compression function of MD5 [9,11]. These attacks are enough to give up on MD4 and MD5 from the point of view of ACR. No collisions have been found for the compression functions of SHA-1 and RIPEMD-160, and these may well be stronger.

KEYING. In the above popular hash functions there is no explicit key. But Damgård [6,7] defines ACR via keyed functions, and it is in this setting that he proves the MD construction correct [7]. Keying hash functions seems essential for a meaningful formalization of security. Thus, in truth, a hash function F does not have the signature above: it must have two arguments, one for the key K and one for the message M. To use F one selects a random key K, publishes it, and from then on one hashes according to F_K. In effect, the key amounts to the description of a particular hash function.

1.2 Target Collision-Resistance

With an ACR hash function F the key K is published and the adversary wins if she manages to find *any* collision x, y for the F_K. The points x and y may depend arbitrarily on K; any pair of distinct points will do. In the notion of Naor and Yung [16] the adversary no longer wins by finding just any collision. The adversary must choose one point, say x, in a way which does *not* depend on K, and then, later, given K, find another point y, this time allowed to depend on K, such that x, y is a collision for F_K. Thus, although it might be easy to find a collision x, y in F_K by making both x, y depend on K, the adversary may be unable to find collisions if she must "commit" to x before seeing K. We call this weakened notion of security *target collision-resistance* (TCR). (In the terminology of [16] it is universal one-wayness.)

Naor and Yung [16] formalize this via the standard "polynomial-time adversaries achieve negligible success probability" approach of asymptotic cryptography. In order to get results which are more directly meaningful for practice, our formalization is non-asymptotic. See Section 2.

NO BIRTHDAYS! Besides being a weaker notion (and hence easier to achieve) we wish to stress one important practical advantage of TCR over ACR: because x must be specified before K is known, birthday attacks to find collisions are not possible. This means the hash length k can be small, like 64 or 80 bits, as compared to 128 or 160 bits for an ACR hash function. This is important to us for several reasons and we will appeal to it later.

GOOD ENOUGH FOR SIGNING. In weakening the security requirement on hash functions we might risk reducing their utility. But TCR is strong enough for the major applications, *if appropriately used*. In particular, it is possible to use TCR hash functions for hashing a message before signing. Due to page limits, the constructions are omitted here: they can be found in [4]. The idea is to pick a new key K for each message M and then sign the pair $(K, F_K(M))$, where F is TCR. This works best for short keys. When they are long some extra tricks can be used, as described in [4], but we are better off with small keys. Thus there is a strong motivation for keeping keys short.

1.3 Making TCR Functions out of Standard Hash Functions

The most direct way to make a TCR hash function is to appropriately key an existing hash function such as MD5 or SHA-1. We caution that one must be careful in how this keying is done. If not, making a TCR assumption about the keyed function may really be no weaker than making an ACR assumption about the original hash function. See Section 4.

1.4 Extending TCR Compression Functions to TCR Hash Functions

Instead of trying to key an entire hash function, as above, a good strategy might be to implement a TCR compression function (perhaps by keying an existing compression function) and then transform it into a (full-fledged) TCR hash function using some simple construction. The question we investigate is how to do this transformation. This turns out to be quite interesting.

THE MD METHOD DOES NOT WORK FOR TCR. Suppose we are given a TCR compression function H in which each κ-bit key specifies a map $H_K\colon \{0,1\}^{k+b} \to \{0,1\}^k$. We want to build a TCR hash function H' in which each key K specifies a map H'_K on arbitrary strings. The obvious thought is to apply the MD method to H_K. However, we show in Section 5.1 that this does not work. We give an example of a compression functions which is secure in the TCR sense but for which the resulting hash function is not.

LINEAR ITERATION: BASIC AND XOR. The most direct extension we found to the MD construction is use a different key at each stage. This works, and its exact security is analyzed in Section 5.2. But the method needs a long key.

We provide a variant of the above scheme which uses only one key for the compression function, but also uses a number of auxiliary keys, which are XORed in at the various stages. This can slightly reduce key sizes, and it also has some advantages from a key-scheduling point of view (eg., it may be slow to "set up" the key of a compression function, so it's best if this not be changed too often).

THE BASIC TREE SCHEME. To get major reductions in key size we turn to trees.[1] Wegman and Carter [25] give a tree-based construction of universal hash functions that reduces key sizes, and Naor and Yung have already pointed out that key lengths for UOWHFs can be reduced by the same method [16, Section 2.3]. We recall this basic tree construction in Section 5.4 and provide a concrete analysis of its security. Then we look at key sizes. Suppose we start with a compression function with key length κ, mapping $2k$ bits to k bits, and we want to hash nk bits to k bits. The basic tree construction yields a hash function with a key size of $\kappa \lg n$ bits.[2] Key lengths have been reduced, but one can reduce them more.

THE XOR TREE SCHEME. Our main construction is the XOR tree scheme. Here, the hash function uses only one key for the compression function and some auxiliary keys. If we start with a compression function with key length κ, mapping $2k$ bits to k bits, and we want to hash nk bits to k bits, the XOR tree construction yields a hash function with a key size of $\kappa + 2k \lg n$ bits.

Recall that k is short, like 64 bits, since we do not need to worry about birthday attacks for TCR functions. On the other hand, κ can be quite large (and in many constructions, it is). So $\kappa + 2k \lg n$ may be much less than $\kappa \lg n$.

---

[1] It may be worth remarking that the obvious idea for reducing key size is to let the key be a seed to a pseudorandom number generator and specify longer keys by stretching the seed to any desired length. The problem is that our keys are public (they are available to the adversary) and pseudorandom generators are of no apparent use in such a context.

[2] This corresponds to a binary tree construction. In Section 5.4 we consider the more general case of starting with a compression function of mk bits to k bits and building an m-ary tree.

1.5 Other Results

REDUCING KEY SIZES FOR OTHER CONSTRUCTIONS. Our main motivation has been building TCR hash functions from primitives underlying popular cryptographic hash functions. But XOR trees can also be used to reduce key sizes for TCR hash functions built from combinatorial or algebraic primitives. For example, the subset sum based construction of [13] uses a key of size sl bits to hash s bits to l bits, where l is a security parameter which controls subset sum instance sizes. (Think of l as a few hundred.) So the size of the key is even longer than the size of the data. The basic (binary) tree scheme can be applied to reduce this: starting with a compression function taking $2l$ bits to l bits (it has key length $\kappa = 2l^2$) the key size of the resulting hash function is $\kappa \lg n = 2l^2 \lg(n)$, where n is the number of (l-bit) blocks hashed. With our (binary) XOR tree scheme, the key size of the resulting function is $\kappa + 2l \lg(n) = 2l(l + \lg(n))$. The latter can be quite a bit smaller. For example if $l = 300$ and $n = 10$ KBytes, the key length for the basic scheme is about 15 times larger than that for the XOR scheme.

INPUT-LENGTH VARIABILITY. The proofs for our constructions rule out adversaries who can find collisions x, y for equal-length strings. In practice, collisions between strings of unequal length must be prevented. To handle this we provide a general mechanism for turning hash functions secure against equal-length collisions into hash functions secure also against collisions of possibly unequal length. This needs just one extra application of the compression function. See Section 6. Given this, we can concentrate on designing functions that resist equal-length collisions.

1.6 Related Work

The general approach to concrete, quantitative security that we are following began with [3].

Keying hash functions has arisen in other contexts. A good deal of work has gone into keying hash functions for message authentication [1,23,14,18]. But that is a very different problem from what we look at here; there, the key is secret, and one is trying to achieve a particular goal of private key cryptography. In another direction for keying a hash function Bellare, Canetti and Krawczyk [2] considered keyed compression functions as pseudorandom functions, and showed that applying the MD construction then yields a pseudorandom function.

A weaker-than-standard notion of hashing is considered in [1], but that notion is based on a hidden key and those hash functions don't suffice for signatures.

2 Notions of Hashing

When one looks at hash functions like MD5 or SHA-1 there is no explicit key. However, no notion of collision-freeness can be properly formalized in such a setting. The first step is thus to talk of families of functions.

FAMILIES OF HASH FUNCTIONS. In a family of hash functions F, each key K specifies a particular hash function F_K in the family. Each such function maps D to $\{0,1\}^k$ where D is some domain associated to the family, and k is the *hash length*, or *output length*, also depending on the family. The key K will be drawn at random from some key space $\{0,1\}^\kappa$, and κ will be called the *key length*. If $D = \{0,1\}^l$ for some l then l is called the *input length*.

Formally, a family F of (keyed) hash functions is a map $F: \{0,1\}^\kappa \times D \to \{0,1\}^k$. We define $F_K: D \to \{0,1\}^k$ by $F_K(x) = F(K,x)$ for each $K \in \{0,1\}^\kappa$ and each $x \in D$. We use either the notation $F_K(x)$ or $F(K,x)$, as convenient.

The hash family F is a *compression function* if the domain is $D = \{0,1\}^m$ for some (small) constant m (eg., $m = 512$). It is an *extended hash function* if $D = \{0,1\}^M$ or $\{0,1\}^{\leq M}$ for some (large) constant M (eg., $M = 2^{64} - 1$).

We say F is *length consistent* if for every $K \in \{0,1\}^\kappa$ and every $x,y \in D$ it is the case that if $|x| = |y|$ then $|F(K,x)| = |F(K,y)|$.

COLLISIONS. A *collision* for a function f defined on a domain D is a pair of strings $x,y \in D$ such that $x \neq y$ but $f(x) = f(y)$. In our setting the function of interest is $f = F_K$ for a randomly chosen key K. Security of a hash family talks about the difficulty of finding collisions in F_K. There are two notions of security. The stronger one we call here *any collision-resistance* (ACR). The weaker one, which is due to Naor and Yung [16], we call *target collision-resistance* (TCR). We now define both notions. First some technicalities.

PROGRAMS AND TIMING. A model of computation is fixed, and the execution time of a program is discussed and analyzed as in any algorithms text, eg. [5]. An *adversary* is a program for our model, written in some fixed programming language. Any program is allowed randomness: the programming language supports the operation FLIPCOIN() which returns a random bit. By convention, when we speak of the running time of an adversary we mean the actual execution time in the fixed model of computation plus the length of the description of the program. For F a family of hash functions we let T_F denote the worst-case time to compute F in the underlying model of computation. Namely, given $K \in \{0,1\}^\kappa$ and $x \in D$, this is the time to output $F(K,x)$.

2.1 Any Collision-Resistance — ACR

The "standard" notion of collision resistance for a function f is that given f it is hard to find a collision x,y for f. In the keyed setting, it can be formalized like this (eg. [6,7]). The adversary C, called a *collision-finder*, is given K chosen at random from $\{0,1\}^\kappa$, and is said to *succeed* if it outputs a collision x,y for F_K. We measure the quality of the hash family by the success probability of the adversary as a function of the time it invests. Formally, a collision-finder C is said to (t, ϵ)-break the family of hash functions $F : \{0,1\}^\kappa \times D \to \{0,1\}^k$ if the running time of the adversary is at most t and the probability that C, on input K, outputs a collision x,y for F_K is at least ϵ. Here the probability is take over K (a random point in $\{0,1\}^\kappa$) and C's random coins. Informally we say F is "any collision-resistant" (ACR) if for every collision-finder who (t, ϵ)-breaks F, the ratio t/ϵ is large.

Note that the adversary is given the (random) point K (the key is "announced") and only then is the adversary asked to find a collisions for F_K. So the adversary may employ a strategy in which the collision which is found depends on K. This makes the notion very strong.

2.2 Target Collision-Resistance — TCR

In the notion of [16] the adversary does not get credit for finding any old collision. The adversary must still find a collision x, y, but now x is not allowed to depend on the key: the adversary must choose it before the key K is known. Only after "committing" to x does the adversary get K. Then she must find y.

Formally, the adversary $C = (C_1, C_2)$ (called a target collision finder) consists of two algorithms, C_1 and C_2. First, C_1 is run, to produce x and possibly some extra "state information" σ that C_1 wants to pass to C_2. We call x the target message. Now, a random key K is chosen and C_2 is run. Algorithm C_2 is given K, x, σ and must find y different from x such that $F_K(x) = F_K(y)$.

The formalization of [16] was asymptotic. Here we provide a concrete one, and call this version of the notion target collision-resistance (TCR).

We begin with some special cases. A target collision finder $C = (C_1, C_2)$ is called an equal-length target collision finder if the messages x, y which C_1 outputs always satisfy $|x| = |y|$. It is called a variable-length target collision finder when no restriction is made on the relative lengths of x, y.

Let $C = (C_1, C_2)$ be a target-collision finder. We say it (t, ϵ)-breaks F if its running time is at most t and it finds a collision with probability at least ϵ. The running time is the sum of the running times of the two algorithms and the probability is over the coins of C_1 and C_2 and the choice of K. We say that F is (t, ϵ)-resistant to equal-length target collisions if there is no equal-length target collision finder which (t, ϵ)-breaks F. F is (t, ϵ)-resistant to variable-length target collisions if there is no variable-length target collision finder which (t, ϵ)-breaks F. If we just say F is (t, ϵ)-TCR, or (t, ϵ)-resistant to target collisions, we mean it is (t, ϵ)-resistant to variable-length target collisions.

Resistance to equal-length target collisions is a weaker notion than resistance to variable-length target collisions: in the former, the adversary is only being given credit if it finds collisions where the messages are of the same length. In practice, we want resistance to variable-length target resistance. However, it turns out the convenient design approach is to focus on resistance to equal-length target collisions and then achieve resistance to variable-length target collisions via a general transformations we present in Section 6.

Informally, we say F is "target collision-resistant" (TCR) (or, resp., TCR to equal-length collisions) if it for every (resp., equal-length) target collision-finder who (t, e)-breaks F, the ratio t/ϵ is large.

3   Composition Lemmas

It is useful to hash a long string in stages, first cutting down its length via one hash function, then applying another to this output to cut it down further. Naor and Yung [16] considered this kind of composition in the context of TCR hash functions. We first state a concrete version of their lemma and then extend it to an equal-length collision analogue which is in fact what we will use.

Let H_1: $\{0,1\}^{\kappa_1} \times \{0,1\}^{l_1} \rightarrow \{0,1\}^{l_2}$ and H_2: $\{0,1\}^{\kappa_2} \times \{0,1\}^{l_2} \rightarrow \{0,1\}^k$ be families of hash functions. The *composition* $H_2 \circ H_1$: $\{0,1\}^{\kappa_1+\kappa_2} \times \{0,1\}^{l_1} \rightarrow \{0,1\}^k$ is the family defined by $(H_2 \circ H_1)(K_2 K_1, M) = H_2(K_2, H_1(K_1, M))$ for all $K_1 \in \{0,1\}^{\kappa_1}$, $K_2 \in \{0,1\}^{\kappa_2}$, and $M \in \{0,1\}^{l_1}$. From the proof of Naor and Yung's composition lemma [16] we extract the concrete security parameters to get the following.

Lemma 1. (TCR composition lemma) *Let H_1: $\{0,1\}^{\kappa_1} \times \{0,1\}^{l_1} \rightarrow \{0,1\}^{l_2}$ and H_2: $\{0,1\}^{\kappa_2} \times \{0,1\}^{l_2} \rightarrow \{0,1\}^k$ be families of hash functions. Assume the first is (t_1, ϵ_1)-secure against target collisions and the second is (t_2, ϵ_2)-secure against target collisions. Then the composition $H_2 \circ H_1$ is (t, ϵ)-secure against target collisions, where $t = \Theta(\min(t_1 - k_2, t_2 - 2T_{H_1} - k_1))$ and $\epsilon = \epsilon_1 + \epsilon_2$.*

In this paper we need such a lemma for the case of equal-length TCR. This requires an extra condition on the first family of hash functions, namely that it be length consistent. See [4] for a proof.

Lemma 2. (TCR composition lemma for equal-length collisions) *Let H_1: $\{0,1\}^{\kappa_1} \times \{0,1\}^{l_1} \rightarrow \{0,1\}^{l_2}$ and H_2: $\{0,1\}^{\kappa_2} \times \{0,1\}^{l_2} \rightarrow \{0,1\}^k$ be families of hash functions. Assume the first is length consistent and (t_1, ϵ_1)-resistant to equal-length target collisions. Assume the second is (t_2, ϵ_2)-resistant to equal-length target collisions. Then the composition $H_2 \circ H_1$ is (t, ϵ)-resistant to equal-length target collisions, where $t = \Theta(\min(t_1 - k_2, t_2 - 2T_{H_1} - k_1))$ and $\epsilon = \epsilon_1 + \epsilon_2$.*

4   TCR Hash Functions from Standard Hash Functions

The most direct way to construct a TCR hash function is to key a function like MD5 or SHA-1. We point out the importance of doing this keying with care.

Suppose, for example, that one keys MD5 through its 128-bit initial chaining value, IV. Denote the resulting hash function family by MD5*. Then breaking MD5* (in the sense of violating TCR) amounts to finding collisions in an algorithm which is identical to MD5 except that it begins with a random, known IV (as opposed to the published one). It seems unlikely that this task would be harder than finding collisions in MD5 itself. It could even be easier!

Alternatively, suppose one tries to use the well-known "envelope" method, setting $\text{MD5}^{**}_K(M) = \text{MD5}(K\|M\|K)$. It seems likely that any extension of Dobbertin's attack [11] which finds collisions in MD5 would also defeat MD5**. Letting md5 denote the compression function of MD5, note that if for any c you can find m, m' such that $\text{md5}(c\|m) = \text{md5}(c\|m')$, then you have broken MD5**.

A safer approach might be to incorporate key bits throughout the message being hashed. For example, with $|K| = 128$ one might intertwine 128 bits of key and the next 384 bits of message into every 512-bit block. (For example, every fourth byte might consist of key.) Now the cryptanalyst's job amounts to finding a collision M, M' in MD5 where we have pre-specified a large number of (random) values to be sprinkled in particular places throughout M and M'. This would seem to be very hard.

Note that the approach above (shuffling key and message bits) is equally at home in defining a TCR compression function based on the compression function underlying a map like MD5 or SHA-1. The resulting compression keyed compression function can then be extended to a full-fledged keyed hash function using the constructs of this paper. Doing this one will gain in provable-security but lose out in increased key length.

5 TCR Hashing based on TCR Compression Functions

Throughout this section messages will be viewed as sequences of blocks of some length l, with l depending on the context. For notational simplicity, let $\Sigma_l = \{0,1\}^l$.

We are given a TCR compression function H. We wish to extend it to a full-fledged hash function H'. We begin with a method which does *not* work.

5.1 The MD Construction Doesn't Propagate TCR

Suppose $H: \{0,1\}^\kappa \times \{0,1\}^{k+b} \to \{0,1\}^k$ and we want to hash messages of nb-bits. The MD method gives a keyed family of functions $MDH^n: \{0,1\}^\kappa \times \{0,1\}^{nb} \to \{0,1\}^k$. To define it, fix some k-bit initial vector, say IV $= 0^k$. Now view the message $M = M[1]\ldots M[n]$ as divided into n blocks, each of b bits. Let $MDH^n(K,M) = C_n$ where $C_0 =$ IV and, for $i \geq 1$, $C_i = H_K(C_{i-1} \parallel M[i])$.

Damgård [7] shows that if H is ACR then so is MDH^n. It would be nice if this worked for TCR too. But it does not. The reason is a little subtle. If H is TCR it still might be easy to find collisions in H_K if we knew K in advance (meaning we were allowed to see K before specifying any point for the collision). However, a few MD iterations of H on a fixed point can effectively surface the key K, causing subsequent iterations to misbehave. This intuition can be formalized by giving an example of a compression function H which is TCR but for which MDH^n is not. To give such an example we must first assume that some TCR compression function exists (else the question is moot).

Proposition 3. *Suppose there exists a TCR compression function* $F: \{0,1\}^\kappa \times \{0,1\}^{k+b} \to \{0,1\}^k$. *Then there exists a TCR compression function H and an integer $n \leq \max(2, \lceil (k + \kappa + b)/b \rceil)$ such that MDH^n is not TCR.*

The statement of the above proposition is "informal" insofar as we have pushed the numerical bounds into the proof.

5.2 The Basic Linear Hash

Though the MD construction doesn't propagate TCR, a natural approach is to iterate just in MDH^n but with a different key at each round. We will show that this does preserve TCR.

We let $H\colon \{0,1\}^\kappa \times \{0,1\}^{k+b} \to \{0,1\}^k$ be the given TCR compression function. The message $M = M[1]\ldots M[s]$ to be hashed is viewed as divided into $s \leq n$ blocks, each of size b.

In the basic linear scheme a hash function is specified by n different keys, $K_1,\ldots,K_n \in \{0,1\}^\kappa$, one key for each application of the underlying compression function. We hash as in the MD method but with a different key at each stage. Again we set the IV to a constant, say 0^k. Formally, $LH^n(K_1\cdots K_n, M) = C_s$ where $C_0 = 0^k$ and for $i = 1,\ldots,s\colon C_i = H(K_i, C_{i-1} \parallel M[i])$. The family of hash functions is $LH^n\colon \{0,1\}^{n\kappa} \times \Sigma_b^{\leq n} \to \{0,1\}^k$. The following theorem says that if the compression function H is resistant to target collisions then so is the extended hash function LH^n. The proof is in [4].

Theorem 4. *Suppose $H\colon \{0,1\}^\kappa \times \{0,1\}^{k+b} \to \{0,1\}^k$ is (t',ϵ')-resistant to target collisions. Suppose $n \geq 1$. Then $LH^n\colon \{0,1\}^{n\kappa} \times \Sigma_b^{\leq n} \to \{0,1\}^k$ is (t,ϵ)-resistant to equal-length target collisions, where $\epsilon = n\epsilon'$ and $t = t' - \Theta(n\cdot[T_H + \kappa])$.*

5.3 The XOR Linear Hash

We present a variant of the above in which the compression function uses the same key K in each stage, but an auxiliary "mask" key K_i, depending on the stage number i, is XORed to the chaining variable in the i-th stage. One advantage is that the key size is reduced compared to the basic scheme for some choices of the parameters. Another advantage is in key scheduling. If the compression function is being computed in hardware it may be preferable to fix the key for the compression function. In software too there can be a penalty for key "setup."

We now describe the scheme properly. Recall the message is $M = M[1]\ldots M[s]$ where $s \leq n$. The function is specified by $n + 1$ different keys $K, K_1,\ldots,K_n$ as indicated above. Let $XLH^n(KK_1\ldots K_n, M) = C_s$ where $C_0 = 0^k$ and for $i = 1,\ldots,s$ we set

$$C'_{i-1} = K_i \oplus C_{i-1} \quad \text{and} \quad C_i = H(K, C'_{i-1} \parallel M[i]).$$

The corresponding family is $XLH^n\colon \{0,1\}^{\kappa+nk} \times \Sigma_b^{\leq n} \to \{0,1\}^k$. The proof of the following can be found in [4].

Theorem 5. *Suppose $H\colon \{0,1\}^\kappa \times \{0,1\}^{k+b} \to \{0,1\}^k$ is (t',ϵ')-resistant to target collisions. Suppose $n \geq 1$. Then $XLH^n\colon \{0,1\}^{\kappa+nk} \times \Sigma_b^{\leq n} \to \{0,1\}^k$ is (t,ϵ)-resistant to equal-length target collisions, where $\epsilon = n\epsilon'$ and $t = t' - \Theta(n\cdot[T_H + \kappa])$.*

5.4 The Basic Tree Scheme

A tree can be used to reduce the key size. A tree scheme described by Naor and Yung [16] uses $\lg n$ keys (for the compression function) to hash an n-block message (using a binary tree). Later we will do better, but first let us provide a description and concrete security analysis of the basic scheme.

We are slightly more general. First, we consider m-ary trees for some $m \geq 2$. This means we start with a compression function $H \colon \{0,1\}^\kappa \times \{0,1\}^{mk} \to \{0,1\}^k$ and want to use it to hash messages of length much more than mk. Second, we do not wish to fix the number of blocks that messages will be.

PARALLEL HASH. We first describe a primitive we will use. We are given a message M which is either a single k-bit block, or else consists of s blocks, each mk bits. In the former case we leave the message unchanged. In the latter case view the message as $M = M[1] \ldots M[s]$ and hash each mk-bit block into a k-bit block via the compression function. We call this procedure "parallel hash."

Formally, for any integer $l \geq 1$ we define $PH^{lm} \colon \{0,1\}^\kappa \times (\Sigma_k \cup \Sigma_{mk}^{\leq l}) \to \Sigma_k^{\leq l}$ like this. If $M \in \Sigma_k$ set $PH^{lm}(K, M) = M$. Else write $M = M[1] \ldots M[s] \in \Sigma_{mk}^s$ ($s \leq l$) and set

$$PH^{lm}(K, M) = H(K, M[1]) \parallel H(K, M[2]) \parallel \cdots \parallel H(K, M[s]).$$

Notice that only one key is used. Notice that $PH^m = H$ is the original compression function augmented to be the identity on messages of size k.

Lemma 6. *Suppose $H \colon \{0,1\}^\kappa \times \{0,1\}^{mk} \to \{0,1\}^k$ is (t', ϵ')-resistant to target collisions. Suppose $l \geq 1$ and let $n = lm$. Then $PH^n \colon \{0,1\}^\kappa \times \Sigma_k \cup \Sigma_{mk}^{\leq l} \to \Sigma_k^{\leq l}$ is (t, ϵ)-resistant to equal-length target collisions, where $\epsilon = l\epsilon'$ and $t = t' - \Theta(nk)$.*

Proof. The proof is a slight extension and "concretization" of the proof sketch in [16, Section 2.3], and for completeness is given in [4]. ∎

BASIC TREE HASH. We assume that the messages we will hash have length bounded by km^d, for some constants m and d. To simplify our exposition we further insist that any message M to be hashed has exactly m^i k-bit blocks, for some $i \in \{0, 1, \ldots, d\}$. The hash function uses d keys $K_1, \ldots, K_d$, and the hash value is denoted $TH(K_1 \ldots K_d, M)$. It is computed on $M = M[1] \cdots M[m^i]$ by building an m-ary tree of depth i. The leaves correspond to the message blocks and the root corresponds to the final hash value. Group the nodes at level 1 (the leaves) into runs of size m and hash each group via $H(K_1, \cdot)$. This yields m^{i-1} values, which form the nodes at level 2 of the tree. Now continue the process. At each level we use a different key. Thus $H(K_j, \cdot)$ is the function used to hash the nodes at level j of the tree. At level i we have m nodes, which are hashed under $H(K_i, \cdot)$ to yield the root, which, at level $i + 1$, is the final hash value.

Formally, the hash function can be defined by compositions of the parallel hashes we described above. Namely

$$TH^{m^d} = PH^m \circ PH^{m^2} \circ \ldots \circ PH^{m^{d-1}} \circ PH^{m^d}. \tag{1}$$

By our notion of composition each parallel hash uses a different key. Thus one key is used for every hash of a given level, but the key changes across levels. So there are d keys in all. We can assess the security by applying the composition lemma and the analysis of the security of the parallel hash.

Theorem 7. *Suppose* H: $\{0,1\}^\kappa \times \{0,1\}^{mk} \to \{0,1\}^k$ *is* (t', ϵ')-*resistant to target collisions. Suppose* $n = m^d$ *where* $d \geq 1$. *Then* TH^n: $\{0,1\}^{d\kappa} \times \bigcup_{i=0}^{d} \Sigma_k^{m^i} \to \{0,1\}^k$ *is* (t, ϵ)-*resistant to equal-length target collisions, where* $\epsilon = n\epsilon'/(m-1)$ *and* $t = t' - \Theta(n \cdot [T_H + \kappa])$.

Proof. For each $i = 1, \ldots, d$, Lemma 6 says that PH^{m^i} is (t_i, ϵ_i)-secure against equal-length collisions, where $t_i = t - \Theta(m^i)$ and $\epsilon_i = m^{i-1}\epsilon'$. Note that PH^{m^i} is length consistent for any $i = 1, \ldots, d$. Now look at Equation (1) and apply Lemma 2 d times. This gives us $\epsilon = (m^0 + \ldots + m^{d-1})\epsilon' = (m^d - 1)\epsilon'/(m-1) \geq n\epsilon'/(m-1)$. The time estimates can be checked similarly. Details omitted. ∎

5.5 The XOR Tree Hash

In the basic tree hash we key the compression function anew at each level of the tree. Thus the key length is $\kappa \cdot \log_m(n)$, which can be large, because κ may be large. In the XOR variant there is one key K defining $H(K, \cdot)$ and this is used at all levels. However, there are auxiliary keys $K_1, \ldots, K_d$, one per level. These are not keys for the compression function: they are just XORed to the data at each stage. As described in Section 5.2, the motivation is that we can get shorter keys, and also better key scheduling.

We now describe the hash function. As before, assume that the messages we will hash have m^i blocks, each k-bits, for some $i \in \{0, \ldots, d\}$. The hash function is described by a "primary key" $K \in \{0,1\}^\kappa$ and d "auxiliary keys" $K_1, \ldots, K_d \in \{0,1\}^{mk}$. The hash value is denoted $XTH(KK_1 \ldots K_d, M)$. It is computed by building an m-ary tree of depth i whose root is the hash value. Level 1 consists of the leaves, which are the m^d individual blocks of the message. Group them into runs of m blocks each. Before hashing each sequence of blocks, however, XOR key K_1 to each group. Now hash each masked group via $H(K, \cdot)$. This yields m^{d-1} values, which form the nodes at level 2 of the tree. Now continue the process. At each level we use a different key for the masking, but the same key K for the hashing.

Here is a more formal description. Let $M = M[1] \ldots M[s]$ be the message, consisting of k-bit blocks, and suppose $s = m^i$. We define $XTH(KK_1 \cdots K_d, M)$ recursively. First assume $i = 0$ so that $M = M[1]$ consists of 1 block. We set $XTH(K, M) = M[1]$. Now suppose the message M is longer, namely $i \geq 1$. Let $N_j = M[(j-1)m^{i-1} + 1] \ldots M[jm^{i-1}]$ for $j = 1, \ldots, m$. Let

$$XTHI(KK_1 \ldots K_{i-1}, M) = $$
$$XTH(KK_1 \ldots K_{i-1}, N_1) \parallel \cdots \parallel XTH(KK_1 \cdots K_{i-1}, N_m) .$$

Then let

$$XTH(KK_1 \ldots K_i, M) = H(K, K_i \oplus XTHI(KK_1 \ldots K_{i-1}, M)) .$$

The first function here computes the input for the next compression stage. Before applying the compression function to it, however, we XOR in the auxiliary key for this stage. We also set $XTH(KK_1 \ldots K_d, M) = XTH(KK_1 \ldots K_i, M)$. We let XTH^n: $\{0,1\}^{\kappa+dmk} \times \bigcup_{i=0}^{d} \Sigma_k^{m^i} \to \{0,1\}^k$ denote the XOR tree hash family for messages of at most $n = m^d$ blocks.

The key length is $\kappa + mk \cdot \log_m(n)$. For $m = 2$, $k = 64$, the resulting key length of $\kappa + 128 \lg n$ is significantly smaller than for the basic tree scheme in the case where the key size of the compression function is quite big, as happens for examples in the constructions of [13].

We can no longer appeal to the composition lemma in proving security, because the hashing at the different levels of the tree involve a common key K. Instead we give a direct proof of security, which can be found in [4].

Theorem 8. *Suppose* H: $\{0,1\}^{\kappa} \times \{0,1\}^{mk} \to \{0,1\}^k$ *is* (t', ϵ')-*resistant to target collisions. Suppose* $n = m^d$ *where* $d \geq 1$. *Then* XTH^n: $\{0,1\}^{\kappa+dmk} \times \bigcup_{i=0}^{d} \Sigma_k^{m^i} \to \{0,1\}^k$ *is* (t, ϵ)-*resistant to equal-length target collisions, where* $\epsilon = n\epsilon'/(m-1)$ *and* $t = t' - \Theta(n \cdot [T_H + mk])$.

6 Length Variability

In Section 5 we proved security against equal-length target collisions. In practice one requires security against variable-length target collisions.

It is often assumed that input-length variability can be handled by padding the final block of a message M to be hashed so that it unambiguously encodes $|M|$. Let $pad(\cdot)$ denote such a padding function (eg., that of [20]). If H is secure against equal-length target collisions is is $H \circ pad$ secure against variable-length target collisions? Not necessarily. And the same applies to ACR. It is easy to construct such examples.

Here, instead, is a general technique to achieve length-variability. It requires one extra application of the compression function. See [4] for a proof.

Theorem 9. *Fix* $m > 0$ *and let* D_1 *be a set of strings each of length less than* 2^m. *Let* H_1: $\{0,1\}^{\kappa_1} \times D_1 \to \{0,1\}^{l_1}$ *and* H_2: $\{0,1\}^{\kappa_2} \times \{0,1\}^{l_1+m} \to \{0,1\}^k$ *be families of hash functions. Assume* H_1 *is* (t_1, ϵ_1)-*secure against equal-length target collisions and* H_2 *is* (t_2, ϵ_2)-*secure against equal-length target collisions. Define* H: $\{0,1\}^{\kappa_1+\kappa_2} \times D_1 \to \{0,1\}^k$ *by*

$$H(K_1 K_2, M) = H_2(K_2, H_1(K_1, M) \parallel |M|_m)$$

where $|M|_m$ *is the length of* M *written as a string of exactly* m *bits,* $M \in D_1$, $K_1 \in \{0,1\}^{\kappa_1}$, *and* $K_2 \in \{0,1\}^{\kappa_2}$. *Then* H *is* (t, ϵ)-*secure against variable-length target collisions, where* $t = \min(t_1 - \Theta(\kappa_2),\ t_2 - \Theta(\kappa_1 - 2T_{H_1} - 2l_1 - 2m))$ *and* $\epsilon = \epsilon_1 + \epsilon_2$.

While length-indicating padding doesn't work in general, does it work for the schemes of Section 5? For LH the answer is *no*: starting with an arbitrary TCR compression function H_0 one can construct a TCR compression function H for which $LH \circ pad$ is insecure against variable-length target collisions. For XLH the answer is *yes*: if H is a TCR compression function then $XLH \circ pad$ is guaranteed to be secure against target collisions; one can appropriately modify the proof of Theorem 5 to show this.

Acknowledgments

The first author is supported in part by a 1996 Packard Foundation Fellowship in Science and Engineering, and NSF CAREER award CCR-9624439. The second author is supported in part by NSF CAREER Award CCR-9624560.

Thanks to the Crypto 97 referees for their comments on this paper.

References

1. M. BELLARE, R. CANETTI AND H. KRAWCZYK, Keying hash functions for message authentication. *Advances in Cryptology – Crypto 96 Proceedings*, Lecture Notes in Computer Science Vol. 1109, N. Koblitz ed., Springer-Verlag, 1996.
2. M. BELLARE, R. CANETTI AND H. KRAWCZYK, Pseudorandom functions revisited: the cascade construction and its concrete security. *Proceedings of the 37th Symposium on Foundations of Computer Science*, IEEE, 1996.
3. M. BELLARE, J. KILIAN AND P. ROGAWAY, The security of cipher block chaining. *Advances in Cryptology – Crypto 94 Proceedings*, Lecture Notes in Computer Science Vol. 839, Y. Desmedt ed., Springer-Verlag, 1994.
4. M. BELLARE AND P. ROGAWAY, Collision-Resistant Hashing: Towards Making UOWHFs Practical. Full version of this paper, available via http://www-cse.ucsd.edu/users/mihir.
5. T. CORMEN, C. LEISERSON AND R. RIVEST, Introduction to Algorithms. McGraw-Hill, 1992.
6. I. DAMGÅRD, Collision Free Hash Functions and Public Key Signature Schemes. *Advances in Cryptology – Eurocrypt 87 Proceedings*, Lecture Notes in Computer Science Vol. 304, D. Chaum ed., Springer-Verlag, 1987.
7. I. DAMGÅRD, A Design Principle for Hash Functions. *Advances in Cryptology – Crypto 89 Proceedings*, Lecture Notes in Computer Science Vol. 435, G. Brassard ed., Springer-Verlag, 1989.
8. B. DEN BOER AND A. BOSSELAERS, An attack on the last two rounds of MD4. *Advances in Cryptology – Crypto 91 Proceedings*, Lecture Notes in Computer Science Vol. 576, J. Feigenbaum ed., Springer-Verlag, 1991.
9. B. DEN BOER AND A. BOSSELAERS, Collisions for the compression function of MD5. *Advances in Cryptology – Eurocrypt 93 Proceedings*, Lecture Notes in Computer Science Vol. 765, T. Helleseth ed., Springer-Verlag, 1993.
10. H. DOBBERTIN, Cryptanalysis of MD4. *Fast Software Encryption—Cambridge Workshop*, Lecture Notes in Computer Science, vol. 1039, D. Gollman, ed., Springer-Verlag, 1996.
11. H. DOBBERTIN, Cryptanalysis of MD5. Rump Session of Eurocrypt 96, May 1996, http://www.iacr.org/conferences/ec96/rump/index.html.

12. H. DOBBERTIN, A. BOSSELAERS AND B. PRENEEL, RIPEMD-160: A strengthened version of RIPEMD, *Fast Software Encryption*, Lecture Notes in Computer Science 1039, D. Gollmann, ed., Springer-Verlag, 1996.

13. R. IMPAGLIAZZO AND M. NAOR, Efficient cryptographic schemes provably as secure as subset sum. *Journal of Cryptology*, Vol. 9, No. 4, Autumn 1996.

14. B. KALISKI AND M. ROBSHAW, Message Authentication with MD5. *RSA Labs' CryptoBytes*, Vol. 1 No. 1, Spring 1995.

15. R. MERKLE, One way hash functions and DES. *Advances in Cryptology – Crypto 89 Proceedings*, Lecture Notes in Computer Science Vol. 435, G. Brassard ed., Springer-Verlag, 1989

16. M. NAOR AND M. YUNG, Universal one-way hash functions and their cryptographic applications. *Proceedings of the 21st Annual Symposium on Theory of Computing*, ACM, 1989.

17. National Institute of Standards, FIPS 180-1, Secure hash standard. April 1995.

18. B. PRENEEL AND P. VAN OORSCHOT, MD-x MAC and building fast MACs from hash functions. *Advances in Cryptology – Crypto 95 Proceedings*, Lecture Notes in Computer Science Vol. 963, D. Coppersmith ed., Springer-Verlag, 1995.

19. RIPE Consortium, Ripe Integrity primitives — Final report of RACE integrity primitives evaluation (R1040). Lecture Notes in Computer Science, vol. 1007, Springer-Verlag, 1995.

20. R. RIVEST, The MD4 message-digest algorithm, *Advances in Cryptology – Crypto 90 Proceedings*, Lecture Notes in Computer Science Vol. 537, A. J. Menezes and S. Vanstone ed., Springer-Verlag, 1990, pp. 303–311. Also IETF RFC 1320 (April 1992).

21. R. RIVEST, The MD5 message-digest algorithm. IETF RFC 1321 (April 1992).

22. J. ROMPEL, One-way functions are necessary and sufficient for digital signatures. *Proceedings of the 22nd Annual Symposium on Theory of Computing*, ACM, 1990.

23. G. Tsudik, Message authentication with one-way hash functions, *Proceedings of Infocom 92*, IEEE Press, 1992.

24. S. VAUDENAY, On the need for multipermutations: cryptanalysis of MD4 and SAFER. *Fast Software Encryption — Leuven Workshop*, Lecture Notes in Computer Science, vol. 1008, Springer-Verlag, 1995, 286–297.

25. WEGMAN AND CARTER, New hash functions and their use in authentication and set equality, *Journal of Computer and System Sciences*, Vol. 22, 1981, pp. 265–279.

Fast and Secure Hashing Based on Codes

Lars Knudsen and Bart Preneel*

Katholieke Universiteit Leuven, Dept. Electrical Engineering-ESAT,
Kardinaal Mercierlaan 94, B–3001 Heverlee, Belgium
{lars.knudsen,bart.preneel}@esat.kuleuven.ac.be

Abstract. This paper considers hash functions based on block ciphers. It presents a new attack on the compression function of the 128-bit hash function MDC-4 using DES with a complexity far less that one would expect, and proposes new constructions of fast and secure compression functions based on error-correcting codes and m-bit block ciphers with an m-bit key. This leads to simple and practical hash function constructions based on block ciphers such as DES, where the key size is slightly smaller than the block size, IDEA, where the key size is twice the block size and to MD4-like hash functions. Under reasonable assumptions about the underlying block cipher, we obtain collision resistant compression functions. Finally we provide examples of hashing constructions based on both DES and IDEA more efficient than previous proposals and discuss applications of our approach for MD4-like hash functions.

1 Introduction

Cryptographic *hash functions* map a string of arbitrary size to a short string of fixed length, typically 128 or 160 bits. Hash functions are used in cryptographic applications such as digital signatures, password protection schemes, and conventional message authentication. For these applications one requires that it is hard to find an input corresponding to a given hash result (preimage resistance) or a second input with the same hash result as a given input (2nd preimage resistance). Moreover, many applications also require that it is hard to find two inputs with the same hash result (collision resistance). For the remainder of this paper we consider hash functions satisfying these three properties.

All important hash functions are *iterated hash functions* based on an easily computable compression function $h(\cdot, \cdot)$ from two binary sequences of respective lengths m and l to a binary sequence of length m. The message M is split into blocks M_i of l bits, $M = (M_1, M_2, \ldots, M_n)$. If the length of M is not a multiple of l, M is padded using an unambiguous padding rule. The *hash result* $\text{Hash}(IV, M) = H = H_t$ is obtained by computing iteratively

$$H_i = h(H_{i-1}, M_i) \qquad i = 1, 2, \ldots, t, \tag{1}$$

* F.W.O. postdoctoral researcher, sponsored by the Fund for Scientific Research, Flanders (Belgium).

where $H_0 = IV$ is a specified *initial value*. Collision attacks, 2nd preimage attacks, and preimage attacks can be applied to both the compression function and the hash function. For the remainder of this paper we make no distinction between the latter two attacks and refer to both of them as preimage attacks.

It is possible to relate the security of Hash($\cdot$) to that of $h(\cdot,\cdot)$ in several models [2, 12, 15, 17]. To do this, one needs to append an additional block at the end of the input string containing its length, known as MD-strengthening (after Merkle [15] and Damgård [2]), leading to the following result.

Theorem 1. *Let* Hash($\cdot$) *be an iterated hash function with MD-strengthening. Then preimage and collision attacks on* Hash($\cdot,\cdot$) *(where an attacker can choose IV freely) have roughly the same complexity as the corresponding attacks on* $h(\cdot,\cdot)$.

Theorem 1 gives a lower bound on the security of Hash($IV, \cdot$). Most practical hash functions do not have a collision resistant compression function and do not treat the two inputs of the compression function in the same way; exceptions are the DES based hash functions of Merkle [15] and those of the authors [11]. As an example, collisions for the compression function of MD5 have been presented in [3, 6].

We consider hash functions based on block ciphers with block length m bits and key length k bits. For convenience we will assume that k is an integer multiple of m. Such a block cipher defines, for each k-bit key, a permutation on m-bit strings. The advantage of constructing hash functions based on block ciphers is the minimization of the design and implementation effort. Moreover, the security of the block cipher can, to a certain extent, be transferred to hash functions. One difference is that the block cipher has to satisfy certain properties even if the key is known (see for example [19]). The main disadvantage of the block cipher approach is that customized hash functions are likely to be more efficient. However, Dobbertin's attacks [4, 6] on specific hash functions such as MD4 [21] and MD5 [22] have shown that this efficiency can come at a high cost, which makes the block cipher constructions more attractive.

We define the *hash rate* of a hash function based on an m-bit block cipher as the number of m-bit message blocks processed per encryption or decryption. The *complexity* of an attack is the total number of operations, i.e., encryptions or decryptions, required for an attacker to succeed with a high probability.

The first secure constructions for a hash function based on a block cipher were the scheme by Matyas, Meyer, and Oseas [14], which has been specified in ISO/IEC 10118 [9], and its dual, widely known as the Davies-Meyer scheme.[2] Both schemes are *single block length hash functions* giving a hash result of only m bits with hash rate 1. A classification of these schemes has been presented by Preneel et al. in [18]. Using a birthday attack collisions for such a scheme can be found in about $2^{m/2}$ operations and a preimage in about 2^m operations. However, since most block ciphers have a block length of $m = 64$ bits, collisions can be found in only 2^{32} operations and hash functions with a larger hash result

---

[2] The real inventors are probably S.M. Matyas and C.H. Meyer.

are needed. The aim of *double block length* hash functions is first of all to achieve a security level against collision attacks of at least 2^m encryptions. For all constructions with rate 1 of a large class the authors have shown that collisions (for H) can be found in no more than $2^{3m/4}$ operations [11]. Using DES the best known constructions are MDC-2 of rate 1/2 [1, 9], and MDC-4 of rate 1/4 [1]; for both a collision attack is believed to require at least 2^m operations. Another important class of constructions are those of Merkle [15]. In [12], Lai and Massey have proposed two constructions for hash functions based on their block cipher IDEA (with $m = 64$ and $k = 128$): Abreast-DM and Tandem-DM have hash rate 1/2 and a claimed security level against collision attacks of 2^m operations.

However, for all these constructions, except for those of Merkle, there is no proof of security and for MDC-2 and MDC-4 collision for the compression function can be found faster than by a brute force attack. Moreover, for block ciphers with $m = 64$ (e.g., DES and IDEA), the double block constructions do not provide an acceptable security level against parallel brute force collision attacks: it follows from [20, 23] that a security level of at least $2^{75} \ldots 2^{80}$ encryptions is required, which is not offered by any of the current proposals. In [11] we developed a new framework for constructing hash functions based on m-bit block ciphers with m-bit keys using linear codes over $GF(2^2)$. Constructions were shown for which finding a collision requires at least $2^{qm/2}$ encryptions (with $q \geq 2$), and finding a preimage requires at least 2^{qm} encryptions at the cost of more internal memory. For $q = 2$, the constructions are more efficient than existing proposals with the same security level.

In this paper we first show that collisions for the compression function of MDC-4 can be found in time $2^{3m/4}$. For DES some key bits have to be fixed to avoid weak keys and the complementation property, hence the complexity of our attack is only 2^{41}, which is quite feasible today. Since the hash rate of MDC-4 is 1/4, i.e., only one block is processed per four encryptions, one should expect a higher level of security for the compression function. Second we generalize the approach of [11]. We propose a method for constructing hash functions based on block ciphers with larger keys and working on smaller blocks. More precisely, we consider m-bit block ciphers with tm-bit keys, $t \geq 1$, using linear codes over $GF(2^s)$, $s \geq 2$. Our constructions result in more efficient hash functions than those of [11]. Using DES and IDEA we show that it is possible to obtain hash rates close to 1 respectively 2 with a high security level against collision attacks.

In the following $E_K(X)$ and $D_K(Y)$ denote the encryption and decryption of the m-bit plaintext X respectively ciphertext Y under the tm-bit key K. It will be assumed that the block cipher has no weaknesses, i.e., that in attacks on the hash functions based on the block cipher, no short-cut attacks on the block cipher will help an attacker.

The remainder of this paper is organized as follows. In §2 we analyze the compression function of MDC-4. Our new constructions are presented in §3, and in §4 we provide examples of efficient constructions using DES, IDEA, and the MD4-like functions, hereafter denoted the *MDx family*.

2 The Compression Function of MDC-4

MDC-2 and MDC-4 [1] are hash functions based on a block cipher; they are also known as the Meyer-Schilling hash functions after the authors of the first paper describing these schemes [16]. MDC-2 is included in ISO/IEC 10118 Part 2, standardizing a hash function based on a block cipher [9]. MDC-2 can be described as follows, where $h(X,Y) = E_X(Y) \oplus Y$ and $E_K(\cdot)$ denotes encryption with an m-bit block cipher with key K:

$$T1_i = h(u(H1_{i-1}), X_i) = LT1_i \parallel RT1_i$$
$$T2_i = h(v(H2_{i-1}), X_i) = LT2_i \parallel RT2_i$$

$$H1_i = LT1_i \parallel RT2_i$$
$$H2_i = LT2_i \parallel RT1_i.$$

The variables $H1_0$ and $H2_0$ are initialized with the values IV_1 and IV_2 respectively, and the hash result is equal to the concatenation of $H1_t$ and $H2_t$. The rate of this scheme is equal to 1/2. The ISO/IEC standard does not specify any particular block cipher; it also requires the specification of two mappings u, v from the ciphertext space to the key space such that $u(IV_1) \neq v(IV_2)$. For DES, u and v omit the parity bits and fix the second and third bit to 10 and 01 respectively, to preclude attacks based on (semi-)weak keys. The compression function of MDC-2 is certainly not collision resistant: for any fixed choice of X_i, one can find collisions for both $H1_i$ and $H2_i$ independently with a simple birthday attack requiring about $2^{m/2}$ operations.

One iteration of MDC-4 [1] is defined as a concatenation of two MDC-2 steps, where the plaintexts in the second step are equal to $H2_{i-1}$ and $H1_{i-1}$:

$$T1_i = h(u(H1_{i-1}), X_i) = LT1_i \parallel RT1_i$$
$$T2_i = h(v(H2_{i-1}), X_i) = LT2_i \parallel RT2_i$$

$$U1_i = LT1_i \parallel RT2_i$$
$$U2_i = LT2_i \parallel RT1_i$$

$$V1_i = h(u(U1_i), H2_{i-1}) = LV1_i \parallel RV1_i$$
$$V2_i = h(v(U2_i), H1_{i-1}) = LV2_i \parallel RV2_i$$

$$H1_i = LV1_i \parallel RV2_i$$
$$H2_i = LV2_i \parallel RV1_i.$$

The rate of MDC-4 is equal to 1/4. It is clear that the "exchange" of $H1_{i-1}$ and $H2_{i-1}$ in the second step does not improve the algorithm: after the exchange of right halves, $U1_i$ and $U2_i$ are symmetric with respect to $H1_{i-1}$ and $H2_{i-1}$.

Finding a collision for the compression function is harder than in the case of MDC-2. On the other hand, collisions for the compression function of MDC-2 with different values of X_i and with the same value of $(H1_{i-1}, H2_{i-1})$ will also yield collisions for MDC-4, but generally this property does not hold for other collisions for the basic function like pseudo-collisions, i.e., collisions where all the inputs to the compression function are varied.

In the following we will show a collision attack on the compression function with a complexity smaller than a brute-force attack:

1. Choose a random value of X_i and of $H2_{i-1}$ (with $i = 1$ this can be the specified initial value).
2. Calculate $T1_i$ for $2^{3m/4}$ values of $H1_{i-1}$. One expects to find an $m/2$-bit string S such that for a set of $2^{m/4}$ values the relevant bits of LT_i are equal to S (in fact for 50% of the strings S there will be $2^{m/4}$ cases or more).
3. Compute $V2_i$ for the $2^{m/4}$ inputs in this set. The probability of obtaining a collision for $V2_i$ is equal to $\left(2^{m/4}\right)^2 /2^{m+1} = 2^{-m/2+1}$.
4. If no match is obtained, one chooses a new value for S; one can avoid re-computing $T1_i$ if one stores $2^{m/4} \cdot 2^{m/2} = 2^{3m/4}$ m-bit quantities.

The expected effort to find a collision for the compression function of MDC-4 is $2^{3m/4}$ encryptions and the storage of $2^{3m/4}$ m-bit quantities. The keys of MDC-4 using DES are effectively only 54 bits long [1]. In that case the collision attack requires about 2^{41} DES encryptions and a storage of $2^{30.5}$ 54-bit quantities (about 10 Gigabyte).

3 Constructions with Linear Codes

In this section we present new constructions for collision resistant hash functions based on an m-bit block cipher and linear codes. These constructions extend a simple hash mode which is believed to be secure to a multiple hash mode. As the simple hash mode we will use the Davies–Meyer hash function:

$$h_i(M_i, H_{i-1}) = E_{M_i}(H_{i-1}) \oplus H_{i-1} . \tag{2}$$

Any of the 12 secure single block length hash functions described in [18] can be used. The advantage of using the Davies-Meyer hash function is that it is defined for block ciphers with different block and key sizes. The following assumption is standard in cryptography today.

Assumption 1 *Let $E_K(.)$ be an m-bit block cipher with a tm-bit key K for an integer $t > 0$. Define the compression function h to be the Davies-Meyer function (2). Then finding collisions for h requires about $2^{m/2}$ encryptions (of an m-bit block), and finding a preimage for h requires about 2^m encryptions.*

Definition 2 (Multiple Davies-Meyer). Let $E_K(.)$ be an m-bit block cipher with a tm-bit key K for integer $t > 0$. Let $h_1, h_2, \ldots h_n$ be different instantiations of the Davies-Meyer function, that is, $h_i(X_i, Y_i) = E_{X_i}(Y_i) \oplus Y_i$, obtained by fixing $\lceil \log_2 n \rceil$ key (or plaintext) bits to different values, where the X_i's are tm-bit string and the Y_i's are m-bit strings. The compression function of a multiple Davies-Meyer scheme takes r m-bit input blocks, which are expanded by an affine mapping to the n pairs (X_i, Y_i). The output is the concatenation of the outputs of the function $h_1, \ldots, h_n$. The output of the compression function shall depend on all r input blocks, in other words, the matrix of the affine mapping has full rank.

A subfunction $h_i(X_i, Y_i)$ is called *active*, if in a collision or preimage attack the input blocks forming (X_i, Y_i) are different. Two subfunctions $h_i(X_i, Y_i)$ and $h_j(X_j, Y_j)$ can be attacked *independently*, if one can vary one (or more) input blocks to the compression function, such that the arguments (X_i, Y_i) of h_i vary, while the arguments (X_j, Y_j) of h_j do not.

We make the following assumption, which is a slightly improved version of the assumption in [11].

Assumption 2 *Assume that a collision or preimage for the compression function of a multiple Davies-Meyer scheme has been found, that is, simultaneously for $h_1, h_2, \ldots h_n$. Let N be the number of active subfunctions and let $N - v$ be the maximum number of the N subfunctions that can be attacked independently. Then it is assumed that obtaining this collision or preimage must have required at least $2^{vm/2}$ respectively 2^{vm} encryptions.*

Note that in an attempt to find collisions or preimages for a multiple Davies-Meyer scheme it will always be possible to fix some input blocks and thereby fix the outputs. Let N denote the number of active subfunctions. What Assumption 2 says is, that if $N - v$ of these N functions can be attacked independently (separately), then there exists no better attack than a brute force attack on the remaining v subfunctions. Note that in the overall complexity of the collision (or the preimage) attack we do not consider the complexity of the attack on the $N - v$ functions, which makes our assumption strong and plausible.

3.1 The new constructions

The following theorem shows how hash functions based on block ciphers can be constructed using non-binary linear error correcting codes. It extends the main theorem of [11].

Theorem 3. *If there exists an $[n, k, d]$ code over $GF(2^{t+1})$ of length n, dimension k, and minimum distance d, with $(t+1)k > n$, for $t \geq 1$ and $m \gg \log_2 n$, then there exists a parallel hash function based on an m-bit block cipher with a tm-bit key for which finding a collision for the compression function requires at least $2^{(d-1)m/2}$ encryptions and finding a preimage requires at least $2^{(d-1)m}$ encryptions provided that Assumption 2 holds. The hash function has an internal memory of $n \cdot m$ bits, and a rate of $(t+1)\frac{k}{n} - 1$.*

Proof: The compression function consists of n different functions h_i with $1 \leq i \leq n$, see Definition 2. The input to the compression function consists of $(t+1)k$ m-bit blocks: the n variables H_{i-1}^1 through H_{i-1}^n (the output of the n functions of the previous iteration) and r message blocks M_i^1 through M_i^r, with $r = (t+1)k - n > 0$. In the following, every individual bit of these m-bit blocks is treated in the same way. The bits of $(t + 1)$ consecutive input blocks are concatenated yielding k elements of $GF(2^{t+1})$. These elements are encoded using the $[n, k, d]$ code, resulting in n elements of $GF(2^{t+1})$. Each of these elements represents the $(t + 1)$-bit inputs to one of the n functions, that is, one bit represents the plaintext block input and the remaining t bits represent the key input to the block cipher. The individual input bits are obtained by representing the elements of $GF(2^{t+1})$ as a vector space over $GF(2)$. This construction guarantees that the conditions for Assumption 2 are satisfied for the value $v = d - 1$. It follows from the minimum distance of the code that at least d subfunctions are active in a collision. Also, k is the maximum number of subfunctions, which can be attacked independently. But since $n - k \geq d - 1$ by the Singleton bound [13, p. 33] the result follows. ∎

The existence of efficient constructions for $d = 3$ and $d = 4$ follows from the existence of perfect Hamming codes over $GF(2^t)$ (see e.g., [13]) with parameters $n = (q^s - 1)/(q - 1)$, $k = n - s$, $d = 3$ for a prime power q, and from the existence of triply extended MDS (maximum distance separable) codes [13, Ch. 11, Th. 10] with parameters $n = q^s + 2$, $k = q^s - 1$, $d = 4$ for an even prime power q.

Corollary 4. *Provided that Assumption 2 holds, there exist parallel hash functions based on an m-bit block cipher with a tm-bit key of rate close to t for which finding a collision (a preimage) takes at least 2^m (2^{2m}) operations respectively at least $2^{3m/2}$ (2^{3m}) operations.*

Proof: From Theorem 3 it follows that there exist hash functions with rates $(t + 1)k/n - 1$. But since for Hamming codes $n/k \rightarrow 1$ for large values of n, the result follows. ∎

This result implies that at the cost of a larger internal memory, using DES one can obtain hash functions of rate 1, and using IDEA one can obtain hash functions of rate 2. For comparison MDC-2 and MDC-4 have rates 1/2 respectively 1/4, and Abreast-DM and Tandem-DM developed for IDEA [12] have rates 1/2.

In [11] we gave an example of a hash function based on an m-bit block cipher with an m-bit key using the code $[8, 5, 3]$ over $GF(2^2)$. The code is obtained by shortening the Hamming code $[21, 18, 3]$. The hash function has rate 1/4 and an internal memory of $8 \cdot m$ bits. In the following we show that one can improve this result.

The idea is to divide the m-bit words into smaller blocks and to use codes over bigger fields. As an example, assume we have a block cipher with m-bit blocks and m-bit keys for even m. In Theorem 3 codes over $GF(2^2)$ are used, where the two bits of the code words represent the plaintext respectively the key inputs to the block ciphers. An alternative method is to divide all m-bit blocks into blocks of $m/2$ bits and use codes over $GF(2^4)$. The advantage of this approach, which will be illustrated later, is that better bounds exist for such codes. This leads to the following generalization.

Theorem 5. *Let m be a multiple of b, i.e., $m = b \cdot m_b$. If there exists an $[n, k, d]$ code over $GF(2^{b(t+1)})$ of length n, dimension k, and minimum distance d, with $(t + 1)k > n$, and $m \gg \log_2 n$, then there exists a parallel hash function based on an m-bit block cipher with a tm-bit key for which finding a collision for the compression function requires at least $2^{(d-1)m/2}$ encryptions provided that Assumption 2 holds. The hash function has an internal memory of $n \cdot m$ bits, has a rate of $(t + 1)\frac{k}{n} - 1$ and works on m_b-bit blocks.*

Proof: Similar to that of Theorem 3. ∎

As an example, we can construct a hash function based on an m-bit block cipher with an m-bit key by using the code $[8, 6, 3]$ over $GF(2^4)$, which is obtained by shortening the Hamming code $[17, 15, 3]$. The hash function has rate $1/2$ and an internal memory of $8 \cdot m$ bits, and is thus twice as fast as the example using the code $[8, 5, 3]$ mentioned above. With $m = 64$ this construction operates on 32-bit words. One can extend this approach to construct hash functions with codes over $GF(2^8)$, i.e., operating on 16-bit words, for example by shortening the $[257, 255, 3]$ Hamming code ($s = 2$). However, since a $[17, 15, 3]$ over $GF(2^4)$ exists, the construction over $GF(2^8)$ is only more efficient for $n > 17$. Using an $[n, k, d]$-code requires $n \cdot m$ bits of internal memory, which make the constructions less attractive for larger n. Moreover, the codes obtained from splitting the blocks into smaller words result in a more complex implementation.

Notes:

1. Apart from the simple security proof and the relatively high rates, the schemes have the advantage that the n encryptions can be carried out in parallel.
2. The disadvantages of the schemes are the increased amount of internal memory and the cost of the code implementation (mainly some exclusive ors).
3. For the preimage attack, the security bounds assume that the entropy of the unknown part of the input is at least $(d - 1)m$ bits.

As for (3), it is clear that if $D < (d - 1)m$ bits are unknown to the attacker, a brute force preimage attack can be done in about 2^D encryptions.

3.2 Output transformation

The constructions presented above have the following problems:

1. since every output bit does not depend on all input bits of the compression function, it is relatively easy to find many inputs for which several output blocks of the compression function are equal,
2. the number of output blocks is typically much larger than the security level suggests.

The solution is to apply an output transformation to the outputs of the compression function. This transformation can be slow, since it has to be applied only once. Therefore there are many possible constructions.

First we present an approach that does not affect the provable security of the compression functions. One encrypts the n output blocks of the compression function using the block cipher with a fixed key, such that all output blocks of the encryption depend on all input blocks in a complicated way. Subsequently, the n blocks concatenated with the n encrypted blocks are hashed using the following construction. Denote with $n_{\min}$ the smallest possible value of n for a given value of d, such that Theorem 5 holds. Compress the $2n$ blocks to $n_{\min}$ blocks using the new construction with $n_{\min}$ parallel blocks (if $n_{\min} < n$, this hash function will have a lower rate than the original one). This approach solves the first problem and partly the second problem. However, if a further reduction to less than $n_{\min}$ blocks is required, other approaches are necessary.

We present next a generic approach for all values of n, which can be used instead of or in conjunction with the first approach. First one constructs from the m-bit block cipher a large, strong block cipher with block length $n \cdot m$ bits. This block cipher can be slow, since it is applied only once. Subsequently, the n ($n_{\min}$) blocks from the compression function are input to a Davies-Meyer construction where the block cipher key is randomly chosen and fixed (and part of the hash function description). Under Assumption 1 this is a secure hash function. The output can be truncated to any s blocks, where $s \geq d - 1$.

We give here an example of such a construction. One iteration consists of the following two steps. First, permute the blocks, such that block i becomes block $i+1$ and the last block becomes the first block. Second, encrypt the input blocks using the small block cipher in CBC mode. More precisely, denote the output of the compression function with $H^1, \ldots, H^n$. Let K_i for $i = 1, \ldots, n$ be n randomly chosen and fixed keys, and let $C^i = H^i$ for $i = 1, \ldots, n$. Repeat the following procedure r times:

$$C^0 = C^n$$
$$C^i = C^{i-1} \text{ for } i = 1, \ldots, n,$$
$$C^i = E_{K_i}(C^i \oplus C^{i-1}) \text{ for } i = 1, \ldots, n$$

(Here we use the same n keys in every iteration. Alternatively, different keys can be used in all rounds. Also the block permutation in the first iteration can be omitted.) After one iteration the last block will depend on all input blocks and in general, block i depends on i input blocks. After two iterations all blocks depend on all input blocks. However, two rounds are insufficient to make a strong block cipher. Therefore we recommend that this block cipher is used with at least $r = n$ rounds. Finally, the result of the output transformation is defined as the concatenation of the blocks $H^i \oplus C^i$ for $i = 1, \ldots, n$. If the output is truncated we recommend that the final output is formed from the blocks with the highest indices. With r rounds the output transformation requires r^2 encryptions of the small block cipher.

In the next section we give some practical examples.

4 Some Practical Examples

This section contains some examples of new constructions for different parameters of the underlying block cipher. We will use the notation (m, k) for an m-bit block cipher with a k-bit key. In the examples to come we use Hamming codes with $d = 3$ and MDS codes with $d = 4$. The existence of the latter codes follows from [13, Ch. 11, Th. 10].

4.1 Using an (m, m)-block cipher

Table 1. Rates and complexities of previous proposals for (m, m)-block ciphers.

Scheme	Rate	Collision h	Collision H	Reference
MDC-2	1/2	$2^{m/2}$	2^m	[1]
MDC-4	1/4	$2^{3m/4}$	2^m	[1]
Merkle	0.27	2^m	2^m	[15]

Table 2. Comparison of constructions based on codes over $GF(2^2)$ and over $GF(2^4)$ for (m, m)-block ciphers.

$GF(2^2)$		$GF(2^4)$		Collision
Code	Rate	Code	Rate	
$[5, 3, 3]$	1/5	$[6, 4, 3]$	1/4	$\geq 2^m$
$[8, 5, 3]$	1/4	$[8, 6, 3]$	1/2	$\geq 2^m$
$[12, 9, 3]$	1/2	$[12, 10, 3]$	2/3	$\geq 2^m$
$[9, 5, 4]$	1/9	$[9, 6, 4]$	1/3	$\geq 2^{3m/2}$
$[16, 12, 4]$	1/2	$[16, 13, 4]$	5/8	$\geq 2^{3m/2}$

Note that the 2^m complexities of MDC-2 and MDC-4 are the best *known* attacks against the hash function, while against Merkle's scheme and the schemes of Table 2 the 2^m complexity is against the compression functions and is a lower bound for the complexity of an attack on the hash function.

In the following we show an implementation of the construction using the code $[9, 6, 4]$. We define $GF(2^4)$ as the extension field $GF(2)[x]/(x^4 + x + 1)$. There are many generator matrices for a $[9, 6, 4]$ linear code over $GF(2^4)$. We have searched for a generator matrix which leads to a simple and efficient compression function, as explained below. The generator matrix has the following form:

$$
\begin{bmatrix}
1 & 0 & 0 & 0 & 0 & 0 & 1 & 1 & 1 \\
0 & 1 & 0 & 0 & 0 & 0 & 1 & \alpha & \beta \\
0 & 0 & 1 & 0 & 0 & 0 & 1 & \beta & \alpha \\
0 & 0 & 0 & 1 & 0 & 0 & \alpha & 1 & \beta \\
0 & 0 & 0 & 0 & 1 & 0 & \alpha & \beta & 1 \\
0 & 0 & 0 & 0 & 0 & 1 & \beta & 1 & \alpha
\end{bmatrix}
\tag{3}
$$

Here 0 and 1 are the additive and multiplicative neutral elements in $GF(2^4)$ and $\alpha = x$, and $\beta = x^3+x^2+1$. The motivation for the choice of the generator matrix is as follows. In an implementation of the compression functions the elements of $GF(2^4)$ are represented as elements of a vector space over $GF(2)$. Clearly, multiplications with 0 and 1 are the easiest to implement. A closer analysis shows that multiplication with α and β in the above example can be implemented with one respectively two exclusive-ors. An exhaustive search for the matrix with the easiest implementation is clearly not feasible, but by restricting ourselves to using the elements 0, 1, α, and β we obtain a solution close to the optimal one.

Let $f_i(X, Y)$ be different instantiations of the function $E_X(Y) \oplus Y$, let X_L and X_R denote the leftmost respectively rightmost $m/2$ bits of X and let $\|$ denote concatenation of $m/2$ bit blocks. Furthermore, let $G^1, \ldots, G^9$ be the 9 input blocks coming from the compression function in the previous iteration and let M^1, M^2, M^3 be the 3 message block inputs. This results in the following compression function:

$$H^1 = f_1(G^1, G^2)$$
$$H^2 = f_2(G^3, G^4)$$
$$H^3 = f_3(G^5, G^6)$$
$$H^4 = f_4(G^7, G^8)$$
$$H^5 = f_5(G^9, M^1)$$
$$H^6 = f_6(M^2, M^3)$$
$$H^7 = f_7(G^1 \oplus G^3 \oplus G^5 \oplus (G_R^7 \| G_L^8) \oplus (G_R^9 \| M_L^1) \oplus (M_{LR}^3 \| M_R^3),$$
$$\quad G^2 \oplus G^4 \oplus G^6 \oplus (G_L^7 \oplus G_R^8 \| G_L^7) \oplus (G_L^9 \oplus M_R^1 \| G_L^9) \oplus (M_L^2 \| M_R^2 \oplus M_{LR}^3))$$
$$H^8 = f_8(G^1 \oplus G^7 \oplus M^2 \oplus (G_R^3 \| G_L^4) \oplus (G_{LR}^5 \| G_L^6) \oplus (M_{LR}^1 \| M_R^1),$$
$$\quad G^2 \oplus G^8 \oplus M^3 \oplus (G_R^4 \oplus G_L^3 \| G_L^3) \oplus \mathcal{G}^5 \oplus (G_L^9 \| G_R^9 \oplus M_{LR}^1))$$
$$H^9 = f_9(G^1 \oplus G^9 \oplus (G_{LR}^4 \| G_R^4) \oplus (G_R^5 \| G_L^6) \oplus (G_{LR}^8 \| G_R^8) \oplus (M_R^2 \| M_L^3),$$
$$\quad G^2 \oplus M^1 \oplus \mathcal{G}^3 \oplus (G_R^6 \oplus G_L^5 \| G_L^5) \oplus \mathcal{G}^7 \oplus (M_L^2 \oplus M_R^3 \| M_L^2)),$$

where $\mathcal{G}^t = (G_L^t \| G_R^t \oplus G_L^{t+1} \oplus G_R^{t+1})$ and $X_{LR} = X_L \oplus X_R$.

As an output transformation we suggest to first hash the 9 blocks to 7 blocks via the compression function using the code $[7, 4, 4]$ ($n_{\min} = 7$ for $d = 4$) and then to hash the 7 blocks to 3 blocks using the output transformation with 7 rounds described in the previous section.

4.2 Using an $(m, 2m)$-block cipher

The only known hash functions based on an $(m, 2m)$-block cipher with a $2m$-bit hash result are the Abreast-DM and the Tandem-DM from [12]. The compression functions of both hash functions process an m-bit block using two encryptions, i.e., with rate $1/2$. Let H_i and G_i denote the intermediate hash values. The Abreast-DM is defined as follows

$$H_i = H_{i-1} \oplus E_{G_{i-1}, M_i}(H_{i-1})$$
$$G_i = G_{i-1} \oplus E_{M_i, H_{i-1}}(\overline{G}_{i-1}),$$

where $\overline{G}$ is the bitwise complemented value of G. The Tandem-DM is defined as follows

$$W_i = E_{G_{i-1},M_i}(H_{i-1})$$
$$H_i = W_i \oplus H_{i-1}$$
$$G_i = G_{i-1} \oplus E_{M_i,W_i}(G_{i-1}).$$

Table 3 lists the rates and complexities of the best known attacks on the two constructions. However, as already indicated before, there exist more efficient

Table 3. Rates and complexities of previous proposal for $(m, 2m)$-block ciphers [12].

Scheme	Rate	Collision h	Collision H
Abreast-DM	1/2	2^m	2^m
Tandem-DM	1/2	2^m	2^m

constructions with a higher security level. Table 4 lists the rates and complexities of such constructions. As before, it is possible divide the m-bit blocks into smaller subblocks. E.g., the blocks can be divided into halves and expanded with a code over $GF(2^6)$, such as $[65, 63, 3]$.

Table 4. Rates and complexities of our proposals for $(m, 2m)$-block ciphers using codes over $GF(2^3)$.

Code	Rate	Collision
$[4, 2, 3]$	1/2	$\geq 2^m$
$[6, 4, 3]$	1	$\geq 2^m$
$[9, 7, 3]$	4/3	$\geq 2^m$
$[5, 2, 4]$	1/5	$\geq 2^{3m/2}$
$[7, 4, 4]$	5/7	$\geq 2^{3m/2}$
$[10, 7, 4]$	11/10	$\geq 2^{3m/2}$

4.3 The MDx family

Both MD4 [21] and MD5 [22] can be viewed as a Davies-Meyer construction with an underlying m-bit block cipher with a $4m$-bit "key." From this perspective, both constructions have rate 4.

However, Dobbertin's attacks [4, 6] on MD4 and MD5 show that the compression functions are not collision resistant, and his attack on the extended MD4 [5] shows that for MD4 even two dependent runs of the compression function are not collision resistant. However, it seems unlikely that Dobbertin's attacks extend to compression functions consisting of two or more instantiations of MD5. We can apply our methods to construct parallel MD5 hash functions based on linear codes over $GF(2^5)$. In Table 5 we list possible constructions.

Table 5. Rates and complexities of our proposals for the MDx family using codes over $GF(2^5)$.

Scheme	Rate	Scheme	Rate	Scheme	Rate
MD4	4	[5, 3, 3]	2	[5, 2, 4]	1
MD5	4	[10, 8, 3]	3	[10, 7, 4]	2.5
		[20, 18, 3]	3.5	[20, 17, 4]	3.25

Since the assumption for our constructions, i.e., that the basic components are secure, does not hold for MD4 and MD5, we do not specify explicit bounds for the complexities of collision attacks on the compression functions. However, we conjecture that for the constructions using MD5 and codes of minimum distance 4, a collision attack is infeasible. The attack requires a simultaneous collision for at least 3 different instantiations with dependent inputs.

5   Conclusion

We have demonstrated that finding collisions of the compression function of MDC-4 takes 2^{41} operations when used with DES. This casts some doubts on the security of MDC-4. We have presented a new method for construction of hash functions based on block ciphers such as DES which is faster and more secure than existing proposals. Also, our method extends to block ciphers such as IDEA where the block size and key size are different. For large values of the internal memory, constructions using IDEA exist with rates close to two, which is a factor of four faster than existing proposals. Finally, we discussed the applications of our result to the MDx family. We show constructions using MD5, almost as fast as MD5, but (conjectured) much more secure than MD5 itself.

References

1. B.O. Brachtl, D. Coppersmith, M.M. Hyden, S.M. Matyas, C.H. Meyer, J. Oseas, S. Pilpel, M. Schilling, *"Data Authentication Using Modification Detection Codes Based on a Public One Way Encryption Function,"* U.S. Patent Number 4,908,861, March 13, 1990.
2. I.B. Damgård, "A design principle for hash functions," *Advances in Cryptology, Proc. Crypto'89, LNCS 435,* G. Brassard, Ed., Springer-Verlag, 1990, pp. 416–427.
3. B. den Boer, A. Bosselaers, "Collisions for the compression function of MD5," *Advances in Cryptology, Proc. Eurocrypt'93, LNCS 765,* T. Helleseth, Ed., Springer-Verlag, 1994, pp. 293–304.
4. H. Dobbertin, "Cryptanalysis of MD4," *Fast Software Encryption, LNCS 1039,* D. Gollmann, Ed., Springer-Verlag, 1996. pp. 53–69.

5. H. Dobbertin, "Extended MD4 Compress is not Collision-free," Unpublished, October 1995.

6. H. Dobbertin, "Cryptanalysis of MD5 compress," *Presented at the rump session of Eurocrypt'96*, May 1996.

7. FIPS 46, *"Data Encryption Standard,"* Federal Information Processing Standard (FIPS), Publication 46, National Bureau of Standards, U.S. Department of Commerce, Washington D.C., January 1977.

8. W. Hohl, X. Lai, T. Meier, C. Waldvogel, "Security of iterated hash functions based on block ciphers," *Advances in Cryptology, Proc. Crypto'93, LNCS 773*, D. Stinson, Ed., Springer-Verlag, 1994, pp. 379–390.

9. ISO/IEC 10118, *"Information technology – Security techniques – Hash-functions, Part 1: General and Part 2: Hash-functions using an n-bit block cipher algorithm,"* 1994.

10. L.R. Knudsen, X. Lai, "New attacks on all double block length hash functions of hash rate 1, including the parallel-DM," *Advances in Cryptology, Proc. Eurocrypt'94, LNCS 959*, A. De Santis, Ed., Springer-Verlag, 1995, pp. 410–418.

11. L.R. Knudsen, B. Preneel, "Hash functions based on block ciphers and quaternary codes," *Advances in Cryptology, Proc. Asiacrypt'96, LNCS 1163*, K. Kim, T. Matsumoto, Eds., Springer-Verlag, 1996, pp. 77–90.

12. X. Lai, *"On the Design and Security of Block Ciphers,"* ETH Series in Information Processing, Vol. 1, J.L. Massey, Ed., Hartung-Gorre Verlag, Konstanz, 1992.

13. F.J. MacWilliams, N.J.A. Sloane, *"The Theory of Error-Correcting Codes,"* North-Holland Publishing Company, Amsterdam, 1978.

14. S.M. Matyas, C.H. Meyer, J. Oseas, "Generating strong one-way functions with cryptographic algorithm," *IBM Techn. Disclosure Bull.*, Vol. 27, No. 10A, 1985, pp. 5658–5659.

15. R. Merkle, "One way hash functions and DES," *Advances in Cryptology, Proc. Crypto'89, LNCS 435*, G. Brassard, Ed., Springer-Verlag, 1990, pp. 428–446.

16. C.H. Meyer, M. Schilling, "Secure program load with Manipulation Detection Code," *Proc. Securicom 1988*, pp. 111–130.

17. M. Naor, M. Yung, "Universal one-way hash functions and their cryptographic applications," *Proc. 21st ACM Symposium on the Theory of Computing*, ACM, 1989, pp. 387–394.

18. B. Preneel, R. Govaerts, J. Vandewalle, "Hash functions based on block ciphers: a synthetic approach," *Advances in Cryptology, Proc. Crypto'93, LNCS 773*, D. Stinson, Ed., Springer-Verlag, 1994, pp. 368–378.

19. V. Rijmen, B. Preneel, "Improved characteristics for differential cryptanalysis of hash functions based on block ciphers," *Fast Software Encryption, LNCS 1008*, B. Preneel, Ed., Springer-Verlag, 1995, pp. 242–248.

20. J.-J. Quisquater, J.-P. Delescaille, "How easy is collision search? Application to DES," *Advances in Cryptology, Proc. Eurocrypt'89, LNCS 434*, J.-J. Quisquater, J. Vandewalle, Eds., Springer-Verlag, 1990, pp. 429–434.

21. R.L. Rivest, "The MD4 message digest algorithm," *Advances in Cryptology, Proc. Crypto'90, LNCS 537*. S. Vanstone, Ed., Springer-Verlag, 1991, pp. 303–311.

22. R.L. Rivest, "The MD5 message-digest algorithm," *Request for Comments (RFC) 1321*, Internet Activities Board, Internet Privacy Task Force, April 1992.

23. P.C. van Oorschot, M.J. Wiener, "Parallel collision search with application to hash functions and discrete logarithms," *Proc. 2nd ACM Conference on Computer and Communications Security*, ACM, 1994, pp. 210–218.

Edit Distance Correlation Attack on the Alternating Step Generator

Jovan Dj. Golić[1] * and Renato Menicocci[2]

[1] School of Electrical Engineering, University of Belgrade
Bulevar Revolucije 73, 11001 Beograd, Yugoslavia
Email: golic@galeb.etf.bg.ac.yu
[2] Fondazione Ugo Bordoni
Via B. Castiglione 59, 00142 Roma, Italy
Email: rmenic@fub.it

Abstract. A novel edit distance between two binary input strings and one binary output string of appropriate lengths which incorporates the stop/go clocking in the alternating step generator is introduced. An efficient recursive algorithm for the edit distance computation is derived. The corresponding correlation attack on the two stop/go clocked shift registers is then proposed. By systematic computer simulations, it is shown that the minimum output segment length required for a successful attack is linear in the total length of the two stop/go clocked shift registers. This is verified by experimental attacks on relatively short shift registers.

Key words. Stream ciphers, clock-controlled shift registers, alternating step generator, edit distance, cryptanalysis, correlation attacks.

1 Introduction

Keystream generators for stream cipher applications consisting of a small number of clock-controlled shift registers combined by a linear function seem to provide an efficient means for producing sequences with long period, high linear complexity, and good statistical properties, see [7]. The stop-and-go clocking is particularly popular for high speed applications. At any time, a stop/go shift register is clocked once if the clock-control input bit is equal to 1 and is not clocked at all otherwise. The clock-control sequence can be generated by another, regularly clocked shift register, whereas the inherent autocorrelation weakness due to the repetition of bits in a stop/go shift register can be overcome by linearly combining its output with the output of an additional regularly clocked shift

* This work was done while the first author was with the Information Security Research Centre, Queensland University of Technology, Brisbane, Australia. Part of this work was carried out while the first author was on leave at the Isaac Newton Institute for Mathematical Sciences, Cambridge, United Kingdom. This research was supported in part by the Science Fund of Serbia, grant #04M02, through the Mathematical Institute, Serbian Academy of Science and Arts.

register [1], which may be the same as the clock-control one, as is the case in the well-known stop/go cascades [7]. Typically, all the shift registers have linear feedback. However, the additional linear feedback shift register (LFSR) introduced to improve the statistics is then vulnerable to a fast correlation attack based on the repetition weakness [12]. If the clock-control LFSR is itself linearly combined with the stop/go one to produce the output, then it succumbs to a specific, conditional correlation attack [10] which exploits the same repetition weakness, but in a different, perhaps unexpected way.

The alternating step generator (ASG) [8] is an interesting combination of three binary LFSRs, two of which, $LFSR_1$ and $LFSR_2$, are stop/go clocked in a special way by the third one, $LFSR_3$, which is regularly clocked. Instead of $LFSR_3$, one may also use any binary keystream generator. More precisely, if the clock-control bit is equal to 1, then $LFSR_1$ is clocked and $LFSR_2$ is not, and if the clock-control bit is equal to 0, then $LFSR_2$ is clocked and $LFSR_1$ is not. The output sequence is formed as the bitwise sum of the two stop/go clocked LFSR sequences. Although it was proposed before the appearance of [12] and [10], the ASG is not vulnerable to the attacks there introduced. It is shown in [8] that the initial state of the clock-control $LFSR_3$ can be recovered via a divide-and-conquer attack which can be regarded as a special kind of the linear consistency attack [11], which appeared later. Namely, if and only if the guess about the initial state of $LFSR_3$ is correct, then the first binary derivative of the output sequence gives rise to the first binary derivatives of both the regularly clocked $LFSR_1$ and $LFSR_2$ sequences which are then easily tested for linear complexity by the Berlekamp-Massey algorithm [9].

The objective of this paper is to investigate whether a divide-and-conquer attack on $LFSR_1$ and $LFSR_2$ is possible. A way of doing this is to take the edit distance [2] or edit probability [3] approaches developed for memoryless combiners (for combiners with memory, see [6]) and adapt them to deal with the stop/go clocking. The main idea is to define the edit distance between two binary input strings and one binary output string of appropriate lengths as the minimum possible number of effective substitutions (complementations) needed in the combination string, produced from the two input strings by the stop/go clocking in the ASG manner, to obtain the output string, where the minimum is taken over all binary clock-control strings. Our first result is to prove that this unusual edit distance can be computed efficiently by a recursive algorithm whose computational complexity is quadratic in the output string length. Our second contribution is to show by systematic experiments obtained by computer simulations that the minimum output sequence length required for the success of the corresponding edit distance correlation attack is linear in the total length of $LFSR_1$ and $LFSR_2$. The reconstruction of the $LFSR_3$ initial state is also discussed.

In Section 2, a more detailed description of the ASG along with its basic properties is provided. The edit distance and the recursive algorithm for its efficient computation are presented in Section 3, and the experimental results on the underlying embedding probabilities are shown in Section 4. The corresponding correlation attack is explained and experimentally verified in Section 5. Conclusions and some open problems are given in Section 6.

2 Alternating Step Generator

In this section, we recall the structure and basic properties of the alternating step generator (ASG), as presented in [8]. As shown in Fig. 1, the output of the ASG is obtained by bitwise addition (modulo two) of the output sequences of two binary linear feedback shift registers, LFSR$_1$ and LFSR$_2$, whose stop/go clocking is defined by a binary clock-control generator (CCG), which is typically another, but regularly clocked LFSR, denoted as LFSR$_3$. It is assumed that LFSR$_1$ and LFSR$_2$ have different irreducible feedback polynomials of respective degrees r_1 and r_2 and respective coprime periods P_1 and P_2. At every step, only one LFSR is stepped and the output bit is assumed to be produced in the step-then-add manner. Let c_t denote the output bit of the CCG at step $t \geq 1$. Then, in order to obtain the ASG output bit z_t at step t, we first step LFSR$_1$ or LFSR$_2$ depending on whether $c_t = 1$ or $c_t = 0$, respectively, and then we add modulo two the output bits of LFSR$_1$ and LFSR$_2$. Observe that LFSR$_1$ is stop/go clocked, whereas LFSR$_2$ is go/stop clocked. In [8], some good cryptographic properties of the ASG, such as a long period (P), a high linear complexity (L), and approximately uniform relative frequency of short output patterns on a period are established, under the assumption that the clock-control sequence is a de Bruijn sequence of period 2^k. More precisely, it is proven that $P = P_1 P_2 2^k$ and $(r_1 + r_2) 2^{k-1} < L \leq (r_1 + r_2) 2^k$, whereas for approximately uniform distribution of the output patterns of length not bigger than $\min(r_1, r_2)$, it is in addition required that the feedback polynomials of LFSR$_1$ and LFSR$_2$ be primitive as well. It is expected that similar results also hold if the CCG is a LFSR with a primitive feedback polynomial whose period is coprime to $P_1 P_2$.

As mentioned in the previous section, it is also shown in [8] that there exists a (linear consistency) divide-and-conquer attack on the clock-control generator.

3 Edit Distance

Let $X^{n+1} = x_1, x_2, \ldots, x_{n+1}$ and $Y^{n+1} = y_1, y_2, \ldots, y_{n+1}$ denote two binary input strings and let $Z^n = z_1, z_2, \ldots, z_n$ denote a binary output string. Given a binary clock-control string $C^n = c_1, c_2, \ldots, c_n$, let $\hat{Z}^n = \hat{z}_1, \hat{z}_2, \ldots, \hat{z}_n$ denote the combination string produced from X^{n+1} and Y^{n+1} by the step-then-add alternating stepping according to C^n (X^{n+1} and Y^{n+1} correspond to the regularly clocked LFSR$_1$ and LFSR$_2$ sequences of length $n + 1$, respectively). Accordingly, we initially have $\hat{z}_1 = x_1 \oplus y_2$ if $c_1 = 0$ and $\hat{z}_1 = x_2 \oplus y_1$ if $c_1 = 1$, whereas for any $1 \leq s \leq n - 1$ and $0 \leq l \leq s$, if l denotes the number of ones in C^s, then $\hat{z}_s = x_{l+1} \oplus y_{s+1-l}$ and we have $\hat{z}_{s+1} = x_{l+1} \oplus y_{s+2-l}$ if $c_{s+1} = 0$ and $\hat{z}_{s+1} = x_{l+2} \oplus y_{s+1-l}$ if $c_{s+1} = 1$.

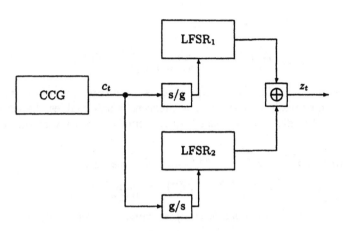

Fig. 1. The alternating step generator.

The *edit distance* between a given pair of strings (X^{n+1}, Y^{n+1}) and a given string Z^n, denoted as $D(X^{n+1}, Y^{n+1}; Z^n)$, is then defined by

$$D(X^{n+1}, Y^{n+1}; Z^n) = \min_{C^n \in \{0,1\}^n} d_H(Z^n, \hat{Z}^n) \qquad (1)$$

where $d_H(Z^n, \hat{Z}^n)$ denotes the Hamming distance between Z^n and $\hat{Z}^n$. In other words, the edit distance is defined as the minimum number of effective substitutions needed to obtain Z^n from $\hat{Z}^n$, where the minimum is over all 2^n binary clock-control strings C^n. Apart from the substitutions in the combination string $\hat{Z}^n$, the edit transformation of the input strings (X^{n+1}, Y^{n+1}) into the output string Z^n also contains one deletion of the first bit from one of the input strings and exactly $n - 1$ repetitions of bits in the input strings, regardless of the clock-control string C^n. Consequently, the only informative part of the edit transformation are effective substitutions, as reflected in our definition of the edit distance.

Our basic objective is to examine whether the defined edit distance can be computed efficiently by an algorithm whose computational complexity is significantly smaller than 2^n, which corresponds to the computation of (1) by the exhaustive search over all C^n. To this end, for any $1 \le s \le n$ and $0 \le l \le s$, we define the *partial edit distance* $W(l, s)$ as $D(X^{s+1}, Y^{s+1}; Z^s)$ under an additional constraint that the binary clock-control string contains exactly l ones, that is,

$$W(l, s) = \min_{C^s : w_H(C^s) = l} d_H(Z^s, \hat{Z}^s) \qquad (2)$$

where $w_H(C^s)$ is the Hamming weight of C^s. Accordingly, the last bit $\hat{z}_s$ is in (2) always produced from the input bits x_{l+1} and y_{s+1-l}, so that the edit transformation involves the prefixes X^{l+1} and Y^{s+1-l} only. The partial edit distance can then be represented as

$$W(l,s) = \min_{C^s:w_H(C^s)=l} \sum_{k=1}^{s} (z_k \oplus \hat{z}_k) \tag{3}$$

where the Hamming distance is expressed as the integer sum of binary variables.

We are now ready to formulate a theorem which enables the efficient computation of the edit distance based on a recursive property of the partial edit distance.

Theorem 1. *For any X^{n+1}, Y^{n+1}, and Z^n, we have*

$$D(X^{n+1}, Y^{n+1}; Z^n) = \min_{0 \le l \le n} W(l,n) \tag{4}$$

where the partial edit distance $W(l,n)$ is computed recursively by

$$W(l,s) = (x_{l+1} \oplus y_{s+1-l} \oplus z_s) + \min(W(l-1, s-1), W(l, s-1)) \tag{5}$$

for $1 \le s \le n$ and $0 \le l \le s$, with the initial values $W(-1, s) = W(s+1, s) = \infty$ and $W(0,0) = 0$.

Proof. First observe that (4) is an immediate consequence of the definition of the partial edit distance. Second, for $s = 1$, (3) directly implies that $W(0,1) = x_1 \oplus y_2 \oplus z_1$ and $W(1,1) = x_2 \oplus y_1 \oplus z_1$ which can also be obtained by (5) from the given initial values.

Now, assume that $s > 1$. Since by definition, $\hat{z}_s = x_{l+1} \oplus y_{s+1-l}$, (3) can then be put into the form

$$W(l,s) = (x_{l+1} \oplus y_{s+1-l} \oplus z_s)$$
$$+ \min \left(\min_{C^{s-1}:w_H(C^{s-1})=l-1} \sum_{k=1}^{s-1} (z_k \oplus \hat{z}_k), \min_{C^{s-1}:w_H(C^{s-1})=l} \sum_{k=1}^{s-1} (z_k \oplus \hat{z}_k) \right) \tag{6}$$

where the first and the second minima correspond to clock-control strings whose last bit c_s is equal to one and zero, respectively (the given initial values take care of the case when $l = 0$ or $l = s$). Equation (5) then follows directly in view of (3). □

The time and space complexities of the recursive algorithm corresponding to Theorem 1 are clearly $O(n^2)$ and $O(n)$, since only the values of the partial edit distance for the current and the preceding value of s have to be stored at a time. The algorithm is thus feasible even if n is very large. One can also store the whole matrix of the partial edit distances, indexed by s and l, in $O(n^2)$ space along with the value(s) of the clock-control bit c_s for which the minimum in (5) is achieved. By backtracking through the matrix, one can then recover all possible clock-control strings giving rise to the minimum number of effective substitutions representing the edit distance.

Some basic symmetry properties of the defined edit distance are captured by the following two propositions. For an arbitrary binary sequence or string $A = a_1, a_2, a_3, a_4, \ldots$, let $\bar{A} = \bar{a}_1, \bar{a}_2, \bar{a}_3, \bar{a}_4, \ldots$, $\hat{A} = a_1, \bar{a}_2, a_3, \bar{a}_4, \ldots$, and $\bar{\hat{A}} = \bar{a}_1, a_2, \bar{a}_3, a_4, \ldots$, where $\bar{a}$ denotes the complement of a bit a.

Proposition 2.

$$D(X^{n+1}, Y^{n+1}; Z^n) = D(\bar{X}^{n+1}, \bar{Y}^{n+1}; Z^n) = D(\bar{X}^{n+1}, Y^{n+1}; \bar{Z}^n). \quad (7)$$

Proof. Given a clock-control string C^n, if $\hat{Z}^n$ is the combination string produced from (X^{n+1}, Y^{n+1}), then $\hat{Z}^n$ is also the combination string produced from $(\bar{X}^{n+1}, \bar{Y}^{n+1})$ and $\bar{\hat{Z}}^n$ is the combination string produced from $(\bar{X}^{n+1}, Y^{n+1})$. The proposition then directly follows from (1). □

Proposition 3.

$$D(X^{n+1}, Y^{n+1}; Z^n) = D(\tilde{X}^{n+1}, \tilde{Y}^{n+1}; \tilde{Z}^n). \quad (8)$$

Proof. Given a clock-control string C^n, if $\hat{Z}^n$ is the combination string produced from (X^{n+1}, Y^{n+1}), then $\tilde{\hat{Z}}^n$ is the combination string produced from $(\tilde{X}^{n+1}, \tilde{Y}^{n+1})$. The proposition then directly follows from (1). □

4 Embedding Probabilities

Let $P_n(D|Z^n)$ denote the probability that the edit distance $D(X^{n+1}, Y^{n+1}; Z^n)$ is equal to D when Z^n is fixed and the pair (X^{n+1}, Y^{n+1}) is randomly chosen according to the uniform probability distribution. Also, let $P_n(D)$ denote the expected value of $P_n(D|Z^n)$ over a uniformly distributed Z^n, that is, the probability that the edit distance is equal to D when the triple (X^{n+1}, Y^{n+1}, Z^n) is randomly chosen according to the uniform probability distribution. The underlying uniform probability distributions over the string pairs and triples mean that the defined probabilities are in fact the fractions of the string pairs and triples such that the edit distance is equal to D, respectively.

In particular, let $P_n(Z^n)$ denote the probability, $P_n(0|Z^n)$, that the edit distance is equal to zero. Since the zero edit distance means that there exists a clock-control sequence such that Z^n is produced from (X^{n+1}, Y^{n+1}) by step-then-add alternating stepping, $P_n(Z^n)$ is also called the embedding probability, given an output string Z^n of length n, see [4] and [5]. Further, let $\bar{P}_n$, P_n^{max}, and P_n^{min} denote the average, the maximum, and the minimum values of $P_n(Z^n)$ over a uniformly distributed Z^n, respectively. All of these embedding probabilities are related to the success of the edit distance correlation attack to be described in the next section. To this end, it is critical that $\bar{P}_n$ decreases with n and it is desirable that this decrease is exponentially fast. Apart from that, it will be nice if P_n^{max} has a similar behavior itself too.

The evaluation of the embedding probabilities defined above seems to be a difficult combinatorial problem, related to the problems investigated in [13], [4], and [5]. The desired exponential decrease with the output string length n is not apparent at all. The approach taken in this section is essentially to compute the embedding probabilities for smaller values of n by exhaustive counting and to estimate their values for larger n by counting on a random sample of suitable input and output strings.

Our first objective is to estimate the edit distance probability distribution $P_n(D)$, $0 \leq D \leq n$. Tables 1 and 2 show the observed minimum, maximum, mean, and median values along with the standard deviation of D in a random sample of 1000 triples (X^{n+1}, Y^{n+1}, Z^n) for $n = 10, (10), 100$ and $n = 200, (100), 1000$, respectively. It appears that the expected value of D increases approximately linearly with n, around $n/10 + 1$, whereas the standard deviation is small and increases very slowly with n.

Table 1: Statistics of D on 1000 random triples (X^{n+1}, Y^{n+1}, Z^n).

n	10	20	30	40	50	60	70	80	90	100
Min	0	0	0	0	1	2	2	3	4	5
Max	5	6	9	10	11	12	14	15	16	18
Mean	1.611	2.783	3.973	5.125	6.147	7.243	8.293	9.336	10.393	11.542
Median	2	3	4	5	6	7	8	9	10	12
Std Dev	.996	1.194	1.373	1.508	1.570	1.715	1.822	1.859	2.002	1.917

Table 2: Statistics of D on 1000 random triples (X^{n+1}, Y^{n+1}, Z^n).

n	200	300	400	500	600	700	800	900	1000
Min	13	24	32	41	51	59	70	77	81
Max	29	40	56	63	72	86	96	104	117
Mean	21.593	31.918	42.016	52.071	61.932	72.518	81.925	92.035	102.15
Median	22	32	42	52	62	72	82	92	102
Std Dev	2.626	2.823	3.137	3.550	3.707	3.767	3.942	4.304	4.562

Our main objective is to compute or estimate the embedding probabilities $\bar{P}_n$, $P_n^{\max}$, and $P_n^{\min}$. Table 3 displays the computed values of these probabilities obtained by exhaustive counting for $n = 5, (1), 10$. For each such n, $P_n^{\min}$ is achieved if Z^n is the constant zero string and $P_n^{\max}$ is achieved if Z^n is the prefix of the sequence $Z = 0, 0, 1, 1, 0, 0, 1, 1, 0, 0, \ldots$. It is conjectured that this is also the case for every n (a related result is proven in [5]). In fact, the values of $P_n^{\max}$ and $P_n^{\min}$ given in Tables 4 and 5 to follow correspond to such strings, since the exhaustive counting is not feasible for larger n. As well, according to Propositions 2 and 3, if the minimum or the maximum is obtained for a string Z^n, then it is also obtained for $\bar{Z}^n$, $\tilde{Z}^n$, and $\bar{\tilde{Z}}^n$.

Table 3: $\bar{P}_n$, $P_n^{\max}$, and $P_n^{\min}$ determined by exhaustive computation.

n	5	6	7	8	9	10
$\bar{P}_n$	.3165	.2598	.2138	.1763	.1455	.1203
$P_n^{\max}$	.4546	.4136	.3766	.3440	.3137	.2869
$P_n^{\min}$	.1479	.08581	.04880	.02734	.01513	.008300

Table 4 gives the estimated values of $\bar{P}_n$, $P_n^{\max}$, and $P_n^{\min}$ for $n = 11, (1), 20$ on a random sample of 10000 pairs (X^{n+1}, Y^{n+1}) for $P_n^{\max}$ and $P_n^{\min}$ and 10000 triples (X^{n+1}, Y^{n+1}, Z^n) for $\bar{P}_n$, except for $n = 11, 12, 13$ where $P_n^{\max}$ and $P_n^{\min}$ are obtained by exhaustive counting.

Table 4: $\bar{P}_n$, $P_n^{\max}$, and $P_n^{\min}$ determined either by a 10000 points estimation or by exhaustive computation*.									
n	11	12	13	14	15	16	17	18	19	20
$\bar{P}_n$	.1026	.0820	.0641	.0587	.0479	.0395	.0349	.0274	.0223	.0188
$P_n^{\max}$	.2621*	.2398*	.2194*	.2045	.1825	.1613	.1532	.1383	.1314	.1176
$P_n^{\min}$	.004516*	.002441*	.001312*	.0007	.0007	.0004	0	0	0	0

Table 5 gives the estimated values of $\bar{P}_n$ and $P_n^{\max}$ for $n = 25, (5), 100$ on a random sample of 30000 triples (X^{n+1}, Y^{n+1}, Z^n) and 30000 pairs (X^{n+1}, Y^{n+1}), respectively. All the obtained estimates of $P_n^{\min}$ are equal to zero, since this probability is then very small.

Table 5a: $\bar{P}_n$ and $P_n^{\max}$ determined by a 30000 points estimation.								
n	25	30	35	40	45	50	55	60
$\bar{P}_n$	.0071	.003033	.0009333	.0004	.0001667	.0000667	.0000667	0
$P_n^{\max}$	.0776	.05127	.0335	.02137	.01293	.007867	.005133	.003233

Table 5b: $\bar{P}_n$ and $P_n^{\max}$ determined by a 30000 points estimation.								
n	65	70	75	80	85	90	95	100
$\bar{P}_n$	0	0	0	0	0	0	0	0
$P_n^{\max}$	.002467	.001833	.001067	.0006	.0006333	.0003667	.0000667	.0000667

Tables 3-5 are consistent with the exponential decrease with n of all the probabilities $\bar{P}_n$, $P_n^{\max}$, and $P_n^{\min}$. Namely, each of them seems to have the form $a\,b^n$, $b < 1$, for large n. The corresponding estimates of the parameters a and b are presented in Table 6. Since not all of them are equally reliable, especially if a is concerned, the derived estimates are shown for each of the tables separately.

Table 6: Estimation of a and b based on Tables 3*, 4**, and 5***.						
	a^*	a^{**}	a^{***}	b^*	b^{**}	b^{***}
$\bar{P}_n$	.8303	.7926	-	.8241	.8292	.7357
$P_n^{\max}$	.7186	.6930	-	.9121	.9151	.9064
$P_n^{\min}$	2.695	-	-	.5622	.5305	-

To be on the conservative side, the probabilities $\bar{P}_n$, $P_n^{\max}$, and $P_n^{\min}$ are approximated for large n by using the maximum values for a and b as

$$\bar{P}_n \approx 0.83 \cdot 0.83^n, \quad P_n^{\max} \approx 0.72 \cdot 0.915^n, \quad P_n^{\min} \approx 2.7 \cdot 0.562^n. \quad (9)$$

5 Correlation Attack

Assume that the feedback polynomials of LFSR_1 and LFSR_2 are known. The objective of the edit distance correlation attack proposed in this section is to reconstruct the secret key dependent initial states of LFSR_1 and LFSR_2 from a known, sufficiently long segment of the output sequence, in the known plaintext scenario. The main point is to measure the statistical dependence or the correlation between the output sequence and the regularly clocked LFSR_1 and LFSR_2 sequences by the edit distance defined in Section 3.

A segment Z^n of the first n successive output bits is produced from the output segments of the regularly clocked LFSR_1 and LFSR_2, called here the input segments, whose lengths are variable depending on the clock-control sequence. If the unknown clock-control sequence is assumed to be purely random, that is, a sequence of independent uniformly distributed binary random variables, then the average length of the input LFSR segments is $n/2 + 1$, whereas their maximum possible length is $n + 1$. Accordingly, for any assumed initial states of LFSR_1 and LFSR_2, one can by their linear recursions generate the two input segments of the same length $n + 1$, X^{n+1} and Y^{n+1}, respectively. The edit distance $D(X^{n+1}, Y^{n+1}; Z^n)$ is then efficiently computed by the recursive algorithm introduced in Section 3. This is repeated for every possible pair of the assumed LFSR_1 and LFSR_2 initial states, altogether $2^{r_1+r_2}$ of them. Since for the correct LFSR initial states the edit distance is clearly equal to zero, all the obtained initial state pairs with the zero edit distance represent the candidates for the correct initial state pairs. The zero edit distance indicates that there exists a clock-control sequence such that Z^n is produced from (X^{n+1}, Y^{n+1}) by step-then-add alternating stepping. In this sense, this edit distance correlation attack can be viewed as a specific embedding attack, just like the Levenshtein-like edit distance attack [2] reduces to the embedding attack [13] in case of a single clock-controlled shift register. Recall that the irregular clocking considered in [2] and [13] is constrained in that the number of clocks per each output bit is an upper-bounded positive integer, whereas the unconstrained clocking is analyzed in [4].

5.1 Theoretical analysis

Ideally, if n is large enough, then there will remain only one candidate for the initial states of LFSR_1 and LFSR_2. This can happen only if the embedding probability $P_n(Z^n)$ defined in Section 4 is sufficiently small. Recall that $P_n(Z^n)$ is the fraction of string pairs (X^{n+1}, Y^{n+1}) for which $D(X^{n+1}, Y^{n+1}; Z^n) = 0$. Namely, the expected number of candidates is clearly $2^{r_1+r_2} P_n(Z_n)$, so that the edit distance correlation attack is deemed successful if and only if, approximately,

$$2^{r_1+r_2} P_n(Z_n) \leq 1, \quad (10)$$

see [4]. According to Section 4, let $\bar{P}_n$, $P_n^{\max}$, and $P_n^{\min}$ denote the average, the maximum, and the minimum values of $P_n(Z^n)$ over a uniformly distributed Z^n, respectively. Then, by substituting these probabilities for $P_n(Z^n)$ in the condition (10), we obtain the minimum output segment length n required for success for the average, for the worst, and for the best Z^n, respectively. More precisely, if any of these probabilities has the exponential form $a\,b^n$, where $b < 1$, then (10) reduces to

$$n \geq \frac{r_1 + r_2 + \log a}{-\log b} \tag{11}$$

which means that the required output segment length is essentially linear in the total length of $LFSR_1$ and $LFSR_2$. In view of (9), the required output segment length is then approximately given as

$$n \geq 3.72\,(r_1 + r_2) - 1 \tag{12}$$

$$n \geq 7.8\,(r_1 + r_2) - 3.7 \tag{13}$$

$$n \geq 1.2\,(r_1 + r_2) + 1.72 \tag{14}$$

in the average, the worst, and the best case, respectively.

The number of the surviving candidate initial state pairs for a chosen n can further be reduced to just one or a very small number by increasing the length n. More precisely, we show that the number of candidate initial state pairs is at least two regardless of how large n is. Moreover, with a certain probability it can also be bigger than two, depending on some linear equations among the candidate pair bits being satisfied or not.

Proposition 4. *The number of candidate initial state pairs for $LFSR_1$ and $LFSR_2$ selected by the zero edit distance criterion is at least equal to two, provided that $n + 1 > \max(r_1, r_2)$.*

Proof. Suppose that the clocking string $C^n = c_1, c_2, \ldots, c_n$, with $c_1 = 1$, is able to transform a pair (X^{n+1}, Y^{n+1}) into Z^n. Suppose that X^{n+1} and Y^{n+1} are generated by $LFSR_1$ and $LFSR_2$ from initial states $X^{r_1} = x_1, x_2, \ldots, x_{r_1}$ and $Y^{r_2} = y_1, y_2, \ldots, y_{r_2}$, respectively, where $n+1 > \max(r_1, r_2)$. Then, there always exists another pair $(\hat{X}^{n+1}, \hat{Y}^{n+1})$ produced by $LFSR_1$ and $LFSR_2$ from appropriate initial states $\hat{X}^{r_1}$ and $\hat{Y}^{r_2}$, respectively, such that $D(\hat{X}^{n+1}, \hat{Y}^{n+1}; Z^n) = 0$. Namely, it suffices to use $\hat{x}_i = x_{i+1}$, $i = 1, 2, \ldots, r_1$, and $\hat{y}_i = y_{i-1}$, $i = 1, 2, \ldots, r_2$, (y_0 is obtained by backward clocking of $LFSR_2$) along with the clocking string $\hat{C}^n$ with $\hat{c}_1 = 0$ and $\hat{c}_i = c_i$, $i = 2, 3, \ldots, n$. An analogous proof is readily obtained for the case where $c_1 = 0$. □

For each obtained candidate initial state pair, one can also store the whole matrix of the partial edit distances and then by backtracking recover all possible clock-control strings C^n giving rise to the zero edit distance. The average number of such clock-control strings per candidate pair can be estimated as

$$m_n = \frac{2^n}{2^n \bar{P}_n} = \frac{1}{\bar{P}_n} \approx 1.2 \cdot 2^{0.269\,n}. \tag{15}$$

Note that if n is chosen so that $2^{r_1+r_2}\bar{P}_n \approx 1$, then $m_n \approx 2^{r_1+r_2}$. If the clock-control sequence is generated by another known LFSR, LFSR$_3$, of length r_3 and if $n > r_3$, then the number of C^n can be reduced by checking the LFSR$_3$ recursion which can be performed sequentially, one bit at a time, by backtracking through the matrix of partial edit distances. In fact, starting from m_{r_3} possible strings C_{r_3} for $n = r_3$, each new bit examined is expected to halve the number of the surviving clock-control strings which is therefore reduced to only one if $n - r_3 \geq \log m_n$, that is, if, approximately,

$$n \geq 1.37\, r_3. \tag{16}$$

The complexity of this search is upper-bounded by m_n, which is close to $2^{0.37\, r_3}$ if $n \approx 1.37\, r_3$ and is close to $2^{r_1+r_2}$ if $n \approx 3.72\,(r_1 + r_2)$. In addition, it may happen that some clock-control bits are uniquely determined without exploiting the LFSR$_3$ recursion which can be used to reduce the search effort. So, if apart from (12), the condition (16) is also satisfied, then the described search is likely to reduce the number of candidate initial state pairs for LFSR$_1$ and LFSR$_2$ to only one, correct pair and also to uniquely determine the initial state of LFSR$_3$. The obtained candidate initial state triples for all the LFSRs are then tested for correctness on a longer output sequence. Note that in view of the structure of the ASG generator, one may expect that different LFSR initial state triples necessarily give rise to different output sequences, so that the solution for the LFSR initial states is very likely to be unique.

Finally, one may observe that if one of the initial states, for LFSR$_1$ or LFSR$_2$, is guessed correctly, then, instead of the embedding probability as such, one should in fact consider the fraction of the input and output string triples giving rise to the zero edit distance provided that one of the input strings is guessed correctly. If this, conditional embedding probability was bigger than the embedding probability defined before, then the number of the obtained candidate pairs would very likely be bigger than just two, see Proposition 4. Note that this is not a problem, since the correct individual initial states can then easily be recovered as the ones that appear in most the obtained candidate pairs. Note that the computational complexity of the proposed edit distance correlation attack remains $O(2^{r_1+r_2})$ anyway. However, for the sake of completeness, we have also experimentally obtained an estimate of this conditional embedding probability, $\approx 0.85 \cdot 0.85^n$, which is very close to the estimate of $\bar{P}_n$ given in (9).

5.2 Experimental attacks

A number of computer simulations were conducted to show that the above edit distance correlation attack can work in practice. The clock-control generator was assumed to be another linear feedback shift register, LFSR$_3$, of length r_3. Only primitive feedback polynomials were used for all the LFSRs. The correlation attack was performed in the way explained in Subsection 5.1. The feedback polynomials were assumed to be known and the objective was to reconstruct the initial states of LFSR$_1$ and LFSR$_2$ along with the initial state of LFSR$_3$ from a sufficiently long segment of the ASG output sequence.

Some examples of the experimental results obtained are shown in Table 7, where $N_{1,2}$ denotes the number of candidate initial state pairs for LFSR$_1$ and LFSR$_2$ satisfying the zero edit distance criterion and N_3 denotes the number of candidate initial states for LFSR$_3$ obtained from the found $N_{1,2}$ pairs by examining the corresponding clock-control strings. As different LFSR initial state triples very likely yield different output sequences, then $N_3 = 1$ effectively reduces the number of candidate initial state pairs for LFSR$_1$ and LFSR$_2$ to only one. Observe that the same ASG was considered in the experiments 1 and 1', the only difference being the keystream sequence length n, which was increased in the experiment 1' to test the ability of the attack to reduce the number of candidate initial states for LFSR$_3$ when $N_{1,2}$ is already at its minimum. Accordingly, in each of the experiments a unique solution for the LFSR initial states was obtained. Similar results were also obtained in a number of other experiments where the shift register lengths r_1 and r_2 were smaller than those from Table 7.

Notice that multiple candidates for the initial states of LSFR$_1$ and LFSR$_2$ can appear and that their number, $N_{1,2}$, which is expected to be a small positive integer, can not be reduced to one by increasing n, see Proposition 4. In practice, as indicated by (12), it was observed that $N_{1,2}$ is minimized by using $n \approx 4(r_1 + r_2)$. Multiple candidates for the initial states of LSFR$_1$ and LFSR$_2$ can effectively be reduced to only one candidate by reconstructing the LFSR$_3$ initial state from all possible clock-control strings, provided that n is sufficiently long, see (16).

Table 7: Experimental results.

Experiment	1	1'	2	3	4	5
n	180	200	120	150	180	160
r_1, r_2, r_3	14, 14, 64	14, 14, 64	15, 15, 48	15, 15, 64	15, 15, 64	16, 16, 64
$N_{1,2}, N_3$	2, 2	2, 1	6, 1	4, 2	2, 1	3, 1

6 Conclusions

A novel edit distance between two binary input strings and one binary output string of appropriate lengths which reflects the specific stop/go clocking in the ASG generator is introduced. An efficient recursive algorithm for the edit distance computation whose time complexity is quadratic in the output string length is derived. By systematic computer simulations, the underlying embedding probabilities are shown to exponentially decrease with the output string length. The corresponding edit distance correlation attack on the two stop/go clocked shift registers in the ASG generator is then proposed. The attack essentially consists in computing the edit distance for every possible pair of the initial states of the two shift registers and in finding all the pairs with the zero edit distance. By using the evaluated embedding probabilities, it is then established that the minimum output segment length required for a successful attack is linear in the total length of the two stop/go clocked shift registers. More precisely, only about four total lengths on average and about eight total lengths in the worst case are sufficient for the success. The reconstruction of the initial state of the clock-control shift register is also discussed. The theory is illustrated by successful experimental attacks conducted on relatively short shift registers. From the practical cryptographic standpoint, the results show that the total length of the two stop/go clocked shift registers should be sufficiently large in order to prevent the exhaustive search over their initial states. On the other hand, note that the clock-control shift register itself should be long enough to prevent the divide-and-conquer attack from [8].

Finding analytical expressions for the embedding probabilities is an interesting, but difficult combinatorial problem, see [13], [5], and [4] for related problems regarding a single clock-controlled shift register. Defining an appropriate edit probability instead of the edit distance is an approach which may possibly reduce the required output segment length, see [3] and [4] for related previous work in this direction. Finally, it remains to be investigated whether a correlation attack on individual stop/go clocked shift registers in the ASG generator based on another special edit distance or edit probability is also possible.

References

1. T. Beth and F. C. Piper, "The stop-and-go generator," Advances in Cryptology - EUROCRYPT '84, *Lecture Notes in Computer Science*, vol. 209, T. Beth, N. Cot, and I. Ingemarsson eds., Springer-Verlag, pp. 88-92, 1985.
2. J. Dj. Golić and M. Mihaljević, "A generalized correlation attack on a class of stream ciphers based on the Levenshtein distance," *Journal of Cryptology*, vol. 3(3), pp. 201-212, 1991.
3. J. Dj. Golić and S. Petrović, "A generalized correlation attack with a probabilistic constrained edit distance," Advances in Cryptology - EUROCRYPT '92, *Lecture Notes in Computer Science*, vol. 658, R. A. Rueppel ed., Springer-Verlag, pp. 472-476, 1993.
4. J. Dj. Golić and L. O'Connor, "Embedding and probabilistic correlation attacks on clock-controlled shift registers," Advances in Cryptology - EUROCRYPT '94, *Lecture Notes in Computer Science*, vol. 950, A. De Santis ed., Springer-Verlag, pp. 230-243, 1995.
5. J. Dj. Golić, "Constrained embedding probability for two binary strings," *SIAM Journal on Discrete Mathematics*, vol. 9(3), pp. 360-364, 1996.
6. J. Dj. Golić, "Edit distance correlation attacks on clock-controlled combiners with memory," Information Security and Privacy, *Lecture Notes in Computer Science*, vol. 1172, J. Pieprzyk ed., Springer-Verlag, pp. 169-181, 1996.
7. D. Gollmann and W. G. Chambers, "Clock-controlled shift registers: a review," *IEEE Journal on Selected Areas in Communications*, vol. 7, pp. 525-533, May 1989.
8. C. G. Günther, "Alternating step generators controlled by de Bruijn sequences," Advances in Cryptology - EUROCRYPT '87, *Lecture Notes in Computer Science*, vol. 304, D. Chaum and W. L. Price eds., Springer-Verlag, pp. 5-14, 1988.
9. J. L. Massey, "Shift-register synthesis and BCH decoding," *IEEE Trans. Inform. Theory*, vol. IT-15, pp. 122-127, Jan. 1969.

10. R. Menicocci, "Cryptanalysis of a two-stage Gollmann cascade generator," *Proceedings of SPRC '93*, Rome, Italy, pp. 62-69, 1993.
11. K. Zeng, C. H. Yang, and T. R. N. Rao, "On the linear consistency test (LCT) in cryptanalysis with applications," Advances in Cryptology - CRYPTO '89, *Lecture Notes in Computer Science*, vol. 435, G. Brassard ed., Springer-Verlag, pp. 164-174, 1990.
12. K. Zeng, C. H. Yang, and T. R. N. Rao, "An improved linear syndrome algorithm in cryptanalysis with applications," Advances in Cryptology - CRYPTO '90, *Lecture Notes in Computer Science*, vol. 537, A. J. Menezes and S. A. Vanstone eds., Springer-Verlag, pp. 34-47, 1991.
13. M. V. Živković, "An algorithm for the initial state reconstruction of the clock-controlled shift register," *IEEE Trans. Inform. Theory*, vol. IT-37, pp. 1488-1490, Sept. 1991.

Differential Fault Analysis
of Secret Key Cryptosystems

Eli Biham

Computer Science Department

Technion – Israel Institute of Technology

Haifa 32000, Israel

biham@cs.technion.ac.il

http://www.cs.technion.ac.il/~biham/

Adi Shamir

Applied Math. and Comp. Sci. Department

The Weizmann Institute of Science

Rehovot 76100, Israel

shamir@wisdom.weizmann.ac.il

Abstract

In September 1996 Boneh, Demillo, and Lipton from Bellcore announced a new type of cryptanalytic attack which exploits computational errors to find cryptographic keys. Their attack is based on algebraic properties of modular arithmetic, and thus it is applicable only to public key cryptosystems such as RSA, and not to secret key algorithms such as the Data Encryption Standard (DES).

In this paper, we describe a related attack, which we call *Differential Fault Analysis*, or *DFA*, and show that it is applicable to almost any secret key cryptosystem proposed so far in the open literature. Our DFA attack can use various fault models and various cryptanalytic techniques to recover the cryptographic secrets hidden in the tamper-resistant device. In particular, we have demonstrated that under the same hardware fault model used by the Bellcore researchers, we can extract the full DES key from a sealed tamper-resistant DES encryptor by analyzing between 50 and 200 ciphertexts generated from unknown but related plaintexts.

In the second part of the paper we develop techniques to identify the keys of completely unknown ciphers (such as SkipJack) sealed in tamper-resistant devices, and to reconstruct the complete specification of DES-like unknown ciphers.

In the last part of the paper, we consider a different fault model, based on permanent hardware faults, and show that it can be used to break DES by analyzing a small number of ciphertexts generated from completely unknown and unrelated plaintexts.

1 Introduction

In September 1996 Boneh, Demillo, and Lipton from Bellcore announced an ingenious new type of cryptanalytic attack which received widespread attention[11,5]. This attack is applicable only to public key cryptosystems such as RSA, and not to secret key algorithms such as the Data Encryption Standard (DES)[17]. According to Boneh, "The algorithm that we apply to the device's faulty computations works against the algebraic structure used in public key cryptography, and another algorithm will have to be devised to work against the non-algebraic operations that are used in secret key techniques". In particular, the original Bellcore attack is based on specific algebraic properties of modular arithmetic, and cannot handle the complex bit manipulations which underly most secret key algorithms.

This type of attack on a tamper-resistant device shows that even cryptographic schemes sealed inside such devices might leak information about the secret key. Earlier papers on this subject, including the papers of Anderson and Kuhn[1], and of Kocher[8], had shown that tamper-resistant devices are vulnerable to several types of attacks including attacks against the protocols, attacks using carelessness of the device's programmers, and timing attacks.

In this paper, we describe a new attack, related to Boneh, Demillo, and Liptons' attack, which we call *Differential Fault Analysis*, or *DFA*, and show that it is applicable to almost any secret key cryptosystem proposed so far in the open literature. In particular, we have actually implemented DFA in the case of DES, and demonstrated that under the same hardware fault model used by the Bellcore researchers, we can extract the full DES key from a sealed tamper-resistant DES encryptor by analyzing between 50 and 200 ciphertexts generated from unknown but related plaintexts. In more specialized cases, as few as five ciphertexts are sufficient to completely reveal the key. The power of Differential Fault Analysis is demonstrated by the fact that even if DES is replaced by triple DES (whose 168 bits of key were assumed to make it practically invulnerable), essentially the same attack can break it with essentially the same number of given ciphertexts.

Differential Fault Analysis can break many additional secret key cryptosystems, including IDEA[9], RC5[19] and Feal[21,16,14,15]. Some ciphers, such as Khufu[13], Khafre[13] and Blowfish[20] compute their S boxes from the key material. In such ciphers, it may be even possible to extract the S boxes themselves, and the keys, using the techniques of Differential Fault Analysis. Differential Fault Analysis can also be applied against stream ciphers, but the implementation might differ in some technical details from the implementation described above. At this point we should note that small differences in the fault models might crucially affect the capabilities and the complexities of the attacks.

Differential fault analysis is not limited to finding the keys of known ciphers: We introduce an asymmetric fault model which makes it possible to find the secret key stored in a tamper-resistant cryptographic device even when nothing is known about the structure and operation of the cryptosystem. A prime example of such a scenario is the SkipJack cryptosystem, which was developed by the NSA, has unknown design, and is embedded as a tamper-resistant chip inside the commercially available Fortezza PC cards. We have not tested this attack on SkipJack, but we believe that it is a realistic threat against some smartcard applications which were not specifically designed to counter it.

Moreover, we show that in most interesting cases we can extract the exact structure of an unknown DES-like cipher sealed in the tamper-resistant device, including the identification of its round functions, S boxes, and subkeys. If the attacker can only encrypt with the tamper-resistant device with a fixed key (e.g., in PIN verifying devices), still the attacker can identify the operation of the cipher with this particular key, which lets him encrypt and decrypt under this key.

The transient fault model used in all these attacks has not been demonstrated in physical experiments, and was questioned by representatives of the smartcard industry. We thus suggest a practical attack based on a different fault model which will hopefully be less controversial. In this model we cut one wire or permanently destroy a single memory cell in the smartcard using a narrow laser-beam. This model allows us to mount a *pure* ciphertext only attack which finds the key with only a few ciphertexts generated from random unrelated unknown plaintexts. The attack is thus applicable even when the smartcard chooses the message it wants to encrypt, and even it if uses counters or random bits to foil differential attacks on related plaintexts.

2   The Attack on DES

Todays' computers are extremely reliable. Still, it is possible for interested parties to intentionally induce faults into some kinds of computations. Although this is almost impossible to do to a remote computer or to a mainframe, it can be done to smartcards.

In many applications (like electronic money, identification systems, or access control) smartcards are used as secure extensions of the host, and enable their owners to apply cryptographic computations without knowing the hosts' secret keys. It is assumed that the smartcards are tamper-resistant, and thus even the owner of a smartcard cannot open it or reverse-engineer it in order to reveal the secret keys kept inside the smartcard.

However, due to the simplicity of the smartcards, and the ability of its owner to control the environment, the owner of the smartcard can force the smartcard to malfunction in many ways, such as changing the power supply voltage, changing the frequency of the (external) clock, and applying radiation of many kinds.

Boneh, Demillo, and Lipton suggested using faults induced by the card owner in order to deduce the private keys in number theoretic public key cryptosystems.

We use the following fault model, similar to the model used by Boneh, Demillo, and Lipton: The smartcard is assumed to have random transient faults in its registers, with some small probability of occurrence in each bit, so that during each encryption/decryption there appears a small number of faults (typically one) during the computation, and each such fault inverts the value of one of the bits, either from zero to one or from one to zero.

More accurately, we assume that during each faulty computation there occurs one (or a few) faults, at random times during the computations, and with random choice of the registers and positions in the registers. Both the bit location and the exact timing of the fault are unknown to the attacker.

This model lets the attacker induce errors at some random position during the DES encryption, i.e., at some random bit of one of the registers in a random round. For

- The CaveTable has a very skewed statistical distribution. It is not a permutation; 92 of the 256 possible 8-bit values never appear; some values appear as many as four times. The distribution appears to be consistent with that of a random function.

The skew in the CaveTable means that the T-box values are skewed, too: we know $T(i) - i$ must appear in the CaveTable, so for any input to the T-box, we can immediately rule out 92 possibilities for the corresponding T-box output without needing any knowledge of the CMEA key.

3.1 A chosen-plaintext attack

CMEA is weak against chosen-plaintext attacks: one can recover all of the T-box entries with about 338 chosen texts (on average) and very little work. This attack works on any fixed block length $n > 2$; the attacker is not assumed to have control over n. We have implemented the attack to empirically verify it for correctness; the attack is extremely successful in our tests[2].

The attack proceeds in two stages, first recovering $T(0)$, and then recovering the remainder of the T-box entries; the CMEA key itself is never identified. First, one learns $T(0)$ with $(256 - 92)/2 = 82$ chosen plaintexts (on average). For each guess x at the value of $T(0)$, obtain the encryption of the message $P = (1 - x, 1 - x, 1 - x, \ldots)$, e.g. the message P where each byte has the value $1 - x$; if the result is of the form $C = (-x, \ldots)$ then we can conclude with high probability that indeed $T(0) = x$. False alarms occasionally occur, but they can be ruled out quickly in the second phase because of the skewed CaveTable distribution. Note that there are only $256 - 92 = 164$ possible values of $T(0)$, since $T(0)$ must appear in the CaveTable, and therefore we expect to identify the correct value after about $164/2 = 82$ trials, on average.

In the second phase of the attack, one learns all of the remaining T-box entries with 256 more chosen plaintexts. For each byte j, to learn the value of $T(j)$, let $k = ((n-1) \oplus j) - (n-2)$, where the desired blocks are n bytes long. Obtain the encryption of the message $P = (1 - T(0), 1 - T(0), \ldots, 1 - T(0), k - T(0), 0)$; if the result is of the form $C = (t - T(0), \ldots)$, then we may conclude that $T(j) = t$, except for a possible error in the LSB. A more sophisticated analysis can resolve the uncertainty in the LSB of the T-box entries.[3]

In practice, chosen-plaintext queries may be available in some special situations. Suppose the targeted cellphone user can be persuaded to a call a phone

[2] M. Bannert has independent implemented our attack, and also reports success [Ban97]; his manuscript also documents some aspects of the chosen-plaintext attack in greater detail than is possible here.
[3] Use the skewed CaveTable to reduce the number of ambiguous CaveTable entries to 164 possibilities. Now for each known text obtained in the second phase, we know both the input P'' and the output C to the third CMEA layer; simulate that layer without the derived T-box values, using trial-and-error for each ambiguous T-box value: one needs at most 2^n trials per text (and in practice far fewer), and wrong trials are quickly eliminated.

the same attack. This latter approach makes it possible to attack triple DES (with 168 bit keys), or DES with independent subkeys (with 768 bit keys).

This attack still works even under more general assumptions on the fault locations, such as faults inside the function F, more than one fault during encryption, or even faults in the key or the key scheduling algorithm.

To check these claims, we have implemented a variant of this attack in which the faults may occur also in the inputs of the F function, rather than only in the right half of the data (i.e., the faults do not affect the following rounds directly), and found that the number of ciphertexts required to find the key is about the same as in the original attack, although the faults may occur in twice as many positions.

If the attacker can induce the faults in a chosen position or a chosen time during encryption, he can improve his results by a large factor. For example, if the attacker can cause the faults to appear uniformly in the last two, three, or four rounds of the DES encryption (rather than in all the 16 rounds), our attack requires only about 10 ciphertexts! If the attacker can choose the exact position of the fault, this number can be further reduced to about 3 ciphertexts.

2.1 Discussion

We described a new attack on ciphers using transient hardware faults. Smartcard designers can try to counter this attack by computing the encryption function twice, and outputting the ciphertext only if the results are identical. This solution is however insufficient: the probability that the same fault occurs during both encryptions may not be sufficiently small. In the attack there are only 512 possible positions for a fault. When computing twice with a single random fault in each of the two encryptions, there is a probability of $1/512$ to generate the same fault in both computations. Since the device outputs the doubly-corrupted results, the attacker receives the same data as in the original attack, if he tries 512 as many encryptions. Thus, instead of generating about 200 ciphertexts in the original attack, the device performs about 100000 encryptions.

In many cryptographic implementations, the key scheduling algorithm precomputes all the subkeys in advance, or computes the subkeys from the key every single encryption. In such cases, it is easy to find ciphertext pairs whose differences result only from one faulty subkey bit, when the faults affect the subkeys. In the attack we should assume that the difference is caused by one faulty subkey bit, or by several subkey bits caused by one fault during the key scheduling algorithm. The number of ciphertexts required for such analysis is expected to remain about the same as in the attack we described.

DFA can also be combined with other types of cryptanalytic attacks: for example, when the cipher is computed by software in the device, the faults might affect the program counter or the loop index. In such cases, we can apply more or fewer rounds than required.

In some implementations, the DES key scheduling is applied with two 28-bit shift registers, C and D, as in its original definition. The key rotates every round, and is restored to its original state at the end of encryption, since the total number of shifts during the 16 rounds is 28. If the faults affect the shifts of these registers, then in the following encryptions the key is changed to a *related* key. Related key cryptanalysis[3], or differential related key cryptanalysis[7] might be applied with DFA in such cases. We expect that linear cryptanalysis[12] can also be combined with DFA in some cases (in a similar way to differential-linear cryptanalysis[10]), especially when the identification of the fault position is highly reliable (or when the fault positions might be chosen by the attacker).

Variants of DFA attacks can in some cases also derive the keys of modes of operation in which only part of the ciphertext is known to the attacker. This is similar to the situation studied in [18] for differential cryptanalysis of the CFB mode of operation.

3   Breaking Unknown Cryptosystems

In this section, we introduce a variant of DFA that can find the secret keys of unknown cryptosystems, even if they are sealed inside tamper-resistant devices, and nothing is known about their design. In this attack, we assume a slightly different fault model: the main assumption behind this fault model is that the cryptographic key is stored in an asymmetric type of memory, in which induced faults are much more likely to change a one bit into a zero than to change a zero bit into a one (or the other way around). CMOS registers seem to be quite symmetric, but most types of non-volatile memory exhibit some degree of asymmetry. For example, a one bit in an EEPROM cell is stored as a small charge on an electrically isolated gate. If the fault is induced by external radiation (e.g., ultraviolet light), then the charges are more likely to leak out of the gate than to be forced into the gate.

To make the analysis simpler, we assume that we can apply a low level physical stress to the tamper-resistant device when it is disconnected from power, whose only possible effect is to occasionally flip one of the one bits in the key register to a zero. The plausibility of this assumption depends on numerous physical and technical considerations, which are beyond the scope of this paper.

We further assume that we are allowed to apply two types of cryptographic functions to the given tamper-resistant device: We can supply a plaintext m and use the current key k stored in the non-volatile memory of the device to get a ciphertext c, or we can supply a new n-bit key k' which replaces k in the non-volatile memory.

The cryptanalytic attack has two stages: In the first stage, we keep the original unknown secret key k stored in the tamper-resistant device, and use it to repeatedly encrypt a fixed plaintext m_0. After each encryption, we disconnect the device from power and apply a gentle physical stress. The resultant stream of ciphertexts is likely to consist of several copies of c_0, followed by several copies of a different c_1, followed by several copies of yet another c_2, until the sequence stabilizes on c_f. Since each change is likely to be the result of one more key bit flipping from one to zero (thus changing the current key k_i into a new variant k_{i+1}), and since there are about $n/2$ one bits in the original unknown key k, we expect f to be about n/2, and c_f to be the result of encrypting m_0 under the all-zero key k_f.

In the second stage of the attack, we work our way backwards from the known all-zero key k_f to the unknown original key k_0. Assuming that we already know some intermediate key k_{i+1}, we assume that k_i differs from k_{i+1} in a single bit position. If we knew the cryptographic algorithm involved, we could easily try all the possible single bit changes in a simple software simulation on a personal computer, and find the (almost certainly unique) change which would give rise to the observed ciphertext c_i. However, we do not need either a simulator or knowledge of the cryptographic algorithm, since we are given the real thing in the form of a tamper-resistant device into which we can load any key we wish, to test out whether it produces the desired ciphertext c_i. We can thus proceed deterministically from the known k_f to the desired k_0 in $O(n)$ stages, trying $O(n)$ keys at each stage. The attack is guaranteed to succeed if the fault model is satisfied, and its total complexity is at most $O(n^2)$ encryptions.

This seems to be the first cryptanalytic attack which makes it possible to find the secret key of a completely unknown cryptosystem in polynomial time (quadratic time in our case). It relies on a particular fault model which is stronger than the transient fault model described above, and which has not been experimentally verified so far. In the full version of this paper we'll discuss numerous extensions of the attack, including the analysis of more complicated fault models in which the sequence of corrupted keys forms a biased random walk in the space of 2^n possible keys.

4 Reconstructing Unknown Ciphers

In this section we show how to reconstruct the full structure of unknown ciphers hidden in tamper-resistant devices. We assume that the attacked cipher is a DES-like cipher, which encrypts by applying some initial permutation and then applying a round function several times. The round function applies some function F to the right half of the data, XORs the result to the left half, and exchanges the roles of the halves. Finally, some final permutation is performed.

In our attack we reconstruct the cipher in several steps. In each step we receive some additional information on the unknown structure of the cipher: we start from the final permutation and continue backwards through the rounds.

Note that the representation of the ciphers is not unique, and thus we cannot identify the exact original definition of the cipher. Instead, we actually find a family of representations, of which the original representation is a member. We can then choose any member of the family as an equivalent representation of the cipher.

In the first step of the attack we study some information on the final permutation: We identify which ciphertext bits come from the right half and which from the left half of the last round, i.e., which bits affect which in the last round. We encrypt several plaintexts several times, and collect pairs of ciphertexts consisting of the real ciphertext of the plaintext and a faulty ciphertext resulting from some fault during encryption of the same plaintext. We identify the faults which occur in the last round (or two) by counting the bits differing between the real ciphertext and the faulty ciphertext. We use the pairs in which the number of differing bits in the ciphertexts is smaller than a threshold (typically about a quarter of the blocksize), in which case, it is almost certain that the fault occurred in the last round, or in the preceding one.

Each fault in the last round differs in one bit in the right half and several bits in the left half. Therefore, the bits of the left half differ more frequently than the bits of the right half. We can thus identify the bits of the right half as those which differ in the least number of pairs.

Once we identify the bits of the right half and the bits of the left half, we can observe which bits of the left half are affected by each bit of the right half via the function F in the last round. In DES-like ciphers the F function is composed of S boxes, and each S box takes a few of the bits of the right half, and affects a few of the bits of the left half. We can easily identify the number of S boxes, and the input and output bits of each S box: we just choose all the pairs which differ by only one bit of the right half, and for each such bit we find the set of all the bits of the left half which differ in those pairs. This information suffices to find the number of S boxes, their sizes, and to identify the input and output bits of each S box.

Then, we reconstruct the content of the S boxes, with the specific unknown key mixed into the input of the S boxes and some unknown value mixed into their output (i.e., we can identify the table $T(x)$, where $T(x) = S(x \oplus k) \oplus u$, where k is the (actual) subkey, and u is derived from the right half of the preceding round). This can be done since the pairs give us the difference $T(x) \oplus T(y)$, for two known values x and y. This reconstruction is very effective. For example, if the unknown cipher is DES, we miss information on only $6 + 4 = 10$ bits out of the $64 * 4 = 256$ unknown bits of each S box, and if the unknown cipher is LOKI[6], we miss information on only $12 + 8 = 20$ bits out of the $2^{12} * 8 = 32768$ bits of each S box of LOKI. These missing bits do not reduce the success of the attack since we actually find all the information we need for peeling off the last round: the subkeys are already mixed into the S boxes and the extra constants are counted naturally as parts of the subkeys when analyzing the preceding rounds. This way we can fully analyze the whole cipher, and receive its full description with the specific key mixed into the S boxes.

Till now we identified the S boxes up to XOR with some unknown constants, some of which are subkeys. We can further identify the S boxes and the subkeys by analyzing encryptions under several keys and comparing the differences between the retrieved tables T. In DES and LOKI the key scheduling algorithm and the S boxes can be easily identified by such comparisons. In these ciphers, where the key scheduling actually select key bits into the subkey (rather than making a more complex computation), the selection pattern of the key bits can be identified by analyzing the effect of different keys on the T boxes. Then, the effect of the subkeys on the T boxes can be removed, resulting with the original S boxes and the key scheduling algorithm.

In some ciphers, the function F is not a composition of several S boxes as in DES and LOKI, but may compute more complex operations. In this case we can model the function F by a most general structure: for each possible subkey, the F function is modeled as an arbitrary function. In this case we can view the F function as applying one huge S box. The number of input and output bits of this huge S box is only half the block size, and thus even though the identification of this whole S box requires much work (about 2^{36} in the case of a 16-round 64-bit DES-like cipher), it can still be done for any particular key. The effect of the key on this S box can be removed in most cases after repeating this computation for several keys.

We have implemented major parts of this attack on a personal computer, using DES as the unknown cipher. Our implementation required only about 500 faulty ciphertexts to identify the bits of the right half, and up to 5000 faulty ciphertexts to identify the input and output bits of the S boxes, without reconstructing their actual entries. Their complete reconstruction would require about 10000 faulty ciphertexts. Note that the complexity of this S box reconstruction crucially depends on the size of the S boxes: in DES there are 64 entries in each S box, and thus about 64 faults in the input of an S box in different ciphertexts suffice to reconstruct the S box (except the 10 missing bits). In LOKI there are 4096 entries in each S box, and thus the number of required faulty ciphertexts is much larger. If we view the F function as one large S box, the number of entries of this huge S box is 2^{32}, and thus the number of required faulty ciphertexts in this case is huge, but still practical.

5 Non-Differential Fault Analysis

The main criticism against differential fault analysis was the transient fault model that was claimed to be unrealistic. Although we still believe that an attack based on radiation-induced faults is possible, we decided to devise a more practical fault model that will hopefully be less controversial.

The fault model considered in this section assumes that we can cut a wire or destroy a memory cell in a register using a narrow laser beam directed into a carefully chosen location in the silicon. As a result, during computation the values entering the affected location can be considered to be permanently stuck at a fixed value, and can no longer affect the rest of the computation. A similar fault model is considered in [2] where it is used in conjunction with micro-probing to mount other attacks on smartcard-based cryptosystems.

5.1 The Basic Attack

Based on this fault model we develop a *pure* ciphertext only attack, which, unlike all other ciphertext only attacks, does not make any assumption on the statistical distribution of the plaintexts. Moreover, the ciphertext should only be received after the permanent fault had already been generated; non-faulty ciphertexts are not required at all in this attack. Therefore, this attack will still work if faulty smartcards, which already have the appropriate faults due to natural reasons, are given to the attacker.

We describe this attack against a smartcard implementation of DES. We assume that DES is implemented in hardware as a single round, which is used 16 times as is described in Figure 1. We generate a permanent fault in the least significant bit of the left-half register, either by destroying the bit or by cutting the wire which enters or leaves this cell, and assume that the value of this bit is permanently stuck at zero. Figure 2 describes the values computed by this iterated implementation in the last round.

For any ciphertext the least significant bit (LSB) of the right half of the ciphertext (L_{16}) is zero, and the LSB of L_{15} is also zero. The LSB of the output of the F-function

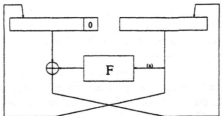

Figure 1. An Iterated Hardware Implementation of DES.

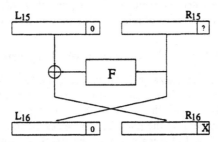

Figure 2. The Last Round in the Iterated Implementation of DES.

of the last round equals the LSB of the left half of the ciphertext (X, since the LSB of L_{15} is zero). This bit is the output of S7 in the last round. The input of S7 is formed by XORing 6 known ciphertext bits with six key bits. We can try all the 64 possible values of the six key bits, and discard any value which does not predict the LSB of the right half of the ciphertext. Each ciphertext is used to discard about half of the remaining key values. Thus, given about six ciphertexts (on whose plaintexts the attacker has no information of any kind) the right value of the six key bits can be identified. The other bits of the last subkey can be found by inducing additional faults, or by enumerating and verifying additional key bits.

5.2 Attacks on Unrolled Implementations

The attack becomes easier if DES is implemented as 16 separate hardware rounds. In this case, a permanent fault induced in one round would not directly affect the other rounds. In this attack, it suffices to destroy the LSB of register L_{15}. In particular, if we destroy the whole L_{15} in this case, we can find the key with only two ciphertexts. We can even reduce the number of ciphertexts required to find 32 bits of the last subkey uniquely to exactly one (rather than only in average) by also destroying the first and sixth input bits entering the S boxes (after they are XORed with the subkey). In this case, the F function becomes reversible, and thus its remaining inputs can be computed uniquely from the outputs.

In an alternative attack, we can destroy all the bits of the subkey registers, except in the first two rounds. Given the ciphertext, remove the last 14 rounds by decrypting with zero subkeys. Equivalently, we can destroy the outputs of the F-function in rounds 3-16, so that the outputs become zero, or even cut the cipher after the 14'th round. In all these cases, the problem is reduced to attacking a two-round DES, which can be done using one ciphertext.

We can even attack the independent-key variant of any iterated cipher. We simply remove a round at a time and prepare the required encrypted data for later analysis. After all the rounds are removed, we find the subkey of the last removed round using the prepared data, and use this found subkey in attacking the second-to-last removed round, etc., till we find all the subkeys. Note that we get two equations on each F-function from two consecutive removed rounds, and thus we get sufficient information for finding the 48-bit subkeys, although each equation leaks only 32 bits.

5.3 Additional Attacks on Iterated Implementations

All the above attacks find the key bits *horizontally*, i.e., a subkey at a time. In iterated hardware implementations we can do the following *vertical* attack which finds the first bits of all the subkeys, then the second bits of all the subkeys, etc. In this attack we first permanently destroy the plaintext to be zero, and encrypt the zero plaintext under the unknown key. In the main step of the attack, we cut the wire leaving the last bit of the subkey in the iterated hardware implementation of the rounds, and encrypt the zero plaintext under the resultant subkeys. Then, we cut the second-to-last wire, etc. When all the subkey bits had been cut, we get the zero plaintext encrypted under the zero subkeys. We can now find the first bits of the subkeys (total of 16 bits in a 16-round cipher) by trying all their possible values (2^{16} trials), and comparing the resulting ciphertexts to the last received ciphertext. Then, we try the values of the second bits of the subkeys, etc, till we find all the bits of the subkeys. For DES in particular, we can use the fact that the key is divided into two halves. We can cut the first 23 bits of the subkey before they are XORed to the right half of the data. Then, we cut the 24'th bit. Then again we cut the next 23 bits. Given the ciphertexts of the destroyed (i.e., 0) plaintext before and after each of these events (four ciphertexts), we can exhaustively search for the 16 key bits which still affect the computation after the last event. When found, we can complete the 12 key bits of the right half of the key, and then search for the 16 key bits of the right half which affect the encryption after the first event. Finally, we complete the remaining 12 key bits.

An additional attack against iterated implementations of DES ignores the data rather than the key. In this attack we cut the 32-bit data input of the F-function (this position is denoted by (x) in Figure 1), such that the output of the F-function becomes dependent only of the subkey and independent of the plaintext. If the actual plaintext is unknown, we also destroy the plaintext register, i.e., set it to zero permanently. The key scheduling algorithm of DES divides the key into two halves, each of them affects only four S boxes in each round (S1-S4, and S5-S8). As a result, 32 bits of the ciphertext depend only on one 28-bit half of the key, and the other 32 bits of the ciphertext depend only on the other 28-bit half of the key. Thus, given the ciphertext of the zero plaintext (or similarly of any plaintext) under the modified cipher, we can easily search for the two halves of the key (2^{28} trials for each, each trial costs about 1/4 encryption).

5.4 Remark

Note that in this model it is easy to apply a chosen plaintext attack without choosing any plaintext — we just destroy the plaintext register. In such a case, even a fault tolerant design in which the smartcard encrypts the plaintext several times and compares the results, will not detect any difference between the ciphertexts.[2] Moreover, if the verification is done by decrypting the ciphertext, the result register can also be destroyed, and thus will always be the same as the plaintext register.

6 Acknowledgements

We would like to gratefully acknowledge the pioneering contribution of Boneh, Demillo, and Lipton, whose ideas were the starting point of our new attack.

References

[1] Ross Anderson, Markus Kuhn, *Tamper Resistance - a Cautionary Note*, proceedings of the Second Usenix Workshop on Electronic Commerce, pp. 1–11, November 1996.

[2] Ross Anderson, Markus Kuhn, *Low Cost Attacks on Tamper Resistant Devices*, proceedings of the 1997 Security Protocols Workshop, Paris, April 7–9, 1997.

[3] Eli Biham, *New Types of Cryptanalytic Attacks Using Related Keys*, Journal of Cryptology, Vol. 7, No. 4, pp. 229–246, 1994.

[4] Eli Biham, Adi Shamir, *Differential Cryptanalysis of the Data Encryption Standard*, Springer-Verlag, 1993.

[5] Dan Boneh, Richard A. Demillo, Richard J. Lipton, *On the Importance of Checking Cryptographic Protocols for Faults*, Lecture Notes in Computer Science, Advances in Cryptology, proceedings of EUROCRYPT'97, pp. 37–51, 1997.

[6] Lawrence Brown, Josef Pieprzyk, Jennifer Seberry, *LOKI - A Cryptographic Primitive for Authentication and Secrecy Applications*, Lecture Notes in Computer Science, Advances in Cryptology, proceedings of AUSCRYPT'90, pp. 229–236, 1990.

[7] John Kelsey, Bruce Schneier, David Wagner, *Key-Schedule Cryptanalysis of IDEA, G-DES, GOST, SAFER, and Triple-DES*, Lecture Notes in Computer Science, Advances in Cryptology, proceedings of CRYPTO'96, pp. 237–251, 1996.

---

[2] Designers should be very careful with such designs, since under some fault models, the comparison of the results might leaks information about the key, which wouldn't be leaked otherwise.

[8] Paul C. Kocher, *Timing Attacks on Implementations of Diffie-Hellman, RSA, DSS, and Other Systems*, Lecture Notes in Computer Science, Advances in Cryptology, proceedings of CRYPTO'96, pp. 104–113, 1996.

[9] Xuejia Lai, James L. Massey, Sean Murphy, *Markov Ciphers and Differential Cryptanalysis*, Lecture Notes in Computer Science, Advances in Cryptology, proceedings of EUROCRYPT'91, pp. 17–38, 1991.

[10] Susan K. Langford, Martin E. Hellman, *Differential-linear cryptanalysis*, Lecture Notes in Computer Science, Advances in Cryptology, proceedings of CRYPTO'94, pp. 17–25, 1994.

[11] John Markoff, *Potential Flaw Seen in Cash Card Security*, New York Times, September 26, 1996.

[12] Mitsuru Matsui, *Linear Cryptanalysis Method for DES Cipher*, Lecture Notes in Computer Science, Advances in Cryptology, proceedings of EUROCRYPT'93, pp. 386–397, 1993.

[13] Ralph C. Merkle, *Fast Software Encryption Functions*, Lecture Notes in Computer Science, Advances in Cryptology, proceedings of CRYPTO'90, pp. 476–501, 1990.

[14] Shoji Miyaguchi, *FEAL-N specifications*, technical note, NTT, 1989.

[15] Shoji Miyaguchi, *The FEAL cipher family*, Lecture Notes in Computer Science, Advances in Cryptology, proceedings of CRYPTO'90, pp. 627–638, 1990.

[16] Shoji Miyaguchi, Akira Shiraishi, Akihiro Shimizu, *Fast Data Encryption Algorithm FEAL-8*, Review of electrical communications laboratories, Vol. 36, No. 4, pp. 433–437, 1988.

[17] National Bureau of Standards, *Data Encryption Standard*, U.S. Department of Commerce, FIPS pub. 46, January 1977.

[18] Bart Preneel, Marnix Nuttin, Vincent Rijmen, Johan Buelens, *Cryptanalysis of the CFB Mode of the DES with a Reduced Number of Rounds*, Lecture Notes in Computer Science, Advances in Cryptology, proceedings of CRYPTO'93, pp. 212–223, 1993.

[19] Ronald L. Rivest, *The RC5 Encryption Algorithm*, proceedings of Fast Software Encryption, Leuven, Lecture Notes in Computer Science, pp. 86–96, 1994.

[20] Bruce Schneier, *Description of a New Variable-Length Key, 64-Bit Block Cipher (Blowfish)*, proceedings of Fast Software Encryption, Cambridge, Lecture Notes in Computer Science, pp. 191–204, 1993.

[21] Akihiro Shimizu, Shoji Miyaguchi, *Fast Data Encryption Algorithm FEAL*, Lecture Notes in Computer Science, Advances in Cryptology, proceedings of EUROCRYPT'87, pp. 267–278. 1987.

Cryptanalysis of the Cellular Message Encryption Algorithm

David Wagner
University of California, Berkeley
daw@cs.berkeley.edu

Bruce Schneier   John Kelsey
Counterpane Systems
{schneier,kelsey}@counterpane.com

Abstract. This paper analyzes the Telecommunications Industry Association's Cellular Message Encryption Algorithm (CMEA), which is used for confidentiality of the control channel in the most recent American digital cellular telephony systems. We describe an attack on CMEA which requires 40–80 known plaintexts, has time complexity about 2^{24}–2^{32}, and finishes in minutes or hours of computation on a standard workstation. This demonstrates that CMEA is deeply flawed.

Keywords: cryptanalysis, block ciphers, cellular telephone

1 Introduction

As the US cellular telephony industry has boomed, the need for security has increased: both for privacy and fraud prevention. Because all cellular communications are sent over a radio link, anyone with the appropriate receiver can passively eavesdrop on all cellphone transmissions in the area without fear of detection. The earliest U.S. cellular telephony systems relied on the high cost of cellular-capable receivers (or scanners) for security. When such scanners become affordable and widely available, the cellphone industry lobbied for protective legislation. But these legal prohibitions have failed to solve the problem, and systems architects have been forced to turn increasingly to cryptography for more robust security.

The cellular telephony industry players are especially concerned with fraud prevention. The FCC estimates that the cellular industry loses more than \$400 million per year to fraud [FCC97]. Cellphone cloning is probably the foremost form of this problem. Because most of today's cellphones identify themselves over public radio links by sending their identity information in the clear, eavesdroppers can (and do) easily misappropriate others' identity information to make fraudulent phone calls. While the latest digital cellphones currently offer some weak protection against casual eavesdroppers because digital technology is so new that inexpensive digital scanners have not yet become widely available, the president of the Cellular Telecommunications Industry Association testified in recent Congressional hearings [Whe97] that "history will likely repeat itself as digital scanners and decoders, though expensive now, drop in price in the future."

Cryptographic mechanisms are one obvious way to combat cloning fraud, and indeed, the industry is turning to cryptography for protection. In 1992, the TR-45 working group within the Telecommunications Industry Association (TIA) developed a standard for integration of cryptographic technology into tomorrow's digital cellular systems [TIA92], which has been updated at least once [TIA95]. Some of the most recent cellphones to hit the market already include these cryptographic protection mechanisms [Nok96].

The TIA standard [TIA95] describes four cryptographic primitives for use in North American digital cellular systems:

- CAVE, a mixing function, is intended for challenge-response authentication protocols and for key generation.
- A repeated XOR mask is applied to voice data for voice privacy[1].
- ORYX, a LSFR-based stream cipher intended for wireless data services.
- CMEA (Control Message Encryption Algorithm), a simple block cipher, is used to encrypt the control channel [Ree91].

The voice privacy algorithms has long been known to be insecure [Bar92, CFP93]. Recent work by the authors has shown that ORYX is insecure as well [WSK97]. This paper focuses on the security of CMEA.

Note that CMEA is not used to protect voice communications. Instead, it is intended to protect sensitive control data, such as the digits dialed by the cellphone user. A successful break of CMEA might reveal user calling patterns. Also sent CMEA-encrypted are digits dialed (all DTMF tones) by the remote endpoint and alphanumeric personal pages recieved by the cellphone user. Finally, compromise of the control channel contents could lead to any confidential data the user types on the keypad: calling card PIN numbers may be an especially widespread concern, and credit card numbers, bank account numbers, and voicemail PIN numbers are also at risk.

This paper is organized as follows. We describe CMEA in Section 2 for reference. Next, Section 3 lists some observations that form a foundation for our later analysis. Then we give effective chosen- and known-plaintext attacks on CMEA in Sections 4 and 5. Finally, Section 6 concludes.

2 A description of CMEA

We describe the CMEA specification fully here for reference. CMEA is a byte-oriented variable-width block cipher with a 64 bit key. Block sizes may be any number of bytes; in practice, US cellular telephony systems typically apply CMEA to 2–6 byte blocks, with the block size potentially varying without any key changes. CMEA is quite simple, and appears to be optimized for 8-bit microprocessors with severe resource limitations.

CMEA consists of three layers. The first layer performs one non-linear pass on the block; this effects left-to-right diffusion. The second layer is a purely linear, unkeyed operation intended to make changes propagate in the opposite direction.

---

[1] The situation is more complicated: time-division multiple access (TDMA) systems use a straight XOR mask, while code-division multiple access (CDMA) systems instead use keyed spread spectrum techniques for security.

One can think of the second step as (roughly speaking) XORing the right half of the block onto the left half. The third layer performs a final non-linear pass on the block from left to right; in fact, it is the inverse of the first layer.

CMEA obtains the non-linearity in the first and third layer from a 8-bit keyed lookup table known as the T-box. The T-box calculates its 8-bit output as

$$T(x) = C((((C((((C(((C((x \oplus K_0) + K_1) + x) \oplus K_2) + K_3) + x) \oplus K_4) + K_5) + x) \oplus K_6) + K_7) + x$$

given input byte x and 8-byte key $K_{0...7}$. In this equation C is an unkeyed 8-bit lookup table known as the CaveTable; all operations are performed using 8-bit arithmetic. The CaveTable is given in Figure 1,

We now provide a specification of CMEA. The algorithm encrypts a n-byte message $P_{0,...,n-1}$ to a ciphertext $C_{0,...,n-1}$ under the key $K_{0...7}$ as follows:

$$y_0 \leftarrow 0$$
$$\text{for } i \leftarrow 0, \ldots, n-1$$
$$\qquad P_i' \leftarrow P_i + T(y_i \oplus i)$$
$$\qquad y_{i+1} \leftarrow y_i + P_i'$$

$$\text{for } i \leftarrow 0, \ldots, \lfloor \tfrac{n}{2} \rfloor - 1$$
$$\qquad P_i'' \leftarrow P_i' \oplus (P_{n-1-i}' \vee 1)$$

$$z_0 \leftarrow 0$$
$$\text{for } i \leftarrow 0, \ldots, n-1$$
$$\qquad z_{i+1} \leftarrow z_i + P_i''$$
$$\qquad C_i \leftarrow P_i'' - T(z_i \oplus i)$$

Here all operations are byte-wide arithmetic: $+$ and $-$ are addition and subtraction modulo 256, $\oplus$ stands for a logical bitwise exclusive or, $\vee$ represents a logical bitwise or, and the keyed T function is as described previously.

CMEA is specified in [TIA92, TIA95]; it is also described in U.S. Patent 5,159,634 [Ree91], though a different T-box method is listed.

3   Preliminaries

First, we list some preliminary observations:

- CMEA is it's own inverse. In other words, every key is a "weak key" (in the strict sense, from the DES nomenclature, of being self-inverse). This was apparently originally a design goal, for unknown reasons.
- CMEA is typically used to encrypt short blocks. Because the cellular telephony specification does not use random IVs, does not use block chaining modes, and encrypts short blocks under CMEA, codebook attacks could be a threat. On the other hand, the cellphone specifications require the CMEA key to be re-derived (using CAVE as a pseudo-random generator) for every

Fig. 1. The CaveTable

hi\lo	.0 .1 .2 .3 .4 .5 .6 .7 .8 .9 .a .b .c .d .e .f
0.	d9 23 5f e6 ca 68 97 b0 7b f2 0c 34 11 a5 8d 4e
1.	0a 46 77 8d 10 9f 5e 62 f1 34 ec a5 c9 b3 d8 2b
2.	59 47 e3 d2 ff ae 64 ca 15 8b 7d 38 21 bc 96 00
3.	49 56 23 15 97 e4 cb 6f f2 70 3c 88 ba d1 0d ae
4.	e2 38 ba 44 9f 83 5d 1c de ab c7 65 f1 76 09 20
5.	86 bd 0a f1 3c a7 29 93 cb 45 5f e8 10 74 62 de
6.	b8 77 80 d1 12 26 ac 6d e9 cf f3 54 3a 0b 95 4e
7.	b1 30 a4 96 f8 57 49 8e 05 1f 62 7c c3 2b da ed
8.	bb 86 0d 7a 97 13 6c 4e 51 30 e5 f2 2f d8 c4 a9
9.	91 76 f0 17 43 38 29 84 a2 db ef 65 5e ca 0d bc
a.	e7 fa d8 81 6f 00 14 42 25 7c 5d c9 9e b6 33 ab
b.	5a 6f 9b d9 fe 71 44 c5 37 a2 88 2d 00 b6 13 ec
c.	4e 96 a8 5a b5 d7 c3 8d 3f f2 ec 04 60 71 1b 29
d.	04 79 e3 c7 1b 66 81 4a 25 9d dc 5f 3e b0 f8 a2
e.	91 34 f6 5c 67 89 73 05 22 aa cb ee bf 18 d0 4d
f.	f5 36 ae 01 2f 94 c3 49 8b bd 58 12 e0 77 6c da

call, so the amount of text required for a codebook attack may often be un-available. (In a codebook attack, one obtains the encryption of every possible plaintext, records those pairs in a lookup table, and uses it to completely decrypt future messages without needing to know the key.)

J. Hillyard [Hil97] has noted that codebook attacks may still be possible in practice. In some contexts, each digit dialed will be encrypted in a separate CMEA block (with fixed padding); because CMEA is used in ECB mode, the result is a simple substitution cipher on the digits 0–9. Techniques from classical cryptography may well suffice to recover useful information about the dialed digits, especially when side information is available.

– One bit of the plaintext leaks. The LSB (least-significant bit) of the ciphertext is the complement of the LSB of the plaintext.

– The T-box has some key equivalence classes. Simultaneously complementing the MSB (most significant bit) of K_0 and K_1 leaves the action of the T-box unchanged; the same holds for K_{2i} and K_{2i+1} for $i = 0, 1, 2, 3$. Therefore for the rest of the paper we take the MSBs of K_0, K_2, K_4, and K_6 to all be 0, without loss of generality, and we see that the effective key length of CMEA is at most 60 bits.

– Recovering the value of all 256 of the T-box entries suffices to break CMEA, even if the key $K_{0...7}$ is never recovered.

– The value of $T(0)$ occupies a position of special importance. $T(0)$ is always used to obtain C_0 from P_0; one cannot trivially predict where other T-box entries are likely to be used. Knowing $T(0)$ lets one learn the inputs to the T-box lookups that modify the second byte in the message.

– The CaveTable has a very skewed statistical distribution. It is not a permutation; 92 of the 256 possible 8-bit values never appear; some values appear as many as four times. The distribution appears to be consistent with that of a random function.

The skew in the CaveTable means that the T-box values are skewed, too: we know $T(i) - i$ must appear in the CaveTable, so for any input to the T-box, we can immediately rule out 92 possibilities for the corresponding T-box output without needing any knowledge of the CMEA key.

3.1 A chosen-plaintext attack

CMEA is weak against chosen-plaintext attacks: one can recover all of the T-box entries with about 338 chosen texts (on average) and very little work. This attack works on any fixed block length $n > 2$; the attacker is not assumed to have control over n. We have implemented the attack to empirically verify it for correctness; the attack is extremely successful in our tests[2].

The attack proceeds in two stages, first recovering $T(0)$, and then recovering the remainder of the T-box entries; the CMEA key itself is never identified. First, one learns $T(0)$ with $(256 - 92)/2 = 82$ chosen plaintexts (on average). For each guess x at the value of $T(0)$, obtain the encryption of the message $P = (1 - x, 1 - x, 1 - x, \ldots)$, e.g. the message P where each byte has the value $1 - x$; if the result is of the form $C = (-x, \ldots)$ then we can conclude with high probability that indeed $T(0) = x$. False alarms occasionally occur, but they can be ruled out quickly in the second phase because of the skewed CaveTable distribution. Note that there are only $256 - 92 = 164$ possible values of $T(0)$, since $T(0)$ must appear in the CaveTable, and therefore we expect to identify the correct value after about $164/2 = 82$ trials, on average.

In the second phase of the attack, one learns all of the remaining T-box entries with 256 more chosen plaintexts. For each byte j, to learn the value of $T(j)$, let $k = ((n-1) \oplus j) - (n-2)$, where the desired blocks are n bytes long. Obtain the encryption of the message $P = (1 - T(0), 1 - T(0), \ldots, 1 - T(0), k - T(0), 0)$; if the result is of the form $C = (t - T(0), \ldots)$, then we may conclude that $T(j) = t$, except for a possible error in the LSB. A more sophisticated analysis can resolve the uncertainty in the LSB of the T-box entries.[3]

In practice, chosen-plaintext queries may be available in some special situations. Suppose the targeted cellphone user can be persuaded to a call a phone

---

[2] M. Bannert has independent implemented our attack, and also reports success [Ban97]; his manuscript also documents some aspects of the chosen-plaintext attack in greater detail than is possible here.

[3] Use the skewed CaveTable to reduce the number of ambiguous CaveTable entries to 164 possibilities. Now for each known text obtained in the second phase, we know both the input P'' and the output C to the third CMEA layer; simulate that layer without the derived T-box values, using trial-and-error for each ambiguous T-box value: one needs at most 2^n trials per text (and in practice far fewer), and wrong trials are quickly eliminated.

number under the attacker's control—perhaps a menuized survey, answering machine, or operator. The phone message the user receives might prompt the user to enter digits (chosen in advance by the attacker), thus silently enabling a chosen-plaintext attack on CMEA. Alternatively, the phone message might send chosen DTMF tones to the targetted cellphone user, thus mounting chosen-plaintext queries at will.

4 A known-plaintext attack on 3-byte blocks

We now describe a known plaintext attack on CMEA needing about 40–80 known texts. The attack assumes that each known plaintext is enciphered with a 3-byte block width. Our (unoptimized) implementation has a time complexity of 2^{24} to 2^{32}, and can be easily parallelized.

Our cryptanalysis has two phases. The first phase gathers information about the T-box entries from the known CMEA encryptions, eliminating many possibilities for the values of each T-box output. In this way we reduce the problem to that of cryptanalysis of the T-box algorithm, given some partial information about T-box input/output pairs. In the second phase, we take advantage of the statistical biases in the CaveTable to cryptanalyze the T-box and recover the CMEA key $K_{0...7}$, using pruned search and meet-in-the-middle techniques to enhance performance.

The first phase is implemented as follows. Because $T(0)$ occupies a position of special importance, we exhaustively search over the 164 possibilities for $T(0)$. (Remember that $T(0)$ must appear in the CaveTable, and so there are only $256 - 92 = 164$ possibilities for it.) For each guess at $T(0)$, we set up a 256×256 array $p_{i,j}$ which records for each i, j whether $T(i) = j$ is possible. All values for $T(i), i > 0$ are initially listed as possible. Since $T(i) - i$ is a CaveTable output and the CaveTable has an uneven distribution, we can immediately rule out 92 values for $T(i)$.

Next, we gradually eliminate impossible values using the known texts as follows. The general idea is that each known plaintext/ciphertext pair lets us establish several implications of the form

$$T(0) = t_0, T(i) = j \quad \Rightarrow \quad T(i') = j'. \tag{1}$$

If we have already eliminated $T(i') = j'$ as impossible, then we can conclude that $T(i) = j$ is also impossible via the contrapositive of (1). In this way, we successively rule out more and more possibilities in the $p_{i,j}$ array, until we either reach a contradiction (in which case we start over with another guess at $T(0)$) or until we run out of logical deductions to make (in which case we proceed to the second phase).

The second phase recovers the CMEA key from the information about T previously accumulated in the $p_{i,j}$ array. Our simplest key recovery algorithm is based on pruned search. First, one guesses K_6 and K_7. Then, we peel off the effect of the last 1/4 of the T-box, and check whether the intermediate value is a possible CaveTable output. The intermediate value must always be one of the 164 possible CaveTable outputs when we find the correct K_6, K_7; because the CaveTable is so heavily skewed, incorrect K_6, K_7 guesses will usually be quickly identified by this test, if we have knowledge about a number of T-box entries. Next, one continues by guessing K_4, K_5, pruning the search as before, and continuing the pruned search until the entire key is recovered. This technique is very effective if enough information is available in the $p_{i,j}$ array.

Unfortunately, pruned search very quickly becomes extremely computationally intensive if too few known texts are available: at each stage, too many candidates survive the pruning, and the search complexity grows exponentially. We have a more sophisticated key recovery algorithm which can reduce the computation workload dramatically in these instances. The basic idea is that the T-box is subject to a classic meet-in-the-middle optimization: one can work halfway through the T-box given only $K_{0...3}$, and one can work backwards up to the middle given just $K_{4...7}$. This enables us to precompute a lookup table that contains the intermediate value corresponding to each $K_{0...3}$ value. Then, we try each possible $K_{4...7}$ value, work backwards through some known T-box outputs, and look for a match in the precomputed lookup table. Of course the search pruning techniques can be applied to $K_{4...7}$ to further reduce the complexity of the meet-in-the-middle algorithm. The combination of pruned search and meet-in-the-middle cryptanalysis allows us to efficiently recover the entire CMEA key with as few as 40–80 known plaintexts.

4.1 The first phase: more details

We describe how to derive implications of the form (1) from some known CMEA encryptions for the first phase. Knowing $T(0)$ lets us recover (for each plaintext/ciphertext pair P, C) y_1, z_1 and thus we learn the inputs to the two T-boxes lookups used to modify C_1. We make a guess (e.g. $T(i) = j$) about the output of the first aforementioned T-box lookup. We can derive the (implied) output of the second T-box lookup by using the known text pair. Then we deduce the (implied) values of y_2, z_2 and thus the inputs to the two T-box lookups used to modify C_2. Next we derive the quantity XORed into C_0 in the second CMEA layer, which lets us calculate the (implied) outputs of the two T-box lookups that modify C_2[4]. Therefore our assumption $T(i) = j$ implies three other derived equations of the form $T(i') = j'$; if any of those three derived input/output pairs i', j' is listed as impossible in $p_{i',j'}$, then we have found a contradiction, and we may conclude that our original assumption was wrong—namely, that the assumed value of the T-box entry was in fact impossible, and that value may be marked as impossible in $p_{i,j}$.

In this way, we can gradually rule out many entries $p_{i,j}$ as impossible. We loop over all i, j and all known texts, until no more deductions can be made. If our guess at $T(0)$ was incorrect, then there will probably be a T-box input

---

[4] The true situation is slightly more complicated. The LSB remains unknown, so we have to try two possibilities; only if both possibilities lead to a contradiction can we rule out the equation $T(i) = j$ as impossible.

for which no possible output values remain, and in this case we will be able to discard our incorrect guess at $T(0)$. Otherwise, we tentatively conclude that our guess at $T(0)$ was correct, and we can usually identify several other known T-box input/output pairs; with this information in hand, we proceed to the second phase. Typically the first phase will identify $T(0)$ uniquely when sufficiently many known plaintexts (about 50 or more) are available[5]; if more possibilities for $T(0)$ are found, the second phase will be invoked for such possibility.

4.2 The second phase: more details

First, we describe how to prune key trials during the key recovery search. Note that a T-box output is of the form

$$T(i) = C(((O + i) \oplus K_6) + K_7) + i$$

for some unknown CaveTable output O. We can calculate $j = C(((O + i) \oplus K_6) + K_7) + i$ for all CaveTable outputs and check whether each such j is listed as possible in $p_{i,j}$; if every such j is listed as impossible, then we can recognize our guess at K_6, K_7 as incorrect. Because there are only 164 possible CaveTable outputs, incorrect guesses at K_6, K_7 will usually be ruled out by some i as long as there is enough information in the $p_{i,j}$ array. These incorrect guesses at K_6, K_7 can thus be pruned from the search tree without any further work.

Next, we give some more details on the meet-in-the-middle approach. This approach is only applicable when we have enough known plaintexts to identify 4 known T-box input/output values $(a, T(a)), (b, T(b)), (c, T(c)), (d, T(d))$ from the $p_{i,j}$ array. For each K_0, K_1, K_2, we compute the intermediate values a', b', c', d' formed after computing T through the known key bytes; for example, $a' = C((a \oplus K_0) + K_1) + a) \oplus K_2$. Next we form the 24-bit index $n = (a' - d', b' - d', c' - d')$, and insert the pair $(n, K_{0...2})$ into a large hash table keyed on n. After repeating for all 2^{22} possible $K_{0...2}$ values, we have built a precomputed lookup table suitable for use in the meet-in-the-middle optimization. To check a trial $K_{4...7}$ value, we work backwards from $T(a), T(b), T(c), T(d)$ as far possible given only $K_{4...7}$ and identify the intermediate values a'', b'', c'', d''. The intermediate values reflect the values of the T-box computations just after addition of K_3: for example,

$$C(((C(((C(a'') + a) \oplus K_4) + K_5) + a) \oplus K_6) + K_7) + a = T(a).$$

We see that a'' can be identified from $a, T(a)$ by working backwards through the T-box computation and inverting the CaveTable where necessary[6], and b'', c'', d''

---

[5] The density of $p_{.,.}$ after all deductions turns out to be a poor estimator for success. For any fixed number of known texts, the density seems to be quite constant—hovering around 0.5 for 40 texts and around 0.35 for 80 texts—and variations don't seem to be very strongly correlated to success in either phase of the attack.

[6] Collisions in the CaveTable may cause multiple possibilities for a'', b'', c'', d'' to be identified; we simply search through them all exhaustively. On the other hand, because some outputs never appear in the CaveTable, sometimes no possibilities will be identified, which lets us immediately prune away $K_{4...7}$. In practice, the number of possibilities is usually small.

can be found similarly. Then we form the 24-bit index $m = (a'' - d'', b'' - d'', c'' - d'')$, search in the precomputed hash table for a matching entry $(n, K_{0...2})$ with $n = m$, and use trial encryption to check the resulting $K_{0...7}$ value. Note that if our guess at $K_{4...7}$ was correct, we have $a'' = a' + K_3$ etc., so that the correct value of $K_{0...2}$ will show up in our search of the precomputed hash table and the correct value of K_3 can be derived as $a'' - a'$; this ensures that we will identify $K_{0...7}$ correctly.

Pruned search lets us dramatically reduce the number of key candidates tried, if there is enough information in the $p_{.,.}$ array. The meet-in-the-middle optimization is a time-space tradeoff that further reduces the computational workload when 4 known T-box input/output values are available. Combining the two approaches yields a key recovery algorithm for the second phase that is very efficient on a standard 100 MHz Pentium with 40 Mb of memory. Furthermore, the search algorithm can easily be parallelized for even greater performance if necessary. Note that we make heavy use of the non-uniform output distribution of the CaveTable, and these analysis techniques would not work if the CaveTable were unbiased.

4.3   Discussion

This known plaintext attack is much more devastating than the chosen plaintext attack described in Section 3.1. Chosen plaintext may be difficult to obtain in practice, but known plaintext is likely to be much easier to acquire.

There are a number of realistic ways that the required known plaintext can be collected in practice. Dialed digits are typically CMEA-encrypted with 3-byte blocks; typically each block will contain only one digit, and often the telephone number dialed will be known. DTMF tones sent on the line will usually be CMEA-encrypted. If the user can be persuaded to dial a number under adversarial control, using their calling card, then the DTMF tones and user-dialed digits will be known to the attacker, providing a ready source of known plaintext; after recovering the CMEA key in a known-plaintext attack, the attacker could decrypt the calling card number and make false calls billed to the victim's name. Furthermore, alphanumeric pages sent to cellular phones are becoming increasingly common, and alphanumeric pages are sent over the control channel. These pages may have a large known component, which will provide some known plaintext. It should be clear that known plaintext may be available from a number of potential sources.

In this section, we have discussed cryptanalysis of CMEA with 3-byte block widths. A block width of 3 bytes is a natural choice to examine. Known plaintext with 3-byte block widths is often readily available in practice; for instance, dialed digits are typically encrypted and transmitted using 3-byte block widths in nearly all digital cellular architectures. Moreover, CMEA appears to be easiest to analyze for short block widths, and most cellular standards avoid block widths shorter than 3 bytes[7]. Therefore, 3-byte blocks are a good indicator of the

---

[7] IS-95 is a notable exception; see Section 5 for a better attack on the 2-byte block widths that are used in some IS-95 messages.

strength of CMEA as used in phone systems; by giving a known-plaintext attack on CMEA with 3-byte blocks, we show that the control channel is not protected adequately in nearly all of the North American digital cellular phone systems.

5 A known-plaintext attack on 2-byte blocks

We saw above that CMEA is insecure when used with a 3-byte block width; now we show that the situation is even worse for 2-byte blocks. In this section, we present an attack on CMEA needing just 4 known plaintexts when 2-byte blocks are in use. Most cellular standards avoid using CMEA with 2-byte blocks. However, this is not just a theoretical attack: a few cellular systems, such as IS-95 (CDMA), do apply CMEA with a 2-byte block width to protect dialed digits, and they will be vulnerable to the improved attack.

The known-plaintext attack on 2-byte blocks follows immediately from our earlier discussion. First, we guess $T(0)$; that lets us recover 4 more T-box values from the first two known texts. (There is no need for a stage corresponding to the first phase of the attack on 3-byte blocks, as we can trivially derive 4 known input/output pairs for the T-box from the known texts.) With those known T-box input/output pairs, we perform a pruned meet-in-the-middle search to derive a number of possibilities for the full CMEA key, as described in Section 4.2. The correct CMEA key can be quickly recognized by trial decryption. The pruned meet-in-the-middle search has work factor 2^{24}–2^{32}, and we will need to do about 30 iterations of the search to handle each of the possibilities for $T(0)$. In sum, this attack requires just 4 known 2-byte plaintexts and has time complexity about 2^{29}–2^{37}.

In fact, the plaintext requirements can be reduced even further, to just two known 2-byte plaintexts and some extra ciphertexts. We don't need to know the decryption of the extra ciphertexts: the extra ciphertexts must merely be enough to information-theoretically determine the CMEA key, so that all incorrect key trials can be recognized and discarded. Note the plaintext often contains redundancy—for instance, when it contains dialed digits, there are only 10 possible values for each nibble, and often much of the input is a public fixed value—so in practice obtaining the necessary extra ciphertexts should be very easy.

6 Conclusions

We have presented several attacks on CMEA, and some of them may be realistically exploitable in practice. We described several possible ways to obtain known plaintext information. One attack that applies to nearly all North American digital cellular standards needs about 40–80 known plaintexts; that many known texts may be available in some situations, although availability is likely to depend on subtleties of the cellular phone system implementation. Though it does not apply to most digital cellphone standards, another attack needs just 4 known plaintexts, which is a much more realistic assumption. At a minimum, these attacks illuminate fundamental certificational weaknesses in CMEA. At worst, widespread attacks on CMEA might be possible in practice.

Our cryptanalysis of CMEA underscores the need for an open cryptographic review process. Betting on new algorithms is always dangerous, and closed-door design and proprietary standards are not conducive to the best odds.

Since being exposed to public scrutiny, three of the four proprietary TIA cryptographic algorithms have been broken: the voice privacy protection was shown to be insecure as early as 1992 [Bar92, CFP93], this paper cryptanalyzes CMEA, and ORYX was recently broken by the authors [WSK97]. This poor success rate provides a strong argument against closed-door design.

In addition, our analysis also shows the importance of explicitly stating security assumptions during every step of the design and development process, and of not reusing security components without throroughly examining the implications of reuse. The CaveTable was designed to have the security properties CAVE needed. Designers reused it for CMEA because they were low on space; this turned out to be a bad idea. CMEA requires different properties from the CaveTable than CAVE does.

In short, CMEA is deeply flawed, and should be carefully reconsidered.

7 Acknowledgements

Greg Rose first pointed out the insecurity of CMEA, and he deserves the credit for that discovery. We were not aware of serious flaws in CMEA until we heard over the grapevine that he had found an effective known-plaintext attack on CMEA; this tip provided the motivation to look more closely at CMEA until we managed to independently re-derive the attack described in this paper. Unfortunately he is not free to publish his analysis, so we offer ours instead. We are extremely grateful to Greg Rose for his immeasurable help.

Also, we thank an anonymous party (for scanning the cellphone cryptography standard and posting it to the Internet [TIA92]), John Young (for acting as a clearinghouse for resources on cellphone crypto), Ron Rivest (for many helpful comments on the presentation of our results), Steve Schear (for some assistance navigating the maze of cellular standards), Niels Ferguson (for useful feedback), and all those early readers who independently pointed out that the number of possibilities for $T(0)$ could be reduced from 256 to 164 in the known plaintext attack.

References

[Ban97] M. Bannert, "Cryptanalysis of the Cellular Message Encryption Algorithm," unpublished manuscript, 1 May 1997.

[Bar92] J.P. Barlow, "Decrypting the Puzzle Palace," *Communications of the ACM*, July 1992.

[CFP93] R. Mechaley, Speaker, Digital telephony and cryptography policy session. *The Third Conference on Computers, Freedom and Privacy*, Burlingame, CA, 1993, Bruce Koball, General Chair.

[FCC97] FCC Wireless Telecommunications Bureau, "FCC-WTB Information of Cellular Fraud," http://www.fcc.gov/wtb/cellfrd.html, Feb 1997.

[Hil97] J. Hillyard, personal communication, 21 May 1997.

[Nok96] Nokia Mobile Phones, "Nokia Announces Anti-Fraud Protection Option for All Models Marketed in 1996," 5 Jan 1996, Tampa Fla., http://www.nokia.com/news/news_htmls/nmp_960105b.html, press release.

[Ree91] J.A. Reeds III, "Cryptosystem for Cellular Telephony," U.S. Patent 5,159,634, Sep 1991.

[TIA92] TIA IS-54 Appendix A, "Dual-mode Cellular System: Authentication, Message Encryption, Voice Privacy Mask Generation, Shared Secret Data Generation, A-Key Verification, and Test Data," Feb 1992, Rev B.

[TIA95] TIA TR45.0.A, "Common Cryptographic Algorithms," June 1995, Rev B.

[Whe97] "Summary of Testimony of Thomas E. Wheeler," Oversight Hearing on Cellular Privacy, 5 Feb 1997, House Commerce Committee, Subcommittee on Telecommunications, Trade, and Consumer Protection. http://www.house.gov/commerce/telecom/hearings/020597/wheeler.pdf

[WSK97] D. Wagner, B. Schneier, J. Kelsey, "Cryptanalysis of ORYX," unpublished manuscript, 4 May 1997.

Author Index

Erratum

C.P. Schnorr: Security of 2^t–Root Identification and Signatures,
Proceedings CRYPTO'96, Springer LNCS 1109, (1996), pp. 143–156
page 148, section 3, line 5 of the proof of Theorem 3.
Correction. The proposed factoring method

Check whether $\{\gcd(Y^{2^i} \pm Z^{2^{i+\ell}}, N)\} = \{p, q\}$ holds for some i with $0 \leq i < t$

fails if $Y^{2^i} = -Z^{2^{i+\ell}}$ holds for some i with $0 \leq i < t$, otherwise it factors N
with probability $\frac{1}{2}$. In the first case continue the factoring algorithm as follows
until it factors N with probability $\frac{1}{2}$:

Supplemental steps to the factoring algorithm. Repeat the entire algorithm using independent coin flips and construct independent pairs (Y, Z) with
$Y^{2^t} = Z^{2^{t+\ell}}$ mod N until either of the following two cases arises.

Case I. $Y^{2^i} \neq -Z^{2^{i+\ell}}$ for all i with $0 \leq i < t$ holds for some (Y, Z). Then terminate as the proposed factoring method succeeds using Y, Z with probability
$\frac{1}{2}$.

Case II. $Y^{2^i} = -Z^{2^{i+\ell}}$ holds for two independent pairs $(Y, Z), (Y', Z')$. Then
replace these pairs by $(Y_{\text{new}}, Z_{\text{new}})$ with $Y_{\text{new}} := YY'$, $Z_{\text{new}} := ZZ'$. If
$Y_{\text{new}}^{2^{i_{\text{new}}}} = -Z_{\text{new}}^{2^{i_{\text{new}}+\ell}}$ holds for some i_{new} then we have $i_{\text{new}} < i$, otherwise
terminate (as the proposed factoring method succeeds using $Y_{\text{new}}, Z_{\text{new}}$ with
probability $\frac{1}{2}$).
Continue the repetitions of the entire algorithm using idependent coin flips
and continue to decrease i until the algorithm either terminates in Case I or
enters Case II with $i = 1$. In the latter case the proposed factoring method
succeeds using $Y_{\text{new}}, Z_{\text{new}}$ with probability $\frac{1}{2}$, in particular $\{\gcd(Y_{\text{new}} \pm Z_{\text{new}}, N)\} = \{p, q\}$ holds with probability $\frac{1}{2}$.

With the supplemental steps the algorithm factorizes N with probability $\frac{1}{2}$. The
supplemental steps increase the time bound for factoring by a factor $O(\ell)$. The
correctness proof of the amended factoring method uses the following observation

We see from $Y^{2^t} = Z^{2^{t+\ell}}$ mod N that Z^{2^t}/Y is a 2^t-root of 1 mod N. This
root is not necessarily uniformly distributed over all 2^t-roots of 1 mod N. But
it is uniformly distributed within certain cosets.

Fact. Let $Y = Y(Z^{2^t})$ *be a function of* Z^{2^t} *that solves* $Y^{2^t} = Z^{2^{t+\ell}}$ *mod* N *with*
$\ell < t$. *Then* Z^{2^t}/Y *takes the roots in* $c_0 R_N(2^t)^{2^\ell}$ *with equal probability for*
all $c_0 \in R_N(2^t)$, *where* $R_N(2^t)$ *denotes the group of* 2^t-*roots of* 1 *mod* N *and*
$R_N(2^t)^{2^\ell} \subset R_N(2^t)$ *denotes the subgroup of* 2^ℓ-*powers.*

All subsequent factoring algorithms in the paper have to be amended in the
same way.

Springer
and the
environment

At Springer we firmly believe that an international science publisher has a special obligation to the environment, and our corporate policies consistently reflect this conviction.
We also expect our business partners – paper mills, printers, packaging manufacturers, etc. – to commit themselves to using materials and production processes that do not harm the environment. The paper in this book is made from low- or no-chlorine pulp and is acid free, in conformance with international standards for paper permanency.

 Springer

Lecture Notes in Computer Science

For information about Vols. 1–1211

please contact your bookseller or Springer-Verlag